Linear Algebra

Linear Algebra

Step by Step

Kuldeep Singh
Senior Lecturer in Mathematics
University of Hertfordshire

OXFORD
UNIVERSITY PRESS

OXFORD
UNIVERSITY PRESS

Great Clarendon Street, Oxford, OX2 6DP,
United Kingdom

Oxford University Press is a department of the University of Oxford.
It furthers the University's objective of excellence in research, scholarship,
and education by publishing worldwide. Oxford is a registered trade mark of
Oxford University Press in the UK and in certain other countries

First Edition published in 2014

Reprinted with corrections 2015

Published in the United States of America by Oxford University Press
198 Madison Avenue, New York, NY 10016, United States of America

British Library Cataloguing in Publication Data
Data available

Library of Congress Control Number: 2013947001

ISBN 978-0-19-965444-4

Printed in the UK by
Ashford Colour Press Ltd

Preface

My interest in mathematics began at school. I am originally of Sikh descent, and as a young child often found English difficult to comprehend, but I discovered an affinity with mathematics, a universal language that I could begin to learn from the same start point as my peers.

Linear algebra is a fundamental area of mathematics, and is arguably the most powerful mathematical tool ever developed. It is a core topic of study within fields as diverse as business, economics, engineering, physics, computer science, ecology, sociology, demography and genetics. For an example of linear algebra at work, one need look no further than the Google search engine, which relies on linear algebra to rank the results of a search with respect to relevance.

My passion has always been to teach, and I have held the position of Senior Lecturer in Mathematics at the University of Hertfordshire for over twenty years, where I teach linear algebra to entry level undergraduates. I am also the author of *Engineering Mathematics Through Applications*, a book that I am proud to say is used widely as the basis for undergraduate studies in many different countries. I also host and regularly update a website dedicated to mathematics.

At the University of Hertfordshire we have over one hundred mathematics undergraduates. In the past we have based our linear algebra courses on various existing textbooks, but in general students have found them hard to digest; one of my primary concerns has been in finding rigorous, yet accessible textbooks to recommend to my students. Because of the popularity of my previously published book, I have felt compelled to construct a book on linear algebra that bridges the considerable divide between school and undergraduate mathematics.

I am somewhat fortunate in that I have had so many students to assist me in evaluating each chapter. In response to their reactions, I have modified, expanded and added sections to ensure that its content entirely encompasses the ability of students with a limited mathematical background, as well as the more advanced scholars under my tutelage. I believe that this has allowed me to create a book that is unparalleled in the simplicity of its explanation, yet comprehensive in its approach to even the most challenging aspects of this topic.

Level

This book is intended for first- and second-year undergraduates arriving with average mathematics grades. Many students find the transition between school and undergraduate mathematics difficult, and this book specifically addresses that gap and allows seamless progression. It assumes limited prior mathematical knowledge, yet also covers difficult material and answers tough questions through the use of clear explanation and a wealth of illustrations. The emphasis of the book is on students learning for themselves by gradually absorbing clearly presented text, supported by patterns, graphs and associated questions. The text allows the student to gradually develop an understanding of a topic, without the need for constant additional support from a tutor.

Pedagogical Issues

The strength of the text is in the large number of examples and the step-by-step explanation of each topic as it is introduced. It is compiled in a way that allows distance learning, with explicit solutions to all of the set problems freely available online <http://www.oup.co.uk/companion/singh>. The miscellaneous exercises at the end of each chapter comprise questions from past exam papers from various universities, helping to reinforce the reader's confidence. Also included are short historical biographies of the leading players in the field of linear algebra. These are generally placed at the beginning of a section to engage the interest of the student from the outset.

Published textbooks on this subject tend to be rather static in their presentation. By contrast, my book strives to be significantly more dynamic, and encourages the engagement of the reader with frequent question and answer sections. The question–answer element is sprinkled liberally throughout the text, consistently testing the student's understanding of the methods introduced, rather than requiring them to remember by rote.

The simple yet concise nature of its content is specifically designed to aid the weaker student, but its rigorous approach and comprehensive manner make it entirely appropriate reference material for mathematicians at every level. Included in the online resource will be a selection of MATLAB scripts, provided for those students who wish to process their work using a computer.

Finally, it must be acknowledged that linear algebra can appear abstract when first encountered by a student. To show off some of its possibilities and potential, interviews with leading academics and practitioners have been placed between chapters, giving readers a taste of what may be to come once they have mastered this powerful mathematical tool.

Acknowledgements

I would particularly like to thank Timothy Peacock for his signficant help in improving this text. In addition I want to thank Sandra Starke for her considerable contribution in making this text accessible.

Thanks too to the OUP team, in particular Keith Mansfield, Viki Mortimer, Smita Gupta and Clare Charles.

Dedication

To Shaheed Bibi Paramjit Kaur

Contents

1 Linear Equations and Matrices **1**

1.1 Systems of Linear Equations 1
1.2 Gaussian Elimination 12
1.3 Vector Arithmetic 27
1.4 Arithmetic of Matrices 41
1.5 Matrix Algebra 57
1.6 The Transpose and Inverse of a Matrix 75
1.7 Types of Solutions 91
1.8 The Inverse Matrix Method 105

Des Higham Interview **127**

2 Euclidean Space **129**

2.1 Properties of Vectors 129
2.2 Further Properties of Vectors 143
2.3 Linear Independence 159
2.4 Basis and Spanning Set 171

Chao Yang Interview **190**

3 General Vector Spaces **191**

3.1 Introduction to General Vector Spaces 191
3.2 Subspace of a Vector Space 202
3.3 Linear Independence and Basis 216
3.4 Dimension 229
3.5 Properties of a Matrix 239
3.6 Linear Systems Revisited 254

Janet Drew Interview **275**

4 Inner Product Spaces **277**

4.1 Introduction to Inner Product Spaces 277
4.2 Inequalities and Orthogonality 290
4.3 Orthonormal Bases 306
4.4 Orthogonal Matrices 321

Anshul Gupta Interview **338**

5 Linear Transformations **339**

5.1 Introduction to Linear Transformations 339
5.2 Kernel and Range of a Linear Transformation 352
5.3 Rank and Nullity 364
5.4 Inverse Linear Transformations 372
5.5 The Matrix of a Linear Transformation 389
5.6 Composition and Inverse Linear Transformations 407

Petros Drineas Interview **429**

6 Determinants and the Inverse Matrix **431**

6.1 Determinant of a Matrix 431
6.2 Determinant of Other Matrices 439
6.3 Properties of Determinants 455
6.4 LU Factorization 472

Françoise Tisseur Interview **490**

7 Eigenvalues and Eigenvectors **491**

7.1 Introduction to Eigenvalues and Eigenvectors 491
7.2 Properties of Eigenvalues and Eigenvectors 503
7.3 Diagonalization 518
7.4 Diagonalization of Symmetric Matrices 533
7.5 Singular Value Decomposition 547

Brief Solutions 567

Index 605

1 Linear Equations and Matrices

SECTION 1.1 Systems of Linear Equations

By the end of this section you will be able to

- solve a linear system of equations
- plot linear graphs and determine the type of solutions

1.1.1 Introduction to linear algebra

We are all familiar with simple one-line equations. An equation is where two mathematical expressions are defined as being equal. Given $3x = 6$, we can almost intuitively see that x must equal 2.

However, the solution isn't always this easy to find, and the following example demonstrates how we can extract information embedded in more than one line of information.

Imagine for a moment that John has bought two ice creams and two drinks for £3.00.

How much did John pay for each item?

Let $x =$ cost of ice cream and $y =$ cost of drink, then the problem can be written as

$$2x + 2y = 3$$

At this point, it is impossible to find a unique value for the cost of each item. However, you are then told that Jane bought two ice creams and one drink for £2.50. With this additional information, we can model the problem as a **system of equations** and look for unique values for the cost of ice creams and drinks. The problem can now be written as

$$\begin{aligned} 2x + 2y &= 3 \\ 2x + y &= 2.5 \end{aligned}$$

Using a bit of guesswork, we can see that the only sensible values for x and y that satisfy both equations are $x = 1$ and $y = 0.5$. Therefore an ice cream must have cost £1.00 and a drink £0.50.

Of course, this is an extremely simple example, the solution to which can be found with a minimum of calculation, but larger systems of equations occur in areas like engineering, science and finance. In order to reliably extract information from multiple linear equations, we need linear algebra. Generally, the complex scientific, or engineering problem can be solved by using linear algebra on linear equations.

*What does the term **linear equation** mean?*
An equation is where two mathematical expressions are defined as being equal.

A linear equation is one where all the variables such as x, y, z have index (power) of 1 or 0 only, for example

$$x + 2y + z = 5$$

is a linear equation. The following are also linear equations:

$$x = 3;\ x + 2y = 5;\ 3x + y + z + w = -8$$

The following are *not* linear equations:

1. $x^2 - 1 = 0$
2. $x + y^4 + \sqrt{z} = 9$
3. $\sin(x) - y + z = 3$

Why not?
In equation (1) the index (power) of the variable x is 2, so this is actually a quadratic equation.
In equation (2) the index of y is 4 and z is 1/2. Remember, $\sqrt{z} = z^{1/2}$.
In equation (3) the variable x is an argument of the trigonometric function sine.

Note that if an equation contains an **argument** of trigonometric, exponential, logarithmic or hyperbolic functions then the equation is not linear.

A set of linear equations is called a **linear system**.

In this first course on linear algebra we examine the following questions regarding linear systems:

- Are there any solutions?
- Does the system have no solution, a unique solution or an infinite number of solutions?
- How can we find all the solutions, if they exist?
- Is there some sort of structure to the solutions?

Linear algebra is a systematic exploration of linear equations and is related to 'a new kind of arithmetic' called the **arithmetic of matrices** which we will discuss later in the chapter.

However, linear algebra isn't exclusively about solving linear systems. The tools of matrices and vectors have a whole wealth of applications in the fields of functional analysis and quantum mechanics, where inner product spaces are important. Other applications include optimization and approximation where the critical questions are:

1. Given a set of points, what's the best linear model for them?
2. Given a function, what's the best polynomial approximation to it?

To solve these problems we need to use the concepts of eigenvalues and eigenvectors and orthonormal bases which are discussed in later chapters.

In all of mathematics, the concept of linearization is critical because linear problems are very well understood and we can say a lot about them. For this reason we try to convert many areas of mathematics to linear problems so that we can solve them.

1.1.2 System of linear equations

We can plot linear equations on a graph. Figure 1.1 shows the example of the two linear equations we discussed earlier.

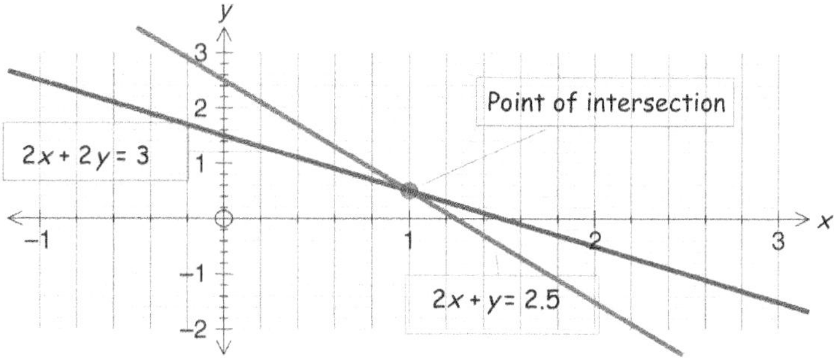

Figure 1.1

Figure 1.2 is an example of the linear equation, $x + y + 2z = 0$, in a 3d coordinate system.

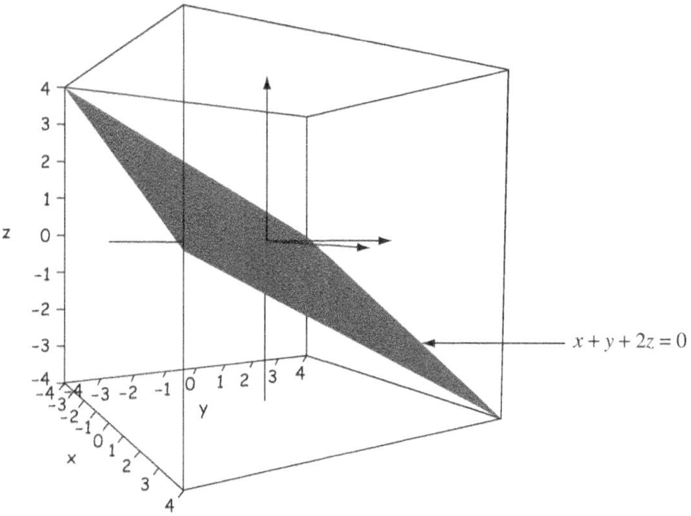

Figure 1.2

What do you notice about the graphs of linear equations?
They are straight lines in 2d and a plane in 3d. This is why they are called linear equations and the study of such equations is called **linear algebra**.

What does the term system of linear equations mean?
Generally a finite number of linear equations with a finite number of unknowns $x, y, z, w, \ldots$ is called a **system of linear equations** or just a **linear system**.

For example, the following is a linear system of three simultaneous equations with three unknowns x, y and z:

$$x + 2y - 3z = 3$$
$$2x - y - z = 11$$
$$3x + 2y + z = -5$$

In general, a linear system of m equations in n unknowns $x_1, x_2, x_3, \ldots, x_n$ is written mathematically as

$$\begin{aligned} a_{11}x_1 + a_{12}x_2 + \cdots + a_{1n}x_n &= b_1 \\ a_{21}x_1 + a_{22}x_2 + \cdots + a_{2n}x_n &= b_2 \\ \vdots \qquad \vdots \qquad\qquad \vdots \quad &\quad \vdots \\ a_{m1}x_1 + a_{m2}x_2 + \cdots + a_{mn}x_n &= b_m \end{aligned} \qquad (*)$$

where the coefficients a_{ij} and b_j represent real numbers. The unknowns $x_1, x_2, \ldots, x_n$ are **placeholders** for real numbers.

Linear algebra involves using a variety of methods for finding solutions to linear systems such as (*).

Example 1.1

Solve the equations about the cost of ice creams and drinks by algebraic means

$$2x + 2y = 3 \qquad (1)$$
$$2x + y = 2.5 \qquad (2)$$

Solution

How do we solve these linear simultaneous equations, (1) and (2)?

Let's think about the information contained in these equations. The x in the first line represents the cost of an ice cream, so must have the same value as the x in the second line. Similarly, the y in the first line that represents the cost of a drink must have the same value as the y in the second line.

It follows that we can combine the two equations to see if together they offer any useful information.

How?

In this case, we subtract equation (2) from equation (1):

$$\begin{array}{rl} 2x + 2y = 3 & (1) \\ -(2x + y = 2.5) & (2) \\ \hline 0 + y = 0.5 & \end{array}$$

Note that the unknown x is eliminated in the last line which leaves $y = 0.5$.

What else do we need to find?

The other unknown x.

How?
By substituting $y = 0.5$ into equation (1):

$$2x + 2(0.5) = 3 \quad \text{implies that} \quad 2x + 1 = 3 \quad \text{gives} \quad x = 1$$

Hence the cost of an ice cream is £1 because $x = 1$ and the cost of a drink is £0.50 because $y = 0.5$; this is the solution to the given simultaneous equations (1) and (2).

This is also the **point of intersection**, (1, 0.5), of the graphs in Fig. 1.1. The procedure outlined in Example 1.1 is called the method of **elimination**. The values $x = 1$ and $y = 0.5$ is the solution of equations (1) and (2). In general, values which satisfy the above linear system are called the **solution** or the **solution set** of the linear system. Here is another example.

Example 1.2

Solve

$$\begin{aligned} 9x + 3y &= 6 \qquad (1) \\ 2x - 7y &= 9 \qquad (2) \end{aligned}$$

Solution

We need to find the values of x and y which satisfy both equations.

How?

Taking one equation from the other doesn't help us here, but we can multiply through either or both equations by a non-zero constant.

If we multiply equation (1) by 2 and (2) by 9 then in both cases the x coefficient becomes 18. Carrying out this operation we have

$$\begin{aligned} 18x + 6y &= 12 \quad [\text{multiplying equation (1) by 2}] \\ 18x - 63y &= 81 \quad [\text{multiplying equation (2) by 9}] \end{aligned}$$

How do we eliminate x from these equations?

To eliminate the unknown x we subtract these equations:

$$\begin{array}{r} 18x + 6y = 12 \\ -(18x - 63y = 81) \\ \hline \end{array}$$

$$\begin{aligned} 0 + [6 - (-63)]\,y &= 12 - 81 \quad [\text{subtracting}] \\ 69y &= -69 \quad \text{which gives } y = -1 \end{aligned}$$

We have $y = -1$.

What else do we need to find?

The value of the placeholder x.

How?

By substituting $y = -1$ into the given equation $9x + 3y = 6$:

$$\begin{aligned} 9x + 3(-1) &= 6 \\ 9x - 3 &= 6 \\ 9x &= 9 \text{ which gives } x = 1 \end{aligned}$$

(continued...)

Hence our solution to the linear system of (1) and (2) is

$$x = 1 \text{ and } y = -1$$

We can check that this is the solution to the given system, (1) and (2), by substituting these values, $x = 1$ and $y = -1$, into the equations (1) and (2).

Note that we can carry out the following operations on a linear system of equations:

1. Interchange any pair of equations.
2. Multiply an equation by a non-zero constant.
3. Add or subtract one equation from another.

By carrying out these steps 1, 2 and 3 we end up with a simpler linear system to solve, but with the same solution set as the original linear system. In the above case we had

$$\begin{aligned} 9x + 3y &= 6 \\ 2x - 7y &= 9 \end{aligned} \quad \Rightarrow \quad \begin{aligned} 9x + 3y &= 6 \\ 69y &= -69 \end{aligned}$$

Of course, the system on the right hand side was much easier to solve. We can also use this method of elimination to solve three simultaneous linear equations with three unknowns, such as the one in the next example.

Example 1.3

Solve the linear system

$$\begin{aligned} x + 2y + 4z &= 7 && (1) \\ 3x + 7y + 2z &= -11 && (2) \\ 2x + 3y + 3z &= 1 && (3) \end{aligned}$$

Solution

What are we trying to find?

The values of x, y and z that satisfy all three equations (1), (2) and (3).

How do we find the values of x, y and z?

By elimination. To eliminate one of these unknowns, we first need to make the coefficients of x (or y or z) equal.

Which one?

There are three choices but we select so that the arithmetic is made easier, in this case it is x. Multiply equation (1) by 2 and then subtract the bottom equation (3):

$$\begin{aligned} 2x + 4y + 8z &= 14 && \text{[multiplying (1) by 2]} \\ -(2x + 3y + 3z &= 1) && (3) \\ \hline 0 + y + 5z &= 13 && \text{[subtracting]} \end{aligned}$$

Note that we have eliminated x and have the equation $y + 5z = 13$.
How can we determine the values of y and z from this equation?
We need another equation with only y and z.
How can we get this?
Multiply equation (1) by 3 and then subtract the second equation (2):

$$\begin{array}{rl} 3x + 6y + 12z = 21 & \text{[multiplying (1) by 3]} \\ -(3x + 7y + 2z = -11) & (2) \\ \hline 0 - y + 10z = 32 & \text{[subtracting]} \end{array}$$

Again there is no x and we have the equation $-y + 10z = 32$.
How can we find y and z?
We now solve the two simultaneous equations that we have obtained

$$\begin{aligned} y + 5z &= 13 \qquad (4) \\ -y + 10z &= 32 \qquad (5) \end{aligned}$$

We add equations (4) and (5), because $y + (-y) = 0$, which eliminates y.

$$0 + 15z = 45 \text{ gives } z = \frac{45}{15} = 3$$

Hence $z = 3$, but how do we find the other two unknowns x and y?
We first determine y by substituting $z = 3$ into equation (4) $y + 5z = 13$:

$$\begin{aligned} y + (5 \times 3) &= 13 \\ y + 15 &= 13 \qquad \text{which gives } y = -2 \end{aligned}$$

We have $y = -2$ and $z = 3$. We still need to find the value of last unknown x.
How do we find the value of x?
By substituting the values we have already found, $y = -2$ and $z = 3$, into the given equation $x + 2y + 4z = 7$ (1):

$$x + (2 \times -2) + (4 \times 3) = 7 \quad \text{gives} \quad x = -1$$

Hence the solution of the given three linear equations is $x = -1, y = -2$ and $z = 3$.
We can illustrate the given equations in a three-dimensional coordinate system as shown in Fig. 1.3.

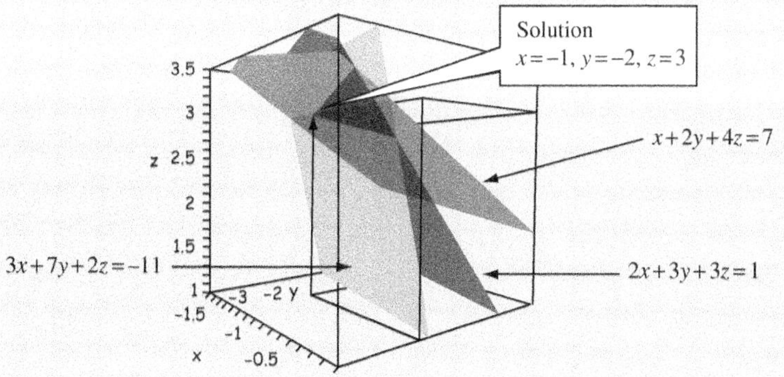

Figure 1.3

Each of the equations (1), (2) and (3) are represented by a plane in a three-dimensional system. The computer generated image above allows us to see where these planes lie with respect to each other. The coordinates of these planes is the solution of the system.

The aim of the above problem was to convert the given system into something simpler that could be solved. We had

$$\begin{aligned} x + 2y + 4z &= 7 \\ 3x + 7y + 2z &= -11 \\ 2x + 3y + 3z &= 1 \end{aligned} \quad \Rightarrow \quad \begin{aligned} x + 2y + 4z &= 7 \\ -y + 10z &= 32 \\ 15z &= 45 \end{aligned}$$

We will examine in detail m equations with n unknowns and develop a more efficient way of solving these later in this chapter.

1.1.3 Types of solutions

We now go back to evaluating a simple system of two linear simultaneous equations and discuss the case where we have no, or an infinite number of solutions. As stated earlier, one of the fundamental questions of linear algebra is how many solutions do we have of a given linear system.

Example 1.4

Solve the linear system

$$\begin{aligned} 2x + 3y &= 6 \quad &(1) \\ 4x + 6y &= 9 \quad &(2) \end{aligned}$$

Solution

How do we solve these equations?

Multiply (1) by 2 and then subtract equation (2):

$$\begin{aligned} 4x + 6y &= 12 \quad &[\text{Multiplying (1) by 2}] \\ -(4x + 6y &= 9) \quad &(2) \\ \hline 0 + 0 &= 3 \end{aligned}$$

But how can we have $0 = 3$*?*

A plot of the graphs of the given equations is shown in Fig. 1.4.

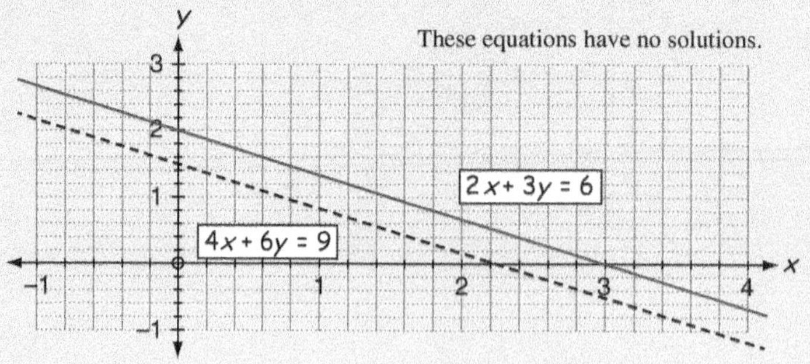

Figure 1.4

Can you see why there is no common solution to these equations?
The solution of the given equations would be the intersection of the lines shown in Fig. 1.4, but these lines are parallel so there is no intersection, therefore no solution.

By examining the given equations,

$$2x + 3y = 6 \qquad (1)$$
$$4x + 6y = 9 \qquad (2)$$

can you see why there is no solution?
If you multiply the first equation (1) by 2 we have

$$4x + 6y = 12$$

This is a contradiction.
Why?
Because we have

$$4x + 6y = 12$$
$$4x + 6y = 9 \qquad (2)$$

that is, $4x + 6y$ equals both 9 and 12. This is clearly impossible. Hence the given linear system has no solution.

A system that has no solution is called **inconsistent**. If the linear system has at least one solution then we say the system is **consistent**.

Can we have more than one solution?
Consider the following example.

Example 1.5

Graph the equations and determine the solution of this system:

$$2x + 3y = 6 \qquad (1)$$
$$4x + 6y = 12 \qquad (2)$$

Solution
The graph of the given equations is shown in Fig. 1.5.

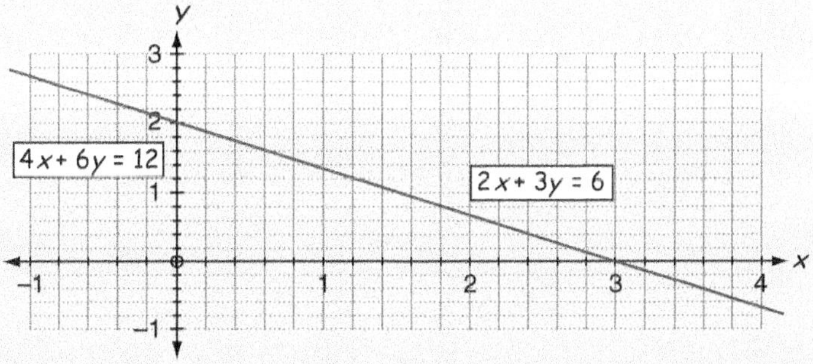

Figure 1.5

(continued...)

What do you notice?
Both the given equations produce exactly the same line; that is they coincide.
How many solutions do these equations have?
An infinite number of solutions, as you can see on the graph of Fig. 1.5. Any point on the line is a solution, and since there are an infinite number of points on the line we have an infinite number of solutions.
How can we write these solutions?
Let $x = a$ - where a is any real number - be a solution.
What then is y equal to?
Substituting $x = a$ into the given equation (1) yields

$$\begin{aligned} 2a + 3y &= 6 \qquad [2x + 3y = 6] \\ 3y &= 6 - 2a \\ y &= \frac{6-2a}{3} = \frac{6}{3} - \frac{2}{3}a = 2 - \frac{2}{3}a \end{aligned}$$

Hence if $x = a$ then $y = 2 - \frac{2}{3}a$.
The solution of the given linear system, (1) and (2), is $x = a$ and $y = 2 - 2a/3$ where a is any real number. You can check this by substituting various values of a. For example, if $a = 1$ then

$$x = 1, \quad y = 2 - 2(1)/3 = 4/3$$

We can check that this answer is correct by substituting these values, $x = 1$ and $y = 4/3$, into equations (1) and (2):

$$\begin{aligned} 2(1) + 3(4/3) &= 2 + 4 = 6 \\ 4(1) + 6(4/3) &= 4 + 8 = 12 \end{aligned} \qquad \begin{bmatrix} 2x + 3y = 6 & (1) \\ 4x + 6y = 12 & (2) \end{bmatrix}$$

Hence our solution works. This solution $x = a$ and $y = 2 - 2a/3$ will satisfy the given equations for any real value of a.

The graphs in Fig. 1.6 represent the three possible solutions to a linear system with two unknowns.

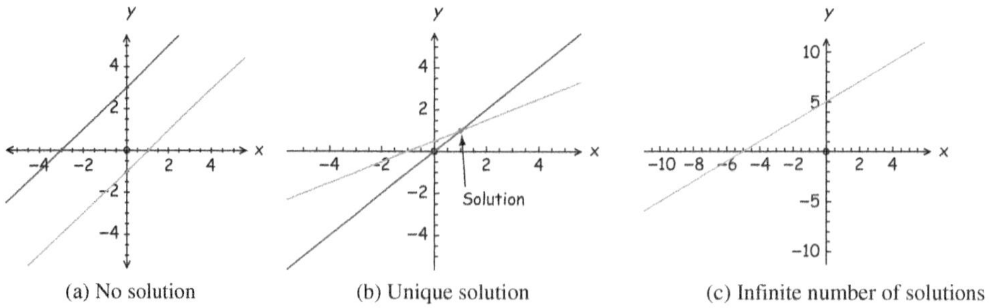
y
x
Solution
(a) No solution
(b) Unique solution
(c) Infinite number of solutions

Figure 1.6

The graphs in Fig. 1.7 illustrate solutions arising from three linear equations and three unknowns.

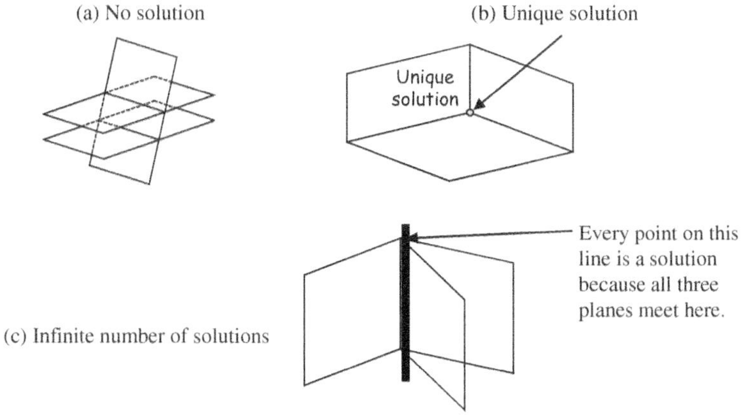

Figure 1.7

Fig. 1.7(a) shows three planes (equations) which have no point in common, hence no solution.

Fig. 1.7(b) shows the three planes (equations) with a unique point in common.

Fig. 1.7(c) shows three planes (equations) with a line in common. Every point on this line is a solution, which means we have an infinite number of solutions.

Summary

A linear equation is an equation in which the unknowns have an index of 1 or 0.
The procedure outlined in this section to solve such systems is the method of elimination.

EXERCISES 1.1

(Brief solutions at end of book. Full solutions available at <http://www.oup.co.uk/companion/singh>.)

1. Which of the following equations are linear equations in x, y and z? If they are not linear explain why not.

(a) $x - y - z = 3$ (b) $\sqrt{x} + y + z = 6$ (c) $\cos(x) + \sin(y) = 1$

(d) $e^{x+y+z} = 1$ (e) $x - 2y + 5z = \sqrt{3}$ (f) $x = -3y$

(g) $x = \dfrac{-b \pm \sqrt{b^2 - 4ac}}{2a}$ (h) $\pi x + y + ez = 5$ (i) $\sqrt{2}x + \frac{1}{2}y + z = 0$

(j) $\sinh^{-1}(x) = \ln\left|x + \sqrt{x^2 + 1}\right|$ (k) $\frac{\pi}{2}x - \sqrt{2}y + z\sin(\pi) = 0$

(l) $x^{2^0} + y^{3^0} + z^{3^0} = 0$ (m) $y^{\cos^2(x)+\sin^2(x)} + x - z = 9$

2. Solve the following linear system by the elimination process discussed in this section.

(a) $\begin{aligned} x + y &= 2 \\ x - y &= 0 \end{aligned}$ (b) $\begin{aligned} 2x - 3y &= 5 \\ x - y &= 2 \end{aligned}$ (c) $\begin{aligned} 2x - 3y &= 35 \\ x - y &= 2 \end{aligned}$

(d) $\begin{aligned} 5x - 7y &= 2 \\ 9x - 3y &= 6 \end{aligned}$ (e) $\begin{aligned} \pi x - 5y &= 2 \\ \pi x - y &= 1 \end{aligned}$ (f) $\begin{aligned} ex - ey &= 2 \\ ex + ey &= 0 \end{aligned}$

(In part (f), e is the irrational number $e = 2.71828182846\cdots$)

3. Solve the following linear systems by the elimination process discussed in this section.

(a) $\begin{aligned} x + y + z &= 3 \\ x - y - z &= -1 \\ 2x + y + 5z &= 8 \end{aligned}$ (b) $\begin{aligned} x + 2y - 2z &= 6 \\ 2x - 3y + z &= -10 \\ 3x - y + 3z &= -16 \end{aligned}$

(c) $\begin{aligned} 3x + y - 2z &= 4 \\ 5x - 3y + 10z &= 32 \\ 7x + 4y + 16z &= 13 \end{aligned}$ (d) $\begin{aligned} 6x - 3y + 2z &= 31 \\ 5x + y + 12z &= 36 \\ 8x + 5y + z &= 11 \end{aligned}$

4. Plot the graphs of these linear equations and decide on the number of solutions of each linear system. If there are any solutions, find them.

(a) $\begin{aligned} 2x + y &= 3 \\ x - y &= 7 \end{aligned}$ (b) $\begin{aligned} 2x + y &= 3 \\ 8x + 4y &= 12 \end{aligned}$ (c) $\begin{aligned} 2x + y &= 3 \\ 2x + y &= 5 \end{aligned}$

(d) $\begin{aligned} 3x - 2y &= 3 \\ 3x - 2y &= 6 \end{aligned}$ (e) $\begin{aligned} 3x - 2y - 3 &= 0 \\ 3x - 2y - 5 &= 0 \end{aligned}$

(f) $\begin{aligned} 5x - 2y - 5 &= 0 \\ 3x - 2y - 3 &= 0 \end{aligned}$

5. Without evaluating, decide on the number of solutions of each of these linear systems.

(a) $\begin{aligned} 7x + y &= 10 \\ x - y &= 7 \end{aligned}$ (b) $\begin{aligned} 12x + 4y &= 16 \\ 8x + 4y &= 16 \end{aligned}$

(c) $\begin{aligned} 2x - y - z &= 3 \\ 4x - 2y - 2z &= 3 \end{aligned}$

SECTION 1.2 Gaussian Elimination

By the end of this section you will be able to

- understand what is meant by a matrix and an augmented matrix
- solve a linear system using Gaussian elimination
- extend row operations so that the solution can be found by inspection

In the previous section we solved a linear system by a process of elimination and substitution. In this section we define this process in a systematic way. In order to do this we need to describe what is meant by a **matrix** and **reduced row echelon form**. Just by examining the matrix of a given linear system in reduced row echelon form we can say a lot of things about the solutions:

1. Are there any solutions?
2. Is the solution unique?
3. Are there an infinite number of solutions?

1.2.1 Introduction to matrices

In our introductory example from the previous section, we formed two equations that described the cost of ice creams x and drinks y:

$$\begin{aligned} 2x + 2y &= 3 \\ 2x + y &= 2.5 \end{aligned}$$

Note that the first column of values only contains coefficients of x, and the second column only contains coefficients of y. So we can write

$$\begin{matrix} x & y \end{matrix}$$
$$\begin{pmatrix} 2 & 2 \\ 2 & 1 \end{pmatrix}, \quad \begin{pmatrix} 3 \\ 2.5 \end{pmatrix}$$

The brackets on the left contain the coefficients from the problem, and is referred to as the **matrix** of coefficients. The brackets on the right hand side contain the total cost in a single column, and is referred to as a **vector**.

Matrices are used in various fields of engineering, science and economics to solve real-life problems such as those in control theory, electrical principles, structural analysis, string theory, quantitative finance and many others. Problems in these areas can often be written as a set of linear simultaneous equations. It is easier to solve these equations by using matrices.

Table 1.1 shows the sales of three different ice cream flavours during the week.

Table 1.1 Ice cream sales

	Monday	Tuesday	Wednesday	Thursday	Friday
Strawberry	12	15	10	16	12
Vanilla	5	9	14	7	10
Chocolate	8	12	10	9	15

We can represent the information in the table in matrix form as

$$\begin{pmatrix} 12 & 15 & 10 & 16 & 12 \\ 5 & 9 & 14 & 7 & 10 \\ 8 & 12 & 10 & 9 & 15 \end{pmatrix}$$

The first column represents the ice cream sales for Monday, second column for Tuesday, ... and the last column for Friday.

Matrices are an efficient way of storing data. In the next section we look formally at the methods used to extract information from the data.

1.2.2 Elementary row operations

Suppose we have a general linear system of m equations with n unknowns labelled $x_1, x_2, x_3, \ldots$ and x_n given by:

$$\begin{aligned} a_{11}x_1 + a_{12}x_2 + \cdots + a_{1n}x_n &= b_1 \\ a_{21}x_1 + a_{22}x_2 + \cdots + a_{2n}x_n &= b_2 \\ \vdots \qquad \vdots \qquad \qquad \vdots \quad &\quad \vdots \\ a_{m1}x_1 + a_{m2}x_2 + \cdots + a_{mn}x_n &= b_m \end{aligned}$$

where the coefficients a_{ij}, b_i are real numbers and $x_1, x_2, x_3, \ldots$ and x_n are placeholders for real numbers that satisfy the equations. This general system can be stored in matrix form as

$$\left(\begin{array}{ccc|c} a_{11} & \cdots & a_{1n} & b_1 \\ \vdots & \ddots & \vdots & \\ a_{m1} & \cdots & a_{mn} & b_m \end{array}\right)$$

This is an **augmented matrix**, which is a matrix containing the coefficients of the unknowns x_1, x_2, $x_3, \ldots, x_n$ and the constant values on the right hand side of the equations. In everyday English language, augmented means 'to increase'. Augmenting a matrix means adding one or more columns to the original matrix. In this case, we have added the b's column to the matrix. These are divided by a vertical line as shown above.

Example 1.6

Consider Example 1.3 section 1.1, where we had to solve the following linear system:

$$\begin{aligned} x + 2y + 4z &= 7 \\ 3x + 7y + 2z &= -11 \\ 2x + 3y + 3z &= 1 \end{aligned}$$

Write the augmented matrix of this linear system.

Solution

An augmented matrix is simply a shorthand way of representing a linear system of equations. Rather than write x, y and z after each coefficient, recognize that the first column contains the coefficients of x, the second column the coefficients of y and so on. Placing the coefficients of x, y and z on the left hand side of the vertical line in the augmented matrix and the constant values 7, -11 and 1 on the right hand side we have

$$\left(\begin{array}{ccc|c} 1 & 2 & 4 & 7 \\ 3 & 7 & 2 & -11 \\ 2 & 3 & 3 & 1 \end{array}\right)$$

In the previous section 1.1 we solved these equations by an elimination process which involved carrying out the following operations:

1. Multiply an equation by a non-zero constant.
2. Add or subtract a multiple of one equation to another.
3. Interchange equations.

Because each row of the augmented matrix corresponds to one equation of the linear system, we can carry out analogous operations such as:

1. Multiply a row by a non-zero constant.
2. Add or subtract a multiple of one row to another.
3. Interchange rows.

We refer to these operations as **elementary row operations**.

1.2.3 Gaussian elimination

Figure 1.8

Gauss (1777–1855) (Fig. 1.8) is widely regarded as one of the three greatest mathematicians of all time, the others being Archimedes and Newton. By the age of 11, Gauss could prove that $\sqrt{2}$ is irrational. At the age of 18, he constructed a regular 17-sided polygon with a compass and unmarked straight edge only. Gauss went as a student to the world-renowned centre for mathematics - Göttingen. Later in life, Gauss took up a post at Göttingen and published papers in number theory, infinite series, algebra, astronomy and optics. The unit of magnetic induction is named after Gauss.

A linear system of equations is solved by carrying out the above elementary row operations 1, 2 and 3 to find the values of the unknowns $x, y, z, w \ldots$ This method saves time because we do not need to write out the unknowns $x, y, z, w \ldots$ each time, and it is more methodical. In general, you will find there is less likelihood of making a mistake by using this Gaussian elimination process.

Example 1.7

Solve the following linear system by using the Gaussian elimination procedure:

$$\begin{aligned} x - 3y + 5z &= -9 \\ 2x - y - 3z &= 19 \\ 3x + y + 4z &= -13 \end{aligned}$$

Solution

What is the augmented matrix in this case?

Let R_1, R_2 and R_3 represent rows 1, 2 and 3 respectively. We have

$$\begin{aligned} x - 3y + 5z &= -9 \\ 2x - y - 3z &= 19 \\ 3x + y + 4z &= -13 \end{aligned} \quad \text{and} \quad \begin{matrix} \text{Row 1} \\ \text{Row 2} \\ \text{Row 3} \end{matrix} \quad \begin{matrix} R_1 \\ R_2 \\ R_3 \end{matrix} \left(\begin{array}{ccc|c} 1 & -3 & 5 & -9 \\ 2 & -1 & -3 & 19 \\ 3 & 1 & 4 & -13 \end{array} \right)$$

Note that each row represents an equation.

How can we find the unknowns x, y and z?

The columns in the matrix represent the x, y and z coefficients respectively. If we can transform this augmented matrix into

$$\begin{matrix} \\ \\ \\ \text{Final row} \end{matrix} \begin{matrix} x & y & z & \\ \downarrow & \downarrow & \downarrow & \\ \left(\begin{array}{ccc|c} * & * & * & * \\ 0 & * & * & * \\ 0 & 0 & A & B \end{array} \right) \end{matrix}$$

where A, B and $*$ represents any real number

then we can find z.

How?

Look at the final row.

What does this represent?

$(0 \times x) + (0 \times y) + (A \times z) = B$

$$Az = B \text{ which gives } z = \frac{B}{A} \text{ provided } A \neq 0$$

Hence we have a value for $z = B/A$.

But how do we find the other two unknowns x and y?

Now we can use a method called **back substitution**. Examine the second row of the above matrix:

$$\begin{matrix} & x \;\; y \;\; z \\ \text{Second row} & \left(\begin{array}{ccc|c} * & * & * & * \\ 0 & * & * & * \\ 0 & 0 & A & B \end{array} \right) \end{matrix}$$

By expanding the second row we get an equation in terms of y and z. From above we already know the value of $z = B/A$, so we can substitute $z = B/A$ and obtain y. Similarly from the first row we can find x by substituting the values of y and z.

We need to perform row operations on the augmented matrix to transform it from:

$$\begin{matrix} R_1 \\ R_2 \\ R_3 \end{matrix} \left(\begin{array}{ccc|c} 1 & -3 & 5 & -9 \\ 2 & -1 & -3 & 19 \\ 3 & 1 & 4 & -13 \end{array}\right) \text{ to } \left(\begin{array}{ccc|c} * & * & * & * \\ 0 & * & * & * \\ 0 & 0 & A & B \end{array}\right)$$

We need to convert this augmented matrix to an equivalent matrix with zeros in the bottom left hand corner. That is 0 in place of 2, 3 and 1.

How do we get 0 in place of 2?

Remember, we can multiply an equation by a non-zero constant, and take one equation away from another. In terms of matrices, this means that we can multiply a row and take one row away from another because each row represents an equation.

To get 0 in place of 2 we multiple row 1, R_1, by 2 and subtract the result from row 2, R_2; that is, we carry out the row operation $R_2 - 2R_1$:

$$\begin{matrix} R_1 \\ R_2^* = R_2 - 2R_1 \\ R_3 \end{matrix} \left(\begin{array}{ccc|c} 1 & -3 & 5 & -9 \\ 2-2(1) & -1-(2\times(-3)) & -3-(2\times 5) & 19-(2\times(-9)) \\ 3 & 1 & 4 & -13 \end{array}\right)$$

We call the new middle row R_2^*. Completing the arithmetic, the middle row becomes

$$\begin{matrix} R_1 \\ R_2^* \\ R_3 \end{matrix} \left(\begin{array}{ccc|c} 1 & -3 & 5 & -9 \\ 0 & 5 & -13 & 37 \\ 3 & 1 & 4 & -13 \end{array}\right)$$

Where else do we need a zero?

Need to get a 0 in place of 3 in the bottom row.

How?

We multiply the top row R_1 by 3 and subtract the result from the bottom row R_3; that is, we carry out the row operation, $R_3 - 3R_1$:

$$\begin{matrix} R_1 \\ R_2^* \\ R_3^* = R_3 - 3R_1 \end{matrix} \left(\begin{array}{ccc|c} 1 & -3 & 5 & -9 \\ 0 & 5 & -13 & 37 \\ 3-3(1) & 1-(3\times(-3)) & 4-(3\times 5) & -13-(3\times(-9)) \end{array}\right)$$

We can call R_3^* the new bottom row of this matrix. Simplifying the arithmetic in the entries gives:

$$\begin{matrix} R_1 \\ R_2^* \\ R_3^* \end{matrix} \left(\begin{array}{ccc|c} 1 & -3 & 5 & -9 \\ 0 & 5 & -13 & 37 \\ 0 & 10 & -11 & 14 \end{array}\right)$$

Note that we now only need to convert the 10 into zero in the bottom row.

How do we get a zero in place of 10?

We can only make use of the bottom two rows, R_2^* and R_3^*.

Why?

Looking at the first column, it is clear that taking any multiple of R_1 away from R_3 will interfere with the zero that we have just worked to establish.

(continued...)

We execute $R_3^* - 2R_2^*$ because

$$10 - (2 \times 5) = 0 \quad \text{(gives a 0 in place of 10)}$$

Therefore we have

$$\begin{array}{r} R_1 \\ R_2^* \\ R_3^{**} = R_3^* - 2R_2^* \end{array} \left(\begin{array}{ccc|c} 1 & -3 & 5 & -9 \\ 0 & 5 & -13 & 37 \\ 0 - (2 \times 0) & 10 - (2 \times 5) & -11 - [2 \times (-13)] & 14 - (2 \times 37) \end{array} \right)$$

which simplifies to

$$\begin{array}{r} \\ R_1 \\ R_2^* \\ R_3^{**} \end{array} \begin{array}{c} \begin{array}{ccc} x & y & z \end{array} \\ \left(\begin{array}{ccc|c} 1 & -3 & 5 & -9 \\ 0 & 5 & -13 & 37 \\ 0 & 0 & 15 & -60 \end{array} \right) \end{array} \quad (\dagger)$$

From the bottom row R_3^{**} we have

$$15z = -60 \text{ which gives } z = -\frac{60}{15} = -4$$

How do we find the other two unknowns x and y?
By expanding the middle row R_2^* of (†) we have:

$$5y - 13z = 37$$

We can find y by substituting $z = -4$ into this

$$5y - 13(-4) = 37 \qquad [\text{Substituting } z = -4]$$

$$5y + 52 = 37 \text{ which implies } 5y = -15 \text{ therefore } y = -\frac{15}{5} = -3$$

How can we find the last unknown x?
By expanding the first row R_1 of (†) we have:

$$x - 3y + 5z = -9$$

Substituting $y = -3$ and $z = -4$ into this:

$$x - (3 \times (-3)) + (5 \times (-4)) = -9$$

$$x + 9 - 20 = -9 \text{ which gives } x = 2$$

Hence our solution to the linear system is $x = 2, y = -3$ and $z = -4$.

Remember, each of the given equations can also be graphically represented by planes in a 3d coordinate system.

The solution is where all three planes (equations) meet. The equations are illustrated in Fig. 1.9.

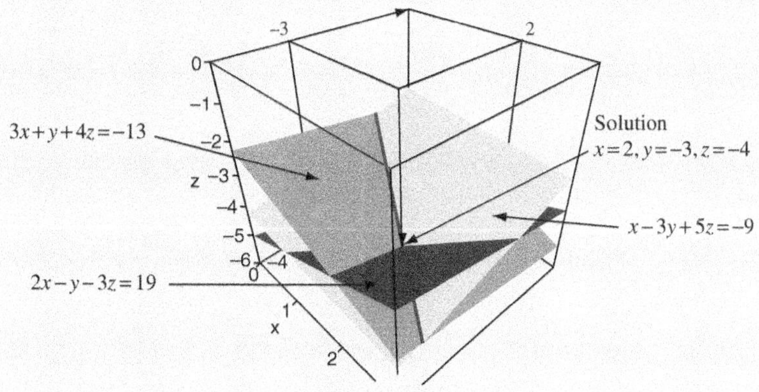

Figure 1.9

In the above example we transformed the given system:

$$\begin{aligned} x - 3y + 5z &= -9 \\ 2x - y - 3z &= 19 \\ 3x + y + 4z &= -13 \end{aligned} \quad \Rightarrow \quad \begin{aligned} x - 3y + 5z &= -9 \\ 5y - 13z &= 37 \\ 15z &= -60 \end{aligned}$$

The system on the right hand side is much easier to solve.

The above process is called **Gaussian elimination** with back substitution. The aim of Gaussian elimination is to produce a 'triangular' matrix with zeros in the bottom left corner of the matrix. This is achieved by the elementary row operations:

1. Multiply a row by a non-zero constant.
2. Add or subtract a multiple of one row from another.
3. Interchange rows.

We say two matrices are **row equivalent** if one matrix is derived from the other by using these three operations.

If augmented matrices of two linear systems are row equivalent then the two systems have the same solution set. You may like to check this for the above Example 1.7, to see that the solution $x = 2$, $y = -3$ and $z = -4$ satisfies the given equations.

In summary, the given linear system of equations is written in an augmented matrix, which is then transformed into a much simpler equivalent augmented matrix, which then allows us to use back substitution to find the solution of the linear system.

Example 1.8

Solve the linear system:

$$\begin{aligned} x + 3y + 2z &= 13 \\ 4x + 4y - 3z &= 3 \\ 5x + y + 2z &= 13 \end{aligned}$$

(continued...)

Solution

How can we find the unknowns x, y and z?

We use Gaussian elimination with back substitution.

The augmented matrix is:

$$\begin{array}{ll} \text{Row 1} & R_1 \\ \text{Row 2} & R_2 \\ \text{Row 3} & R_3 \end{array} \left(\begin{array}{ccc|c} 1 & 3 & 2 & 13 \\ 4 & 4 & -3 & 3 \\ 5 & 1 & 2 & 13 \end{array}\right)$$

Need to convert the entries in this triangle to zeros.

Our aim is to convert this augmented matrix so that there are 0's in the bottom left hand corner, that is; the first 4 in the second row reduces to zero, and the 5 and 1 from the bottom row reduce to zero. Hence $4 \to 0$, $5 \to 0$ and $1 \to 0$.

To get 0 in place of the first 4 in the middle row we multiply row 1, R_1, by 4 and take the result away from row 2, R_2, that is $R_2 - 4R_1$. To get 0 in place of 5 in the bottom row we multiply row 1, R_1, by 5 and take the result away from row 3, R_3, that is $R_3 - 5R_1$. Combining the two row operations, $R_2 - 4R_1$ and $R_3 - 5R_1$, we have

$$\begin{array}{r} R_1 \\ R_2^\dagger = R_2 - 4R_1 \\ R_3^\dagger = R_3 - 5R_1 \end{array} \left(\begin{array}{ccc|c} 1 & 3 & 2 & 13 \\ 4-(4\times 1) & 4-(4\times 3) & -3-(4\times 2) & 3-(4\times 13) \\ 5-(5\times 1) & 1-(5\times 3) & 2-(5\times 2) & 13-(5\times 13) \end{array}\right)$$

We call the new row 2 and 3 – $R_2^\dagger$ and $R_3^\dagger$ respectively. This simplifies to:

$$\begin{array}{r} R_1 \\ R_2^\dagger \\ R_3^\dagger \end{array} \left(\begin{array}{ccc|c} 1 & 3 & 2 & 13 \\ 0 & -8 & -11 & -49 \\ 0 & -14 & -8 & -52 \end{array}\right)$$

We have nearly obtained the required matrix with zeros in the bottom left hand corner. We need a 0 in place of -14 in the new bottom row, $R_3^\dagger$. We can only use the second and third rows, $R_2^\dagger$ and $R_3^\dagger$.

Why?

Because if we use first row, R_1, we will get a non-zero number in place of the zero already established in $R_3^\dagger$.

How can we obtain a 0 in place of −14?

$$R_3^\dagger - \frac{14}{8}R_2^\dagger \text{ because } -14 - \left[\frac{14}{8} \times (-8)\right] = 0 \text{ (gives 0 in place of } -14)$$

Therefore

$$\begin{array}{r} R_1 \\ R_2^\dagger \\ R_3^{\dagger\dagger} = R_3^\dagger - \frac{14}{8}R_2^\dagger \end{array} \left(\begin{array}{ccc|c} 1 & 3 & 2 & 13 \\ 0 & -8 & -11 & -49 \\ 0 & -14 - \left[\frac{14}{8} \times (-8)\right] & -8 - \left[\frac{14}{8} \times (-11)\right] & -52 - \left(\frac{14}{8} \times (-49)\right) \end{array}\right)$$

which simplifies to

$$\begin{array}{r} \\ R_1 \\ R_2^\dagger \\ R_3^{\dagger\dagger} \end{array} \begin{array}{c} \begin{array}{ccc} x & y & z \end{array} \\ \left(\begin{array}{ccc|c} 1 & 3 & 2 & 13 \\ 0 & -8 & -11 & -49 \\ 0 & 0 & 45/4 & 135/4 \end{array}\right) \end{array} \quad (*)$$

This time we have called the bottom row $R_3^{\dagger\dagger}$. From this row, $R_3^{\dagger\dagger}$, we have

$$\frac{45}{4}z = \frac{135}{4} \text{ which gives } z = \frac{135}{45} = 3$$

How can we find the unknown, y?
By expanding the second row $R_2^{\dagger}$ in (*) we have

$$-8y - 11z = -49$$

We know from above that $z = 3$, therefore, substituting $z = 3$ gives

$$\begin{aligned} -8y - 11(3) &= -49 \\ -8y - 33 &= -49 \\ -8y &= -49 + 33 = -16 \text{ which yields } y = (-16)/(-8) = 2 \end{aligned}$$

So far we have $y = 2$ and $z = 3$.
How can we find the last unknown, x?
By expanding the first row, R_1, of (*) we have

$$x + 3y + 2z = 13$$

Substituting our values already found, $y = 2$ and $z = 3$, we have

$$\begin{aligned} x + (3 \times 2) + (2 \times 3) &= 13 \\ x + 6 + 6 &= 13 \text{ which gives } x = 1 \end{aligned}$$

Hence $x = 1$, $y = 2$ and $z = 3$ is our solution. We can illustrate these equations as shown in Fig. 1.10.

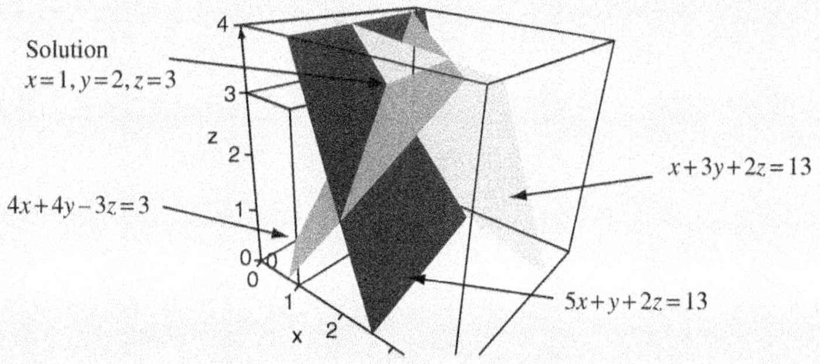

Figure 1.10

We can check that this solution is correct by substituting these $x = 1$, $y = 2$ and $z = 3$ into the given equations.

In the above example we carried out row operations so that:

$$\begin{aligned} x + 3y + 2z &= 13 \\ 4x + 4y - 3z &= 3 \\ 5x + y + 2z &= 13 \end{aligned} \quad \Rightarrow \quad \begin{aligned} x + 3y + 2z &= 13 \\ -8y - 11z &= -49 \\ 45z/4 &= 135/4 \end{aligned}$$

Note that the right hand system is much easier to solve.

1.2.4 Extending row operations

The Gaussian elimination process can be extended in the above example so that the first non-zero number in the bottom row of (*) is 1, that is

$$\begin{array}{c} R_1 \\ R_2^{\dagger} \\ R_3^{\dagger\dagger} \end{array} \left(\begin{array}{ccc|c} 1 & 3 & 2 & 13 \\ 0 & -8 & -11 & -49 \\ 0 & 0 & 45/4 & 135/4 \end{array}\right)$$

Convert this into 1

How do we convert 45/4 into 1?
Multiply the bottom row $R_3^{\dagger\dagger}$ by $\frac{4}{45}$.

$$\begin{array}{c} R_1 \\ R_2^{\dagger} \\ R_3' = \dfrac{4R_3^{\dagger\dagger}}{45} \end{array} \left(\begin{array}{ccc|c} 1 & 3 & 2 & 13 \\ 0 & -8 & -11 & -49 \\ 0 & 0 & \frac{4}{45}\left(\frac{45}{4}\right) & \frac{4}{45}\left(\frac{135}{4}\right) \end{array}\right)$$

which simplifies to

$$\begin{array}{c} R_1 \\ R_2^{\dagger} \\ R_3' \end{array} \left(\begin{array}{ccc|c} 1 & 3 & 2 & 13 \\ 0 & -8 & -11 & -49 \\ 0 & 0 & 1 & 3 \end{array}\right)$$

The advantage of this is that we get the z value directly. From the bottom row, R_3', we have $z = 3$. We can extend these row operations further and obtain the following matrix:

$$\left(\begin{array}{ccc|c} 1 & 0 & 0 & * \\ 0 & 1 & 0 & * \\ 0 & 0 & 1 & * \end{array}\right) \qquad (*)$$

Why would we want to achieve this sort of augmented matrix?
Because we can read off the x, y and z values directly from this augmented matrix. The only problem is in doing the arithmetic, because achieving this sort of matrix can be a laborious process.

This augmented matrix (*) is said to be in reduced row echelon form.

A matrix is in **reduced row echelon form**, normally abbreviated to **rref**, if it satisfies all the following conditions:

1. If there are any rows containing only zero entries then they are located in the bottom part of the matrix.
2. If a row contains non-zero entries then the first non-zero entry is a 1. This 1 is called a **leading 1**.
3. The leading 1's of two consecutive non-zero rows go strictly from top left to bottom right of the matrix.
4. The only non-zero entry in a column containing a leading 1 is the leading 1.

If condition (4) is *not* satisfied then we say that the matrix is in **row echelon form** and drop the qualification 'reduced'. In some linear algebra literature the leading 1 condition is relaxed and it is enough to say that any non-zero number is the **leading coefficient**.

For example, the following are all in reduced row echelon form:

$$\begin{pmatrix} 0 & \boxed{1} & 0 & 8 & 0 \\ 0 & 0 & \boxed{1} & 0 & 4 \\ 0 & 0 & 0 & 0 & 0 \\ 0 & 0 & 0 & 0 & 0 \end{pmatrix}, \quad \begin{pmatrix} \boxed{1} & 0 & 0 & 5 \\ 0 & \boxed{1} & 0 & -6 \\ 0 & 0 & \boxed{1} & 9 \end{pmatrix} \text{ and } \begin{pmatrix} 0 & \boxed{1} & 3 & 0 & 4 & 0 & 6 \\ 0 & 0 & 0 & \boxed{1} & 5 & 0 & -1 \\ 0 & 0 & 0 & 0 & 0 & \boxed{1} & 9 \\ 0 & 0 & 0 & 0 & 0 & 0 & 0 \end{pmatrix}$$

The following matrices are *not* in reduced row echelon form:

$$\mathbf{A} = \begin{pmatrix} 0 & \boxed{1} & 5 & 0 & 0 \\ 0 & 0 & \boxed{1} & 0 & 0 \\ 0 & 0 & 0 & \boxed{1} & 0 \\ 0 & 0 & 0 & 0 & \boxed{1} \end{pmatrix}, \mathbf{B} = \begin{pmatrix} \boxed{1} & 0 & 0 \\ 0 & \boxed{1} & 0 \\ \boxed{1} & 0 & 0 \end{pmatrix} \text{ and } \mathbf{C} = \begin{pmatrix} 0 & 0 & 0 & 0 & 0 & 0 & 0 \\ 0 & 0 & 0 & \boxed{1} & 5 & 8 & -1 \\ 0 & 0 & 0 & 0 & 0 & \boxed{1} & 9 \\ 0 & 0 & 0 & 0 & 0 & 0 & 0 \end{pmatrix}$$

Why not?

In matrix **A** the third column contains a leading one but has a non-zero entry, 5.

In matrix **B** the leading ones do not go from top left to bottom right.

In matrix **C** the top row of zeros should be relegated to the bottom of the matrix as stated in condition (1) above.

However, matrix **A** is in row echelon form but not in reduced row echelon form. Matrices **B** and **C** are not in row echelon form.

The procedure which places an augmented matrix into row echelon form is called Gaussian elimination and the algorithm which places an augmented matrix into a reduced row echelon form is called **Gauss–Jordan** elimination.

Example 1.9

Place the augmented matrix
$$\begin{array}{ccc} x & y & z \end{array}$$
$$\left(\begin{array}{ccc|c} 1 & 5 & -3 & -9 \\ 0 & -13 & 5 & 37 \\ 0 & 0 & 5 & -15 \end{array} \right)$$
into reduced row echelon form.

Solution

Why should we want to place this matrix into reduced row echelon form?
In a nutshell, it's to avoid back substitution. If we look at the bottom row of the given augmented matrix we have $5z = -15$.

We need to divide by 5 in order to find the z value.

The reduced row echelon form, rref, gives us the values of the unknowns directly, and we do not need to carry out further manipulation or elimination.

What does reduced row echelon form mean in this case?
It means convert the given augmented matrix

$$\begin{array}{c} R_1 \\ R_2 \\ R_3 \end{array} \left(\begin{array}{ccc|c} 1 & 5 & -3 & -9 \\ 0 & -13 & 5 & 37 \\ 0 & 0 & 5 & -15 \end{array} \right) \text{ into something like } \left(\begin{array}{ccc|c} 1 & 0 & 0 & * \\ 0 & 1 & 0 & * \\ 0 & 0 & 1 & * \end{array} \right)$$

This means that we need to get 0 in place of the 5 in the second row, and 0's in place of the 5 and -3 in the first row. We also need a 1 in place of -13 in the middle row and 1 in place of the 5 in the bottom row.

How do we convert the 5 in the bottom row into 1?
Divide the last row by 5 (remember, this is the same as multiplying by $1/5$):

$$\begin{array}{c} R_1 \\ R_2 \\ R_3' = R_3/5 \end{array} \left(\begin{array}{ccc|c} 1 & 5 & -3 & -9 \\ 0 & -13 & 5 & 37 \\ 0 & 0 & 1 & -3 \end{array} \right)$$

How do we get 0 in place of -3 in the first row and the 5 in the second row?
We execute the row operations $R_1 + 3R_3'$ and $R_2 - 5R_3'$:

$$\begin{array}{c} R_1^* = R_1 + 3R_3' \\ R_2^* = R_2 - 5R_3' \\ R_3' \end{array} \left(\begin{array}{ccc|c} 1 + 3\,(0) & 5 + 3\,(0) & -3 + 3\,(1) & -9 + 3\,(-3) \\ 0 - 5\,(0) & -13 - 5\,(0) & 5 - 5\,(1) & 37 - 5\,(-3) \\ 0 & 0 & 1 & -3 \end{array} \right)$$

Simplifying the entries gives

$$\begin{array}{c} R_1^* \\ R_2^* \\ R_3' \end{array} \left(\begin{array}{ccc|c} 1 & 5 & 0 & -18 \\ 0 & -13 & 0 & 52 \\ 0 & 0 & 1 & -3 \end{array} \right)$$

How do we get a 1 in place of -13?
Divide the middle row by -13 (remember, this is the same as multiplying by $-1/13$):

$$\begin{matrix} R_1^* \\ R_2^{**} = R_2^*/(-13) \\ R_3' \end{matrix} \left(\begin{array}{ccc|c} 1 & 5 & 0 & -18 \\ 0 & -13/(-13) & 0 & 52/(-13) \\ 0 & 0 & 1 & -3 \end{array} \right)$$

Simplifying the second row gives

$$\begin{matrix} R_1^* \\ R_2^{**} \\ R_3' \end{matrix} \left(\begin{array}{ccc|c} 1 & 5 & 0 & -18 \\ 0 & 1 & 0 & -4 \\ 0 & 0 & 1 & -3 \end{array} \right)$$

This matrix is now in row echelon form but not in reduced row echelon form. We need to convert the 5 in the top row into 0 to get it into reduced row echelon form.

How?

We carry out the row operation, $R_1^* - 5R_2^{**}$:

$$\begin{matrix} R_1^{**} = R_1^* - 5R_2^{**} \\ R_2^{**} \\ R_3' \end{matrix} \left(\begin{array}{ccc|c} 1-5(0) & 5-5(1) & 0-5(0) & -18-5(-4) \\ 0 & 1 & 0 & -4 \\ 0 & 0 & 1 & -3 \end{array} \right)$$

Simplifying the top row entries gives

$$\begin{matrix} \\ R_1^{**} \\ R_2^{**} \\ R_3' \end{matrix} \begin{matrix} \begin{matrix} x & y & z & \end{matrix} \\ \left(\begin{array}{ccc|c} 1 & 0 & 0 & 2 \\ 0 & 1 & 0 & -4 \\ 0 & 0 & 1 & -3 \end{array} \right) \end{matrix}$$

Hence we have placed the given augmented matrix into reduced row echelon form.

Now we can read off the x, y and z values, that is $x = 2$, $y = -4$ and $z = -3$.

We will prove later on that the matrix in reduced row echelon form is unique. This means that however we vary our row operations we will always end up with the same matrix in reduced row echelon form. However, the matrix in row echelon form is *not* unique.

Summary

To find the solution to a linear system of m equations by n unknowns we aim to produce:

$$\begin{array}{ccccccc} a_{11}x_1 & + & \cdots & + & a_{1n}x_n & = & b_1 \\ a_{21}x_1 & + & \cdots & + & a_{2n}x_n & = & b_2 \\ \vdots & & & & \vdots & = & \vdots \\ a_{m1}x_1 & + & \cdots & + & a_{mn}x_n & = & b_m \end{array}$$

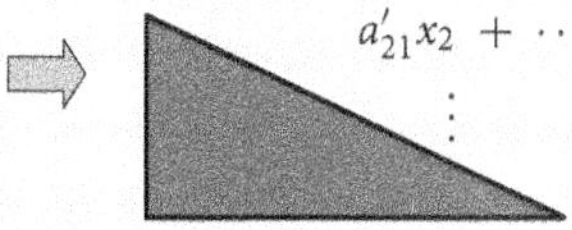

$$\begin{array}{rcl} a_{11}x_1 + a_{12}x_2 + \cdots + a_{1n}x_n & = & b_1 \\ a'_{21}x_2 + \cdots + a'_{2n}x_n & = & b'_2 \\ \vdots \qquad\qquad \vdots & & \vdots \\ a'_{mn}x_n & = & b'_m \end{array}$$

EXERCISES 1.2

(Brief solutions at end of book. Full solutions available at <http://www.oup.co.uk/companion/singh>.)

1. Solve the following linear system by applying Gaussian elimination with back substitution:

(a) $\begin{aligned} x + y &= 7 \\ x - 2y &= 4 \end{aligned}$

(b) $\begin{aligned} x + 2y - 3z &= 3 \\ 2x - y - z &= 11 \\ 3x + 2y + z &= -5 \end{aligned}$

(c) $\begin{aligned} 2x + 2y + z &= 10 \\ x - 3y + 4z &= 0 \\ 3x - y + 6z &= 12 \end{aligned}$

(d) $\begin{aligned} x + 2y + z &= 1 \\ 2x + 2y + 3z &= 2 \\ 5x + 8y + 2z &= 4 \end{aligned}$

(e) $\begin{aligned} 10x + y - 5z &= 18 \\ -20x + 3y + 20z &= 14 \\ 5x + 3y + 5z &= 9 \end{aligned}$

2. Solve the following linear system by placing the augmented matrix in row echelon form.

(a) $\begin{aligned} x + 2y + 3z &= 12 \\ 2x - y + 5z &= 3 \\ 3x + 3y + 6z &= 21 \end{aligned}$

(b) $\begin{aligned} 2x - y - 4z &= 0 \\ 3x + 5y + 2z &= 5 \\ 4x - 3y + 6z &= -16 \end{aligned}$

(c) $\begin{aligned} 3x - y + 7z &= 9 \\ 5x + 3y + 2z &= 10 \\ 9x + 2y - 5z &= 6 \end{aligned}$

3. Solve the following linear system by placing the augmented matrix in reduced row echelon form.

(a) $\begin{aligned} x + y + 2z &= 9 \\ 4x + 4y - 3z &= 3 \\ 5x + y + 2z &= 13 \end{aligned}$

(b) $\begin{aligned} x + y + z &= -2 \\ 2x - y - z &= -4 \\ 4x + 2y - 3z &= -3 \end{aligned}$

(c) $\begin{aligned} 2x + y - z &= 2 \\ 4x + 3y + 2z &= -3 \\ 6x - 5y + 3z &= -14 \end{aligned}$

(d) $\begin{aligned} -2x + 3y - 2z &= 8 \\ -x + 2y - 10z &= 0 \\ 5x - 7y + 4z &= -20 \end{aligned}$

SECTION 1.3 Vector Arithmetic

By the end of this section you will be able to

- understand what is meant by vectors and scalars
- use vector arithmetic such as addition, scalar multiplication and dot product

As shown in the previous section, matrix notation provides a systematic way of both analysing and solving linear systems. In section 1.2 we came across a single column matrix that we referred to as a vector. Vectors, being the simplest of matrices, are also the most frequently used.

However, to understand the 'how and why' of matrices we need to grasp the underlying concept of vectors. Vectors are the gateway to comprehending matrices. By analysing the properties of vectors we shall come to a fuller understanding of matrices in general.

Well, what are vectors?

In many situations answering the question 'how much?' is enough. Sometimes, one needs to know not only 'how much', but also 'in which direction?' Vectors are the natural type of thing to answer such questions with, for they are capable of expressing geometry, not just 'size'.

For instance, physicists rely upon vectors to mathematically express the motion of an object in terms of its direction, as well as the rate that it is travelling. Engineers will use a vector to express the magnitude of a force and the direction in which it is acting. Each additional component to a problem simply requires an additional entry in the vector that describes it. Using two- and three-dimensional vectors to express physical problems also allows for a geometric interpretation, so data can be plotted and visually compared.

In more abstract problems, *many* more dimensions might be employed, meaning that the resultant vector is impossible to plot in more than three-dimensions but astonishingly the mathematics works in this space too, and the resulting solutions are no less valid.

First we need to formalize some of the mathematics so that we can work with vectors.

1.3.1 Vectors and Scalars

The physical interpretation of a vector is a quantity that has size (magnitude) and direction. The instruction '*walk due north for 5 kilometres*' can be expressed as a vector; its magnitude is 5 km and its direction is due north.

Velocity, acceleration, force and displacement are all vector quantities.

The instruction '*Go on a 5 km walk*' is not a vector because it has no direction; all that is specified is the *length* of the walk, but we don't know where to start or where to head.

We shall now start referring to this as a scalar.

So what are scalars?

A scalar is a number that measures the size of a particular quantity.

Length, area, volume, mass and temperature are all scalar quantities.

How do we write down vectors and scalars and how can we distinguish between them?

A vector from O to A is denoted by $\overrightarrow{OA}$, or written in bold typeface **a** and can be represented geometrically as shown in Fig. 1.11.

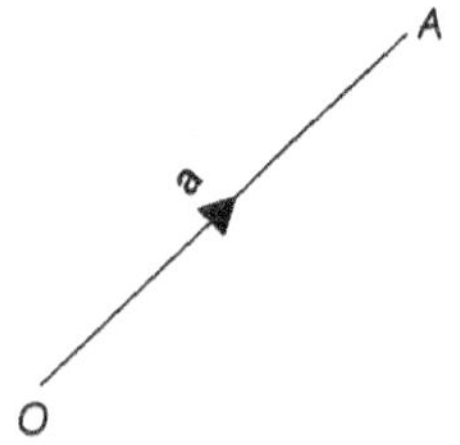

Figure 1.11

A scalar is denoted by a, not in bold, so that we can distinguish between vectors and scalars.

Two vectors are equivalent if they have the same direction and magnitude. For example, the vectors **d** and **e** in Fig. 1.12 are equivalent, that is $\mathbf{d} = \mathbf{e}$.

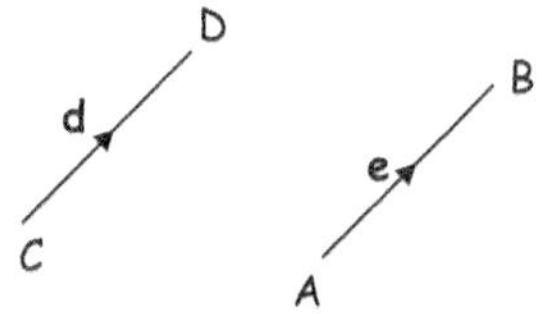

Figure 1.12

The vectors **d** and **e** have the same direction and magnitude but only differ in position. There are many examples of vectors in the real world:

(a) A displacement of 20 m to the horizontal, right of an object from O to A (Fig. 1.13).

Figure 1.13

(b) A force on an object acting vertically downwards (Fig. 1.14).

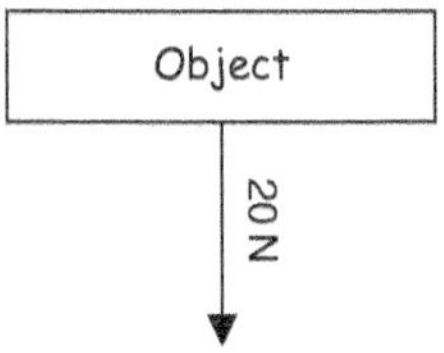

Figure 1.14

(c) The velocity and acceleration of a ball thrown vertically upwards are vectors (Fig. 1.15).

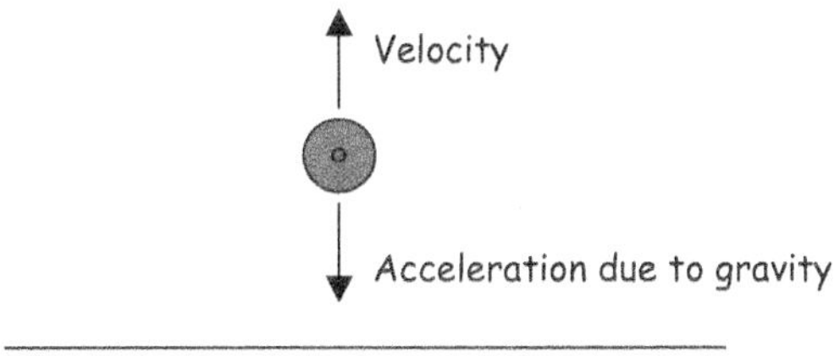

Figure 1.15

1.3.2 Vector addition and scalar multiplication

The result of adding two vectors such as **a** and **b** in Fig. 1.16 is the diagonal of the parallelogram, **a** + **b**, as shown.

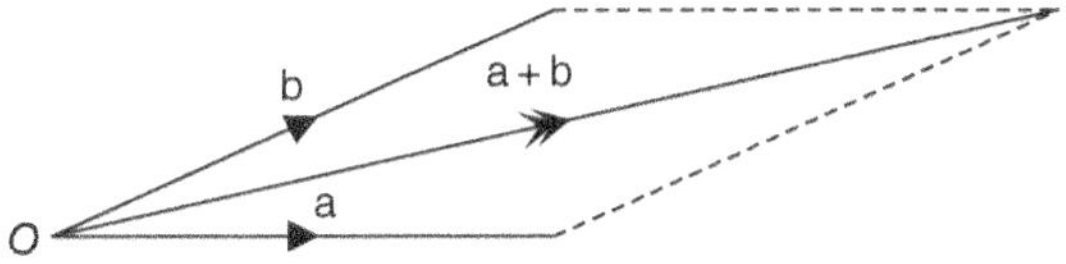

Figure 1.16

The multiplication $k\mathbf{a}$ of a real number k with a vector **a** is the product of the size of **a** with the number k. For example, 2**a** is the vector in the same direction as vector **a** but the magnitude is twice as long (Fig. 1.17).

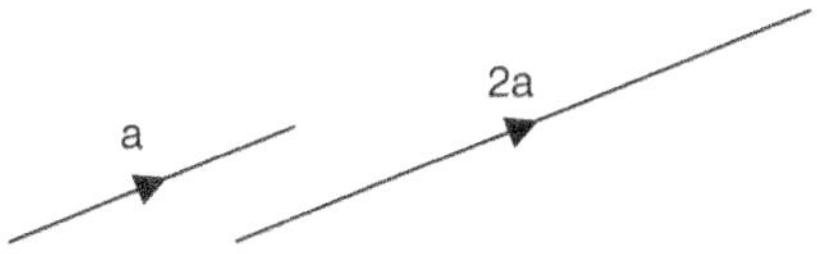

Figure 1.17

What does the vector $\frac{1}{2}\mathbf{a}$ look like?

Figure 1.18

It's the same direction as vector **a** but half the magnitude (Fig. 1.18).

*What effect does a negative k have on a vector such as k**a**?*

If $k = -2$ then $-2\mathbf{a}$ is the vector **a** but in the opposite direction and the magnitude is multiplied by 2 (Fig. 1.19):

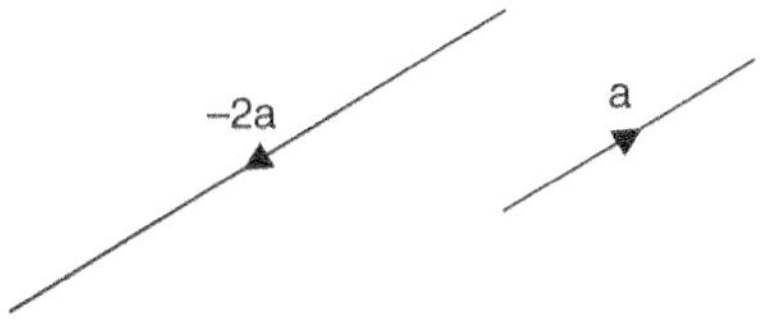

Figure 1.19

A vector $-\mathbf{a}$ is the vector $\mathbf{a}$ but in the opposite direction. We can define this as

$$-\mathbf{a} = (-1)\,\mathbf{a}$$

We call the product $k\mathbf{a}$ scalar multiplication.
We can also subtract vectors as shown in Fig. 1.20.

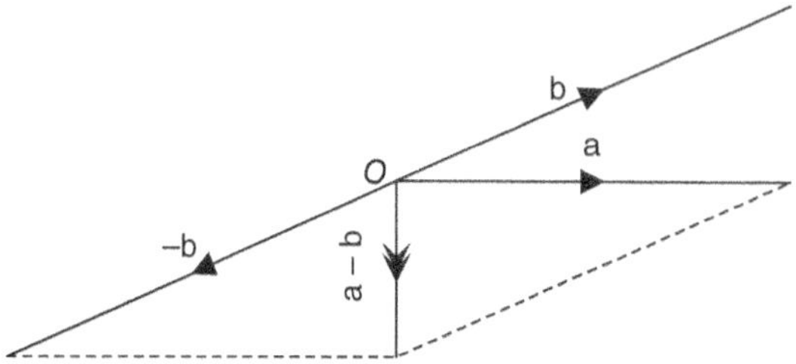

Figure 1.20

The vector subtraction of two vectors $\mathbf{a}$ and $\mathbf{b}$ is defined by

$$\mathbf{a} - \mathbf{b} = \mathbf{a} + (-\mathbf{b}) \tag{1.1}$$

1.3.3 Vectors in $\mathbb{R}^2$

What is meant by $\mathbb{R}^2$?
$\mathbb{R}^2$ is the $x - y$ plane representing the Cartesian coordinate system named after the French mathematician and philosopher Rene Descartes.

Figure 1.21 Rene Descartes 1596–1650

Rene Descartes (Fig. 1.21) was a French philosopher born in 1596. He attended a Jesuit college and, because of his poor health, he was allowed to remain in bed until 11 o'clock in the morning, a habit he continued until his death in 1650.

After graduating in 1618, he went to Holland to study mathematics. Here Descartes lived a solitary life, concentrating only on mathematics and philosophy. His main contribution to mathematics was analytical geometry, which includes our present x-y plane and the three-dimensional space with x, y and z axes.

This coordinate system cemented algebra with geometry. Prior to the coordinate system, geometry and algebra were two different subjects.

In 1649, Descartes left Holland to tutor Queen Christina of Sweden. However, she wanted to study mathematics at dawn, which did not suit Descartes who never rose before 11 am. As a combination of the early starts and the harsh Swedish winter, Descartes died of pneumonia in 1650.

The points in the Cartesian plane are ordered pairs with reference to the origin, which is denoted by O.

What does the term 'ordered pair' mean?

The order of the entries matters, that is the coordinate (a, b) is different from (b, a), provided $a \neq b$.

The coordinate (a, b) can be written as a column $\begin{pmatrix} a \\ b \end{pmatrix}$ and is called a **column vector** or simply a **vector**. For example, the following are all vectors in the plane $\mathbb{R}^2$ (Fig. 1.22):

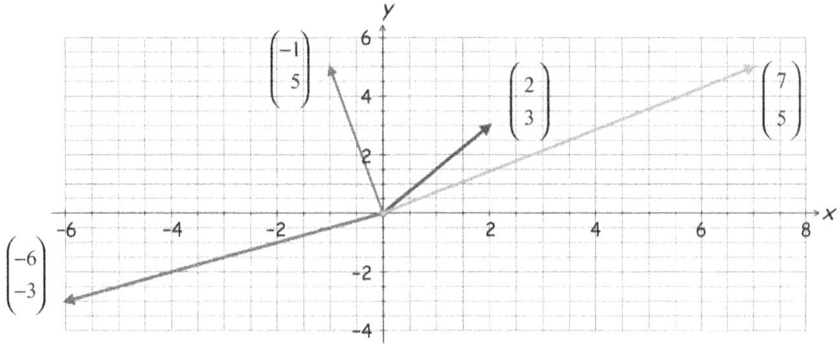

Figure 1.22

These are examples of vectors with two entries, $\begin{pmatrix} -6 \\ -3 \end{pmatrix}, \begin{pmatrix} 7 \\ 5 \end{pmatrix}, \begin{pmatrix} 2 \\ 3 \end{pmatrix}$ and $\begin{pmatrix} -1 \\ 5 \end{pmatrix}$.

The set of all vectors with two entries is denoted by $\mathbb{R}^2$ and pronounced 'r two'. The $\mathbb{R}$ indicates that the entries are real numbers.

We can add and subtract vectors in $\mathbb{R}^2$ as stated above, that is we apply the parallelogram law on the vectors (Fig. 1.23).

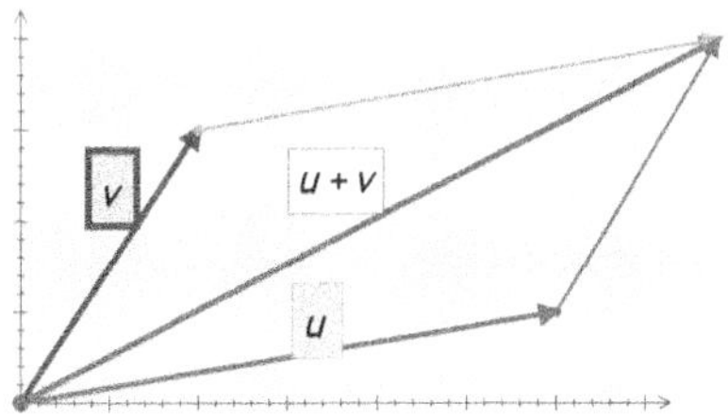

Figure 1.23

Example 1.10

Let $\mathbf{u} = \begin{pmatrix} 3 \\ -1 \end{pmatrix}$ and $\mathbf{v} = \begin{pmatrix} -2 \\ 3 \end{pmatrix}$. Plot $\mathbf{u} + \mathbf{v}$ and write down $\mathbf{u} + \mathbf{v}$ as a column vector.

Solution

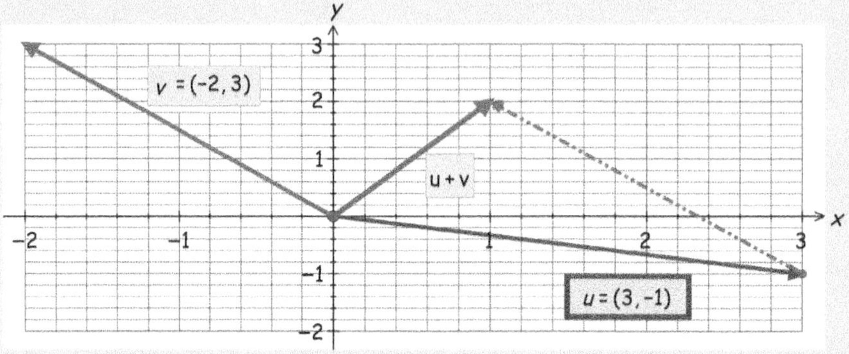

Figure 1.24

By examining Fig. 1.24 we see that the coordinates of $\mathbf{u} + \mathbf{v}$ are $(1, 2)$ and this is written as a column vector $\begin{pmatrix} 1 \\ 2 \end{pmatrix}$.

What do you notice about your result?

If we add x and y coordinates in the vectors separately then we obtain the resultant vector.

That is, we evaluate $\mathbf{u} + \mathbf{v} = \begin{pmatrix} 3 \\ -1 \end{pmatrix} + \begin{pmatrix} -2 \\ 3 \end{pmatrix} = \begin{pmatrix} 3-2 \\ -1+3 \end{pmatrix} = \begin{pmatrix} 1 \\ 2 \end{pmatrix}$, which means that we can simply add the corresponding entries of the vector to find $\mathbf{u} + \mathbf{v}$.

In general, if $\mathbf{u} = \begin{pmatrix} a \\ b \end{pmatrix}$ and $\mathbf{v} = \begin{pmatrix} c \\ d \end{pmatrix}$ then we add corresponding entries:

$$\mathbf{u} + \mathbf{v} = \begin{pmatrix} a \\ b \end{pmatrix} + \begin{pmatrix} c \\ d \end{pmatrix} = \begin{pmatrix} a+c \\ b+d \end{pmatrix}$$

Example 1.11

Let $\mathbf{v} = \begin{pmatrix} 3 \\ 1 \end{pmatrix}$. Plot the vectors $\frac{1}{2}\mathbf{v}$, $2\mathbf{v}$, $3\mathbf{v}$ and $-\mathbf{v}$ on the same axes.

Solution

Plotting each of these vectors on $\mathbb{R}^2$ is shown in Fig. 1.25.

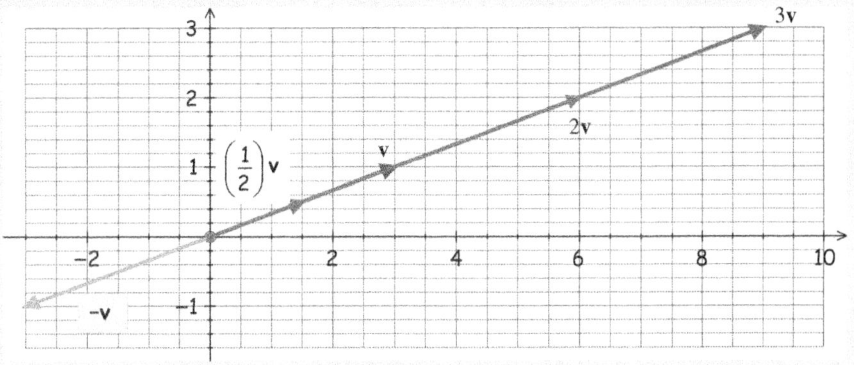

Figure 1.25

Note that by reading off the coordinates of each vector:

$$\frac{1}{2}\mathbf{v} = \frac{1}{2}\begin{pmatrix}3\\1\end{pmatrix} = \begin{pmatrix}1.5\\0.5\end{pmatrix},\ 2\mathbf{v} = 2\begin{pmatrix}3\\1\end{pmatrix} = \begin{pmatrix}6\\2\end{pmatrix},\ 3\mathbf{v} = 3\begin{pmatrix}3\\1\end{pmatrix} = \begin{pmatrix}9\\3\end{pmatrix} \text{ and}$$

$$-\mathbf{v} = -\begin{pmatrix}3\\1\end{pmatrix} = \begin{pmatrix}-3\\-1\end{pmatrix}$$

Remember, the product $k\mathbf{v}$ is called scalar multiplication. The term **scalar** comes from the Latin word **scala** meaning ladder. Scalar multiplication changes the length or the scale of the vector as you can see in Fig. 1.25.

In general, if $\mathbf{v} = \begin{pmatrix}a\\b\end{pmatrix}$ then the scalar multiplication

$$k\mathbf{v} = k\begin{pmatrix}a\\b\end{pmatrix} = \begin{pmatrix}ka\\kb\end{pmatrix}$$

1.3.4 Vectors in $\mathbb{R}^3$

What does the notation $\mathbb{R}^3$ mean?

$\mathbb{R}^3$ is the set of all ordered triples of real numbers and is also called 3-space.

We can extend the vector properties in $\mathbb{R}^2$ mentioned above to three dimensions $\mathbb{R}^3$ pronounced 'r three'.

The $x - y$ plane can be extended to cover three dimensions by including a third axis called the z axis. This axis is at right angles to the other two, x and y, axes. The position of a vector in three dimensions is given by three coordinates (x, y, z).

For example, the following vector $\begin{pmatrix}1\\2\\5\end{pmatrix}$ in $\mathbb{R}^3$ is represented geometrically in Fig. 1.26:

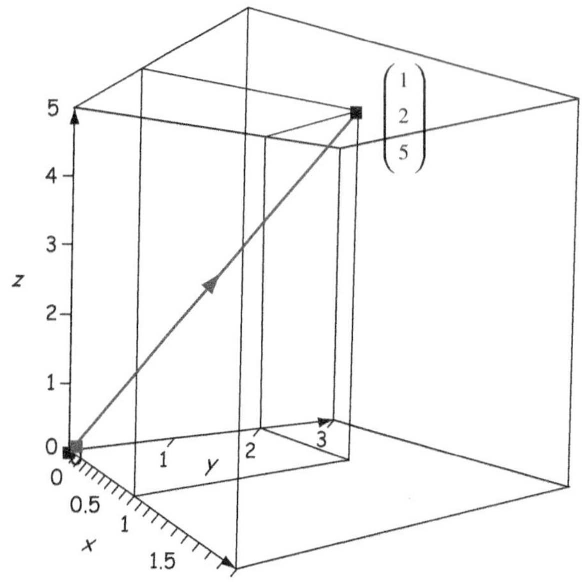

Figure 1.26

Vector addition and scalar multiplication are carried out the same way as in the plane $\mathbb{R}^2$. That is, if $\mathbf{u} = \begin{pmatrix} a \\ b \\ c \end{pmatrix}$ and $\mathbf{v} = \begin{pmatrix} d \\ e \\ f \end{pmatrix}$ then vector addition and scalar multiplication are defined as

$$\mathbf{u} + \mathbf{v} = \begin{pmatrix} a \\ b \\ c \end{pmatrix} + \begin{pmatrix} d \\ e \\ f \end{pmatrix} = \begin{pmatrix} a+d \\ b+e \\ c+f \end{pmatrix}, \quad k\mathbf{u} = k\begin{pmatrix} a \\ b \\ c \end{pmatrix} = \begin{pmatrix} ka \\ kb \\ kc \end{pmatrix}$$

1.3.5 Vectors in $\mathbb{R}^n$

What does $\mathbb{R}^n$ represent?

In the 17th century Rene Descartes used ordered pairs of real numbers, $\mathbf{v} = \begin{pmatrix} a \\ b \end{pmatrix}$, to describe vectors in a plane and extended it to ordered triples of real numbers, $\mathbf{v} = \begin{pmatrix} a \\ b \\ c \end{pmatrix}$, to describe vectors in three-dimensional space.

Can we extend this to an ordered quadruple of real numbers, $\mathbf{v} = \begin{pmatrix} a \\ b \\ c \\ d \end{pmatrix}$, or n-tuples of real numbers, $\mathbf{v} = \begin{pmatrix} v_1 \\ v_2 \\ \vdots \\ v_n \end{pmatrix}$?

Yes. In the 17th century vectors were defined as geometric objects and there was no geometric interpretation of $\mathbb{R}^n$ for n greater than three dimensions. However, in the 19th century vectors began to be seen as mathematical objects that can be added, subtracted and scalar multiplied that allowed us to extend the vector definition.

There are many real life situations where more than three variables are involved. For example

1. weather conditions
2. performance of companies in a stock market
3. mortality rates in a population

In 1905 the great German mathematician Hilbert (1862–1943) produced his famous theory of a vector space of infinitely many variables known as Hilbert space.

A vector $\mathbf{v} = \begin{pmatrix} v_1 \\ v_2 \\ \vdots \\ v_n \end{pmatrix}$ is called an n-dimensional vector. An example is $\mathbf{v} = \begin{pmatrix} 1 \\ -2 \\ \vdots \\ 8 \end{pmatrix}$ where 8 is the nth entry.

Hence $\mathbb{R}^n$ is the set of all n-dimensional vectors where $\mathbb{R}$ signifies that the entries of the vector are real numbers, that is, $v_1, v_2, v_3, \ldots$ and v_n are all real numbers. The real number v_j of the vector $\mathbf{v}$ is called the component or more precisely the jth component of the vector $\mathbf{v}$.

This $\mathbb{R}^n$ is also called n-space or the vector space of n-tuples.

Note that the vectors are ordered n-tuples.

What does this mean?

The vector $\mathbf{v} = \begin{pmatrix} 1 \\ -2 \\ \vdots \\ 8 \end{pmatrix}$ is different from $\begin{pmatrix} -2 \\ 1 \\ \vdots \\ 8 \end{pmatrix}$; that is, the order of the entries matters.

How do we draw vectors in $\mathbb{R}^n$ for $n \geq 4$?

We cannot draw pictures of vectors in $\mathbb{R}^4, \mathbb{R}^5, \mathbb{R}^6 \ldots$ but we *can* carry out arithmetic in this multidimensional space.

Two vectors $\mathbf{u}$ and $\mathbf{v}$ are equal if they have the same number of components and all the corresponding components are equal.

How can we write this in mathematical notation?

Let $\mathbf{u} = \begin{pmatrix} u_1 \\ \vdots \\ u_n \end{pmatrix}$ and $\mathbf{v} = \begin{pmatrix} v_1 \\ \vdots \\ v_n \end{pmatrix}$ then

(1.2) $\mathbf{u} = \mathbf{v}$ if and only if entries $u_j = v_j$ for $j = 1, 2, 3, \ldots, n$.

For example, the vectors $\begin{pmatrix} 1 \\ 5 \\ 7 \end{pmatrix}$ and $\begin{pmatrix} 1 \\ 7 \\ 5 \end{pmatrix}$ are not equal because the corresponding components are not equal.

Example 1.12

Let $\mathbf{u} = \begin{pmatrix} x-3 \\ y+1 \\ z+x \end{pmatrix}$ and $\mathbf{v} = \begin{pmatrix} 1 \\ 2 \\ 3 \end{pmatrix}$. If $\mathbf{u} = \mathbf{v}$ then determine the real numbers x, y and z.

Solution

Since $\mathbf{u} = \mathbf{v}$ we have

$$\begin{aligned} x-3=1 &\quad \text{gives} \quad x=4 \\ y+1=2 &\quad \text{gives} \quad y=1 \\ z+x=3 &\quad \text{gives} \quad z+4=3 \quad \text{implies} \quad z=-1 \end{aligned}$$

Our solution is $x = 4, y = 1$ and $z = -1$.

Vector addition in $\mathbb{R}^n$ is defined as

$$\mathbf{u}+\mathbf{v} = \begin{pmatrix} u_1 \\ \vdots \\ u_n \end{pmatrix} + \begin{pmatrix} v_1 \\ \vdots \\ v_n \end{pmatrix} = \begin{pmatrix} u_1+v_1 \\ \vdots \\ u_n+v_n \end{pmatrix} \tag{1.3}$$

The sum of the vectors $\mathbf{u}$ and $\mathbf{v}$ is calculated by adding the corresponding components. Note that $\mathbf{u}+\mathbf{v}$ is also a vector in $\mathbb{R}^n$.

Scalar multiplication $k\mathbf{v}$ is carried out by multiplying each component of the vector $\mathbf{v}$ by the scalar k:

$$k\mathbf{v} = k\begin{pmatrix} v_1 \\ \vdots \\ v_n \end{pmatrix} = \begin{pmatrix} kv_1 \\ \vdots \\ kv_n \end{pmatrix} \tag{1.4}$$

Again $k\mathbf{v}$ is a vector in $\mathbb{R}^n$. This $\mathbb{R}^n$ is called **Euclidean space.**

Euclidean space is the space of all n-tuples of real numbers. Here n is any natural number $1, 2, 3, 4, \ldots$ and is called the **dimension** of $\mathbb{R}^n$.

1.3.6 Introduction to the dot (inner) product

So far we have looked at the two fundamental operations of linear algebra – vector addition and scalar multiplication of vectors in $\mathbb{R}^n$.

How do we multiply vectors?

One way is to take the **dot product** which we define next.

Let $\mathbf{u} = \begin{pmatrix} u_1 \\ u_2 \\ \vdots \\ u_n \end{pmatrix}$ and $\mathbf{v} = \begin{pmatrix} v_1 \\ v_2 \\ \vdots \\ v_n \end{pmatrix}$ be vectors in $\mathbb{R}^n$, then the dot product of $\mathbf{u}$ and $\mathbf{v}$ which is denoted by $\mathbf{u} \cdot \mathbf{v}$ is the quantity given by

$$\mathbf{u} \cdot \mathbf{v} = u_1 v_1 + u_2 v_2 + u_3 v_3 + \cdots + u_n v_n \tag{1.5}$$

This multiplication is called the **dot** or **inner product** of the vectors $\mathbf{u}$ and $\mathbf{v}$.

Also note that the dot product of two vectors $\mathbf{u}$ and $\mathbf{v}$ is obtained by multiplying each component u_j with its corresponding component v_j and then adding the results.

Example 1.13

Let $\mathbf{u} = \begin{pmatrix} -3 \\ 1 \\ 7 \end{pmatrix}$ and $\mathbf{v} = \begin{pmatrix} 9 \\ 2 \\ -4 \end{pmatrix}$. Find $\mathbf{u} \cdot \mathbf{v}$.

Solution

Applying the above formula (1.5) gives

$$\mathbf{u} \cdot \mathbf{v} = \begin{pmatrix} -3 \\ 1 \\ 7 \end{pmatrix} \cdot \begin{pmatrix} 9 \\ 2 \\ -4 \end{pmatrix} = (-3 \times 9) + (1 \times 2) + (7 \times (-4)) = -27 + 2 - 28 = -53$$

Hence $\mathbf{u} \cdot \mathbf{v} = -53$.

1.3.7 Linear combination of vectors

Let $\mathbf{v}_1, \mathbf{v}_2, \ldots, \mathbf{v}_n$ be vectors in $\mathbb{R}^n$ and $k_1, k_2, \ldots, k_n$ be scalars, then the dot product

$$\begin{pmatrix} k_1 \\ \vdots \\ k_n \end{pmatrix} \cdot \begin{pmatrix} \mathbf{v}_1 \\ \vdots \\ \mathbf{v}_n \end{pmatrix} = k_1\mathbf{v}_1 + k_2\mathbf{v}_2 + k_3\mathbf{v}_3 + \cdots + k_n\mathbf{v}_n \text{ is a } \textbf{linear combination} \tag{1.6}$$

The entries in one of the vectors in formula (1.6) is another set of vectors $\mathbf{v}_1, \ldots, \mathbf{v}_n$, so the dot product is a linear combination of vectors.

The dot product combines the fundamental operations – scalar multiplication and vector addition by linear combination.

Example 1.14

Let $\mathbf{u} = \begin{pmatrix} -3 \\ 1 \end{pmatrix}$ and $\mathbf{v} = \begin{pmatrix} 1 \\ 3 \end{pmatrix}$. Find the following linear combinations:

(a) $\mathbf{u} + \mathbf{v}$ **(b)** $3\mathbf{u} + 2\mathbf{v}$ **(c)** $\mathbf{u} - \mathbf{u}$

(d) Determine the values of scalars x and y such that the linear combination $x\mathbf{u} + y\mathbf{v} = \mathbf{O}$.

Solution

(a) By adding the corresponding entries we have

$$\mathbf{u} + \mathbf{v} = \begin{pmatrix} -3 \\ 1 \end{pmatrix} + \begin{pmatrix} 1 \\ 3 \end{pmatrix} = \begin{pmatrix} -2 \\ 4 \end{pmatrix}$$

(b) By applying the rules of scalar multiplication and vector addition we have

$$3\mathbf{u} + 2\mathbf{v} = 3\begin{pmatrix} -3 \\ 1 \end{pmatrix} + 2\begin{pmatrix} 1 \\ 3 \end{pmatrix} = \begin{pmatrix} -3 \times 3 \\ 1 \times 3 \end{pmatrix} + \begin{pmatrix} 2 \times 1 \\ 2 \times 3 \end{pmatrix} = \begin{pmatrix} -9 + 2 \\ 3 + 6 \end{pmatrix} = \begin{pmatrix} -7 \\ 9 \end{pmatrix}$$

(c) We have

$$\mathbf{u} - \mathbf{u} = \begin{pmatrix} -3 \\ 1 \end{pmatrix} - \begin{pmatrix} -3 \\ 1 \end{pmatrix} = \begin{pmatrix} -3 + 3 \\ 1 - 1 \end{pmatrix} = \begin{pmatrix} 0 \\ 0 \end{pmatrix} = \mathbf{O}$$

Hence $\mathbf{u} - \mathbf{u}$ gives the zero vector $\mathbf{O}$.

(d) We have the linear combination $x\mathbf{u} + y\mathbf{v} = \mathbf{O}$:

$$x\mathbf{u} + y\mathbf{v} = x\begin{pmatrix} -3 \\ 1 \end{pmatrix} + y\begin{pmatrix} 1 \\ 3 \end{pmatrix} = \begin{pmatrix} -3x + y \\ x + 3y \end{pmatrix} = \begin{pmatrix} 0 \\ 0 \end{pmatrix}$$

We need to solve the simultaneous equations:

$$\left.\begin{aligned} -3x + y &= 0 \\ x + 3y &= 0 \end{aligned}\right\} \text{ implies that } x = y = 0$$

Hence the linear combination $0\mathbf{u} + 0\mathbf{v}$ gives the zero vector, that is $0\mathbf{u} + 0\mathbf{v} = \mathbf{O}$.

Linear combinations are sprinkled throughout this book so it is important that you understand that it combines scalars and vectors.

Note that for any vector $\mathbf{v}$ we have

$$\mathbf{v} - \mathbf{v} = \mathbf{O} \text{ where } \mathbf{O} = \begin{pmatrix} 0 \\ \vdots \end{pmatrix}$$

The **zero vector** in $\mathbb{R}^n$ is denoted by $\mathbf{O}$ and is defined as being 'non-empty', although all entries are zero.

Summary

Physically, vectors express magnitude as well as direction. Scalars only define magnitude.

$\mathbb{R}^n$ is n-space where n is a natural number such as 1, 2, 3, 4, 5, ... Vectors in $\mathbb{R}^n$ are denoted by

$$\mathbf{v} = \begin{pmatrix} v_1 \\ \vdots \\ v_n \end{pmatrix}.$$

We can linearly combine vectors by applying the two fundamental operations of linear algebra - scalar multiplication and vector addition.

EXERCISES 1.3

(Brief solutions at end of book. Full solutions available at <http://www.oup.co.uk/companion/singh>.)

1. Consider the following vectors **a** and **b** as shown in Fig. 1.27.

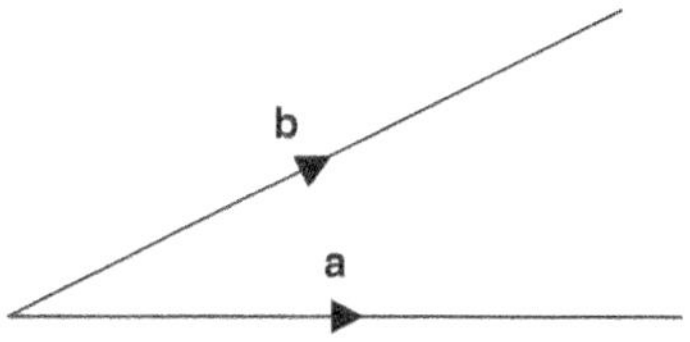

Figure 1.27

Sketch the following vectors:

(a) $\mathbf{a}+\mathbf{b}$ (b) $\mathbf{a}-\mathbf{b}$ (c) $3\mathbf{a}$ (d) $-\frac{1}{2}\mathbf{b}$ (e) $3\mathbf{a}-\frac{1}{2}\mathbf{b}$

2. Let $\mathbf{u} = \begin{pmatrix} 1 \\ -1 \end{pmatrix}$ and $\mathbf{v} = \begin{pmatrix} 2 \\ 1 \end{pmatrix}$. Plot the following vectors on the same axes in $\mathbb{R}^2$:

(a) $\mathbf{u}$ (b) $\mathbf{v}$ (c) $\mathbf{u}+\mathbf{v}$ (d) $\mathbf{u}-\mathbf{v}$

Determine

(e) $\mathbf{u}\cdot\mathbf{v}$ (f) $\mathbf{v}\cdot\mathbf{u}$ (g) $\mathbf{u}\cdot\mathbf{u}$ (h) $\mathbf{v}\cdot\mathbf{v}$

3. Let $\mathbf{u} = \begin{pmatrix} 2 \\ 3 \\ -1 \end{pmatrix}$ and $\mathbf{v} = \begin{pmatrix} 5 \\ 1 \\ -2 \end{pmatrix}$ be vectors in $\mathbb{R}^3$. Evaluate the following:

(a) $\mathbf{u}\cdot\mathbf{v}$ (b) $\mathbf{v}\cdot\mathbf{u}$ (c) $\mathbf{u}\cdot\mathbf{u}$ (d) $\mathbf{v}\cdot\mathbf{v}$

4. Let $\mathbf{u} = \begin{pmatrix} 2 \\ -1 \end{pmatrix}$. Plot the following vectors on the same axes in $\mathbb{R}^2$:

(a) $\mathbf{u}$ (b) $-\mathbf{u}$ (c) $2\mathbf{u}$ (d) $3\mathbf{u}$ (e) $-2\mathbf{u}$

5. Let $\mathbf{u} = \begin{pmatrix} -1 \\ 1 \end{pmatrix}$ and $\mathbf{v} = \begin{pmatrix} 3 \\ -1 \end{pmatrix}$. Let $\mathbf{w}$ be the linear combination $\mathbf{w} = \begin{pmatrix} 1 \\ \lambda \end{pmatrix} \cdot \begin{pmatrix} \mathbf{u} \\ \mathbf{v} \end{pmatrix}$ where λ is a scalar. Find the linear combination for the following values of λ:

(a) $\lambda = 1$ (b) $\lambda = -1$ (c) $\lambda = \frac{1}{2}$ (d) $\lambda = -\frac{1}{2}$ (e) $\lambda = \frac{1}{3}$

Sketch on the same axes the vectors $\mathbf{w}$ obtained in parts (a), (b), (c), (d) and (e).

6. Let $\mathbf{u} = \begin{pmatrix} -2 \\ 2 \end{pmatrix}$, $\mathbf{v} = \begin{pmatrix} 2 \\ 1 \end{pmatrix}$ and $\mathbf{w} = \begin{pmatrix} k \\ c \end{pmatrix} \cdot \begin{pmatrix} \mathbf{u} \\ \mathbf{v} \end{pmatrix}$ where k and c are scalars.
Write out the vector $\mathbf{w}$ as a linear combination for the following values of k and c:

(a) $k = 1, c = 1$ (b) $k = \frac{1}{2}, c = \frac{1}{2}$ (c) $k = -\frac{1}{2}, c = \frac{1}{2}$ (d) $k = \frac{1}{2}, c = -\frac{1}{2}$

Sketch on the same axes the vectors $\mathbf{w}$ obtained in parts (a), (b), (c) and (d).

7. Let $\mathbf{u} = \begin{pmatrix} x+3 \\ y-2 \end{pmatrix}$ and $\mathbf{v} = \begin{pmatrix} x-2 \\ y+11 \end{pmatrix}$. Given that $\mathbf{u} + \mathbf{v} = \mathbf{O}$ determine the values of the real numbers x and y.

8. Let $\mathbf{u} = \begin{pmatrix} 1 \\ 0 \end{pmatrix}$, $\mathbf{v} = \begin{pmatrix} 0 \\ 1 \end{pmatrix}$ and $\mathbf{w} = \begin{pmatrix} x \\ y \end{pmatrix}$ be vectors in $\mathbb{R}^2$. Show that $x\mathbf{u} + y\mathbf{v} = \mathbf{w}$.

9. Let $\mathbf{u} = \begin{pmatrix} -3 \\ 5 \\ 8 \end{pmatrix}$ and $\mathbf{v} = \begin{pmatrix} 7 \\ -1 \\ 2 \end{pmatrix}$. Find the linear combinations:

(a) $\mathbf{u} + \mathbf{v}$ (b) $5\mathbf{u}$ (c) $2\mathbf{u} + 6\mathbf{v}$ (d) $\mathbf{u} - 3\mathbf{v}$ (e) $-5\mathbf{u} - 4\mathbf{v}$

10. Let $\mathbf{u} = \begin{pmatrix} -9 \\ 2 \\ 4 \end{pmatrix}$, $\mathbf{v} = \begin{pmatrix} -2 \\ 1 \\ 3 \end{pmatrix}$ and $\mathbf{w} = \begin{pmatrix} 1 \\ -2 \\ 5 \end{pmatrix}$. Find the linear combinations:

(a) $\mathbf{u} + \mathbf{v} + \mathbf{w}$ (b) $\mathbf{u} - \mathbf{v} - \mathbf{w}$ (c) $2\mathbf{u} + \mathbf{v} - \mathbf{w}$ (d) $-2\mathbf{u} + 3\mathbf{v} + 5\mathbf{w}$

11. Let $\mathbf{u} = \begin{pmatrix} 1 \\ 0 \\ 0 \end{pmatrix}$, $\mathbf{v} = \begin{pmatrix} 0 \\ 1 \\ 0 \end{pmatrix}$, $\mathbf{w} = \begin{pmatrix} 0 \\ 0 \\ 1 \end{pmatrix}$ and $\mathbf{x} = \begin{pmatrix} x \\ y \\ z \end{pmatrix}$ be vectors in $\mathbb{R}^3$. Show that

$$x\mathbf{u} + y\mathbf{v} + z\mathbf{w} = \mathbf{x}$$

12. Find the real numbers x, y and z, if

$$x\begin{pmatrix} 1 \\ 2 \\ 0 \end{pmatrix} + y\begin{pmatrix} 0 \\ 1 \\ -1 \end{pmatrix} + z\begin{pmatrix} -2 \\ 0 \\ 6 \end{pmatrix} = \begin{pmatrix} 5 \\ 3 \\ 17 \end{pmatrix}$$

13. Let $\mathbf{u} = \begin{pmatrix} -1 \\ 3 \\ 2 \\ 0 \end{pmatrix}$, $\mathbf{v} = \begin{pmatrix} 3 \\ -2 \\ 5 \\ 1 \end{pmatrix}$, $\mathbf{w} = \begin{pmatrix} 0 \\ -1 \\ 1 \\ 2 \end{pmatrix}$ and $\mathbf{x} = \begin{pmatrix} x \\ y \\ z \\ a \end{pmatrix}$ be vectors in $\mathbb{R}^4$.

    Determine the following linear combinations:

    (a) $\mathbf{u} + \mathbf{v} + \mathbf{w}$ (b) $\mathbf{u} - \mathbf{v} - \mathbf{w}$ (c) $\mathbf{u} - 2\mathbf{v} + 3\mathbf{w}$ (d) $\mathbf{u} - 3\mathbf{w} + \mathbf{x}$

    If $\mathbf{u} + \mathbf{v} + \mathbf{w} + \mathbf{x} = \mathbf{O}$ determine the values of x, y, z and a.

14. Let $\mathbf{e}_k$ be a vector in $\mathbb{R}^n$ which has a 1 in the kth component and zeros elsewhere, that is:

$$\mathbf{e}_1 = \begin{pmatrix} 1 \\ 0 \\ 0 \\ \vdots \\ 0 \end{pmatrix}, \mathbf{e}_2 = \begin{pmatrix} 0 \\ 1 \\ 0 \\ \vdots \\ 0 \end{pmatrix}, \ldots, \mathbf{e}_k = \begin{pmatrix} 0 \\ \vdots \\ 1 \\ 0 \\ \vdots \end{pmatrix}, \ldots, \mathbf{e}_n = \begin{pmatrix} 0 \\ 0 \\ \vdots \\ 0 \\ 1 \end{pmatrix}$$

    Let $\mathbf{u}$ be a vector given by $\mathbf{u} = \begin{pmatrix} x_1 \\ \vdots \\ x_k \\ \vdots \\ x_n \end{pmatrix}$. Show that

$$\mathbf{u} = x_1\mathbf{e}_1 + x_2\mathbf{e}_2 + \cdots + x_k\mathbf{e}_k + \cdots + x_n\mathbf{e}_n$$

15. Let $\mathbf{u}$ and $\mathbf{v}$ be vectors in $\mathbb{R}^n$. **Disprove** the following proposition:
    If $\mathbf{u} \cdot \mathbf{v} = 0$ then $\mathbf{u} = \mathbf{O}$ or $\mathbf{v} = \mathbf{O}$.

SECTION 1.4 Arithmetic of Matrices

By the end of this section you will be able to

- execute arithmetic operations with matrices
- use matrix theory to perform transformations

1.4.1 Matrices revisited

*What does the term **matrix** mean?*

A matrix is an array of numbers enclosed in brackets. The numbers in the array are called the entries or the elements of the matrix. The term **matrices** is the plural of matrix.

Figure 1.28 Arthur Cayley, 1821–1895.

Arthur Cayley (Fig. 1.28), an English lawyer who became a mathematician, was the first person to develop matrices as we know them today. He graduated from Trinity College, Cambridge, in 1842 and taught there for four years. In 1849, he took up the profession of lawyer and worked at the courts of Lincoln's Inn in London for 14 years. He returned to mathematics in 1863 and was appointed Professor of Pure Mathematics at Cambridge, with a substantial decrease in salary but was nonetheless happy to pursue his interest in mathematics. In 1858, he published 'Memoir on the Theory of Matrices' which contained the definition of a matrix. Cayley thought that matrices were of no practical use whatsoever, just a convenient mathematical notation. He could not have been more wrong. Linear algebra is used today in engineering, the sciences, medicine, statistics and economics.

The **size** of a matrix is given by the number of rows and columns. For example, $\begin{pmatrix} 12 & 6 \\ 3 & 7 \end{pmatrix}$ is a 2×2 matrix (2 rows by 2 columns) and is called a **square** matrix. The size of this matrix is verbally stated as being '2 by 2'.

$\begin{pmatrix} 3 & 2 \\ 7 & 6 \\ 5 & 9 \\ 2 & 1 \end{pmatrix}$ is a 4×2 matrix (4 rows by 2 columns) and is not a square matrix. An example of a 2×4 matrix is $\begin{pmatrix} 1 & 2 & 3 & 4 \\ 5 & 6 & 7 & 8 \end{pmatrix}$. The size of this matrix is '2 by 4'.

Note that we state the number of rows first and then the number of columns. Hence 2×4 and 4×2 are different size matrices.

A common notation for general matrices is:

$$\begin{array}{c c} & \text{Column 1} \quad \text{Column 2} \\ \begin{array}{c}\text{Row 1}\\ \text{Row 2}\end{array} & \begin{pmatrix} a_{11} & a_{12} \\ a_{21} & a_{22} \end{pmatrix} \end{array}$$

where a_{12} is the entry in the first row and second column.

What row and column is the element a_{21} in?

Element a_{21} is in the second row and first column. Note that the subscript of each element states row number first and then the column number. The position of a_{12} in a matrix is different from a_{21}.

What is the position of an element a_{23} in a 3 by 4 matrix?

Element a_{23} is in the second row and third column of a 3 by 4 matrix:

$$\mathbf{A} = \begin{pmatrix} a_{11} & a_{12} & \overset{\text{Col. 3}}{a_{13}} & a_{14} \\ a_{21} & a_{22} & \boxed{a_{23}} & a_{24} \\ a_{31} & a_{32} & a_{33} & a_{34} \end{pmatrix} \text{Row 2}$$

Generally lowercase letters represent the elements (or entries) of the matrices and bold capital represent the matrix itself.

The columns of a matrix are called the **column** vectors of the matrix. The rows of a matrix are called the **row** vectors of the matrix.

1.4.2 Transformation and scalar multiplication

Computer graphics are essentially created using high-speed transformations. For example, a computer animated sequence is based on modelling surfaces of connecting triangles. The computer stores the vertices of the triangle in its memory and then certain operations such as rotations, translations, reflections and enlargements are carried out by (transformation) matrices.

In this context, we apply a (transformation) matrix in order to perform a function such as rotation, reflection or translation, as shown in Fig. 1.29:

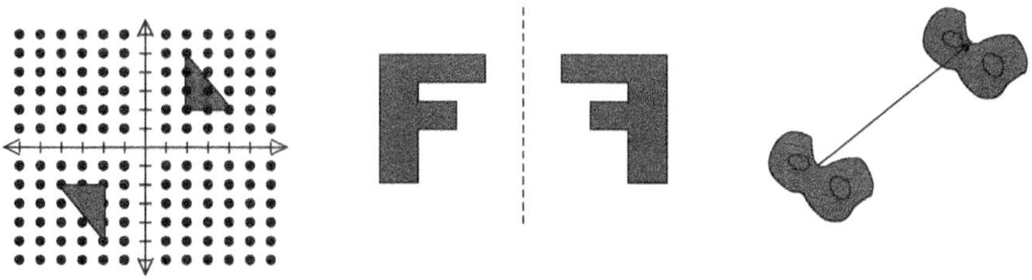

Rotation Reflection Translation

Figure 1.29

First we examine scalar multiplication of matrices.

*What does the term **scalar** mean?*

A scalar is a number which is used to multiply the entries of a matrix.

What does scalar multiplication mean?

Let $\mathbf{A}$ be a matrix and k be a scalar, then the scalar multiplication $k\mathbf{A}$ is the matrix constructed by multiplying each entry of $\mathbf{A}$ by k.

$$\text{If } \mathbf{A} = \begin{pmatrix} a_{11} & \cdots & a_{1n} \\ \vdots & \vdots & \vdots \\ a_{m1} & \cdots & a_{mn} \end{pmatrix} \text{ then } k\mathbf{A} = \begin{pmatrix} ka_{11} & \cdots & ka_{1n} \\ \vdots & \vdots & \vdots \\ ka_{m1} & \cdots & ka_{mn} \end{pmatrix} \tag{1.7}$$

Note that scalar multiplication of a matrix results in a matrix of the same size. Next we show what scalar multiplication means in terms of transformations.

Example 1.15

Consider the vertices of a triangle (Fig. 1.30) given by $P(2, 0)$, $Q(2, 3)$ and $R(0, 0)$:

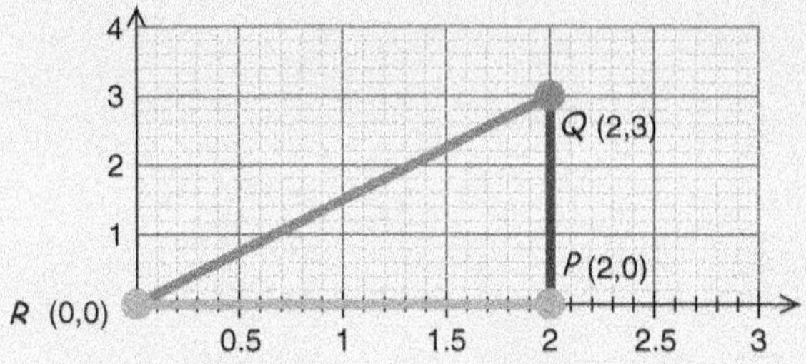

Figure 1.30

$$A = \begin{pmatrix} \overset{P}{2} & \overset{Q}{2} & \overset{R}{0} \\ 0 & 3 & 0 \end{pmatrix}$$

This can be represented in matrix form $\mathbf{A}$ with the coordinates of the point P as the entries in the first column, the coordinates of the point Q as entries in the second column and the coordinates of the point R as entries in the last column. Each of these vertices is represented by a column vector in the matrix $\mathbf{A}$.
Determine the image of the triangle under the transformation performed by $2\mathbf{A}$.

Solution
Carrying out the scalar multiplication we have

$$2\mathbf{A} = 2 \begin{pmatrix} \overset{P}{2} & \overset{Q}{2} & \overset{R}{0} \\ 0 & 3 & 0 \end{pmatrix} = \begin{pmatrix} \overset{P'}{4} & \overset{Q'}{4} & \overset{R'}{0} \\ 0 & 6 & 0 \end{pmatrix} \quad \text{[Doubling each vector]}$$

Plotting the given triangle PQR (Fig. 1.31), and the transformed triangle $P'Q'R'$ we get:

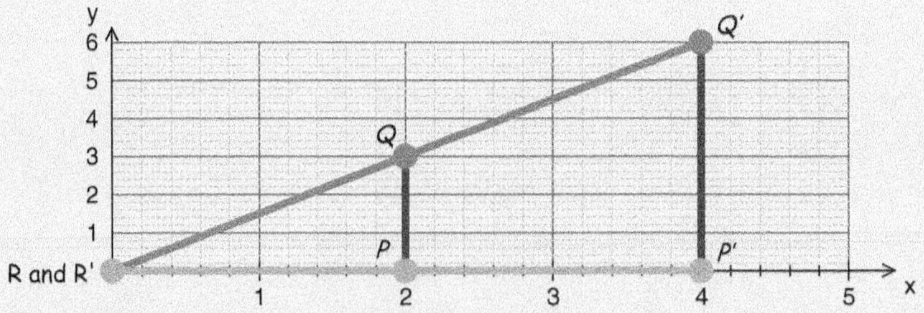

Figure 1.31

What effect does scalar multiplication 2A have on the initial triangle PQR?
This scalar multiplication increases the length of sides of the triangle by a factor of 2. This means that $2\mathbf{A}$ *doubles* the length of each side of the initial triangle.

What would be the image of the triangle PQR under the scalar multiplication of $\frac{1}{2}$A?
It would scale each side of the triangle to *half* its initial size.

Example 1.16

Determine the following:

(a) $10\begin{pmatrix} 1 & 2 \\ 3 & 4 \end{pmatrix}$ **(b)** $\frac{1}{2}\begin{pmatrix} 1 & 2 & 5 & 6 & 1 \\ 7 & 9 & 11 & 6 & 3 \end{pmatrix}$

Solution

(a) Multiplying each entry of the matrix by 10 gives

$$10\begin{pmatrix} 1 & 2 \\ 3 & 4 \end{pmatrix} = \begin{pmatrix} 1\times 10 & 2\times 10 \\ 3\times 10 & 4\times 10 \end{pmatrix} = \begin{pmatrix} 10 & 20 \\ 30 & 40 \end{pmatrix}$$

(b) Multiplying each element in the second matrix by 1/2 (or dividing by 2) gives

$$\frac{1}{2}\begin{pmatrix} 1 & 2 & 5 & 6 & 1 \\ 7 & 9 & 11 & 6 & 3 \end{pmatrix} = \begin{pmatrix} 0.5 & 1 & 2.5 & 3 & 0.5 \\ 3.5 & 4.5 & 5.5 & 3 & 1.5 \end{pmatrix}$$

We can also go the other way, that is factorize out a common term. For example

$$\begin{pmatrix} 10 & 15 \\ 35 & 40 \end{pmatrix} = \begin{pmatrix} 5\times 2 & 5\times 3 \\ 5\times 7 & 5\times 8 \end{pmatrix} = 5\begin{pmatrix} 2 & 3 \\ 7 & 8 \end{pmatrix}$$

1.4.3 Matrix addition

We add matrices in the same way that we add vectors. To add or subtract two matrices we add or subtract the corresponding locations in each matrix respectively. For example

$$\begin{pmatrix} 2 & 9 \\ 4 & 1 \end{pmatrix} + \begin{pmatrix} 3 & 8 \\ 5 & 7 \end{pmatrix} = \begin{pmatrix} 2+3 & 9+8 \\ 4+5 & 1+7 \end{pmatrix} = \begin{pmatrix} 5 & 17 \\ 9 & 8 \end{pmatrix}$$

Also

$$\begin{pmatrix} 2 & 9 \\ 4 & 1 \end{pmatrix} - \begin{pmatrix} 3 & 8 \\ 5 & 7 \end{pmatrix} = \begin{pmatrix} 2-3 & 9-8 \\ 4-5 & 1-7 \end{pmatrix} = \begin{pmatrix} -1 & 1 \\ -1 & -6 \end{pmatrix}$$

The sum of matrices of size m by n (m rows by n columns) is defined by:

$$\begin{pmatrix} a_{11} & \cdots & a_{1n} \\ \vdots & \vdots & \vdots \\ a_{m1} & \cdots & a_{mn} \end{pmatrix} + \begin{pmatrix} b_{11} & \cdots & b_{1n} \\ \vdots & \vdots & \vdots \\ b_{m1} & \cdots & b_{mn} \end{pmatrix} = \begin{pmatrix} a_{11}+b_{11} & \cdots & a_{1n}+b_{1n} \\ \vdots & \vdots & \vdots \\ a_{m1}+b_{m1} & \cdots & a_{mn}+b_{mn} \end{pmatrix} \tag{1.8}$$

Similarly the subtraction of these matrices is:

$$\begin{pmatrix} a_{11} & \cdots & a_{1n} \\ \vdots & \vdots & \vdots \\ a_{m1} & \cdots & a_{mn} \end{pmatrix} - \begin{pmatrix} b_{11} & \cdots & b_{1n} \\ \vdots & \vdots & \vdots \\ b_{m1} & \cdots & b_{mn} \end{pmatrix} = \begin{pmatrix} a_{11}-b_{11} & \cdots & a_{1n}-b_{1n} \\ \vdots & \vdots & \vdots \\ a_{m1}-b_{m1} & \cdots & a_{mn}-b_{mn} \end{pmatrix} \tag{1.9}$$

Note that for addition (or subtraction) we add (or subtract) the elements in corresponding locations. The result of this is another matrix of the same size as the initial matrices.

Example 1.17

Determine the following:

(a) $\begin{pmatrix} 1 & 2 \\ 3 & 4 \end{pmatrix} + \begin{pmatrix} 5 & 6 \\ 7 & 8 \end{pmatrix}$ **(b)** $\begin{pmatrix} 1 & 2 & 5 & 6 & 1 \\ 7 & 9 & 11 & 6 & 3 \end{pmatrix} + \begin{pmatrix} 9 & 8 & 5 & 7 & 6 \\ 2 & 13 & 7 & 2 & 3 \end{pmatrix}$

(c) $\begin{pmatrix} 1 & 2 & 3 \\ 4 & 5 & 6 \\ 7 & 8 & 9 \end{pmatrix} - \begin{pmatrix} 1 & 1 & 1 \\ 3 & 3 & 3 \\ 4 & 4 & 4 \end{pmatrix}$ **(d)** $\begin{pmatrix} 2 & 3 \\ 5 & 4 \end{pmatrix} - \begin{pmatrix} 1 & 2 & 4 \\ 8 & 4 & 0 \end{pmatrix}$

Solution

(a) Adding the corresponding elements of each matrix gives

$$\begin{pmatrix} 1 & 2 \\ 3 & 4 \end{pmatrix} + \begin{pmatrix} 5 & 6 \\ 7 & 8 \end{pmatrix} = \begin{pmatrix} 1+5 & 2+6 \\ 3+7 & 4+8 \end{pmatrix} = \begin{pmatrix} 6 & 8 \\ 10 & 12 \end{pmatrix}$$

(b) Again adding the corresponding entries of the given matrices we have

$$\begin{pmatrix} 1 & 2 & 5 & 6 & 1 \\ 7 & 9 & 11 & 6 & 3 \end{pmatrix} + \begin{pmatrix} 9 & 8 & 5 & 7 & 6 \\ 2 & 13 & 7 & 2 & 3 \end{pmatrix} = \begin{pmatrix} 10 & 10 & 10 & 13 & 7 \\ 9 & 22 & 18 & 8 & 6 \end{pmatrix}$$

(c) Similarly we have

$$\begin{pmatrix} 1 & 2 & 3 \\ 4 & 5 & 6 \\ 7 & 8 & 9 \end{pmatrix} - \begin{pmatrix} 1 & 1 & 1 \\ 3 & 3 & 3 \\ 4 & 4 & 4 \end{pmatrix} = \begin{pmatrix} 1-1 & 2-1 & 3-1 \\ 4-3 & 5-3 & 6-3 \\ 7-4 & 8-4 & 9-4 \end{pmatrix} = \begin{pmatrix} 0 & 1 & 2 \\ 1 & 2 & 3 \\ 3 & 4 & 5 \end{pmatrix}$$

(d) We cannot evaluate the given matrices $\begin{pmatrix} 2 & 3 \\ 5 & 4 \end{pmatrix} - \begin{pmatrix} 1 & 2 & 4 \\ 8 & 4 & 0 \end{pmatrix}$.

Why not?

Because they are of different size. The first matrix, $\begin{pmatrix} 2 & 3 \\ 5 & 4 \end{pmatrix}$, is a 2 by 2 matrix and the second matrix, $\begin{pmatrix} 1 & 2 & 4 \\ 8 & 4 & 0 \end{pmatrix}$, is a 2 by 3 matrix, therefore we cannot subtract these matrices.

Remember, when we add or subtract matrices the resulting matrix is always the same size as the given matrices. This is also the case for vector addition and subtraction, because a vector is simply a particular type of matrix.

1.4.4 Matrix vector product

Let us reconsider our example of simultaneous equations from section 1.1. We formed two equations that described the cost of ice creams x and drinks y:

$$\begin{aligned} 2x + 2y &= 3 \\ 2x + y &= 2.5 \end{aligned}$$

We can write these equations as a linear combination of the vectors:

$$\begin{pmatrix} 2 \\ 2 \end{pmatrix} x + \begin{pmatrix} 2 \\ 1 \end{pmatrix} y = \begin{pmatrix} 3 \\ 2.5 \end{pmatrix}$$

We can separate out the coefficients and the unknowns x and y of the left hand side:

$$\begin{pmatrix} 2 & 2 \\ 2 & 1 \end{pmatrix} \begin{pmatrix} x \\ y \end{pmatrix}$$

This is a matrix times a vector. In order to use this method, we need to define matrix times a vector as the product of row by column:

$$\begin{pmatrix} \text{Row 1} \\ \text{Row 2} \end{pmatrix} \times (\text{Column}) = \begin{pmatrix} (\text{Row 1}) \times (\text{Column}) \\ (\text{Row 2}) \times (\text{Column}) \end{pmatrix}$$

We have

$$(\text{Row 1}) \times (\text{Col}) = \begin{pmatrix} 2 & 2 \end{pmatrix} \begin{pmatrix} x \\ y \end{pmatrix} = 2x + 2y$$

$$(\text{Row 2}) \times (\text{Col}) = \begin{pmatrix} 2 & 1 \end{pmatrix} \begin{pmatrix} x \\ y \end{pmatrix} = 2x + y$$

Hence the above equations can be written in matrix form as

$$\begin{pmatrix} 2 & 2 \\ 2 & 1 \end{pmatrix} \begin{pmatrix} x \\ y \end{pmatrix} = \begin{pmatrix} 3 \\ 2.5 \end{pmatrix}$$

We can write any m by n linear equations in matrix form. The following simultaneous equations:

$$\begin{aligned} x + y + 3z &= 5 \\ -2x - y + 5z &= 6 \end{aligned}$$

are written in matrix form as $\mathbf{Ax} = \mathbf{b}$, where

$$\mathbf{A} = \begin{pmatrix} 1 & 1 & 3 \\ -2 & -1 & 5 \end{pmatrix}, \ \mathbf{x} = \begin{pmatrix} x \\ y \\ z \end{pmatrix} \text{ and } \mathbf{b} = \begin{pmatrix} 5 \\ 6 \end{pmatrix}$$

Linear equations are generally written in compact form as $\mathbf{Ax} = \mathbf{b}$, where $\mathbf{A}$ is the matrix of coefficients of the unknowns, $\mathbf{x}$ is a vector of the unknowns and $\mathbf{b}$ is the vector containing the constant values.

Clearly working with two or three simple simultaneous equations does not require matrix intervention, but mathematicians, scientists and engineers frequently work with much more complex systems, which would be virtually impossible to solve without matrices.

Matrices are a much more efficient and systematic way of solving linear equations.

We can examine the geometric interpretation of systems of equations by looking at transformations. For example

$$\begin{aligned} x - 3y &= -1 \\ x + y &= 2 \end{aligned}$$

These equations can be written as $\mathbf{Ax} = \mathbf{b}$ where $\mathbf{A} = \begin{pmatrix} 1 & -3 \\ 1 & 1 \end{pmatrix}$, $\mathbf{x} = \begin{pmatrix} x \\ y \end{pmatrix}$ and $\mathbf{b} = \begin{pmatrix} -1 \\ 2 \end{pmatrix}$, which means the matrix $\mathbf{A}$ transforms the vector $\mathbf{x} = \begin{pmatrix} x \\ y \end{pmatrix}$ to $\begin{pmatrix} -1 \\ 2 \end{pmatrix}$ as shown in Fig. 1.32.

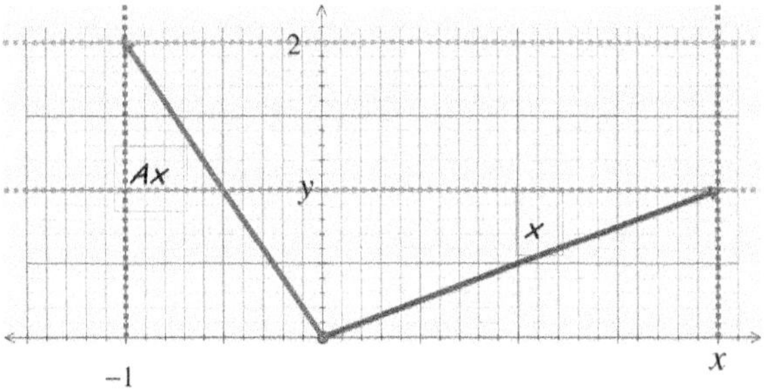

Figure 1.32

We say that the matrix $\mathbf{A}$ transforms the vector $\mathbf{x}$ to another vector $\mathbf{b}$, say $\mathbf{A}$ acts on $\mathbf{x}$ to give $\mathbf{b}$. This $\mathbf{Ax} = \mathbf{b}$ is like a function f acting on the argument x to give a value $f(x) = y$. $\mathbf{A}$ acts on $\mathbf{x}$ to give $\mathbf{b}$.

In general, if matrix $\mathbf{A} = \begin{pmatrix} \mathbf{v}_1 & \mathbf{v}_2 & \ldots & \mathbf{v}_n \end{pmatrix}$ where $\mathbf{v}_1, \mathbf{v}_2, \ldots, \mathbf{v}_n$ are the column vectors of matrix $\mathbf{A}$ and $\mathbf{x}$ is the vector of the unknowns $x_1, x_2, \ldots, x_n$ then we define matrix vector product $\mathbf{Ax}$ as the following linear combination:

$$\text{(1.10)} \qquad \mathbf{Ax} = \begin{pmatrix} \mathbf{v}_1 & \mathbf{v}_2 & \cdots & \mathbf{v}_n \end{pmatrix} \begin{pmatrix} x_1 \\ \vdots \\ x_n \end{pmatrix} = x_1\mathbf{v}_1 + x_2\mathbf{v}_2 + \cdots + x_n\mathbf{v}_n$$

Note that the number of columns of matrix $\mathbf{A}$ must equal the number of entries of the vector $\mathbf{x}$. The product $\mathbf{Ax}$ is the linear combination of the column vectors of matrix $\mathbf{A}$.

Example 1.18

Calculate $\mathbf{Ax}$ for the following:

(a) $\mathbf{A} = \begin{pmatrix} 2 & 3 \\ -1 & 5 \end{pmatrix}, \mathbf{x} = \begin{pmatrix} 2 \\ 1 \end{pmatrix}$ **(b)** $\mathbf{A} = \begin{pmatrix} 1 & 2 \\ 3 & 4 \\ 5 & 6 \end{pmatrix}, \mathbf{x} = \begin{pmatrix} 3 \\ 4 \end{pmatrix}$ **(c)** $\mathbf{A} = \begin{pmatrix} 1 & 0 \end{pmatrix}, \mathbf{x} = \begin{pmatrix} 3 \\ 4 \end{pmatrix}$

Solution

(a) Writing this as the linear combination of columns of the matrix:

$$\mathbf{Ax} = \begin{pmatrix} 2 & 3 \\ -1 & 5 \end{pmatrix}\begin{pmatrix} 2 \\ 1 \end{pmatrix} = 2\begin{pmatrix} 2 \\ -1 \end{pmatrix} + 1\begin{pmatrix} 3 \\ 5 \end{pmatrix} = \begin{pmatrix} 4+3 \\ -2+5 \end{pmatrix} = \begin{pmatrix} 7 \\ 3 \end{pmatrix}$$

The given matrix $\mathbf{A}$ acting on the vector $\begin{pmatrix} 2 \\ 1 \end{pmatrix}$ yields the vector $\begin{pmatrix} 7 \\ 3 \end{pmatrix}$.

(b) Similarly we have

$$\mathbf{Ax} = \begin{pmatrix} 1 & 2 \\ 3 & 4 \\ 5 & 6 \end{pmatrix}\begin{pmatrix} 3 \\ 4 \end{pmatrix} = 3\begin{pmatrix} 1 \\ 3 \\ 5 \end{pmatrix} + 4\begin{pmatrix} 2 \\ 4 \\ 6 \end{pmatrix} = \begin{pmatrix} 3+8 \\ 9+16 \\ 15+24 \end{pmatrix} = \begin{pmatrix} 11 \\ 25 \\ 39 \end{pmatrix}$$

(c) We have

$$\mathbf{Ax} = \begin{pmatrix} 1 & 0 \end{pmatrix}\begin{pmatrix} 3 \\ 4 \end{pmatrix} = (3\,(1) + 4\,(0)) = (3) = 3$$

Note that a 1×1 matrix such as (3) does not require brackets.

It is important to note that $\mathbf{Ax}$ means the matrix $\mathbf{A}$ acts on the vector $\mathbf{x}$.

1.4.5 Matrix times matrix

Multiplication of matrices is not quite as straightforward as addition and subtraction, but the method is not difficult to grasp. However, matrix multiplication is the most important arithmetic operation.

How do we define a matrix times a matrix?

Exactly as we defined a matrix times a vector. Consider the matrix multiplication $\mathbf{AB}$ where

$$\mathbf{A} = \begin{pmatrix} 1 & -3 \\ 1 & 0 \end{pmatrix} \text{ and } \mathbf{B} = \begin{matrix} \mathbf{c}_1 & \mathbf{c}_2 & \mathbf{c}_3 \end{matrix} \atop \begin{pmatrix} 1 & 2 & 3 \\ 4 & 5 & 6 \end{pmatrix}$$

We denote the columns of matrix $\mathbf{B}$ as vectors $\mathbf{c}_1, \mathbf{c}_2$ and $\mathbf{c}_3$. Next we see how matrix $\mathbf{A}$ acts on each of these vectors $\mathbf{c}_1, \mathbf{c}_2$ and $\mathbf{c}_3$:

$$\mathbf{Ac}_1 = \begin{pmatrix} 1 & -3 \\ 1 & 0 \end{pmatrix}\begin{pmatrix} 1 \\ 4 \end{pmatrix} = 1\begin{pmatrix} 1 \\ 1 \end{pmatrix} + 4\begin{pmatrix} -3 \\ 0 \end{pmatrix} = \begin{pmatrix} -11 \\ 1 \end{pmatrix}$$

$$\mathbf{Ac}_2 = \begin{pmatrix} 1 & -3 \\ 1 & 0 \end{pmatrix}\begin{pmatrix} 2 \\ 5 \end{pmatrix} = 2\begin{pmatrix} 1 \\ 1 \end{pmatrix} + 5\begin{pmatrix} -3 \\ 0 \end{pmatrix} = \begin{pmatrix} -13 \\ 2 \end{pmatrix}$$

$$\mathbf{A}\mathbf{c}_3 = \begin{pmatrix} 1 & -3 \\ 1 & 0 \end{pmatrix}\begin{pmatrix} 3 \\ 6 \end{pmatrix} = 3\begin{pmatrix} 1 \\ 1 \end{pmatrix} + 6\begin{pmatrix} -3 \\ 0 \end{pmatrix} = \begin{pmatrix} -15 \\ 3 \end{pmatrix}$$

We define matrix multiplication **AB** as

$$\mathbf{AB} = \begin{pmatrix} 1 & -3 \\ 1 & 0 \end{pmatrix}\overset{\begin{matrix}\mathbf{c}_1 & \mathbf{c}_2 & \mathbf{c}_3\end{matrix}}{\begin{pmatrix} 1 & 2 & 3 \\ 4 & 5 & 6 \end{pmatrix}} = \overset{\begin{matrix}\mathbf{Ac}_1 & \mathbf{Ac}_2 & \mathbf{Ac}_3\end{matrix}}{\begin{pmatrix} -11 & -13 & -15 \\ 1 & 2 & 3 \end{pmatrix}} \quad \text{[From above]}$$

Generally, matrix multiplication is carried out as follows:

Matrix multiplication of matrix **A** with n columns, and matrix $\mathbf{B} = (\mathbf{c}_1 \ \ \mathbf{c}_2 \ \cdots \ \mathbf{c}_k)$ with n rows and k columns, is defined as

(1.11) $$\mathbf{AB} = \mathbf{A}(\mathbf{c}_1 \ \ \mathbf{c}_2 \ \cdots \ \mathbf{c}_k) = (\mathbf{Ac}_1 \ \ \mathbf{Ac}_2 \ \cdots \ \mathbf{Ac}_k)$$

Matrix multiplication **AB** means that the matrix **A** acts on each of the column vectors of **B**.

For **AB** we must have

Number of columns of matrix **A** = Number of rows of matrix **B**.

We can only multiply matrices if the number of columns of the left hand matrix equals the number of rows of the right hand matrix.

Remember, we defined a matrix times a vector as row by column multiplication. Applying this to two lots of 2 by 2 matrices gives:

$$\begin{pmatrix} a & b \\ c & d \end{pmatrix} \times \begin{pmatrix} e & f \\ g & h \end{pmatrix} = \begin{pmatrix} ae + bg & af + bh \\ ce + dg & cf + dh \end{pmatrix}$$

For example, applying this to the matrices:

$$\begin{pmatrix} 2 & 1 \\ 3 & 7 \end{pmatrix} \times \begin{pmatrix} 4 & 9 \\ 5 & 8 \end{pmatrix} = \begin{pmatrix} (2\times 4) + (1\times 5) & (2\times 9) + (1\times 8) \\ (3\times 4) + (7\times 5) & (3\times 9) + (7\times 8) \end{pmatrix} = \begin{pmatrix} 13 & 26 \\ 47 & 83 \end{pmatrix}$$

Matrix multiplication can be carried out row by column:

$$\begin{matrix}\text{Row 1} \\ \text{Row 2}\end{matrix}\begin{pmatrix} 2 & 1 \\ 3 & 7 \end{pmatrix} \times \overset{\begin{matrix}\text{Col. 1} & \text{Col. 2}\end{matrix}}{\begin{pmatrix} 4 & 9 \\ 5 & 8 \end{pmatrix}} = \begin{pmatrix} \text{Row 1} \times \text{Col. 1} & \text{Row 1} \times \text{Col. 2} \\ \text{Row 2} \times \text{Col. 1} & \text{Row 2} \times \text{Col. 2} \end{pmatrix}$$

We can only multiply matrices if the number of columns of the left hand matrix equals the number of rows of the right hand matrix.

In general, if **A** is a $m \times r$ (m rows by r columns) matrix and **B** is a $r \times n$ (r rows by n columns) matrix then the multiplication **AB** results in a $m \times n$ matrix.

We can illustrate the size of the matrix multiplication as shown in Fig. 1.33.

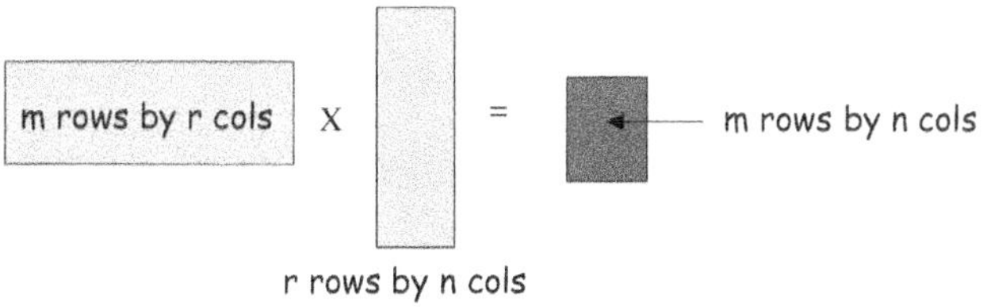

Figure 1.33

Example 1.19

Calculate the following, if possible:

(a) $(3\ \ 4)\begin{pmatrix} 2 & 3 \\ -1 & 5 \end{pmatrix}$ **(b)** $\begin{pmatrix} 2 & 3 & 6 \\ 1 & 5 & 7 \end{pmatrix}\begin{pmatrix} 3 & 7 \\ 4 & 2 \\ 1 & 3 \end{pmatrix}$ **(c)** $\begin{pmatrix} 2 & 3 & 6 \\ 1 & 5 & 7 \end{pmatrix}\begin{pmatrix} 3 & 7 \\ 4 & 2 \end{pmatrix}$ **(d)** $\begin{pmatrix} 1 \\ 2 \end{pmatrix}(x\ \ y)$

Solution

(a) Row by column multiplication gives:

$$\begin{aligned}(3\ \ 4)\begin{pmatrix} 2 & 3 \\ -1 & 5 \end{pmatrix} &= \big((3\times 2)+(4\times -1)\quad (3\times 3)+(4\times 5)\big) \\ &= (2\ \ 29)\end{aligned}$$

(b) Again using row by column gives:

$$\begin{aligned}\begin{pmatrix} 2 & 3 & 6 \\ 1 & 5 & 7 \end{pmatrix}\begin{pmatrix} 3 & 7 \\ 4 & 2 \\ 1 & 3 \end{pmatrix} &= \begin{pmatrix} (2\times 3)+(3\times 4)+(6\times 1) & (2\times 7)+(3\times 2)+(6\times 3) \\ (1\times 3)+(5\times 4)+(7\times 1) & (1\times 7)+(5\times 2)+(7\times 3) \end{pmatrix} \\ &= \begin{pmatrix} 24 & 38 \\ 30 & 38 \end{pmatrix}\end{aligned}$$

(c) $\begin{pmatrix} 2 & 3 & 6 \\ 1 & 5 & 7 \end{pmatrix}\begin{pmatrix} 3 & 7 \\ 4 & 2 \end{pmatrix}$. Since the number of columns in the left hand matrix is three and the number of rows in the right hand matrix is two, we cannot multiply these matrices together.

(d) Applying the row by column technique we have

$$\begin{pmatrix} 1 \\ 2 \end{pmatrix}(x\ \ y) = \begin{pmatrix} 1\times x & 1\times y \\ 2\times x & 2\times y \end{pmatrix} = \begin{pmatrix} x & y \\ 2x & 2y \end{pmatrix}$$

Note, the size of this matrix multiplication is 2 by 2.

A matrix with just one row such as $(3\ \ 4)$ is called a **row matrix** or **row vector**. In general, a matrix of size $1 \times n$ (one row by n columns) is called a row vector.

Are there any other row vectors in Example 1.19?

Yes, $(x\ \ y)$.

*What do you think the term **column matrix** means?*

A matrix with just one column such as $\begin{pmatrix} 1 \\ 2 \end{pmatrix}$. Again, a matrix of size $n \times 1$ (n rows by one column) is generally called a **column vector**.

The following example demonstrates the practical application of matrix multiplication.

A local shop sells three types of ice cream flavours: strawberry, vanilla and chocolate. Strawberry costs £2 each, vanilla £1 each and chocolate £3 each. The sales of each ice cream are as shown in Table 1.2.

Table 1.2

	Monday	Tuesday	Wednesday	Thursday	Friday
Strawberry (S)	12	15	10	16	12
Vanilla (V)	5	9	14	7	10
Chocolate (C)	8	12	10	9	15

What are the sales takings for each of these days?

Of course, you can solve this problem without matrices, but using matrix notation provides a systematic way of evaluating the sales for each day.

Let $S =$ Strawberry, $V =$ Vanilla and $C =$ Chocolate. Writing out the matrices and carrying out the matrix multiplication row by column gives:

$$\overset{\displaystyle S\ V\ C}{\begin{pmatrix} 2 & 1 & 3 \end{pmatrix}} \overset{\displaystyle M \quad T \quad W \quad TH \quad F}{\begin{pmatrix} 12 & 15 & 10 & 16 & 12 \\ 5 & 9 & 14 & 7 & 10 \\ 8 & 12 & 10 & 9 & 15 \end{pmatrix}} = \begin{pmatrix} 53 & 75 & 64 & 66 & 79 \end{pmatrix}$$

Hence the takings for Monday are:

$$(2 \times 12) + (1 \times 5) + (3 \times 8) = £53$$

Similarly for the other days, we have Tuesday £75, Wednesday £64, Thursday £66 and Friday £79. The matrix on the right hand side gives the takings for each weekday.

1.4.6 Computation of matrices

We can write the computation of matrices in compact form. For example, let **A** and **B** be matrices then **A** + **B** means add the matrices **A** and **B**. The computation 2**A** means multiply every entry of the matrix **A** by 2.

A software package such as MATLAB can be used for computing matrices. In fact, MATLAB is short for 'Matrix Laboratory'. It is a very useful tool which can be used to eliminate the drudgery from lengthy calculations.

There are many mathematical software packages such as MAPLE and MATHEMATICA, but MATLAB is particularly useful for linear algebra. MATLAB commands and other details are given on the website. In the present release of MATLAB, the matrices are entered with square brackets rather than round ones.

For example, in MATLAB to write the matrix $\mathbf{A} = \begin{pmatrix} 1 & 2 \\ 3 & 4 \end{pmatrix}$ we enter A=[1 2; 3 4] after the command prompt ≫. The semicolon indicates the end of the row.

Example 1.20

Let $\mathbf{A} = \begin{pmatrix} 1 & 2 & 5 \\ 4 & 6 & 9 \end{pmatrix}$, $\mathbf{B} = \begin{pmatrix} 2 & 5 & 6 \\ 1 & 7 & 2 \\ 9 & 6 & 1 \end{pmatrix}$ and $\mathbf{C} = \begin{pmatrix} 5 & 4 & 9 \\ 7 & 4 & 0 \\ 6 & 9 & 8 \end{pmatrix}$

Compute the following:

(a) $\mathbf{B} + \mathbf{C}$ **(b)** $\mathbf{A} + \mathbf{B}$ **(c)** $3\mathbf{B}+2\mathbf{C}$ **(d)** $\mathbf{AB}$ **(e)** $\mathbf{BA}$

What do you notice about your results to parts (d) and (e)?

Solution

(a) Adding the corresponding entries by hand gives

$$\mathbf{B} + \mathbf{C} = \begin{pmatrix} 2 & 5 & 6 \\ 1 & 7 & 2 \\ 9 & 6 & 1 \end{pmatrix} + \begin{pmatrix} 5 & 4 & 9 \\ 7 & 4 & 0 \\ 6 & 9 & 8 \end{pmatrix} = \begin{pmatrix} 2+5 & 5+4 & 6+9 \\ 1+7 & 7+4 & 2+0 \\ 9+6 & 6+9 & 1+8 \end{pmatrix} = \begin{pmatrix} 7 & 9 & 15 \\ 8 & 11 & 2 \\ 15 & 15 & 9 \end{pmatrix}$$

To use MATLAB, enter the matrices $\mathbf{B}$ and $\mathbf{C}$ after the command prompt >>. Separate each matrix by a comma. Then enter $\mathbf{B} + \mathbf{C}$ and the output should give the same result as above.

(b) $\mathbf{A} + \mathbf{B}$ is impossible because matrices are of different size.

What sizes are matrix $\mathbf{A}$ *and matrix* $\mathbf{B}$*?*

$\mathbf{A}$ is a 2×3 matrix and $\mathbf{B}$ is a 3×3 matrix, so we cannot add (or subtract) these matrices.

If we try this in MATLAB, we receive a message saying 'Matrix dimensions must agree'.

(c) We have

$$3\mathbf{B} + 2\mathbf{C} = 3\begin{pmatrix} 2 & 5 & 6 \\ 1 & 7 & 2 \\ 9 & 6 & 1 \end{pmatrix} + 2\begin{pmatrix} 5 & 4 & 9 \\ 7 & 4 & 0 \\ 6 & 9 & 8 \end{pmatrix}$$

$$= \begin{pmatrix} 6 & 15 & 18 \\ 3 & 21 & 6 \\ 27 & 18 & 3 \end{pmatrix} + \begin{pmatrix} 10 & 8 & 18 \\ 14 & 8 & 0 \\ 12 & 18 & 16 \end{pmatrix} = \begin{pmatrix} 16 & 23 & 36 \\ 17 & 29 & 6 \\ 39 & 36 & 19 \end{pmatrix} \left[\begin{array}{l}\text{Adding the} \\ \text{corresponding} \\ \text{entries}\end{array}\right]$$

In MATLAB, multiplication is carried out by using the command *. Once the matrices $\mathbf{B}$ and $\mathbf{C}$ have been entered, you do not need to enter them again. Enter 3*$\mathbf{B}$ + 2*$\mathbf{C}$ to check the above evaluation.

(d) Using row by column multiplication we have

$$\mathbf{AB} = \begin{pmatrix} 1 & 2 & 5 \\ 4 & 6 & 9 \end{pmatrix} \times \begin{pmatrix} 2 & 5 & 6 \\ 1 & 7 & 2 \\ 9 & 6 & 1 \end{pmatrix}$$

$$= \begin{pmatrix} (1\times 2)+(2\times 1)+(5\times 9) & (1\times 5)+(2\times 7)+(5\times 6) & (1\times 6)+(2\times 2)+(5\times 1) \\ (4\times 2)+(6\times 1)+(9\times 9) & (4\times 5)+(6\times 7)+(9\times 6) & (4\times 6)+(6\times 2)+(9\times 1) \end{pmatrix}$$

$$= \begin{pmatrix} 2+2+45 & 5+14+30 & 6+4+5 \\ 8+6+81 & 20+42+54 & 24+12+9 \end{pmatrix} = \begin{pmatrix} 49 & 49 & 15 \\ 95 & 116 & 45 \end{pmatrix}$$

(continued...)

(e) **BA** is impossible to calculate because the number of columns in the left hand matrix, **B**, is not equal to the number of rows in the right hand matrix, **A**.

*How many columns does the matrix **B** have?*
Matrix **B** has three columns but matrix **A** has two rows. Since the number of columns of matrix **B**, three, does not match the number of rows of matrix **A**, two, we cannot evaluate **BA**.

Notice that the matrix multiplication **AB** does not equal the matrix multiplication **BA**. In fact we have just demonstrated that **BA** cannot be evaluated while **AB** is computed, as shown in part (d) above.

It is important to note that matrix multiplication is not the same as multiplying two real numbers. In matrix multiplication the order of the multiplication does matter.

We will discuss important arithmetic properties of matrices in the next section but next we apply this matrix multiplication to transformations.

The six vertices of the letter L shown in Fig. 1.34 can be represented by the matrix **L** given by:

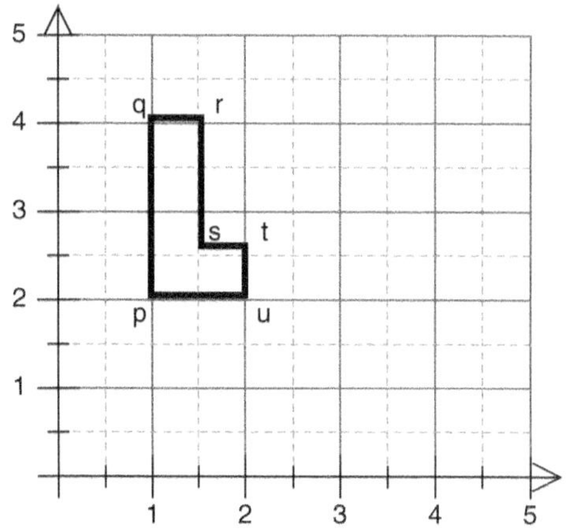

$$\mathbf{L} = \begin{pmatrix} \overset{p}{1} & \overset{q}{1} & \overset{r}{1.5} & \overset{s}{1.5} & \overset{t}{2} & \overset{u}{2} \\ 2 & 4 & 4 & 2.5 & 2.5 & 2 \end{pmatrix}$$

Figure 1.34

Example 1.21

Let $\mathbf{A} = \begin{pmatrix} 1 & 0.5 \\ 0 & 1 \end{pmatrix}$ and $\mathbf{L} = \begin{pmatrix} 1 & 1 & 1.5 & 1.5 & 2 & 2 \\ 2 & 4 & 4 & 2.5 & 2.5 & 2 \end{pmatrix}$ be the matrix representing the points *pqrstu* shown in Fig. 1.34. Determine the image of the letter L under the transformation carried out by the matrix multiplication **AL**.

Solution

Let the image of the corners *p*, *q*, *r*, *s*, *t* and *u* be denoted by p', q', r', s', t' and u' respectively. The six vertices *p*, *q*, *r*, *s*, *t* and *u* of letter L are vectors, and the matrix **A** transforms each of these to new vectors p', q', r', s', t' and u' respectively.

Carrying out the matrix multiplication $\mathbf{AL}$:

$$\mathbf{AL} = \begin{pmatrix} 1 & 0.5 \\ 0 & 1 \end{pmatrix} \begin{pmatrix} \overset{p}{1} & \overset{q}{1} & \overset{r}{1.5} & \overset{s}{1.5} & \overset{t}{2} & \overset{u}{2} \\ 2 & 4 & 4 & 2.5 & 2.5 & 2 \end{pmatrix} = \begin{pmatrix} \overset{p'}{2} & \overset{q'}{3} & \overset{r'}{3.5} & \overset{s'}{2.75} & \overset{t'}{3.25} & \overset{u'}{3} \\ 2 & 4 & 4 & 2.5 & 2.5 & 2 \end{pmatrix}$$

Plotting the transformed letter $p'q'r's't'u'$ as shown in Fig. 1.35.

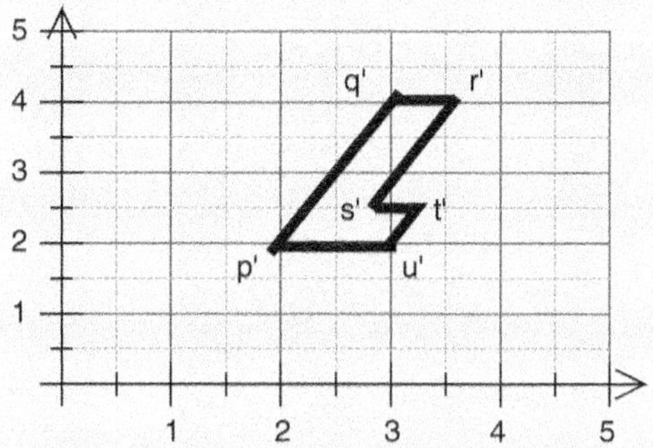

Figure 1.35

The transformation represented by matrix multiplication $\mathbf{AL}$ *italicizes* the letter L. The columns of our transformed matrix are the coordinates of the italicized letter L.

How does matrix $\mathbf{A}$ in Example 1.21 produce this italics of the letter L?

The matrix $\mathbf{A} = \begin{pmatrix} 1 & \boxed{0.5} \\ 0 & 1 \end{pmatrix}$ shifts the coordinates p, q, r, s, t and u horizontally by half or 0.5 of their height because of the 0.5 entry in the first row. Matrix $\mathbf{A}$ transforms vectors as follows:

$$\begin{pmatrix} 1 \\ 2 \end{pmatrix} \to \begin{pmatrix} 2 \\ 2 \end{pmatrix}, \begin{pmatrix} 1 \\ 4 \end{pmatrix} \to \begin{pmatrix} 3 \\ 4 \end{pmatrix}, \begin{pmatrix} 1.5 \\ 4 \end{pmatrix} \to \begin{pmatrix} 3.5 \\ 4 \end{pmatrix}, \ldots, \begin{pmatrix} 2 \\ 2 \end{pmatrix} \to \begin{pmatrix} 3 \\ 2 \end{pmatrix}$$

Summary

A matrix is an array of numbers. We can add or subtract matrices of the same size.
Scalar multiplication scales the matrix.
Matrix multiplication $\mathbf{AB}$ is carried out by multiplying the column vectors of matrix $\mathbf{B}$ by the matrix $\mathbf{A}$. $\mathbf{AB}$ means that the matrix $\mathbf{A}$ acts on each of the column vectors of matrix $\mathbf{B}$.

EXERCISES 1.4

(Brief solutions at end of book. Full solutions available at <http://www.oup.co.uk/companion/singh>.)

You may like to check your numerical solutions by using MATLAB.

1. For $\mathbf{A} = \begin{pmatrix} 1 & 2 \\ 3 & -1 \end{pmatrix}$, $\mathbf{B} = \begin{pmatrix} 6 & -1 \\ 5 & 3 \end{pmatrix}$ and $\mathbf{C} = \begin{pmatrix} -1 \\ 1 \end{pmatrix}$, (uppercase $\mathbf{C}$ is a vector) evaluate, where possible, the following:

(a) **A** + **B** (b) **B** + **A** (c) **B** + **B** + **B** (d) 3**B**
(e) 3**A** + 2**B** (f) **A** + **C** (g) **B** + **C** (h) **AC**
(i) **BC** (j) 5**A** − 7**BC** (k) 3**AC** − 2**BC**

What do you notice about your answers to parts (c) and (d)?

2. Consider the following matrices:

$$\mathbf{A}=\begin{pmatrix}1 & -1 & 7\\ 2 & 9 & 6\end{pmatrix},\ \mathbf{B}=\begin{pmatrix}5 & 1 & 4\\ 8 & 2 & 7\\ 1 & 4 & 9\end{pmatrix},\ \mathbf{C}=\begin{pmatrix}2\\ 4\\ 7\end{pmatrix}\ \text{and}\ \mathbf{D}=\begin{pmatrix}7 & 9 & 4\\ 1 & 3 & -5\\ 2 & -1 & -3\end{pmatrix}$$

Compute the following where possible:

(a) **A** − **A** (b) 3**A** − 2**A** (c) **BC** (d) **CB** (e) **B** + **D** (f) **D** + **B**

(g) **A** − **C** (h) $\frac{1}{2}\mathbf{C}$ (i) **BD** (j) **DB** (k) **BD** − **DB** (l) **CD**

3. Evaluate the following (a, b, c and d are any real numbers):

(a) $\begin{pmatrix}2 & 4\\ 3 & 9\end{pmatrix}\begin{pmatrix}1 & 0\\ 0 & 1\end{pmatrix}$ (b) $\begin{pmatrix}6 & 7\\ 2 & 3\end{pmatrix}\begin{pmatrix}1 & 0\\ 0 & 1\end{pmatrix}$ (c) $\begin{pmatrix}a & b\\ c & d\end{pmatrix}\begin{pmatrix}1 & 0\\ 0 & 1\end{pmatrix}$

(d) $\begin{pmatrix}2 & 3 & 6\\ 1 & 4 & 5\\ 0 & 9 & 7\end{pmatrix}\begin{pmatrix}1 & 0 & 0\\ 0 & 1 & 0\\ 0 & 0 & 1\end{pmatrix}$

What do you notice about your results?

4. Determine the following:

(a) $\begin{pmatrix}3 & 7\\ 2 & 5\end{pmatrix}\begin{pmatrix}5 & -7\\ -2 & 3\end{pmatrix}$ (b) $\frac{1}{5}\begin{pmatrix}3 & -4\\ -7 & 11\end{pmatrix}\begin{pmatrix}11 & 4\\ 7 & 3\end{pmatrix}$ (c) $\frac{1}{4}\begin{pmatrix}7 & -9\\ -5 & 7\end{pmatrix}\begin{pmatrix}7 & 9\\ 5 & 7\end{pmatrix}$

5. Let $\mathbf{A}=\begin{pmatrix}1 & -1\\ 1 & -1\end{pmatrix}$ and find $\mathbf{A}^2=\mathbf{A}\times\mathbf{A}$. [$\mathbf{A}^2$ in MATLAB is evaluated by the command A^2]. Comment upon your result.

6. Evaluate $\begin{pmatrix}5 & -1 & -2\\ 10 & -2 & -4\\ 15 & -3 & -6\end{pmatrix}\begin{pmatrix}1 & 1 & 3\\ 1 & -1 & -1\\ 2 & 3 & 8\end{pmatrix}$. Comment upon your result.

7. Complete the subscript numbers in the following matrix:

$$\begin{pmatrix}a_{\square} & a_{\square} & a_{\square} & a_{\square}\\ a_{\square} & a_{\square} & a_{\square} & a_{\square}\end{pmatrix}$$

8. Determine $\mathbf{A}^2=\mathbf{A}\times\mathbf{A}$, $\mathbf{A}^3=\mathbf{A}\times\mathbf{A}\times\mathbf{A}$ and $\mathbf{A}^4=\mathbf{A}\times\mathbf{A}\times\mathbf{A}\times\mathbf{A}$ for

(a) $\mathbf{A}=\begin{pmatrix}1 & 0\\ 0 & 1\end{pmatrix}$ (b) $\mathbf{A}=\begin{pmatrix}0 & 1\\ -1 & 0\end{pmatrix}$ (c) $\mathbf{A}=\begin{pmatrix}1/2 & 1/2\\ 1/2 & 1/2\end{pmatrix}$

(The matrix in part (c) is called a **Markov matrix** because the numbers in the columns total 1.)

*A **discrete dynamical system** is described by the formula $\mathbf{x}_n=\mathbf{A}^n\mathbf{x}$. Write down the formula for $\mathbf{x}_n=\mathbf{A}^n\mathbf{x}$ with the matrices in parts (a), (b) and (c).

9. Let the top of a table be given by the coordinates $P(1,2)$, $Q(2,2)$, $R(2,4)$ and $S(1,4)$. Write down the matrix **A** which represents this table top $PQRS$.

Let

$$\mathbf{B} = \begin{pmatrix} 4 & 4 & 4 & 4 \\ 0 & 0 & 0 & 0 \end{pmatrix}, \quad \mathbf{C} = \begin{pmatrix} 0 & 1 \\ -1 & 0 \end{pmatrix} \text{ and } \mathbf{D} = \begin{pmatrix} -1 & 0 \\ 0 & 1 \end{pmatrix}$$

Determine the image of the table top under the following transformation and illustrate what effect each transformation has.

(a) $\mathbf{A} - \mathbf{B}$ (b) $3\mathbf{A}$ (c) $\mathbf{CA}$ (d) $\mathbf{DA}$

10. Determine the image of the matrix **F** under the transformation **AF** where

$$\mathbf{A} = \begin{pmatrix} 1 & 0.2 \\ 0 & 1 \end{pmatrix} \text{ and } \mathbf{F} = \begin{pmatrix} 1 & 1 & 2 & 2 & 1.4 & 1.4 & 2 & 2 & 1.4 & 1.4 \\ 1 & 3 & 3 & 2.6 & 2.6 & 2 & 2 & 1.6 & 1.6 & 1 \end{pmatrix}$$

Plot your image.

11. Determine the vector $\mathbf{x} = \begin{pmatrix} x \\ y \end{pmatrix}$ such that $\mathbf{Ax} = \mathbf{O}$ where $\mathbf{O} = \begin{pmatrix} 0 \\ 0 \end{pmatrix}$ for:

(a) $\mathbf{A} = \begin{pmatrix} 1 & 2 \\ 3 & 5 \end{pmatrix}$ (b) $\mathbf{A} = \begin{pmatrix} 2 & 7 \\ 3 & 15 \end{pmatrix}$ *(c) $\mathbf{A} = \begin{pmatrix} 1 & 4 \\ 3 & 12 \end{pmatrix}$

12. *Determine whether the vector **w** is a linear combination of vectors **u** and **v** for:

(i) $\mathbf{w} = \begin{pmatrix} 1 \\ 0 \end{pmatrix}$; $\mathbf{u} = \begin{pmatrix} 5 \\ 8 \end{pmatrix}$ and $\mathbf{v} = \begin{pmatrix} 2 \\ 4 \end{pmatrix}$

(ii) $\mathbf{w} = \begin{pmatrix} 0 \\ 1 \end{pmatrix}$; $\mathbf{u} = \begin{pmatrix} 5 \\ 8 \end{pmatrix}$ and $\mathbf{v} = \begin{pmatrix} 2 \\ 4 \end{pmatrix}$

(iii) $\mathbf{w} = \begin{pmatrix} 1 \\ 2 \\ 3 \end{pmatrix}$; $\mathbf{u} = \begin{pmatrix} 1 \\ 0 \\ 0 \end{pmatrix}$ and $\mathbf{v} = \begin{pmatrix} 0 \\ 1 \\ 0 \end{pmatrix}$

(iv) $\mathbf{w} = \begin{pmatrix} 1 \\ 2 \\ 3 \end{pmatrix}$; $\mathbf{u} = \begin{pmatrix} 4 \\ 8 \\ 0 \end{pmatrix}$ and $\mathbf{v} = \begin{pmatrix} 1 \\ 2 \\ -3/7 \end{pmatrix}$

SECTION 1.5 Matrix Algebra

By the end of this section you will be able to

- simplify matrices using algebra
- prove and use results of matrix arithmetic
- compare and contrast between matrix and real number algebra

This section is a little different from previous sections in that it requires us to provide justification for the solutions we arrive at.

Why do we need to prove results?

Figure 1.36 Andrew Wiles 1953–present © C. J. Mozzochi, Princeton, NJ.

When the British mathematician Sir Andrew Wiles (born 11 April 1953) – see Fig. 1.36 – was a young boy, he stumbled upon one of the most enduring mathematics problems in history.

In 1637, lawyer and amateur mathematician Pierre de Fermat stated that '*there are no positive integers a, b and c, that satisfy the equation* $a^n + b^n = c^n$, *for any integer value of n greater than two.*' Unfortunately, the truth of this statement is not self-evident, even if it was to Fermat. If we start by trying $n = 3$, there are an infinite number of values to try for a, b and c. Because Fermat left no proof to back up his claim, the statement could never be verified, although huge prizes were offered for a solution.

Wiles dedicated most of his life to the seemingly unfathomable problem, and after 32 years of studying, finally struck upon a series of complicated steps that unquestionably underpinned Fermat's claim.

From this example, it is easy to see why getting into the habit of proving your results is a good idea. We want to prove our results because:

1. We want to make sure that our mathematical statements are logically correct.
2. Proofs lead to a deeper understanding of concepts.
3. Proofs explain not only 'how', but 'why' a particular statement is unquestionably true.
4. Proofs generalize mathematical concepts.

Mathematical proof is a lot more powerful than a scientific proof because it is not subject to experimental data, so once we have carried out a mathematical proof we know that our result is absolutely correct and permanent. Pythagoras' theorem is as true today as it was over two thousand years ago. Pythagoras died knowing that no one could ever dispute his theorem, because it was developed on absolute truth.

For example, we often use the identity $(x + y)^2 = x^2 + 2xy + y^2$ but this is not true simply because it works for every example we try but because we can prove this result. Once you have proven the general result $(x + y)^2 = x^2 + 2xy + y^2$ then you don't need to worry about trying it for particular numbers because you have no doubt that the result is correct.

A lot of students have difficulties with this abstract mathematics of proving statements because it is initially very hard to understand what is going on. If you can push yourself to think logically then after a while you will enjoy this kind of mathematics a lot more than the techniques or methods of approach that we have been using up to now. It is better to understand mathematics rather than just learn to follow certain rules.

Proofs are a critical part of the mathematical thought process but it does take time to digest and understand proofs.

1.5.1 Properties of matrix addition

First, we discuss the zero matrix.

What do you think the term zero matrix means?

A zero matrix is a matrix which has all zero entries and it is denoted by $\mathbf{O}$. The 2×2, 2×4 and 3×3 zero matrices are

$$\begin{pmatrix} 0 & 0 \\ 0 & 0 \end{pmatrix}, \begin{pmatrix} 0 & 0 & 0 & 0 \\ 0 & 0 & 0 & 0 \end{pmatrix} \text{ and } \begin{pmatrix} 0 & 0 & 0 \\ 0 & 0 & 0 \\ 0 & 0 & 0 \end{pmatrix} \text{ respectively.}$$

Sometimes the zero matrix is denoted by $\mathbf{O}_{mn}$, meaning that it is a zero matrix of size $m \times n$ (m rows by n columns). The above matrices are denoted $\mathbf{O}_{22}$, $\mathbf{O}_{24}$ and $\mathbf{O}_{33}$ respectively.

In Exercises 1.4, question 1 (a) and (b), we showed that for particular matrices $\mathbf{A}$ and $\mathbf{B}$

$$\mathbf{A} + \mathbf{B} = \mathbf{B} + \mathbf{A}.$$

This is true for all matrices that can be added.

In Exercises 1.4, question 1 (c) and (d) we also showed that for the particular matrix B, $\mathbf{B} + \mathbf{B} + \mathbf{B} = 3\mathbf{B}$. In general,

$$\underbrace{\mathbf{B} + \mathbf{B} + \mathbf{B} + \cdots + \mathbf{B}}_{n \text{ copies}} = n\mathbf{B}$$

Most of the laws of algebra of real numbers can also be extended to algebra of matrices. These laws, and the above properties, are summarized in the next theorem.

Theorem (1.12). Properties of Matrix Addition.
Let $\mathbf{A}$, $\mathbf{B}$ and $\mathbf{C}$ be matrices of the same size $m \times n$ (m rows by n columns). Then

(a) $\mathbf{A} + \mathbf{B} = \mathbf{B} + \mathbf{A}$

(b) $(\mathbf{A} + \mathbf{B}) + \mathbf{C} = \mathbf{A} + (\mathbf{B} + \mathbf{C})$ (Associative law for addition)

(c) $\underbrace{\mathbf{A} + \mathbf{A} + \mathbf{A} + \cdots + \mathbf{A}}_{k \text{ copies}} = k\mathbf{A}$

(d) There is an $m \times n$ matrix called the zero matrix, denoted $\mathbf{O}$, which has the property

$$\mathbf{A} + \mathbf{O} = \mathbf{A}$$

(e) There is a matrix denoted $-\mathbf{A}$ such that

$$\mathbf{A} + (-\mathbf{A}) = \mathbf{A} - \mathbf{A} = \mathbf{O}$$

This matrix $-\mathbf{A}$ is called the **additive inverse** of $\mathbf{A}$.

Of course, having worked through the previous sections, we feel that the statements above must be true. But we need to prove these statements for *all* matrices and not only the ones in the previous sections.

Proof.

We will prove parts (a) and (c) only. The rest are left as questions in Exercises 1.5.

Let $\mathbf{A} = \begin{pmatrix} a_{11} & \cdots & a_{1n} \\ \vdots & \vdots & \vdots \\ a_{m1} & \cdots & a_{mn} \end{pmatrix}$ and $\mathbf{B} = \begin{pmatrix} b_{11} & \cdots & b_{1n} \\ \vdots & \vdots & \vdots \\ b_{m1} & \cdots & b_{mn} \end{pmatrix}$.

(a) Remember, the order of adding real numbers does not matter, that is for real numbers a and b we have $a + b = b + a$.

Adding the matrices **A** and **B** given above yields

$$\begin{aligned}
\mathbf{A}+\mathbf{B} &= \begin{pmatrix} a_{11} & \cdots & a_{1n} \\ \vdots & \vdots & \vdots \\ a_{m1} & \cdots & a_{mn} \end{pmatrix} + \begin{pmatrix} b_{11} & \cdots & b_{1n} \\ \vdots & \vdots & \vdots \\ b_{m1} & \cdots & b_{mn} \end{pmatrix} \\
&= \begin{pmatrix} a_{11}+b_{11} & \cdots & a_{1n}+b_{1n} \\ \vdots & \vdots & \vdots \\ a_{m1}+b_{m1} & \cdots & a_{mn}+b_{mn} \end{pmatrix} && \left[\begin{array}{l}\text{Adding the} \\ \text{corresponding} \\ \text{entries}\end{array}\right] \\
&= \begin{pmatrix} b_{11}+a_{11} & \cdots & b_{1n}+a_{1n} \\ \vdots & \vdots & \vdots \\ b_{m1}+a_{m1} & \cdots & b_{mn}+a_{mn} \end{pmatrix} && \left[\begin{array}{l}\text{Changing the} \\ \text{order of} \\ \text{addition}\end{array}\right] \\
&= \underbrace{\begin{pmatrix} b_{11} & \cdots & b_{1n} \\ \vdots & \vdots & \vdots \\ b_{m1} & \cdots & b_{mn} \end{pmatrix}}_{=\mathbf{B}} + \underbrace{\begin{pmatrix} a_{11} & \cdots & a_{1n} \\ \vdots & \vdots & \vdots \\ a_{m1} & \cdots & a_{mn} \end{pmatrix}}_{=\mathbf{A}} \\
&= \mathbf{B}+\mathbf{A}
\end{aligned}$$

Hence we have proved the statement, $\mathbf{A}+\mathbf{B} = \mathbf{B}+\mathbf{A}$. ■

(c) Here we are required to prove that $\underbrace{\mathbf{A}+\mathbf{A}+\mathbf{A}+\cdots+\mathbf{A}}_{k \text{ copies}} = k\mathbf{A}$. We have

$$\begin{aligned}
\underbrace{\mathbf{A}+\mathbf{A}+\mathbf{A}+\cdots+\mathbf{A}}_{k \text{ copies}} &= \begin{pmatrix} a_{11} & \cdots & a_{1n} \\ \vdots & \vdots & \vdots \\ a_{m1} & \cdots & a_{mn} \end{pmatrix} + \begin{pmatrix} a_{11} & \cdots & a_{1n} \\ \vdots & \vdots & \vdots \\ a_{m1} & \cdots & a_{mn} \end{pmatrix} + \cdots + \begin{pmatrix} a_{11} & \cdots & a_{1n} \\ \vdots & \vdots & \vdots \\ a_{m1} & \cdots & a_{mn} \end{pmatrix} \\
&= \begin{pmatrix} \underbrace{a_{11}+a_{11}+\cdots+a_{11}}_{k \text{ copies}} & \cdots & \underbrace{a_{1n}+a_{1n}+\cdots+a_{1n}}_{k \text{ copies}} \\ \vdots & \vdots & \vdots \\ \underbrace{a_{m1}+a_{m1}+\cdots+a_{m1}}_{k \text{ copies}} & \cdots & \underbrace{a_{mn}+a_{mn}+\cdots+a_{mn}}_{k \text{ copies}} \end{pmatrix} \\
&= \begin{pmatrix} ka_{11} & \cdots & ka_{1n} \\ \vdots & \vdots & \vdots \\ ka_{m1} & \cdots & ka_{mn} \end{pmatrix} = \underbrace{k}_{\substack{\text{factorizing} \\ \text{out } k}} \underbrace{\begin{pmatrix} a_{11} & \cdots & a_{1n} \\ \vdots & \vdots & \vdots \\ a_{m1} & \cdots & a_{mn} \end{pmatrix}}_{=\mathbf{A}} = k\mathbf{A}
\end{aligned}$$

Hence we have proved that $\underbrace{\mathbf{A}+\mathbf{A}+\mathbf{A}+\cdots+\mathbf{A}}_{k \text{ copies}} = k\mathbf{A}$.

■

The symbol ■ signifies the end of a proof. We will use this symbol ■ throughout the book.

We can use these results to simplify matrices as the next example demonstrates.

Example 1.22

Let $\mathbf{A} = \begin{pmatrix} 1 & -1 & 3 & 7 \\ 2 & 9 & 5 & -6 \end{pmatrix}$, $\mathbf{B} = \begin{pmatrix} 7 & 6 & -3 & 2 \\ 1 & 4 & 5 & 3 \end{pmatrix}$ and $\mathbf{C} = \begin{pmatrix} -2 & -7 & 8 & 6 \\ 3 & -9 & 2 & 1 \end{pmatrix}$. Determine

(a) $\mathbf{A}+\mathbf{B}$ **(b)** $\mathbf{B}+\mathbf{A}$ **(c)** $(\mathbf{A}+\mathbf{B})+\mathbf{C}$ **(d)** $\mathbf{A}+(\mathbf{B}+\mathbf{C})$

(e) $\mathbf{A}+\mathbf{O}$ **(f)** $\mathbf{A}+\mathbf{A}+\mathbf{A}+\mathbf{A}+\mathbf{A}$ **(g)** $\mathbf{C}+(-\mathbf{C})$

Solution

(a) We have

$$\mathbf{A}+\mathbf{B} = \begin{pmatrix} 1 & -1 & 3 & 7 \\ 2 & 9 & 5 & -6 \end{pmatrix} + \begin{pmatrix} 7 & 6 & -3 & 2 \\ 1 & 4 & 5 & 3 \end{pmatrix}$$

$$= \begin{pmatrix} 1+7 & -1+6 & 3+(-3) & 7+2 \\ 2+1 & 9+4 & 5+5 & -6+3 \end{pmatrix} = \begin{pmatrix} 8 & 5 & 0 & 9 \\ 3 & 13 & 10 & -3 \end{pmatrix}$$

(b) Clearly by Theorem (1.12) (a) we have

$$\mathbf{B}+\mathbf{A} = \mathbf{A}+\mathbf{B} = \begin{pmatrix} 8 & 5 & 0 & 9 \\ 3 & 13 & 10 & -3 \end{pmatrix} \quad \text{[By part (a)]}$$

(c) We have already evaluated $\mathbf{A}+\mathbf{B}$ in part (a). Therefore we have

$$(\mathbf{A}+\mathbf{B})+\mathbf{C} = \underbrace{\begin{pmatrix} 8 & 5 & 0 & 9 \\ 3 & 13 & 10 & -3 \end{pmatrix}}_{\text{By Part (a)}} + \begin{pmatrix} -2 & -7 & 8 & 6 \\ 3 & -9 & 2 & 1 \end{pmatrix}$$

$$= \begin{pmatrix} 8-2 & 5-7 & 0+8 & 9+6 \\ 3+3 & 13-9 & 10+2 & -3+1 \end{pmatrix} = \begin{pmatrix} 6 & -2 & 8 & 15 \\ 6 & 4 & 12 & -2 \end{pmatrix}$$

(d) By Theorem (1.12) (b) we have

$$\mathbf{A}+(\mathbf{B}+\mathbf{C}) = (\mathbf{A}+\mathbf{B})+\mathbf{C} \underset{\text{By Part (c)}}{=} \begin{pmatrix} 6 & -2 & 8 & 15 \\ 6 & 4 & 12 & -2 \end{pmatrix}$$

(e) By Theorem (1.12) (d) we have $\mathbf{A}+\mathbf{O} = \mathbf{A} = \begin{pmatrix} 1 & -1 & 3 & 7 \\ 2 & 9 & 5 & -6 \end{pmatrix}$.

(continued...)

(f) By Theorem (1.12) (c) we have

$$\underbrace{\mathbf{A}+\mathbf{A}+\mathbf{A}+\mathbf{A}+\mathbf{A}}_{=5\mathbf{A}} = 5\mathbf{A} = 5\begin{pmatrix} 1 & -1 & 3 & 7 \\ 2 & 9 & 5 & -6 \end{pmatrix}$$

$$= \begin{pmatrix} 5 & -5 & 15 & 35 \\ 10 & 45 & 25 & -30 \end{pmatrix} \quad \begin{bmatrix}\text{Multiplying each} \\ \text{entry by 5}\end{bmatrix}$$

(g) By Theorem (1.12) (e) we have $\mathbf{C} + (-\mathbf{C}) = \mathbf{C} - \mathbf{C} = \begin{pmatrix} 0 & 0 & 0 & 0 \\ 0 & 0 & 0 & 0 \end{pmatrix} = \mathbf{O}_{24}$.

1.5.2 Properties of scalar multiplication

We now examine some properties of another fundamental linear algebra operation – scalar multiplication.

Theorem (1.13). Let $\mathbf{A}$ and $\mathbf{B}$ both be $m \times n$ (m rows by n columns) matrices and c, k be scalars. Then

(a) $(ck)\,\mathbf{A} = c\,(k\mathbf{A})$
(b) $k\,(\mathbf{A}+\mathbf{B}) = k\mathbf{A} + k\mathbf{B}$
(c) $(c+k)\,\mathbf{A} = c\mathbf{A} + k\mathbf{A}$

Proof.
We will only prove part (c) here. You are asked to prove the remaining results in Exercises 1.5.

Let $\mathbf{A} = \begin{pmatrix} a_{11} & \cdots & a_{1n} \\ \vdots & \vdots & \vdots \\ a_{m1} & \cdots & a_{mn} \end{pmatrix}$ then

$$(c+k)\,\mathbf{A} = (c+k)\begin{pmatrix} a_{11} & \cdots & a_{1n} \\ \vdots & \vdots & \vdots \\ a_{m1} & \cdots & a_{mn} \end{pmatrix}$$

$$= \begin{pmatrix} (c+k)\,a_{11} & \cdots & (c+k)\,a_{1n} \\ \vdots & \vdots & \vdots \\ (c+k)\,a_{m1} & \cdots & (c+k)\,a_{mn} \end{pmatrix} \quad \begin{bmatrix}\text{Multiplying each entry} \\ \text{by } (c+k)\end{bmatrix}$$

$$= \begin{pmatrix} ca_{11}+ka_{11} & \cdots & ca_{1n}+ka_{1n} \\ \vdots & \vdots & \vdots \\ ca_{m1}+ka_{m1} & \cdots & ca_{mn}+ka_{mn} \end{pmatrix} \quad \begin{bmatrix}\text{Opening up the} \\ \text{brackets}\end{bmatrix}$$

$$= \begin{pmatrix} ca_{11} & \cdots & ca_{1n} \\ \vdots & \vdots & \vdots \\ ca_{m1} & \cdots & ca_{mn} \end{pmatrix} + \begin{pmatrix} ka_{11} & \cdots & ka_{1n} \\ \vdots & \vdots & \vdots \\ ka_{m1} & \cdots & ka_{mn} \end{pmatrix} \quad \left[\begin{array}{l}\text{Separating out} \\ \text{terms into } c\text{'s and } k\text{'s}\end{array}\right]$$

$$= \underbrace{c}_{\substack{\text{Factorizing} \\ \text{out } c}} \underbrace{\begin{pmatrix} a_{11} & \cdots & a_{1n} \\ \vdots & \vdots & \vdots \\ a_{m1} & \cdots & a_{mn} \end{pmatrix}}_{=\mathbf{A}} + \underbrace{k}_{\substack{\text{Factorizing} \\ \text{out } k}} \underbrace{\begin{pmatrix} a_{11} & \cdots & a_{1n} \\ \vdots & \vdots & \vdots \\ a_{m1} & \cdots & a_{mn} \end{pmatrix}}_{=\mathbf{A}} = c\mathbf{A} + k\mathbf{A}$$

■

Example 1.23

Let $\mathbf{A} = \begin{pmatrix} -1 & 3 & 4 \\ 9 & 6 & 1 \end{pmatrix}$, $\mathbf{B} = \begin{pmatrix} 1 & 5 & 8 \\ -2 & 3 & 2 \end{pmatrix}$, $c = -2, k = 5$ and $\mathbf{x} = \begin{pmatrix} x \\ y \\ z \end{pmatrix}$. Determine:

(a) $(c+k)\,\mathbf{B}$ **(b)** $c\mathbf{B} + k\mathbf{B}$ **(c)** $-10\mathbf{A}$ **(d)** $c\,(k\mathbf{A})$ **(e)** $k\,(\mathbf{A}+\mathbf{B})\,\mathbf{x}$

(f) Write the following linear system in terms of $\mathbf{A}$, $\mathbf{B}$, c, k, $\mathbf{x}$ and $\mathbf{O}$:

$$\begin{aligned} 14x + 14y + 28z &= 0 \\ -77x - 21y + 7z &= 0 \end{aligned}$$

Solution

(a) We first evaluate $(c+k)$ and then scalar multiply this result by matrix $\mathbf{B}$:

$$(c+k)\,\mathbf{B} = (-2+5)\,\mathbf{B} \quad [\text{Substituting } c = -2 \text{ and } k = 5]$$

$$= 3\mathbf{B} = 3\begin{pmatrix} 1 & 5 & 8 \\ -2 & 3 & 2 \end{pmatrix} = \begin{pmatrix} 3 & 15 & 24 \\ -6 & 9 & 6 \end{pmatrix} \quad \left[\begin{array}{l}\text{Multiplying each} \\ \text{entry by 3}\end{array}\right]$$

(b) By Theorem (1.13) (c) we have $(c+k)\,\mathbf{B} = c\mathbf{B} + k\mathbf{B}$, therefore we have the same solution as part (a).

(c) *How do we evaluate $-10\mathbf{A}$?*
Multiply each entry of matrix $\mathbf{A}$ by -10:

$$-10\mathbf{A} = -10\begin{pmatrix} -1 & 3 & 4 \\ 9 & 6 & 1 \end{pmatrix} = \begin{pmatrix} 10 & -30 & -40 \\ -90 & -60 & -10 \end{pmatrix}$$

(d) *What is $c(k\mathbf{A})$ equal to?*
By Theorem (1.13) (a) we have $(ck)\,\mathbf{A} = c\,(k\mathbf{A})$ and $ck = -2 \times 5 = -10$, therefore $c\,(k\mathbf{A}) = (ck)\,\mathbf{A} = -10\mathbf{A}$. The evaluation of $-10\mathbf{A}$ was done in part (c) above.

(e) Adding the matrices $\mathbf{A}$ and $\mathbf{B}$ and multiplying the result by $k = 5$ and $\mathbf{x}$ we have

$$k\,(\mathbf{A}+\mathbf{B})\,\mathbf{x} = 5\left[\begin{pmatrix} -1 & 3 & 4 \\ 9 & 6 & 1 \end{pmatrix} + \begin{pmatrix} 1 & 5 & 8 \\ -2 & 3 & 2 \end{pmatrix}\right]\mathbf{x}$$

$$= 5\begin{pmatrix} 0 & 8 & 12 \\ 7 & 9 & 3 \end{pmatrix}\begin{pmatrix} x \\ y \\ z \end{pmatrix} = \begin{pmatrix} 0 & 40 & 60 \\ 35 & 45 & 15 \end{pmatrix}\begin{pmatrix} x \\ y \\ z \end{pmatrix} = \begin{pmatrix} 40y + 60z \\ 35x + 45y + 15z \end{pmatrix}$$

(continued...)

(f) We can write the linear system as

$$\begin{aligned} 14x + 14y + 28z &= 0 \\ -77x - 21y + 7z &= 0 \end{aligned} \quad \Rightarrow \quad \begin{pmatrix} 14 & 14 & 28 \\ -77 & -21 & 7 \end{pmatrix} \begin{pmatrix} x \\ y \\ z \end{pmatrix} = \begin{pmatrix} 0 \\ 0 \end{pmatrix}$$

The 2 by 3 matrix has a common factor of 7. Taking this out, we have

$$7\begin{pmatrix} 2 & 2 & 4 \\ -11 & -3 & 1 \end{pmatrix} \qquad (*)$$

Since we are given $c = -2$, $k = 5$, we can write $7 = 5 - (-2) = k - c$.

Can we find any relationship between the matrix in (*) and the given matrices $\mathbf{A} = \begin{pmatrix} -1 & 3 & 4 \\ 9 & 6 & 1 \end{pmatrix}$ *and* $\mathbf{B} = \begin{pmatrix} 1 & 5 & 8 \\ -2 & 3 & 2 \end{pmatrix}$*?*

Adding matrices **A** and **B** does not give us the coefficients in (*). However, **B** − **A** works:

$$\mathbf{B} - \mathbf{A} = \begin{pmatrix} 1 & 5 & 8 \\ -2 & 3 & 2 \end{pmatrix} - \begin{pmatrix} -1 & 3 & 4 \\ 9 & 6 & 1 \end{pmatrix} = \begin{pmatrix} 2 & 2 & 4 \\ -11 & -3 & 1 \end{pmatrix} \quad \left[\begin{array}{l}\text{Subtracting the} \\ \text{corresponding entries}\end{array}\right]$$

Hence we can write the given linear system as

$$(k - c)(\mathbf{B} - \mathbf{A})\mathbf{x} = \mathbf{O}$$

1.5.3 Properties of matrix multiplication

The algebraic rules of matrices are closely related to the algebraic rules of real numbers, but there are some differences which you need to be aware of.

Can you remember one difference between matrix and real number algebraic rules from the previous section?

Matrix multiplication is not commutative, which means that the *order in which* we multiply matrices matters. If **A** and **B** are matrices then in general

(1.14) $$\mathbf{AB} \neq \mathbf{BA} \quad [\text{Not equal}]$$

In Example 1.20 from the previous section we were given the matrices

$$\mathbf{A} = \begin{pmatrix} 1 & 2 & 5 \\ 4 & 6 & 9 \end{pmatrix} \text{ and } \mathbf{B} = \begin{pmatrix} 2 & 5 & 6 \\ 1 & 7 & 2 \\ 9 & 6 & 1 \end{pmatrix}$$

and we found that $\mathbf{AB} = \begin{pmatrix} 49 & 49 & 15 \\ 95 & 116 & 45 \end{pmatrix}$, but **BA** could not be evaluated because the number of columns (3) of the left hand matrix, **B**, is not equal to the number of rows (2) of the right hand matrix, **A**; that is, we cannot evaluate **BA** while **AB** is as given above.

You need to remember this important difference between matrix multiplication and multiplication of real numbers. We say that matrix multiplication is *not commutative*.

Consider the following matrix multiplication **AB**:

$$\begin{array}{c} \\ \\ \\ i\text{th row} \\ \\ \\ \end{array}\begin{pmatrix} a_{11} & a_{12} & \cdots & a_{1r} \\ a_{21} & a_{22} & & a_{2r} \\ \vdots & \vdots & \vdots & \vdots \\ \boxed{a_{i1}} & a_{i2} & & \cdots a_{ir} \\ \vdots & \vdots & \vdots & \vdots \\ a_{m1} & a_{m2} & \cdots & a_{mr} \end{pmatrix}\overset{j\text{th column}}{\begin{pmatrix} b_{11} & \cdots & \boxed{b_{1j}} & \cdots b_{1n} \\ b_{21} & \cdots & b_{2j} & \cdots b_{2n} \\ \vdots & \vdots & \vdots & \vdots \\ b_{r1} & \cdots & b_{rj} & b_{rn} \end{pmatrix}} = \begin{pmatrix} (\mathbf{AB})_{11} & \cdots & (\mathbf{AB})_{1j} & \cdots (\mathbf{AB})_{1n} \\ (\mathbf{AB})_{12} & \cdots & (\mathbf{AB})_{2j} & \cdots (\mathbf{AB})_{2n} \\ \vdots & \vdots & \vdots & \vdots \\ (\mathbf{AB})_{i1} & \cdots & \boxed{(\mathbf{AB})_{ij}} & \cdots (\mathbf{AB})_{in} \\ \vdots & \vdots & \vdots & \vdots \\ (\mathbf{AB})_{m1} & \cdots & (\mathbf{AB})_{mj} & \cdots (\mathbf{AB})_{mn} \end{pmatrix}$$

We can calculate specific entries of the matrix multiplication **AB** by using the notation $(\mathbf{AB})_{ij}$. Applying the row by column multiplication to the ith row and jth column yields:

$$a_{i1}b_{1j} + a_{i2}b_{2j} + \cdots + a_{ir}b_{rj} = (\mathbf{AB})_{ij}$$

The entry in the ith row and jth column of a matrix **AB** is denoted by $(\mathbf{AB})_{ij}$. Hence

$$(\mathbf{AB})_{ij} = \text{Row}_{\text{i}}\,(\mathbf{A}) \times \text{Column}_{\text{j}}\,(\mathbf{B})$$

where $i = 1, 2, 3, \ldots, m$ and $j = 1, 2, 3, \ldots, n$.

Writing out the complete row and column gives

$$(1.15) \qquad (\mathbf{AB})_{ij} = \begin{pmatrix} a_{i1} & a_{i2} & a_{i3} & \cdots & a_{ir} \end{pmatrix}\begin{pmatrix} b_{1j} \\ b_{2j} \\ \vdots \\ b_{rj} \end{pmatrix}$$

$$= \left(a_{i1}b_{1j}\right) + \left(a_{i2}b_{2j}\right) + \left(a_{i3}b_{3j}\right) + \cdots + \left(a_{ir}b_{rj}\right) = \sum_{K=1}^{r} a_{iK}b_{Kj}$$

The sigma notation $\sum_{K=1}^{r} a_{iK}b_{Kj}$ means summing the term $a_{iK}b_{Kj}$ from $K = 1$ to $K = r$.

For example, (i) $\sum_{k=1}^{10} k = 1 + 2 + 3 + 4 + \cdots + 10 = 55$

(ii) $\sum_{k=1}^{10} 2^k = 2^1 + 2^2 + 2^3 + \cdots + 2^{10} = 2046$

This sigma notation can be difficult to get accustomed to but this compact way of writing a sum is common in mathematics so you will see this in mathematical literature.

Note, that $(\mathbf{AB})_{ij}$ is the ij entry of matrix **AB**. We can evaluate specific entries of the matrix multiplication **AB** by using this formula as the next example demonstrates.

Example 1.24

Let $\mathbf{A} = \begin{pmatrix} 1 & 3 & 6 \\ 4 & 7 & -1 \\ 5 & 3 & 2 \end{pmatrix}$ and $\mathbf{B} = \begin{pmatrix} -1 & 3 & -5 & -6 \\ 2 & 1 & -7 & 2 \\ 6 & 4 & 9 & 1 \end{pmatrix}$.

Determine the following elements $(\mathbf{AB})_{12}$, $(\mathbf{AB})_{23}$, $(\mathbf{AB})_{31}$, $(\mathbf{AB})_{43}$ and $(\mathbf{AB})_{34}$.

Solution

What does the notation $(\mathbf{AB})_{12}$ mean?

$(\mathbf{AB})_{12}$ is the entry in the first row and the second column of matrix $\mathbf{AB}$ which is equal to

$$(\mathbf{AB})_{12} = (1 \;\; 3 \;\; 6)\begin{pmatrix} 3 \\ 1 \\ 4 \end{pmatrix} = (1 \times 3) + (3 \times 1) + (6 \times 4) = 30$$

What does the notation $(\mathbf{AB})_{23}$ mean?

$(\mathbf{AB})_{23}$ is the entry in the second row and the third column of matrix $\mathbf{AB}$ which is equal to

$$(\mathbf{AB})_{23} = (4 \;\; 7 \;\; -1)\begin{pmatrix} -5 \\ -7 \\ 9 \end{pmatrix} = (4 \times (-5)) + (7 \times (-7)) + (-1 \times 9) = -78$$

Similarly, we have $(\mathbf{AB})_{31}$ is the third row of matrix $\mathbf{A}$ times the first column of matrix $\mathbf{B}$.

$$(\mathbf{AB})_{31} = (5 \;\; 3 \;\; 2)\begin{pmatrix} -1 \\ 2 \\ 6 \end{pmatrix} = (5 \times (-1)) + (3 \times 2) + (2 \times 6) = 13$$

What does the notation $(\mathbf{AB})_{43}$ mean?

$(\mathbf{AB})_{43}$ is the fourth row of matrix $\mathbf{A}$ times the third column of matrix $\mathbf{B}$, but matrix $\mathbf{A}$ only has three rows so we cannot evaluate $(\mathbf{AB})_{43}$.

$(\mathbf{AB})_{34}$ is the third row of matrix $\mathbf{A}$ times the fourth column of matrix $\mathbf{B}$

$$(\mathbf{AB})_{34} = (5 \;\; 3 \;\; 2)\begin{pmatrix} -6 \\ 2 \\ 1 \end{pmatrix} = (5 \times (-6)) + (3 \times 2) + (2 \times 1) = -22$$

Hence in summary we have $(\mathbf{AB})_{12} = 30$, $(\mathbf{AB})_{23} = -78$, $(\mathbf{AB})_{31} = 13$, $(\mathbf{AB})_{34} = -22$ but we cannot evaluate $(\mathbf{AB})_{43}$ because matrix $\mathbf{A}$ does not have four rows.

Note, that we can work out $(\mathbf{AB})_{34} = -22$ but not $(\mathbf{AB})_{43}$. Be careful in evaluating $(\mathbf{AB})_{ij}$ because the orders of the row i and column j matters; that is, for a general matrix $\mathbf{C}$ we have $\mathbf{C}_{ij} \neq \mathbf{C}_{ji}$ [not equal].

Let us examine other properties of matrix multiplication.

Example 1.25

Let $\mathbf{A} = \begin{pmatrix} 2 & 3 \\ 6 & 7 \\ 1 & 2 \end{pmatrix}$, $\mathbf{B} = \begin{pmatrix} 1 & 2 \\ 3 & 4 \end{pmatrix}$ and $\mathbf{C} = \begin{pmatrix} 5 & 6 \\ 7 & 8 \end{pmatrix}$. Determine $\mathbf{A}(\mathbf{B} + \mathbf{C})$ and $\mathbf{AB} + \mathbf{AC}$.

Solution

First we find $\mathbf{B}+\mathbf{C}=\begin{pmatrix}1 & 2\\ 3 & 4\end{pmatrix}+\begin{pmatrix}5 & 6\\ 7 & 8\end{pmatrix}=\begin{pmatrix}6 & 8\\ 10 & 12\end{pmatrix}$. Next we left multiply this by matrix $\mathbf{A}$:

$$\mathbf{A}(\mathbf{B}+\mathbf{C})=\begin{pmatrix}2 & 3\\ 6 & 7\\ 1 & 2\end{pmatrix}\begin{pmatrix}6 & 8\\ 10 & 12\end{pmatrix}=\begin{pmatrix}42 & 52\\ 106 & 132\\ 26 & 32\end{pmatrix}$$

Similarly, we can show $\mathbf{AB}+\mathbf{AC}$ gives the same result. In this case, we have

$$\mathbf{A}(\mathbf{B}+\mathbf{C})=\mathbf{AB}+\mathbf{AC}$$

Proposition (1.16). Properties of matrix multiplication.
Let $\mathbf{A}$, $\mathbf{B}$ and $\mathbf{C}$ be matrices of an appropriate size so that the following arithmetic operation can be carried out. Then

(a) $(\mathbf{AB})\,\mathbf{C}=\mathbf{A}(\mathbf{BC})$ (associative law for multiplication).
(b) $\mathbf{A}(\mathbf{B}+\mathbf{C})=\mathbf{AB}+\mathbf{AC}$ (left distributive law).
(c) $(\mathbf{B}+\mathbf{C})\,\mathbf{A}=\mathbf{BA}+\mathbf{CA}$ (right distributive law).
(d) $\mathbf{A}\times\mathbf{O}=\mathbf{O}\times\mathbf{A}=\mathbf{O}$. Note that multiplication is commutative in this case.

Here we only prove part (b). You are asked to prove the remaining parts in Exercises 1.5.

The proof of part (b) might be challenging to follow because of the notation involved. In fact, it is a lot less complicated than it appears at first. All we are really doing is looking at the individual elements of the matrix formed by $\mathbf{A}(\mathbf{B}+\mathbf{C})$, and comparing them to the individual elements of the matrix formed by $\mathbf{AB}+\mathbf{AC}$.

Proof of (b).
Let $\mathbf{A}$ be a matrix of size $m\times r$ (m rows by r columns), $\mathbf{B}$ and $\mathbf{C}$ be matrices of size $r\times n$ (r rows by n columns). Also let the entries in matrices $\mathbf{A}$, $\mathbf{B}$ and $\mathbf{C}$ be denoted by a_{ij}, b_{jk} and c_{jk} respectively. [The subscripts ij and jk gives the position of the entry.]

If we can prove this result for an arbitrary row and column then we can apply our result to all the rows and columns.

We need to prove $\mathbf{A}(\mathbf{B}+\mathbf{C})=\mathbf{AB}+\mathbf{AC}$. First we look at the left hand side of this identity and then we examine the right and show that they are equal.

Adding the two matrices inside the brackets on the left hand side gives:

$$\mathbf{B}+\mathbf{C}=\begin{pmatrix}b_{11}+c_{11} & b_{12}+c_{12}\cdots & b_{1j}+c_{1j} & \cdots b_{1n}+c_{1n}\\ b_{21}+c_{21} & b_{22}+c_{22}\cdots & b_{2j}+c_{2j} & \cdots b_{2n}+c_{2n}\\ \vdots & \vdots & \vdots & \vdots\\ b_{r1}+c_{r1} & b_{r2}+c_{r2}\cdots & b_{rj}+c_{rj} & \cdots b_{rn}+c_{rn}\end{pmatrix}$$

Writing out the left hand side **A** (**B** + **C**) we have

$$\begin{array}{c} \\ \\ \\ i\text{th Row} \\ \\ \\ \end{array}\begin{pmatrix} a_{11} & a_{12} & \cdots & a_{1r} \\ a_{21} & a_{22} & & a_{2r} \\ \vdots & \vdots & \vdots & \vdots \\ \boxed{a_{i1}} & a_{i2} & & \cdots a_{ir} \\ \vdots & \vdots & \vdots & \vdots \\ a_{m1} & a_{m2} & & \cdots a_{mr} \end{pmatrix}\begin{pmatrix} b_{11}+c_{11} & b_{12}+c_{12}\cdots & \overset{j\text{th Col}}{b_{1j}+c_{1j}} & \cdots b_{1n}+c_{1n} \\ b_{21}+c_{21} & b_{22}+c_{22}\cdots & b_{2j}+c_{2j} & \cdots b_{2n}+c_{2n} \\ \vdots & \vdots & \vdots & \vdots \\ b_{r1}+c_{r1} & b_{r2}+c_{r2}\cdots & b_{rj}+c_{rj} & \cdots b_{rn}+c_{rn} \end{pmatrix}$$

Expanding the ith row and jth column entry of $[\mathbf{A}(\mathbf{B}+\mathbf{C})]_{ij}$ we have

$$[\mathbf{A}(\mathbf{B}+\mathbf{C})]_{ij} = a_{i1}\left(b_{1j}+c_{1j}\right) + a_{i2}\left(b_{2j}+c_{2j}\right) + \cdots + a_{ir}\left(b_{rj}+c_{rj}\right) \quad (\dagger)$$

Now we examine the right hand side, **AB** + **AC**. By using the previous formula (1.15):

$$(\mathbf{AB})_{ij} = \left(a_{i1}b_{1j}\right) + \left(a_{i2}b_{2j}\right) + \left(a_{i3}b_{3j}\right) + \cdots + \left(a_{ir}b_{rj}\right)$$

For the ith row and jth column entry we have

$$(\mathbf{AB})_{ij} = a_{i1}b_{1j} + a_{i2}b_{2j} + \cdots + a_{ir}b_{rj}$$

Similarly

$$(\mathbf{AC})_{ij} = a_{i1}c_{1j} + a_{i2}c_{2j} + \cdots + a_{ir}c_{rj}$$

Adding these gives

$$\begin{aligned} (\mathbf{AB})_{ij} + (\mathbf{AC})_{ij} &= a_{i1}b_{1j} + a_{i1}c_{1j} + a_{i2}b_{2j} + a_{i2}c_{2j} + \cdots + a_{ir}b_{rj} + a_{ir}c_{rj} \\ &= a_{i1}\left(b_{1j}+c_{1j}\right) + a_{i2}\left(b_{2j}+c_{2j}\right) + \cdots + a_{ir}\left(b_{rj}+c_{rj}\right) && \left[\begin{array}{l}\text{Factorizing} \\ \text{out the } a\text{'s}\end{array}\right] \\ &= [\mathbf{A}(\mathbf{B}+\mathbf{C})]_{ij} && [\text{By } (\dagger)] \end{aligned}$$

Since $(\mathbf{AB})_{ij} + (\mathbf{AC})_{ij} = [\mathbf{A}(\mathbf{B}+\mathbf{C})]_{ij}$ for an arbitrary row and column entry, we have our general result, **AB** + **AC** = **A** (**B** + **C**).

■

Example 1.26

Let $\mathbf{A} = \begin{pmatrix} 1 & 2 \\ 9 & 7 \end{pmatrix}$, $\mathbf{B} = \begin{pmatrix} 2 & 3 & -1 \\ 5 & -7 & 6 \end{pmatrix}$, $\mathbf{C} = \begin{pmatrix} 2 \\ -5 \\ 9 \end{pmatrix}$ and **O** be the zero matrix. Evaluate:

(a) (**AB**) **C** **(b)** **A** (**BC**) **(c)** **A** (**B** + **C**) **(d)** **AB** + **AC**

(e) (**B** + **C**) **A** **(f)** $\mathbf{AO}_{22}$ **(g)** $\mathbf{BO}_{33}$ **(h)** $\mathbf{CO}_{13}$

Solution

(a) Multiplying the bracket term first:

$$(\mathbf{AB}) = \begin{pmatrix} 1 & 2 \\ 9 & 7 \end{pmatrix}\begin{pmatrix} 2 & 3 & -1 \\ 5 & -7 & 6 \end{pmatrix}$$
$$= \begin{pmatrix} 12 & -11 & 11 \\ 53 & -22 & 33 \end{pmatrix}$$

$$(\mathbf{AB})\,\mathbf{C} = \begin{pmatrix} 12 & -11 & 11 \\ 53 & -22 & 33 \end{pmatrix}\begin{pmatrix} 2 \\ -5 \\ 9 \end{pmatrix} = \begin{pmatrix} 178 \\ 513 \end{pmatrix}$$

(b) By Theorem (1.16) (a) we have $(\mathbf{AB})\,\mathbf{C} = \mathbf{A}\,(\mathbf{BC})$, therefore we have the answer of part (a).

(c) Matrices $\mathbf{B}$ and $\mathbf{C}$ are of different size, therefore we cannot add them. That is we cannot evaluate $\mathbf{B} + \mathbf{C}$ which means $\mathbf{A}\,(\mathbf{B} + \mathbf{C})$ is impossible.

(d) By Theorem (1.16) (b) we have $\mathbf{AB} + \mathbf{AC} = \mathbf{A}\,(\mathbf{B} + \mathbf{C})$, therefore we also cannot evaluate $\mathbf{AB} + \mathbf{AC}$ because we cannot add $\mathbf{B}$ and $\mathbf{C}$.

(e) As part (c) above, we cannot add matrices $\mathbf{B}$ and $\mathbf{C}$. Hence we cannot evaluate $(\mathbf{B} + \mathbf{C})\,\mathbf{A}$.
For the remaining parts, we use Theorem (1.16) (d) which says that

$$\mathbf{A} \times \mathbf{O} = \mathbf{O} \times \mathbf{A} = \mathbf{O}$$

For parts **(f)** and **(g)** we have

$$\mathbf{AO}_{22} = \begin{pmatrix} 1 & 2 \\ 9 & 7 \end{pmatrix}\begin{pmatrix} 0 & 0 \\ 0 & 0 \end{pmatrix} = \begin{pmatrix} 0 & 0 \\ 0 & 0 \end{pmatrix},$$

$$\mathbf{BO}_{33} = \begin{pmatrix} 2 & 3 & -1 \\ 5 & -7 & 6 \end{pmatrix}\begin{pmatrix} 0 & 0 & 0 \\ 0 & 0 & 0 \\ 0 & 0 & 0 \end{pmatrix} = \begin{pmatrix} 0 & 0 & 0 \\ 0 & 0 & 0 \end{pmatrix}$$

(h) We have $\mathbf{CO}_{13} = \begin{pmatrix} 2 \\ -5 \\ 9 \end{pmatrix}\begin{pmatrix} 0 & 0 & 0 \end{pmatrix} = (0) = 0.$

1.5.4 Matrix powers

Matrix powers are particularly useful in Markov chains – these are based on matrices whose entries are probabilities. Many real life systems have an element of uncertainty and develop over time and this can be explained through Markov chains.

A Markov chain is a sequence of random variables with the property that given the present state, the future and past states are independent. For example, the game Monopoly where the states are determined entirely by dice is a Markov chain. However, games like poker are not a Markov chain because what is displayed depends on past moves.

See question 10 of Exercises 1.5 for a concrete example.

We define matrix powers for square matrices only.

What is a square matrix?
A square matrix has an equal number of rows and columns, so any n by n matrix is a square matrix.

Let **A** be a square matrix then we define

$$\mathbf{A}^k = \underbrace{\mathbf{A} \times \mathbf{A} \times \mathbf{A} \times \cdots \times \mathbf{A}}_{k \text{ copies}} \quad \text{where } k \text{ is a positive integer} \tag{1.17}$$

Why can't we define $\mathbf{A}^k$ for non-square matrices?
Suppose matrix $\mathbf{A}$ is $m \times n$ (m rows by n columns) and m does not equal n ($m \neq n$).

What is $\mathbf{A}^2$ equal to?

$$\mathbf{A}^2 = \overbrace{\begin{pmatrix} a_{11} & \cdots & a_{1n} \\ \vdots & \vdots & \vdots \\ a_{m1} & \cdots & a_{mn} \end{pmatrix}}^{n \text{ columns}} \begin{pmatrix} a_{11} & \cdots & a_{1n} \\ \vdots & \vdots & \vdots \\ a_{m1} & \cdots & a_{mn} \end{pmatrix} \; m \text{ rows}$$

Well, we cannot carry out this matrix multiplication because the number of columns of the left hand matrix is n but the number of rows of the right hand matrix is m and these are *not* equal ($m \neq n$), therefore we cannot multiply. If we cannot find $\mathbf{A}^2$ then it follows that we cannot evaluate $\mathbf{A}^k$ for $k \geq 2$. Hence formula (1.17) is only valid if $\mathbf{A}$ is a square matrix, that is $m = n$.

A square matrix to the power 0 is the identity matrix (to be defined in the next section), that is:

$$\mathbf{A}^0 = \mathbf{I} \text{ (identity matrix)}$$

Example 1.27

Let $\mathbf{A} = \begin{pmatrix} -3 & 7 \\ 1 & 8 \end{pmatrix}$ and $\mathbf{B} = \begin{pmatrix} 5 & 9 \\ 2 & 3 \end{pmatrix}$. Determine:

(a) $\mathbf{A}^2$ **(b)** $\mathbf{B}^2$ **(c)** $2\mathbf{AB}$ **(d)** $(\mathbf{A} + \mathbf{B})^2$ **(e)** $\mathbf{A}^2 + 2\mathbf{AB} + \mathbf{B}^2$
(f) *Why is the following result not valid* $(\mathbf{A} + \mathbf{B})^2 = \mathbf{A}^2 + 2\mathbf{AB} + \mathbf{B}^2$?
(g) *What does* $(\mathbf{A} + \mathbf{B})^2$ *equal?*

Solution

(a) Remember, $\mathbf{A}^2 = \mathbf{A} \times \mathbf{A}$ so we have

$$\mathbf{A}^2 = \begin{pmatrix} -3 & 7 \\ 1 & 8 \end{pmatrix} \times \begin{pmatrix} -3 & 7 \\ 1 & 8 \end{pmatrix} = \begin{pmatrix} 16 & 35 \\ 5 & 71 \end{pmatrix}$$

(b) Similarly we have

$$\mathbf{B}^2 = \begin{pmatrix} 5 & 9 \\ 2 & 3 \end{pmatrix} \times \begin{pmatrix} 5 & 9 \\ 2 & 3 \end{pmatrix} = \begin{pmatrix} 43 & 72 \\ 16 & 27 \end{pmatrix}$$

(c) *How do we work out 2**AB**?*
Multiply the matrices $\mathbf{A}$ and $\mathbf{B}$ and then multiply our result by 2:

$$2\mathbf{AB} = 2\begin{pmatrix} -3 & 7 \\ 1 & 8 \end{pmatrix}\begin{pmatrix} 5 & 9 \\ 2 & 3 \end{pmatrix} = 2\begin{pmatrix} -1 & -6 \\ 21 & 33 \end{pmatrix} = \begin{pmatrix} -2 & -12 \\ 42 & 66 \end{pmatrix}$$

(d) *How do we find* $(\mathbf{A}+\mathbf{B})^2$*?*
Remember, $(\mathbf{A}+\mathbf{B})^2 = (\mathbf{A}+\mathbf{B}) \times (\mathbf{A}+\mathbf{B})$. First, we add the matrices $\mathbf{A}$ and $\mathbf{B}$ and then square the result:

$$\mathbf{A}+\mathbf{B} = \begin{pmatrix} -3 & 7 \\ 1 & 8 \end{pmatrix} + \begin{pmatrix} 5 & 9 \\ 2 & 3 \end{pmatrix} = \begin{pmatrix} 2 & 16 \\ 3 & 11 \end{pmatrix}$$

$$(\mathbf{A}+\mathbf{B})^2 = \begin{pmatrix} 2 & 16 \\ 3 & 11 \end{pmatrix} \times \begin{pmatrix} 2 & 16 \\ 3 & 11 \end{pmatrix} = \begin{pmatrix} 52 & 208 \\ 39 & 169 \end{pmatrix}$$

(e) To find $\mathbf{A}^2 + 2\mathbf{AB} + \mathbf{B}^2$ we add the above parts **(a)**, **(b)** and **(c)**:

$$\mathbf{A}^2 + 2\mathbf{AB} + \mathbf{B}^2 = \underbrace{\begin{pmatrix} 16 & 35 \\ 5 & 71 \end{pmatrix}}_{=\mathbf{A}^2} + \underbrace{\begin{pmatrix} -2 & -12 \\ 42 & 66 \end{pmatrix}}_{=2\mathbf{AB}} + \underbrace{\begin{pmatrix} 43 & 72 \\ 16 & 27 \end{pmatrix}}_{=\mathbf{B}^2}$$

$$= \begin{pmatrix} 16-2+43 & 35-12+72 \\ 5+42+16 & 71+66+27 \end{pmatrix} = \begin{pmatrix} 57 & 95 \\ 63 & 164 \end{pmatrix}$$

Note, that parts (d) and (e) are **not** equal. That is

$$(\mathbf{A}+\mathbf{B})^2 \neq \mathbf{A}^2 + 2\mathbf{AB} + \mathbf{B}^2 \ \left[\text{not equal}\right]$$

Why not?

(f) Because matrix multiplication is not commutative that means order of multiplication matters. Remember, the matrix multiplication $\mathbf{AB}$ does not equal $\mathbf{BA}$.

What is $(\mathbf{A}+\mathbf{B})^2$ *equal to?*

$$\begin{aligned}(\mathbf{A}+\mathbf{B})^2 &= (\mathbf{A}+\mathbf{B}) \times (\mathbf{A}+\mathbf{B}) \\ &= \mathbf{A}^2 + \mathbf{AB} + \mathbf{BA} + \mathbf{B}^2\end{aligned}$$

$\mathbf{AB} \neq \mathbf{BA}$ [not equal], therefore $\mathbf{AB} + \mathbf{BA} \neq 2\mathbf{AB}$ [not equal].
Let's check the result for this particular example.
We know what $\mathbf{AB}$ *is by part (c) above but what is* $\mathbf{BA}$ *equal to?*

$$\mathbf{BA} = \begin{pmatrix} 5 & 9 \\ 2 & 3 \end{pmatrix}\begin{pmatrix} -3 & 7 \\ 1 & 8 \end{pmatrix} = \begin{pmatrix} -6 & 107 \\ -3 & 38 \end{pmatrix}$$

(continued...)

(g) Next we work out $\mathbf{A}^2 + \mathbf{AB} + \mathbf{BA} + \mathbf{B}^2$. By the above parts we have

$$\mathbf{A}^2 + \mathbf{AB} + \mathbf{BA} + \mathbf{B}^2 = \underbrace{\begin{pmatrix} 16 & 35 \\ 5 & 71 \end{pmatrix}}_{=\mathbf{A}^2} + \underbrace{\begin{pmatrix} -1 & -6 \\ 21 & 33 \end{pmatrix}}_{=\mathbf{AB}} + \underbrace{\begin{pmatrix} -6 & 107 \\ -3 & 38 \end{pmatrix}}_{=\mathbf{BA}} + \underbrace{\begin{pmatrix} 43 & 72 \\ 16 & 27 \end{pmatrix}}_{=\mathbf{B}^2}$$

$$= \begin{pmatrix} 16 - 1 - 6 + 43 & 35 - 6 + 107 + 72 \\ 5 + 21 - 3 + 16 & 71 + 33 + 38 + 27 \end{pmatrix}$$

$$= \begin{pmatrix} 52 & 208 \\ 39 & 169 \end{pmatrix}$$

This is the same answer as to part (d). Therefore we have $(\mathbf{A} + \mathbf{B})^2 = \mathbf{A}^2 + \mathbf{AB} + \mathbf{BA} + \mathbf{B}^2$.

As seen in the last example, we need to be very careful in carrying over our real number algebra to matrix algebra. It does not always work. We can prove the general rule:

$$(\mathbf{A} + \mathbf{B})^2 = \mathbf{A}^2 + \mathbf{AB} + \mathbf{BA} + \mathbf{B}^2$$

provided **A** and **B** are square matrices of the same size.

In mathematics, if you are asked to show a result is false then you only need to produce one counter example to the result.

Example 1.28

Show the following result is false by giving an example:

$$(\mathbf{A} - \mathbf{B})^2 = \mathbf{O} \text{ implies } \mathbf{A} = \mathbf{B}$$

Solution

We need to give an example for which this result does not work.

Since we have $\mathbf{A} - \mathbf{B}$, both matrices **A** and **B** must be of the same size. Let's take the simplest case of size 2 by 2. We need $(\mathbf{A} - \mathbf{B})^2 = \mathbf{O}$ so let $(\mathbf{A} - \mathbf{B}) = \begin{pmatrix} a & b \\ c & d \end{pmatrix}$ then

$$(\mathbf{A} - \mathbf{B})^2 = \begin{pmatrix} a & b \\ c & d \end{pmatrix}\begin{pmatrix} a & b \\ c & d \end{pmatrix} = \begin{pmatrix} a^2 + bc & ab + bd \\ ca + dc & bc + d^2 \end{pmatrix} = \begin{pmatrix} 0 & 0 \\ 0 & 0 \end{pmatrix}$$

Equating the first entry gives

$$a^2 + bc = 0 \Rightarrow a^2 = -bc$$

We can select any numbers which satisfy this $a^2 = -bc$. We like to deal with whole numbers because they are easier, so let $b = -1$ and $c = 4$ then $a = \sqrt{4} = \pm 2$. Choose a to be the positive root, so let $a = 2$. We need to find a value for d so that the above is satisfied.

How?

Equating the last entry

$$bc + d^2 = 0 \Rightarrow d^2 = -bc$$

We already have $b = -1$ and $c = 4$ so $d = \sqrt{4} = \pm 2$. This time select $d = -2$. Hence

$$(\mathbf{A} - \mathbf{B}) = \begin{pmatrix} a & b \\ c & d \end{pmatrix} = \begin{pmatrix} 2 & -1 \\ 4 & -2 \end{pmatrix} \quad [\text{because } a = 2, b = -1, c = 4, d = -2]$$

Check that $(\mathbf{A} - \mathbf{B})^2 = \mathbf{O}$. Now we just need to find matrices $\mathbf{A}$ and $\mathbf{B}$ such that

$$\mathbf{A} - \mathbf{B} = \begin{pmatrix} 2 & -1 \\ 4 & -2 \end{pmatrix} \quad (*)$$

There are an infinite number of matrices $\mathbf{A}$ and $\mathbf{B}$ which give the above result (*). Here is one choice, but you may have chosen others:

$$\mathbf{A} = \begin{pmatrix} 3 & -1 \\ 2 & 1 \end{pmatrix} \text{ and } \mathbf{B} = \begin{pmatrix} 1 & 0 \\ -2 & 3 \end{pmatrix}$$

Hence we have found matrices $\mathbf{A}$ and $\mathbf{B}$ such that

$$(\mathbf{A} - \mathbf{B})^2 = \mathbf{O} \text{ does not imply } \mathbf{A} = \mathbf{B}$$

Summary

Generally, if $\mathbf{A}$ and $\mathbf{B}$ are matrices for which multiplication is a valid operation then

$$\mathbf{AB} \neq \mathbf{BA} \quad [\text{not equal}]$$

We cannot blindly apply the properties of real number algebra to matrix algebra.
Let $\mathbf{A}$ be a square matrix then

(1.17) $$\mathbf{A}^k = \underbrace{\mathbf{A} \times \mathbf{A} \times \mathbf{A} \times \cdots \times \mathbf{A}}_{k \text{ copies}}$$

EXERCISES 1.5

(Brief solutions at end of book. Full solutions available at <http://www.oup.co.uk/companion/singh>.)

You may like to check your numerical solutions by using MATLAB.

1. Let $\mathbf{A} = \begin{pmatrix} 2 & -2 & 5 \\ 0 & -1 & 7 \end{pmatrix}$, $\mathbf{B} = \begin{pmatrix} -1 & 3 & 7 \\ 2 & -9 & 6 \end{pmatrix}$ and $\mathbf{C} = \begin{pmatrix} 9 & 5 & 8 \\ -6 & -1 & 6 \end{pmatrix}$. Determine

   (a) $\mathbf{A} + \mathbf{B}$ (b) $(\mathbf{A} + \mathbf{B}) + \mathbf{C}$ (c) $\mathbf{A} + (\mathbf{B} + \mathbf{C})$ (d) $\mathbf{B} + (\mathbf{C} + \mathbf{A})$
   (e) $\mathbf{C} + \mathbf{O}$ (f) $\mathbf{B} + \mathbf{B} + \mathbf{B} + \mathbf{B} + \mathbf{B}$ (g) $5\mathbf{B}$ (h) $\mathbf{C} + (-\mathbf{C})$

2. Let $\mathbf{A} = \begin{pmatrix} 5 & -1 & -2 \\ 1 & -3 & 2 \end{pmatrix}$.

   (a) Find a matrix **B** such that $\mathbf{A} + \mathbf{B} = \mathbf{O}$ where **O** is the 2×3 zero matrix.

   (b) Let $\mathbf{C} = \begin{pmatrix} 1 & 2 & 3 \\ 4 & 5 & 6 \end{pmatrix}$. Determine $\mathbf{A} + (-\mathbf{C})$ and $\mathbf{A} - \mathbf{C}$.

   (c) Determine $\mathbf{A} - \mathbf{B} - \mathbf{C}$ and $\mathbf{A} - (\mathbf{B} - \mathbf{C})$.

3. Let $\mathbf{A} = \begin{pmatrix} 1 & 5 & -2 \\ 3 & 7 & -9 \end{pmatrix}$, $\mathbf{B} = \begin{pmatrix} -1 & 0 & 7 \\ -3 & 2 & 1 \\ -4 & 1 & 6 \end{pmatrix}$, $\mathbf{C} = \begin{pmatrix} 7 & 8 & -1 \\ -8 & 6 & -6 \\ 6 & -3 & 5 \end{pmatrix}$ and $\mathbf{O} = \mathbf{O}_{33}$.

   Determine, if possible, the following:

   (a) **AB** (b) **BA** (c) $\mathbf{A}(\mathbf{B} + \mathbf{C})$ (d) $\mathbf{AB} + \mathbf{AC}$
   (e) $(\mathbf{B} + \mathbf{C})\mathbf{A}$ (f) $\mathbf{BA} + \mathbf{CA}$ (g) **CO** (h) $\mathbf{BO} + \mathbf{CO}$
   (i) $\mathbf{OB} + \mathbf{OC}$ (j) $(\mathbf{A} + \mathbf{B})\mathbf{C}$ (k) $\mathbf{AC} + \mathbf{BC}$

4. Let $\mathbf{A} = \begin{pmatrix} 1 & 2 & 3 \\ 4 & 5 & 6 \\ 7 & 8 & 9 \end{pmatrix}$ and $\mathbf{I} = \begin{pmatrix} 1 & 0 & 0 \\ 0 & 1 & 0 \\ 0 & 0 & 1 \end{pmatrix}$. Determine (a) **AI** (b) **IA**

   What do you notice about your results?

5. Let $\mathbf{A} = \begin{pmatrix} 1 & 2 & 3 \\ 4 & 5 & 6 \end{pmatrix}$ and $\mathbf{B} = \begin{pmatrix} 2 & -1 \\ 2 & 8 \\ 7 & 5 \end{pmatrix}$. Evaluate the following:

$$(\mathbf{AB})_{11}, (\mathbf{AB})_{12}, (\mathbf{AB})_{21}, (\mathbf{AB})_{22} \text{ and } \mathbf{AB}$$

6. For $\mathbf{A} = \begin{pmatrix} -1 & 3 & 5 \\ 4 & 1 & -7 \end{pmatrix}$, $\mathbf{B} = \begin{pmatrix} 1 & -3 & -5 \\ -4 & -1 & 7 \end{pmatrix}$, $c = -9$ and $k = 8$ determine

   (a) $(ck)\mathbf{A}$ (b) $c(k\mathbf{A})$ (c) $k(\mathbf{A} + \mathbf{B})$ (d) $k\mathbf{A} + k\mathbf{B}$
   (e) $(c + k)\mathbf{A}$ (f) $c\mathbf{A} + k\mathbf{A}$ (g) $(c + k)\mathbf{B}$ (h) $c\mathbf{B} + k\mathbf{B}$

7. Determine the scalar λ (Greek letter 'lambda') which satisfies $\mathbf{Ax} = \lambda\mathbf{x}$ where

$$\mathbf{A} = \begin{pmatrix} 2 & 1 \\ 1 & 2 \end{pmatrix} \text{ and } \mathbf{x} = \begin{pmatrix} 1 \\ 1 \end{pmatrix}$$

8. Let $\mathbf{A} = \begin{pmatrix} 1/3 & 1/3 & 1/3 \\ 1/3 & 1/3 & 1/3 \\ 1/3 & 1/3 & 1/3 \end{pmatrix}$. Find $\mathbf{A}^2, \mathbf{A}^3$ and $\mathbf{A}^4$.

   What do you think $\mathbf{A}^n$ *will be for any natural number* n*?*

   A discrete dynamical system is described by the formula $\mathbf{x}_n = \mathbf{A}^n\mathbf{x}$. Write the formula for the given matrix $\mathbf{x}_n = \mathbf{A}^n\mathbf{x}$ for the given matrix **A**.

9. Prove that the following results are false:

   (a) $\mathbf{AB} = \mathbf{O} \Rightarrow \mathbf{A} = \mathbf{O}$ or $\mathbf{B} = \mathbf{O}$.

   (b) $\mathbf{AB} - \mathbf{BA} = \mathbf{O} \Rightarrow \mathbf{A} = \mathbf{B}$.

10. Let $\mathbf{T}$ be a (transition) matrix of a Markov chain and $\mathbf{p}$ be a probability vector. Then the probability that the chain is in a particular state after k steps is given by the vector $\mathbf{p}_k$:

$$\mathbf{p}_k = \mathbf{T}^k\mathbf{p}$$

Determine $\mathbf{p}_k$ by using MATLAB or any appropriate software for

$$\mathbf{T} = \begin{pmatrix} 0.6 & 0.7 \\ 0.4 & 0.3 \end{pmatrix}, \quad \mathbf{p} = \begin{pmatrix} 0.5 \\ 0.5 \end{pmatrix} \text{ and } k = 1, 2, 10, 100 \text{ and } 100\,000$$

What do you notice about your results?

11. Prove property (d) of Proposition (1.16).
12. Prove properties (b), (d) and (e) of Theorem (1.12).
13. Prove properties (a) and (b) of Theorem (1.13).
*15. Prove property (a) of Theorem (1.16).

SECTION 1.6 The Transpose and Inverse of a Matrix

By the end of this section you will be able to

- prove properties of the matrix transpose
- define and prove properties of the inverse matrix

We should now be familiar with matrix operations such as addition, subtraction and multiplication.

What about division of matrices?
You cannot divide matrices. The nearest operation to division of matrices is the inverse matrix which we discuss in this section. We can use the inverse matrix to find solutions of linear systems.

There are other important operations of matrices such as transpose.

1.6.1 Transpose of a matrix

What do you think the ***transpose*** *of a matrix might be?*
It is a new matrix which is made by rotating the rows of the given matrix. For example, if

$$\begin{pmatrix} 1 & 2 \\ 3 & 4 \\ 5 & 6 \end{pmatrix} \text{ then the transpose of this matrix is } \begin{pmatrix} 1 & 3 & 5 \\ 2 & 4 & 6 \end{pmatrix}$$

Column 1 becomes row 1 and column 2 becomes row 2. If $\mathbf{A}$ is the given matrix then the transpose of $\mathbf{A}$ is denoted by $\mathbf{A}^T$. In the above case, we have $\mathbf{A}^T = \begin{pmatrix} 1 & 3 & 5 \\ 2 & 4 & 6 \end{pmatrix}$.

In general, the entry a_{ij} (ith row by jth column) of matrix $\mathbf{A}$ is transposed to a_{ji} (jth row by ith column) in $\mathbf{A}^T$. We define the transpose of a matrix as

(1.18) $$\mathbf{A} = (a_{ij}) \text{ implies } \mathbf{A}^T = (a_{ji}) \quad \text{[The subscript } ij \text{ changes to } ji\text{]}$$

Example 1.29

Find the transpose of the following matrices:

(i) $\mathbf{A} = \begin{pmatrix} -9 & 2 & 3 \\ 7 & -2 & 9 \\ 6 & -1 & 5 \end{pmatrix}$ **(ii)** $\mathbf{B} = \begin{pmatrix} 1 & 0 & 0 \\ 0 & 2 & 0 \\ 0 & 0 & 3 \end{pmatrix}$ **(iii)** $\mathbf{C} = \begin{pmatrix} -1 & 3 & 4 \\ 7 & 9 & 0 \end{pmatrix}$ **(iv)** $\mathbf{D} = \begin{pmatrix} 1 \\ 2 \\ 3 \end{pmatrix}$

Solution

(i) *Swapping* rows and columns we have

$$\mathbf{A}^T = \begin{pmatrix} -9 & 7 & 6 \\ 2 & -2 & -1 \\ 3 & 9 & 5 \end{pmatrix}$$

(ii) Transposing $\mathbf{B} = \begin{pmatrix} 1 & 0 & 0 \\ 0 & 2 & 0 \\ 0 & 0 & 3 \end{pmatrix}$ gives the same matrix, that is $\mathbf{B}^T = \begin{pmatrix} 1 & 0 & 0 \\ 0 & 2 & 0 \\ 0 & 0 & 3 \end{pmatrix}$.

Transposing this matrix does not change the matrix, that is $\mathbf{B}^T = \mathbf{B}$.

(iii) We have

$$\mathbf{C}^T = \begin{pmatrix} -1 & 7 \\ 3 & 9 \\ 4 & 0 \end{pmatrix} \qquad \left[\text{because } \mathbf{C} = \begin{pmatrix} -1 & 3 & 4 \\ 7 & 9 & 0 \end{pmatrix}\right]$$

(iv) *What sort of matrix is* $\mathbf{D} = \begin{pmatrix} 1 \\ 2 \\ 3 \end{pmatrix}$?

D is a *column* matrix (vector) and transposing makes it into a *row* matrix $\mathbf{D}^T = (1 \ \ 2 \ \ 3)$.

A column vector such as $\mathbf{x} = \begin{pmatrix} x_1 \\ \vdots \\ x_n \end{pmatrix}$ can be written as the transpose of a row vector, that is $\mathbf{x} = \begin{pmatrix} x_1 & \cdots & x_n \end{pmatrix}^T$. This approach of writing a column vector $\mathbf{x}$ as the transpose of a row vector saves space. In this case, column vector $\mathbf{x}$ takes up n lines while $\mathbf{x} = (x_1 \ \cdots \ x_n)^T$ can be written on a single line.

Note, that transposing a matrix of size $m \times n$ (m rows by n columns) gives a matrix of size $n \times m$ (n rows by m columns). If $m \neq n$ then transposing changes the shape of the matrix.

James Sylvester and Arthur Cayley are considered by many to be the fathers of linear algebra.

Figure 1.37 James Sylvester 1814 to 1897.

James Sylvester (Fig. 1.37) was born to a London Jewish family in 1814. He went to school in London and in 1837 took the Mathematical Tripos exam at Cambridge. The Mathematical Tripos is an undergraduate mathematics exam which used to distinguish students by ranking them in order of merit. The word tripos is derived from a three-legged stool which students used to sit on in order to take the examination.

Sylvester did not graduate until 1841 and even then from Trinity College, Dublin rather than Cambridge because he refused to take the oath of the Church of England.

In 1841, aged 27, he become Professor of Mathematics at the University of Virginia, USA, for six months. He resigned his post in March 1842 because of the lenient attitude of the university towards disruptive students.

He returned to England in 1843 and began to study law and met Arthur Cayley, also a mathematics graduate studying law. During this time, they made a significant contribution to what would became matrix theory.

In 1877 he went back to America to become Professor of Mathematics at the new Johns Hopkins University. A year later he founded the *American Journal of Mathematics*, the first mathematics journal to be distributed in the USA.

In 1883, he returned to England to become Professor of Geometry at Oxford University, where he remained until his death in 1897.

Sylvester gave us the term **matrix**. He also adopted other mathematical terms such as invariant (something that does not vary) and discriminant.

1.6.2 Properties of matrix transpose

Next we prove some of the results concerning the transpose of a matrix.

Theorem (1.19). Properties of matrix transpose.

Let **A** and **B** be matrices of appropriate size so that the operations below can be carried out. We have the following properties (k is a scalar):

(a) $\left(\mathbf{A}^T\right)^T = \mathbf{A}$ (b) $\left(k\mathbf{A}\right)^T = k\mathbf{A}^T$

(c) $(\mathbf{A}+\mathbf{B})^T = \mathbf{A}^T + \mathbf{B}^T$ (d) $(\mathbf{AB})^T = \mathbf{B}^T\mathbf{A}^T$

We only prove properties (a) and (c). You are asked to prove the remaining in Exercises 1.6.

(a) What does $\left(\mathbf{A}^T\right)^T = \mathbf{A}$ *mean?*

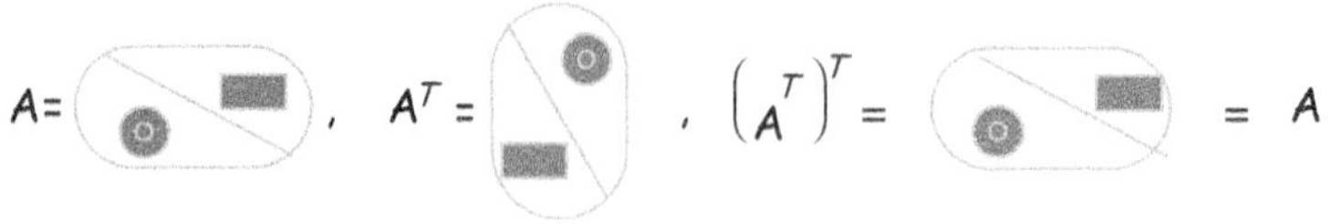

It means that the transpose of a transposed matrix is the matrix you started with. You could think of it as flipping the matrix across the diagonal from top left to bottom right and then flipping it again which gets you back to where you started.

Proof of (a).

Let a_{ij} be the entry in the ith row and jth column of the matrix $\mathbf{A}$. Then

$$\mathbf{A}^T = \left(a_{ij}\right)^T = \left(a_{ji}\right)$$

Executing another transpose on this yields $\left(\mathbf{A}^T\right)^T = \left(a_{ji}\right)^T = \left(a_{ij}\right) = \mathbf{A}$.

Notice that the entries have swapped twice, for example $a_{21} \to a_{12} \to a_{21}$, $a_{31} \to a_{13} \to a_{31}$, and so on. Hence $\left(\mathbf{A}^T\right)^T = \mathbf{A}$.

■

Proof of (c).

$$(\mathbf{A}+\mathbf{B})^T = \left[\begin{pmatrix} a_{11} & a_{12} & \cdots & a_{1n} \\ a_{21} & a_{22} & \cdots & a_{2n} \\ \vdots & \vdots & \vdots & \vdots \\ a_{m1} & a_{m2} & \cdots & a_{mn} \end{pmatrix} + \begin{pmatrix} b_{11} & b_{12} & \cdots & b_{1n} \\ b_{21} & b_{22} & \cdots & b_{2n} \\ \vdots & \vdots & \vdots & \vdots \\ b_{m1} & b_{m2} & \cdots & b_{mn} \end{pmatrix}\right]^T$$

$$= \begin{pmatrix} a_{11}+b_{11} & a_{12}+b_{12} & \cdots & a_{1n}+b_{1n} \\ a_{21}+b_{21} & a_{22}+b_{22} & \cdots & a_{2n}+b_{2n} \\ \vdots & \vdots & \vdots & \vdots \\ a_{m1}+b_{m1} & a_{m2}+b_{m2} & \cdots & a_{mn}+b_{mn} \end{pmatrix}^T \quad \left[\text{adding the corresponding entries}\right]$$

$$= \begin{pmatrix} a_{11}+b_{11} & a_{21}+b_{21} & \cdots & a_{m1}+b_{m1} \\ a_{12}+b_{12} & a_{22}+b_{22} & \cdots & a_{m2}+b_{m2} \\ \vdots & \vdots & \vdots & \vdots \\ a_{1n}+b_{1n} & a_{2n}+b_{2n} & \cdots & a_{mn}+b_{mn} \end{pmatrix} \quad \left[\text{taking the transpose, that is interchanging rows and columns}\right]$$

$$= \underbrace{\begin{pmatrix} a_{11} & a_{21} & \cdots & a_{m1} \\ a_{12} & a_{22} & \cdots & a_{m2} \\ \vdots & \vdots & \vdots & \vdots \\ a_{1n} & a_{2n} & \cdots & a_{mn} \end{pmatrix}}_{=\mathbf{A}^T} + \underbrace{\begin{pmatrix} b_{11} & b_{21} & \cdots & b_{m1} \\ b_{12} & b_{22} & \cdots & b_{m2} \\ \vdots & \vdots & \vdots & \vdots \\ b_{1n} & b_{2n} & \cdots & b_{mn} \end{pmatrix}}_{=\mathbf{B}^T} = \mathbf{A}^T + \mathbf{B}^T$$

Hence we have our result $(\mathbf{A}+\mathbf{B})^T = \mathbf{A}^T + \mathbf{B}^T$.

■

The MATLAB command for transpose is an apostrophe. For example, to find the transpose of $\begin{pmatrix} 1 & 2 & 3 \\ 4 & 5 & 6 \end{pmatrix}$ we enter the following command [1 2 3; 4 5 6]′ after the prompt >>.

Example 1.30

Let $\mathbf{A} = \begin{pmatrix} 3 & -4 & 1 \\ 5 & 2 & 6 \end{pmatrix}$ and $\mathbf{B} = \begin{pmatrix} -2 & 7 & 5 \\ 1 & 3 & -9 \end{pmatrix}$. Determine

(a) $(\mathbf{A}^T)^T$ **(b)** $(2\mathbf{A})^T - (3\mathbf{B})^T$ **(c)** $(\mathbf{A}+\mathbf{B})^T$ **(d)** $\mathbf{A}^T + \mathbf{B}^T$ **(e)** $(\mathbf{AB})^T$

Solution

(a) By Theorem (1.19) property (a) we have $(\mathbf{A}^T)^T = \mathbf{A} = \begin{pmatrix} 3 & -4 & 1 \\ 5 & 2 & 6 \end{pmatrix}$.

(b) Applying the same theorem property (b) we have

$$(2\mathbf{A})^T - (3\mathbf{B})^T = 2\mathbf{A}^T - 3\mathbf{B}^T \quad \left[\text{by Theorem (1.19) (b) } (k\mathbf{A})^T = k\mathbf{A}^T\right]$$

$$= 2\begin{pmatrix} 3 & -4 & 1 \\ 5 & 2 & 6 \end{pmatrix}^T - 3\begin{pmatrix} -2 & 7 & 5 \\ 1 & 3 & -9 \end{pmatrix}^T$$

$$= 2\begin{pmatrix} 3 & 5 \\ -4 & 2 \\ 1 & 6 \end{pmatrix} - 3\begin{pmatrix} -2 & 1 \\ 7 & 3 \\ 5 & -9 \end{pmatrix} \underset{\text{multiplying by scalars}}{=} \begin{pmatrix} 6 & 10 \\ -8 & 4 \\ 2 & 12 \end{pmatrix} - \begin{pmatrix} -6 & 3 \\ 21 & 9 \\ 15 & -27 \end{pmatrix}$$

$$= \begin{pmatrix} 12 & 7 \\ -29 & -5 \\ -13 & 39 \end{pmatrix}$$

(c) Substituting matrices **A** and **B** we have

$$(\mathbf{A}+\mathbf{B})^T = \left[\begin{pmatrix} 3 & -4 & 1 \\ 5 & 2 & 6 \end{pmatrix} + \begin{pmatrix} -2 & 7 & 5 \\ 1 & 3 & -9 \end{pmatrix}\right]^T$$

$$\underset{\text{adding the entries}}{=} \begin{pmatrix} 1 & 3 & 6 \\ 6 & 5 & -3 \end{pmatrix}^T \underset{\text{taking transpose}}{=} \begin{pmatrix} 1 & 6 \\ 3 & 5 \\ 6 & -3 \end{pmatrix}$$

(d) *What is* $\mathbf{A}^T + \mathbf{B}^T$ *equal to?*
By Theorem (1.19) property (c) we have $\mathbf{A}^T + \mathbf{B}^T = (\mathbf{A}+\mathbf{B})^T$. This was evaluated in part (c) above.

(e) *What is* $(\mathbf{AB})^T$ *equal to?*
We cannot evaluate $(\mathbf{AB})^T$ because we cannot work out **AB**. Matrix multiplication **AB** is impossible because the number of columns of matrix **A** does not equal the number of rows of matrix **B**.

It is not enough that matrices **A**, **B** are of the same size in order to execute matrix multiplication as you can observe in the final part of the above example.

1.6.3 Identity matrix

*What does the term **identity matrix** mean?*

The identity matrix is a matrix denoted by $\mathbf{I}$ such that

$$\mathbf{AI} = \mathbf{A} \text{ for any matrix } \mathbf{A} \tag{1.20}$$

This is similar to real numbers where the number 1 is the identity element which satisfies $x1 = x$ for any real number x.

What does the identity matrix look like?

It is a matrix with 1's along the leading diagonal (top left to bottom right) and zeros elsewhere.

For example, $\begin{pmatrix} 1 & 0 \\ 0 & 1 \end{pmatrix}$, $\begin{pmatrix} 1 & 0 & 0 \\ 0 & 1 & 0 \\ 0 & 0 & 1 \end{pmatrix}$ and $\begin{pmatrix} 1 & 0 & 0 & 0 \\ 0 & 1 & 0 & 0 \\ 0 & 0 & 1 & 0 \\ 0 & 0 & 0 & 1 \end{pmatrix}$ are all identity matrices.

How can we distinguish between these 2 by 2, 3 by 3 and 4 by 4 identity matrices?

We can denote the size in the subscript of $\mathbf{I}$ as $\mathbf{I}_2$, $\mathbf{I}_3$ and $\mathbf{I}_4$ respectively.

*Is the identity matrix, **I**, a square matrix?*

Yes the identity must be a square matrix.

How can we write the formal definition of the identity matrix?

Definition (1.21). An identity matrix is a square matrix denoted by $\mathbf{I}$ and defined by

$$\mathbf{I} = (i_{kj}) = \begin{cases} 1 & \text{if} \quad k = j \\ 0 & \text{if} \quad k \neq j \end{cases}$$

What does definition (1.21) mean?

All the entries in the leading diagonal (top left to bottom right) of a matrix $\mathbf{I}$ are 1, that is

$$i_{11} = i_{22} = i_{33} = i_{44} = \cdots = 1$$

and all the other entries away from the leading diagonal are zero.

For a 2 by 2 matrix we have:

$$\begin{pmatrix} a & b \\ c & d \end{pmatrix}\begin{pmatrix} 1 & 0 \\ 0 & 1 \end{pmatrix} = \begin{pmatrix} a & b \\ c & d \end{pmatrix} \text{ where } a, b, c \text{ and } d \text{ are real numbers.}$$

The identity matrix is illustrated in the next example.

Example 1.31

Let $P(1,1)$, $Q(3,1)$, $R(1,3)$ and matrix $\mathbf{A}$ represent the vertices of this triangle. Determine the image of the triangle PQR under the transformation $\mathbf{IA}$.

Solution

Carrying out the matrix multiplication **IA** gives

$$\mathbf{IA} = \begin{pmatrix} 1 & 0 \\ 0 & 1 \end{pmatrix} \overset{\begin{matrix} P & Q & R \end{matrix}}{\begin{pmatrix} 1 & 3 & 1 \\ 1 & 1 & 3 \end{pmatrix}} = \overset{\begin{matrix} P' & Q' & R' \end{matrix}}{\begin{pmatrix} 1 & 3 & 1 \\ 1 & 1 & 3 \end{pmatrix}}$$

We can plot this as shown in Fig. 1.38.

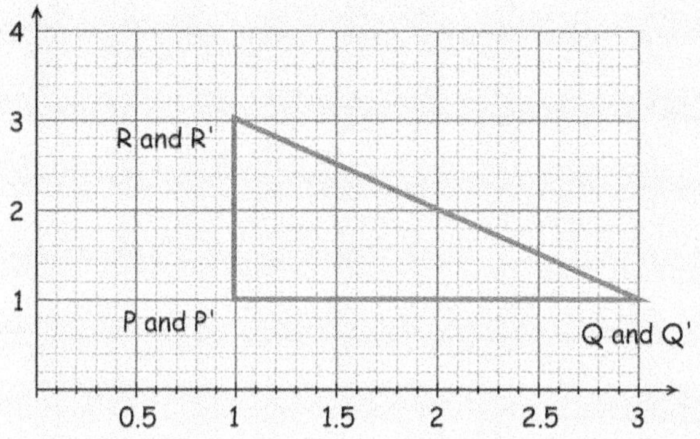

Figure 1.38

The transformation given by the matrix multiplication **IA** does *not* change the given triangle *PQR*. This means that the triangle *PQR* remains fixed under this transformation **IA**.

Note the different size identity matrices required for the following matrix multiplication:

$$\begin{pmatrix} -1 & 4 & 7 \\ 3 & -6 & 9 \end{pmatrix} \begin{pmatrix} 1 & 0 & 0 \\ 0 & 1 & 0 \\ 0 & 0 & 1 \end{pmatrix} = \begin{pmatrix} -1 & 4 & 7 \\ 3 & -6 & 9 \end{pmatrix}$$

$$\begin{pmatrix} 1 & 0 \\ 0 & 1 \end{pmatrix} \begin{pmatrix} -1 & 4 & 7 \\ 3 & -6 & 9 \end{pmatrix} = \begin{pmatrix} -1 & 4 & 7 \\ 3 & -6 & 9 \end{pmatrix}$$

If **A** is not a square matrix then the identity matrix is of different size, depending on pre- or post-multiplying by the identity matrix.

*What does pre-multiplying by **I** mean?*

It means the left hand matrix in a matrix multiplication is **I**.

*What does post-multiplying by **I** mean?*

It means the right hand matrix in a matrix multiplication is **I**.

If we post-multiply matrix $\mathbf{A} = \begin{pmatrix} -1 & 4 & 7 \\ 3 & -6 & 9 \end{pmatrix}$ by the identity matrix we use a 3 by 3 identity matrix $\mathbf{I}_3$, that is $\mathbf{A} \times \mathbf{I}_3 = \mathbf{A}$. If we pre-multiply **A** by the identity matrix we use the 2 by 2 identity matrix, that is $\mathbf{I}_2 \times \mathbf{A} = \mathbf{A}$.

In general, if **A** is a m by n (m rows by n columns) matrix then $\mathbf{I}_m\mathbf{A} = \mathbf{A}\mathbf{I}_n$.

A property of the identity is:

Proposition (1.22). Let $\mathbf{I}$ be the identity matrix. Then $\mathbf{I}^T = \mathbf{I}$

Proof – See Exercises 1.6.

■

What use is the identity matrix if the transformation remains fixed?
We need the identity matrix in order to define and explain the inverse matrix.

1.6.4 Inverse matrix

Let $x \neq 0$ be a real number then the inverse of x is a real number x^{-1} such that

$$x\left(x^{-1}\right) = 1$$

1 is the identity element of real numbers.

What do you think an inverse matrix is?
Given a square matrix $\mathbf{A}$ then the inverse of $\mathbf{A}$ is a square matrix $\mathbf{B}$ such that

$$\mathbf{AB} = \mathbf{I}$$

where $\mathbf{I}$ is the identity matrix defined earlier.

The inverse matrix of $\mathbf{A}$ is denoted by $\mathbf{A}^{-1}$ where $\mathbf{A}^{-1} = \mathbf{B}$ in the above case. Note, that

$$\mathbf{A}^{-1} \neq \frac{1}{\mathbf{A}}$$

The inverse matrix is not equal to 1 over $\mathbf{A}$. Actually 1 over $\mathbf{A}$ is not defined for a matrix $\mathbf{A}$.

We will define the process of finding $\mathbf{A}^{-1}$ later in this chapter.

Graphically, the inverse matrix $\mathbf{A}^{-1}$ performs the transformation shown in Fig. 1.39.

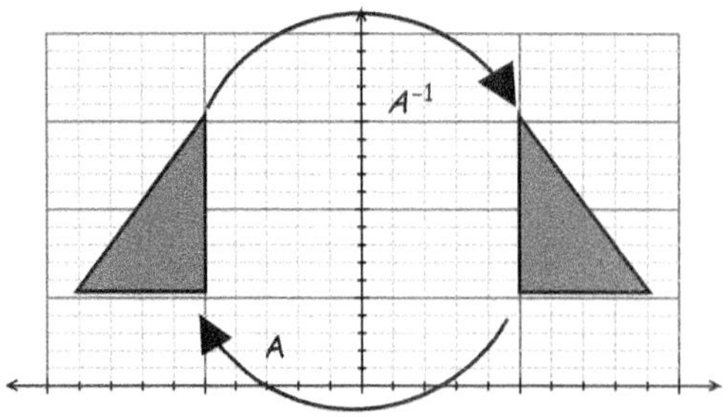

Figure 1.39

If a transformation $\mathbf{A}$ is applied to an object as shown in Fig. 1.39 then the transformation $\mathbf{A}^{-1}$ undoes $\mathbf{A}$ so the net result of $\mathbf{A}^{-1}\mathbf{A} = \mathbf{I}$ is to leave the object unchanged.

Definition (1.23). A square matrix $\mathbf{A}$ is said to be **invertible** or **non-singular** if there is a matrix $\mathbf{B}$ of the same size such that

$$\mathbf{AB} = \mathbf{BA} = \mathbf{I}$$

Matrix $\mathbf{B}$ is called the (multiplicative) **inverse** of $\mathbf{A}$ and is denoted by $\mathbf{A}^{-1}$.

Normally we say matrices $\mathbf{A}$ and $\mathbf{B}$ are the inverse of each other. If matrix $\mathbf{A}$ is the inverse of matrix $\mathbf{B}$ then $\mathbf{B}$ is the inverse of $\mathbf{A}$.

If matrices $\mathbf{A}$ and $\mathbf{B}$ are of the same size then it can be shown that $\mathbf{AB} = \mathbf{I}$ if and only if $\mathbf{BA} = \mathbf{I}$. In this case, matrix multiplication is commutative, which means it does not matter about the order of multiplication.

If $\mathbf{BA} = \mathbf{I}$ then we call matrix $\mathbf{B}$ the **left inverse** of $\mathbf{A}$. Also, if $\mathbf{AC} = \mathbf{I}$ then we call matrix $\mathbf{C}$ the **right inverse** of $\mathbf{A}$. We will show in Exercises 1.6 that the left inverse is equal to the right inverse of an invertible matrix, that is $\mathbf{B} = \mathbf{C}$.

Example 1.32

Show that matrices $\mathbf{A}$ and $\mathbf{B}$ are the inverse of each other, where

$$\mathbf{A} = \begin{pmatrix} 3 & 7 \\ 2 & 5 \end{pmatrix} \text{ and } \mathbf{B} = \begin{pmatrix} 5 & -7 \\ -2 & 3 \end{pmatrix}$$

Solution

What do we need to show?

Need to show that the matrix multiplications $\mathbf{AB}$ and $\mathbf{BA}$ give the identity matrix $\mathbf{I}$.

$$\mathbf{AB} = \begin{pmatrix} 3 & 7 \\ 2 & 5 \end{pmatrix}\begin{pmatrix} 5 & -7 \\ -2 & 3 \end{pmatrix} = \begin{pmatrix} 1 & 0 \\ 0 & 1 \end{pmatrix} = \mathbf{I}$$

Similarly, we have $\mathbf{BA} = \mathbf{I}$. Therefore, matrices $\mathbf{A}$ and $\mathbf{B}$ are the inverse of each other.

Example 1.33

Show that matrix $\mathbf{B}$ is the inverse of matrix $\mathbf{A}$ given that

$$\mathbf{A} = \begin{pmatrix} 1 & 2 & 0 \\ 2 & 5 & -1 \\ 4 & 10 & -1 \end{pmatrix} \text{ and } \mathbf{B} = \begin{pmatrix} 5 & 2 & -2 \\ -2 & -1 & 1 \\ 0 & -2 & 1 \end{pmatrix}$$

(continued...)

Solution

Similar to Example 1.32 but the size of the matrix is 3 by 3. Multiplying the matrices:

$$\mathbf{AB} = \begin{pmatrix} 1 & 2 & 0 \\ 2 & 5 & -1 \\ 4 & 10 & -1 \end{pmatrix} \begin{pmatrix} 5 & 2 & -2 \\ -2 & -1 & 1 \\ 0 & -2 & 1 \end{pmatrix} = \begin{pmatrix} 1 & 0 & 0 \\ 0 & 1 & 0 \\ 0 & 0 & 1 \end{pmatrix} = \mathbf{I}$$

Hence the matrix $\mathbf{B}$ is the inverse of matrix $\mathbf{A}$ because $\mathbf{AB} = \mathbf{I}$. You can also show that $\mathbf{A}$ is the inverse matrix of $\mathbf{B}$ by proving $\mathbf{BA}$ is also equal to $\mathbf{I}$.

Why is the inverse matrix important?

Remember, we want to use matrices to solve a system of linear equations which we can generally write as $\mathbf{Ax} = \mathbf{b}$ where $\mathbf{x}$ is the vector of unknowns that we need to find. If we multiply both sides of this $\mathbf{Ax} = \mathbf{b}$ by the inverse matrix $\mathbf{A}^{-1}$ we obtain:

$$\mathbf{A}^{-1}(\mathbf{Ax}) = \mathbf{A}^{-1}\mathbf{b}$$

$$\underbrace{\left(\mathbf{A}^{-1}\mathbf{A}\right)}_{\mathbf{I}}\mathbf{x} = \mathbf{Ix} = \mathbf{x} = \mathbf{A}^{-1}\mathbf{b}$$

Hence we can find the unknowns by finding the inverse matrix, because $\mathbf{x} = \mathbf{A}^{-1}\mathbf{b}$.

Example 1.34

Solve the following linear system:

$$\begin{aligned} x + 2y \quad\;\; &= 1 \\ 2x + 5y - z &= 2 \\ 4x + 10y - z &= 3 \end{aligned}$$

Solution

Let $\mathbf{A}$ be the matrix of coefficients and $\mathbf{x}$ be the vector of unknowns. Then the above system can be written as $\mathbf{Ax} = \mathbf{b}$ where

$$\mathbf{A} = \begin{pmatrix} 1 & 2 & 0 \\ 2 & 5 & -1 \\ 4 & 10 & -1 \end{pmatrix}, \mathbf{x} = \begin{pmatrix} x \\ y \\ z \end{pmatrix} \text{ and } \mathbf{b} = \begin{pmatrix} 1 \\ 2 \\ 3 \end{pmatrix}$$

By the above we have $\mathbf{x} = \mathbf{A}^{-1}\mathbf{b}$.

What is $\mathbf{A}^{-1}$ equal to?

$\mathbf{A}^{-1} = \mathbf{B} = \begin{pmatrix} 5 & 2 & -2 \\ -2 & -1 & 1 \\ 0 & -2 & 1 \end{pmatrix}$ which we found in Example 1.33 above. Therefore

$$\mathbf{x} = \mathbf{A}^{-1}\mathbf{b} = \begin{pmatrix} 5 & 2 & -2 \\ -2 & -1 & 1 \\ 0 & -2 & 1 \end{pmatrix} \begin{pmatrix} 1 \\ 2 \\ 3 \end{pmatrix} = \begin{pmatrix} 3 \\ -1 \\ -1 \end{pmatrix} = \begin{pmatrix} x \\ y \\ z \end{pmatrix}$$

Hence $x = 3, y = -1$ and $z = -1$. Check that this solution satisfies the given linear system.

However, not all matrices have an inverse as we describe next.

Example 1.35

Show that matrix $\mathbf{A}$ does not have an inverse, where $\mathbf{A} = \begin{pmatrix} 3 & 2 \\ 6 & 4 \end{pmatrix}$.

Solution

Let $\mathbf{B} = \begin{pmatrix} a & b \\ c & d \end{pmatrix}$ be a general 2 by 2 matrix. If we multiply out $\mathbf{AB}$ we have

$$\mathbf{AB} = \begin{pmatrix} 3 & 2 \\ 6 & 4 \end{pmatrix}\begin{pmatrix} a & b \\ c & d \end{pmatrix} = \begin{pmatrix} \boxed{3a+2c} & 3b+2d \\ \boxed{6a+4c} & 6b+4d \end{pmatrix} = \begin{pmatrix} \boxed{1} & 0 \\ \boxed{0} & 1 \end{pmatrix}$$

Equating the entries of the first column gives:

$$3a + 2c = 1 \quad (1)$$

$$6a + 4c = 0 \quad (2)$$

These equations are inconsistent because equation (2) is double the first on the left hand side but is not doubled on the right. There are no real numbers a, b, c and d which satisfy both equations (1) and (2). Hence the matrix $\mathbf{A}$ does not have an inverse.

Definition (1.24). A square matrix $\mathbf{A}$ is said to be **non-invertible** or **singular** if there is no matrix $\mathbf{B}$ such that $\mathbf{AB} = \mathbf{BA} = \mathbf{I}$.

The zero matrix $\mathbf{O}$ has no inverse. We say the zero matrix is non-invertible.

1.6.5 Properties of the inverse matrix

In this subsection, we discuss various properties of the inverse matrix.

You may find the remaining parts of this section challenging to follow because they deal with proving results. Generally, if not universally, students find understanding and constructing their own proofs very demanding. This is because a thorough understanding of each step is needed. However, you will thoroughly enjoy linear algebra if you can follow and construct your own proofs.

For the remainder of this section you will need to know the definition of an invertible matrix, which is definition (1.23).

Proposition (1.25). The inverse of an invertible (non-singular) matrix is **unique**.

What does this proposition mean?

There is *only* one inverse matrix of an invertible matrix, or mathematically there is only one matrix $\mathbf{B}$, such that $\mathbf{AB} = \mathbf{BA} = \mathbf{I}$.

Proof.

How can we prove this proposition?

We suppose there are two inverse matrices associated with an invertible matrix $\mathbf{A}$, and then show that these are in fact equal. Let's nominate these inverse matrices as $\mathbf{B}$ and $\mathbf{C}$ and show that $\mathbf{B} = \mathbf{C}$. These matrices, $\mathbf{B}$ and $\mathbf{C}$, must satisfy $\mathbf{AB} = \mathbf{I}$ and $\mathbf{AC} = \mathbf{I}$ (because $\mathbf{B}$ and $\mathbf{C}$ are inverses of $\mathbf{A}$).

Since both, $\mathbf{AB}$ and $\mathbf{AC}$, are equal to the identity matrix $\mathbf{I}$ we can equate them:

$$\mathbf{AB} = \mathbf{AC}$$

Left multiply this equation $\mathbf{AB} = \mathbf{AC}$ by matrix $\mathbf{B}$:

$$\begin{aligned}
\mathbf{B}(\mathbf{AB}) &= \mathbf{B}(\mathbf{AC}) \\
\underbrace{(\mathbf{BA})}_{=\mathbf{I}}\mathbf{B} &= \underbrace{(\mathbf{BA})}_{=\mathbf{I}}\mathbf{C} && \left[\begin{array}{l}\text{Since matrices } \mathbf{A} \text{ and } \mathbf{B} \text{ are inverse} \\ \text{of each other, } \mathbf{BA} = \mathbf{AB} = \mathbf{I}\end{array}\right] \\
\mathbf{IB} &= \mathbf{IC} \\
\mathbf{B} &= \mathbf{C} && \left[\text{Remember } \mathbf{IB} = \mathbf{B} \text{ and } \mathbf{IC} = \mathbf{C}\right]
\end{aligned}$$

Hence we have proven our proposition that the inverse matrix is unique.

■

Proposition (1.26). If $\mathbf{A}$ is an invertible matrix then $\mathbf{A}^{-1}$ is invertible and $\left(\mathbf{A}^{-1}\right)^{-1} = \mathbf{A}$.

This proposition means that the inverse of the inverse matrix is the matrix we started with.

This is like taking an ice cube, melting it, and then refreezing it into an ice cube again.

Proof.

Let $\mathbf{B} = \mathbf{A}^{-1}$ then by definition of the inverse matrix (1.23):

$\mathbf{A}$ is said to be invertible if there is a $\mathbf{B}$ such that $\mathbf{AB} = \mathbf{BA} = \mathbf{I}$.

We know that $\mathbf{B}$ is an invertible matrix because $\mathbf{BA} = \mathbf{I}$. Required to prove $\mathbf{B}^{-1} = \mathbf{A}$.

Why?

Because at the beginning of the proof we stated $\mathbf{B} = \mathbf{A}^{-1}$.

By the definition of the inverse matrix we have

$$\mathbf{BB}^{-1} = \mathbf{I} \text{ and } \mathbf{BA} = \mathbf{I}$$

Equating these gives

$$\mathbf{BB}^{-1} = \mathbf{BA}$$

Left multiplying both sides by $\mathbf{B}^{-1}$ yields:

$$\begin{aligned}
\mathbf{B}^{-1}\left(\mathbf{B}\mathbf{B}^{-1}\right) &= \mathbf{B}^{-1}\left(\mathbf{B}\mathbf{A}\right) \\
\left(\mathbf{B}^{-1}\mathbf{B}\right)\mathbf{B}^{-1} &= \left(\mathbf{B}^{-1}\mathbf{B}\right)\mathbf{A} \\
\mathbf{I}\mathbf{B}^{-1} &= \mathbf{I}\mathbf{A} \qquad \left[\text{remember } \mathbf{B}^{-1}\mathbf{B} = \mathbf{I}\right] \\
\mathbf{B}^{-1} &= \mathbf{A} \qquad \left[\text{because } \mathbf{I}\mathbf{X} = \mathbf{X} \text{ for any matrix } \mathbf{X}\right]
\end{aligned}$$

Hence $\mathbf{B}^{-1} = \mathbf{A}$, which means $\left(\mathbf{A}^{-1}\right)^{-1} = \mathbf{A}$ because $\mathbf{B} = \mathbf{A}^{-1}$. Thus we have our result.

■

Proposition (1.27). Let $\mathbf{A}$ and $\mathbf{B}$ be invertible matrices, then $\mathbf{AB}$ is invertible and $(\mathbf{AB})^{-1} = \mathbf{B}^{-1}\mathbf{A}^{-1}$

What does this proposition mean?

It means that the product of two invertible matrices is also invertible. Note that the inverse of $\mathbf{A} \times \mathbf{B}$ is equal to the inverse of $\mathbf{B}$ multiplied by the inverse of $\mathbf{A}$. You undo the last operation first. Compare this with putting on your shirt and then your tie. You remove your tie first and then your shirt.

We can illustrate this by means of a flow chart:

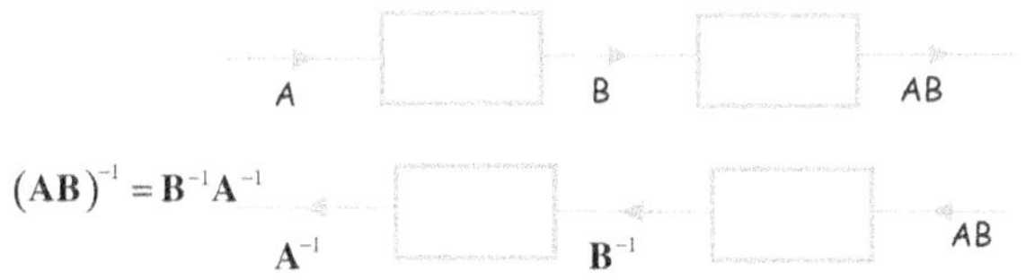

Proof.

The inverse of $\mathbf{AB}$ is given by $(\mathbf{AB})^{-1}$.

What do we need to show?

Required to prove $\mathbf{B}^{-1}\mathbf{A}^{-1}$ is the inverse of $\mathbf{AB}$; that is, we need to show

$$(\mathbf{AB})\,\mathbf{B}^{-1}\mathbf{A}^{-1} = \mathbf{I} \text{ and } \mathbf{B}^{-1}\mathbf{A}^{-1}\,(\mathbf{AB}) = \mathbf{I}$$

Let's consider the first of these, $(\mathbf{AB})\,\mathbf{B}^{-1}\mathbf{A}^{-1}$:

$$\begin{aligned}
(\mathbf{AB})\,\mathbf{B}^{-1}\mathbf{A}^{-1} &= \mathbf{A}\underbrace{\left(\mathbf{B}\mathbf{B}^{-1}\right)}_{=\mathbf{I}}\mathbf{A}^{-1} \\
&= \mathbf{A}\mathbf{I}\mathbf{A}^{-1} = \mathbf{A}\mathbf{A}^{-1} = \mathbf{I}
\end{aligned}$$

Similarly we have $\mathbf{B}^{-1}\mathbf{A}^{-1}(\mathbf{AB}) = \mathbf{I}$.

Hence by definition of the inverse matrix (1.23), we conclude that $\mathbf{B}^{-1}\mathbf{A}^{-1}$ is the inverse of $\mathbf{AB}$ which is denoted by $(\mathbf{AB})^{-1}$. We have our result $(\mathbf{AB})^{-1} = \mathbf{B}^{-1}\mathbf{A}^{-1}$.

■

Proposition (1.28). Let $\mathbf{A}$ be an invertible (non-singular) matrix and k be a non-zero scalar then $(k\mathbf{A})^{-1}$ is invertible (non-singular) and $(k\mathbf{A})^{-1} = \frac{1}{k}\mathbf{A}^{-1}$.

Proof – See Exercises 1.6 question 5.

■

Proposition (1.29). Let $\mathbf{A}$ be an invertible (non-singular) matrix then the transpose of the matrix, $\mathbf{A}^T$, is also invertible and $(\mathbf{A}^T)^{-1} = (\mathbf{A}^{-1})^T$.

What does this proposition mean?
For an invertible matrix, you can change the order of the inverse and transpose.

Proof.
Similar to the proof of Proposition (1.26).

The inverse matrix of $\mathbf{A}^T$ is denoted by $(\mathbf{A}^T)^{-1}$. We need to show that $(\mathbf{A}^{-1})^T$ is equal to the inverse of $\mathbf{A}^T$; that is, we need to show that

$$(\mathbf{A}^{-1})^T \mathbf{A}^T = \mathbf{I} \text{ and } \mathbf{A}^T (\mathbf{A}^{-1})^T = \mathbf{I}$$

Examining the first of these $(\mathbf{A}^{-1})^T \mathbf{A}^T$ we have

$$\begin{aligned} (\mathbf{A}^{-1})^T \mathbf{A}^T &= (\mathbf{A}\mathbf{A}^{-1})^T && \left[\text{using Theorem (1.12) (d) } \mathbf{Y}^T\mathbf{X}^T = (\mathbf{X}\mathbf{Y})^T\right] \\ &= \mathbf{I}^T && \left[\text{remember } \mathbf{A}\mathbf{A}^{-1} = \mathbf{I}\right] \\ &= \mathbf{I} && \left[\text{remember } \mathbf{I}^T = \mathbf{I} \text{ because } \mathbf{I} \text{ is the identity}\right] \end{aligned}$$

Similarly we have $\mathbf{A}^T (\mathbf{A}^{-1})^T = \mathbf{I}$.

In summary we have $(\mathbf{A}^{-1})^T \mathbf{A}^T = \mathbf{A}^T (\mathbf{A}^{-1})^T = \mathbf{I}$. Hence $(\mathbf{A}^{-1})^T$ is the inverse matrix of $\mathbf{A}^T$. We have our required result $(\mathbf{A}^T)^{-1} = (\mathbf{A}^{-1})^T$.

■

In Exercises 1.6, there are questions on **proof by induction**. The method is fully explained on the book's website. Normally if a statement to be proved is valid for positive integers n then we can apply mathematical induction. For example, say we want to prove

$$1 + 2 + 3 + \cdots + n = \frac{n(n+1)}{2} \quad (\dagger)$$

then we would use proof by induction. Let $P(n)$ represent this statement $(\dagger)$. The three steps of mathematical induction are:

Step 1 Check the result is true for some base case such as $n = 1$.

Step 2 Assume the result is true for $n = k$.

Step 3 Prove the result for $n = k + 1$ by using steps 1 and 2.

If you are not familiar with proof by induction then it is important that you go over the relevant material on the website. Some examples of proof by induction are also on the website.

Summary

The properties of the transpose matrices are given by:

$$\left(\mathbf{A}^T\right)^T = \mathbf{A}; \quad (k\mathbf{A})^T = k\mathbf{A}^T; \quad (\mathbf{A}+\mathbf{B})^T = \mathbf{A}^T + \mathbf{B}^T; \quad (\mathbf{AB})^T = \mathbf{B}^T\mathbf{A}^T$$

A square matrix $\mathbf{A}$ is said to be invertible if there is a matrix $\mathbf{B}$ of the same size such that

$$\mathbf{AB} = \mathbf{BA} = \mathbf{I}$$

If $\mathbf{A}$ and $\mathbf{B}$ are invertible matrices then we have

$$\left(\mathbf{A}^{-1}\right)^{-1} = \mathbf{A}; \quad (\mathbf{AB})^{-1} = \mathbf{B}^{-1}\mathbf{A}^{-1}; \quad \left(\mathbf{A}^T\right)^{-1} = \left(\mathbf{A}^{-1}\right)^T$$

EXERCISES 1.6

(Brief solutions at end of book. Full solutions available at <http://www.oup.co.uk/companion/singh>.)

1. Find $\mathbf{A}^T$ in each of the following cases:

(a) $\mathbf{A} = \begin{pmatrix} 1 & 2 \\ 3 & 4 \end{pmatrix}$ (b) $\mathbf{A} = \begin{pmatrix} 1 & 2 & 3 \\ -1 & -2 & -3 \end{pmatrix}$ (c) $\mathbf{A} = \begin{pmatrix} -1 & 5 & 9 & 100 \end{pmatrix}$

(d) $\mathbf{A} = \begin{pmatrix} a & b \\ c & d \end{pmatrix}$ (e) $\mathbf{A} = \begin{pmatrix} 0 & 0 \\ 0 & 0 \\ 0 & 0 \end{pmatrix}$ (f) $\mathbf{A} = \begin{pmatrix} 0 & 1 & 0 & 0 \\ 0 & 0 & 1 & 0 \\ 0 & 0 & 0 & 1 \\ 0 & 0 & 0 & 0 \end{pmatrix}$

2. Let $\mathbf{A} = \begin{pmatrix} -1 & 4 & 8 \\ -9 & 1 & 2 \end{pmatrix}$, $\mathbf{B} = \begin{pmatrix} 5 & 8 \\ 0 & -6 \\ 5 & 6 \end{pmatrix}$, $\mathbf{C} = \begin{pmatrix} -4 & 1 \\ 6 & 5 \end{pmatrix}$ and

$\mathbf{D} = \begin{pmatrix} -6 & 3 & 1 \\ 8 & 9 & -2 \\ 6 & -1 & 5 \end{pmatrix}$

Compute the following, if possible. *(You can check your answers using MATLAB.)*

(a) $(\mathbf{AB})^T$ (b) $(\mathbf{BC})^T$ (c) $\mathbf{C} - \mathbf{C}^T$ (d) $\mathbf{D} - \mathbf{D}^T$ (e) $\left(\mathbf{D}^T\right)^T$

(f) $(2\mathbf{C})^T$ (g) $\mathbf{A}^T + \mathbf{B}$ (h) $\mathbf{A} + \mathbf{B}^T$ (i) $\left(\mathbf{A}^T + \mathbf{B}\right)^T$ (j) $\left(2\mathbf{A}^T - 5\mathbf{B}\right)^T$

(k) $(-\mathbf{D})^T$ (l) $-\left(\mathbf{D}^T\right)$ (m) $\left(\mathbf{C}^2\right)^T$ (n) $\left(\mathbf{C}^T\right)^2$

3. Let $\mathbf{u} = \begin{pmatrix} 1 \\ 2 \\ 3 \end{pmatrix}$ and $\mathbf{v} = \begin{pmatrix} 4 \\ 5 \\ 6 \end{pmatrix}$. Determine (a) $\mathbf{u}^T\mathbf{v}$ (b) $\mathbf{v}^T\mathbf{u}$ (c) $\mathbf{u} \cdot \mathbf{v}$

4. Prove $\mathbf{I}^T = \mathbf{I}$ where $\mathbf{I}$ is the identity matrix.

5. Prove Proposition (1.28).

6. Let $\mathbf{A}$ be any matrix and k be a scalar. Prove that $(k\mathbf{A})^T = k\mathbf{A}^T$.

7. Let $\mathbf{A}$ be a square matrix. Prove that $(\mathbf{A}^2)^T = (\mathbf{A}^T)^2$.

8. Prove the following:
   If an invertible matrix $\mathbf{A}$ has a left inverse $\mathbf{B}$ and a right inverse $\mathbf{C}$, then $\mathbf{B} = \mathbf{C}$.

9. Let matrices $\mathbf{A}$ and $\mathbf{B}$ be invertible. Prove that $\mathbf{A} + \mathbf{B}$ may not be invertible.

10. * Prove Proposition (1.19) part (d). (This is a difficult result to establish, but try to prove this without looking at the solutions. It will bring you great pleasure if you can prove this result.)

11. Let $\mathbf{A}$, $\mathbf{B}$ and $\mathbf{C}$ be matrices of the appropriate size so that they can be multiplied. Prove that $(\mathbf{ABC})^T = \mathbf{C}^T\mathbf{B}^T\mathbf{A}^T$.

12. * (If you do not know the method of proof by induction then you will need to read up on 'Proof by Induction' on the web to do the following question.)
    Let $\mathbf{A}_1, \mathbf{A}_2, \ldots, \mathbf{A}_{n-1}$ and $\mathbf{A}_n$ be matrices of the appropriate size so that the matrix multiplication, $\mathbf{A}_1 \times \mathbf{A}_2 \times \cdots \times \mathbf{A}_{n-1} \times \mathbf{A}_n$, is valid.
    Prove that for any natural number n (a positive integer) we have

$$(\mathbf{A}_1\mathbf{A}_2 \cdots \mathbf{A}_{n-1}\mathbf{A}_n)^T = \mathbf{A}_n^T\mathbf{A}_{n-1}^T \cdots \mathbf{A}_2^T\mathbf{A}_1^T.$$

13. Let $\mathbf{A}$ be a square matrix. Prove that $(\mathbf{A}^n)^T = (\mathbf{A}^T)^n$, where n is a natural number.

14. Let $\mathbf{A} = \begin{pmatrix} 0 & 0 \\ 0 & 0 \end{pmatrix}$, $\mathbf{B} = \begin{pmatrix} 1 & 1 \\ 1 & 1 \end{pmatrix}$ and $\mathbf{C} = \begin{pmatrix} 1 & 3 \\ 2 & 6 \end{pmatrix}$. Show that matrices $\mathbf{A}$, $\mathbf{B}$ and $\mathbf{C}$ are singular, that is non-invertible.

15. Show that matrices $\mathbf{A}$ and $\mathbf{B}$ are inverses of each other:
    (a) $\mathbf{A} = \begin{pmatrix} 9 & 2 \\ 13 & 3 \end{pmatrix}$, $\mathbf{B} = \begin{pmatrix} 3 & -2 \\ -13 & 9 \end{pmatrix}$ (b) $\mathbf{A} = \begin{pmatrix} 1 & 2 & 3 \\ 0 & 1 & 4 \\ 0 & 0 & 1 \end{pmatrix}$, $\mathbf{B} = \begin{pmatrix} 1 & -2 & 5 \\ 0 & 1 & -4 \\ 0 & 0 & 1 \end{pmatrix}$
    (c) $\mathbf{A} = \begin{pmatrix} 1 & 0 & 2 \\ 2 & -1 & 3 \\ 4 & 1 & 8 \end{pmatrix}$, $\mathbf{B} = \begin{pmatrix} -11 & 2 & 2 \\ -4 & 0 & 1 \\ 6 & -1 & -1 \end{pmatrix}$
    Hence, or otherwise, solve the following linear system:

$$\begin{aligned} x \quad\;\; + 2z &= 1 \\ 2x - y + 3z &= 2 \\ 4x + y + 8z &= 3 \end{aligned}$$

16. Let $\mathbf{A} = \begin{pmatrix} \sin(\theta) & \cos(\theta) \\ -\cos(\theta) & \sin(\theta) \end{pmatrix}$ and $\mathbf{B} = \begin{pmatrix} \sin(\theta) & -\cos(\theta) \\ \cos(\theta) & \sin(\theta) \end{pmatrix}$. Show that $\mathbf{A}$ and $\mathbf{B}$ are the inverse of each other. Solve the linear system

$$\sin(\theta)\,x + \cos(\theta)\,y = 1$$
$$-\cos(\theta)\,x + \sin(\theta)\,y = -1$$

17. Explain why an invertible matrix $\mathbf{A}$ has to be a square matrix.
18. Let $\mathbf{A}$ and $\mathbf{B}$ be invertible matrices of the appropriate size so that matrix multiplication is a valid operation. Prove that $\left((\mathbf{AB})^{-1}\right)^T = \left(\mathbf{A}^T\right)^{-1}\left(\mathbf{B}^T\right)^{-1}$.
19. Let $\mathbf{A}$, $\mathbf{B}$ and $\mathbf{C}$ be invertible matrices of the appropriate size so that the following matrix multiplication is valid: $\mathbf{ABC}$. Prove that $(\mathbf{ABC})^{-1} = \mathbf{C}^{-1}\mathbf{B}^{-1}\mathbf{A}^{-1}$.
20. * (You will need to read up on 'Proof by Induction' on the web to do the following question.)

    Let $\mathbf{A}_1, \mathbf{A}_2, \ldots, \mathbf{A}_{n-1}$ and $\mathbf{A}_n$ be matrices of the appropriate size so that

    $$\mathbf{A}_1 \times \mathbf{A}_2 \times \cdots \times \mathbf{A}_{n-1} \times \mathbf{A}_n$$

    is a valid operation. Prove that

    $$(\mathbf{A}_1 \times \mathbf{A}_2 \times \cdots \times \mathbf{A}_{n-1} \times \mathbf{A}_n)^{-1} = \mathbf{A}_n^{-1} \times \mathbf{A}_{n-1}^{-1} \times \cdots \times \mathbf{A}_2^{-1} \times \mathbf{A}_1^{-1}$$

21. In matrix algebra we define the negative indices of an invertible (non-singular) square matrix $\mathbf{A}$ by

    $$\mathbf{A}^{-n} = \left(\mathbf{A}^{-1}\right)^n \text{ where } n \text{ is a natural number (positive integer)}$$

    By using this definition prove that $\mathbf{A}^{-n} = (\mathbf{A}^n)^{-1}$.
22. Let $\mathbf{C}$ be an invertible matrix and $\mathbf{A}$ and $\mathbf{B}$ be matrices of appropriate size so that matrix multiplications below are valid. Prove the following:

    (a) If $\mathbf{AC} = \mathbf{BC}$ then $\mathbf{A} = \mathbf{B}$. (This is called the **right cancellation property**.)

    (b) If $\mathbf{CA} = \mathbf{CB}$ then $\mathbf{A} = \mathbf{B}$. (This is called the **left cancellation property**.)
23. Prove that if $\mathbf{A}$ is an invertible (non-singular) matrix then $\mathbf{AB} = \mathbf{O} \Rightarrow \mathbf{B} = \mathbf{O}$.
24. Let $\mathbf{P}$ and $\mathbf{A}$ be square matrices and matrix $\mathbf{P}$ be invertible. Prove the following results:
    (a) $\left(\mathbf{P}^{-1}\mathbf{AP}\right)^2 = \mathbf{P}^{-1}\mathbf{A}^2\mathbf{P}$ (b) $\left(\mathbf{P}^{-1}\mathbf{AP}\right)^3 = \mathbf{P}^{-1}\mathbf{A}^3\mathbf{P}$ (c) $\left(\mathbf{P}^{-1}\mathbf{AP}\right)^n = \mathbf{P}^{-1}\mathbf{A}^n\mathbf{P}$
    (Hint: Use mathematical induction.)

SECTION 1.7 Types of Solutions

By the end of this section you will be able to

- find a unique solution, infinitely many solutions or establish no solutions to a linear system
- solve homogeneous and non-homogeneous linear systems

Remember, linear algebra is the study of linear systems of equations.

In this section, we examine more complex linear systems which involve many equations in many variables. We try to reduce these linear systems to their simplest possible form by breaking them down to reduced row echelon form.

Here, we apply row operations to determine if there are no, unique or an infinite number of solutions to a linear system.

1.7.1 Equivalent systems

In section 1.2 we found the solutions to simultaneous equations by writing them in matrix form, and then converting to a row equivalent matrix.

What does row equivalent mean?
Two matrices are row equivalent if one is derived from the other by applying elementary row operations as described on page 19.

Proposition (1.30). If a linear system is described by the augmented matrix $(\mathbf{A} \mid \mathbf{b})$ and it is row equivalent to $(\mathbf{R} \mid \mathbf{b}')$ then both linear systems have the same solution set.

Proof.
This follows from section 1.1 because the augmented matrix $(\mathbf{R} \mid \mathbf{b}')$ is derived from $(\mathbf{A} \mid \mathbf{b})$ by elementary row operations which are:

1. Multiply a row by a non-zero constant.
2. Add a multiple of one row to another.
3. Interchange rows.

These are equivalent to:

1. Multiply an equation by a non-zero constant.
2. Add a multiple of one equation to another.
3. Interchange equations.

From section 1.1, we know that carrying out the bottom three operations on a linear system yields the same solution as the initial linear system. The two sets of operations are equivalent, therefore $(\mathbf{A} \mid \mathbf{b})$ and $(\mathbf{R} \mid \mathbf{b}')$ have the same solution set.

■

This proposition means that when solving a linear system $\mathbf{Ax} = \mathbf{b}$, it can be simplified to an easier problem $\mathbf{Rx} = \mathbf{b}'$ which has the same solution as $\mathbf{Ax} = \mathbf{b}$.

1.7.2 Types of solutions

As discussed in section 1.1 there are three types of solutions to a linear system:

1. No solution
2. Unique solution
3. Infinite number of solutions

All the examples of section 1.2 had unique solutions. If the system has no solution, we say it is **inconsistent**. We can show that a system is inconsistent by using the above row operations, as the next example illustrates.

Example 1.36

Show that the following linear system is inconsistent:

$$\begin{aligned} x + y + 2z &= 3 \\ -x + 3y - 5z &= 7 \\ 2x - 2y + 7z &= 1 \end{aligned}$$

Solution

The augmented matrix with each row labelled is given by

$$\begin{matrix} R_1 \\ R_2 \\ R_3 \end{matrix} \left(\begin{array}{ccc|c} 1 & 1 & 2 & 3 \\ -1 & 3 & -5 & 7 \\ 2 & -2 & 7 & 1 \end{array} \right)$$

We use Gaussian elimination to simplify the problem. To get a 0 in place of -1 in the second row and the 2 in the bottom row we execute operations, $R_2 + R_1$ and $R_3 - 2R_1$, respectively.

$$\begin{matrix} R_1 \\ R_2^* = R_2 + R_1 \\ R_3^* = R_3 - 2R_1 \end{matrix} \left(\begin{array}{ccc|c} 1 & 1 & 2 & 3 \\ -1+1 & 3+1 & -5+2 & 7+3 \\ 2-(2\times 1) & -2-(2\times 1) & 7-(2\times 2) & 1-(2\times 3) \end{array} \right)$$

Simplifying the arithmetic in the entries gives

$$\begin{matrix} R_1 \\ R_2^* \\ R_3^* \end{matrix} \left(\begin{array}{ccc|c} 1 & 1 & 2 & 3 \\ 0 & 4 & -3 & 10 \\ 0 & -4 & 3 & -5 \end{array} \right)$$

Adding the bottom two rows, $R_3^* + R_2^*$, and simplifying gives:

$$\begin{matrix} \\ R_1 \\ R_2^* \\ R_3^{**} = R_3^* + R_2^* \end{matrix} \begin{matrix} \begin{matrix} x & y & z & \end{matrix} \\ \left(\begin{array}{ccc|c} 1 & 1 & 2 & 3 \\ 0 & 4 & -3 & 10 \\ 0 & 0 & 0 & 5 \end{array} \right) \end{matrix}$$

Expanding the bottom row, R_3^{**}, we have $0x + 0y + 0z = 5$ which means that $0 = 5$.
Clearly 0 cannot equal 5, therefore the given linear system is inconsistent.
The given equations can be illustrated as shown in Fig. 1.40.

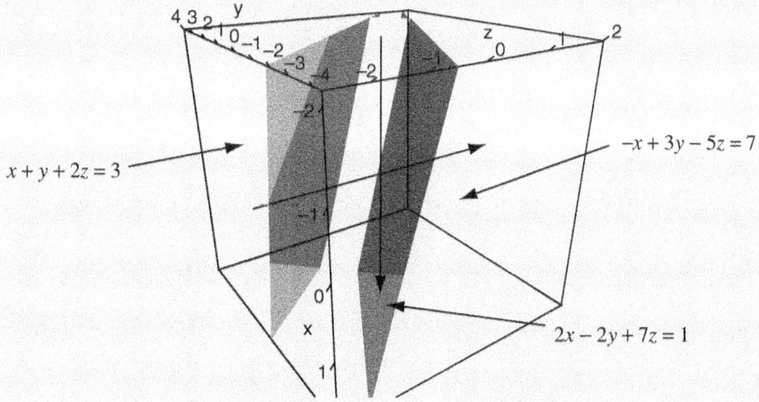

Figure 1.40

(continued...)

These equations are inconsistent because the three planes (equations) do not meet, which means there is no point common to all three. There is no solution to the equations.

For a 3 by 3 linear system Fig. 1.41 illustrates situations where there are no solutions.

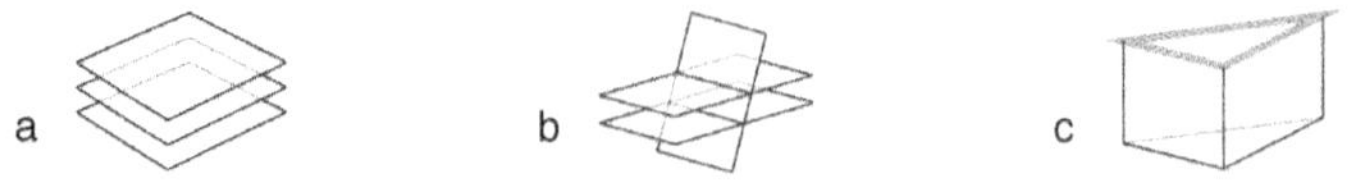

Equations with three parallel planes. Equations with two parallel planes. Three equations form a prism.

Figure 1.41

In each of these cases there is no point which is common to all three planes, and this means that there is no solution to the linear system.

In general, if a linear system leads to

$$0x_1 + 0x_2 + 0x_3 + \cdots 0x_n = b$$

where $b \neq 0$ (b is not zero) then the system is inconsistent and does not have a solution.

A linear system $(\mathbf{A} \mid \mathbf{b})$ has no solution if and only if the row equivalent $\left(\mathbf{R} \mid \mathbf{b}'\right)$ has no solution.

1.7.3 Homogeneous systems

A **homogeneous** linear system is defined as $\mathbf{Ax} = \mathbf{O}$ where $\mathbf{A}$ is an m by n matrix and $\mathbf{x}$ is a vector.

Example 1.37

Solve the following homogeneous linear system:

$$\begin{aligned} -x + 2y + 3z &= 0 \\ x - 4y - 13z &= 0 \\ -3x + 5y + 4z &= 0 \end{aligned} \quad \text{which can be written as} \quad \begin{pmatrix} -1 & 2 & 3 \\ 1 & -4 & -13 \\ -3 & 5 & 4 \end{pmatrix}\begin{pmatrix} x \\ y \\ z \end{pmatrix} = \begin{pmatrix} 0 \\ 0 \\ 0 \end{pmatrix}$$

Solution

What is the augmented matrix equal to?

$$\begin{matrix} R_1 \\ R_2 \\ R_3 \end{matrix} \left(\begin{array}{ccc|c} -1 & 2 & 3 & 0 \\ 1 & -4 & -13 & 0 \\ -3 & 5 & 4 & 0 \end{array}\right)$$

We need to get 0's in the bottom left hand corner of the matrix, that is in place of 1, -3 and 5.

How?

The row operations $R_2 + R_1$ and $R_3 - 3R_1$ give 0 in place of 1 and -3 respectively:

$$\begin{matrix} R_1 \\ R_2^* = R_2 + R_1 \\ R_3^* = R_3 - 3R_1 \end{matrix} \left(\begin{array}{ccc|c} -1 & 2 & 3 & 0 \\ 1-1 & -4+2 & -13+3 & 0+0 \\ -3-3(-1) & 5-3(2) & 4-3(3) & 0-3(0) \end{array} \right)$$

Simplifying the entries gives

$$\begin{matrix} R_1 \\ R_2^* \\ R_3^* \end{matrix} \left(\begin{array}{ccc|c} -1 & 2 & 3 & 0 \\ 0 & -2 & -10 & 0 \\ 0 & -1 & -5 & 0 \end{array} \right)$$

Executing the row operation $2R_3^* - R_2^*$, and simplifying, yields:

$$\begin{matrix} R_1 \\ R_2^* \\ R_3^{**} = 2R_3^* - R_2^* \end{matrix} \left(\begin{array}{ccc|c} -1 & 2 & 3 & 0 \\ 0 & -2 & -10 & 0 \\ 0 & 0 & 0 & 0 \end{array} \right)$$

Multiplying the first row by -1 and the second row by $-1/2$ we have

$$\begin{matrix} \\ R_1^* = -R_1 \\ R_2^{**} = -R_2^*/2 \\ R_3^{**} \end{matrix} \begin{matrix} \begin{matrix} x & y & z & \end{matrix} \\ \left(\begin{array}{ccc|c} 1 & -2 & -3 & 0 \\ 0 & 1 & 5 & 0 \\ 0 & 0 & 0 & 0 \end{array} \right) \end{matrix} \qquad (\dagger)$$

This augmented matrix is now in row echelon form.

By expanding the middle row, R_2^{**}, of (†) we have

$$y + 5z = 0 \text{ which gives } y = -5z$$

How can we find x?

Expanding the first row R_1^* of (†):

$$x - 2y - 3z = 0 \text{ which gives } x = 2y + 3z \qquad (*)$$

As long as these equations, $y = -5z$ and $x = 2y + 3z$, are satisfied, we conclude that we have a solution to the given linear system. In fact, there are an infinite number of solutions to this system because we can let z be any real number and then write x and y in terms of z. Let $z = t$, where t is any real number, then $y = -5z = -5t$. Rewriting (*) using these values:

$$\begin{aligned} x = 2y + 3z &= 2(-5t) + 3t \\ &= -10t + 3t = -7t \end{aligned}$$

Hence our solution is $x = -7t$, $y = -5t$ and $z = t$ where t is any real number. This means that we have an infinite number of solutions, which are in the proportion; x is -7 times z, y is -5 times z and z can take any value t.

(continued...)

The given equations are illustrated in Fig. 1.42

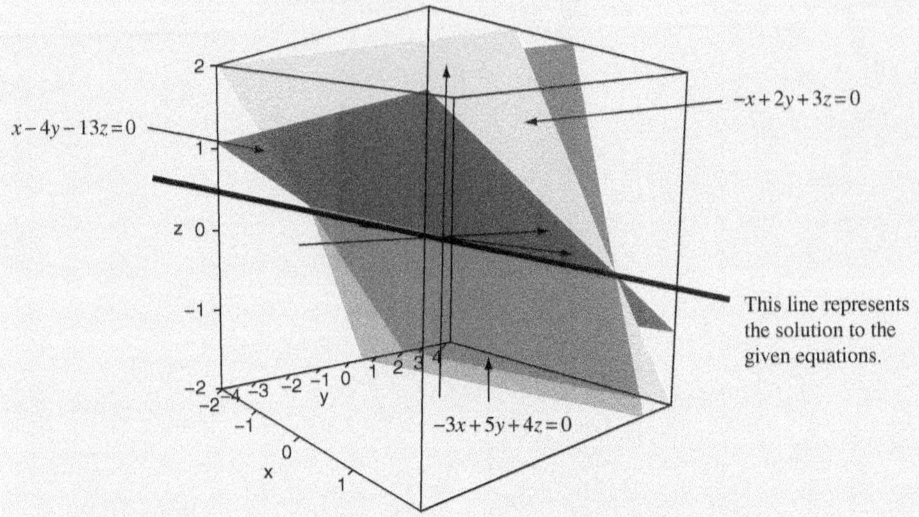

Figure 1.42

In this case the three given equations have a line in common, and any point on this line is a solution; that is why we have an infinite number of solutions.

As shown above, we can write the general solution as $x = -7t$, $y = -5t$ and $z = t$, where t is an arbitrary number called a **parameter**. A parameter is something that can be varied to give a family of solutions.

Particular solutions are given by substituting real numbers for the parameter t. For example, substituting $t = 1$ gives the particular solution $x = -7t = -7$, $y = -5t = -5$ and $z = t = 1$. Check that this is actually a solution to the above linear system.

What is the particular solution if $t = 2$?

$$x = -7t = -14, y = -5t = -10 \text{ and } z = t = 2$$

The solution for $t = \pi$ is $x = -7t = -7\pi$, $y = -5t = -5\pi$ and $z = t = \pi$.

What is the particular solution if $t = 0$?

$$x = 0, y = 0 \text{ and } z = 0$$

All these solutions:

$$(-7, -5, 1), (-14, -10, 2), (-7\pi, -5\pi, \pi), (0, 0, 0)$$

lie on the straight line shown in Fig. 1.42.

The solution $x = 0$, $y = 0$ and $z = 0$ is called the **trivial solution** to the homogeneous system.

What does the term homogeneous mean?

A linear system is called **homogeneous** if all the constant terms on the right hand side are zero; that is, $\mathbf{Ax} = \mathbf{O}$:

$$\begin{aligned} a_{11}x_1 + a_{12}x_2 + \cdots + a_{1n}x_n &= 0 \\ a_{21}x_1 + a_{22}x_2 + \cdots + a_{2n}x_n &= 0 \\ \vdots \quad\quad \vdots \quad\quad\quad\quad \vdots \quad &\quad \vdots \\ a_{m1}x_1 + a_{m2}x_2 + \cdots + a_{mn}x_n &= 0 \end{aligned}$$

In general, if a linear system of equations contains

$$0x_1 + 0x_2 + 0x_3 + \cdots + 0x_n = 0 \qquad (*)$$

then this equation can be removed without affecting the solution. A row with all zero entries is called a **zero row**.

Note, that carrying out elementary row operations on zero rows gives another zero row. You can verify this by applying the three elementary row operations on a zero row.

To understand the proof of the next Proposition (1.31) you will need to be sure that you can recall the definition of reduced row echelon form, which was given in section 1.2.

Proposition (1.31). Let the above homogeneous linear system $\mathbf{Ax} = \mathbf{O}$, whose augmented matrix is $(\mathbf{A}\,|\mathbf{O})$, be row equivalent to $(\mathbf{R}\,|\mathbf{O})$, where $\mathbf{R}$ is an equivalent matrix in reduced row echelon form. Let there be n unknowns and r non-zero rows in $\mathbf{R}$. If $r < n$ (the number of non-zero equations (rows) in $\mathbf{R}$ is less than the number of unknowns) then the linear system $(\mathbf{A}\,|\mathbf{O})$ has an infinite number of solutions.

By the above Proposition (1.30), $(\mathbf{A}\,|\mathbf{O})$ and $(\mathbf{R}\,|\mathbf{O})$ have the same solution set, therefore in the following proof we consider the augmented matrix in reduced row echelon form $(\mathbf{R}\,|\mathbf{O})$.

Proof.

(This might be a bit hard to grasp on first reading.)

We are given that the number of non-zero equations (rows) r is less than the number of unknowns n. The non-zero equations of $\mathbf{R}$ is rectangular in shape (Fig. 1.43):

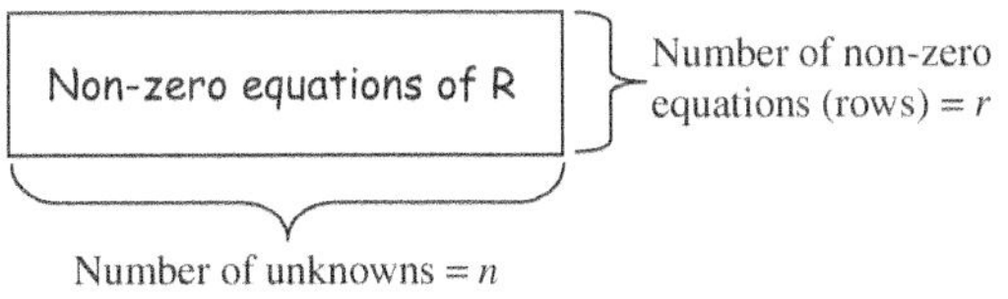

Figure 1.43

The reduced row echelon form matrix $\mathbf{R}$ has r non-zero rows. This means that there are exactly r leading 1's.

Why?

Because **R** is in reduced row echelon form, which means that every non-zero row has a leading 1. However, because the matrix is rectangular, $r < n$, there must be at least one column, call it x_j (free variable), which has no leading 1:

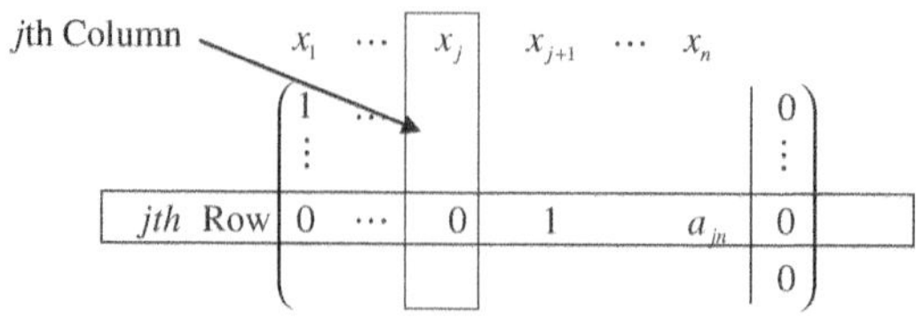

This means that the variable x_j can take any value:

$x_j = 1, 2, 3, \pi, -1, \ldots$ which we can write as $x_j = s$ where s is any real number.

Hence we have infinite number of solutions.

■

Proposition (1.31) means that if a homogeneous linear system $\mathbf{Ax} = \mathbf{O}$ has more unknowns than (non-zero) equations then this system has an infinite number of solutions.

In Example 1.37 the row echelon form was given by (†) on page 95.

Non-zero equations of **R**. → R_1^*, R_2^{**}

$$\begin{array}{c} \\ R_1^* \\ R_2^{**} \\ R_3^{**} \end{array} \begin{array}{c} \begin{array}{ccc} x & y & z \end{array} \leftarrow 3 \text{ unknowns} \\ \left(\begin{array}{ccc|c} 1 & -2 & -3 & 0 \\ 0 & 1 & 5 & 0 \\ 0 & 0 & 0 & 0 \end{array}\right) \end{array}$$

We had three unknowns ($n = 3$) but only two non-zero rows ($r = 2$) in row echelon form therefore there must be an infinite number of solutions because $r = 2 < 3 = n$.

If the number of unknowns n is greater than the number of non-zero equations in row echelon form, $n > r$, then we say that there are $n - r$ **free variables.** We normally assign parameters such as s, t etc. to these free variables. In Example 1.37 we had $3 - 2 = 1$ free variable which was z and we assigned the parameter t to it. The free variables are the unknowns which do not start any equation (row) in row echelon form. In the above, none of the equations (rows) begin with a value for z, therefore z is the free variable; that is $z = t$, where t is any real number and we can find x and y in terms of t, $x = -7t$, $y = -5t$.

Applying Proposition (1.31) to a general 3 by 3 (three equations with three unknowns) system we can state the following:

Case I If there are three non-zero equations (rows) in row echelon form, then we have a unique solution $x = 0, y = 0$ and $z = 0$ – Fig. 1.44(a).

Case II If there are less than three non-zero equations (rows) in row echelon form then the solution is not unique and forms a line or a plane – Fig. 1.44(b) and (c).

Note, that for Fig. 1.44(b) and (c) we have an infinite number of solutions.

Since we have a **homogeneous** system, we must have a solution even if it is only the trivial solution $x = 0, y = 0$ and $z = 0$.

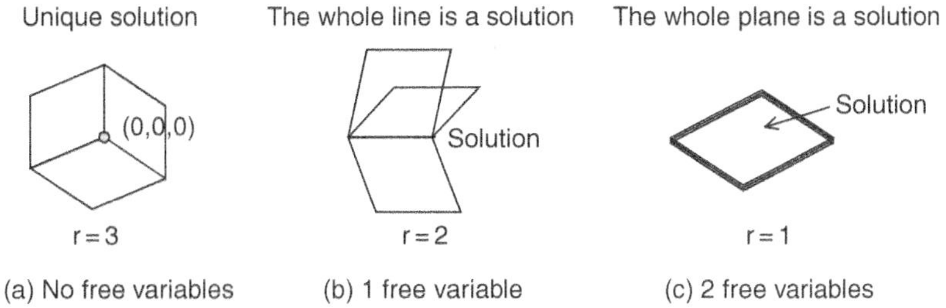

Figure 1.44

1.7.4 Non-homogeneous systems

A *non*-homogeneous linear system is defined as $\mathbf{Ax} = \mathbf{b}$ where $\mathbf{b} \neq \mathbf{O}$:

$$\begin{array}{ccccccccc} a_{11}x_1 & + & a_{12}x_2 & + & \cdots & + & a_{1n}x_n & = & b_1 \\ a_{21}x_1 & + & a_{22}x_2 & + & \cdots & + & a_{2n}x_n & = & b_2 \\ \vdots & & \vdots & & & & \vdots & & \vdots \\ a_{m1}x_1 & + & a_{m2}x_2 & + & \cdots & + & a_{mn}x_n & = & b_m \end{array}$$

Here, at least one of the *b*'s must take a value other than zero.

Proposition (1.32). Let a consistent non-homogeneous linear system, $\mathbf{Ax} = \mathbf{b}$ where $\mathbf{b} \neq \mathbf{O}$, be row equivalent to the augmented matrix $(\mathbf{R} \mid \mathbf{b}')$ where $\mathbf{R}$ is in reduced row echelon form and there are n unknowns and r non-zero rows in $\mathbf{R}$.

If $r < n$ then the linear system $\mathbf{Ax} = \mathbf{b}$ has an infinite number of solutions.

Proof – See Exercises 1.7.

■

Proposition (1.32) means that if the consistent linear system $\mathbf{Ax} = \mathbf{b}$ has more unknowns than non-zero equations in reduced row echelon form then this system has an infinite number of solutions. Again, the non-zero equations of $\mathbf{R}$ form a rectangular matrix (more columns than rows):

R

Example 1.38

Solve the following non-homogeneous linear system using Gaussian elimination:

$$\begin{array}{rrrrrcr} x & -y & +2z & & +3u & = & 1 \\ -x & +y & & +2w & -5u & = & 5 \\ x & -y & +4z & +2w & +4u & = & 13 \\ -2x & +2y & -5z & -w & -3u & = & -1 \end{array}$$

(continued...)

Solution

The augmented matrix is

$$\begin{array}{c} R_1 \\ R_2 \\ R_3 \\ R_4 \end{array}\left(\begin{array}{ccccc|c} 1 & -1 & 2 & 0 & 3 & 1 \\ -1 & 1 & 0 & 2 & -5 & 5 \\ 1 & -1 & 4 & 2 & 4 & 13 \\ -2 & 2 & -5 & -1 & -3 & -1 \end{array}\right)$$

Note, that when a particular unknown does not exist in the equation, we place a 0 in the coefficient part. To transform this matrix into (reduced) row echelon form, we execute the following elementary row operations:

$$\begin{array}{r} R_1 \\ R_2^* = R_2 + R_1 \\ R_3^* = R_3 - R_1 \\ R_4^* = R_4 + 2R_1 \end{array}\left(\begin{array}{ccccc|c} 1 & -1 & 2 & 0 & 3 & 1 \\ -1+1 & 1-1 & 0+2 & 2+0 & -5+3 & 5+1 \\ 1-1 & -1-(-1) & 4-2 & 2-0 & 4-3 & 13-1 \\ -2+2(1) & 2+2(-1) & -5+2(2) & -1+2(0) & -3+2(3) & -1+2(1) \end{array}\right)$$

Simplifying this gives

$$\begin{array}{c} R_1 \\ R_2^* \\ R_3^* \\ R_4^* \end{array}\left(\begin{array}{ccccc|c} 1 & -1 & 2 & 0 & 3 & 1 \\ 0 & 0 & 2 & 2 & -2 & 6 \\ 0 & 0 & 2 & 2 & 1 & 12 \\ 0 & 0 & -1 & -1 & 3 & 1 \end{array}\right)$$

To obtain 0's in the bottom rows we have to carry out the following row operations:

$$\begin{array}{r} R_1 \\ R_2^* \\ R_3^{**} = R_3^* - R_2^* \\ R_4^{**} = 2R_4^* + R_2^* \end{array}\left(\begin{array}{ccccc|c} 1 & -1 & 2 & 0 & 3 & 1 \\ 0 & 0 & 2 & 2 & -2 & 6 \\ 0-0 & 0-0 & 2-2 & 2-2 & 1-(-2) & 12-6 \\ 2(0)+0 & 2(0)+0 & 2(-1)+2 & 2(-1)+2 & 2(3)+(-2) & 2(1)+6 \end{array}\right)$$

Simplifying the entries gives

$$\begin{array}{c} R_1 \\ R_2^* \\ R_3^{**} \\ R_4^{**} \end{array}\left(\begin{array}{ccccc|c} 1 & -1 & 2 & 0 & 3 & 1 \\ 0 & 0 & 2 & 2 & -2 & 6 \\ 0 & 0 & 0 & 0 & 3 & 6 \\ 0 & 0 & 0 & 0 & 4 & 8 \end{array}\right)$$

Dividing the third row by 3 and the fourth row by 4 (or multiplying by 1/3 and 1/4 respectively) gives

$$\begin{array}{r} R_1 \\ R_2^* \\ R_3^\dagger = R_3^{**}/3 \\ R_4^\dagger = R_4^{**}/4 \end{array}\left(\begin{array}{ccccc|c} 1 & -1 & 2 & 0 & 3 & 1 \\ 0 & 0 & 2 & 2 & -2 & 6 \\ 0 & 0 & 0 & 0 & 1 & 2 \\ 0 & 0 & 0 & 0 & 1 & 2 \end{array}\right)$$

Subtracting the last two rows from each other and dividing the second row R_2^* by 2 we have

$$\begin{array}{r} \\ R_1 \\ R_2^*/2 \\ R_3^\dagger \\ R_4^\dagger - R_3^\dagger \end{array}\begin{array}{c} \begin{array}{ccccc} x & y & z & w & u \end{array} \\ \left(\begin{array}{ccccc|c} \boxed{1} & -1 & 2 & 0 & 3 & 1 \\ 0 & 0 & \boxed{1} & 1 & -1 & 3 \\ 0 & 0 & 0 & 0 & \boxed{1} & 2 \\ 0 & 0 & 0 & 0 & 0 & 0 \end{array}\right) \end{array} \qquad (\dagger\dagger)$$

This matrix is now in row echelon form. We have three non-zero rows with five unknowns, therefore by the above Proposition (1.32) we conclude that we have an infinite number of solutions and there are $5 - 3 = 2$ free variables. In (††) none of the equations (rows) start with y and w, therefore these are the free variables.

From the third row $R_3^\dagger$ we have $u = 2$. By expanding the second row $R_2^*/2$ we have

$$\begin{aligned} z + w - u &= 3 \\ z + w - 2 &= 3 \qquad \left[\text{substituting } u = 2\right] \\ z + w &= 5 \text{ which gives } z = 5 - w \end{aligned}$$

Let $w = t$ where t is an arbitrary real number then $z = 5 - t$.

From the first row R_1 we have

$$\begin{aligned} x - y + 2z + 3u &= 1 \\ x - y + 2(5 - t) + 3(2) &= 1 \qquad \left[\text{substituting } z = 5 - t \text{ and } u = 2\right] \\ x - y + 10 - 2t + 6 &= 1 \qquad \left[\text{simplifying}\right] \\ x &= y + 2t - 15 \end{aligned}$$

Let $y = s$, where s is an any real number, then

$$x = s + 2t - 15$$

Hence our solution to the given linear system is

$$x = s + 2t - 15, \quad y = s, \quad z = 5 - t, \quad w = t \text{ and } u = 2$$

This is the general solution and you can find particular values by substituting any real numbers for s and t. We have an infinite number of solutions.

1.7.5 Applications to puzzles

We can use Gaussian elimination to solve puzzles. We can solve Sudoku puzzles by using this technique. Many of these have an infinite number of solutions but the puzzles in national newspapers are for the general audience, so the answers are generally limited to the positive whole numbers below 10 or 25.

Most people think that they have no choice but to solve these puzzles by trial and error, but applying Gaussian elimination provides a general solution and is done systematically.

Example 1.39

In the following puzzle the row and columns add up to the numbers which are in the bold font. Find all the unknowns x's. (This is a much tougher example than you would find in a national newspaper.)

x_1	x_2	x_3	**15**
x_4	x_5	x_6	**6**
8	**7**	**6**	

(continued...)

Solution

Forming the equations by summing up the rows and columns we have

$$\begin{array}{llllllll} x_1 & +x_2 & +x_3 & & & & = 15 \\ & & & x_4 & + x_5 & + x_6 & = 6 \\ x_1 & & & +x_4 & & & = 8 \\ & x_2 & & & + x_5 & & = 7 \\ & & x_3 & & & + x_6 & = 6 \end{array} \quad \begin{bmatrix} \text{first row} \\ \text{second row} \\ \text{first column} \\ \text{second column} \\ \text{third column} \end{bmatrix}$$

As long as this system is consistent, we can be sure that we have an infinite number of solutions, because we have more unknowns (6) than equations (5).

The augmented matrix and reduced row echelon form which is evaluated by using mathematical software such as MAPLE or MATLAB is given by:

$$\left(\begin{array}{cccccc|c} 1 & 1 & 1 & 0 & 0 & 0 & 15 \\ 0 & 0 & 0 & 1 & 1 & 1 & 6 \\ 1 & 0 & 0 & 1 & 0 & 0 & 8 \\ 0 & 1 & 0 & 0 & 1 & 0 & 7 \\ 0 & 0 & 1 & 0 & 0 & 1 & 6 \end{array}\right) \Rightarrow \begin{array}{c} \\ \text{row 1} \\ \text{row 2} \\ \text{row 3} \\ \text{row 4} \\ \text{row 5} \end{array} \left(\begin{array}{cccccc|c} x_1 & x_2 & x_3 & x_4 & x_5 & x_6 & \\ 1 & 0 & 0 & 0 & -1 & -1 & 2 \\ 0 & 1 & 0 & 0 & 1 & 0 & 7 \\ 0 & 0 & 1 & 0 & 0 & 1 & 6 \\ 0 & 0 & 0 & 1 & 1 & 1 & 6 \\ 0 & 0 & 0 & 0 & 0 & 0 & 0 \end{array}\right)$$

In reduced row echelon form we have four non-zero equations and six unknowns, therefore there are $6 - 4 = 2$ free variables. None of the equations start with x_5 and x_6, so these are our free variables. Let $x_5 = s$ and $x_6 = t$, where s and t are any real numbers.

Expanding row 4, we have $x_4 + x_5 + x_6 = 6$ implies $x_4 = 6 - x_5 - x_6 = 6 - s - t$

Expanding row 3, we have $x_3 + x_6 = 6$ implies $x_3 = 6 - x_6 = 6 - t$.

Expanding row 2, we have $x_2 + x_5 = 7$ implies $x_2 = 7 - x_5 = 7 - s$.

Expanding row 1, we have $x_1 - x_5 - x_6 = 2$ implies $x_1 = 2 + x_5 + x_6 = 2 + s + t$.

Our solution is

$$x_1 = 2 + s + t,\ x_2 = 7 - s,\ x_3 = 6 - t,\ x_4 = 6 - s - t,\ x_5 = s \text{ and } x_6 = t \qquad (*)$$

where s and t are any real numbers. A particular solution can be found by letting s and t take on particular values. For instance, let $s = 1, t = 2$, choosing these values so that they are different values which might be one of the requirements of the puzzle and the numbers are nice and easy. Substituting $s = 1, t = 2$ into (*) gives

$$x_1 = 5,\ x_2 = 6,\ x_3 = 4,\ x_4 = 3,\ x_5 = 1 \text{ and } x_6 = 2$$

Check that this is actually a solution to the above puzzle.

Of course, there are an infinite number of solutions to above puzzle - you could take

$$s = 1,\ t = 1 \text{ or } s = -1,\ t = 5 \text{ or } s = \pi,\ t = \sqrt{2}$$

and then feed these numbers into (*), which will give other values for the unknowns.

In the above example, the solution:

$$x_1 = 2 + s + t, \quad x_2 = 7 - s, \quad x_3 = 6 - t, \quad x_4 = 6 - s - t, \quad x_5 = s \text{ and } x_6 = t \qquad (*)$$

is the **general solution**, and can be written in vector form as

$$\mathbf{x} = \begin{pmatrix} x_1 \\ x_2 \\ x_3 \\ x_4 \\ x_5 \\ x_6 \end{pmatrix} = \begin{pmatrix} 2+s+t \\ 7-s \\ 6-t \\ 6-s-t \\ s \\ t \end{pmatrix} = s\begin{pmatrix} 1 \\ -1 \\ 0 \\ -1 \\ 1 \\ 0 \end{pmatrix} + t\begin{pmatrix} 1 \\ 0 \\ -1 \\ -1 \\ 0 \\ 1 \end{pmatrix} + \begin{pmatrix} 2 \\ 7 \\ 6 \\ 6 \\ 0 \\ 0 \end{pmatrix}$$

Another application of Gaussian elimination is analysing a mechanical device such as a robotic arm. In an engineering context **degrees of freedom** are important.

If there are d free variables in a linear system then we have d degrees of freedom.

Summary

Let $(\mathbf{A} \mid \mathbf{b})$ and $(\mathbf{R} \mid \mathbf{b}')$ be row equivalent where $\mathbf{R}$ is in reduced row echelon form.
Let n be the number of unknowns and r be the number of non-zero equations in $\mathbf{R}$.
If $r < n$ then the linear system $(\mathbf{A} \mid \mathbf{b})$ has an infinite number of solutions.

EXERCISES 1.7

(Brief solutions at end of book. Full solutions available at <http://www.oup.co.uk/companion/singh>.)

1. Determine whether the following linear systems have a unique, infinitely many or no solutions. Also determine the solutions in the case of the system consisting of infinitely many and unique solutions.

(a) $\begin{aligned} x+3y+2z &= 5 \\ 2x-y-z &= 1 \\ -x+2y+z &= 3 \end{aligned}$

(b) $\begin{aligned} -x+y+z &= 0 \\ 3x-2y+5z &= 0 \\ 4x-y-2z &= 0 \end{aligned}$

(c) $\begin{aligned} -x+y+z &= 2 \\ 2x+2y+3z &= 5 \\ 6x+6y+9z &= 7 \end{aligned}$

(d) $\begin{aligned} x+y-z &= 2 \\ x+2y+z &= 4 \\ 3x+3y-3z &= 6 \end{aligned}$

(e) $\begin{aligned} 3x-3y-z+2w &= 0 \\ 6x-7y+z+w &= 0 \\ x-y-2z-w &= 0 \\ 2x-2y+6z+8w &= 0 \end{aligned}$

(f) $\begin{aligned} 2x+3y+5z+2w &= 6 \\ 2x+3y+2z+2w &= 7 \\ 8x+12y+20z+8w &= 24 \\ x+2y+4z+5w &= 6 \end{aligned}$

(g) $\begin{aligned} -10y \quad +38z &= 6 \\ 5x+6y-8z+4w &= 3 \\ 10x+7y+3z+8w &= 9 \end{aligned}$

(h) $\begin{aligned} -y \qquad\qquad\quad +5u &= 3 \\ 3x-4y+z+6w+7u &= 5 \\ 15x-20y+2z+30w+3u &= -1 \\ 12x-16y+7z+24w+60u &= 10 \end{aligned}$

(i) $\begin{aligned} -y \qquad\qquad\quad + 5u &= 3 \\ 3x - 4y + z + 6w + 7u &= 5 \\ 15x - 20y + 2z + 30w + 3u &= -1 \\ 12x - 16y + 7z + 24w + 60u &= 46 \end{aligned}$

(j) $\begin{aligned} 2x - y + 3z + 5w + 5u &= 1 \\ x + 7y + 6z - 11w + 7u &= -8 \\ x - 2y + 6z + 4w + u &= 5 \\ 2x - 6y \qquad + 14w + 2u &= \frac{20}{3} \end{aligned}$

2. The reduced row echelon form of the corresponding equations have been evaluated using software. Determine the solutions in vector form:

(a) $\begin{aligned} x + y - 2z + 4w &= 5 \\ 2x + 2y - 3z + w &= 3 \\ 3x + 3y - 4z - 2w &= 1 \end{aligned} \quad \Rightarrow \quad \left(\begin{array}{cccc|c} 1 & 1 & 0 & -10 & -9 \\ 0 & 0 & 1 & -7 & -7 \\ 0 & 0 & 0 & 0 & 0 \end{array}\right)$

(b) $\begin{aligned} x_1 + x_2 + x_3 + x_4 + 2x_5 + 3x_6 &= 0 \\ 2x_1 + 2x_2 + 2x_3 + 2x_4 + 4x_5 + 5x_6 &= 0 \\ 3x_1 + 3x_2 + 3x_3 + 3x_4 + 6x_5 + 7x_6 &= 0 \end{aligned} \quad \Rightarrow \quad \left(\begin{array}{cccccc|c} 1 & 1 & 1 & 1 & 2 & 0 & 0 \\ 0 & 0 & 0 & 0 & 1 & 0 & 0 \\ 0 & 0 & 0 & 0 & 0 & 0 & 0 \end{array}\right)$

(c) $\begin{aligned} x_1 + 3x_2 - 2x_3 + x_4 + 2x_6 &= 0 \\ 2x_1 + 6x_2 - 5x_3 - 2x_4 - 3x_5 + 4x_6 &= -1 \\ x_3 + 2x_4 + 3x_5 &= 1 \\ x_1 + 3x_2 + 4x_4 + 6x_5 + 2x_6 &= 3 \end{aligned} \quad \Rightarrow \quad \left(\begin{array}{cccccc|c} 1 & 3 & 0 & 0 & 6 & 2 & 2 \\ 0 & 0 & 1 & 0 & 3 & 0 & 1 \\ 0 & 0 & 0 & 1 & 0 & 0 & 0 \\ 0 & 0 & 0 & 0 & 0 & 0 & 0 \end{array}\right)$

3. Determine the value of k in the following, so that the system where z is not zero is consistent, and solve it:

$$\begin{aligned} 2x - y - 4z &= k \\ -x + y + 2z &= k \\ -x + y + kz &= k \end{aligned}$$

4. Let $\mathbf{u}$ and $\mathbf{v}$ where $\mathbf{u}$ is not a multiple of $\mathbf{v}$ be solutions of a linear homogeneous system $\mathbf{Ax} = \mathbf{O}$. Prove that for any scalars such as k and c the vector $k\mathbf{u} + c\mathbf{v}$ is also a solution of $\mathbf{Ax} = \mathbf{O}$. This means that $\mathbf{Ax} = \mathbf{O}$ has an infinite number of solutions, provided $\mathbf{u}$ and $\mathbf{v}$ are solutions.

5. Consider the non-homogeneous linear system $\mathbf{Ax} = \mathbf{b}$ where $\mathbf{b} \neq \mathbf{O}$. If $\mathbf{x}_p$ is a particular solution to this $\mathbf{Ax} = \mathbf{b}$ and $\mathbf{x}_h$ is the solution to the associated homogeneous system $\mathbf{Ax} = \mathbf{O}$, then prove that $\mathbf{x}_p + \mathbf{x}_h$ is a solution to $\mathbf{Ax} = \mathbf{b}$.

6. Prove Proposition (1.32).

7. *Prove that the reduced row echelon form (rref) of an n by n matrix either is the identity matrix $\mathbf{I}$ or contains at least one row of zeros.

You may use software for the remaining questions.

8. The numbers highlighted with bold are totals for each column and row. Solve the following puzzle and write the general solution in vector form:

x_1	x_2	x_3	**6**
x_4	x_5	x_6	**15**
5	**7**	**9**	

9. Solve the following puzzle and write the general solution in vector form:

x_1	x_2	x_3	**16**
x_4	x_5	x_6	**21**
x_7	x_8	x_9	**8**
17	**15**	**13**	

10. The number in each circle is the sum of the numbers in the four squares surrounding it. Solve the following Sujiko puzzle and write the general solution in vector form:

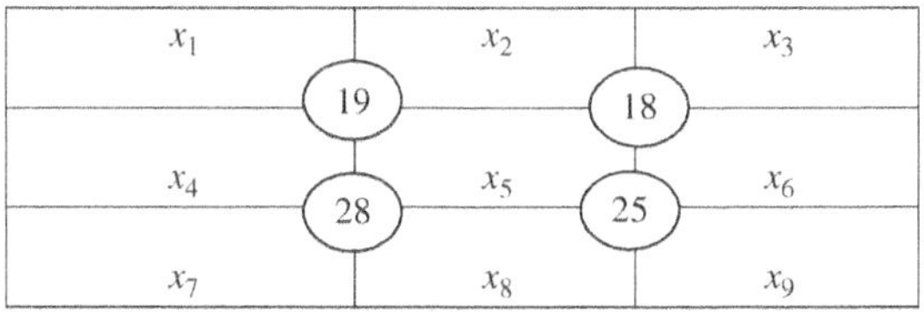

11. Global positioning system (GPS) is used to find our location. Satellites send signals to a receiver on earth with information about the location of the satellite in xyz- coordinate system and the time t when the signal was transmitted. After some calculations we obtain the following three linear equations. Solve these equations.

$$\begin{aligned} 2x - 4y + 4z + 0.077t &= 3.86 \\ -2y + 2z - 0.056t &= -3.47 \\ 2x - 2y \qquad\qquad &= 0 \end{aligned}$$

SECTION 1.8 The Inverse Matrix Method

By the end of this section you will be able to

- understand what is meant by an elementary matrix
- determine the inverse of a given matrix
- apply the inverse matrix method to encryption of messages

Another way to solve the linear system $\mathbf{Ax} = \mathbf{b}$ where $\mathbf{A}$ is invertible is to use the inverse matrix, because $\mathbf{x} = \mathbf{A}^{-1}\mathbf{b}$. However, finding the inverse of a matrix is a long process and we can often solve linear systems quickly using Gaussian elimination.

So why do we want to find the inverse matrix?

The inverse matrix is not useful for solving a single linear system $\mathbf{Ax} = \mathbf{b}$, but it is useful if one needs to solve many systems that have the same coefficient matrix $\mathbf{A}$. For example, suppose you have 100 linear systems $\mathbf{Ax}_k = \mathbf{b}_k$, where the coefficients are fixed but the right hand side vector $\mathbf{b}_k$ varies and you want to solve these. In this case, you would invert the coefficient matrix $\mathbf{A}$ only once and then find $\mathbf{x}_k = \mathbf{A}^{-1}\mathbf{b}_k$ for each $\mathbf{b}_k$.

There are many real life applications where you might need to solve the linear system, $\mathbf{Ax} = \mathbf{b}$, in which matrix $\mathbf{A}$ remains the same while vector $\mathbf{b}$ changes. In that case, it is more efficient to solve for the inverse of $\mathbf{A}$ once, then multiply that inverse by the various $\mathbf{b}$ vectors. This is because matrix $\mathbf{A}$ is generally a common property but vector $\mathbf{b}$ a particular property of the system.

In order to evaluate the inverse matrix we need to first describe elementary matrices and show how they are related to row operations.

1.8.1 Elementary matrices

An **elementary matrix** is a matrix obtained by a **single** row operation on the identity matrix $\mathbf{I}$.

For example, $\begin{pmatrix} 1 & 0 & 0 \\ 0 & \sqrt{2} & 0 \\ 0 & 0 & 1 \end{pmatrix}$ is an elementary matrix because we have multiplied the middle row of the identity matrix $\begin{pmatrix} 1 & 0 & 0 \\ 0 & 1 & 0 \\ 0 & 0 & 1 \end{pmatrix}$ by $\sqrt{2}$. We carried out a single row operation.

Example 1.40

Which of the following are elementary matrices?

$$\mathbf{A} = \begin{pmatrix} 1 & 0 & 0 \\ 0 & 1 & 0 \\ 0 & 0 & 23 \end{pmatrix}, \quad \mathbf{B} = \begin{pmatrix} 1 & 0 & -5 & 0 \\ 0 & 1 & 0 & 0 \\ 0 & 0 & 1 & 0 \\ 0 & 0 & 0 & 1 \end{pmatrix}, \quad \mathbf{C} = \begin{pmatrix} 1 & 0 & 3 \\ 0 & 4 & 0 \\ 0 & 0 & 1 \end{pmatrix} \text{ and } \mathbf{D} = \begin{pmatrix} 0 & 1 \\ 1 & 0 \end{pmatrix}$$

Solution

$\mathbf{A}$, $\mathbf{B}$ and $\mathbf{D}$ are elementary matrices.

Why?

Matrix $\mathbf{A}$ is an elementary matrix because we can obtain this matrix by multiplying the bottom row of the identity matrix by 23. Matrix $\mathbf{B}$ is an elementary matrix because we can obtain this matrix by adding -5 times the third row to the first row of the identity matrix $\mathbf{I}_4$. Matrix $\mathbf{D}$ is clearly an elementary matrix because we have just interchanged the rows of the identity matrix $\begin{pmatrix} 1 & 0 \\ 0 & 1 \end{pmatrix}$. Remember, swapping rows is a single row operation.

Matrix **C** is not an elementary matrix because we cannot obtain the matrix **C** by a single row operation on the identity matrix. Two row operations on the identity matrix are required to obtain matrix **C**. The two row operations are:

1. Add 3 times the bottom row of the identity matrix to the top row.
2. Multiply the middle row of the identity matrix by 4.

If an elementary matrix, **E**, is obtained from the identity matrix by a certain row operation then the matrix multiplication **EA** produces the same operation on the matrix **A**. For example, consider the general 3 by 3 matrix and the elementary matrix obtained from the identity matrix by interchanging the bottom two rows:

$$\mathbf{A} = \begin{pmatrix} a & b & c \\ d & e & f \\ g & h & i \end{pmatrix} \text{ and } \mathbf{E} = \begin{pmatrix} 1 & 0 & 0 \\ 0 & 0 & 1 \\ 0 & 1 & 0 \end{pmatrix} \text{ interchanged}$$

Then the matrix multiplication $\mathbf{EA} = \begin{pmatrix} 1 & 0 & 0 \\ 0 & 0 & 1 \\ 0 & 1 & 0 \end{pmatrix} \begin{pmatrix} a & b & c \\ d & e & f \\ g & h & i \end{pmatrix} = \begin{pmatrix} a & b & c \\ g & h & i \\ d & e & f \end{pmatrix}$ interchanged

What does the given elementary matrix **E** *do to the matrix* **A**?

It performs the same row operation of interchanging the bottom two rows.

What effect does the matrix multiplication **EA**, *where* $\mathbf{E} = \begin{pmatrix} 1 & 0 & 0 \\ 0 & -2 & 0 \\ 0 & 0 & 1 \end{pmatrix}$, *have on the general 3 by 3 matrix* **A**?

The elementary matrix **E** in this case is obtained from the identity matrix by multiplying the middle row by -2. Carrying out the matrix multiplication **EA** gives

$$\mathbf{EA} = \begin{pmatrix} 1 & 0 & 0 \\ 0 & -2 & 0 \\ 0 & 0 & 1 \end{pmatrix} \begin{pmatrix} a & b & c \\ d & e & f \\ g & h & i \end{pmatrix} = \begin{pmatrix} a & b & c \\ -2d & -2e & -2f \\ g & h & i \end{pmatrix}$$

Hence the matrix multiplication **EA** performs the same elementary row operation of multiplying the middle row by -2.

In general, if **E** is a n by n elementary matrix obtained by performing a particular row operation on the identity matrix $\mathbf{I}_n$ and **A** is an n by r matrix then the matrix multiplication **EA** performs the same row operation on **A**. We can use matrix **E** to execute a row operation.

Can you remember what is meant by the term row equivalent from the previous section?

Matrices that are obtained from one another by a finite number of elementary row operations are said to be row equivalent.

Definition (1.33). A matrix **B** is row equivalent to a matrix **A** if and only if there are a finite number of elementary matrices $\mathbf{E}_1, \mathbf{E}_2, \mathbf{E}_3, \ldots$ and $\mathbf{E}_n$ such that $\mathbf{B} = \mathbf{E}_n\mathbf{E}_{n-1} \cdots \mathbf{E}_2\mathbf{E}_1\mathbf{A}$.

This definition means that the matrix **B** can be obtained from matrix **A** by a finite number of row operations. Note, that **A** is pre-multiplied (left multiplied) by elementary matrices.

Note, that if matrix **B** is row equivalent to matrix **A** then the reverse is also true, matrix **A** is row equivalent to matrix **B**. You are asked to show this result in Exercises 1.8.

Proposition (1.34). An elementary matrix is invertible (non-singular) and its inverse is also an elementary matrix.

Proof – See Exercises 1.8.

■

1.8.2 Equivalent statements

You will need to look up what is meant by reduced row echelon form (rref), which was defined in subsection 1.2.4 to understand this subsection.

In the remaining subsections, we find the inverse matrix by placing the given matrix into reduced row echelon form, but first we establish an important theorem showing equivalence.

*What does the term **equivalence** mean?*

Two statements P and Q are equivalent if both P and Q are true or both are false. Another way of saying this is 'P if and only if Q' denoted by $P \Leftrightarrow Q$. This means that P implies Q and Q implies P, - it goes both ways, $P \Rightarrow Q$ and $Q \Rightarrow P$.

We prove propositions of the form $P \Rightarrow Q$ by assuming statement P to be true and then deduce Q.

In general, we say four statements P, Q, R and S are equivalent if $P \Rightarrow Q \Rightarrow R \Rightarrow S \Rightarrow P$.

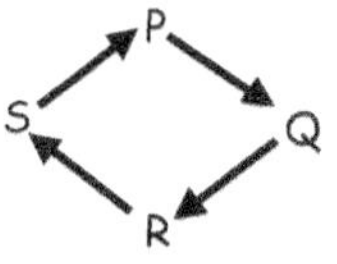

Next we prove an important theorem regarding invertible matrices.

Why do we bother with proofs?

One of the fundamental aims of a linear algebra course is to learn reasoning. Many people see mathematics as just a tool used by engineers and scientists, but this misses the depth of reasoning and the inherent beauty of it. Proofs explain why mathematics works and not just how.

Theorem (1.35). Let **A** be a n by n matrix, then the following 4 statements are equivalent:

(a) The matrix **A** is invertible (non-singular).

(b) The linear system $\mathbf{Ax} = \mathbf{O}$ only has the trivial solution $\mathbf{x} = \mathbf{O}$.

(c) The reduced row echelon form of the matrix **A** is the identity matrix **I**.

(d) **A** is a product of elementary matrices.

Proof – How can we prove this theorem?

We show (a) implies (b) and (b) implies (c) and (c) implies (d) and (d) implies (a). In symbolic notation, this is (a) $\Rightarrow$ (b) $\Rightarrow$ (c) $\Rightarrow$ (d) $\Rightarrow$ (a). We first prove (a) implies (b).

For (a) $\Rightarrow$ (b):

We assume statement (a) to be true and deduce statement (b).

By assuming **A** is an invertible matrix we need to show that the linear system $\mathbf{Ax} = \mathbf{O}$ only has the trivial solution $\mathbf{x} = \mathbf{O}$. Consider the linear system $\mathbf{Ax} = \mathbf{O}$, because **A** is invertible, therefore there is a unique matrix $\mathbf{A}^{-1}$ such that $\mathbf{A}^{-1}\mathbf{A} = \mathbf{I}$. Multiplying both sides of $\mathbf{Ax} = \mathbf{O}$ by $\mathbf{A}^{-1}$ yields

$$\underbrace{\mathbf{A}^{-1}\mathbf{A}}_{=\mathbf{I}}\mathbf{x} = \mathbf{A}^{-1}\mathbf{O}$$

$$\mathbf{Ix} = \mathbf{A}^{-1}\mathbf{O} = \mathbf{O}$$

$$\mathbf{Ix} = \mathbf{O} \text{ which gives } \mathbf{x} = \mathbf{O} \qquad [\text{because } \mathbf{Ix} = \mathbf{x}]$$

The answer $\mathbf{x} = \mathbf{O}$ means that we only have the trivial solution $\mathbf{x} = \mathbf{O}$. Hence we have shown (a) $\Rightarrow$ (b). Next we prove (b) implies (c).

For (b) $\Rightarrow$ (c):

The procedure to prove (b) $\Rightarrow$ (c) is to assume statement (b) and deduce (c). This time we assume that $\mathbf{Ax} = \mathbf{O}$ only has the trivial solution $\mathbf{x} = \mathbf{O}$ and, by using this, prove that the reduced row echelon form of the matrix **A** is the identity matrix **I**.

The reduced row echelon form of matrix **A** cannot have a row (equation) of zeros, otherwise we would have n unknowns but less than n non-zero rows (equations), which means that by Proposition (1.31): if $r < n$ then the linear system $\mathbf{Ax} = \mathbf{O}$ has an infinite number of solutions.

We would have an infinite number of solutions. However, we only have a unique solution to $\mathbf{Ax} = \mathbf{O}$, therefore there are no zero rows.

Question (7) of Exercises 1.7 claims: the reduced row echelon form of a matrix is either the identity **I** or it contains a row of zeros.

Hence the reduced row echelon form of **A** is the identity matrix. We have (b) $\Rightarrow$ (c).

For (c) $\Rightarrow$ (d):

In this case, we assume part (c), that is 'the reduced row echelon form of the matrix **A** is the identity matrix **I**', which means that the matrix **A** is row equivalent to the identity matrix **I**. By definition (1.33):

B is row equivalent to a matrix **A** if and only if $\mathbf{B} = \mathbf{E}_n\mathbf{E}_{n-1}\cdots\mathbf{E}_2\mathbf{E}_1\mathbf{A}$.

There are elementary matrices $\mathbf{E}_1, \mathbf{E}_2, \mathbf{E}_3, \ldots$ and $\mathbf{E}_k$ such that

$$\begin{aligned}\mathbf{A} &= \mathbf{E}_k\mathbf{E}_{k-1}\cdots\mathbf{E}_2\mathbf{E}_1\mathbf{I} && [\text{because } \mathbf{A} \text{ is row equivalent to } \mathbf{I}]\\ &= \mathbf{E}_k\mathbf{E}_{k-1}\cdots\mathbf{E}_2\mathbf{E}_1 && [\text{because } \mathbf{I} \text{ is the identity matrix}]\end{aligned}$$

This shows that matrix **A** is a product of elementary matrices. We have (c) $\Rightarrow$ (d).

For (d) $\Rightarrow$ (a):

In this last case, we assume that matrix **A** is a product of elementary matrices and deduce that matrix **A** is invertible. By Proposition (1.34): an elementary matrix is invertible.

We know that elementary matrices are invertible (have an inverse) and therefore the matrix multiplication $\mathbf{E}_k\mathbf{E}_{k-1}\cdots\mathbf{E}_2\mathbf{E}_1$ is invertible. In fact, we have

$$\begin{aligned}\mathbf{A}^{-1} &= \left(\mathbf{E}_k\mathbf{E}_{k-1}\cdots\mathbf{E}_2\mathbf{E}_1\right)^{-1} \\ &= \mathbf{E}_1^{-1}\mathbf{E}_2^{-1}\cdots\mathbf{E}_{k-1}^{-1}\mathbf{E}_k^{-1} \qquad \left[\text{because } (\mathbf{XYZ})^{-1} = \mathbf{Z}^{-1}\mathbf{Y}^{-1}\mathbf{X}^{-1}\right]\end{aligned}$$

Hence the matrix **A** is invertible, which means that we have proven (d) $\Rightarrow$ (a).

We have shown (a) $\Rightarrow$ (b) $\Rightarrow$ (c) $\Rightarrow$ (d) $\Rightarrow$ (a) which means that the four statements (a), (b), (c) and (d) are equivalent.

■

1.8.3 Determining the inverse matrix

Any two statements out of four in the above Theorem (1.35) are equivalent. That is, (a) $\Rightarrow$ (b) and (b) $\Rightarrow$ (a), which is normally written as (a) $\Leftrightarrow$ (b).

By the above Theorem (1.35), we also have (a) $\Leftrightarrow$ (c) which means that the (n by n) matrix **A** is invertible $\Leftrightarrow$ **A is row equivalent to the identity matrix I**. There are elementary matrices $\mathbf{E}_1, \mathbf{E}_2, \mathbf{E}_3, \ldots$ and $\mathbf{E}_k$ such that

$$\mathbf{I} = \mathbf{E}_k\mathbf{E}_{k-1}\cdots\mathbf{E}_2\mathbf{E}_1\mathbf{A} \qquad (^*)$$

Right multiplying both sides of (*) by $\mathbf{A}^{-1}$ gives

$$\begin{aligned}\underbrace{\mathbf{IA}^{-1}}_{=\mathbf{A}^{-1}} &= \mathbf{E}_k\mathbf{E}_{k-1}\cdots\mathbf{E}_2\mathbf{E}_1\,\underbrace{\mathbf{AA}^{-1}}_{=\mathbf{I}} \\ \mathbf{A}^{-1} &= \mathbf{E}_k\mathbf{E}_{k-1}\cdots\mathbf{E}_2\mathbf{E}_1\mathbf{I}\end{aligned}$$

Hence from $\mathbf{I} = \mathbf{E}_k\mathbf{E}_{k-1}\cdots\mathbf{E}_2\mathbf{E}_1\mathbf{A}$ we deduced that $\mathbf{A}^{-1} = \mathbf{E}_k\mathbf{E}_{k-1}\cdots\mathbf{E}_2\mathbf{E}_1\mathbf{I}$.

What does this mean?

It means that the same elementary matrices $\mathbf{E}_1, \mathbf{E}_2, \mathbf{E}_3, \ldots$ and $\mathbf{E}_k$ transform the identity matrix **I** into the invertible matrix $\mathbf{A}^{-1}$. This actually implies that we have to perform the same row operations that transform the matrix **A** to the identity matrix **I** and also transform **I** to the inverse matrix $\mathbf{A}^{-1}$. Summarizing this:

$$(\mathbf{A}\,|\,\mathbf{I}) \times \mathbf{A}^{-1} = \left(\mathbf{I}\,\middle|\,\mathbf{A}^{-1}\right)$$

This means that we convert $(\mathbf{A}\,|\,\mathbf{I})$ to $\left(\mathbf{I}\,\middle|\,\mathbf{A}^{-1}\right)$. Hence the row operations that transform matrix **A** into **I**, also transform **I** into $\mathbf{A}^{-1}$.

Example 1.41

Determine the inverse matrix $\mathbf{A}^{-1}$ given that $\mathbf{A} = \begin{pmatrix} 1 & 0 & 2 \\ 2 & -1 & 3 \\ 4 & 1 & 8 \end{pmatrix}$.

Solution

Above we established that the row operations for transforming $\mathbf{A}$ into the identity matrix $\mathbf{I}$ are the same as for transforming the identity matrix $\mathbf{I}$ into the inverse matrix $\mathbf{A}^{-1}$, therefore we carry out the row operations simultaneously. This is achieved by transforming the augmented matrix $(\mathbf{A} \mid \mathbf{I})$ into the augmented matrix $(\mathbf{I} \mid \mathbf{A}^{-1})$. Labelling the rows:

$$\begin{matrix} R_1 \\ R_2 \\ R_3 \end{matrix} \left(\begin{array}{ccc|ccc} 1 & 0 & 2 & 1 & 0 & 0 \\ 2 & -1 & 3 & 0 & 1 & 0 \\ 4 & 1 & 8 & 0 & 0 & 1 \end{array}\right) = (\mathbf{A} \mid \mathbf{I})$$

Remember that R_1, R_2 and R_3 represent the first, second and third rows respectively.

Our aim is to convert the given matrix $\mathbf{A}$ (left) into the identity matrix $\mathbf{I}$.

How?

Need to convert the 2 and 4 in the bottom two rows into zeros by executing the following row operations:

$$\begin{matrix} R_1 \\ R_2' = R_2 - 2R_1 \\ R_3' = R_3 - 4R_1 \end{matrix} \left(\begin{array}{ccc|ccc} 1 & 0 & 2 & 1 & 0 & 0 \\ 2-2(1) & -1-2(0) & 3-2(2) & 0-2(1) & 1-2(0) & 0-2(0) \\ 4-4(1) & 1-4(0) & 8-4(2) & 0-4(1) & 0-4(0) & 1-4(0) \end{array}\right)$$

Simplifying the entries gives

$$\begin{matrix} R_1 \\ R_2' \\ R_3' \end{matrix} \left(\begin{array}{ccc|ccc} 1 & 0 & 2 & 1 & 0 & 0 \\ 0 & -1 & -1 & -2 & 1 & 0 \\ 0 & 1 & 0 & -4 & 0 & 1 \end{array}\right)$$

We interchange the bottom two rows, R_2' and R_3':

$$\begin{matrix} R_1 \\ R_2^* = R_3' \\ R_3^* = R_2' \end{matrix} \left(\begin{array}{ccc|ccc} 1 & 0 & 2 & 1 & 0 & 0 \\ 0 & 1 & 0 & -4 & 0 & 1 \\ 0 & -1 & -1 & -2 & 1 & 0 \end{array}\right)$$

Note that we have swapped the rows around so that the second row is in the correct format for the identity matrix on the *left hand side* of the vertical line.

What else do we need for the identity matrix?

Convert the first -1 in the bottom row into zero.

How?

Add the bottom two rows and simplify the entries:

$$\begin{matrix} R_1 \\ R_2^* \\ R_3^{**} = R_3^* + R_2^* \end{matrix} \left(\begin{array}{ccc|ccc} 1 & 0 & 2 & 1 & 0 & 0 \\ 0 & 1 & 0 & -4 & 0 & 1 \\ 0 & 0 & -1 & -6 & 1 & 1 \end{array}\right)$$

The -1 in the bottom row needs to be $+1$. Multiplying the bottom row by -1 yields

$$\begin{matrix} R_1 \\ R_2^* \\ R_3^{\dagger} = -R_3^{**} \end{matrix} \left(\begin{array}{ccc|ccc} 1 & 0 & 2 & 1 & 0 & 0 \\ 0 & 1 & 0 & -4 & 0 & 1 \\ 0 & 0 & 1 & 6 & -1 & -1 \end{array}\right)$$

We have nearly arrived at our destination, which is the identity matrix on the left.

(continued...)

We only need to convert the 2 in the top row into 0. Carry out the row operation $R_1 - 2R_3^\dagger$:

$$\begin{matrix} R_1^* = R_1 - 2R_3^\dagger \\ R_2^* \\ R_3^\dagger \end{matrix} \left(\begin{array}{ccc|ccc} 1-2(0) & 0-2(0) & 2-2(1) & 1-2(6) & 0-2(-1) & 0-2(-1) \\ 0 & 1 & 0 & -4 & 0 & 1 \\ 0 & 0 & 1 & 6 & -1 & -1 \end{array}\right)$$

Simplifying the arithmetic in the top row gives the identity on the left:

$$\begin{matrix} R_1^* \\ R_2^* \\ R_3^\dagger \end{matrix} \left(\begin{array}{ccc|ccc} 1 & 0 & 0 & -11 & 2 & 2 \\ 0 & 1 & 0 & -4 & 0 & 1 \\ 0 & 0 & 1 & 6 & -1 & -1 \end{array}\right) = \left(\mathbf{I} \middle| \mathbf{A}^{-1}\right)$$

Hence the inverse matrix, $\mathbf{A}^{-1}$, is the matrix with the entries on the *right hand side* of the vertical line,

$$\mathbf{A}^{-1} = \begin{pmatrix} -11 & 2 & 2 \\ -4 & 0 & 1 \\ 6 & -1 & -1 \end{pmatrix}.$$

This might seem like a tedious way of finding the inverse of a 3×3 matrix, but there is no easy method to find $\mathbf{A}^{-1}$ (if it exists) for any given 3×3 or larger matrix $\mathbf{A}$.

1.8.4 Solving linear equations

Consider a general linear system which is written in matrix form as

$$\mathbf{Ax} = \mathbf{b} \qquad [\mathbf{A} \text{ is invertible}]$$

Multiplying this by the inverse matrix $\mathbf{A}^{-1}$ gives

$$\underbrace{\mathbf{A}^{-1}\mathbf{A}}_{=\mathbf{I}}\mathbf{x} = \mathbf{A}^{-1}\mathbf{b}$$

$$\mathbf{Ix} = \mathbf{A}^{-1}\mathbf{b} \text{ implies } \mathbf{x} = \mathbf{A}^{-1}\mathbf{b} \quad [\text{because } \mathbf{Ix} = \mathbf{x}]$$

Hence the solution of a linear system $\mathbf{Ax} = \mathbf{b}$ where $\mathbf{A}$ is invertible is given by

$$\mathbf{x} = \mathbf{A}^{-1}\mathbf{b} \tag{1.36}$$

This is why we need to determine the inverse matrix $\mathbf{A}^{-1}$. The next example demonstrates how this works for particular linear systems of equations.

Example 1.42

Solve both the linear systems:

(a) $\begin{aligned} x \quad\;\; + 2z &= 5 \\ 2x - y + 3z &= 7 \\ 4x + y + 8z &= 10 \end{aligned}$

(b) $\begin{aligned} x \quad\;\; + 2z &= 1 \\ 2x - y + 3z &= 2 \\ 4x + y + 8z &= 3 \end{aligned}$

Solution

(a) Writing this in matrix form $\mathbf{Ax} = \mathbf{b}$ where

$$\mathbf{A} = \begin{pmatrix} 1 & 0 & 2 \\ 2 & -1 & 3 \\ 4 & 1 & 8 \end{pmatrix}, \quad \mathbf{x} = \begin{pmatrix} x \\ y \\ z \end{pmatrix} \text{ and } \mathbf{b} = \begin{pmatrix} 5 \\ 7 \\ 10 \end{pmatrix}$$

By the above result (1.36) we have $\mathbf{x} = \mathbf{A}^{-1}\mathbf{b}$, provided $\mathbf{A}$ is invertible.
What is the inverse matrix $\mathbf{A}^{-1}$ equal to?
The matrix $\mathbf{A}$ is the same matrix as the previous example, therefore the inverse matrix is already evaluated:

$$\mathbf{A}^{-1} = \begin{pmatrix} -11 & 2 & 2 \\ -4 & 0 & 1 \\ 6 & -1 & -1 \end{pmatrix}$$

Hence

$$\mathbf{x} = \begin{pmatrix} x \\ y \\ z \end{pmatrix} = \begin{pmatrix} -11 & 2 & 2 \\ -4 & 0 & 1 \\ 6 & -1 & -1 \end{pmatrix}\begin{pmatrix} 5 \\ 7 \\ 10 \end{pmatrix} \quad \left[\mathbf{x} = \mathbf{A}^{-1}\mathbf{b}\right]$$

$$= \begin{pmatrix} -(11 \times 5) + (2 \times 7) + (2 \times 10) \\ -(4 \times 5) + (0 \times 7) + (1 \times 10) \\ (6 \times 5) - (1 \times 7) - (1 \times 10) \end{pmatrix} = \begin{pmatrix} -21 \\ -10 \\ 13 \end{pmatrix}$$

We have $x = -21$, $y = -10$ and $z = 13$. You can check this is the solution to the given linear system by substituting these values into the system.

(b) The coefficient matrix is the same and we have only changed the vector $\mathbf{b}$. Hence

$$\mathbf{x} = \begin{pmatrix} x \\ y \\ z \end{pmatrix} = \begin{pmatrix} -11 & 2 & 2 \\ -4 & 0 & 1 \\ 6 & -1 & -1 \end{pmatrix}\begin{pmatrix} 1 \\ 2 \\ 3 \end{pmatrix} = \begin{pmatrix} -1 \\ -1 \\ 1 \end{pmatrix} \left[\text{because } \mathbf{x} = \mathbf{A}^{-1}\mathbf{b}\right]$$

We have $x = -1$, $y = -1$ and $z = 1$.

In the above example, we have solved two linear systems in one go because we had $\mathbf{A}^{-1}$. Once we have the inverse matrix $\mathbf{A}^{-1}$, we can solve a whole sequence of systems such as

$$\mathbf{Ax}_1 = \mathbf{b}_1, \quad \mathbf{Ax}_2 = \mathbf{b}_2, \quad \mathbf{Ax}_3 = \mathbf{b}_3, \ldots, \quad \mathbf{Ax}_k = \mathbf{b}_k \text{ by using } \mathbf{x}_j = \mathbf{A}^{-1}\mathbf{b}_j.$$

Proposition (1.37). The linear system $\mathbf{Ax} = \mathbf{b}$ has a unique solution $\Leftrightarrow$ $\mathbf{A}$ is invertible.

Proof – Exercises 1.8.

■

We can add this statement (1.37) to the main Theorem (1.35) to get:

Theorem (1.38). Let **A** be an n by n matrix, then the following five statements are equivalent:

(a) The matrix **A** is invertible (non-singular).

(b) The linear system $\mathbf{Ax} = \mathbf{O}$ only has the trivial solution $\mathbf{x} = \mathbf{O}$.

(c) The reduced row echelon form of the matrix **A** is the identity matrix **I**.

(d) **A** is a product of elementary matrices.

(e) $\mathbf{Ax} = \mathbf{b}$ has a unique solution.

1.8.5 Non-invertible (singular) matrices

If the above approach of trying to convert $(\mathbf{A} \mid \mathbf{I})$ into the augmented matrix $\left(\mathbf{I} \mid \mathbf{A}^{-1}\right)$ cannot be achieved then the matrix **A** is non-invertible (singular). This means that the matrix **A** does not have an inverse. By Theorem (1.38), we know that the matrix **A** is invertible (has an inverse) $\Leftrightarrow$ the reduced row echelon form of **A** is the identity matrix **I**. Remember, the reduced row echelon form of an n by n matrix can only: (1) be an identity matrix or (2) have a row of zeros.

If we end up with a row of zeros then the given matrix is non-invertible.

Proposition (1.39). Let **A** be a square matrix and **R** be the reduced row echelon form of **A**. Then **R** has at least one row of zeros $\Leftrightarrow$ **A** is non-invertible (singular).

Proof.
See Exercises 1.8.

■

The proof of this result can be made a lot easier if we understand some mathematical logic. Generally to prove a statement of the type $P \Leftrightarrow Q$ we assume P to be true and then deduce Q. Then we assume Q to be true and deduce P.

However, in mathematical logic this can also be proven by showing:

$$(\text{Not } P) \Leftrightarrow (\text{Not } Q)$$

Means that $(\text{Not } P) \Rightarrow (\text{Not } Q)$ and $(\text{Not } Q) \Rightarrow (\text{Not } P)$. This is because statements $P \Leftrightarrow Q$ and $(\text{Not } P) \Leftrightarrow (\text{Not } Q)$ are equivalent. See website for more details.
The following demonstrates this proposition.

Example 1.43

Show that the matrix **A** is non-invertible where

$$\mathbf{A} = \begin{pmatrix} 1 & -2 & 3 & 5 \\ 2 & 5 & 6 & 9 \\ -3 & 1 & 2 & 3 \\ 1 & 13 & -30 & -49 \end{pmatrix}$$

Solution

Writing this in the augmented matrix form $(\mathbf{A} \mid \mathbf{I})$ we have

$$\begin{matrix} R_1 \\ R_2 \\ R_3 \\ R_4 \end{matrix} \left(\begin{array}{cccc|cccc} 1 & -2 & 3 & 5 & 1 & 0 & 0 & 0 \\ 2 & 5 & 6 & 9 & 0 & 1 & 0 & 0 \\ -3 & 1 & 2 & 3 & 0 & 0 & 1 & 0 \\ 1 & 13 & -30 & -49 & 0 & 0 & 0 & 1 \end{array}\right)$$

We need to convert the 2, −3 and 1 (bottom row) in the first column into 0's.

How?

By executing the following row operations:

$$\begin{matrix} R_1 \\ R_2^* = R_2 - 2R_1 \\ R_3^* = R_3 + 3R_1 \\ R_4^* = R_4 - R_1 \end{matrix} \left(\begin{array}{cccc|cccc} 1 & -2 & 3 & 5 & 1 & 0 & 0 & 0 \\ 2-2(1) & 5-2(-2) & 6-2(3) & 9-2(5) & -2 & 1 & 0 & 0 \\ -3+3(1) & 1+3(-2) & 2+3(3) & 3+3(5) & 3 & 0 & 1 & 0 \\ 1-1 & 13-(-2) & -30-3 & -49-5 & -1 & 0 & 0 & 1 \end{array}\right)$$

Simplifying the arithmetic gives

$$\begin{matrix} R_1 \\ R_2^* \\ R_3^* \\ R_4^* \end{matrix} \left(\begin{array}{cccc|cccc} 1 & -2 & 3 & 5 & 1 & 0 & 0 & 0 \\ 0 & 9 & 0 & -1 & -2 & 1 & 0 & 0 \\ 0 & -5 & 11 & 18 & 3 & 0 & 1 & 0 \\ 0 & 15 & -33 & -54 & -1 & 0 & 0 & 1 \end{array}\right)$$

Note, that the bottom row is −3 times the third row on the left. Executing $R_4^* + 3R_3^*$:

$$\begin{matrix} R_1 \\ R_2^* \\ R_3^* \\ R_4^{**} = R_4^* + 3R_3^* \end{matrix} \left(\begin{array}{cccc|cccc} 1 & -2 & 3 & 5 & 1 & 0 & 0 & 0 \\ 0 & 9 & 0 & -1 & -2 & 1 & 0 & 0 \\ 0 & -5 & 11 & 18 & 3 & 0 & 1 & 0 \\ 0 & 0 & 0 & 0 & 8 & 0 & 3 & 1 \end{array}\right)$$

Since we have a zero row on the *left hand side* of the vertical line, therefore the reduced row echelon form will have a zero row. We conclude by Proposition (1.37):

$$rref(\mathbf{A}) = \text{R has at least one row of zeros} \Leftrightarrow \mathbf{A} \text{ is non-invertible.}$$

The matrix $\mathbf{A}$ is non-invertible (singular). That is the matrix $\mathbf{A}$ does not have an inverse.

If the matrix $\mathbf{A}$ does not have an inverse then the linear system $\mathbf{Ax} = \mathbf{b}$ cannot have a unique solution. We must have an infinite number or no solution.

We normally use the compact notation $rref(\mathbf{A}) = \mathbf{R}$ to mean that the reduced row echelon form of $\mathbf{A}$ is $\mathbf{R}$. This notation $rref(A)$ is the command in MATLAB to find the rref of $\mathbf{A}$.

1.8.6 Applications to cryptography

Cryptography is the study of communication by stealth. It involves the coding and decoding of messages. This is a growing area of linear algebra applications because agencies such as the CIA use cryptography to encode and decode information.

One way to code a message is to use matrices.

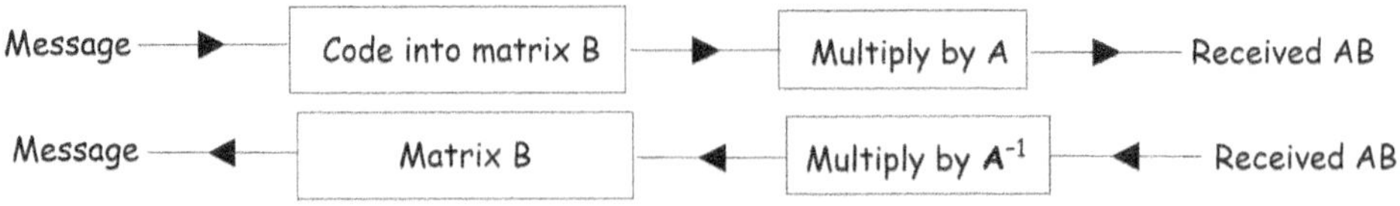

For example, let **A** be an invertible matrix. The message is encrypted into a matrix **B** such that the matrix multiplication **AB** is a valid operation. Send the message generated by the matrix multiplication **AB**. At the other end, they will need to know the inverse matrix $\mathbf{A}^{-1}$ in order to decode the message because $\mathbf{A}^{-1}(\mathbf{AB}) = \mathbf{B}$. Remember that the matrix **B** contains the message.

A simple way of encoding messages is to represent each letter of the alphabet by its position in the alphabet and then add 3 to this. For example, we can create Table 1.3.

Table 1.3

Alphabet	A	B	C	D	...	W	X	Y	Z	
Position	1	2	3	4	...	23	24	25	26	27
Position +3	4	5	6	7	...	26	27	28	29	30

The final column represents space and we nominate this by a value of $27 + 3 = 30$.

To eliminate tedium from calculations, we use appropriate software to carry out the following example.

Example 1.44

Encode the message 'OPERATION BLUESTAR' by using matrix **A** where

$$\mathbf{A} = \begin{pmatrix} 1 & 2 & 3 \\ 2 & 1 & 2 \\ 3 & 2 & 4 \end{pmatrix}$$

Solution

Using the above table we have the numerical values as shown in Table 1.4.

Table 1.4

O	P	E	R	A	T	I	O	N		B	L	U	E	S	T	A	R
18	19	8	21	4	23	12	18	17	30	5	15	24	8	22	23	4	21

Since the matrix **A** is a 3 by 3 matrix, we write these numbers in 3 by 1 vectors so that matrix multiplication is a valid operation. Hence

$$\begin{pmatrix} 18 \\ 19 \\ 8 \end{pmatrix}, \begin{pmatrix} 21 \\ 4 \\ 23 \end{pmatrix}, \begin{pmatrix} 12 \\ 18 \\ 17 \end{pmatrix}, \begin{pmatrix} 30 \\ 5 \\ 15 \end{pmatrix}, \begin{pmatrix} 24 \\ 8 \\ 22 \end{pmatrix} \text{ and } \begin{pmatrix} 23 \\ 4 \\ 21 \end{pmatrix}$$

Putting these together as column vectors of matrix $\mathbf{B}$ so that we can multiply by $\mathbf{A}$ in one go:

$$\mathbf{B} = \begin{pmatrix} 18 & 21 & 12 & 30 & 24 & 23 \\ 19 & 4 & 18 & 5 & 8 & 4 \\ 8 & 23 & 17 & 15 & 22 & 21 \end{pmatrix}$$

Multiplying this matrix by $\mathbf{A}$ (use software to do this):

$$\mathbf{AB} = \begin{pmatrix} 1 & 2 & 3 \\ 2 & 1 & 2 \\ 3 & 2 & 4 \end{pmatrix} \begin{pmatrix} 18 & 21 & 12 & 30 & 24 & 23 \\ 19 & 4 & 18 & 5 & 8 & 4 \\ 8 & 23 & 17 & 15 & 22 & 21 \end{pmatrix} = \begin{pmatrix} 80 & 98 & 99 & 85 & 106 & 94 \\ 71 & 92 & 76 & 95 & 100 & 92 \\ 124 & 163 & 140 & 160 & 176 & 161 \end{pmatrix}$$

These numbers in the columns of matrix $\mathbf{AB}$ are transmitted, that is the encoded message is:

$$80, 71, 124, 98, 92, 163, 99, 76, 140, 85, 95, 160, 106, 100, 176, 94, 92, 161$$

By examining these numbers you cannot say they are related to the alphabet.
How do we decode this message?
We need to multiply by the inverse matrix. By using software the inverse matrix $\mathbf{A}^{-1}$ is

$$\mathbf{A}^{-1} = \begin{pmatrix} 0 & 2 & -1 \\ 2 & 5 & -4 \\ -1 & -4 & 3 \end{pmatrix}$$

To decode the message we need to find $\mathbf{A}^{-1}(\mathbf{AB})$, which is evaluated by using software:

$$\mathbf{A}^{-1}(\mathbf{AB}) = \begin{pmatrix} 0 & 2 & -1 \\ 2 & 5 & -4 \\ -1 & -4 & 3 \end{pmatrix} \begin{pmatrix} 80 & 98 & 99 & 85 & 106 & 94 \\ 71 & 92 & 76 & 95 & 100 & 92 \\ 124 & 163 & 140 & 160 & 176 & 161 \end{pmatrix} = \begin{pmatrix} 18 & 21 & 12 & 30 & 24 & 23 \\ 19 & 4 & 18 & 5 & 8 & 4 \\ 8 & 23 & 17 & 15 & 22 & 21 \end{pmatrix}$$

The columns of this matrix $\mathbf{A}^{-1}(\mathbf{AB})$ are the entries:

$$18, 19, 8, 21, 4, 23, 12, 18, 17, 30, 5, 15, 24, 8, 22, 23, 4, 21$$

Using the above Table 1.3 backwards, we can read the message as 'OPERATION BLUESTAR'.

We use cryptography all the time. For example, emails, websites, ATM cards and digital passwords are all protected by encryption.

Summary

You can find the inverse matrix of $\mathbf{A}$ by transforming the augmented matrix $(\mathbf{A} \mid \mathbf{I})$ into the augmented matrix $(\mathbf{I} \mid \mathbf{A}^{-1})$ by elementary row operations.
If in the process of this you get a row of zeros then the matrix $\mathbf{A}$ is non-invertible.

EXERCISES 1.8

(Brief solutions at end of book. Full solutions available at <http://www.oup.co.uk/companion/singh>.)

1. *Which of the following are elementary matrices?*

   For the matrices which are elementary, state the row operation on the identity matrix which produced them.

$$\mathbf{A}=\begin{pmatrix}3&0&0\\0&3&0\\0&0&3\end{pmatrix},\ \mathbf{B}=\begin{pmatrix}-1&0&0\\0&1&0\\0&0&1\end{pmatrix},\ \mathbf{C}=\begin{pmatrix}0&1&0\\1&0&0\\0&0&1\end{pmatrix},\ \mathbf{D}=\begin{pmatrix}1&0&0\\3&1&0\\0&0&1\end{pmatrix},$$

$$\mathbf{E}=\begin{pmatrix}0&1&0\\1&0&0\\0&0&2\end{pmatrix},\ \mathbf{F}=\begin{pmatrix}-1&0&-1\\0&1&0\\0&0&1\end{pmatrix},\ \mathbf{G}=\begin{pmatrix}0&0&1&0\\0&1&0&0\\1&0&0&0\\0&0&0&1\end{pmatrix}\text{ and }\mathbf{H}=\begin{pmatrix}1&0&0&0\\0&1&0&0\\0&0&0&0\\0&0&0&1\end{pmatrix}$$

2. Find the inverse of the following elementary matrices:

   (a) $\mathbf{E}_1=\begin{pmatrix}-1&0\\0&1\end{pmatrix}$ (b) $\mathbf{E}_2=\begin{pmatrix}0&1\\1&0\end{pmatrix}$ (c) $\mathbf{E}_3=\begin{pmatrix}1&0\\0&-2\end{pmatrix}$

   (d) $\mathbf{E}_4=\begin{pmatrix}-5&0&0\\0&1&0\\0&0&1\end{pmatrix}$ (e) $\mathbf{E}_5=\begin{pmatrix}1&0&0\\0&-\sqrt{2}&0\\0&0&1\end{pmatrix}$ (f) $\mathbf{E}_6=\begin{pmatrix}1&0&0\\0&1&0\\0&0&\pi\end{pmatrix}$

   Also describe the inverse operation.

3. Find the matrix multiplication **EA** (matrix **A** is the general 3×3 matrix) for the following elementary matrices ($k\neq 0$).

   What effect, if any, does the elementary matrix **E** *have on the matrix* **A**?

   (a) $\mathbf{E}=\begin{pmatrix}1&0&0\\0&-1&0\\0&0&1\end{pmatrix}$ (b) $\mathbf{E}=\begin{pmatrix}0&0&1\\0&1&0\\1&0&0\end{pmatrix}$ (c) $\mathbf{E}=\begin{pmatrix}k&0&0\\0&1&0\\0&0&1\end{pmatrix}$

   (d) $\mathbf{E}=\begin{pmatrix}1&0&0\\0&1&0\\0&0&-1/k\end{pmatrix}$

4. Find the inverse matrix, if it exists, by using row operations:

   (a) $\mathbf{A}=\begin{pmatrix}1&2\\-1&4\end{pmatrix}$ (b) $\mathbf{B}=\begin{pmatrix}2&-5\\-6&1\end{pmatrix}$ (c) $\mathbf{C}=\begin{pmatrix}1&0&2\\2&3&1\\3&6&0\end{pmatrix}$

   (d) $\mathbf{D}=\begin{pmatrix}1&-1&1\\1&0&-1\\0&0&-1\end{pmatrix}$ (e) $\mathbf{E}=\begin{pmatrix}2&-1&0\\-1&2&-1\\0&-1&2\end{pmatrix}$

   (f) $\mathbf{F}=\begin{pmatrix}1&3&4\\-1&1&1\\2&1&-2\end{pmatrix}$ (g) $\mathbf{G}=\begin{pmatrix}-2&5&3&1\\-9&2&-5&6\\2&4&8&16\\4&8&16&32\end{pmatrix}$

(h) $\mathbf{H} = \begin{pmatrix} 1 & 2 & -2 & 3 \\ -2 & 1 & -5 & -6 \\ -5 & -10 & 9 & -15 \\ -6 & -12 & 27 & -18 \end{pmatrix}$ (i) $\mathbf{J} = \begin{pmatrix} 1 & 2 & -2 & 3 \\ -2 & 1 & -5 & -6 \\ -5 & -10 & 9 & -15 \\ -6 & -12 & 12 & -19 \end{pmatrix}$

5. Determine whether the following linear systems have a unique solution by using the inverse matrix method. Also find the solutions in the unique case.

(a) $\begin{aligned} x + 2y &= 3 \\ -x + 4y &= 5 \end{aligned}$ (b) $\begin{aligned} 2x - 5y &= 3 \\ -6x + y &= -1 \end{aligned}$

(c) $\begin{aligned} x \quad\quad + 2z &= -1 \\ 2x + 3y + z &= 1 \\ 3x + 6y \quad\quad &= 9 \end{aligned}$ (d) $\begin{aligned} x - y + z &= 10 \\ x \quad\quad - z &= 3 \\ -z &= 5 \end{aligned}$

(e) $\begin{aligned} 2x - y \quad\quad &= 5 \\ -x + 2y - z &= 7 \\ -y + 2z &= 3 \end{aligned}$ (f) $\begin{aligned} x + 3y + 4z &= -2 \\ -x + y + z &= 3 \\ 2x + y - 2z &= 6 \end{aligned}$

(g) $\begin{aligned} x + 2y - 2z + 3w &= 8 \\ -2x + y - 5z - 6w &= 1 \\ -5x - 10y + 9z - 15w &= -1 \\ -6x - 12y + 12z - 19w &= 5 \end{aligned}$

For the next two questions use any appropriate software to carry out the evaluations.

6. By using Table 1.3 on page 116 encode the following messages using the matrix

$$\mathbf{A} = \begin{pmatrix} 3 & 4 & 5 \\ 1 & 3 & 1 \\ 1 & 1 & 2 \end{pmatrix}$$

(a) ATTACK (b) IS YOUR PARTNER HOME

7. **Leontief input-output model.** The Leontief model represents the economy as a linear system. Consider a particular economy which depends on oil (O), energy (E) and services (S). The **input-output matrix A** of such an economy is given by:

$$\begin{array}{c} \\ \text{O} \\ \text{E} \\ \text{S} \end{array} \begin{array}{c} \begin{array}{ccc} \text{O} & \text{E} & \text{S} \end{array} \\ \begin{pmatrix} 0.25 & 0.15 & 0.1 \\ 0.4 & 0.15 & 0.2 \\ 0.15 & 0.2 & 0.2 \end{pmatrix} \end{array} = \mathbf{A}$$

The numbers in the first row are produced as follows:

To produce one unit of oil the oil industry uses 0.25 units of oil, 0.15 units of energy and 0.1 units of services. Similarly the numbers in the other rows are established. The production vector $\mathbf{p}$ and the demand vector $\mathbf{d}$ satisfies $\mathbf{p} = \mathbf{A}\mathbf{p} + \mathbf{d}$.

Determine the production (the production needed) vector $\mathbf{p}$ if $\mathbf{d} = (100\ 100\ 100)^T$.

8. Let $\mathbf{A}$ be an invertible matrix. Show that $\mathbf{A}^T\mathbf{x} = \mathbf{b}$ has a unique solution.

9. Consider the linear systems $\mathbf{Ax} = \mathbf{b}$ and $\mathbf{Bx} = \mathbf{c}$ where both $\mathbf{A}$ and $\mathbf{B}$ are invertible. Show that $(\mathbf{A} + \mathbf{B})\,\mathbf{x} = \mathbf{b} + \mathbf{c}$ may have an infinite number of solutions.

10. Prove Proposition (1.34).
11. Prove that if a matrix **B** is row equivalent to matrix **A** then there exists an invertible matrix **P** such that $\mathbf{B} = \mathbf{PA}$.
12. (a) Prove that if a matrix **B** is row equivalent to matrix **A** then matrix **A** is row equivalent to matrix **B**.
    (b) Prove that if matrix **A** is row equivalent to matrix **B** and matrix **B** is row equivalent to matrix **C** then **A** is row equivalent to **C**.
13. Prove Proposition (1.37).
14. *Prove Proposition (1.39).
15. Explain why there are only three types of elementary matrices.
16. Prove that the transpose of an elementary matrix is an elementary matrix.

MISCELLANEOUS EXERCISES 1

(Brief solutions at end of book. Full solutions available at <http://www.oup.co.uk/companion/singh>.)

In this exercise you may check your numerical answers using MATLAB.

1.1. Let $\mathbf{A} = \begin{pmatrix} 1 & 2 \\ 3 & 4 \end{pmatrix}$ and $\mathbf{B} = \begin{pmatrix} 5 & 6 \\ 7 & 8 \end{pmatrix}$. Determine (a) $(\mathbf{A} - \mathbf{B})(\mathbf{A} + \mathbf{B})$ (b) $\mathbf{A}^2 - \mathbf{B}^2$
Explain why $\mathbf{A}^2 - \mathbf{B}^2 \neq (\mathbf{A} - \mathbf{B})(\mathbf{A} + \mathbf{B})$.

University of Hertfordshire, UK

1.2. Let **A** and **B** be invertible (non-singular) n by n matrices. Find the errors, if any, in the following derivation:

$$\begin{aligned} \mathbf{AB}(\mathbf{AB})^{-1} &= \mathbf{ABA}^{-1}\mathbf{B}^{-1} \\ &= \mathbf{AA}^{-1}\mathbf{BB}^{-1} \\ &= \mathbf{I} \times \mathbf{I} = \mathbf{I} \end{aligned}$$

You need to explain why you think there is an error.

University of Hertfordshire, UK

1.3. Given the matrix

$$\mathbf{A} = \frac{1}{7}\begin{pmatrix} 3 & -2 & -6 \\ -2 & 6 & -3 \\ -6 & -3 & -2 \end{pmatrix}$$

(a) Compute $\mathbf{A}^2$ and $\mathbf{A}^3$.
(b) Based on these results, determine the matrices $\mathbf{A}^{-1}$ and $\mathbf{A}^{2004}$.

University of Wisconsin, USA

1.4. Let **A**, **B** and **C** be matrices defined by

$$\mathbf{A} = \begin{pmatrix} 1 & 0 & -1 \\ 1 & 1 & 1 \\ 1 & 2 & 3 \end{pmatrix}, \quad \mathbf{B} = \begin{pmatrix} 1 & 1 \\ 1 & 2 \\ 0 & -1 \\ 0 & 1 \end{pmatrix}, \quad \mathbf{C} = \begin{pmatrix} 1 & -1 & 1 & 0 \\ -2 & 1 & 2 & 3 \end{pmatrix}$$

Which of the following are defined?

$$\mathbf{A}^t,\ \mathbf{AB},\ \mathbf{B}+\mathbf{C},\ \mathbf{A}-\mathbf{B},\ \mathbf{CB},\ \mathbf{BC}^t,\ \mathbf{A}^2$$

Compute those matrices which are defined.

Jacobs University, Germany
(part question)

1.5. Show that $\mathbf{A}^T\mathbf{A}$ is symmetric for all matrices **A**.

Memorial University, Canada

1.6. Find conditions on a, b, c, d, such that the matrix $\begin{bmatrix} a & b \\ c & d \end{bmatrix}$ commutes with the matrix $\begin{bmatrix} 1 & 2 \\ 3 & 0 \end{bmatrix}$.

Memorial University, Canada

1.7. Give an example of the following, or state that no such example exists: 2×2 matrix **A** and 2×1 non-zero vectors **u** and **v** such that $\mathbf{Au} = \mathbf{Av}$ yet $\mathbf{u} \neq \mathbf{v}$.

Illinois State University, USA
(part question)

1.8. (a) If $\mathbf{A} = \begin{pmatrix} 1 & 2 \\ 3 & 4 \end{pmatrix}$ and $\mathbf{B} = \begin{pmatrix} 0 & 1 \\ -1 & 0 \end{pmatrix}$, compute $\mathbf{A}^2$, $\mathbf{B}^2$, **AB** and **BA**.

(b) If $\mathbf{A} = \begin{pmatrix} a & b \\ c & d \end{pmatrix}$ and $\mathbf{B} = \begin{pmatrix} e & f \\ g & h \end{pmatrix}$, compute $\mathbf{AB} - \mathbf{BA}$.

Queen Mary, University of London, UK

1.9. Let $\mathbf{M} = \begin{pmatrix} 1 & 1 \\ 1 & 1 \end{pmatrix}$. Compute $\mathbf{M}^n$ for $n = 2, 3, 4$. Find a function $c(n)$ such that $\mathbf{M}^n = c(n)\,\mathbf{M}$ for all $n \in \mathbb{Z}, n \geq 1$. (You are not required to prove any of your results.)

Queen Mary, University of London, UK
(part question)

1.10. Let $\mathbf{A} = \begin{pmatrix} \frac{1}{3} & \frac{1}{3} \\ \frac{1}{3} & \frac{1}{3} \end{pmatrix}$. Determine (i) $\mathbf{A}^2$ (ii) $\mathbf{A}^3$

Prove that $\mathbf{A}^n = \frac{1}{2}\left(\frac{2}{3}\right)^n \mathbf{A}$.

University of Hertfordshire, UK

1.11. How many rows does **B** have if **BC** is a 4×6 matrix? Explain.

Washington State University, USA

1.12. Determine, with explanation, if the following matrices are invertible.

(a) $\begin{bmatrix} 2 & -1 & -1 \\ -1 & 2 & -1 \\ -1 & -1 & 2 \end{bmatrix}$

(b) The $n \times n$ matrix $\mathbf{B}$ if $\mathbf{ABC} = \mathbf{I}_n$ and $\mathbf{A}$ and $\mathbf{C}$ are $n \times n$ invertible matrices.

Illinois State University, USA

1.13. (a) What is meant by saying that the $n \times n$ matrix $\mathbf{A}$ is invertible?

(b) Find the inverse of the matrix,

$$\mathbf{A} = \begin{bmatrix} 2 & 6 & 6 \\ 2 & 7 & 6 \\ 2 & 7 & 7 \end{bmatrix}$$

Check your answer.

University of Sussex, UK

1.14. Obtain a Gaussian array from the following set of equations and transform it to reduced row echelon form to obtain any solutions:

$$\begin{aligned} x + y + 2z &= 8 \\ -x - 2y + 3z &= 1 \\ 3x - 7y + 4z &= 10 \end{aligned}$$

Check your answer!
Any other method will not gain full marks.

University of Sussex, UK

1.15. Let $\mathbf{A} = \begin{bmatrix} 1 & 2 & 7 \\ -2 & 5 & 4 \\ -5 & 6 & -3 \end{bmatrix}$, and let $\mathbf{b} = \begin{bmatrix} 3 \\ 3 \\ 1 \end{bmatrix}$.

(a) Reduce the augmented matrix for the system $\mathbf{Ax} = \mathbf{b}$ to reduced echelon form.

(b) Write the solution set for the system $\mathbf{Ax} = \mathbf{b}$ in parametric vector form.

University of South Carolina, USA
(part question)

1.16. Find the general solution in vector form to

$$\begin{aligned} x_1 - x_2 - 2x_3 - 8x_4 &= -3 \\ -2x_1 + x_2 + 2x_3 + 9x_4 &= 5 \\ 3x_1 - 2x_2 - 3x_3 - 15x_4 &= -9 \end{aligned}$$

using the Gaussian elimination algorithm. (Note: This means that you are allowed to use your calculator only as a check.)

Illinois State University, USA

1.17. Solve the following system of linear equations by Gauss–Jordan elimination:

$$\begin{aligned} x_1 + 2x_2 + x_3 - 4x_4 &= 1 \\ x_1 + 3x_2 + 7x_3 + 2x_4 &= 2 \\ x_1 - 11x_3 - 16x_4 &= -1 \end{aligned}$$

McGill University, Canada

1.18. Reduce the following augmented matrix to its (i) echelon form, (ii) row echelon form and (iii) reduced row echelon form:

$$\begin{bmatrix} 1 & 2 & 5 & 6 \\ 3 & 7 & 3 & 2 \\ 2 & 5 & 11 & 22 \end{bmatrix}$$

National University of Singapore
(part question)

1.19. Show that the following system of linear equations is inconsistent.

$$\begin{aligned} x_1 + 2x_2 - 3x_3 - 5x_4 &= -13 \\ 3x_1 + x_2 + 4x_3 - 4x_4 &= 5 \\ 2x_1 - 2x_2 + 3x_3 - 10x_4 &= 7 \\ x_1 + x_2 + 2x_3 + 2x_4 &= 0 \end{aligned}$$

National University of Ireland, Galway
(part question)

1.20. Let **A** be the 3×3 matrix determined by

$$\mathbf{A}\begin{bmatrix} 0 \\ 1 \\ 1 \end{bmatrix} = \begin{bmatrix} -1 \\ 0 \\ 2 \end{bmatrix}, \quad \mathbf{A}\begin{bmatrix} 1 \\ 0 \\ 1 \end{bmatrix} = \begin{bmatrix} 0 \\ -1 \\ 2 \end{bmatrix}, \quad \mathbf{A}\begin{bmatrix} 1 \\ 1 \\ 0 \end{bmatrix} = \begin{bmatrix} 1 \\ 1 \\ 2 \end{bmatrix}$$

Find **A**.

Columbia University, New York, USA

1.21. Consider the matrix $\mathbf{A} = \begin{pmatrix} 2 & 2 & 0 & 2 \\ -1 & -1 & 2 & 1 \\ 2 & 2 & -1 & 1 \\ -1 & -1 & 1 & 0 \end{pmatrix}$. Find rref(**A**).

Johns Hopkins University, USA
(part question)

1.22. Find all the solutions to the following system of linear equations.

$$\begin{aligned} 2x_1 + 3x_2 + x_3 + 4x_4 - 9x_5 &= 17 \\ x_1 + x_2 + x_3 + x_4 - 3x_5 &= 6 \\ x_1 + x_2 + x_3 + 2x_4 - 5x_5 &= 8 \\ 2x_1 + 2x_2 + 2x_3 + 3x_4 - 8x_5 &= 14 \end{aligned}$$

RWTH Aachen University, Germany

1.23. Let $\mathbf{A} = \begin{bmatrix} 1 & 0 \\ -5 & 2 \end{bmatrix}$.

(a) Write $\mathbf{A}^{-1}$ as a product of two elementary matrices.

(b) Write $\mathbf{A}$ as a product of two elementary matrices.

McGill University, Canada

1.24. Write down the elementary matrix $\mathbf{E}$ that satisfies $\mathbf{EA} = \mathbf{B}$ where

$$\mathbf{A} = \begin{bmatrix} 5 & 1 & 9 \\ 2 & 4 & 0 \\ 1 & 1 & 0 \\ -1 & 3 & -2 \end{bmatrix} \text{ and } \mathbf{B} = \begin{bmatrix} 1 & 1 & 0 \\ 2 & 4 & 0 \\ 5 & 1 & 9 \\ -1 & 3 & -2 \end{bmatrix}$$

University of Western Ontario, Canada

1.25. If

$$\left(\mathbf{A}^T - 3\begin{bmatrix} 1 & 2 \\ -1 & 3 \end{bmatrix}\right)^{-1} = \begin{bmatrix} 2 & 1 \\ 1 & 1 \end{bmatrix}$$

then $\mathbf{A}$=

A. $\begin{bmatrix} 5 & 5 \\ 1 & 0 \end{bmatrix}$ B. $\begin{bmatrix} 7 & 8 \\ 0 & 10 \end{bmatrix}$ C. $\begin{bmatrix} -1 & 5 \\ 7 & 9 \end{bmatrix}$ D. $\begin{bmatrix} 4 & -4 \\ 5 & 11 \end{bmatrix}$

E. $\begin{bmatrix} 2 & -3 \\ -6 & 8 \end{bmatrix}$ F. $\begin{bmatrix} 8 & 2 \\ 7 & -5 \end{bmatrix}$

University of Ottawa, Canada

1.26. Suppose that

$$\mathbf{B} = \begin{bmatrix} 1 & 3 & 5 \\ 0 & 1 & 2 \\ 1 & 3 & 6 \end{bmatrix}, \quad \mathbf{C} = \begin{bmatrix} 3 & 0 & 1 \\ 0 & 2 & 0 \\ 5 & 0 & -2 \end{bmatrix}, \quad \mathbf{D} = \begin{bmatrix} -3 & 1 & 2 \\ 1 & 0 & -1 \\ 0 & -2 & 0 \end{bmatrix}$$

and that the 3×3 matrix $\mathbf{X}$ satisfies $\mathbf{B}(\mathbf{X} + \mathbf{C}) = \mathbf{D}$. Find $\mathbf{X}$.

University of Western Ontario, Canada

1.27. (a) Use Gaussian elimination with back substitution or Gauss–Jordan to solve:

$$\begin{aligned} x + y + z &= 5 \\ -x + 3y + 2z &= -2 \\ 2x + y + z &= 1 \end{aligned}$$

(b) Check your answer using the inverse matrix method.

(c) Can you change the coefficient of the z variable in the third equation so that the resulting system has no solutions? Find the value of the coefficient or explain why such a change would be impossible.

Saint Michael's College Vermont, USA

1.28. Prove (give a clear reason): If $\mathbf{A}$ is a symmetric invertible matrix then $\mathbf{A}^{-1}$ is also symmetric.

Massachusetts Institute of Technology USA

1.29. If $\mathbf{A}$ is a matrix such that $\mathbf{A}^2 - \mathbf{A} + \mathbf{I} = \mathbf{O}$ show that $\mathbf{A}$ is invertible with inverse $\mathbf{I} - \mathbf{A}$.

McGill University Canada 2007
(part question)

1.30. (a) Define what is meant by a square matrix $\mathbf{A}$ being invertible. Show that the inverse of $\mathbf{A}$, if it exists, is unique.

(b) Show that the product of any finite number of invertible matrices is invertible.

(c) Find the inverse of the matrix

$$\mathbf{A} = \begin{bmatrix} 1 & 0 & 1 \\ -1 & 1 & 1 \\ 0 & 1 & 0 \end{bmatrix}$$

University of Sussex, UK

1.31. Let $\mathbf{A}$ and $\mathbf{B}$ be $n \times n$ invertible matrices, with $\mathbf{AXA}^{-1} = \mathbf{B}$. Explain why $\mathbf{X}$ is invertible and calculate $\mathbf{X}^{-1}$ in terms of $\mathbf{A}$ and $\mathbf{B}$.

University of South Carolina, USA
(part question)

1.32. What is the set of all solutions to the following system of equations?

$$\begin{bmatrix} 0 & 1 & 1 & 0 & 2 & 0 & 4 \\ 0 & 0 & 0 & 1 & 3 & 0 & 5 \\ 0 & 1 & 1 & 0 & 2 & 1 & 10 \end{bmatrix} \begin{bmatrix} a \\ b \\ c \\ d \\ e \\ f \\ g \end{bmatrix} = \begin{bmatrix} 7 \\ 8 \\ 16 \end{bmatrix}$$

Columbia University, New York, USA

Sample Questions

1.33. Show that

$$\mathbf{AB} = \mathbf{I} \Leftrightarrow \mathbf{BA} = \mathbf{I}$$

1.34. Prove that: If $\mathbf{AB}$ are square matrices and $\mathbf{AB}$ is invertible (non-singular) then both $\mathbf{A}$ and $\mathbf{B}$ are invertible with $(\mathbf{AB})^{-1} = \mathbf{B}^{-1}\mathbf{A}^{-1}$.

1.35. Let $\mathbf{D}$ be the following n by n matrix

$$\mathbf{D} = \begin{pmatrix} a_{11} & 0 & \cdots & 0 \\ 0 & a_{22} & 0 & \vdots \\ \vdots & 0 & \ddots & 0 \\ 0 & \vdots & 0 & a_{nn} \end{pmatrix}$$

(i) Prove that $\mathbf{D}^p = \begin{pmatrix} a_{11}^p & 0 & \cdots & 0 \\ 0 & a_{22}^p & 0 & \vdots \\ \vdots & 0 & \ddots & 0 \\ 0 & \vdots & 0 & a_{nn}^p \end{pmatrix}$ where p is a positive integer.

(ii) Prove that the inverse matrix $\mathbf{D}^{-1} = \begin{pmatrix} a_{11}^{-1} & 0 & \cdots & 0 \\ 0 & a_{22}^{-1} & 0 & \vdots \\ \vdots & 0 & \ddots & 0 \\ 0 & \vdots & 0 & a_{nn}^{-1} \end{pmatrix}$.

Des Higham

is the 1966 Professor of Numerical Analysis in the Department of Mathematics and Statistics at the University of Strathclyde in Glasgow, UK.

Tell us about yourself and your work.

I develop and apply mathematical models and computational algorithms. As a mathematician, I am mostly interested in proving results, for example, showing that a model can reproduce known behaviour of a physical system and then using it to make predictions that can be tested experimentally. On the algorithmic side, I aim to prove that a computational technique solves the required problem accurately and effectively. However, I like to work alongside colleagues from other areas, including the life sciences and e-business. This is a great way for me to get hold of realistic data sets, and it often throws up challenges that inspire new mathematical ideas. So I find it very rewarding to have my research stimulated and stress-tested by outside applications.

How do you use linear algebra in your job?

A good example is my research in network science. Here we are typically given data about pairwise interactions (who phoned who, who emailed who, who follows who on Twitter). This type of information is naturally represented by a matrix, and many of the interesting questions (who are the key players, where are the bottlenecks, where is the best place to learn the latest rumour, and can we categorize people into groups?) can be thought of as problems in linear algebra.

How important is linear algebra?

Tasks such as solving linear systems and linear least squares problems and computing eigenvalues/eigenvectors are the nuts and bolts that go into most of the algorithms used in scientific computing.

What are the challenges connected with the subject?

In today's high-tech digital world, we are being swamped with data. In fact, a lot of it is about us! There is behavioural information about our on-line activities, our digital communication patterns and our supermarket spending habits, and new experimental techniques are generating more fundamental biological information about our DNA sequences and the behaviour of our genes and our neurons. We need bigger computers and smarter algorithms to keep up with the data deluge.

What are the key issues in your area of linear algebra research?

At the moment, I am focusing on two aspects. First, modelling issues for networks: what 'laws of motion' are responsible for the patterns of interactions that we see in digital social media; i.e. how do matrix entries change over time? Second, what can we usefully do with large scale interaction data; i.e. how do we summarize a giant matrix?

Have you any particular messages that you would like to give to students starting off studying linear algebra?

You only have to look at the success of Google, and its underlying PageRank algorithm, to get a feel for the relevance and power of linear algebra. If you want to understand ideas in mathematics and statistics, and develop a useful set of skills, linear algebra is the place to start.

2 Euclidean Space

SECTION 2.1 Properties of Vectors

By the end of this section you will be able to

- prove some properties of vectors in $\mathbb{R}^n$
- understand what is meant by the dot product and norm
- prove dot product and norm properties of vectors

In chapter 1 we looked at $\mathbb{R}^n$ or Euclidean n-space, named after the Greek mathematician Euclid.

Figure 2.1 Euclid lived around 300 BC © Fondazione Cariplo.

Euclid (Fig. 2.1) is famously known for his work 'The Elements' which is also called 'Euclid's Elements'. This work has been used in mathematics teaching for over two thousand years and was first published as a book in 1482. Only the Bible has been printed more times than 'Euclid's Elements'.

Up until the 1970s, school mathematics in the UK consisted of learning various parts of Euclid's Elements. The concept of mathematical proof and logical reasoning is what made this work survive for so long.

Any element of $\mathbb{R}^n$ is a vector as we discussed in chapter 1.

What are the two attributes of a vector?

Length and direction. For example, navigating in the air is impossible without knowing both length and direction of a vector (Fig. 2.2).

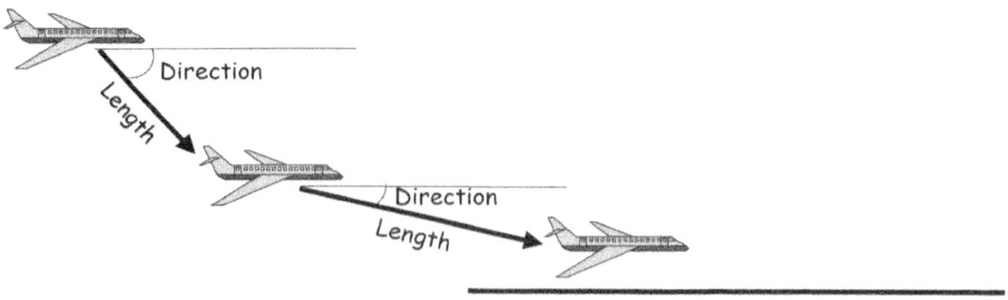

Figure 2.2

We discuss the length and direction of a vector in these next two sections.

Remember, a vector is represented by components. For example, suppose your journey home requires the following directions:

1. 3 blocks east
2. 2 blocks north
3. 4 floors up

We can specify this in three directions:

1. east/west – x axis
2. north/south – y axis
3. up/down – z axis

If we have three independent variables such as x, y and z, we can use $\mathbb{R}^3$ to view a problem geometrically. In fact, we don't have to stop at three, we can consider any number of independent variables, although clearly these do not represent a geometric interpretation that we can plot.

Linear algebra is directly related to the geometry of $\mathbb{R}^n$ because in $\mathbb{R}^n$ we can convert between geometry and algebra.

2.1.1 Vector addition and scalar multiplication properties

Remember, $\mathbb{R}^2$ forms a plane, and $\mathbb{R}^3$ is three-dimensional space.

Recall that a column vector in $\mathbb{R}^n$ is just a special matrix of size n (number of rows) by one (number of columns), therefore all the properties of matrices that we looked at in the last chapter also hold for a vector in $\mathbb{R}^n$.

Next we define the two fundamental operations of linear algebra – **vector addition** and **scalar multiplication**.

Proposition (2.1). Let $\mathbf{u}$, $\mathbf{v}$ and $\mathbf{w}$ be vectors in $\mathbb{R}^n$ and k, c be real numbers (or real scalars). We have the following results:

(i) $\mathbf{u}+\mathbf{v}=\mathbf{v}+\mathbf{u}$ (commutative law)

(ii) $(\mathbf{u}+\mathbf{v})+\mathbf{w}=\mathbf{u}+(\mathbf{v}+\mathbf{w})$ (associative law)

(iii) There exists a zero vector $\mathbf{O}$ such that $\mathbf{u} + \mathbf{O} = \mathbf{u}$ (neutral element)

(iv) For every vector $\mathbf{u}$ there is a vector $-\mathbf{u}$ such that

$$\mathbf{u} + (-\mathbf{u}) = \mathbf{O} \text{ (additive inverse)}$$

(v) $k(\mathbf{u} + \mathbf{v}) = k\mathbf{u} + k\mathbf{v}$ (distributive law)

(vi) $(k + c)\mathbf{u} = k\mathbf{u} + c\mathbf{u}$ (distributive law)

(vii) $(kc)\mathbf{u} = k(c\mathbf{u})$ (associative law)

(viii) For every vector $\mathbf{u}$ we have $1\mathbf{u} = \mathbf{u}$ (neutral element)

Proof.
All of these properties follow from the matrix properties that were discussed in chapter 1.

■

We can use this proposition (2.1) to prove other properties of vectors in $\mathbb{R}^n$.

Proposition (2.2). Let $\mathbf{u}$ be a vector in $\mathbb{R}^n$. Then the vector $-\mathbf{u}$ which satisfies property (iv) in the above Proposition (2.1) is unique:

$$\mathbf{u} + (-\mathbf{u}) = \mathbf{O}$$

Proof.
Let $\mathbf{v}$ be a vector in $\mathbb{R}^n$ such that

$$\mathbf{u} + \mathbf{v} = \mathbf{O} \quad (*)$$

What do we need to show?
Show that $\mathbf{v} = -\mathbf{u}$. Adding $-\mathbf{u}$ to both sides of (*) gives

$$-\mathbf{u} + (\mathbf{u} + \mathbf{v}) = (-\mathbf{u}) + \mathbf{O}$$
$$(-\mathbf{u} + \mathbf{u}) + \mathbf{v} = (-\mathbf{u}) + \mathbf{O}$$
$$\underbrace{(\mathbf{u} + (-\mathbf{u}))}_{=\mathbf{O}} + \mathbf{v} = -\mathbf{u}$$
$$\mathbf{O} + \mathbf{v} = -\mathbf{u}$$
$$\mathbf{v} = -\mathbf{u}$$

■

What does Proposition (2.2) mean?
It means that for every vector $\mathbf{u}$ in $\mathbb{R}^n$ there is only one vector $-\mathbf{u}$ in $\mathbb{R}^n$ such that

$$\mathbf{u} + (-\mathbf{u}) = \mathbf{O}$$

Proposition (2.3). Let $\mathbf{u}$ be a vector in $\mathbb{R}^n$ then $(-1)\mathbf{u} = -\mathbf{u}$.

Proof.

Let $\mathbf{u} = \begin{pmatrix} u_1 \\ \vdots \\ u_n \end{pmatrix}$. We first show that $\mathbf{u} + (-1)\mathbf{u} = \mathbf{O}$ and then use the above Proposition (2.2) which says $-\mathbf{u}$ is unique to deduce $(-1)\mathbf{u} = -\mathbf{u}$:

$$\mathbf{u} + (-1)\,\mathbf{u} = \begin{pmatrix} u_1 \\ \vdots \\ u_n \end{pmatrix} + (-1)\begin{pmatrix} u_1 \\ \vdots \\ u_n \end{pmatrix} = \begin{pmatrix} u_1 \\ \vdots \\ u_n \end{pmatrix} + \begin{pmatrix} -u_1 \\ \vdots \\ -u_n \end{pmatrix} \qquad \left[\text{scalar multiplication}\right]$$

$$= \begin{pmatrix} u_1 & -u_1 \\ \vdots & \vdots \\ u_n & -u_n \end{pmatrix} = \begin{pmatrix} 0 \\ \vdots \\ 0 \end{pmatrix} = \mathbf{O}$$

We have $\mathbf{u} + (-1)\mathbf{u} = \mathbf{O}$ and by the above Proposition (2.2) the vector $-\mathbf{u}$ which satisfies $\mathbf{u} + (-\mathbf{u}) = \mathbf{O}$ is unique, hence

$$(-1)\mathbf{u} = -\mathbf{u}$$

■

So far, we have looked at the two fundamental operations of linear algebra – vector addition and scalar multiplication of vectors in $\mathbb{R}^n$.

Next we look in more detail at the product of two vectors, covered in chapter 1.

2.1.2 Dot (inner) product revisited

Let $\mathbf{u} = \begin{pmatrix} u_1 \\ \vdots \\ u_n \end{pmatrix}$ and $\mathbf{v} = \begin{pmatrix} v_1 \\ \vdots \\ v_n \end{pmatrix}$ be vectors in $\mathbb{R}^n$ then the dot product of $\mathbf{u}$ and $\mathbf{v}$ denoted by $\mathbf{u} \cdot \mathbf{v}$ was defined in section 1.3.6 of the last chapter as

$$\mathbf{u} \cdot \mathbf{v} = u_1 v_1 + u_2 v_2 + u_3 v_3 + \cdots + u_n v_n \tag{1.5}$$

This is the same as matrix multiplication of the transpose of $\mathbf{u}$ by the column vector $\mathbf{v}$:

$$\mathbf{u} \cdot \mathbf{v} = \mathbf{u}^T \mathbf{v} = \begin{pmatrix} u_1 \\ \vdots \\ u_n \end{pmatrix}^T \begin{pmatrix} v_1 \\ \vdots \\ v_n \end{pmatrix}$$

$$= \begin{pmatrix} u_1 & u_2 & \cdots & u_n \end{pmatrix} \begin{pmatrix} v_1 \\ \vdots \\ v_n \end{pmatrix} = u_1 v_1 + u_2 v_2 + u_3 v_3 + \cdots + u_n v_n$$

Note, that this is how we multiply matrices – row by column. Recall that a vector is simply a single column matrix. We transpose the *left hand* vector $\mathbf{u}$ because we can only carry out matrix multiplication if the number of columns of the *left hand* matrix, $\mathbf{u}$, is equal to the number of rows of the *right hand* matrix, $\mathbf{v}$.

This matrix multiplication is called the **dot** or **inner product** of the vectors **u** and **v** and

(2.4) $$\mathbf{u} \cdot \mathbf{v} = \mathbf{u}^T\mathbf{v} = u_1v_1 + u_2v_2 + u_3v_3 + \cdots + u_nv_n$$

What do you notice about your answer?

The result of the dot product of two vectors in $\mathbb{R}^n$ is a real number; that is, carrying out the dot product operation on two vectors gives us a scalar not a vector.

Also, note that the dot product of two vectors **u** and **v** is obtained by multiplying each component u_j with its corresponding component v_j and adding the results.

Example 2.1

Let $\mathbf{u} = (-3 \;\; 1 \;\; 7 \;\; -5)^T$ and $\mathbf{v} = (9 \;\; 2 \;\; -4 \;\; 1)^T$. Find $\mathbf{u} \cdot \mathbf{v}$.

Solution

The given vectors $\mathbf{u} = (-3 \;\; 1 \;\; 7 \;\; -5)^T$ and $\mathbf{v} = (9 \;\; 2 \;\; -4 \;\; 1)^T$ were written in terms of the transpose of the vector so that we can save space.

Applying the above formula (2.4) gives

$$\begin{aligned}\mathbf{u} \cdot \mathbf{v} = \begin{pmatrix} -3 \\ 1 \\ 7 \\ -5 \end{pmatrix} \cdot \begin{pmatrix} 9 \\ 2 \\ -4 \\ 1 \end{pmatrix} &= (-3 \times 9) + (1 \times 2) + (7 \times (-4)) + ((-5) \times 1) \\ &= -27 + 2 - 28 - 5 = -58\end{aligned}$$

Hence $\mathbf{u} \cdot \mathbf{v} = -58$.

You can check this answer, $\mathbf{u} \cdot \mathbf{v} = -58$, by using the numerical software MATLAB. The MATLAB command for the dot product of two vectors **u** and **v** is dot (u, v).

We can use dot product to write an equation. For example

$$3x + 4y = 11 \text{ can be written as } \begin{pmatrix} 3 \\ 4 \end{pmatrix} \cdot \begin{pmatrix} x \\ y \end{pmatrix} = (3 \;\; 4)\begin{pmatrix} x \\ y \end{pmatrix} = 11$$

To solve this equation means 'to determine the vector whose dot product with the vector (3, 4) equals 11'.

The dot product also gives information about the angle between the two vectors but we will discuss this in the next section.

Example 2.2

Let $\mathbf{u} = \begin{pmatrix} 2 \\ 5 \end{pmatrix}$ and $\mathbf{v} = \begin{pmatrix} 5 \\ -2 \end{pmatrix}$. Plot these vectors **u** and **v** in $\mathbb{R}^2$ and find $\mathbf{u} \cdot \mathbf{v}$.

(continued...)

Solution

Plotting the vectors **u** and **v** in $\mathbb{R}^2$ gives (see Fig. 2.3).

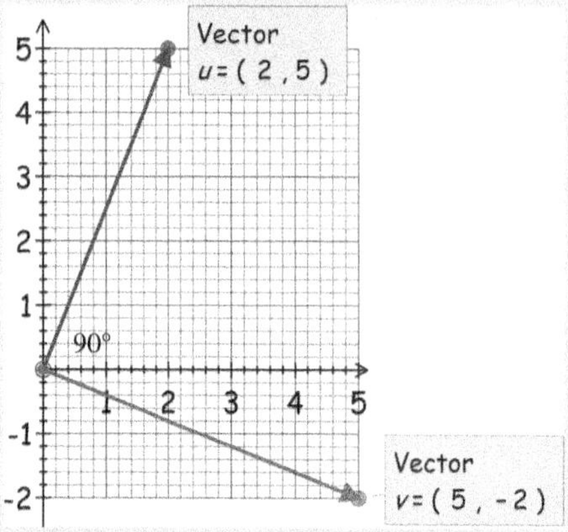

Figure 2.3

$$\mathbf{u}\cdot\mathbf{v}=\begin{pmatrix}2\\5\end{pmatrix}\cdot\begin{pmatrix}5\\-2\end{pmatrix}=(2\times 5)+(5\times(-2))=10-10=0$$

What do you notice about your answer?

Dot product is zero. Also note that the vectors **u** and **v** are perpendicular to each other.

Vectors **u** and **v** are perpendicular or orthogonal if and only if their dot product is zero. Two vectors **u** and **v** in $\mathbb{R}^n$ are said to be **perpendicular** or **orthogonal** ⇔

(2.5) $$\mathbf{u}\cdot\mathbf{v}=0$$

In linear algebra the concept of orthogonality is very important.

Example 2.3

Let $\mathbf{u}=\begin{pmatrix}a\\b\end{pmatrix}$ and $\mathbf{v}=\begin{pmatrix}b\\-a\end{pmatrix}$ be vectors in $\mathbb{R}^2$. Prove that these vectors are orthogonal.

Solution

Carrying out the dot product of the given two vectors by applying the above formula (2.4):

$$\mathbf{u}\cdot\mathbf{v}=\begin{pmatrix}a\\b\end{pmatrix}\cdot\begin{pmatrix}b\\-a\end{pmatrix}=(a\times b)+(b\times(-a))=ab-ba=0$$

The dot product $\mathbf{u}\cdot\mathbf{v}$ is zero, therefore by the above formula (2.5) we conclude that the vectors **u** and **v** are orthogonal (perpendicular) as shown in Fig. 2.4.

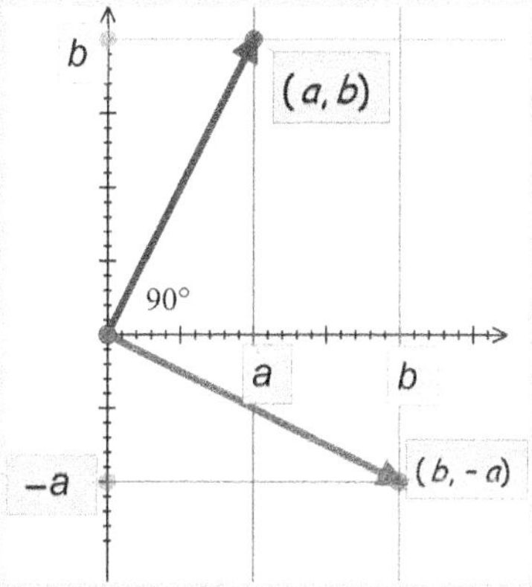

Figure 2.4

We can also apply the above formula (2.5), and use the dot product to solve equations such as $3x + 4y = 0$.

Here, we need to find all the vectors $(x\ \ y)^T$ such that

$$\begin{pmatrix} 3 \\ 4 \end{pmatrix} \cdot \begin{pmatrix} x \\ y \end{pmatrix} = 0 \quad \left[\text{remember } \begin{pmatrix} 3 \\ 4 \end{pmatrix} \cdot \begin{pmatrix} x \\ y \end{pmatrix} = 3x + 4y\right]$$

We know from the above formula (2.5), that if $\mathbf{u} \cdot \mathbf{v} = 0$, then the vectors $\mathbf{u}$ and $\mathbf{v}$ are perpendicular to each other.

We need to find the vectors which are perpendicular (orthogonal) to the vector $\begin{pmatrix} 3 \\ 4 \end{pmatrix}$.

We can illustrate this (Fig. 2.5).

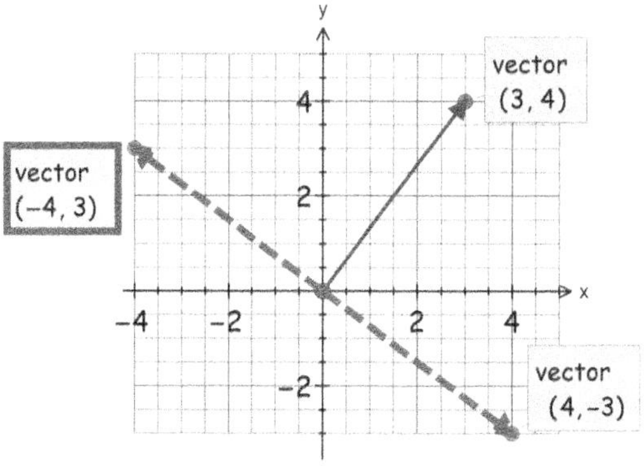

Figure 2.5

By examining Fig. 2.5 we see that every vector on the dashed line is perpendicular to the vector $\begin{pmatrix} 3 \\ 4 \end{pmatrix}$ so we have an infinite number of solutions to the given equation $3x + 4y = 0$:

$$x = 4,\ y = -3 \quad \text{or} \quad x = -4,\ y = 3 \quad \text{or} \quad x = 0,\ y = 0 \ldots$$

We are given one equation, $3x + 4y = 0$, and two unknowns; x and y. From chapter 1 (Proposition (1.31)) we know that there are an infinite number of solutions to this equation, because we have fewer equations (1) than unknowns (2).

2.1.3 Properties of dot (inner) product

Next we state some basic properties of the dot product in $\mathbb{R}^n$.

Proposition (2.6). Let **u**, **v** and **w** be vectors in $\mathbb{R}^n$ and k be a real scalar (real number). We have the following:

(i) $(\mathbf{u}+\mathbf{v})\cdot\mathbf{w} = \mathbf{u}\cdot\mathbf{w} + \mathbf{v}\cdot\mathbf{w}$ [distributive law]

(ii) $\mathbf{u}\cdot\mathbf{v} = \mathbf{v}\cdot\mathbf{u}$ [commutative law]

(iii) $(k\mathbf{u})\cdot\mathbf{v} = k(\mathbf{u}\cdot\mathbf{v}) = (\mathbf{u}\cdot k\mathbf{v})$

(iv) $\mathbf{u}\cdot\mathbf{u} \geq 0$ and we have $\mathbf{u}\cdot\mathbf{u} = 0 \Leftrightarrow \mathbf{u} = \mathbf{O}$
($\mathbf{u}\cdot\mathbf{u}$ is the square of the length of the vector **u**.)

Proof.

Let $\mathbf{u} = (u_1 \ \cdots \ u_n)^T = \begin{pmatrix} u_1 \\ \vdots \\ u_n \end{pmatrix}$ and $\mathbf{v} = (v_1 \ \cdots \ v_n)^T = \begin{pmatrix} v_1 \\ \vdots \\ v_n \end{pmatrix}$ be vectors in $\mathbb{R}^n$.

(i) See Exercises 2.1.

(ii) Required to prove $\mathbf{u}\cdot\mathbf{v} = \mathbf{v}\cdot\mathbf{u}$:

$$\begin{aligned}
\mathbf{u}\cdot\mathbf{v} &= u_1v_1 + u_2v_2 + u_3v_3 + \cdots + u_nv_n && \text{[applying (2.4)]} \\
&= v_1u_1 + v_2u_2 + v_3u_3 + \cdots + v_nu_n && [\text{because } u_j \text{ and } v_j \text{ are real so } u_jv_j = v_ju_j] \\
&= \mathbf{v}\cdot\mathbf{u}
\end{aligned}$$

Note that matrix multiplication is not commutative but the dot product is.

■

(iii) *How do we prove* $(k\mathbf{u})\cdot\mathbf{v} = k(\mathbf{u}\cdot\mathbf{v})$?
By expanding the left hand side and showing that it is equivalent to the right:

$$\begin{aligned}
(k\mathbf{u})\cdot\mathbf{v} &= \left[k\begin{pmatrix} u_1 \\ \vdots \\ u_n \end{pmatrix}\right]\cdot\begin{pmatrix} v_1 \\ \vdots \\ v_n \end{pmatrix} = \begin{pmatrix} ku_1 \\ \vdots \\ ku_n \end{pmatrix}\cdot\begin{pmatrix} v_1 \\ \vdots \\ v_n \end{pmatrix} && \left[\begin{array}{l}\text{scalar multiplication} \\ \text{of the vector } \mathbf{u} \text{ by } k\end{array}\right] \\
&= ku_1v_1 + ku_2v_2 + ku_3v_3 + \cdots + ku_nv_n && \text{[applying (2.4)]} \\
&= k(u_1v_1 + u_2v_2 + u_3v_3 + \cdots + u_nv_n) && \text{[factorizing]} \\
&= k(\mathbf{u}\cdot\mathbf{v}) && [\text{because } u_1v_1 + \cdots + u_nv_n = \mathbf{u}\cdot\mathbf{v}]
\end{aligned}$$

Hence we have shown that $(k\mathbf{u})\cdot\mathbf{v} = k(\mathbf{u}\cdot\mathbf{v})$.

Similarly we can show $k(\mathbf{u}\cdot\mathbf{v}) = (\mathbf{u}\cdot k\mathbf{v})$.

■

(iv) What does $\mathbf{u} \cdot \mathbf{u} \geq 0$ mean?

As a real number u satisfies $u \times u \geq 0$ so similarly for vector $\mathbf{u}$ we have $\mathbf{u} \cdot \mathbf{u} \geq 0$.

How do we prove this result?

Substitute $\mathbf{u} = \begin{pmatrix} u_1 \\ \vdots \\ u_n \end{pmatrix}$ and then evaluate $\mathbf{u} \cdot \mathbf{u}$:

$$\mathbf{u} \cdot \mathbf{u} = \begin{pmatrix} u_1 \\ \vdots \\ u_n \end{pmatrix} \cdot \begin{pmatrix} u_1 \\ \vdots \\ u_n \end{pmatrix} = u_1 u_1 + u_2 u_2 + u_3 u_3 + \cdots + u_n u_n \qquad \text{[applying (2.4)]}$$

$$= (u_1)^2 + (u_2)^2 + (u_3)^2 + \cdots + (u_n)^2 \geq 0$$

Since all the u's are real numbers then each $(u_j)^2 \geq 0$. Hence we have shown $\mathbf{u} \cdot \mathbf{u} \geq 0$.

How do we show the last part: $\mathbf{u} \cdot \mathbf{u} = 0 \Leftrightarrow \mathbf{u} = \mathbf{O}$?

From above, we have $\mathbf{u} \cdot \mathbf{u} = (u_1)^2 + (u_2)^2 + (u_3)^2 + \cdots + (u_n)^2$, therefore

$$\mathbf{u} \cdot \mathbf{u} = 0 \Leftrightarrow (u_1)^2 + (u_2)^2 + (u_3)^2 + \cdots + (u_n)^2 = 0$$

$$\Leftrightarrow u_1 = u_2 = u_3 = \cdots = u_n = 0$$

Remember that, from the definition of the zero vector $u_1 = u_2 = u_3 = \cdots = u_n = 0$, we have $\mathbf{u} = \mathbf{O}$. Hence we have our result $\mathbf{u} \cdot \mathbf{u} = 0 \Leftrightarrow \mathbf{u} = \mathbf{O}$.

■

2.1.4 The norm or length of a vector

Let $\mathbf{u}$ be a vector in $\mathbb{R}^n$. The **length** or **norm** of a vector $\mathbf{u}$ is denoted by $\|\mathbf{u}\|$. We can use Pythagoras' theorem to find a way to define the norm of a vector. Consider a vector $\mathbf{u}$ in $\mathbb{R}^2$ (Fig. 2.6):

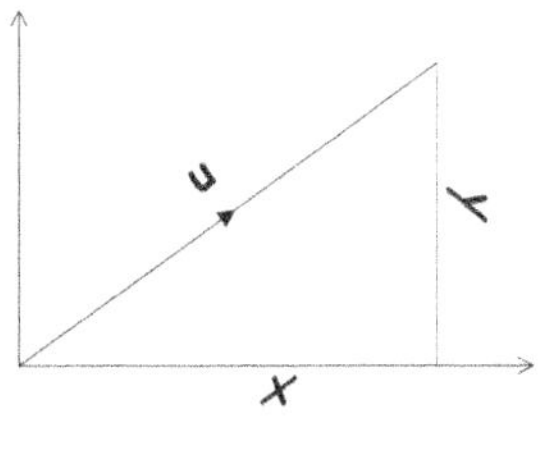

Figure 2.6

The length or norm of a vector $\mathbf{u} = (x \quad y)^T$ in $\mathbb{R}^2$ is given by:

$$\|\mathbf{u}\| = \sqrt{x^2 + y^2}$$

Consider a vector $\mathbf{v}$ in $\mathbb{R}^3$ (Fig. 2.7):

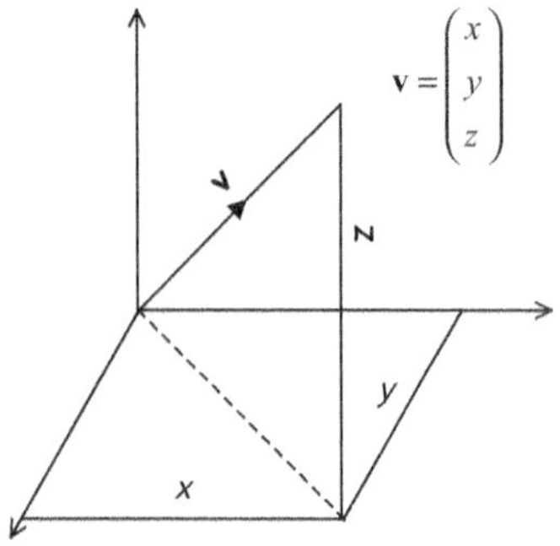

Figure 2.7

The length or norm of a vector $\mathbf{v} = (x \quad y \quad z)^T$ in $\mathbb{R}^3$ is given by

$$\|\mathbf{v}\| = \sqrt{x^2 + y^2 + z^2}$$

Pythagoras' Theorem. Let $\mathbf{u}$ be a vector in $\mathbb{R}^n$ then the length of $\mathbf{u} = \begin{pmatrix} u_1 \\ \vdots \\ u_n \end{pmatrix}$ is given by:

$$\|\mathbf{u}\| = \sqrt{(u_1)^2 + (u_2)^2 + (u_3)^2 + \cdots + (u_n)^2} \tag{2.7}$$

The norm of a vector $\mathbf{u}$ is a real number which gives the length of the vector $\mathbf{u}$.

Example 2.4

Let $\mathbf{u} = \begin{pmatrix} -7 \\ -2 \end{pmatrix}$ and $\mathbf{v} = \begin{pmatrix} 8 \\ 3 \end{pmatrix}$. Plot these vectors $\mathbf{u}$ and $\mathbf{v}$ in $\mathbb{R}^2$ and evaluate the norms $\|\mathbf{u}\|$ and $\|\mathbf{v}\|$.

Solution

Using Pythagoras' theorem (2.7) gives the lengths:

$$\|\mathbf{u}\| = \sqrt{(-7)^2 + (-2)^2} = \sqrt{53} = 7.280 \quad [3\text{ dp}]$$

$$\|\mathbf{v}\| = \sqrt{8^2 + 3^2} = \sqrt{73} = 8.544 \quad [3\text{ dp}]$$

Plotting the vectors $\mathbf{u} = \begin{pmatrix} -7 \\ -2 \end{pmatrix}$ and $\mathbf{v} = \begin{pmatrix} 8 \\ 3 \end{pmatrix}$ and labelling $\|\mathbf{u}\|$ and $\|\mathbf{v}\|$ in $\mathbb{R}^2$ gives (Fig. 2.8):

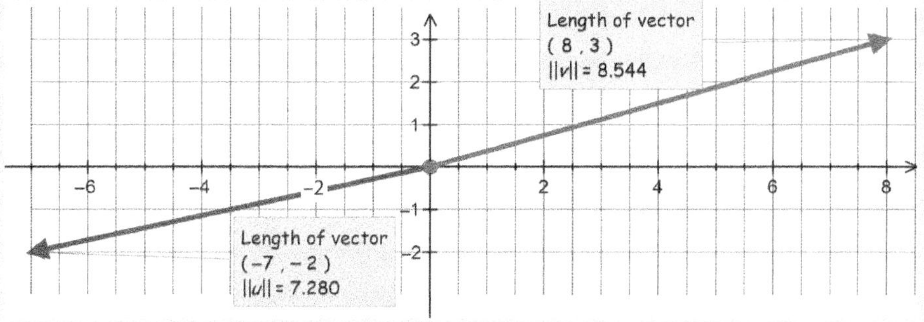

Figure 2.8

We can also write the length of the vector in terms of the dot product:
Let **u** be in $\mathbb{R}^n$ then

(2.8) $$\|\mathbf{u}\| = \sqrt{\mathbf{u}\cdot\mathbf{u}} \quad \text{[positive root]}$$

Proof.
First, by applying formula (2.4) $\mathbf{u}\cdot\mathbf{v} = u_1v_1 + u_2v_2 + \cdots + u_nv_n$ we can find $\mathbf{u}\cdot\mathbf{u}$:

$$\begin{aligned}\mathbf{u}\cdot\mathbf{u} &= u_1u_1 + u_2u_2 + u_3u_3 + \cdots + u_nu_n \\ &= (u_1)^2 + (u_2)^2 + (u_3)^2 + \cdots + (u_n)^2\end{aligned}$$

Second, by taking the square root of this

$$\sqrt{\mathbf{u}\cdot\mathbf{u}} = \sqrt{(u_1)^2 + (u_2)^2 + (u_3)^2 + \cdots + (u_n)^2} \underbrace{=}_{\text{By (2.7)}} \|\mathbf{u}\|$$

which is our required result.

■

This $\|\mathbf{u}\| = \sqrt{\mathbf{u}\cdot\mathbf{u}}$ is called the **Euclidean norm**. There are other types of norms which we will discuss in the next chapter.

The **distance function** (also called the **metric**) written as $d(\mathbf{u}, \mathbf{v})$ is the distance between vectors $\mathbf{u} = (u_1\ u_2\ \cdots\ u_n)^T$ and $\mathbf{v} = (v_1\ v_2\ \cdots\ v_n)^T$ in $\mathbb{R}^n$ and is defined as

(2.9) $$d(\mathbf{u}, \mathbf{v}) = \|\mathbf{u} - \mathbf{v}\|$$

We can use this distance function to find the distance between various satellites for a GPS system.

Example 2.5

Let $\mathbf{s}_1 = (1\ 2\ 3)^T$, $\mathbf{s}_2 = (7\ 4\ 3)^T$ and $\mathbf{s}_3 = (2\ 1\ 9)^T$ be the positions of three satellites as shown in Fig. 2.9. Find the distances between the satellites.

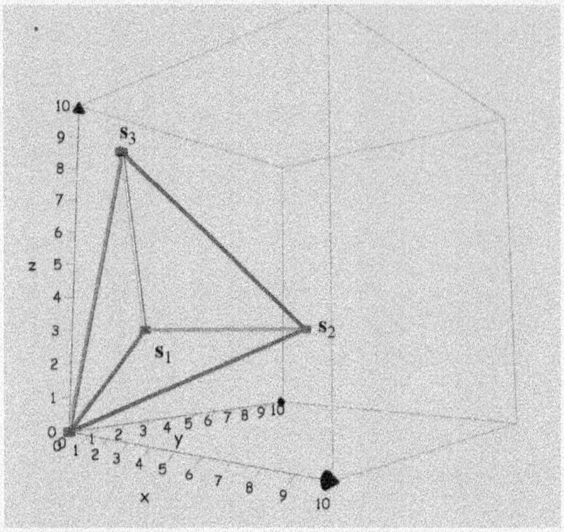

Figure 2.9

Solution

Using the above formula (2.9) we have $d(\mathbf{s}_1, \mathbf{s}_2) = \|\mathbf{s}_1 - \mathbf{s}_2\|$.

What is $\mathbf{s}_1 - \mathbf{s}_2$ *equal to?*

$$\mathbf{s}_1 - \mathbf{s}_2 = \begin{pmatrix} 1 \\ 2 \\ 3 \end{pmatrix} - \begin{pmatrix} 7 \\ 4 \\ 3 \end{pmatrix} = \begin{pmatrix} 1-7 \\ 2-4 \\ 3-3 \end{pmatrix} = \begin{pmatrix} -6 \\ -2 \\ 0 \end{pmatrix}$$

By applying (2.9) we have

$$d(\mathbf{s}_1, \mathbf{s}_2) = \|\mathbf{s}_1 - \mathbf{s}_2\| = \sqrt{(-6)^2 + (-2)^2 + 0^2} = \sqrt{40} = 6.32 \text{ (2dp)}$$

Similarly, the distance between other satellites is

$$d(\mathbf{s}_1, \mathbf{s}_3) = 6.16 \text{ (2dp) and } d(\mathbf{s}_2, \mathbf{s}_3) = 8.37 \text{ (2dp)}$$

Next we state and prove certain properties of the Euclidean norm of a vector.

2.1.5 Properties of the norm of a vector

For a scalar k we define the modulus of k denoted $|k|$ as

$$|k| = \sqrt{k^2}$$

Proposition (2.10). Let $\mathbf{u}$ be a vector in $\mathbb{R}^n$ and k be a real scalar. We have the following:

(i) $\|\mathbf{u}\| \geq 0$ [positive] and $\|\mathbf{u}\| = 0 \Leftrightarrow \mathbf{u} = \mathbf{O}$.

(ii) $\|k\mathbf{u}\| = |k|\,\|\mathbf{u}\|$

Proof.
Let $\mathbf{u}$ be a vector in $\mathbb{R}^n$; therefore we can write this as $\mathbf{u} = (u_1 \cdots u_n)^T$.

(i) Required to prove $\|\mathbf{u}\| \geq 0$. By Pythagoras' theorem (2.7), we have the length of $\mathbf{u}$:

$$\|\mathbf{u}\| = \sqrt{(u_1)^2 + (u_2)^2 + (u_3)^2 + \cdots + (u_n)^2}$$

Since the square root is positive, $\|\mathbf{u}\| \geq 0$.

Next we prove the equality; that is $\|\mathbf{u}\| = 0 \Leftrightarrow \mathbf{u} = \mathbf{O}$. We have $\|\mathbf{u}\| = 0$, which means that

$$\|\mathbf{u}\| = \sqrt{\mathbf{u} \cdot \mathbf{u}} = 0 \Leftrightarrow \mathbf{u} \cdot \mathbf{u} = 0$$

By Proposition (2.6) part (iv), we have $\mathbf{u} \cdot \mathbf{u} = 0 \Leftrightarrow \mathbf{u} = \mathbf{O}$. We have proven our equality.

(ii) Expanding the left hand side of $\|k\mathbf{u}\| = |k|\,\|\mathbf{u}\|$ by applying definition; (2.8) $\|\mathbf{v}\| = \sqrt{\mathbf{v} \cdot \mathbf{v}}$ gives

$$\begin{aligned} \|k\mathbf{u}\| &= \sqrt{k\mathbf{u} \cdot k\mathbf{u}} = \sqrt{k^2(\mathbf{u} \cdot \mathbf{u})} \\ &= \sqrt{k^2}\sqrt{\mathbf{u} \cdot \mathbf{u}} && \left[\text{because } k^2 \text{ and } \mathbf{u} \cdot \mathbf{u} \text{ are real, so} \sqrt{ab} = \sqrt{a}\sqrt{b}\right] \\ &= |k|\,\|\mathbf{u}\| && \left[\text{from above we have } \sqrt{k^2} = |k|\right] \end{aligned}$$

■

Normally, to obtain the length (norm) of a given vector $\mathbf{v}$ you will find it easier to determine $\|\mathbf{v}\|^2 = \mathbf{v} \cdot \mathbf{v}$ and then take the square root of your result to find $\|\mathbf{v}\|$.

Summary

Let $\mathbf{u}$ and $\mathbf{v}$ be vectors in $\mathbb{R}^n$ then the dot product $\mathbf{u} \cdot \mathbf{v}$ is given by (2.4)

$$\mathbf{u} \cdot \mathbf{v} = u_1v_1 + u_2v_2 + u_3v_3 + \cdots + u_nv_n$$

The vectors $\mathbf{u}$ and $\mathbf{v}$ are **orthogonal** or **perpendicular** $\Leftrightarrow \mathbf{u} \cdot \mathbf{v} = 0$
The norm or length of a vector $\mathbf{u}$ in $\mathbb{R}^n$ is defined as

$$\|\mathbf{u}\| = \sqrt{\mathbf{u} \cdot \mathbf{u}} = \sqrt{(u_1)^2 + (u_2)^2 + (u_3)^2 + \cdots + (u_n)^2}$$

EXERCISES 2.1

(Brief solutions at end of book. Full solutions available at <http://www.oup.co.uk/companion/singh>.)

You may like to check your answers using the numerical software MATLAB.
The MATLAB command for norm of a vector $\mathbf{u}$ *is norm*(u).

1. Let $\mathbf{u} = \begin{pmatrix} -1 \\ 3 \end{pmatrix}$ and $\mathbf{v} = \begin{pmatrix} 2 \\ 1 \end{pmatrix}$ be vectors in $\mathbb{R}^2$. Evaluate the following:

   (a) $\mathbf{u} \cdot \mathbf{v}$ (b) $\mathbf{v} \cdot \mathbf{u}$ (c) $\mathbf{u} \cdot \mathbf{u}$ (d) $\mathbf{v} \cdot \mathbf{v}$ (e) $\|\mathbf{u}\|^2$
   (f) $\|\mathbf{v}\|^2$ (g) $\|\mathbf{u}\|$ (h) $\|\mathbf{v}\|$ (i) $\|\mathbf{u} + \mathbf{v}\|^2$ (j) $d(\mathbf{u}, \mathbf{v})$

2. Let $\mathbf{u} = (2 \;\; 3 \;\; -1)^T$ and $\mathbf{v} = (5 \;\; 1 \;\; -2)^T$ be vectors in $\mathbb{R}^3$. Evaluate
   (a) $\mathbf{u} \cdot \mathbf{v}$ (b) $\mathbf{v} \cdot \mathbf{u}$ (c) $\mathbf{u} \cdot \mathbf{u}$ (d) $\mathbf{v} \cdot \mathbf{v}$ (e) $\|\mathbf{u}\|^2$
   (f) $\|\mathbf{v}\|^2$ (g) $\|\mathbf{u}\|$ (h) $\|\mathbf{v}\|$ (i) $\|\mathbf{u} + \mathbf{v}\|^2$ (j) $d(\mathbf{u}, \mathbf{v})$

3. Let $\mathbf{u} = (-1 \;\; 2 \;\; 5 \;\; -3)^T$ and $\mathbf{v} = (2 \;\; -3 \;\; -1 \;\; 5)^T$ be vectors in $\mathbb{R}^4$. Evaluate
   (a) $\mathbf{u} \cdot \mathbf{v}$ (b) $\mathbf{v} \cdot \mathbf{u}$ (c) $\mathbf{u} \cdot \mathbf{u}$ (d) $\mathbf{v} \cdot \mathbf{v}$ (e) $\|\mathbf{u}\|^2$
   (f) $\|\mathbf{v}\|^2$ (g) $\|\mathbf{u}\|$ (h) $\|\mathbf{v}\|$ (i) $\|\mathbf{u} + \mathbf{v}\|^2$ (j) $d(\mathbf{u}, \mathbf{v})$

4. Let $\mathbf{i} = \begin{pmatrix} 1 \\ 0 \\ 0 \end{pmatrix}$, $\mathbf{j} = \begin{pmatrix} 0 \\ 1 \\ 0 \end{pmatrix}$ and $\mathbf{k} = \begin{pmatrix} 0 \\ 0 \\ 1 \end{pmatrix}$. Show that

   (a) $\mathbf{i} \cdot \mathbf{i} = 1$ (b) $\mathbf{j} \cdot \mathbf{j} = 1$ (c) $\mathbf{k} \cdot \mathbf{k} = 1$ (d) $\mathbf{i} \cdot \mathbf{j} = 0$ (e) $\mathbf{i} \cdot \mathbf{k} = 0$ (f) $\mathbf{j} \cdot \mathbf{k} = 0$
   If $\mathbf{u} = a\mathbf{i} + b\mathbf{j} + c\mathbf{k}$ and $\mathbf{v} = d\mathbf{i} + e\mathbf{j} + f\mathbf{k}$ show that $\mathbf{u} \cdot \mathbf{v} = ad + be + cf$.

5. Let $\mathbf{u} = \begin{pmatrix} -3 \\ 2 \end{pmatrix}$ and $\mathbf{v} = \begin{pmatrix} -2 \\ -3 \end{pmatrix}$ be vectors in $\mathbb{R}^2$. Plot these vectors in $\mathbb{R}^2$ and show that the vectors are orthogonal. Hence solve the equation $-3x + 2y = 0$.

6. Let $\mathbf{u} = \begin{pmatrix} 7 \\ -2 \end{pmatrix}$ and $\mathbf{v} = \begin{pmatrix} -5 \\ 3 \end{pmatrix}$. Determine the following norms and plot your result on the same axes:
   (a) $\|\mathbf{u} + \mathbf{v}\|$ (b) $d(\mathbf{u}, \mathbf{v}) = \|\mathbf{u} - \mathbf{v}\|$

7. Prove properties (ii), (v) and (vii) of Proposition (2.1).

8. Prove property (i) of Proposition (2.6).

9. Determine $\frac{1}{\|\mathbf{u}\|}\mathbf{u}$ for the following vectors in $\mathbb{R}^n$:
   (a) $\mathbf{u} = (2 \;\; -7)^T$ (b) $\mathbf{u} = (-9 \;\; 3 \;\; 7)^T$ (c) $\mathbf{u} = (-3 \;\; 5 \;\; 8 \;\; 6)^T$
   (d) $\mathbf{u} = (-6 \;\; 2 \;\; 8 \;\; 3 \;\; 5)^T$

   Determine $\left\| \frac{1}{\|\mathbf{u}\|}\mathbf{u} \right\|$ in each case.

What do you notice about your results?

10. Show that, for any non-zero vector $\mathbf{u}$ in $\mathbb{R}^n$, we have $\left\| \frac{1}{\|\mathbf{u}\|}\mathbf{u} \right\| = 1$.

11. Let $\mathbf{u}$ and $\mathbf{v}$ be vectors in $\mathbb{R}^n$. *Disprove* the following propositions:
    (a) If $\mathbf{u} \cdot \mathbf{v} = 0$ then $\mathbf{u} = \mathbf{O}$ or $\mathbf{v} = \mathbf{O}$.
    (b) $\|\mathbf{u} + \mathbf{v}\| = \|\mathbf{u}\| + \|\mathbf{v}\|$

12. Let $\mathbf{u}_1, \mathbf{u}_2, \mathbf{u}_3, \ldots, \mathbf{u}_n$ be orthogonal vectors in $\mathbb{R}^n$. Prove
    (i) $\|\mathbf{u}_1 + \mathbf{u}_2\|^2 = \|\mathbf{u}_1\|^2 + \|\mathbf{u}_2\|^2$
    *(ii) $\|\mathbf{u}_1 + \mathbf{u}_2 + \cdots + \mathbf{u}_n\|^2 = \|\mathbf{u}_1\|^2 + \|\mathbf{u}_2\|^2 + \cdots + \|\mathbf{u}_n\|^2$
    For part (ii) use mathematical induction.

13. Let $\mathbf{u}$ and $\mathbf{v}$ be vectors in $\mathbb{R}^n$. Prove that
(a) $\|\mathbf{u}+\mathbf{v}\|^2 + \|\mathbf{u}-\mathbf{v}\|^2 = 2\|\mathbf{u}\|^2 + 2\|\mathbf{v}\|^2$ (b) $\|\mathbf{u}+\mathbf{v}\|^2 - \|\mathbf{u}-\mathbf{v}\|^2 = 4(\mathbf{u}\cdot\mathbf{v})$
14. This question is on the properties of the distance function $d(\mathbf{u}\ \mathbf{v})$ where $\mathbf{u}$ and $\mathbf{v}$ are vectors in n-space.
(i) Show that $d(\mathbf{u}, \mathbf{v}) = \sqrt{(u_1 - v_1)^2 + (u_2 - v_2)^2 + \cdots + (u_n - v_n)^2}$.
(ii) Prove that $d(\mathbf{u}, \mathbf{v}) = d(\mathbf{v}, \mathbf{u})$.
(iii) Prove that $d(\mathbf{u}, \mathbf{v}) \geq 0$ and $d(\mathbf{u}, \mathbf{v}) = 0 \Leftrightarrow \mathbf{u} = \mathbf{v}$.

SECTION 2.2 Further Properties of Vectors

By the end of this section you will be able to

- evaluate the angle between two vectors in $\mathbb{R}^n$
- prove inequalities associated with the inner product and norm

In the last section we mentioned that the dot product has a geometric significance in $\mathbb{R}^2$ and $\mathbb{R}^3$, explained in subsection 2.2.2. In this section, we discuss the angle between two vectors. Additionally, we examine normalization, a process that standardizes the length of the vectors, while retaining information about their direction. This simplification is useful when we are interested primarily in the direction of vectors.

Subsection 2.2.3, which deals with inequalities and proofs, requires you to recall some of the properties of dot products and norms of vectors.

2.2.1 Revision of norm and dot product

We first do a numerical example of norms (lengths) and dot product. Remember, the norm or length of a vector $\mathbf{v}$ is denoted by $\|\mathbf{v}\|$ and the dot product with a dot $\cdot$ between the two vectors.

Example 2.6

Let $\mathbf{u} = \begin{pmatrix} 1 \\ 5 \end{pmatrix}$ and $\mathbf{v} = \begin{pmatrix} 4 \\ 1 \end{pmatrix}$ be in $\mathbb{R}^2$. Determine the following:

(i) $|\mathbf{u}\cdot\mathbf{v}|$ **(ii)** $\|\mathbf{u}\|\,\|\mathbf{v}\|$ **(iii)** $\|\mathbf{u}\| + \|\mathbf{v}\|$ **(iv)** $\|\mathbf{u}+\mathbf{v}\|$

Solution

(i) We have $\mathbf{u}\cdot\mathbf{v} = \begin{pmatrix} 1 \\ 5 \end{pmatrix} \cdot \begin{pmatrix} 4 \\ 1 \end{pmatrix} = (1\times 4) + (5 \times 1) = 9$ so $|\mathbf{u}\cdot\mathbf{v}| = |9| = 9$.

(continued...)

(ii) Remember, the symbol $\|\ \|$ means the length of the vector. By Pythagoras (2.7) we have

$$\|\mathbf{u}\| = \sqrt{1^2+5^2} = \sqrt{26} \quad \text{and} \quad \|\mathbf{v}\| = \sqrt{4^2+1^2} = \sqrt{17}$$

Multiplying these we have $\|\mathbf{u}\|\,\|\mathbf{v}\| = \sqrt{26}\times\sqrt{17} = 21.02$ (2dp).

(iii) Adding both the above results in part (ii) we have

$$\|\mathbf{u}\| + \|\mathbf{v}\| = \sqrt{26}+\sqrt{17} = 9.22 \text{ (2dp)}$$

(iv) Similarly, by using Pythagoras:

$$\|\mathbf{u}+\mathbf{v}\| = \left\|\begin{pmatrix}1\\5\end{pmatrix}+\begin{pmatrix}4\\1\end{pmatrix}\right\| = \left\|\begin{pmatrix}5\\6\end{pmatrix}\right\| = \sqrt{5^2+6^2} = \sqrt{61} = 7.81 \text{ (2 dp)}$$

We can illustrate these in $\mathbb{R}^2$ (Fig. 2.10):

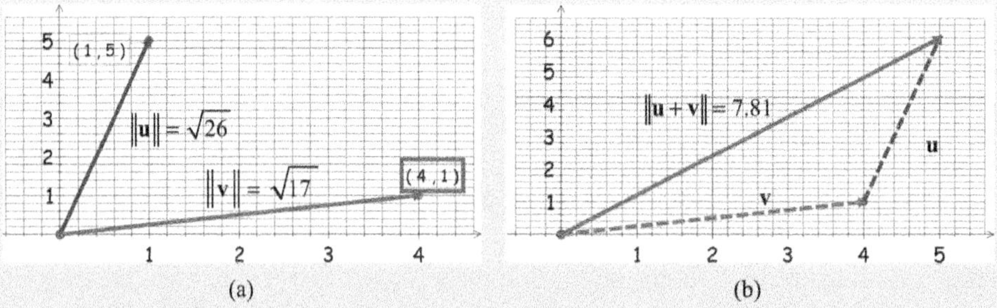

Figure 2.10

2.2.2 Angle between two vectors

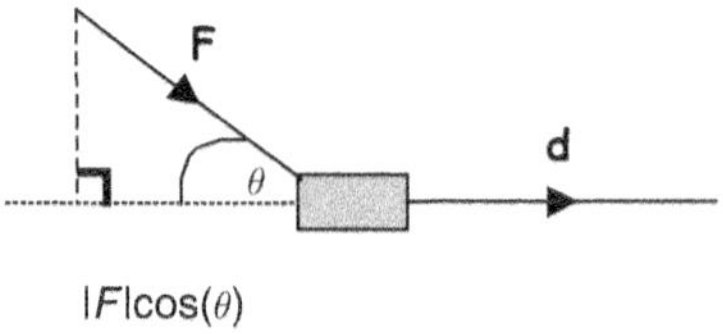

Figure 2.11 shows a constant force **F** applied to an object and, as a result, the object moves a distance $\|\mathbf{d}\|$. The work done by the force is the product of the magnitude of the force in the line of action, $\|\mathbf{F}\|\cos(\theta)$, and the distance an object has moved $\|\mathbf{d}\|$. Hence

$$\text{Work done} = \|\mathbf{F}\|\cos(\theta)\times\|\mathbf{d}\|$$

Work is defined as the product of the force applied in a particular direction and the distance it moves in that direction.

A real-world example of this is to imagine a rope tied to an object, perhaps a barge. You need to move the barge by pulling the rope (Fig. 2.12).

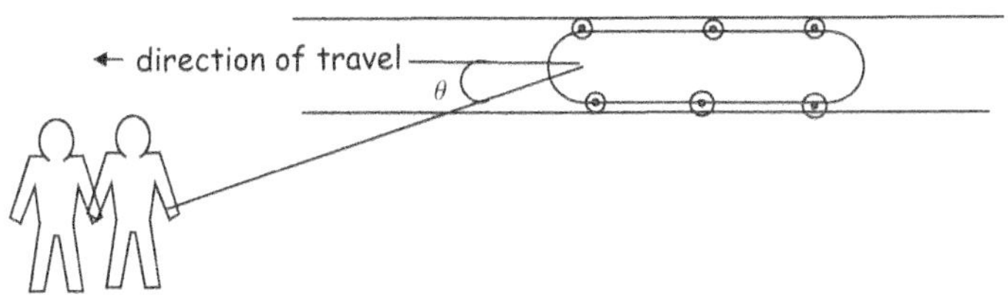

Figure 2.12

If you could stand directly behind the barge and push, then all your force would be in the direction of movement.

In this case, **F** and **d** are parallel and the angle between them is 0° so we have

$$\text{Work done} = \|\mathbf{F}\|\cos(0^\circ) \times \|\mathbf{d}\| = \|\mathbf{F}\| \times \|\mathbf{d}\| \qquad [\text{because } \cos(0^\circ) = 1]$$

This is the least possible force used to push the object because we are pushing in the same direction as we would like the object to move.

If you push the barge in a direction perpendicular (orthogonal) to the canal, it would not move forward at all, so you would do no work because

$$\text{Work done} = \|\mathbf{F}\|\cos(90^\circ) \times \|\mathbf{d}\| = 0 \qquad [\text{because } \cos(90^\circ) = 0]$$

The actual amount of work done in moving the barge along the canal is a value somewhere between these two possibilities, and is given by the angle the rope makes with the direction of the canal.

How can we find the angle θ between two vectors in the above diagrams?

To find the angle between two vectors we need to use the cosine rule from trigonometry (Fig. 2.13):

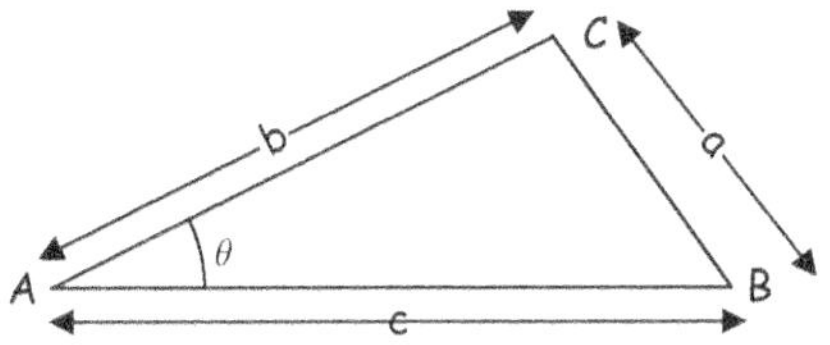

Figure 2.13

With reference to Fig. 2.13, the cosine rule from trigonometry states that

$$a^2 = b^2 + c^2 - 2bc\cos(\theta)$$

Consider two non-zero vectors $\mathbf{u} = (u_1, u_2)$ and $\mathbf{v} = (v_1, v_2)$ in $\mathbb{R}^2$ with an angle θ between them (Fig. 2.14):

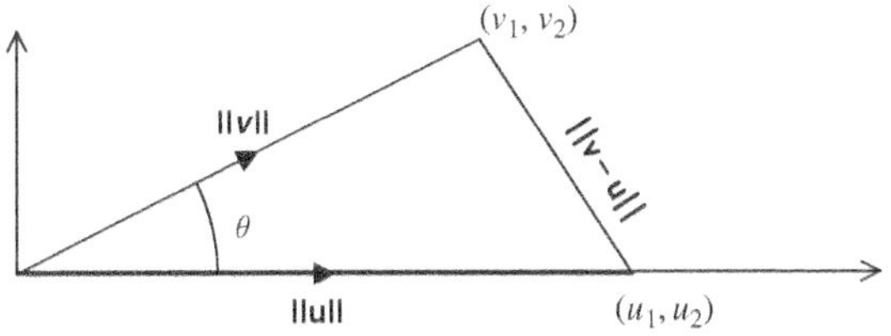

Figure 2.14

By applying the above cosine rule to this triangle we have

$$\|\mathbf{v}-\mathbf{u}\|^2 = \|\mathbf{v}\|^2 + \|\mathbf{u}\|^2 - 2\|\mathbf{v}\|\,\|\mathbf{u}\|\cos(\theta)$$

Rearranging this:

$$\begin{aligned}\|\mathbf{v}\|\,\|\mathbf{u}\|\cos(\theta) &= \frac{1}{2}(\|\mathbf{v}\|^2 + \|\mathbf{u}\|^2 - \|\mathbf{v}-\mathbf{u}\|^2)\\ &= \frac{1}{2}\left[v_1^2+v_2^2+u_1^2+u_2^2-(v_1-u_1)^2-(v_2-u_2)^2\right]\\ &= \frac{1}{2}\left[v_1^2+v_2^2+u_1^2+u_2^2-\left(v_1^2-2u_1v_1+u_1^2\right)-\left(v_2^2-2u_2v_2+u_2^2\right)\right]\\ &= \frac{1}{2}\left[2u_1v_1+2u_2v_2\right] = u_1v_1+u_2v_2\\ \|\mathbf{v}\|\,\|\mathbf{u}\|\cos(\theta) &= u_1v_1+u_2v_2 \underbrace{=}_{\text{By (2.4)}} \mathbf{u}\cdot\mathbf{v}\end{aligned}$$

Therefore the dot product is

$$\mathbf{u}\cdot\mathbf{v} = \|\mathbf{u}\|\,\|\mathbf{v}\|\cos(\theta) \tag{2.11}$$

What does this formula mean?

It means that the dot (inner) product of two vectors $\mathbf{u}$ and $\mathbf{v}$ is the length of vector $\mathbf{u}$ times the length of vector $\mathbf{v}$ times the cosine of the angle θ between them.

In the example of pulling a barge, the work done, $W = \|\mathbf{F}\|\,\|\mathbf{d}\|\cos(\theta)$, in moving the barge a distance $\|\mathbf{d}\|$ is given by the dot product:

$$W = \|\mathbf{F}\|\,\|\mathbf{d}\|\cos(\theta) = \mathbf{F}\cdot\mathbf{d}$$

where $\mathbf{F}$ is the constant (pulling) force applied to the barge. Hence, the work done is given by the dot product of force $\mathbf{F}$ and distance $\mathbf{d}$.

In physical terms, we can define the dot product of two vectors $\mathbf{u}$ and $\mathbf{v}$ as the work done by $\mathbf{v}$ in moving the object a distance of $\|\mathbf{u}\|$ in the direction of $\mathbf{u}$ (Fig. 2.15).

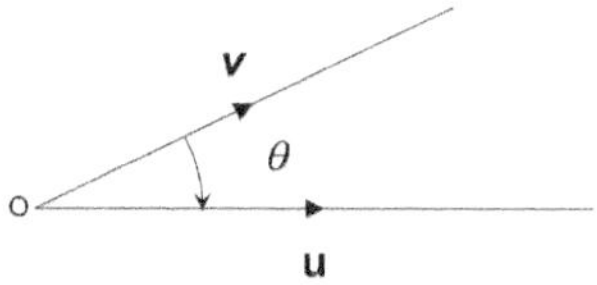

Figure 2.15

Note the following:

(a) If two vectors $\mathbf{u}$ and $\mathbf{v}$ are pointing in same directions then the dot product is positive. The two vectors are working (pushing or pulling) in the same direction (Fig. 2.16(a)).

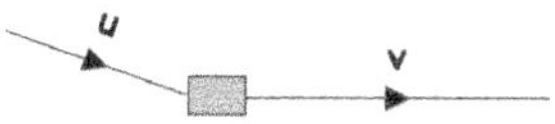

Figure 2.16(a)

(b) If two vectors **u** and **v** are perpendicular (orthogonal) then the dot product is 0 because $\cos(90°) = 0$. (The force applied in the direction of the vector **u** contributes nothing to the motion in the direction of vector **v**.) (Fig. 2.16(b))

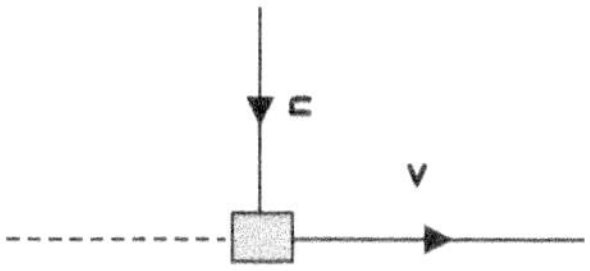

Figure 2.16(b)

(c) If two vectors **u** and **v** are pointing in opposite directions then the dot product is negative (Fig. 2.16(c)).

Figure 2.16(c)

The two vectors are working (pulling) in opposite directions.

When we talk about the angle θ between two vectors we mean the angle which lies between 0° and 180° (or 0 to π radians), as shown in Fig. 2.17:

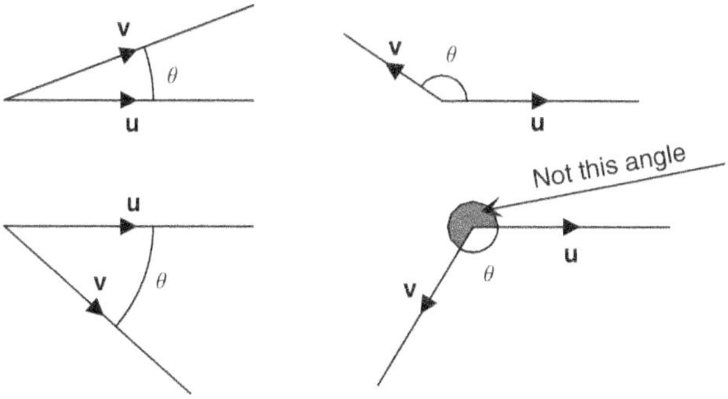

Figure 2.17

Making $\cos(\theta)$ the subject of formula (2.11) we have

(2.12) $$\cos(\theta) = \frac{\mathbf{u}\cdot\mathbf{v}}{\|\mathbf{u}\|\,\|\mathbf{v}\|}$$ provided **u** and **v** are non-zero vectors.

For the above Example 2.6 the angle θ between the vectors $\mathbf{u} = \begin{pmatrix}1\\5\end{pmatrix}$ and $\mathbf{v} = \begin{pmatrix}4\\1\end{pmatrix}$ is

$$\cos(\theta) = \frac{\mathbf{u}\cdot\mathbf{v}}{\|\mathbf{u}\|\,\|\mathbf{v}\|} = \frac{9}{21.02} \quad \left[\text{because } \mathbf{u}\cdot\mathbf{v} \underbrace{=}_{\text{By part (i)}} 9 \text{ and } \|\mathbf{u}\|\,\|\mathbf{v}\| \underbrace{=}_{\text{By part (ii)}} 21.02\right]$$

Taking the inverse cosine of both sides gives $\theta = \cos^{-1}\left(\frac{9}{21.02}\right) = 64.88°$ (see Fig. 2.10(a)).

Example 2.7

Let $\mathbf{u} = \begin{pmatrix} -5 \\ -1 \end{pmatrix}$ and $\mathbf{v} = \begin{pmatrix} 4 \\ 2 \end{pmatrix}$ be vectors in $\mathbb{R}^2$. Determine the angle between these vectors and label the angle and the vectors $\mathbf{u}$ and $\mathbf{v}$ in $\mathbb{R}^2$.

Solution

How do we find the angle between the given vectors?

By using formula (2.12) $\cos(\theta) = \dfrac{\mathbf{u} \cdot \mathbf{v}}{\|\mathbf{u}\| \|\mathbf{v}\|}$.

We need to find $\mathbf{u} \cdot \mathbf{v}$, which is the dot product, and lengths $\|\mathbf{u}\|$, $\|\mathbf{v}\|$ of vectors $\mathbf{u}$ and $\mathbf{v}$.

What is $\mathbf{u} \cdot \mathbf{v}$ equal to?

Using formula (2.4) $\mathbf{u} \cdot \mathbf{v} = u_1 v_1 + u_2 v_2 + \cdots + u_n v_n$ we have

$$\mathbf{u} \cdot \mathbf{v} = \begin{pmatrix} -5 \\ -1 \end{pmatrix} \cdot \begin{pmatrix} 4 \\ 2 \end{pmatrix} = (-5 \times 4) + (-1 \times 2) = -22$$

(The negative dot product means the two vectors are working in opposite directions.)

How can we determine the norm (length) of the vectors $\mathbf{u}$ and $\mathbf{v}$?

By applying Pythagoras' theorem (2.7) from the last section:

$$\|\mathbf{u}\| = \left\| \begin{pmatrix} -5 \\ -1 \end{pmatrix} \right\| = \sqrt{(-5)^2 + (-1)^2} = \sqrt{26}$$

$$\|\mathbf{v}\| = \left\| \begin{pmatrix} 4 \\ 2 \end{pmatrix} \right\| = \sqrt{4^2 + 2^2} = \sqrt{20}$$

Substituting $\mathbf{u} \cdot \mathbf{v} = -22$, $\|\mathbf{u}\| = \sqrt{26}$ and $\|\mathbf{v}\| = \sqrt{20}$ into $\cos(\theta) = \dfrac{\mathbf{u} \cdot \mathbf{v}}{\|\mathbf{u}\| \|\mathbf{v}\|}$ gives

$$\cos(\theta) = \frac{-22}{\sqrt{26}\sqrt{20}} = -\frac{22}{\sqrt{26 \times 20}} = -0.965$$

What is the angle θ equal to?

$$\theta = \cos^{-1}(-0.965) = 164.74^\circ \quad \text{[Taking inverse cos]}$$

Plotting these vectors and labelling the angle between them in $\mathbb{R}^2$ gives (Fig. 2.18):

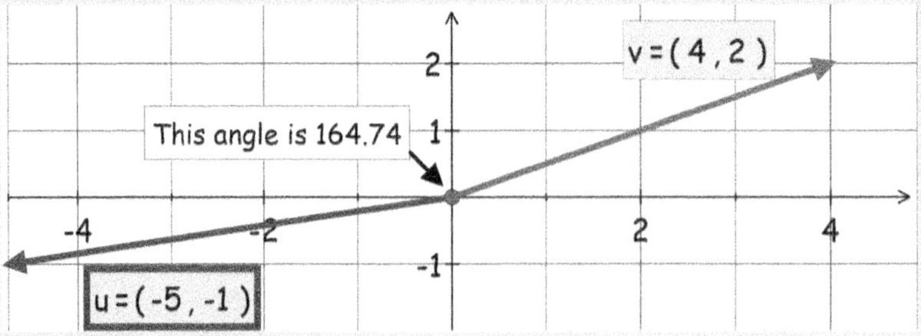

Figure 2.18

(Note, the two vectors are pulling in opposite directions.)

The formula for the angle between two non-zero vectors can be extended to any n-space, that is let $\mathbf{u}$ and $\mathbf{v}$ be any non-zero vectors in $\mathbb{R}^n$ then

$$(2.13) \qquad \cos(\theta) = \frac{\mathbf{u} \cdot \mathbf{v}}{\|\mathbf{u}\| \|\mathbf{v}\|} \quad \text{where } 0 \le \theta \le \pi \text{ (radians)}$$

In the exercises, we show that this angle, θ, between the vectors is unique.

Example 2.8

Let $\mathbf{u} = \begin{pmatrix} 3 \\ -1 \\ 7 \end{pmatrix}$ and $\mathbf{v} = \begin{pmatrix} -2 \\ 1 \\ 9 \end{pmatrix}$ be vectors in $\mathbb{R}^3$. Determine the angle between these vectors.

Solution

The procedure is very similar to Example 2.7. We need to use the above formula (2.13):

$$\cos(\theta) = \frac{\mathbf{u} \cdot \mathbf{v}}{\|\mathbf{u}\| \|\mathbf{v}\|}.$$

What are the dot product $\mathbf{u} \cdot \mathbf{v}$ and the norms $\|\mathbf{u}\|$ and $\|\mathbf{v}\|$ equal to?

Applying dot product formula (2.4) we have

$$\mathbf{u} \cdot \mathbf{v} = \begin{pmatrix} 3 \\ -1 \\ 7 \end{pmatrix} \cdot \begin{pmatrix} -2 \\ 1 \\ 9 \end{pmatrix} = (3 \times (-2)) + (-1 \times 1) + (7 \times 9) = 56$$

To evaluate the lengths we use Pythagoras' theorem (2.7):

$$\|\mathbf{u}\| = \sqrt{3^2 + (-1)^2 + 7^2} = \sqrt{59} \text{ and } \|\mathbf{v}\| = \sqrt{(-2)^2 + 1^2 + 9^2} = \sqrt{86}$$

Substituting $\mathbf{u} \cdot \mathbf{v} = 56$, $\|\mathbf{u}\| = \sqrt{59}$ and $\|\mathbf{v}\| = \sqrt{86}$ into $\cos(\theta) = \frac{\mathbf{u} \cdot \mathbf{v}}{\|\mathbf{u}\| \|\mathbf{v}\|}$ gives

$$\cos(\theta) = \frac{56}{\sqrt{59}\sqrt{86}} = \frac{56}{\sqrt{59 \times 86}} = 0.786$$

How do we find the angle θ?

$$\theta = \cos^{-1}(0.786) = 38.19^\circ \quad \text{[inverse cos]}$$

2.2.3 Inequalities

Next, we prove some inequalities in relation to the dot product and norm (length) of vectors.

Cauchy–Schwarz inequality (2.14). Let $\mathbf{u}$ and $\mathbf{v}$ be vectors in $\mathbb{R}^n$ then

$$|\mathbf{u} \cdot \mathbf{v}| \le \|\mathbf{u}\| \|\mathbf{v}\|$$

For the above Example 2.6 we had

$$|\mathbf{u}\cdot\mathbf{v}| \underbrace{=}_{\text{By part (i)}} 9 \leq 21.02 \underbrace{=}_{\text{By part (ii)}} \|\mathbf{u}\|\,\|\mathbf{v}\|$$

Proof.

How can we prove this inequality for any vectors in $\mathbb{R}^n$?

If the vectors $\mathbf{u}$ and $\mathbf{v}$ are non-zero then we can use the above formula:

(2.11) $$\mathbf{u}\cdot\mathbf{v} = \|\mathbf{u}\|\,\|\mathbf{v}\|\cos(\theta)$$

Taking the modulus of both sides we have

$$|\mathbf{u}\cdot\mathbf{v}| = |\|\mathbf{u}\|\,\|\mathbf{v}\|\cos(\theta)|$$

The lengths are positive or zero, $\|\mathbf{u}\| \geq 0$ and $\|\mathbf{v}\| \geq 0$, therefore the modulus of these is just $\|\mathbf{u}\|$ and $\|\mathbf{v}\|$ respectively.

Why?

Because if $x \geq 0$ then $|x| = x$. Hence we have $|\mathbf{u}\cdot\mathbf{v}| = \|\mathbf{u}\|\,\|\mathbf{v}\|\ |\cos(\theta)|$.

What can we say about the size of $|\cos(\theta)|$ for any real angle θ?

From trigonometry we know that the cosine of any real angle cannot be greater than 1:

$$0 \leq |\cos(\theta)| \leq 1 \qquad \text{[Lies between 0 and 1]}$$

By substituting the right hand inequality $|\cos(\theta)| \leq 1$ into the above derived equation $|\mathbf{u}\cdot\mathbf{v}| = \|\mathbf{u}\|\,\|\mathbf{v}\|\ |\cos(\theta)|$ we have

$$|\mathbf{u}\cdot\mathbf{v}| \leq \|\mathbf{u}\|\,\|\mathbf{v}\|\ (1) = \|\mathbf{u}\|\,\|\mathbf{v}\|$$

Hence we have proven the Cauchy–Schwarz inequality, $|\mathbf{u}\cdot\mathbf{v}| \leq \|\mathbf{u}\|\,\|\mathbf{v}\|$, for non-zero vectors $\mathbf{u}$ and $\mathbf{v}$.

If $\mathbf{u} = \mathbf{O}$ or $\mathbf{v} = \mathbf{O}$ then

$$\mathbf{u}\cdot\mathbf{v} = 0 \text{ and therefore } |\mathbf{u}\cdot\mathbf{v}| = 0 \text{ and } \|\mathbf{u}\|\,\|\mathbf{v}\| = 0$$

Cauchy–Schwarz inequality, $|\mathbf{u}\cdot\mathbf{v}| \leq \|\mathbf{u}\|\,\|\mathbf{v}\|$, holds in this case as well because $0 \leq 0$.

■

The Cauchy–Schwarz inequality claims that the dot product is less than or equal to the multiplication of lengths of the vectors.

Figure 2.19 (a) Cauchy 1789–1857.

Augustin Cauchy (Fig. 2.19(a)) was born in 1789 in Paris, France. He took the entrance exam for the prestigious Ecole Polytechnique in 1805 and graduated in 1807. For the next eight years he had various posts in the field of engineering but he was persuaded by Lagrange and Laplace to convert to mathematics. In 1815, he became Assistant Professor of Mathematics Analysis at the Ecole Polytechnique and he was the first person to make analysis rigorous.

Figure 2.19 (b) Schwarz 1843–1921.

Hermann Schwarz (Fig. 2.19(b)) was born in 1843 in Poland and wanted to study chemistry at university. However, Weierstrass and Kummer convinced him to study mathematics at the Technical University of Berlin.

In 1864, he received his doctorate under the supervision of Weierstrass and completed a teaching qualification in 1867. Schwarz held various posts at the University of Halle, and in 1875 he managed to secure the post of Chair of Mathematics at Göttingen University. At that time, Göttingen University was building a prolific reputation in the field of mathematics but in 1892 he left Göttingen to take up a professorship at the University of Berlin.

Next we prove an inequality dealing with addition of lengths of vectors rather than multiplication.

To establish the next inequality we need to use the following results of inequalities from real numbers:

$$|x + y + z| \le |x| + |y| + |z| \qquad (^*)$$

$$x \le y \Leftrightarrow \sqrt{x} \le \sqrt{y} \text{ provided } x \ge 0,\ y \ge 0 \qquad (^{**})$$

Minkowski (triangular) inequality (2.15). Let $\mathbf{u}$ and $\mathbf{v}$ be vectors in $\mathbb{R}^n$ then

$$\|\mathbf{u}+\mathbf{v}\| \le \|\mathbf{u}\| + \|\mathbf{v}\|$$

What does this mean in $\mathbb{R}^2$?

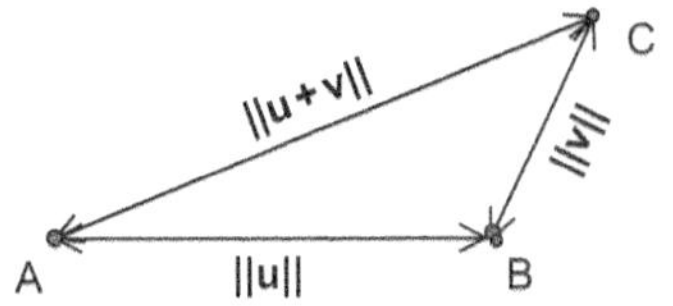

Figure 2.20

The length (norm) $\|\mathbf{u}+\mathbf{v}\|$ is less than or equal to the other two lengths (norms) added together (Fig. 2.20). (If a donkey is at station A and the food is located at station C then the donkey would travel the shortest distance along AC not AB and then BC, because the route AB–BC would always be at least as long, or greater.)

In the above Example 2.6 for Fig. 2.10(b) we had:

$$\|\mathbf{u}+\mathbf{v}\| \underbrace{=}_{\text{By part (iv)}} 7.81 \le 9.22 \underbrace{=}_{\text{By part (iii)}} \|\mathbf{u}\| + \|\mathbf{v}\|$$

Proof.

How can we prove this inequality?

We examine $\|\mathbf{u}+\mathbf{v}\|^2$ and show that this is less than or equal to $(\|\mathbf{u}\|+\|\mathbf{v}\|)^2$ and then take the square root of both sides to get our inequality.

What is $\|\mathbf{u}+\mathbf{v}\|^2$ equal to?

$$\begin{aligned}\|\mathbf{u}+\mathbf{v}\|^2 &= (\mathbf{u}+\mathbf{v})\cdot(\mathbf{u}+\mathbf{v}) \qquad \left[\text{By (2.8)} \quad \|\mathbf{u}\| = \sqrt{\mathbf{u}\cdot\mathbf{u}}\right]\\ &= (\mathbf{u}\cdot\mathbf{u}) + \underbrace{\mathbf{u}\cdot\mathbf{v}+\mathbf{v}\cdot\mathbf{u}}_{=2\mathbf{u}\cdot\mathbf{v}} + (\mathbf{v}\cdot\mathbf{v})\\ &= \|\mathbf{u}\|^2 + 2(\mathbf{u}\cdot\mathbf{v}) + \|\mathbf{v}\|^2 \qquad \left[\text{Because } \mathbf{u}\cdot\mathbf{u} = \|\mathbf{u}\|^2 \text{ and } \mathbf{v}\cdot\mathbf{v} = \|\mathbf{v}\|^2\right]\end{aligned}$$

We can take the modulus of both sides. Since $\|\mathbf{u}+\mathbf{v}\|^2 \ge 0$, the modulus of this is the same, that is $\left|\|\mathbf{u}+\mathbf{v}\|^2\right| = \|\mathbf{u}+\mathbf{v}\|^2$. We have

$$\begin{aligned}\|\mathbf{u}+\mathbf{v}\|^2 &= \left|\|\mathbf{u}\|^2 + 2(\mathbf{u}\cdot\mathbf{v}) + \|\mathbf{v}\|^2\right|\\ &\le \left|\|\mathbf{u}\|^2\right| + |2\mathbf{u}\cdot\mathbf{v}| + \left|\|\mathbf{v}\|^2\right| \qquad \left[\text{applying } |x+y+z| \le |x|+|y|+|z| \quad (*)\right]\end{aligned}$$

Because $\|\mathbf{u}\|^2 \ge 0$ and $\|\mathbf{v}\|^2 \ge 0$, the modulus of these is the same, that is

$$\left|\|\mathbf{u}\|^2\right| = \|\mathbf{u}\|^2 \text{ and } \left|\|\mathbf{v}\|^2\right| = \|\mathbf{v}\|^2$$

The middle term on the right hand side is $|2\mathbf{u}\cdot\mathbf{v}| = 2|\mathbf{u}\cdot\mathbf{v}|$. Hence we have

$$\|\mathbf{u}+\mathbf{v}\|^2 \le \|\mathbf{u}\|^2 + 2|\mathbf{u}\cdot\mathbf{v}| + \|\mathbf{v}\|^2$$

By applying the above Cauchy–Schwarz inequality (2.14) $|\mathbf{u} \cdot \mathbf{v}| \le \|\mathbf{u}\| \, \|\mathbf{v}\|$ to this middle term in the last line we have

$$\begin{aligned}\|\mathbf{u}+\mathbf{v}\|^2 &\le \|\mathbf{u}\|^2 + 2|\mathbf{u}\cdot\mathbf{v}| + \|\mathbf{v}\|^2 \\ &\le \|\mathbf{u}\|^2 + 2\|\mathbf{u}\| \, \|\mathbf{v}\| + \|\mathbf{v}\|^2 \quad \left[\text{applying Cauchy–Schwarz } |\mathbf{u}\cdot\mathbf{v}| \le \|\mathbf{u}\|\|\mathbf{v}\|\right] \\ &= (\|\mathbf{u}\| + \|\mathbf{v}\|)^2 \quad \left[\text{using } a^2+2ab+b^2=(a+b)^2\right]\end{aligned}$$

Taking the square root of this $\|\mathbf{u}+\mathbf{v}\|^2 \le (\|\mathbf{u}\| + \|\mathbf{v}\|)^2$ gives our required inequality

$$\|\mathbf{u}+\mathbf{v}\| \le \|\mathbf{u}\| + \|\mathbf{v}\| \qquad [\text{By } (^{**})]$$

■

Next we give a brief biography of Minkowski (pronounced 'Minkofski').

Figure 2.21 Minkowski.

Hermann Minkowski (Fig. 2.21) was born in 1864 to Jewish parents in Lithuania, but received his education in Germany at the University of Königsberg.

At the age of 19, while still a student, Minkowski won a prize from the French Academy of Sciences. In 1885, he received his doctorate under the supervision of Lindemann. (Lindemann was the first to prove the transcendence of π, in 1882.)

Minkowski taught at the universities of Bonn, Königsberg, Zurich and finally settled in Göttingen in 1902, where he secured a post in the Mathematics Department. He become a close friend of David Hilbert.

Minkowski died at the young age of 44.

Note, that the Cauchy–Schwarz inequality is about the lengths of products of vectors:

$$|\mathbf{u}\cdot\mathbf{v}| \le \|\mathbf{u}\|\|\mathbf{v}\|$$

while the Minkowski inequality is about the lengths of addition of vectors:

$$\|\mathbf{u}+\mathbf{v}\| \le \|\mathbf{u}\| + \|\mathbf{v}\|$$

2.2.4 Unit vectors

A vector of length 1 is called a **unit vector**. In Exercises 2.1, we showed that for any non-zero vector $\mathbf{u}$ in $\mathbb{R}^n$ we have $\left\| \frac{1}{\|\mathbf{u}\|}\mathbf{u} \right\| = 1$.

What does this mean?

It means that we can always find a unit vector in the *direction* of any non-zero vector **u** by dividing the given vector by its length $\|\mathbf{u}\|$ (Fig. 2.22).

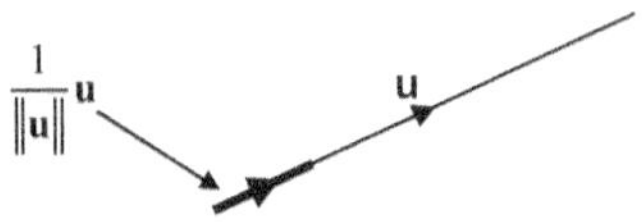

Figure 2.22

For example, a vector in a particular direction of length 5 can be divided by 5 to give a vector in the same direction but length 1 (unit vector).

The process of finding a unit vector in the direction of the given vector **u** is called **normalizing**. The unit vector in the direction of the vector **u** is normally denoted by $\hat{\mathbf{u}}$ (pronounced as 'u hat') meaning it is a vector of length 1, that is

$$\hat{\mathbf{u}} = \frac{1}{\|\mathbf{u}\|}\mathbf{u} \tag{2.16}$$

Later on in this chapter we will see that normalizing vectors simplifies calculations. Examples of unit vectors are shown in Fig. 2.23.

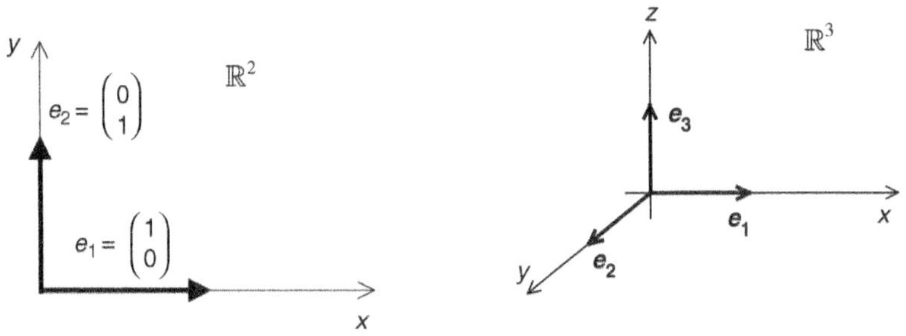

Figure 2.23

The vectors shown in Fig. 2.23, $\mathbf{e}_1 = \begin{pmatrix}1\\0\end{pmatrix}$ and $\mathbf{e}_2 = \begin{pmatrix}0\\1\end{pmatrix}$, are unit vectors in $\mathbb{R}^2$, and $\mathbf{e}_1 = (1 \quad 0 \quad 0)^T$, $\mathbf{e}_2 = (0 \quad 1 \quad 0)^T$ and $\mathbf{e}_3 = (0 \quad 0 \quad 1)^T$ are unit vectors in $\mathbb{R}^3$. These are normally called the **standard** unit vectors.

For any n space, $\mathbb{R}^n$, the standard unit vectors are defined by

$$\mathbf{e}_1 = \begin{pmatrix}1\\0\\\vdots\\\\0\end{pmatrix}, \ \mathbf{e}_2 = \begin{pmatrix}0\\1\\0\\\vdots\\0\end{pmatrix}, \ldots \ \mathbf{e}_k = \begin{pmatrix}0\\\vdots\\1\\0\\\vdots\end{pmatrix}, \ \cdots, \ \mathbf{e}_n = \begin{pmatrix}0\\\vdots\\0\\0\\1\end{pmatrix}$$

That is, we have 1 in the kth position of the vector $\mathbf{e}_k$ and zeros everywhere else.

Actually these are examples of **perpendicular unit** vectors called **orthonormal** vectors, which means that they are normalized and they are orthogonal. Hence orthonormal vectors have two properties:

1. All the vectors are **orthogonal** to each other (perpendicular to each other).
2. All vectors are normalized, that is they have a norm or length of 1 (unit vectors).

Orthonormal (perpendicular unit) vectors are important in linear algebra.

2.2.5 Application of vectors

An application of vectors is the **support vector machine**, which is a computer algorithm. The algorithm produces the best hyperplane which separates data groups (or vectors). Hyperplanes are general planes in $\mathbb{R}^n$. In support vector machines, we are interested in finding the shortest distance between the hyperplane and the vectors.

A hyperplane is a general plane in n-space. In two-space it is a line, as shown in Fig. 2.24.

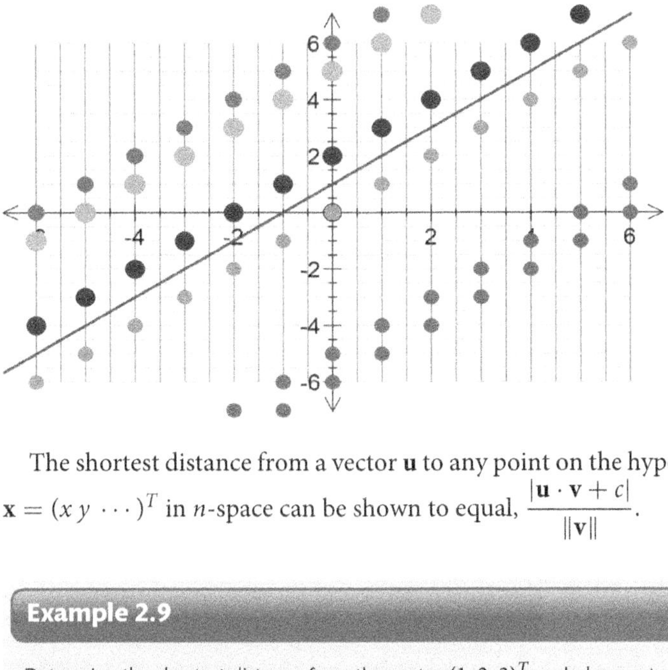

Figure 2.24

The shortest distance from a vector $\mathbf{u}$ to any point on the hyperplane $\mathbf{v} \cdot \mathbf{x} + c = 0$ where $\mathbf{x} = (x\ y\ \cdots)^T$ in n-space can be shown to equal, $\dfrac{|\mathbf{u} \cdot \mathbf{v} + c|}{\|\mathbf{v}\|}$.

Example 2.9

Determine the shortest distance from the vector $(1\ \ 2\ \ 3)^T$ and plane $x + y + z = -1$. This is the distance of the thick line shown in Fig. 2.25.

Solution

We can write the plane $x + y + z = -1$ as $x + y + z + 1 = 0$, and in dot product form:

$$\begin{pmatrix} 1 \\ 1 \\ 1 \end{pmatrix} \cdot \begin{pmatrix} x \\ y \\ z \end{pmatrix} + 1 = 0 \quad [\text{This is of the form } \mathbf{v} \cdot \mathbf{x} + c = 0]$$

(continued...)

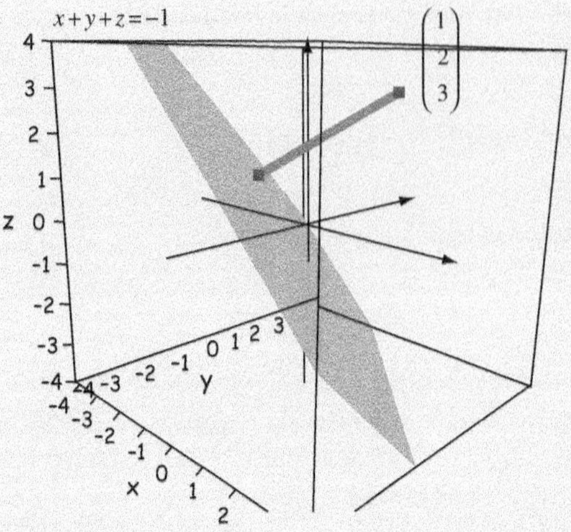

Figure 2.25

Using the above formula with $\mathbf{u} = \begin{pmatrix} 1 \\ 2 \\ 3 \end{pmatrix}$, $\mathbf{v} = \begin{pmatrix} 1 \\ 1 \\ 1 \end{pmatrix}$ and $c = 1$:

$$\frac{|\mathbf{u} \cdot \mathbf{v} + c|}{\|\mathbf{v}\|} = \frac{|(1 \times 1) + (2 \times 1) + (3 \times 1) + 1|}{\sqrt{(1)^2 + 1^2 + 1^2}} = \frac{7}{\sqrt{3}} = 4.04 \text{ (2dp)}$$

The shortest distance between the plane $x + y + z = -1$ and the vector $(1 \;\; 2 \;\; 3)^T$ is 4.04.

Summary

Let $\mathbf{u}$ and $\mathbf{v}$ be non-zero vectors in $\mathbb{R}^n$ and if θ is the angle between the vectors then

$$\cos(\theta) = \frac{\mathbf{u} \cdot \mathbf{v}}{\|\mathbf{u}\| \|\mathbf{v}\|}$$

A unit vector in the direction of the non-zero vector $\mathbf{u}$ is denoted by $\hat{\mathbf{u}}$.

EXERCISES 2.2

(Brief solutions at end of book. Full solutions available at <http://www.oup.co.uk/companion/singh>.)

You may like to check your numerical answers using the numerical software MATLAB.

*To find the angle between two vectors **u** and **v** in MATLAB enter the command:* dot(u,v)/(norm(u)*norm(v)), acosd(ans).

1. Let $\mathbf{u}$ and $\mathbf{v}$ be vectors in $\mathbb{R}^2$. For the following $\mathbf{u}$ and $\mathbf{v}$ determine the angle between the vectors and label this angle and the vectors $\mathbf{u}$ and $\mathbf{v}$ in $\mathbb{R}^2$.

   (a) $\mathbf{u} = \begin{pmatrix} 1 \\ 1 \end{pmatrix}, \mathbf{v} = \begin{pmatrix} 0 \\ 1 \end{pmatrix}$ (b) $\mathbf{u} = \begin{pmatrix} 1 \\ 0 \end{pmatrix}, \mathbf{v} = \begin{pmatrix} 0 \\ 1 \end{pmatrix}$

   (c) $\mathbf{u} = \begin{pmatrix} -2 \\ 3 \end{pmatrix}, \mathbf{v} = \begin{pmatrix} 1/2 \\ -1/2 \end{pmatrix}$

2. For the following vectors $\mathbf{u}$ and $\mathbf{v}$ in $\mathbb{R}^3$ determine the angle between them.

   (a) $\mathbf{u} = \begin{pmatrix} -1 \\ 1 \\ 3 \end{pmatrix}, \mathbf{v} = \begin{pmatrix} 3 \\ -1 \\ 5 \end{pmatrix}$ (b) $\mathbf{u} = \begin{pmatrix} 1 \\ 0 \\ 0 \end{pmatrix}, \mathbf{v} = \begin{pmatrix} 0 \\ 0 \\ 15 \end{pmatrix}$

   (c) $\mathbf{u} = \begin{pmatrix} -1 \\ 2 \\ 3 \end{pmatrix}, \mathbf{v} = \begin{pmatrix} \sqrt{2} \\ 1/\sqrt{2} \\ -1 \end{pmatrix}$

3. Find the angle between the following vectors in $\mathbb{R}^4$:

   (a) $\mathbf{u} = \begin{pmatrix} 2 & 3 & -8 & 1 \end{pmatrix}^T, \mathbf{v} = \begin{pmatrix} -1 & 2 & -5 & -3 \end{pmatrix}^T$

   (b) $\mathbf{u} = \begin{pmatrix} -2 & -3 & -1 & -1 \end{pmatrix}^T, \mathbf{v} = \begin{pmatrix} 1 & 2 & 3 & 4 \end{pmatrix}^T$

   (c) $\mathbf{u} = \begin{pmatrix} \pi & \sqrt{2} & 0 & 1 \end{pmatrix}^T, \mathbf{v} = \begin{pmatrix} 1/\pi & \sqrt{2} & -1 & 1 \end{pmatrix}^T$

4. Determine the value of k so that the following vectors are orthogonal to each other:

   (a) $\mathbf{u} = \begin{pmatrix} -1 \\ 5 \\ k \end{pmatrix}, \mathbf{v} = \begin{pmatrix} -3 \\ 2 \\ 7 \end{pmatrix}$ (b) $\mathbf{u} = \begin{pmatrix} 2 \\ -1 \\ 3 \end{pmatrix}, \mathbf{v} = \begin{pmatrix} 3 \\ 1 \\ k \end{pmatrix}$

   (c) $\mathbf{u} = \begin{pmatrix} 0 \\ -k \\ \sqrt{2} \end{pmatrix}, \mathbf{v} = \begin{pmatrix} -7 \\ 5 \\ k \end{pmatrix}$

5. Determine the unit vector $\hat{\mathbf{u}}$ for each of the following vectors. (Normalize these vectors.)

   (a) $\mathbf{u} = \begin{pmatrix} 2 & 3 \end{pmatrix}^T$ (b) $\mathbf{u} = \begin{pmatrix} 1 & 2 & 3 \end{pmatrix}^T$

   (c) $\mathbf{u} = \begin{pmatrix} 1/2 & -1/2 & 1/4 \end{pmatrix}^T$ (d) $\mathbf{u} = \begin{pmatrix} \sqrt{2} & 2 & -\sqrt{2} & \sqrt{2} \end{pmatrix}^T$

   (e) $\mathbf{u} = \begin{pmatrix} -\pi/5 & \pi & -\pi & \pi/10 & 0 \end{pmatrix}^T$

6. Determine the value(s) of k so that $\hat{\mathbf{u}} = \begin{pmatrix} 1/\sqrt{2} \\ 1/2 \\ k \end{pmatrix}$ is a unit vector.

7. (a) Show that $\mathbf{u} = \begin{pmatrix} \cos(\theta) \\ \sin(\theta) \end{pmatrix}$ is a unit vector.

   (b) Plot this vector $\mathbf{u}$ in $\mathbb{R}^2$ for $\theta = \frac{\pi}{4}$.

   (c) Let $\mathbf{v} = \begin{pmatrix} \cos(\theta) \\ -\sin(\theta) \end{pmatrix}$ be a vector in $\mathbb{R}^2$. On the same axes plot $\mathbf{v}$ for $\theta = \frac{\pi}{4}$.

   (d) Determine the angle between the vectors $\mathbf{u}$ and $\mathbf{v}$.

8. Show that the vectors $\mathbf{u} = \begin{pmatrix} a \\ b \end{pmatrix}$ and $\mathbf{v} = \begin{pmatrix} -b \\ a \end{pmatrix}$ in $\mathbb{R}^2$ are orthogonal.

9. Let vectors $\mathbf{u} = \begin{pmatrix} \cos(A) \\ \sin(A) \end{pmatrix}$ and $\mathbf{v} = \begin{pmatrix} \cos(B) \\ \sin(B) \end{pmatrix}$ be in $\mathbb{R}^2$. Show that

$$\mathbf{u} \cdot \mathbf{v} = \cos(A - B)$$

10. Prove that the angle between a non-zero vector $\mathbf{u}$ and $-\mathbf{u}$ in any n-space, $\mathbb{R}^n$, is π radians or 180°.

11. Find a vector which is orthogonal to $\mathbf{u} = (1 \ \ 1 \ \ 1)^T$. Determine all the vectors orthogonal to $\mathbf{u}$.

12. Find the shortest distance, correct to 2dp, between the vectors and the corresponding hyperplanes:
   (a) $(1 \ \ 1)^T$ and $y = x + 1$ (b) $(0.5 \ \ 2)^T$ and $y = 2x - 1$
   (c) $(-1 \ \ -3 \ \ 3)^T$ and $2x + y - z = 7$
   (d) $(1 \ \ 2 \ \ 3 \ \ 4)^T$ and $x + 2y + z + w = 10$

13. Prove the Cauchy–Schwarz inequality by the following procedure:
   (a) Show that the dot product $(k\mathbf{u} + \mathbf{v}) \cdot (k\mathbf{u} + \mathbf{v}) \geq 0$ where $\mathbf{u}$ and $\mathbf{v}$ are vectors in $\mathbb{R}^n$.

   (b) Show that $(k\mathbf{u} + \mathbf{v}) \cdot (k\mathbf{u} + \mathbf{v}) = k^2(\mathbf{u} \cdot \mathbf{u}) + 2k(\mathbf{u} \cdot \mathbf{v}) + (\mathbf{v} \cdot \mathbf{v})$.

   (c) Equate the right hand side in part (b) to the general quadratic

$$ak^2 + bk + c$$

   (d) Assuming that $ak^2 + bk + c \geq 0$ if and only if $b^2 \leq 4ac$, show the Cauchy–Schwarz inequality $|\mathbf{u} \cdot \mathbf{v}| \leq \|\mathbf{u}\| \, \|\mathbf{v}\|$.
   [Also assume the following inequality from the main text

$$x \leq y \Leftrightarrow \sqrt{x} \leq \sqrt{y} \text{ provided } x \geq 0, \ y \geq 0$$

   and $\sqrt{x^2} = |x|$.]

14. By applying the Cauchy–Schwarz inequality, prove that $-1 \leq \cos(\theta) \leq 1$, where θ is the angle between two given vectors $\mathbf{u}$ and $\mathbf{v}$ in any n-space $\mathbb{R}^n$.

15. Let $\mathbf{u}$ and $\mathbf{v}$ be non-zero vectors in $\mathbb{R}^n$. Prove that the angle θ between the two vectors $\mathbf{u}$ and $\mathbf{v}$ is **unique** in the range $[0, \pi]$.
   [Hint: Consider $\cos(\theta) = \dfrac{\mathbf{u} \cdot \mathbf{v}}{\|\mathbf{u}\| \|\mathbf{v}\|}$ and $\cos(\beta) = \dfrac{\mathbf{u} \cdot \mathbf{v}}{\|\mathbf{u}\| \|\mathbf{v}\|}$. Use the identity
   $\cos(A) - \cos(B) = -2\sin\left(\dfrac{A+B}{2}\right)\sin\left(\dfrac{A-B}{2}\right)$.]

SECTION 2.3 Linear Independence

By the end of this section you will be able to

- understand what is meant by linear independence
- test vectors for linear independence
- prove properties about linear independence

2.3.1 Standard unit vectors in $\mathbb{R}^n$

What does the term standard unit vector mean?

Recall from the last section that unit vectors are of length 1, and **standard unit vectors** in $\mathbb{R}^n$ are column vectors with 1 in the kth position of the vector $\mathbf{e}_k$ and zeros everywhere else (Fig. 2.26).

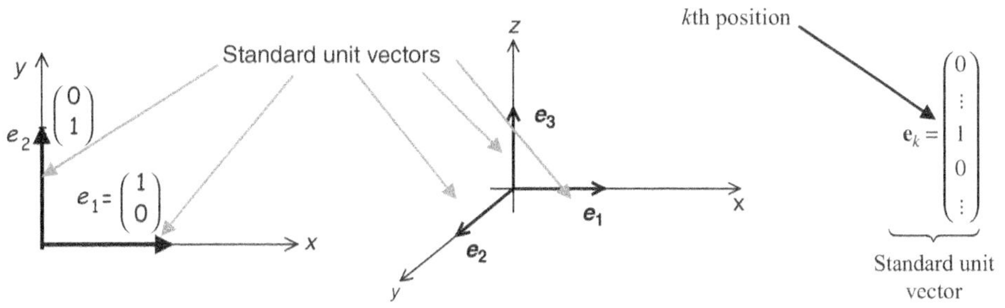

Figure 2.26

Why are these standard unit vectors important?

Because we can write any vector $\mathbf{u}$ of $\mathbb{R}^n$ in terms of scalars and standard unit vectors as we showed in Exercises 1.3, question 14. We proved the following important result:

Proposition (2.17). Let $\mathbf{u} = \begin{pmatrix} x_1 & \cdots & x_k & \cdots & x_n \end{pmatrix}^T$ be any vector in $\mathbb{R}^n$ then

$$\mathbf{u} = \underbrace{x_1}_{\text{scalar}} \underbrace{\mathbf{e}_1}_{\text{unit vector}} + \underbrace{x_2}_{\text{scalar}} \underbrace{\mathbf{e}_2}_{\text{unit vector}} + \cdots + \underbrace{x_k}_{\text{scalar}} \underbrace{\mathbf{e}_k}_{\text{unit vector}} + \cdots + \underbrace{x_n}_{\text{scalar}} \underbrace{\mathbf{e}_n}_{\text{unit vector}}$$

The position of vector $\mathbf{u}$ can be described (uniquely) by these scalars and unit vectors $\mathbf{e}_1, \mathbf{e}_2, \ldots$ and $\mathbf{e}_n$.

For example, the vector $\mathbf{u} = \begin{pmatrix} 2 \\ 3 \end{pmatrix}$ in $\mathbb{R}^2$ can be written as

$$\begin{pmatrix} 2 \\ 3 \end{pmatrix} = 2\begin{pmatrix} 1 \\ 0 \end{pmatrix} + 3\begin{pmatrix} 0 \\ 1 \end{pmatrix} = 2\mathbf{e}_1 + 3\mathbf{e}_2 \qquad \text{[In this case the scalars } x_1 = 2 \text{ and } x_2 = 3.]$$

Note that the scalars $x_1 = 2$ and $x_2 = 3$ are the coordinates of the vector $\mathbf{u}$.

This representation

$$\mathbf{u} = \begin{pmatrix} x_1 \\ \vdots \\ x_n \end{pmatrix} = x_1\mathbf{e}_1 + x_2\mathbf{e}_2 + \cdots + x_n\mathbf{e}_n$$

is a linear combination of the scalars and standard unit vectors $\mathbf{e}_1, \mathbf{e}_2, \ldots$ and $\mathbf{e}_n$. We can write this $\mathbf{u} = x_1\mathbf{e}_1 + x_2\mathbf{e}_2 + \cdots + x_n\mathbf{e}_n$ in matrix form as

$$\mathbf{u} = (\mathbf{e}_1 \;\; \mathbf{e}_2 \;\; \mathbf{e}_3 \;\; \cdots \;\; \mathbf{e}_n) \begin{pmatrix} x_1 \\ \vdots \\ x_n \end{pmatrix} \text{ where } \mathbf{e}_1 = \begin{pmatrix} 1 \\ 0 \\ \vdots \end{pmatrix}, \ldots, \mathbf{e}_n = \begin{pmatrix} 0 \\ \vdots \\ 1 \end{pmatrix}$$

The matrix $(\mathbf{e}_1 \;\; \mathbf{e}_2 \;\; \mathbf{e}_3 \;\; \cdots \;\; \mathbf{e}_n) = \mathbf{I}$ where $\mathbf{I}$ is the identity matrix.

Proposition (2.18). Let $\mathbf{u}$ be any vector in $\mathbb{R}^n$ then the linear combination

$$\mathbf{u} = x_1\mathbf{e}_1 + x_2\mathbf{e}_2 + \cdots + x_k\mathbf{e}_k + \cdots + x_n\mathbf{e}_n$$

is unique.

What does this proposition mean?

It means for any vector $\mathbf{u}$ the *scalars* in the above linear combination are *unique*.

Proof.

Let the vector $\mathbf{u}$ be written as another linear combination:

$$\mathbf{u} = y_1\mathbf{e}_1 + y_2\mathbf{e}_2 + \cdots + y_k\mathbf{e}_k + \cdots + y_n\mathbf{e}_n$$

What do we need to show?

We must show that, in fact, all the scalars are equal: $y_1 = x_1$, $y_2 = x_2, \cdots$ and $y_n = x_n$.

Equate the two linear combinations because both are equal to $\mathbf{u}$:

$$\begin{aligned} x_1\mathbf{e}_1 + x_2\mathbf{e}_2 + \cdots + x_n\mathbf{e}_n &= y_1\mathbf{e}_1 + y_2\mathbf{e}_2 + \cdots + y_n\mathbf{e}_n = \mathbf{u} \\ x_1\mathbf{e}_1 + x_2\mathbf{e}_2 + \cdots + x_n\mathbf{e}_n - y_1\mathbf{e}_1 - y_2\mathbf{e}_2 - \cdots - y_n\mathbf{e}_n &= \mathbf{u} - \mathbf{u} = \mathbf{O} \\ (x_1 - y_1)\,\mathbf{e}_1 + (x_2 - y_2)\,\mathbf{e}_2 + \cdots + (x_n - y_n)\,\mathbf{e}_n &= \mathbf{O} \quad [\text{factorizing}] \end{aligned}$$

Writing out $\mathbf{e}_1, \mathbf{e}_2, \ldots, \mathbf{e}_n$ and $\mathbf{O}$ as column vectors in the last line we have

$$(x_1 - y_1)\begin{pmatrix} 1 \\ 0 \\ \vdots \\ 0 \end{pmatrix} + (x_2 - y_2)\begin{pmatrix} 0 \\ 1 \\ 0 \\ \vdots \end{pmatrix} + \cdots + (x_n - y_n)\begin{pmatrix} 0 \\ 0 \\ \vdots \\ 1 \end{pmatrix} = \begin{pmatrix} 0 \\ 0 \\ \vdots \\ 0 \end{pmatrix}$$

$$\left[\text{remember } \mathbf{e}_k = \begin{pmatrix} \left.\vdots\right\} \text{zeros} \\ 1 \\ 0 \\ \vdots \end{pmatrix}\right]$$

Scalar multiplying and adding these vectors gives

$$\begin{pmatrix} x_1 - y_1 \\ x_2 - y_2 \\ \vdots \\ x_n - y_n \end{pmatrix} = \begin{pmatrix} 0 \\ 0 \\ \vdots \\ 0 \end{pmatrix}$$

Hence we have $x_1 - y_1 = 0,\ x_2 - y_2 = 0,\ x_3 - y_3 = 0, \ldots$ and $x_n - y_n = 0$ which gives:

$$x_1 = y_1,\ x_2 = y_2,\ x_3 = y_3, \ldots \text{ and } x_n = y_n$$

Therefore, the given linear combination, $\mathbf{u} = x_1\mathbf{e}_1 + x_2\mathbf{e}_2 + \cdots + x_n\mathbf{e}_n$, is unique.

■

2.3.2 Linear independence

Example 2.10

Find the values of the scalars $x_1, x_2, x_3, \ldots$ and x_n in the following:

$$x_1\mathbf{e}_1 + x_2\mathbf{e}_2 + \cdots + x_k\mathbf{e}_k + \cdots + x_n\mathbf{e}_n = \mathbf{O}$$

Solution

Substituting $\mathbf{e}_1, \mathbf{e}_2, \ldots$ and $\mathbf{e}_n$ we have

$$x_1 \begin{pmatrix} 1 \\ 0 \\ \vdots \\ 0 \end{pmatrix} + x_2 \begin{pmatrix} 0 \\ 1 \\ 0 \\ \vdots \end{pmatrix} + \cdots + x_n \begin{pmatrix} 0 \\ 0 \\ \vdots \\ 1 \end{pmatrix} = \begin{pmatrix} 0 \\ 0 \\ \vdots \\ 0 \end{pmatrix}$$

We can write this in matrix form as

$$\begin{pmatrix} 1 & & 0 \\ & \ddots & \\ 0 & & 1 \end{pmatrix} \begin{pmatrix} x_1 \\ \vdots \\ x_n \end{pmatrix} = \begin{pmatrix} 0 \\ \vdots \\ 0 \end{pmatrix} \qquad [\text{This is } \mathbf{Ix} = \mathbf{O}]$$

In compact form, we have $\mathbf{Ix} = \mathbf{O}$ where $\mathbf{I}$ is the identity matrix. $\mathbf{Ix} = \mathbf{O}$ gives $\mathbf{x} = \mathbf{O}$. The zero vector $\mathbf{x} = \mathbf{O}$ has entries $x_1 = 0,\ x_2 = 0,\ x_3 = 0, \ldots$ and $x_n = 0$.

We say that the standard unit vectors $\mathbf{e}_1, \mathbf{e}_2, \ldots$ and $\mathbf{e}_n$ are **linearly independent**, when any one of the vectors $\mathbf{e}_k$ *cannot* be made by a linear combination of the others.

One of the reasons we can write any vector of $\mathbb{R}^n$ in terms of $\mathbf{e}_1, \mathbf{e}_2, \ldots$ and $\mathbf{e}_n$ is because these vectors are linearly independent.

Definition (2.19). We say vectors $\mathbf{v}_1,\ \mathbf{v}_2,\ \mathbf{v}_3, \ldots$ and $\mathbf{v}_n$ in $\mathbb{R}^n$ are **linearly independent** $\Leftrightarrow$ the only real scalars $k_1,\ k_2,\ k_3, \ldots$ and k_n which satisfy:

$$k_1\mathbf{v}_1 + k_2\mathbf{v}_2 + k_3\mathbf{v}_3 + \cdots + k_n\mathbf{v}_n = \mathbf{O} \text{ are } k_1 = k_2 = k_3 = \cdots = k_n = 0$$

What does this mean?

The only solution to the linear combination $k_1\mathbf{v}_1 + k_2\mathbf{v}_2 + \cdots + k_n\mathbf{v}_n = \mathbf{O}$ occurs when all the scalars $k_1, k_2, k_3, \ldots$ and k_n are equal to zero. In other words you cannot make any one of the vectors $\mathbf{v}_j$, say, by a linear combination of the others.

We can write the linear combination $k_1\mathbf{v}_1 + k_2\mathbf{v}_2 + k_3\mathbf{v}_3 + \cdots + k_n\mathbf{v}_n = \mathbf{O}$ in matrix form as

$$\begin{pmatrix} \mathbf{v}_1 & \mathbf{v}_2 & \cdots & \mathbf{v}_n \end{pmatrix}\begin{pmatrix} k_1 \\ \vdots \\ k_n \end{pmatrix} = \begin{pmatrix} 0 \\ \vdots \\ 0 \end{pmatrix}$$

The first column of the matrix $(\mathbf{v}_1\ \ \mathbf{v}_2\ \ \cdots\ \ \mathbf{v}_n)$ is given by the entries in $\mathbf{v}_1$, the second column is given by the entries in $\mathbf{v}_2$ and the nth column by entries in $\mathbf{v}_n$.

The standard unit vectors are not the only vectors in $\mathbb{R}^n$ which are linearly independent. In the following example, we show another set of linearly independent vectors.

Example 2.11

Show that $\mathbf{u} = \begin{pmatrix} -1 \\ 1 \end{pmatrix}$ and $\mathbf{v} = \begin{pmatrix} 2 \\ 3 \end{pmatrix}$ are linearly independent in $\mathbb{R}^2$ and plot them.

Solution

Consider the linear combination:

$$k\mathbf{u} + c\mathbf{v} = \mathbf{O} \quad [\text{using } k \text{ and } c \text{ as scalars}]$$

Substituting the given vectors $\mathbf{u}$ and $\mathbf{v}$ into this $k\mathbf{u} + c\mathbf{v} = \mathbf{O}$:

$$k\begin{pmatrix} -1 \\ 1 \end{pmatrix} + c\begin{pmatrix} 2 \\ 3 \end{pmatrix} = \begin{pmatrix} 0 \\ 0 \end{pmatrix}$$

Let $\mathbf{A} = (\mathbf{u}\ \ \mathbf{v}) = \begin{pmatrix} -1 & 2 \\ 1 & 3 \end{pmatrix}$ and $\mathbf{x} = \begin{pmatrix} k \\ c \end{pmatrix}$. We need to solve $\mathbf{Ax} = \mathbf{O}$ where $\mathbf{O} = \begin{pmatrix} 0 \\ 0 \end{pmatrix}$.

Carrying out row operations on the augmented matrix $(\mathbf{A} \mid \mathbf{O})$:

$$\begin{array}{c} \\ R_1 \\ R_2 \end{array}\begin{array}{c} \begin{array}{cc} k & c \end{array} \\ \left(\begin{array}{cc|c} -1 & 2 & 0 \\ 1 & 3 & 0 \end{array}\right) \end{array} \quad \Longrightarrow \quad \begin{array}{c} \\ R_1 \\ R_2 + R_1 \end{array}\begin{array}{c} \begin{array}{cc} k & c \end{array} \\ \left(\begin{array}{cc|c} -1 & 2 & 0 \\ 0 & 5 & 0 \end{array}\right) \end{array}$$

From the bottom row, $R_2 + R_1$, we have $5c = 0$, which gives $c = 0$. Substituting this into the first row yields $k = 0$. This means that the only values of the scalars are $k = 0$ and $c = 0$.

Hence the linear combination $k\mathbf{u} + c\mathbf{v} = \mathbf{O}$ yields $k = 0$ and $c = 0$, therefore the given vectors $\mathbf{u}$ and $\mathbf{v}$ are linearly independent, because all the scalars, k and c, are equal to zero. The plot of the given vectors is shown in Fig. 2.27.

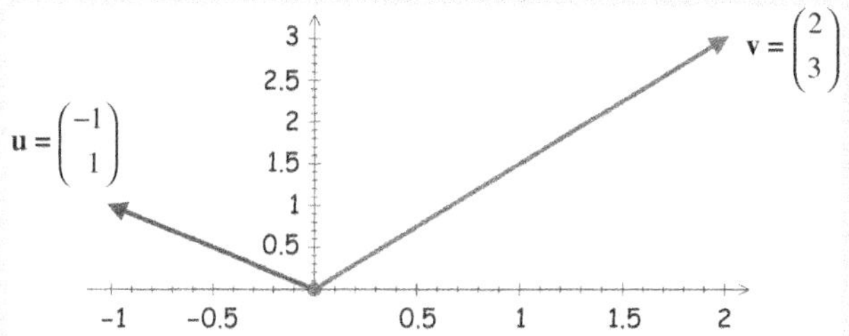

Figure 2.27

When the vectors **u** and **v** are linearly independent, it means that they are not scalar multiples of each other. Arbitrary linear independent vectors **u** and **v** in $\mathbb{R}^2$ can be illustrated as shown in Fig. 2.28:

Figure 2.28

Linearly independent vectors **u** and **v** have different directions.

2.3.3 Linear dependence

What does linear dependence mean?

Definition (2.20). Conversely we have: the vectors $\mathbf{v}_1, \mathbf{v}_2, \mathbf{v}_3, \ldots$ and $\mathbf{v}_n$ in $\mathbb{R}^n$ are linearly **dependent** $\Leftrightarrow$ the scalars $k_1, k_2, k_3, \ldots$ and k_n are not all zero and satisfy

$$k_1\mathbf{v}_1 + k_2\mathbf{v}_2 + k_3\mathbf{v}_3 + \cdots + k_n\mathbf{v}_n = \mathbf{O}$$

Linear dependence of vectors $\mathbf{v}_1, \mathbf{v}_2, \mathbf{v}_3, \ldots$ and $\mathbf{v}_n$ means that there are non-zero scalars k's which satisfy

$$k_1\mathbf{v}_1 + k_2\mathbf{v}_2 + k_3\mathbf{v}_3 + \cdots + k_n\mathbf{v}_n = \mathbf{O}$$

Example 2.12

Show that $\mathbf{u} = \begin{pmatrix} -3 \\ 1 \end{pmatrix}$ and $\mathbf{v} = \begin{pmatrix} 1 \\ -1/3 \end{pmatrix}$ are linearly dependent in $\mathbb{R}^2$ and plot them.

Solution

Consider the linear combination:

$$k\mathbf{u} + c\mathbf{v} = \mathbf{O} \qquad [k \text{ and } c \text{ are scalars}]$$

(continued...)

Substituting the given vectors

$$k\mathbf{u} + c\mathbf{v} = k\begin{pmatrix} -3 \\ 1 \end{pmatrix} + c\begin{pmatrix} 1 \\ -1/3 \end{pmatrix} = \begin{pmatrix} 0 \\ 0 \end{pmatrix}$$

The augmented matrix for this is given by

$$\begin{matrix} R_1 \\ R_2 \end{matrix} \left(\begin{array}{cc|c} -3 & 1 & 0 \\ 1 & -1/3 & 0 \end{array}\right) \qquad \left[\text{because } (\mathbf{u} \;\; \mathbf{v} \mid \mathbf{O})\right]$$

Carrying out the row operation $3R_2 + R_1$:

$$\begin{matrix} \\ R_1 \\ 3R_2 + R_1 \end{matrix} \begin{matrix} \;\; k \;\; c \\ \left(\begin{array}{cc|c} -3 & 1 & 0 \\ 0 & 0 & 0 \end{array}\right) \end{matrix}$$

From the top row we have $-3k + c = 0$, which implies $c = 3k$. Let $k = 1$ (for ease of arithmetic) then

$$c = 3k = 3 \times 1 = 3$$

Substituting our values $k = 1$ and $c = 3$ into $k\mathbf{u} + c\mathbf{v} = \mathbf{O}$ gives

$$\mathbf{u} + 3\mathbf{v} = \mathbf{O} \text{ or } \mathbf{u} = -3\mathbf{v}$$

We have found non-zero scalars, $k = 1$ and $c = 3$, which satisfy $k\mathbf{u} + c\mathbf{v} = \mathbf{O}$, therefore the given vectors $\mathbf{u}$ and $\mathbf{v}$ are linearly dependent, and $\mathbf{u} = -3\mathbf{v}$.

Plotting the given vectors $\mathbf{u}$ and $\mathbf{v}$ we have (Fig. 2.29):

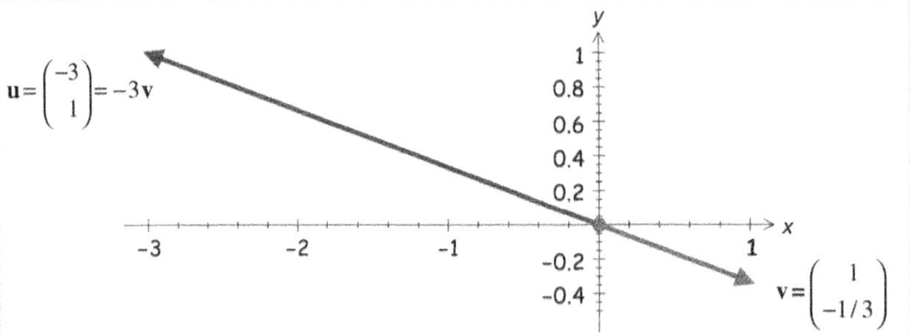

Figure 2.29

Note that $\mathbf{u} = -3\mathbf{v}$ means that the vector $\mathbf{u}$ is a scalar multiple (-3) of the vector $\mathbf{v}$. Hence the size of vector $\mathbf{u}$ is three times the size of vector $\mathbf{v}$, but in the opposite direction. If vectors $\mathbf{u}$ and $\mathbf{v}$ in $\mathbb{R}^2$ are linearly dependent then we have

$$k\mathbf{u} + c\mathbf{v} = \mathbf{O} \text{ where } k \neq 0 \text{ or } c \neq 0$$

That is, at least one of scalars is not zero. Suppose $k \neq 0$ then

$$\begin{aligned} k\mathbf{u} &= -c\mathbf{v} && [\text{transposing } k\mathbf{u} + c\mathbf{v} = \mathbf{O}] \\ \mathbf{u} &= -\frac{c}{k}\mathbf{v} && [\text{dividing by } k] \end{aligned}$$

This means that the vector $\mathbf{u}$ is a scalar multiple of the other vector $\mathbf{v}$, which suggests that $\mathbf{u}$ is in the same (or opposite) direction as vector $\mathbf{v}$. Plotting these we have (Fig. 2.30):

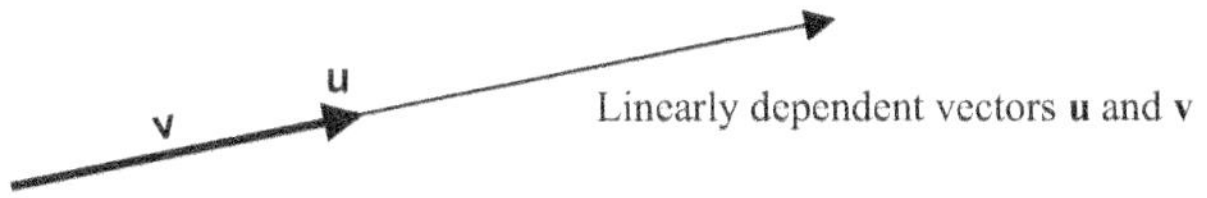

Figure 2.30

Of course, in the previous Example 2.12, we could have let $k = 2, 3, \pi, 666 \ldots$. Any non-zero number will do. Generally, it makes the arithmetic easier if we use $k = 1$.

Example 2.13

Determine whether the vectors $\mathbf{u} = \begin{pmatrix} -3 \\ 1 \\ 0 \end{pmatrix}$, $\mathbf{v} = \begin{pmatrix} 0 \\ 1 \\ -1 \end{pmatrix}$ and $\mathbf{w} = \begin{pmatrix} 2 \\ 0 \\ 0 \end{pmatrix}$ in $\mathbb{R}^3$ are linearly dependent or independent.

Solution

Consider the following linear combination

$$k_1\mathbf{u} + k_2\mathbf{v} + k_3\mathbf{w} = \mathbf{O}$$

where k's are scalars.

Let $\mathbf{A} = (\mathbf{u} \;\; \mathbf{v} \;\; \mathbf{w}) = \begin{pmatrix} -3 & 0 & 2 \\ 1 & 1 & 0 \\ 0 & -1 & 0 \end{pmatrix}$ and $\mathbf{x} = \begin{pmatrix} k_1 \\ k_2 \\ k_3 \end{pmatrix}$. We need to solve $\mathbf{Ax} = \mathbf{O}$.

Carrying out row operations on the augmented matrix:

$$\begin{array}{c} \\ R_1 \\ R_2 \\ R_3 \end{array} \begin{array}{c} \begin{array}{ccc} k_1 & k_2 & k_3 \end{array} \\ \left(\begin{array}{ccc|c} -3 & 0 & 2 & 0 \\ 1 & 1 & 0 & 0 \\ 0 & -1 & 0 & 0 \end{array}\right) \end{array} \Rightarrow \begin{array}{c} \\ R_1 \\ R_2 + R_3 \\ R_3 \end{array} \begin{array}{c} \begin{array}{ccc} k_1 & k_2 & k_3 \end{array} \\ \left(\begin{array}{ccc|c} -3 & 0 & 2 & 0 \\ 1 & 0 & 0 & 0 \\ 0 & -1 & 0 & 0 \end{array}\right) \end{array}$$

From the right hand matrix we have $k_1 = k_2 = k_3 = 0$, that is all the scalars are zero:

$$k_1\mathbf{u} + k_2\mathbf{v} + k_3\mathbf{w} = \mathbf{O} \text{ implies } k_1 = k_2 = k_3 = 0$$

Hence the given vectors $\mathbf{u}$, $\mathbf{v}$ and $\mathbf{w}$ are linearly independent.

2.3.4 Properties of linear dependence

In this subsection we describe easier ways of testing for independence.

Example 2.14

Test the following vectors for linear independence:

$$\mathbf{u} = \begin{pmatrix} 2 \\ 3 \\ 7 \end{pmatrix}, \; \mathbf{v} = \begin{pmatrix} -4 \\ 19 \\ -5 \end{pmatrix}, \; \mathbf{w} = \begin{pmatrix} 0 \\ 0 \\ 0 \end{pmatrix}$$

(continued...)

Solution
Consider the linear combination:

$$k_1\mathbf{u} + k_2\mathbf{v} + k_3\mathbf{w} = \mathbf{O} \quad (\dagger)$$

Let $k_1 = k_2 = 0$ and $k_3 \neq 0$, then for these values of scalars, the above linear combination (†) is satisfied. Hence we have non-zero scalar(s), $k_3 \neq 0$, which implies that the given vectors are linearly dependent.

The presence of the zero vector ensures that the vectors are linearly dependent. We don't need to find the values of the scalars in the case where we have the zero vector.

Proposition (2.21). Let $\mathbf{v}_1, \mathbf{v}_2, \ldots$ and $\mathbf{v}_n$ be vectors in $\mathbb{R}^n$. If at least one of these vectors, $\mathbf{v}_j$ say, is the zero vector then the vectors $\mathbf{v}_1, \mathbf{v}_2, \ldots$ and $\mathbf{v}_n$ are linearly dependent.

Proof.
Consider the linear combination

$$k_1\mathbf{v}_1 + k_2\mathbf{v}_2 + \cdots + k_j\mathbf{v}_j + \cdots + k_n\mathbf{v}_n = \mathbf{O} \quad (*)$$

In (*) take $k_j \neq 0$ [non-zero number] and all the other scalars equal to zero, that is

$$k_1 = k_2 = \cdots = k_{j-1} = k_{j+1} = \cdots = k_n = 0$$

Since $\mathbf{v}_j = \mathbf{O}$ we have $k_j\mathbf{v}_j = \mathbf{O}$, which means that all scalars in (*) are not zero ($k_j \neq 0$). By Definition (2.20) in section 2.3.3:

If non-zero k's satisfy $k_1\mathbf{v}_1 + k_2\mathbf{v}_2 + \cdots + k_n\mathbf{v}_n = \mathbf{O}$ then vectors $\mathbf{v}$'s are dependent.

We have that vectors $\mathbf{v}_1$, $\mathbf{v}_2, \ldots$ and $\mathbf{v}_n$ are linearly dependent because $k_j \neq 0$.

■

What does Proposition (2.21) mean?
If among the vectors $\mathbf{v}_1, \mathbf{v}_2, \mathbf{v}_3, \ldots$ and $\mathbf{v}_n$, one of these is the zero vector then they are linearly dependent.

Example 2.15

Test the vectors $\mathbf{u} = \begin{pmatrix} -3 \\ 1 \\ 0 \end{pmatrix}$, $\mathbf{v} = \begin{pmatrix} 0 \\ 1 \\ -1 \end{pmatrix}$, $\mathbf{w} = \begin{pmatrix} 2 \\ 0 \\ 0 \end{pmatrix}$ and $\mathbf{x} = \begin{pmatrix} 1 \\ 2 \\ 3 \end{pmatrix}$ in $\mathbb{R}^3$ for linear independence.

Solution
Consider the linear combination

$$k_1\mathbf{u} + k_2\mathbf{v} + k_3\mathbf{w} + k_4\mathbf{x} = \mathbf{O}$$

What do we need to find?

We need to determine the values of the scalars, *k*'s.

Substituting the given vectors into this linear combination:

$$k_1\mathbf{u} + k_2\mathbf{v} + k_3\mathbf{w} + k_4\mathbf{x} = k_1\begin{pmatrix}-3\\1\\0\end{pmatrix} + k_2\begin{pmatrix}0\\1\\-1\end{pmatrix} + k_3\begin{pmatrix}2\\0\\0\end{pmatrix} + k_4\begin{pmatrix}1\\2\\3\end{pmatrix} = \begin{pmatrix}0\\0\\0\end{pmatrix}$$

The augmented matrix of this is given by

$$\begin{matrix}R_1\\R_2\\R_3\end{matrix}\left(\begin{array}{cccc|c}-3 & 0 & 2 & 1 & 0\\1 & 1 & 0 & 2 & 0\\0 & -1 & 0 & 3 & 0\end{array}\right) \quad \left[\text{Using } (\mathbf{u}\ \ \mathbf{v}\ \ \mathbf{w}\ \ \mathbf{x}\ \mid\ \mathbf{O})\right]$$

Carrying out the row operation $R_2 + R_3$ gives one extra zero in the middle row:

$$\begin{matrix}\\R_1\\R_2+R_3\\R_3\end{matrix}\begin{matrix}\begin{matrix}k_1 & k_2 & k_3 & k_4\end{matrix}\\\left(\begin{array}{cccc|c}-3 & 0 & 2 & 1 & 0\\1 & 0 & 0 & 5 & 0\\0 & -1 & 0 & 3 & 0\end{array}\right)\end{matrix}$$

From the bottom row, we have $-k_2 + 3k_4 = 0$ which gives $k_2 = 3k_4$. Let $k_4 = 1$:

$$k_2 = 3k_4 = 3 \times 1 = 3$$

From the middle row we have $k_1 + 5k_4 = 0$ implies that $k_1 = -5k_4 = -5 \times 1 = -5$.
The top row gives $-3k_1 + 2k_3 + k_4 = 0$. Substituting $k_1 = -5$ and $k_4 = 1$ into this:

$$-3(-5) + 2k_3 + 1 = 0 \text{ implies } k_3 = -8$$

Our scalars are $k_1 = -5$, $k_2 = 3$, $k_3 = -8$ and $k_4 = 1$. Substituting these into the above linear combination $k_1\mathbf{u} + k_2\mathbf{v} + k_3\mathbf{w} + k_4\mathbf{x} = \mathbf{O}$ gives the relationship between the vectors:

$$-5\mathbf{u} + 3\mathbf{v} - 8\mathbf{w} + \mathbf{x} = \mathbf{O} \text{ or } \mathbf{x} = 5\mathbf{u} - 3\mathbf{v} + 8\mathbf{w}$$

Since we have non-zero scalars (*k*'s) the given vectors are linearly dependent.

The linear combination $\mathbf{x} = 5\mathbf{u} - 3\mathbf{v} + 8\mathbf{w}$ means that we can make the vector $\mathbf{x}$ out of the vectors $\mathbf{u}$, $\mathbf{v}$ and $\mathbf{w}$.

In the next proposition we will prove that if there are more vectors than the value of n in the n-space then the vectors are linearly dependent. In the above Example 2.19 we had four vectors $\mathbf{u}$, $\mathbf{v}$, $\mathbf{w}$ and $\mathbf{x}$ in $\mathbb{R}^3$ and $4 > 3$, therefore the given vectors $\mathbf{u}$, $\mathbf{v}$, $\mathbf{w}$ and $\mathbf{x}$ were linearly dependent.

Proposition (2.22). Let $\mathbf{v}_1, \mathbf{v}_2, \mathbf{v}_3, \ldots$ and $\mathbf{v}_m$ be different vectors in $\mathbb{R}^n$. If $n < m$, that is the value of n in the n-space is less than the number m of vectors, then the vectors $\mathbf{v}_1, \mathbf{v}_2, \mathbf{v}_3, \ldots$ and $\mathbf{v}_m$ are linearly dependent.

Proof.
Consider the linear combination of the given vectors $\mathbf{v}_1$, $\mathbf{v}_2$, $\mathbf{v}_3, \ldots$ and $\mathbf{v}_m$:

$$k_1\mathbf{v}_1 + k_2\mathbf{v}_2 + \cdots + k_n\mathbf{v}_n + k_{n+1}\mathbf{v}_{n+1} + \cdots + k_m\mathbf{v}_m = \mathbf{O} \qquad (*)$$

The number of equations is n because each vector belongs to the n-space $\mathbb{R}^n$ but the number of unknowns $k_1, k_2, k_3, \ldots, k_n, k_{n+1}, \ldots, k_m$ is m. Writing this out we have

$$\underbrace{k_1\begin{pmatrix} v_{11} \\ v_{12} \\ \vdots \\ v_{1n} \end{pmatrix} + k_2\begin{pmatrix} v_{21} \\ v_{22} \\ \vdots \\ v_{2n} \end{pmatrix} + \cdots + k_n\begin{pmatrix} v_{n1} \\ v_{n2} \\ \vdots \\ v_{nn} \end{pmatrix} + k_{n+1}\begin{pmatrix} v_{(n+1)1} \\ v_{(n+1)2} \\ \vdots \\ v_{(n+1)n} \end{pmatrix} + \cdots + k_m\begin{pmatrix} v_{m1} \\ v_{m2} \\ \vdots \\ v_{mn} \end{pmatrix}}_{m \text{ unknowns}}$$

$$= \left.\begin{pmatrix} 0 \\ 0 \\ \vdots \\ 0 \end{pmatrix}\right\} n \text{ equations}$$

By Proposition (1.31) of chapter 1:

In a linear system $\mathbf{Ax} = \mathbf{O}$, if the number of equations is less than the number unknowns then the system has an infinite number of solutions.

In our linear system the number of equations n is less than the number of unknowns m because we are given $n < m$. Therefore we have an infinite number of k's which satisfy $(*)$ and this means all the k's are not zero. By Definition (2.20) in section 2.3.3:

If non-zero k's satisfy $k_1\mathbf{v}_1 + k_2\mathbf{v}_2 + \cdots + k_n\mathbf{v}_n = \mathbf{O}$ then vectors $\mathbf{v}$'s are dependent.

Hence the given vectors $\mathbf{v}_1, \mathbf{v}_2, \mathbf{v}_3, \ldots$ and $\mathbf{v}_m$ are dependent.

■

Normally, we write the vectors as a collection in a set. A set is denoted by { } and is a collection of objects. The objects are called elements or members of the set.

We can write the set of vectors $\mathbf{v}_1, \mathbf{v}_2, \mathbf{v}_3, \ldots$ and $\mathbf{v}_n$ as

$$S = \{\mathbf{v}_1, \mathbf{v}_2, \mathbf{v}_3, \ldots, \mathbf{v}_n\}$$

We use the symbol S for a set.

The following is an important test for linear independence.

Proposition (2.23). Let $S = \{\mathbf{v}_1, \mathbf{v}_2, \mathbf{v}_3, \ldots, \mathbf{v}_n\}$ be n vectors in the n-space $\mathbb{R}^n$. Let $\mathbf{A}$ be the n by n matrix whose columns are given by the vectors $\mathbf{v}_1, \mathbf{v}_2, \mathbf{v}_3, \ldots$ and $\mathbf{v}_n$:

$$\mathbf{A} = (\mathbf{v}_1 \ \ \mathbf{v}_2 \ \ \cdots \ \ \mathbf{v}_n)$$

Then vectors $\mathbf{v}_1, \mathbf{v}_2, \ldots, \mathbf{v}_n$ are linearly independent $\Leftrightarrow$ matrix $\mathbf{A}$ is invertible.

Proof – See Exercises 2.3.

■

This proposition means that the columns of matrix **A** are linearly independent ⇔ **A** is invertible. We can add this to the main theorem of the last chapter – Theorem (1.38):

Theorem (2.24). Let **A** be a n by n matrix, then the following six statements are equivalent:

(a) The matrix **A** is invertible (non-singular).

(b) The linear system $\mathbf{Ax} = \mathbf{O}$ only has the trivial solution $\mathbf{x} = \mathbf{O}$.

(c) The reduced row echelon form of the matrix **A** is the identity matrix **I**.

(d) **A** is a product of elementary matrices.

(e) $\mathbf{Ax} = \mathbf{b}$ has a unique solution.

(f) Columns of matrix **A** are linearly independent.

Summary

Consider the following linear combination:

$$k_1\mathbf{v}_1 + k_2\mathbf{v}_2 + k_3\mathbf{v}_3 + \cdots + k_n\mathbf{v}_n = \mathbf{O}$$

If the only solution to this $k_1 = k_2 = k_3 = \cdots = k_n = 0$ (all scalars are zero) then the vectors $\mathbf{v}_1, \mathbf{v}_2, \mathbf{v}_3, \ldots$ and $\mathbf{v}_n$ are linearly independent. Otherwise the vectors are linearly dependent.

EXERCISES 2.3

(Brief solutions at end of book. Full solutions available at <http://www.oup.co.uk/companion/singh>.)

1. Determine whether the following vectors are linearly dependent in $\mathbb{R}^2$:

(a) $\mathbf{e}_1 = \begin{pmatrix} 1 \\ 0 \end{pmatrix}, \mathbf{e}_2 = \begin{pmatrix} 0 \\ 1 \end{pmatrix}$ (b) $\mathbf{u} = \begin{pmatrix} 3 \\ 4 \end{pmatrix}, \mathbf{v} = \begin{pmatrix} -6 \\ -8 \end{pmatrix}$

(c) $\mathbf{u} = \begin{pmatrix} 6 \\ 10 \end{pmatrix}, \mathbf{v} = \begin{pmatrix} -3 \\ -5 \end{pmatrix}$ (d) $\mathbf{u} = \begin{pmatrix} \pi \\ -2\pi \end{pmatrix}, \mathbf{v} = \begin{pmatrix} -1 \\ 2 \end{pmatrix}$

(e) $\mathbf{u} = \begin{pmatrix} 0 \\ 0 \end{pmatrix}, \mathbf{v} = \begin{pmatrix} -1 \\ 1 \end{pmatrix}$

2. Determine whether the following vectors are linearly dependent in $\mathbb{R}^3$:

(a) $\mathbf{e}_1 = \begin{pmatrix} 1 \\ 0 \\ 0 \end{pmatrix}, \mathbf{e}_2 = \begin{pmatrix} 0 \\ 1 \\ 0 \end{pmatrix}, \mathbf{e}_3 = \begin{pmatrix} 0 \\ 0 \\ 1 \end{pmatrix}$

(b) $\mathbf{u} = \begin{pmatrix} 2 \\ 2 \\ 2 \end{pmatrix}, \mathbf{v} = \begin{pmatrix} 1 \\ 2 \\ -1 \end{pmatrix}, \mathbf{w} = \begin{pmatrix} 0 \\ 0 \\ 1 \end{pmatrix}$ (c) $\mathbf{u} = \begin{pmatrix} 1 \\ 1 \\ 1 \end{pmatrix}, \mathbf{v} = \begin{pmatrix} -2 \\ -2 \\ -2 \end{pmatrix}$

(d) $\mathbf{u} = \begin{pmatrix} -1 \\ 2 \\ -3 \end{pmatrix}$, $\mathbf{v} = \begin{pmatrix} 0 \\ -4 \\ 6 \end{pmatrix}$, $\mathbf{w} = \begin{pmatrix} 2 \\ 0 \\ 6 \end{pmatrix}$

3. Determine whether the following vectors are linearly dependent in $\mathbb{R}^4$:
   (a) $\mathbf{u} = (0\ \ -1\ \ 0\ \ 3)^T$, $\mathbf{v} = (1\ \ 0\ \ 5\ \ 0)^T$, $\mathbf{w} = (2\ \ 1\ \ 0\ \ 0)^T$, $\mathbf{x} = (0\ \ 1\ \ 0\ \ -4)^T$
   (b) $\mathbf{u} = (1\ \ -1\ \ 3\ \ 3)^T$, $\mathbf{v} = (0\ \ 1\ \ 5\ \ 0)^T$, $\mathbf{w} = (-3\ \ -6\ \ -9\ \ -4)^T$, $\mathbf{x} = (-5\ \ 5\ \ -15\ \ -15)^T$
   (c) $\mathbf{u} = (-2\ \ 2\ \ 3\ \ 4)^T$, $\mathbf{v} = (0\ \ 3\ \ -2\ \ -3)^T$, $\mathbf{w} = (2\ \ -2\ \ -1\ \ 0)^T$, $\mathbf{x} = (0\ \ 3\ \ 0\ \ 1)^T$
4. Let $\mathbf{u}$ and $\mathbf{v}$ be non-zero vectors in $\mathbb{R}^n$. Prove that if $\mathbf{u} = k\mathbf{v}$ where k is a real scalar then the vectors $\mathbf{u}$ and $\mathbf{v}$ are linearly dependent. (Vectors $\mathbf{u}$ and $\mathbf{v}$ are scalar multiples of each other.)
5. Let $\mathbf{u}$, $\mathbf{v}$ and $\mathbf{w}$ be any vectors in $\mathbb{R}^n$. Prove that the vectors $\mathbf{u}+\mathbf{v}$, $\mathbf{v}+\mathbf{w}$ and $\mathbf{u}-\mathbf{w}$ are linearly dependent.
6. Let $\mathbf{e}_1$ and $\mathbf{e}_2$ be the standard unit vectors in $\mathbb{R}^2$. Prove that $\mathbf{e}_1$ and $\mathbf{e}_1+\mathbf{e}_2$ are linearly independent.
7. Let $\mathbf{e}_1$, $\mathbf{e}_2$ and $\mathbf{e}_3$ be the standard unit vectors in $\mathbb{R}^3$. Prove that $\mathbf{e}_1$, $\mathbf{e}_1+\mathbf{e}_2$ and $\mathbf{e}_1+\mathbf{e}_2+\mathbf{e}_3$ are linearly independent.
8. Let $\mathbf{u}$, $\mathbf{v}$, $\mathbf{w}$ and $\mathbf{x}$ be linearly independent vectors in $\mathbb{R}^n$. Prove that

$$\mathbf{u}+\mathbf{v},\ \mathbf{v}+\mathbf{w},\ \mathbf{w}+\mathbf{x} \text{ and } \mathbf{u}+\mathbf{x}$$

   are linearly dependent vectors in $\mathbb{R}^n$.
9. Let $\mathbf{u}$, $\mathbf{v}$ and $\mathbf{w}$ be linearly independent vectors in $\mathbb{R}^n$. Let the vector $\mathbf{x}$ be in $\mathbb{R}^n$ such that $\mathbf{x} = k_1\mathbf{u} + k_2\mathbf{v} + k_3\mathbf{w}$, where the k's are real scalars. Prove that this representation of the vector $\mathbf{x}$ is unique.
10. Let $S = \{\mathbf{v}_1, \mathbf{v}_2, \mathbf{v}_3, \ldots, \mathbf{v}_n\}$ be a set of linear independent vectors in $\mathbb{R}^n$. Prove that $T = \{c_1\mathbf{v}_1, c_2\mathbf{v}_2, c_3\mathbf{v}_3, \ldots, c_n\mathbf{v}_n\}$, where the c's are real non-zero scalars, is also a set of linear independent vectors.
11. Prove that if $S = \{\mathbf{v}_1, \mathbf{v}_2, \mathbf{v}_3, \ldots, \mathbf{v}_n\}$ is linearly independent then any subset $T = \{\mathbf{v}_1, \mathbf{v}_2, \mathbf{v}_3, \ldots, \mathbf{v}_m\}$, where $m < n$, is also linearly independent.
12. Determine the real values of t in the following vectors which form linear independence in $\mathbb{R}^3$:

$$\mathbf{u} = \begin{pmatrix} t \\ 1 \\ 1 \end{pmatrix},\ \mathbf{v} = \begin{pmatrix} -1 \\ t \\ 1 \end{pmatrix} \text{ and } \mathbf{w} = \begin{pmatrix} 1 \\ 1 \\ t \end{pmatrix}$$

13. Prove Proposition (2.23).

SECTION 2.4 Basis and Spanning Set

By the end of this section you will be able to

- show that given vectors span $\mathbb{R}^n$
- test whether given vectors are a basis for $\mathbb{R}^n$
- prove properties of basis

To describe a vector in $\mathbb{R}^n$ we need a coordinate system. A basis is a coordinate system or framework which describes the Euclidean n-space.

For example, there are infinitely many vectors in the plane $\mathbb{R}^2$, but we can describe all of these by using the standard unit vectors $\mathbf{e}_1 = (1 \;\; 0)^T$ in the x direction and $\mathbf{e}_2 = (0 \;\; 1)^T$ in the y direction.

We can write a vector $\begin{pmatrix} 2 \\ 3 \end{pmatrix}$, *but what does 2 and 3 represent?*

It indicates two units in the x direction and three units in the y direction which can be written as $2\mathbf{e}_1 + 3\mathbf{e}_2$.

2.4.1 Spanning sets

From the last section we know that we can write any vector in $\mathbb{R}^n$ in terms of the standard unit vectors $\mathbf{e}_1, \mathbf{e}_2, \ldots$ and $\mathbf{e}_n$.

Example 2.16

Let $\mathbf{v} = (a \;\; b \;\; c)^T$ be any vector in $\mathbb{R}^3$. Write this vector $\mathbf{v}$ in terms of the unit vectors:

$$\mathbf{e}_1 = (1 \;\; 0 \;\; 0)^T,\;\; \mathbf{e}_2 = (0 \;\; 1 \;\; 0)^T \text{ and } \mathbf{e}_3 = (0 \;\; 0 \;\; 1)^T$$

Vectors $\mathbf{e}_1, \mathbf{e}_2$ and $\mathbf{e}_3$ specify x, y and z directions respectively. (Illustrated in Fig. 2.26.)

Solution

We have

$$\begin{pmatrix} a \\ b \\ c \end{pmatrix} = a\begin{pmatrix} 1 \\ 0 \\ 0 \end{pmatrix} + b\begin{pmatrix} 0 \\ 1 \\ 0 \end{pmatrix} + c\begin{pmatrix} 0 \\ 0 \\ 1 \end{pmatrix} = a\mathbf{e}_1 + b\mathbf{e}_2 + c\mathbf{e}_3$$

That is, we can write the vector $\mathbf{v}$ as a linear combination of vectors $\mathbf{e}_1, \mathbf{e}_2$ and $\mathbf{e}_3$.

We say that the vectors $\mathbf{e}_1$, $\mathbf{e}_2$ and $\mathbf{e}_3$ **span** or **generate** $\mathbb{R}^3$, because the linear combination:

$$a\mathbf{e}_1 + b\mathbf{e}_2 + c\mathbf{e}_3$$

produces any vector in $\mathbb{R}^3$. We define the term 'span' as follows:

Definition (2.25). Consider the n vectors in the set $S = \{\mathbf{v}_1, \mathbf{v}_2, \mathbf{v}_3, \ldots, \mathbf{v}_n\}$ in the n-space, $\mathbb{R}^n$. If every vector in $\mathbb{R}^n$ can be produced by a linear combination of these vectors $\mathbf{v}_1, \mathbf{v}_2, \mathbf{v}_3, \ldots$ and $\mathbf{v}_n$ then we say these vectors **span** or **generate** the n-space, $\mathbb{R}^n$.

This set $S = \{\mathbf{v}_1, \mathbf{v}_2, \mathbf{v}_3, \ldots, \mathbf{v}_n\}$ is called the **spanning set**. We also say that the set S spans the n-space or S spans $\mathbb{R}^n$.

For example, the standard unit vectors $\mathbf{e}_1 = \begin{pmatrix} 1 \\ 0 \end{pmatrix}$ and $\mathbf{e}_2 = \begin{pmatrix} 0 \\ 1 \end{pmatrix}$ span $\mathbb{R}^2$ because $k\mathbf{e}_1$ spans the x axis and $c\mathbf{e}_2$ spans the y axis. Hence, by introducing scalars, k and c, the linear combination, $k\mathbf{e}_1 + c\mathbf{e}_2$, of these vectors $\mathbf{e}_1$ and $\mathbf{e}_2$ can produce any vector in $\mathbb{R}^2$.

To check that given vectors $\{\mathbf{v}_1, \mathbf{v}_2, \mathbf{v}_3, \ldots, \mathbf{v}_n\}$ span $\mathbb{R}^n$, we carry out the following:

We show that an arbitrary vector $\mathbf{w} = \begin{pmatrix} w_1 & \cdots & w_n \end{pmatrix}^T$ is a linear combination of these vectors:

$$\mathbf{w} = k_1\mathbf{v}_1 + k_2\mathbf{v}_2 + \cdots + k_n\mathbf{v}_n \text{ where } k\text{'s are scalars.}$$

The unit vectors are not the only vectors which span $\mathbb{R}^2$, there are other vectors which also span $\mathbb{R}^2$, as the next example demonstrates.

Example 2.17

Consider the vectors $\mathbf{u} = \begin{pmatrix} 1 \\ 2 \end{pmatrix}$ and $\mathbf{v} = \begin{pmatrix} -1 \\ 1 \end{pmatrix}$ in $\mathbb{R}^2$:

(i) Show that the vectors $\mathbf{u}$ and $\mathbf{v}$ span $\mathbb{R}^2$.

(ii) Write the vector $(3\ 2)^T$ in terms of the given vectors $\mathbf{u}$ and $\mathbf{v}$.

Solution

(i) Let $\mathbf{w} = \begin{pmatrix} a \\ b \end{pmatrix}$ be an arbitrary vector in $\mathbb{R}^2$. Consider the linear combination:

$$k\mathbf{u} + c\mathbf{v} = \mathbf{w} \text{ where } k \text{ and } c \text{ are scalars}$$

$k\mathbf{u}$ specifies vectors in the $\begin{pmatrix} 1 \\ 2 \end{pmatrix}$ direction and $c\mathbf{v}$ specifies vectors in the $\begin{pmatrix} -1 \\ 1 \end{pmatrix}$ direction. We need to show that we can make any vector $\mathbf{w}$ out of $\mathbf{u}$ and $\mathbf{v}$.

We can write this $k\mathbf{u} + c\mathbf{v} = \mathbf{w}$ in matrix form as

$$\begin{matrix} \mathbf{u} & \mathbf{v} \end{matrix} \qquad\qquad \mathbf{w}$$
$$\begin{pmatrix} 1 & -1 \\ 2 & 1 \end{pmatrix}\begin{pmatrix} k \\ c \end{pmatrix} = \begin{pmatrix} a \\ b \end{pmatrix}$$

Writing this as an augmented matrix, we have

$$\begin{matrix} R_1 \\ R_2 \end{matrix}\left(\begin{array}{cc|c} 1 & -1 & a \\ 2 & 1 & b \end{array}\right)$$

We execute row operations so that we can find the values for scalars k and c:

$$\begin{matrix} R_1 \\ R_2 - 2R_1 \end{matrix} \left(\begin{array}{cc|c} 1 & -1 & a \\ 0 & 3 & b-2a \end{array}\right) \quad (\text{columns } k, c)$$

From the bottom row we have

$$3c = b - 2a \text{ implies } c = \frac{b-2a}{3}$$

Substituting this $c = (b-2a)/3$ into the top row gives:

$$k - \frac{b-2a}{3} = a \text{ implies } k = \frac{b-2a}{3} + a = \frac{b-2a+3a}{3} = \frac{b+a}{3}$$

We have found the scalars $k = \frac{a+b}{3}$ and $c = \frac{b-2a}{3}$. Therefore $\mathbf{w} = k\mathbf{u} + c\mathbf{v}$, which means that these vectors $\mathbf{u}$ and $\mathbf{v}$ span or generate $\mathbb{R}^2$. We can illustrate these vectors as shown in Fig. 2.31.

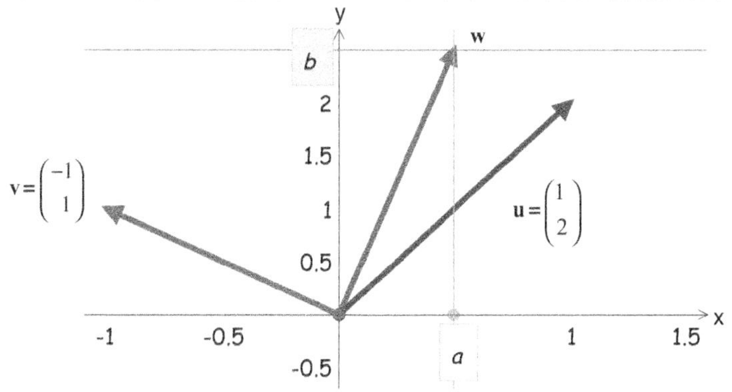

Figure 2.31

Any vector in $\mathbb{R}^2$ can be written as:

$$\begin{pmatrix} a \\ b \end{pmatrix} = \left(\frac{a+b}{3}\right)\mathbf{u} + \left(\frac{b-2a}{3}\right)\mathbf{v}$$

(ii) *How do we write the vector* $\begin{pmatrix} 3 \\ 2 \end{pmatrix}$ *in terms of* $\mathbf{u} = \begin{pmatrix} 1 \\ 2 \end{pmatrix}$ *and* $\mathbf{v} = \begin{pmatrix} -1 \\ 1 \end{pmatrix}$?

We can use part (i) with $a = 3$ and $b = 2$, because we have shown above that any vector in $\mathbb{R}^2$ can be generated by the vectors $\mathbf{u}$ and $\mathbf{v}$. Substituting these, $a = 3$ and $b = 2$, into $\begin{pmatrix} a \\ b \end{pmatrix} = \left(\frac{a+b}{3}\right)\mathbf{u} + \left(\frac{b-2a}{3}\right)\mathbf{v}$ gives

$$\begin{pmatrix} 3 \\ 2 \end{pmatrix} = \left(\frac{3+2}{3}\right)\mathbf{u} + \left(\frac{2-2(3)}{3}\right)\mathbf{v} = \frac{5}{3}\mathbf{u} - \frac{4}{3}\mathbf{v}$$

(continued...)

This is illustrated in Fig. 2.32.

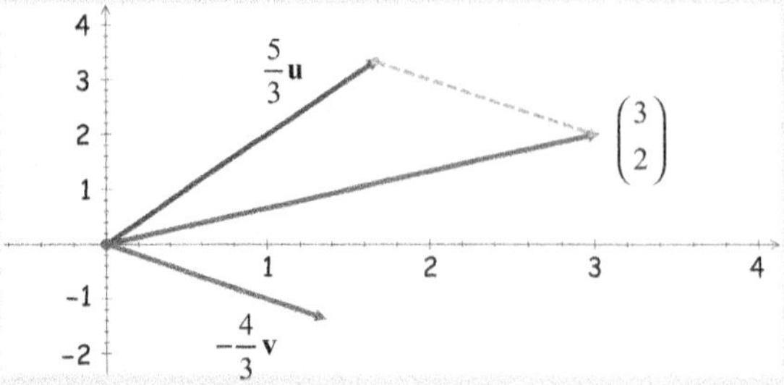

Figure 2.32

The vector $\begin{pmatrix}3\\2\end{pmatrix}$ is made by adding $\frac{5}{3}\mathbf{u}$ and $-\frac{4}{3}\mathbf{v}$.

You may like to check this by arithmetic means: $\frac{5}{3}\mathbf{u}-\frac{4}{3}\mathbf{v}=\begin{pmatrix}3\\2\end{pmatrix}$

If we cannot write an arbitrary vector $\mathbf{w}$ as a linear combination of vectors $\{\mathbf{v}_1, \dots, \mathbf{v}_n\}$ then these vectors do not span $\mathbb{R}^n$.

Example 2.18

Show that the vectors $\mathbf{u}=\begin{pmatrix}1\\-2\end{pmatrix}$ and $\mathbf{v}=\begin{pmatrix}0\\0\end{pmatrix}$ do not span $\mathbb{R}^2$. (Columns of $\begin{pmatrix}1 & 0\\-2 & 0\end{pmatrix}$ do not span $\mathbb{R}^2$.)

Solution

Let $\mathbf{w}=\begin{pmatrix}a\\b\end{pmatrix}$ be an arbitrary vector in $\mathbb{R}^2$. Consider the linear combination

$$k\mathbf{u}+c\mathbf{v}=\mathbf{w}$$

where k and c are real scalars

This linear combination can be written in matrix form as

$$\begin{pmatrix}1 & 0\\-2 & 0\end{pmatrix}\begin{pmatrix}k\\c\end{pmatrix}=\begin{pmatrix}a\\b\end{pmatrix} \Rightarrow \begin{pmatrix}k\\-2k\end{pmatrix}=\begin{pmatrix}a\\b\end{pmatrix}$$

We have the simultaneous equations

$$k=a$$
$$-2k=b \text{ or } b=-2a \qquad [\text{because } k=a]$$

This case only works if $b=-2a$, that is for the vector $\mathbf{w}=\begin{pmatrix}a\\b\end{pmatrix}=\begin{pmatrix}a\\-2a\end{pmatrix}=a\begin{pmatrix}1\\-2\end{pmatrix}$.

Vectors $\mathbf{u}$ and $\mathbf{v}$ only span (generate) vectors in the direction of $\begin{pmatrix}1\\-2\end{pmatrix}$.

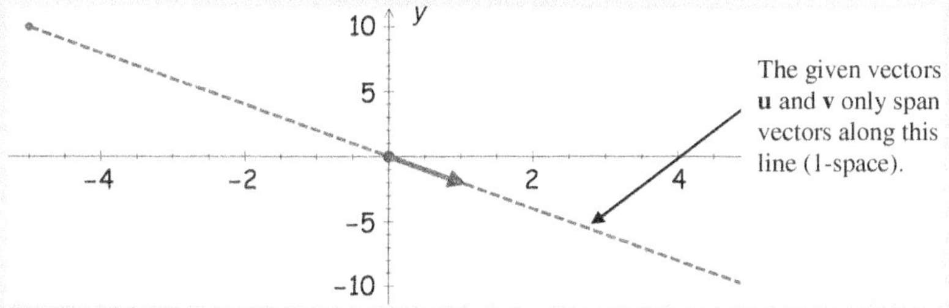

Figure 2.33

Any vector away from the dashed line in Fig. 2.33 cannot be made by a linear combination of the given vectors $\mathbf{u}$ and $\mathbf{v}$.

We conclude that the vectors $\mathbf{u}$ and $\mathbf{v}$ do not span $\mathbb{R}^2$.

2.4.2 Basis

We want a simple way to write down our vectors.

How can we do this?

Given some vectors we can generate others by a linear combination. We need *just* enough vectors to build all other vectors from them through linear combination. This set of just enough vectors is called a basis.

An example is the standard unit vectors $\mathbf{e}_1 = (1 \;\; 0)^T$, $\mathbf{e}_2 = (0 \;\; 1)^T$ for $\mathbb{R}^2$. This is the basis which forms the x and y axes of $\mathbb{R}^2$ because $\mathbf{e}_1 = (1 \;\; 0)^T$ specifies the x direction and $\mathbf{e}_2 = (0 \;\; 1)^T$ specifies the y direction.

Each additional basis vector introduces a new direction.

Definition (2.26). Consider the n vectors $\mathbf{v}_1, \mathbf{v}_2, \mathbf{v}_3, \ldots$ and $\mathbf{v}_n$ in the n space, $\mathbb{R}^n$. These vectors form a **basis** for $\mathbb{R}^n \Leftrightarrow$

(i) $\mathbf{v}_1, \mathbf{v}_2, \mathbf{v}_3, \ldots$ and $\mathbf{v}_n$ span $\mathbb{R}^n$ and

(ii) $\mathbf{v}_1, \mathbf{v}_2, \mathbf{v}_3, \ldots$ and $\mathbf{v}_n$ are linearly independent

We can write the vectors $\mathbf{v}_1, \mathbf{v}_2, \mathbf{v}_3, \ldots$ and $\mathbf{v}_n$ as a set $B = \{\mathbf{v}_1, \mathbf{v}_2, \mathbf{v}_3, \ldots, \mathbf{v}_n\}$. These are called the **basis vectors** – independent vectors which span $\mathbb{R}^n$. Any vector in $\mathbb{R}^n$ can be constructed from the basis vectors.

Bases (plural of basis) are the most efficient spanning sets. There are many sets of vectors that can span a space. However, in these sets some of the vectors might be redundant in spanning the space (because they can be 'made' from the other vectors in the set). A basis has no redundant vectors. This is exactly what is captured by demanding linear independence in the definition.

Example 2.19

Show that the vectors $\mathbf{u} = \begin{pmatrix} 1 \\ 1 \end{pmatrix}$ and $\mathbf{v} = \begin{pmatrix} 1 \\ -1 \end{pmatrix}$ form a basis for $\mathbb{R}^2$.

Solution

We are required to show two things: that **u** and **v** **(i)** span $\mathbb{R}^2$ and **(ii)** are linearly independent.

(i) *How do we verify that vectors* **u** *and* **v** *span* $\mathbb{R}^2$?

Let $\mathbf{w} = \begin{pmatrix} a \\ b \end{pmatrix}$ be an arbitrary vector in $\mathbb{R}^2$ and consider the linear combination:

$$k\mathbf{u} + c\mathbf{v} = \mathbf{w}$$

Substituting the given vectors **u**, **v** and **w** yields:

$$k\begin{pmatrix} 1 \\ 1 \end{pmatrix} + c\begin{pmatrix} 1 \\ -1 \end{pmatrix} = \begin{pmatrix} a \\ b \end{pmatrix} \quad \text{with the augmented matrix} \quad \begin{matrix} R_1 \\ R_2 \end{matrix} \left(\begin{array}{cc|c} 1 & 1 & a \\ 1 & -1 & b \end{array}\right)$$

Carrying out the row operation $R_2 + R_1$:

$$\begin{matrix} \\ R_1 \\ R_2 + R_1 \end{matrix} \begin{matrix} \begin{matrix} k & c & \end{matrix} \\ \left(\begin{array}{cc|c} 1 & 1 & a \\ 2 & 0 & b+a \end{array}\right) \end{matrix}$$

Solving this for scalars gives $k = \frac{1}{2}(a+b)$ and $c = \frac{1}{2}(a-b)$. Since $\mathbf{w} = k\mathbf{u} + c\mathbf{v}$ we can write any vector **w** in $\mathbb{R}^2$ as

$$\mathbf{w} = \begin{pmatrix} a \\ b \end{pmatrix} = \frac{1}{2}(a+b)\begin{pmatrix} 1 \\ 1 \end{pmatrix} + \frac{1}{2}(a-b)\begin{pmatrix} 1 \\ -1 \end{pmatrix}$$

Therefore we conclude that the vectors **u** and **v** span $\mathbb{R}^2$.

(ii) *What else do we need to show for vectors* **u** *and* **v** *to be a basis?*

We need to verify that they are linearly independent. To show linearly independence of two vectors we just need to check that they are not scalar multiples of each other. (See question (4) of Exercises 2.3.)

The given vectors $\begin{pmatrix} 1 \\ 1 \end{pmatrix} \neq m\begin{pmatrix} 1 \\ -1 \end{pmatrix}$ (m is scalar) are not scalar multiples of each other, therefore they are linearly independent.

Vectors **u** and **v** both span $\mathbb{R}^2$ and are linearly independent so they are a basis for $\mathbb{R}^2$.

Equivalently we can say that the columns of matrix $\mathbf{A} = (\mathbf{u} \;\; \mathbf{v})$ form a basis for $\mathbb{R}^2$.

The standard unit vectors $\mathbf{e}_1$ and $\mathbf{e}_2$ (illustrated in Fig. 2.34(b) on the next page) are another basis for $\mathbb{R}^2$. This is generally called the **natural** or **standard basis** for $\mathbb{R}^2$.

Figure 2.34(a) shows some scalar multiples of the vectors **u** and **v** of the above example. These basis vectors **u** and **v** form another coordinate system for $\mathbb{R}^2$ as shown. Figure 2.34(b) shows the natural basis $\mathbf{e}_1$ and $\mathbf{e}_2$, which form our normal x-y coordinate system for $\mathbb{R}^2$.

Why use non-standard basis such as the vectors **u** *and* **v** *shown in Fig. 2.34(a)?*

Examining a vector in a different basis (coordinate system) may bring out structure related to that basis, which is hidden in the standard representation. It may be a relevant and useful structure.

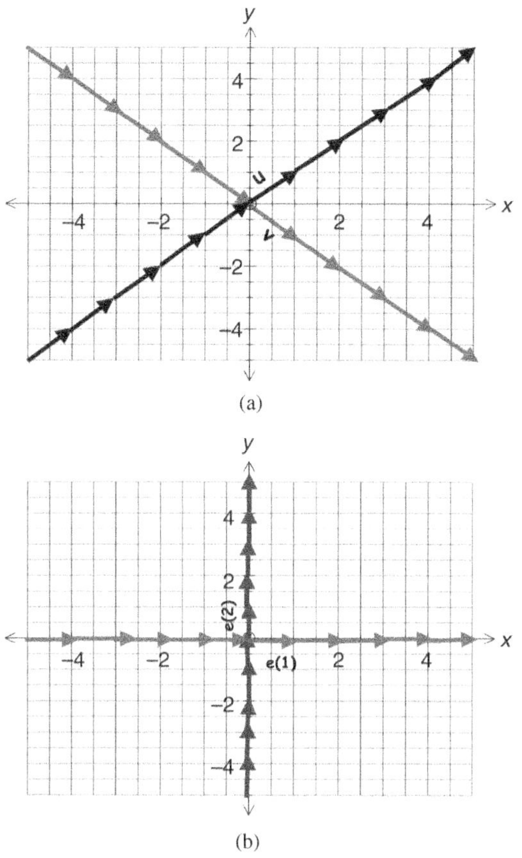

Figure 2.34

For example, looking at a force applied to an object on an inclined plane, the components parallel and perpendicular to the plane are far more useful than the horizontal and vertical components. Choosing a basis wisely can greatly reduce the amount of arithmetic we have to do.

In general, the standard unit vectors $\{\mathbf{e}_1, \mathbf{e}_2, \mathbf{e}_3, \ldots, \mathbf{e}_n\}$ form a **natural basis** for $\mathbb{R}^n$. Try proving this by showing linear independence and span.

This agrees with our usual Cartesian coordinate system on $\mathbb{R}^n$. For example, the vector $\mathbf{u} = (-1 \ \ 7 \ \ 2 \ \ -4 \ \ 9)^T$ in $\mathbb{R}^5$ can be written as

$$\mathbf{u} = (-1)\,\mathbf{e}_1 + 7\mathbf{e}_2 + 2\mathbf{e}_3 + (-4)\,\mathbf{e}_4 + 9\mathbf{e}_5$$

Each vector $\mathbf{x}$ in $\mathbb{R}^n$ is denoted by $\mathbf{x} = (x_1 \ \ x_2 \ \ \cdots \ \ x_n)^T$ which can be written as

$$\mathbf{x} = x_1\mathbf{e}_1 + x_2\mathbf{e}_2 + \cdots + x_n\mathbf{e}_n$$

in the standard basis $\{\mathbf{e}_1, \mathbf{e}_2, \mathbf{e}_3, \ldots, \mathbf{e}_n\}$ for $\mathbb{R}^n$.

2.4.3 Properties of bases

Proposition (2.27). Any n linearly independent vectors in $\mathbb{R}^n$ form a basis for $\mathbb{R}^n$.

Proof – Exercises 2.4.

■

Note that this is an important result because it means that given n vectors in the n-space, $\mathbb{R}^n$, it is enough to show that they are linearly independent to form a basis. We don't need to show that they span $\mathbb{R}^n$ as well. For example, we only need:

(i) 3 linearly independent vectors in $\mathbb{R}^3$ to form a basis for $\mathbb{R}^3$.

(ii) 103 linearly independent vectors in $\mathbb{R}^{103}$ to form a basis for $\mathbb{R}^{103}$.

Proposition (2.28). Any n vectors which span $\mathbb{R}^n$ form a basis for $\mathbb{R}^n$.

Proof – Exercises 2.4.

■

Again, we only need to show that n vectors span $\mathbb{R}^n$ to prove that they are a basis for $\mathbb{R}^n$.

For example, we need 666 vectors that span $\mathbb{R}^{666}$ to form a basis for $\mathbb{R}^{666}$.

Both these results, (2.27) and (2.28), make life a lot easier because if we have n vectors in $\mathbb{R}^n$ then we only need to check one of the conditions, either independence or span.

Example 2.20

Show that the vectors $\mathbf{u} = \begin{pmatrix} 1 \\ 0 \\ 1 \end{pmatrix}$, $\mathbf{v} = \begin{pmatrix} 0 \\ 1 \\ -1 \end{pmatrix}$ and $\mathbf{w} = \begin{pmatrix} -2 \\ 3 \\ 0 \end{pmatrix}$ form a basis for $\mathbb{R}^3$.

Solution

What do we need to show?

Required to prove that the given vectors are linearly independent in $\mathbb{R}^3$.

Why don't we also have to prove that these vectors span $\mathbb{R}^3$?

Because by the above Proposition (2.27):

Any n linearly independent vectors in $\mathbb{R}^n$ form a basis for $\mathbb{R}^n$.

It is enough to show that three vectors are linearly independent for them to be a basis for the 3-space, $\mathbb{R}^3$. Consider the linear combination:

$$k_1\mathbf{u} + k_2\mathbf{v} + k_3\mathbf{w} = \mathbf{O}$$

Using row operations on the augmented matrix $(\mathbf{A}|\mathbf{O})$ with $\mathbf{A} = (\mathbf{u} \;\; \mathbf{v} \;\; \mathbf{w})$:

$$\begin{matrix} \\ \text{R}_1 \\ \text{R}_2 \\ \text{R}_3 \end{matrix} \begin{matrix} \begin{matrix} k_1 & k_2 & k_3 \end{matrix} \\ \left(\begin{array}{ccc|c} 1 & 0 & -2 & 0 \\ 0 & 1 & 3 & 0 \\ 1 & -1 & 0 & 0 \end{array}\right) \end{matrix} \quad \text{gives } k_1 = k_2 = k_3 = 0$$

The given vectors $\mathbf{u}$, $\mathbf{v}$ and $\mathbf{w}$ are linearly independent because all the scalars are zero.

By the above Proposition (2.27) the given three vectors $\mathbf{u}$, $\mathbf{v}$ and $\mathbf{w}$ form a basis for $\mathbb{R}^3$.

Figure 2.35 shows vectors produced by $\mathbf{u}$, $\mathbf{v}$ and $\mathbf{w}$.

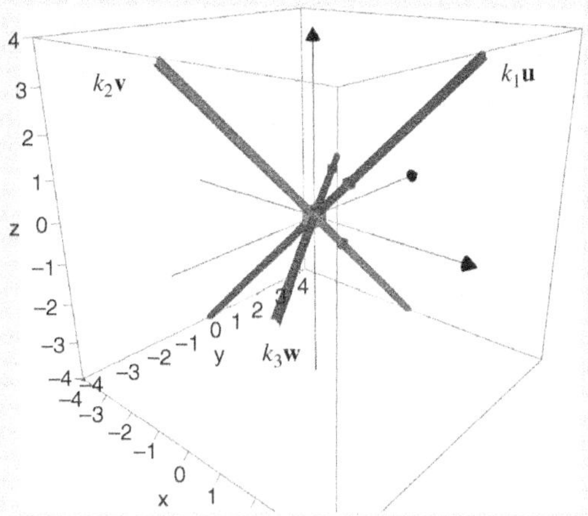

Figure 2.35

By linearly combining these vectors, we can make any vector $\mathbf{x}$ in $\mathbb{R}^3$:

$$\mathbf{x} = k_1\,\mathbf{u} + k_2\,\mathbf{v} + k_3\,\mathbf{w} \quad \text{(k's are scalars)}$$

We can measure vectors such as $\mathbf{x}$ with reference to these basis vectors $\{\mathbf{u},\ \mathbf{v},\ \mathbf{w}\}$.

Example 2.21

Determine whether the following vectors form a basis for $\mathbb{R}^4$:

$$\mathbf{v}_1 = (1\ \ 4\ \ -3\ \ 6)^T,\ \ \mathbf{v}_2 = (9\ \ 3\ \ -1\ \ -6)^T,\ \ \mathbf{v}_3 = (5\ \ 2\ \ 11\ \ -1)^T \text{ and } \mathbf{v}_4 = (0\ \ 0\ \ 0\ \ 0)^T$$

Solution

Since $\mathbf{v}_4 = (0\ \ 0\ \ 0\ \ 0)^T = \mathbf{O}$ is the zero vector, by Proposition (2.21) of the last section:

If at least one of vectors is $\mathbf{O}$ then the vectors are dependent.

Therefore the vectors $\{\mathbf{v}_1, \mathbf{v}_2, \mathbf{v}_3, \mathbf{v}_4\}$ form a linearly dependent set which means that these vectors cannot form a basis for $\mathbb{R}^4$. Remember that for a basis, we need the vectors to be linearly independent.

The remaining work in this section is a little more difficult than the above. You will need to be sure that you understand the definitions of spanning set, linear independence and basis of $\mathbb{R}^n$.

Proposition (2.29). Let the vectors $\{\mathbf{v}_1, \mathbf{v}_2, \mathbf{v}_3, \ldots, \mathbf{v}_n\}$ be a basis for $\mathbb{R}^n$. Every vector in $\mathbb{R}^n$ can be written uniquely as a linear combination of the vectors in this basis.

What does this proposition mean?

There is only one way of writing any vector as a linear combination of the basis vectors.

We have already proven this result for the standard basis in Proposition (2.18) of the last section.

Proof.
Let $\mathbf{u}$ be an arbitrary vector in $\mathbb{R}^n$. We are given that the vectors $\{\mathbf{v}_1, \mathbf{v}_2, \ldots, \mathbf{v}_n\}$ form a basis, so they span $\mathbb{R}^n$ which means that we can write the vector $\mathbf{u}$ in $\mathbb{R}^n$ as a linear combination of $\{\mathbf{v}_1, \mathbf{v}_2, \ldots, \mathbf{v}_n\}$. There exist scalars $k_1, k_2, k_3, \ldots$ and k_n which satisfy

$$\mathbf{u} = k_1\mathbf{v}_1 + k_2\mathbf{v}_2 + k_3\mathbf{v}_3 + \cdots + k_n\mathbf{v}_n$$

Suppose we can write this vector $\mathbf{u}$ as another linear combination of the basis vectors

$$\mathbf{u} = c_1\mathbf{v}_1 + c_2\mathbf{v}_2 + c_3\mathbf{v}_3 + \cdots + c_n\mathbf{v}_n$$

where the c's are scalars.

What do we need to prove?
We need to prove that the two sets of scalars are equal: $k_1 = c_1,\ k_2 = c_2, \ldots$ and $k_n = c_n$. Equating the two linear combinations because both are equal to $\mathbf{u}$ gives

$$\begin{aligned}
k_1\mathbf{v}_1 + k_2\mathbf{v}_2 + \cdots + k_n\mathbf{v}_n &= c_1\mathbf{v}_1 + c_2\mathbf{v}_2 + \cdots + c_n\mathbf{v}_n = \mathbf{u}\\
k_1\mathbf{v}_1 + k_2\mathbf{v}_2 + \cdots + k_n\mathbf{v}_n - c_1\mathbf{v}_1 - c_2\mathbf{v}_2 - \cdots - c_n\mathbf{v}_n &= \mathbf{u} - \mathbf{u} = \mathbf{O}\\
(k_1 - c_1)\mathbf{v}_1 + (k_2 - c_2)\mathbf{v}_2 + \cdots + (k_n - c_n)\mathbf{v}_n &= \mathbf{O} \quad \text{[factorizing]}
\end{aligned}$$

The basis vectors $\{\mathbf{v}_1, \mathbf{v}_2, \mathbf{v}_3, \ldots, \mathbf{v}_n\}$ are linearly independent, therefore all the scalars are equal to zero.

Why?
Because this is the definition of linear independence given in the last section (2.19):
Vectors $\mathbf{v}_1, \mathbf{v}_2, \mathbf{v}_3, \ldots$ and $\mathbf{v}_n$ in $\mathbb{R}^n$ are linearly independent $\Leftrightarrow$

$$m_1\mathbf{v}_1 + m_2\mathbf{v}_2 + m_3\mathbf{v}_3 + \cdots + m_n\mathbf{v}_n = \mathbf{O} \text{ gives } m_1 = m_2 = m_3 = \cdots = m_n = 0$$

Applying this to the above derivation:

$$(k_1 - c_1)\mathbf{v}_1 + (k_2 - c_2)\mathbf{v}_2 + \cdots + (k_n - c_n)\mathbf{v}_n = \mathbf{O}$$

we have

$$\begin{gathered}
k_1 - c_1 = 0,\ k_2 - c_2 = 0,\ k_3 - c_3 = 0, \ldots \text{ and } k_n - c_n = 0\\
k_1 = c_1,\ k_2 = c_2,\ k_3 = c_3, \ldots \text{ and } k_n = c_n
\end{gathered}$$

Hence any arbitrary vector $\mathbf{u}$ can be written uniquely as a linear combination of the basis vectors $\{\mathbf{v}_1, \mathbf{v}_2, \mathbf{v}_3, \ldots, \mathbf{v}_n\}$.

■

Next we prove a lemma. A lemma is a proposition or theorem, the proof of which is used as a stepping stone towards proving something of greater interest. However, there are many lemmas in mathematics which have become important results in themselves, such as Zorn's lemma, Euclid's lemma and Gauss's lemma.

Lemma (2.30). Let $T = \{\mathbf{w}_1, \mathbf{w}_2, \mathbf{w}_3, \ldots, \mathbf{w}_m\}$ be a set of m vectors that are linearly independent in $\mathbb{R}^n$ then $m \leq n$.

What does this lemma mean?

There can be at most n independent vectors in $\mathbb{R}^n$. For example, there can only be three or less independent vectors in $\mathbb{R}^3$.

What is the largest number of linearly independent vectors in $\mathbb{R}^{10}$?

10

Proof.

By Proposition (2.22): In $\mathbb{R}^n$ if $n < m$, then the vectors $\mathbf{v}_1, \mathbf{v}_2, \ldots$ and $\mathbf{v}_m$ are dependent.

We are given that T is a set of vectors that are linearly independent. Therefore $n \geq m$ which is our required result.

■

The maximum number of independent vectors in $\mathbb{R}^n$ form a basis for $\mathbb{R}^n$.

The next proof is a 'proof by contradiction'. Here the procedure to prove that a statement P implies a statement Q (denoted $P \Rightarrow Q$) is to suppose that P *does not* imply Q, and then prove that this can't be the case. Proof by contradiction is a common technique in mathematics and is a powerful mathematical tool often used to prove results.

For example, show that $\dfrac{x^2+2}{x} = x$ has no solution.

Proof – Suppose that the given equation *does have* a solution, call it a:

$$\frac{a^2+2}{a} = a \text{ implies } a^2 + 2 = a^2 \text{ implies } 2 = 0$$

but 2 cannot equal 0, therefore we have arrived at a contradiction. This means that there is something wrong. Since all our derivations are correct, our supposition that there is a solution is incorrect.

■

Another example of proof by contradiction is in the field of number theory: to prove that there are an infinite number of prime numbers, we first suppose that there are only a finite number of primes numbers, and then arrive at a contradiction.

Proposition (2.31). Every basis of $\mathbb{R}^n$ contains exactly n vectors.

Proof.
Let $B = \{\mathbf{v}_1, \mathbf{v}_2, \mathbf{v}_3, \ldots, \mathbf{v}_m\}$ be a basis for $\mathbb{R}^n$. By the above Lemma (2.30) we have

$$m \le n \qquad \text{[because the vectors are linearly independent]}$$

Suppose $m < n$. Let $T = \{\mathbf{e}_1, \mathbf{e}_2, \mathbf{e}_3, \ldots, \mathbf{e}_n\}$ be the natural basis for $\mathbb{R}^n$. We can write each of these vectors in the set T in terms of the basis vectors $B = \{\mathbf{v}_1, \mathbf{v}_2, \ldots, \mathbf{v}_m\}$:

$$\begin{aligned}
\mathbf{e}_1 &= a_{11}\mathbf{v}_1 + a_{21}\mathbf{v}_2 + \cdots + a_{m1}\mathbf{v}_m \\
\mathbf{e}_2 &= a_{12}\mathbf{v}_1 + a_{22}\mathbf{v}_2 + \cdots + a_{m2}\mathbf{v}_m \\
\vdots &\quad\ \vdots \qquad\ \vdots \quad \vdots \qquad \vdots \quad \vdots \\
\mathbf{e}_n &= a_{1n}\mathbf{v}_1 + a_{2n}\mathbf{v}_2 + \cdots + a_{mn}\mathbf{v}_m
\end{aligned}$$

Consider the linear combination

$$c_1\mathbf{e}_1 + c_2\mathbf{e}_2 + \cdots + c_n\mathbf{e}_n = \mathbf{O} \qquad (*)$$

Substituting the above for the vectors $\mathbf{e}_1, \mathbf{e}_2, \mathbf{e}_3, \ldots, \mathbf{e}_n$ into this and rearranging:

$$\begin{aligned}
&c_1\left(a_{11}\mathbf{v}_1 + \cdots + a_{m1}\mathbf{v}_m\right) + c_2\left(a_{12}\mathbf{v}_1 + \cdots + a_{m2}\mathbf{v}_m\right) + \cdots \\
&\qquad\qquad + c_n\left(a_{1n}\mathbf{v}_1 + \cdots + a_{mn}\mathbf{v}_m\right) = \mathbf{O} \\
&\left(c_1a_{11} + c_2a_{12} + \cdots + c_na_{1n}\right)\mathbf{v}_1 + \left(c_1a_{21} + c_2a_{22} + \cdots + c_na_{2n}\right)\mathbf{v}_2 + \cdots \\
&\qquad\qquad + \left(c_1a_{m1} + c_2a_{m2} + \cdots + c_na_{mn}\right)\mathbf{v}_m = \mathbf{O}
\end{aligned}$$

The vectors in $B = \{\mathbf{v}_1, \mathbf{v}_2, \mathbf{v}_3, \ldots \mathbf{v}_m\}$ are linearly independent because they are a basis, so all the scalars (bracketed terms) in the above must be equal to zero:

$$\begin{aligned}
c_1a_{11} + c_2a_{12} + \cdots + c_na_{1n} &= 0 \qquad \text{(Equation 1)} \\
c_1a_{21} + c_2a_{22} + \cdots + c_na_{2n} &= 0 \qquad \text{(Equation 2)} \\
\vdots \qquad\quad \vdots \qquad\qquad \vdots \quad &\ \ \vdots \\
c_1a_{m1} + c_2a_{m2} + \cdots + c_na_{mn} &= 0 \qquad \text{(Equation } m\text{)}
\end{aligned}$$

We can write this in matrix form as $\mathbf{Ax} = \mathbf{O}$ where

$$\mathbf{A} = \begin{pmatrix} a_{11} & \cdots & a_{1n} \\ \vdots & \ddots & \vdots \\ a_{m1} & \cdots & a_{mn} \end{pmatrix}, \quad \mathbf{x} = \begin{pmatrix} c_1 \\ \vdots \\ c_n \end{pmatrix} \text{ and } \mathbf{O} = \begin{pmatrix} 0 \\ \vdots \\ 0 \end{pmatrix}$$

We have m equations and n unknowns. Our supposition is that $m < n$, therefore there are fewer (m) equations than the number (n) of unknowns (c's). By Proposition (1.31):

If the number of equations is less than the number of unknowns then the linear system $\mathbf{Ax} = \mathbf{O}$ has an infinite number of solutions.

We have an infinite number of solutions which means that $\mathbf{Ax} = \mathbf{O}$ has non-trivial solutions, so all the c's are not zero.

By the linear combination in (*) we conclude that the vectors $T = \{\mathbf{e}_1, \mathbf{e}_2, \mathbf{e}_3, \ldots, \mathbf{e}_n\}$ are linear dependent because all the c's are not zero. This cannot be the case because this set T is the standard basis for $\mathbb{R}^n$, which means that these vectors are linearly independent. We have a contradiction, so our supposition that $m < n$ must be wrong, therefore $m = n$.

Hence every basis of $\mathbb{R}^n$ has exactly n vectors.

■

This result means that in a basis, there are enough vectors to span (generate) the whole n-space, $\mathbb{R}^n$, but not too many so that the vectors become dependent. Hence we need exactly n vectors to form a basis for $\mathbb{R}^n$.

Proposition (2.32). Any n non-zero orthogonal vectors in $\mathbb{R}^n$ form a basis for $\mathbb{R}^n$.

Proof – Exercises 2.4.

■

Summary

The set S is a basis for $\mathbb{R}^n$ $\Leftrightarrow$

(i) $S = \{\mathbf{v}_1, \mathbf{v}_2, \mathbf{v}_3, \ldots, \mathbf{v}_n\}$ spans $\mathbb{R}^n$ and

(ii) $S = \{\mathbf{v}_1, \mathbf{v}_2, \mathbf{v}_3, \ldots, \mathbf{v}_n\}$ is linearly independent

Every basis of $\mathbb{R}^n$ contains exactly n vectors.

EXERCISES 2.4

(Brief solutions at end of book. Full solutions available at <http://www.oup.co.uk/companion/singh>.)

1. Determine whether the following vectors span $\mathbb{R}^2$:

   (a) $\mathbf{e}_1 = \begin{pmatrix} 1 \\ 0 \end{pmatrix}$ and $\mathbf{e}_2 = \begin{pmatrix} 0 \\ 1 \end{pmatrix}$ (b) $\mathbf{u} = \begin{pmatrix} 1 \\ 1 \end{pmatrix}$ and $\mathbf{v} = \begin{pmatrix} -1 \\ 1 \end{pmatrix}$

   (c) $\mathbf{u} = \begin{pmatrix} 2 \\ 2 \end{pmatrix}$ and $\mathbf{v} = \begin{pmatrix} -1 \\ -1 \end{pmatrix}$ (d) $\mathbf{u} = \begin{pmatrix} 1 \\ 2 \end{pmatrix}$ and $\mathbf{v} = \begin{pmatrix} -1 \\ 10 \end{pmatrix}$

2. Determine whether the following vectors span $\mathbb{R}^3$:

   (a) $\mathbf{u} = \begin{pmatrix} 2 \\ 0 \\ 0 \end{pmatrix}$, $\mathbf{v} = \begin{pmatrix} 0 \\ 2 \\ 0 \end{pmatrix}$ and $\mathbf{w} = \begin{pmatrix} 0 \\ 0 \\ 2 \end{pmatrix}$

   (b) $\mathbf{u} = \begin{pmatrix} 1 \\ 1 \\ 1 \end{pmatrix}$, $\mathbf{v} = \begin{pmatrix} 2 \\ 2 \\ 2 \end{pmatrix}$ and $\mathbf{w} = \begin{pmatrix} 1 \\ 2 \\ 3 \end{pmatrix}$

(c) $\mathbf{u} = \begin{pmatrix} 1 \\ 1 \\ 1 \end{pmatrix}$, $\mathbf{v} = \begin{pmatrix} 1 \\ 1 \\ 0 \end{pmatrix}$ and $\mathbf{w} = \begin{pmatrix} 1 \\ 0 \\ 0 \end{pmatrix}$

(d) $\mathbf{u} = \begin{pmatrix} 1 \\ 2 \\ 1 \end{pmatrix}$, $\mathbf{v} = \begin{pmatrix} 2 \\ 4 \\ 0 \end{pmatrix}$ and $\mathbf{w} = \begin{pmatrix} -2 \\ -2 \\ 3 \end{pmatrix}$

3. Determine whether the following vectors form a basis for $\mathbb{R}^2$:

(a) $\mathbf{u} = (1 \ \ 2)^T, \ \mathbf{v} = (0 \ \ 1)^T$ (b) $\mathbf{u} = (-2 \ \ -4)^T, \ \mathbf{v} = (1 \ \ 2)^T$

(c) $\mathbf{u} = (4 \ \ 1)^T, \ \mathbf{v} = (1 \ \ 3)^T$ (d) $\mathbf{u} = (3 \ \ 5)^T, \ \mathbf{v} = (2 \ \ 3)^T$

4. Explain why the following vectors do not form a basis for the indicated Euclidean space:

(a) $\mathbf{u} = (1 \ \ 2)^T, \ \mathbf{v} = (0 \ \ 1)^T, \ \mathbf{w} = (1 \ \ 0)^T$ for $\mathbb{R}^2$

(b) $\mathbf{u} = (1 \ \ 1 \ \ 1)^T, \ \mathbf{v} = (1 \ \ 1 \ \ 0)^T, \ \mathbf{w} = (0 \ \ 0 \ \ 0)^T$ for $\mathbb{R}^3$

(c) $\mathbf{u} = (1 \ \ 1 \ \ 1 \ \ 1)^T, \ \mathbf{v} = (1 \ \ 1 \ \ 0 \ \ -1)^T, \ \mathbf{w} = (-1 \ \ -2 \ \ -3 \ \ 4)^T,$
$\mathbf{x} = (2 \ \ 2 \ \ 2 \ \ 2)^T$ for $\mathbb{R}^4$

(d) $\mathbf{u} = (1 \ \ 2 \ \ 3 \ \ 4 \ \ 5)^T, \ \mathbf{v} = (1 \ \ 5 \ \ 0 \ \ 2 \ \ 4)^T, \ \mathbf{w} = (3 \ \ 2 \ \ -5 \ \ 4 \ \ -9)^T,$
$\mathbf{x} = (2 \ \ 9 \ \ 2 \ \ 7 \ \ 7)^T$ for $\mathbb{R}^5$

5. Determine whether vector **b** is in the space spanned by the columns of matrix A:

(a) $\mathbf{A} = \begin{pmatrix} 1 & 1 & 0 \\ 2 & 5 & 0 \\ 3 & 4 & 9 \end{pmatrix}, \mathbf{b} = \begin{pmatrix} 1 \\ 3 \\ 4 \end{pmatrix}$ (b) $\mathbf{A} = \begin{pmatrix} 1 & 2 & 3 \\ 4 & 5 & 6 \\ 7 & 8 & 9 \end{pmatrix}, \mathbf{b} = \begin{pmatrix} 1 \\ 3 \\ 4 \end{pmatrix}$

6. Show that the following vectors form a basis for $\mathbb{R}^3$:

(a) $\mathbf{u} = (5 \ \ 0 \ \ 0)^T, \ \mathbf{v} = (0 \ \ 6 \ \ 0)^T, \ \mathbf{w} = (0 \ \ 0 \ \ 7)^T$

(b) $\mathbf{u} = (\alpha \ \ 0 \ \ 0)^T, \ \mathbf{v} = (0 \ \ \beta \ \ 0)^T, \ \mathbf{w} = (0 \ \ 0 \ \ \lambda)^T$, where α, β and λ are any non-zero real numbers.

7. Show that the vectors $\mathbf{u} = (1 \ \ 1 \ \ 2)^T, \ \mathbf{v} = (-1 \ \ 1 \ \ -2)^T, \ \mathbf{w} = (1 \ \ -5 \ \ 2)^T$ do not form a basis for $\mathbb{R}^3$.

8. Show that the vectors

$$\mathbf{u} = (1 \ \ 1 \ \ 1 \ \ 1)^T, \ \mathbf{v} = (0 \ \ 1 \ \ 1 \ \ 1)^T, \ \mathbf{w} = (0 \ \ 0 \ \ 1 \ \ 1)^T, \ \mathbf{x} = (0 \ \ 0 \ \ 0 \ \ 1)^T$$

form a basis for $\mathbb{R}^4$.

9. Let $\mathbf{u} = (1 \ \ 5 \ \ 0)^T, \quad \mathbf{v} = (0 \ \ 1 \ \ 0)^T$.

(a) Determine the dependency of **u** and **v**.

(b) Find the space spanned by $\{\mathbf{u}, \mathbf{v}\}$.

(c) Write down a general vector **w** such that $\{\mathbf{u}, \mathbf{v}, \mathbf{w}\}$ form a basis for $\mathbb{R}^3$.

10. Let $\mathbf{v}_1, \mathbf{v}_2, \mathbf{v}_3, \ldots, \mathbf{v}_n$ span $\mathbb{R}^n$. Show that $\mathbf{v}_1, \mathbf{v}_2, \mathbf{v}_3, \ldots, \mathbf{v}_n$ and $\mathbf{w}$ also span $\mathbb{R}^n$.

11. Let $S = \{\mathbf{v}_1, \mathbf{v}_2, \mathbf{v}_3, \ldots, \mathbf{v}_n\}$ be a basis for $\mathbb{R}^n$. Prove that

$$T = \left\{k_1\mathbf{v}_1, k_2\mathbf{v}_2, k_3\mathbf{v}_3, \ldots, k_n\mathbf{v}_n\right\}$$

where none of the k's are zero, is also a basis for $\mathbb{R}^n$.

12. Let matrix $\mathbf{A} = (\mathbf{v}_1 \;\; \mathbf{v}_2 \;\; \cdots \;\; \mathbf{v}_n)$. Prove that $\mathbf{A}$ is invertible $\Leftrightarrow$ $\{\mathbf{v}_1, \ldots, \mathbf{v}_n\}$ form a basis for $\mathbb{R}^n$.
13. Prove Proposition (2.27).
14. Prove Proposition (2.28).
15. Prove Proposition (2.32).
16. *Let $\{\mathbf{v}_1, \mathbf{v}_2, \mathbf{v}_3, \ldots, \mathbf{v}_n\}$ be a basis for $\mathbb{R}^n$ and $\mathbf{A}$ be an invertible matrix. Prove that $\{\mathbf{A}\mathbf{v}_1, \mathbf{A}\mathbf{v}_2, \mathbf{A}\mathbf{v}_3, \ldots, \mathbf{A}\mathbf{v}_n\}$ is also a basis for $\mathbb{R}^n$.

MISCELLANEOUS EXERCISES 2

(Brief solutions at end of book. Full solutions available at <http://www.oup.co.uk/companion/singh>.)

2.1. Determine whether the following vectors are linearly independent in the vector space V (**show your working**).

$$V = \mathbb{R}^3, \quad \left\{\begin{bmatrix}1\\0\\0\end{bmatrix}, \begin{bmatrix}1\\1\\1\end{bmatrix}, \begin{bmatrix}1\\0\\0\end{bmatrix}\right\}$$

Purdue University, USA
(part question)

2.2. Give an example or state that no such example exists of :
A set S containing two vectors from $\mathbb{R}^3$ such that Span $S = \mathbb{R}^3$.

Illinois State University, USA
(part question)

2.3. Let $\mathbf{v}_1 = (1,\;\; 1,\;\; 1)$, $\mathbf{v}_2 = (1,\;\; 0,\;\; -1)$ and $\mathbf{v}_3 = (0,\;\; 1,\;\; 1)$.

(a) Are $\mathbf{v}_1$ and $\mathbf{v}_2$ linearly independent? Explain why or why not.

(b) Is $(3,\;\; 2,\;\; 1)$ in the span of $\mathbf{v}_1$ and $\mathbf{v}_2$? Explain why or why not.

(c) Is $\{\mathbf{v}_1,\;\; \mathbf{v}_2,\;\; \mathbf{v}_3\}$ a basis for $\mathbb{R}^3$?

Mount Holyoke College, Massachusetts, USA

2.4. Can the vector $(3,\;\; 1,\;\; 1)$ be expressed as a linear combination of the vectors $(2,\;\; 5,\;\; -1)$, $(1,\;\; 6,\;\; 0)$, $(5,\;\; 2,\;\; -4)$? Justify your answer.

University of Western Ontario, Canada

2.5. (a) Define what is meant by stating that a set of vectors $\{\mathbf{v}_1, \mathbf{v}_2, \ldots, \mathbf{v}_k\}$ in $\mathbb{R}^n$ is *linearly dependent.*

(b) Prove that the following set of vectors in $\mathbb{R}^3$ are linearly dependent.

$$\left\{ \begin{bmatrix} 1 \\ 2 \\ -1 \end{bmatrix}, \begin{bmatrix} 2 \\ -1 \\ 3 \end{bmatrix}, \begin{bmatrix} 1 \\ 7 \\ -6 \end{bmatrix} \right\}$$

University of Manchester, UK

2.6. (a) Find two different bases for the real vector space $\mathbb{R}^2$.

(b) Do the following sets form a basis for $\mathbb{R}^3$? If not, determine whether they are linearly independent, a spanning set for $\mathbb{R}^3$ or neither.

i. $\{(0, 0, 1), (0, 2, 1), (3, 2, 1), (4, 5, 6)\}$

ii. $\{(1, 0, 1), (4, 3, -1), (5, 3, -4)\}$

City University, London, UK
(part question)

2.7. Find a Basis for the following subspace S of the given vector space V. (**Do not show it is a subspace.**)

$$V = \mathbb{R}^5, \quad S = \left\{ \begin{bmatrix} a+b \\ b \\ c \\ 0 \\ c+b \end{bmatrix} \;\middle|\; a, b, c \in \mathbb{R} \right\}$$

Purdue University, USA
(part question)

2.8. Find the largest possible number of linearly independent vectors among

$$\mathbf{v}_1 = \begin{bmatrix} 1 \\ -1 \\ 0 \\ 0 \end{bmatrix} \quad \mathbf{v}_2 = \begin{bmatrix} 1 \\ 0 \\ -1 \\ 0 \end{bmatrix} \quad \mathbf{v}_3 = \begin{bmatrix} 1 \\ 0 \\ 0 \\ -1 \end{bmatrix} \quad \mathbf{v}_4 = \begin{bmatrix} 0 \\ 1 \\ -1 \\ 0 \end{bmatrix} \quad \mathbf{v}_5 = \begin{bmatrix} 0 \\ 1 \\ 0 \\ -1 \end{bmatrix} \quad \mathbf{v}_6 = \begin{bmatrix} 0 \\ 0 \\ 1 \\ -1 \end{bmatrix}$$

Illinois State University, USA

2.9. (a) Let $\mathbf{v}_1 = \begin{pmatrix} 3 \\ -1 \\ 1 \\ 0 \end{pmatrix}, \mathbf{v}_2 = \begin{pmatrix} 4 \\ -7 \\ 3 \\ 2 \end{pmatrix}, \mathbf{v}_3 = \begin{pmatrix} 3 \\ 7 \\ -2 \\ -6 \end{pmatrix}.$

Show that $\{\mathbf{v}_1, \mathbf{v}_2, \mathbf{v}_3\}$ are linearly independent.

(b) For what value(s) of λ is $\left\{\begin{pmatrix}1\\1\\2\end{pmatrix}, \begin{pmatrix}1\\0\\-1\end{pmatrix}, \begin{pmatrix}1\\-1\\\lambda\end{pmatrix}\right\}$ a basis for $\mathbb{R}^3$?

Provide your reasoning.

King's College London, UK
(part question)

2.10. Find a real number c such that the vectors $\begin{pmatrix}1\\2\\4\end{pmatrix}, \begin{pmatrix}2\\3\\5\end{pmatrix}, \begin{pmatrix}2\\7\\c\end{pmatrix}$ do not form a basis of $\mathbb{R}^3$.

University of California Berkeley, USA
(part question)

2.11. Suppose that $\{\mathbf{v}_1, \ \mathbf{v}_2\}$ are linearly independent set in $\mathbb{R}^m$. Show that $\{\mathbf{v}_1, \mathbf{v}_1 + \mathbf{v}_2\}$ is also linearly independent.

University of Maryland, Baltimore County, USA
(part question)

2.12. Which of the following form a basis for $\mathbb{R}^3$? Why?

(a) $\left\{\begin{pmatrix}1\\2\\3\end{pmatrix}, \begin{pmatrix}2\\-1\\5\end{pmatrix}\right\}$ (b) $\left\{\begin{pmatrix}1\\2\\3\end{pmatrix}, \begin{pmatrix}4\\5\\6\end{pmatrix}, \begin{pmatrix}6\\9\\12\end{pmatrix}\right\}$

(c) $\left\{\begin{pmatrix}1\\2\\3\end{pmatrix}, \begin{pmatrix}4\\5\\6\end{pmatrix}, \begin{pmatrix}7\\8\\9\end{pmatrix}, \begin{pmatrix}10\\11\\12\end{pmatrix}\right\}$

University of Maryland, Baltimore County, USA
(part question)

2.13. Consider the three vectors $\mathbf{u} = (1, \ 3, \ 1), \mathbf{v} = (4, \ 2, \ -1)$ and $\mathbf{w} = (-3, \ 1, \ 2)$.

(a) Either prove that the $\mathbf{u}$ is in the span of the vectors $\mathbf{v}$ and $\mathbf{w}$, or prove that it is not.

(b) Are the three vectors $\mathbf{u}$, $\mathbf{v}$ and $\mathbf{w}$ linearly dependent, or linearly independent?

Clark University, USA

2.14. Let $\mathbf{u}_1 = \begin{bmatrix}1\\0\\1\end{bmatrix}, \ \mathbf{u}_2 = \begin{bmatrix}-1\\4\\1\end{bmatrix}, \ \mathbf{u}_3 = \begin{bmatrix}2\\1\\-2\end{bmatrix}$ and $\mathbf{x} = \begin{bmatrix}8\\-4\\-3\end{bmatrix}$.

(a) Determine if the set $\{\mathbf{u}_1, \mathbf{u}_2, \mathbf{u}_3\}$ is orthogonal.

(b) Is $\{\mathbf{u}_1, \mathbf{u}_2, \mathbf{u}_3\}$ a basis for $\mathbb{R}^3$? If yes, express $\mathbf{x}$ as a linear combination of $\mathbf{u}_1, \mathbf{u}_2$ and $\mathbf{u}_3$.

Washington State University, USA

2.15. Let $\mathbf{u} = \begin{bmatrix}2\\0\\3\end{bmatrix}, \ \mathbf{v} = \begin{bmatrix}6\\2\\-4\end{bmatrix}$ and $\mathbf{x} = \begin{bmatrix}2\\4\\0\end{bmatrix}$. Find the following:

(a) $\mathbf{u} \cdot \mathbf{u}$

(b) $\|\mathbf{u}\|$

(c) A unit vector in the direction of $\mathbf{u}$.

(d) $\mathbf{u} \cdot \mathbf{v}$

(e) Are $\mathbf{u}$ and $\mathbf{v}$ orthogonal? Explain.

(f) The distance between $\mathbf{v}$ and $\mathbf{x}$.

Washington State University, USA

2.16. (a) Define the *scalar product* of $\mathbf{x} = (x_1, \ldots, x_n)$ and $\mathbf{y} = (y_1, \ldots, y_n)$ in $\mathbb{R}^n$.

(b) Define the norm of $\mathbf{x} = (x_1, \ldots, x_n)$ in $\mathbb{R}^n$.

(c) In $\mathbb{R}^n$, describe what is meant by the angle θ between the two vectors $\mathbf{u}$ and $\mathbf{v}$. Use this to find the angle between the vectors $\mathbf{u}, \mathbf{v} \in \mathbb{R}^5$ given by

$$\mathbf{u} = (-1,\ 1,\ 1,\ -1,\ 0),\quad \mathbf{v} = (0,\ 2,\ 1,\ 0,\ 2)$$

University of Sussex, UK
(part question)

2.17. (a) Define the standard inner product and the Euclidean norm on the three-dimensional Euclidean vector space $\mathbb{R}^3$, i.e. $\langle \mathbf{x},\ \mathbf{y} \rangle$ and $\|\mathbf{x}\|$ for $\mathbf{x}, \mathbf{y} \in \mathbb{R}^3$.

(b) Define an orthonormal set of vectors.

(c) In $\mathbb{R}^3$, let $\mathbf{u} = (1,\ 1,\ 1)$. By finding the condition for the general vector $(x,\ y,\ z)$ to be orthogonal to $\mathbf{u}$, find a vector $\mathbf{v}$ that is orthogonal to $\mathbf{u}$. Then, by finding the condition for the general vector $(x,\ y,\ z)$ to be orthogonal to $\mathbf{v}$, find a vector $\mathbf{w}$ that is orthogonal to both $\mathbf{u}$ and $\mathbf{v}$. Hence construct an orthonormal set of vectors that contains a scalar multiple of $(1,\ 1,\ 1)$.

Queen Mary, University of London, UK

2.18. (a) Let $\mathbf{u} = (2,\ 2,\ -1,\ 3)$, $\mathbf{v} = (1,\ 3,\ 2,\ -2)$, $\mathbf{w} = (3,\ 1,\ 3,\ 1)$ be vectors in $\mathbb{R}^4$. Compute the following dot products $\mathbf{u} \cdot \mathbf{v}$, $(\mathbf{u} - \mathbf{v}) \cdot (\mathbf{u} - \mathbf{w})$ and $(2\mathbf{u} - 3\mathbf{v}) \cdot (\mathbf{u} + 4\mathbf{w})$.

(b) Verify the Cauchy–Schwarz inequality for $\mathbf{u} = (1, 1, 1)$ and $\mathbf{v} = (-1, 1, 2)$.

Jacobs University, Germany
(part question)

2.19. Let $\mathbf{v}_1, \mathbf{v}_2, \mathbf{v}_3 \in \mathbb{R}^3$ be an orthogonal set of non-zero vectors. Prove that they are linearly independent.

University of Southampton, UK
(part question)

2.20. Let $\{\mathbf{u}_1, \mathbf{u}_2, \ldots, \mathbf{u}_n\}$ be a basis for $\mathbb{R}^n$, and let $\mathbf{A}$ be an invertible $n \times n$ matrix. Prove that $\{\mathbf{A}\mathbf{u}_1, \mathbf{A}\mathbf{u}_2, \ldots, \mathbf{A}\mathbf{u}_n\}$ is a basis for $\mathbb{R}^n$.

Illinois State University, USA

Sample Questions

2.21. Let $\mathbf{u}$ and $\mathbf{v}$ be vectors in $\mathbb{R}^n$. Prove that

$$\mathbf{u} \cdot \mathbf{v} = 0 \quad \Leftrightarrow \quad \|\mathbf{u} + \mathbf{v}\| = \|\mathbf{u} - \mathbf{v}\|$$

2.22. Prove that if for all non-zero vectors $\mathbf{u}$ in $\mathbb{R}^n$ we have $\mathbf{u} \cdot \mathbf{v} = \mathbf{u} \cdot \mathbf{w}$ then $\mathbf{v} = \mathbf{w}$.

2.23. Let $\mathbf{u} = \begin{pmatrix} a \\ b \end{pmatrix}$ and $\mathbf{v} = \begin{pmatrix} c \\ d \end{pmatrix}$ be vectors in $\mathbb{R}^2$. Prove that vectors $\mathbf{u}$ and $\mathbf{v}$ are linearly independent $\Leftrightarrow ad - bc \neq 0$.

2.24. Prove that a set of m vectors where $m < n$ *cannot* span $\mathbb{R}^n$.

2.25. Prove that any n vectors which span $\mathbb{R}^n$ form a basis for $\mathbb{R}^n$.

Chao Yang

is a Staff Scientist in the Scientific Computing Group at the Lawrence Berkeley National Laboratory (LBNL) in California.

Tell us about yourself and your work.

I have been working on numerical linear algebra since graduate school (Rice University). My expertise is in solving large-scale eigenvalue problems. At LBNL, I work on a number of interesting science projects. All of them require using linear algebra to obtain desirable solutions.

How do you use linear algebra in your job?

I use linear algebra to solve quantum mechanics problems. One of the fundamental problems in chemistry and material science is to understand the electronic structure of atoms, molecules and solids. This problem can be solved by computing the solutions to Schrödinger's equation, which is an eigenvalue problem. There are various ways to obtain approximations to the solution. All of them ultimately result in solving a large sparse matrix eigenvalue problem.

How important is linear algebra?

Linear algebra is indispensable in my research. Every problem eventually boils down to a linear algebra problem.

What are the challenges connected with the subject?

The main challenge is the size of the problem, which nowadays can be in the order of millions or even billions. However, these large problems often have structures (e.g. sparsity). Therefore, to solve these problems, one has to take advantage of the structure.

What are the key issues in your area of linear algebra research?

Convergence of iterative solvers for linear equations, least squares problems and eigenvalue problems, methods that can take advantage of the sparsity or other structures of matrices.

3 General Vector Spaces

SECTION 3.1 Introduction to General Vector Spaces

By the end of this section you will be able to

- understand what is meant by a vector space
- give examples of vector spaces
- prove properties of vector spaces

In the last chapter we used vectors to represent a quantity with 'magnitude and direction'. In fact, this 'magnitude–direction' definition relates to a specific application of vectors, which is used in the fields of science and engineering. However, vectors have a much broader range of applications, and they are more generally defined as 'elements of a vector space'. In this chapter we describe what is meant by a vector space and how it is mathematically defined.

We discuss the vector spaces that accommodate matrices, functions and polynomials. You may find this chapter more abstract than the previous chapters, because for Euclidean spaces $\mathbb{R}^2$ and $\mathbb{R}^3$ we were able to visualize and even plot the vectors. It is clearly impossible to draw a vector in four dimensions, but that doesn't make the mathematics carried out in that vector space any less valid.

First, we define what is meant by a general vector space in terms of a set of axioms.

*What does the term **axiom** mean?*

An **axiom** is a statement which is self-evidently true and doesn't have a proof.

We define vector space with 10 axioms, based on the two fundamental operations of linear algebra - vector addition and scalar multiplication.

3.1.1 Vector space

Let V be a non-empty set of elements called vectors. We define two operations on the set V– **vector addition** and **scalar multiplication**. Scalars are real numbers.

Let $\mathbf{u}$, $\mathbf{v}$ and $\mathbf{w}$ be vectors in the set V. The set V is called a **vector space** if it satisfies the following 10 axioms.

1. The vector addition $\mathbf{u} + \mathbf{v}$ is also in the vector space V. Generally in mathematics we say that we have **closure** under vector addition if this property holds.
2. Commutative law: $\mathbf{u} + \mathbf{v} = \mathbf{v} + \mathbf{u}$.

3. Associative law: $(\mathbf{u}+\mathbf{v})+\mathbf{w}=\mathbf{u}+(\mathbf{v}+\mathbf{w})$.
4. Neutral element. There is a vector called the **zero** vector in V denoted by $\mathbf{O}$ which satisfies

$$\mathbf{u}+\mathbf{O}=\mathbf{u} \text{ for every vector } \mathbf{u} \text{ in } V$$

5. Additive inverse. For every vector $\mathbf{u}$ there is a vector $-\mathbf{u}$ (minus $\mathbf{u}$) which satisfies the following:

$$\mathbf{u}+(-\mathbf{u})=\mathbf{O}$$

6. Let k be a real scalar then $k\mathbf{u}$ is also in V. We say that we have **closure** under scalar multiplication if this axiom is satisfied.
7. Associative law for scalar multiplication. Let k and c be real scalars then

$$k(c\mathbf{u})=(kc)\mathbf{u}$$

8. Distributive law for vectors. Let k be a real scalar then

$$k(\mathbf{u}+\mathbf{v})=k\mathbf{u}+k\mathbf{v}$$

9. Distributive law for scalars. Let k and c be real scalars then

$$(k+c)\mathbf{u}=k\mathbf{u}+c\mathbf{u}$$

10. Identity element. For every vector $\mathbf{u}$ in V we have

$$1\mathbf{u}=\mathbf{u}$$

We say that if the elements of the set V satisfy the above 10 axioms then V is called a vector space and the elements are known as vectors. This might seem like a long list to digest, so don't worry if it seems a little intimidating at this point. We will use these axioms frequently in the next few sections, and you will soon become familiar with them.

3.1.2 Examples of vector spaces

Can you think of any examples of vector spaces?

The Euclidean spaces of the last chapter - $V=\mathbb{R}^2,\mathbb{R}^3,\ldots,\mathbb{R}^n$ - are *all* examples of vector spaces.

Are there any other examples of a vector space?

The set of matrices M_{22} that are *all* matrices of size 2 by 2 where matrix addition and scalar multiplication is defined as in chapter 1 form their own vector space.

Let $\mathbf{u}=\begin{pmatrix} a & b \\ c & d \end{pmatrix}$, $\mathbf{v}=\begin{pmatrix} e & f \\ g & h \end{pmatrix}$ and $\mathbf{w}=\begin{pmatrix} i & j \\ k & l \end{pmatrix}$ be matrices in M_{22}.

What is the zero vector in M_{22}?

The zero vector is the zero matrix of size 2 by 2 which is given by $\begin{pmatrix} 0 & 0 \\ 0 & 0 \end{pmatrix}=\mathbf{O}$.

The rules of matrix algebra established in chapter 1 ensure that all 10 axioms are satisfied, defining M_{22} as an example of a vector space. You are asked to check this in Exercises 3.1.

We can show that the set M_{23}, which is the set of matrices of size 2 by 3, also forms a vector space. You are asked to do this in Exercises 3.1.

There also exists vector space which is neither Euclidean space nor formed by a set of matrices. For example, the set of polynomials denoted $P(t)$ whose elements take the form:

$$\mathbf{p}(t) = c_0 + c_1 t + c_2 t^2 + \cdots + c_n t^n$$

where $c_0,\ c_1,\ c_2,\ \ldots$ and c_n are real numbers called the coefficients, form a vector space. The following are examples of polynomials

$$\mathbf{p}(t) = t^2 - 1,\ \mathbf{q}(t) = 1 + 2t + 7t^2 + 12t^3 - 3t^4 \text{ and } \mathbf{r}(t) = 7$$

The **degree** of a polynomial is the highest index (power) which has a non-zero coefficient, that is the maximum n for which $c_n \neq 0$.

What is the degree of $\mathbf{p}(t)$, $\mathbf{q}(t)$ and $\mathbf{r}(t)$?

$\mathbf{p}(t) = t^2 - 1$ is of degree 2, $\mathbf{q}(t) = 1 + 2t + 7t^2 + 12t^3 - 3t^4$ is of degree 4 and $\mathbf{r}(t) = 7$ is of degree 0. Note that the last polynomial $\mathbf{r}(t) = 7 = 7t^0$ is of degree 0 because 0 is the highest index with a non-zero coefficient (c_0).

*How do we define the **zero polynomial**?*

The zero polynomial, denoted by $\mathbf{O}$, is given by

$$\mathbf{O} = 0 + 0t + 0t^2 + \cdots + 0t^n$$

All the coefficient c's are zero. We say that the zero polynomial has *no degree*.

Let $P(t)$ be the set of polynomials of degree less than or equal to n where n is a positive or zero integer. The zero polynomial $\mathbf{O}$ is also in the set $P(t)$.

We define the two operations of vector addition and scalar multiplication as:

1. Vector addition $\mathbf{p}(t) + \mathbf{q}(t)$ is the normal addition of polynomials.
2. Scalar multiplication $k\mathbf{p}(t)$ is the normal multiplication of a constant, k, with the polynomial, $\mathbf{p}(t)$.

In the next example we show that $P(t)$ is also a vector space.

Example 3.1

Show that $P(t)$, the set of polynomials of degree n or less, is indeed a vector space.

Solution

We need to check all 10 axioms with the elements or vectors (polynomials) of this space which have the general polynomial form:

$$\mathbf{p}(t) = c_0 + c_1 t + c_2 t^2 + \cdots + c_n t^n,\ \mathbf{q}(t) = d_0 + d_1 t + d_2 t^2 + \cdots + d_n t^n \text{ and}$$
$$\mathbf{r}(t) = e_0 + e_1 t + e_2 t^2 + \cdots + e_n t^n$$

(continued...)

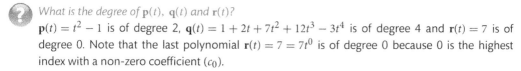

1. Adding two vectors (two polynomials):

$$\mathbf{p}(t)+\mathbf{q}(t)=\underbrace{c_0+c_1t+c_2t^2+\cdots+c_nt^n}_{=\mathbf{p}(t)}+\underbrace{d_0+d_1t+d_2t^2+\cdots+d_nt^n}_{=\mathbf{q}(t)}$$

$$=(c_0+d_0)+(c_1+d_1)t+(c_2+d_2)t^2+\cdots+(c_n+d_n)t^n$$

Hence $\mathbf{p}(t)+\mathbf{q}(t)$ is in $P(t)$ which means that we have *closure* under addition.

2. Commutative law:
We are required to show that $\mathbf{p}(t)+\mathbf{q}(t)=\mathbf{q}(t)+\mathbf{p}(t)$:

$$\mathbf{p}(t)+\mathbf{q}(t)=c_0+c_1t+c_2t^2+\cdots+c_nt^n+d_0+d_1t+d_2t^2+\cdots+d_nt^n$$

$$=\underbrace{d_0+d_1t+d_2t^2+\cdots+d_nt^n}_{=\mathbf{q}(t)}+\underbrace{c_0+c_1t+c_2t^2+\cdots+c_nt^n}_{=\mathbf{p}(t)}$$

$$=\mathbf{q}(t)+\mathbf{p}(t)$$

3. Associative law:
We need to show that $[\mathbf{p}(t)+\mathbf{q}(t)]+\mathbf{r}(t)=\mathbf{p}(t)+[\mathbf{q}(t)+\mathbf{r}(t)]$. Substituting the above polynomials into this yields:

$$[\mathbf{p}(t)+\mathbf{q}(t)]+\mathbf{r}(t)=\left[\underbrace{c_0+c_1t+c_2t^2+\cdots+c_nt^n}_{=\mathbf{p}(t)}+\underbrace{d_0+d_1t+d_2t^2+\cdots+d_nt^n}_{=\mathbf{q}(t)}\right]$$

$$+\underbrace{e_0+e_1t+e_2t^2+\cdots+e_nt^n}_{=\mathbf{r}(t)}$$

$$=c_0+c_1t+c_2t^2+\cdots+c_nt^n+d_0+d_1t+d_2t^2+\cdots+d_nt^n$$

$$+e_0+e_1t+e_2t^2+\cdots+e_nt^n$$

$$=\underbrace{c_0+c_1t+c_2t^2+\cdots+c_nt^n}_{=\mathbf{p}(t)}$$

$$+\left(\underbrace{d_0+d_1t+d_2t^2+\cdots+d_nt^n}_{=\mathbf{q}(t)}+\underbrace{e_0+e_1t+e_2t^2+\cdots+e_nt^n}_{=\mathbf{r}(t)}\right)$$

$$=\mathbf{p}(t)+[\mathbf{q}(t)+\mathbf{r}(t)]$$

4. *What is the zero vector in this case?*
The zero polynomial $\mathbf{O}$ is the real number 0 and is in the set $P(t)$. Clearly

$$\mathbf{p}(t)+\mathbf{O}=\underbrace{c_0+c_1t+c_2t^2+\cdots+c_nt^n}_{=\mathbf{p}(t)}+0=\mathbf{p}(t)$$

5. Additive inverse. For the polynomial $\mathbf{p}(t)$ we have the inverse as $-\mathbf{p}(t)$ and

$$\begin{aligned}\mathbf{p}(t)-\mathbf{p}(t) &= c_0+c_1t+c_2t^2+\cdots+c_nt^n-(c_0+c_1t+c_2t^2+\cdots+c_nt^n)\\ &= c_0+c_1t+c_2t^2+\cdots+c_nt^n-c_0-c_1t-c_2t^2-\cdots-c_nt^n\\ &= (c_0-c_0)+(c_1-c_1)t+(c_2-c_2)t^2+\cdots+(c_n-c_n)t^n\\ &= 0+0+0+\cdots+0=0=\mathbf{O}\end{aligned}$$

6. Let k be a real scalar then

$$\begin{aligned}k\mathbf{p}(t) &= k(c_0+c_1t+c_2t^2+\cdots+c_nt^n)\\ &= kc_0+kc_1t+kc_2t^2+\cdots+kc_nt^n\end{aligned}$$

Hence $k\mathbf{p}(t)$ is also in the space $P(t)$ because it is a polynomial of degree less than or equal to n.

7. Associative law for scalar multiplication. Let k_1 and k_2 be real scalars:

$$\begin{aligned}k_1(k_2\mathbf{p}(t)) &= k_1(k_2[c_0+c_1t+c_2t^2+\cdots+c_nt^n])\\ &= k_1(k_2c_0+k_2c_1t+k_2c_2t^2+\cdots+k_2c_nt^n)\\ &= k_1k_2c_0+k_1k_2c_1t+k_1k_2c_2t^2+\cdots+k_1k_2c_nt^n\\ &= k_1k_2\underbrace{(c_0+c_1t+c_2t^2+\cdots+c_nt^n)}_{=\mathbf{p}(t)} \qquad \text{[factorizing]}\\ &= (k_1k_2)\mathbf{p}(t)\end{aligned}$$

8. Distributive law. Let k be a real scalar, then we have:

$$\begin{aligned}k[\mathbf{p}(t)+\mathbf{q}(t)] &= k\left[\underbrace{c_0+c_1t+c_2t^2+\cdots+c_nt^n}_{=\mathbf{p}(t)}+\underbrace{d_0+d_1t+d_2t^2+\cdots+d_nt^n}_{=\mathbf{q}(t)}\right]\\ &= kc_0+kc_1t+kc_2t^2+\cdots+kc_nt^n+kd_0+kd_1t+kd_2t^2+\cdots+kd_nt^n\\ &\underset{\text{Factorising}}{=} k\underbrace{(c_0+c_1t+c_2t^2+\cdots+c_nt^n)}_{=\mathbf{p}(t)}+k\underbrace{(d_0+d_1t+d_2t^2+\cdots+d_nt^n)}_{=\mathbf{q}(t)}\\ &= k\mathbf{p}(t)+k\mathbf{q}(t)\end{aligned}$$

9. Distributive law for scalars. Let k_1 and k_2 be real scalars:

$$\begin{aligned}(k_1+k_2)\mathbf{p}(t) &= (k_1+k_2)(c_0+c_1t+c_2t^2+\cdots+c_nt^n)\\ &= k_1(c_0+c_1t+\cdots+c_nt^n)+k_2(c_0+c_1t+\cdots+c_nt^n)\\ &= k_1\mathbf{p}(t)+k_2\mathbf{p}(t)\end{aligned}$$

10. The identity element is the real number 1. We have

$$\begin{aligned}1\mathbf{p}(t) &= 1(c_0+c_1t+c_2t^2+\cdots+c_nt^n)\\ &= (1\times c_0)+(1\times c_1t)+(1\times c_2t^2)+\cdots+(1\times c_nt^n)\\ &= c_0+c_1t+c_2t^2+\cdots+c_nt^n=\mathbf{p}(t)\end{aligned}$$

All 10 axioms are satisfied, therefore the set $P(t)$ of polynomials of degree n or less is a vector space.

Example 3.2

Let $P(t)$ be the set of polynomials of degree equal to n where n is a positive integer. Show that this set $P(t)$ is not a vector space.

Remark Note the difference between Example 3.1 and this example. Here the polynomial is of degree n exactly. In Example 3.1 the degree of the polynomial was less than or equal to n.

Solution

If we check the first axiom, that is if we add *any* two vectors in the set $P(t)$, then their addition should also be in the set. Let

$$\mathbf{p}(t) = 4 + 5t^n \text{ and } \mathbf{q}(t) = 3 - 5t^n$$

We have $\mathbf{p}(t) + \mathbf{q}(t) = 4 + 5t^n + 3 - 5t^n = 7$. This result is not inside the set $P(t)$.

Why not?

Because $P(t)$ is the set of polynomials of degree n, but $\mathbf{p}(t) + \mathbf{q}(t) = 7$ is of degree 0, which means that it cannot be a member of this set $P(t)$.

Hence $P(t)$ is not a vector space because axiom 1 fails, that is we do not have closure under vector addition.

Example 3.3

Let V be the set of integers $\mathbb{Z}$. Let vector addition be defined as the normal addition of integers, and scalar multiplication by the usual multiplication of integers by a real scalar, which is any real number.

Show that this set is not a vector space with respect to this definition of vector addition and scalar multiplication.

Solution

The set of integers $\mathbb{Z}$ is not a vector space because when we multiply an integer by a real scalar, which is any real number, then the result may not be an integer.

For example, if we consider the integer 2 and multiply this by the scalar 1/3 then the result is $\frac{1}{3}(2) = \frac{2}{3}$ which is not an integer. Our result after scalar multiplication is not in the set of integers $\mathbb{Z}$. Hence the set of integers fails axiom 6.

We do not have closure under scalar multiplication, therefore the set of integers $\mathbb{Z}$ is not a vector space with respect to vector addition and scalar multiplication as defined above.

In general if a set forms a vector space with the scalars being real numbers then we say we have a vector space over the real numbers.

Another example of vector space is that which contains real-valued functions. You should be familiar with functions.

Examples of functions are $f(x) = x$, $f(x) = x^2 + 1$, $f(x) = \sin(x)$ and $f(x) = e^x$.

Let a and b be real numbers where $a \leq b$, then the notation $[a, b]$ is called the interval and is graphically displayed as shown in Fig. 3.1.

We use square brackets to indicate that $[a, b]$ is inclusive of the end points a and b. ($]a, b[$ contains every value between the end points, but does not contain the end points a and b.)

Figure 3.1

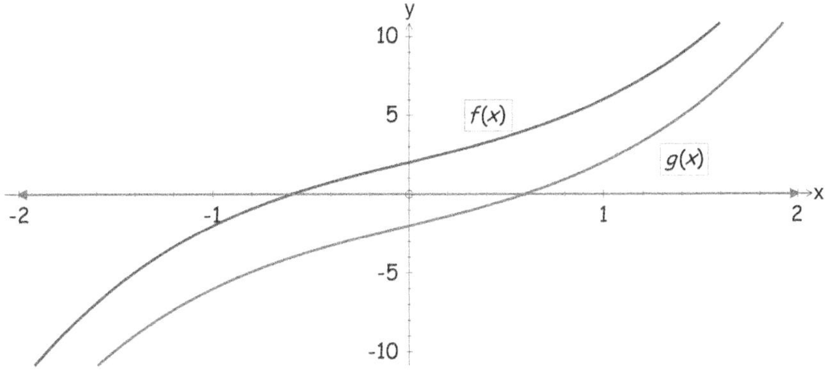

Figure 3.2

The functions in Fig. 3.2 are defined for the interval $[-2, 2]$.

Let $F[a, b]$ be the set of all functions defined in the interval $[a, b]$. Let **f** and **g** be functions in $F[a, b]$. Normally **f** and **g** (being vectors) would be in bold but since we are used to seeing functions without bold symbols we will sometimes write these as f and g respectively. We define:

(i) **Vector addition**. The sum $f + g$ is also in $F[a, b]$ and

$$(f + g)(x) = f(x) + g(x)$$

(ii) **Scalar multiplication**. Let k be a real scalar then for *all* x in $[a, b]$

$$(kf)(x) = kf(x)$$

The zero vector in $F[a, b]$ is the zero function $0(x)$ which is equal to zero for *all* x in the interval $[a, b]$, that is

$$0(x) = 0 \qquad \text{for all } x \in [a, b]$$

For any function f there is a function $-f$ (minus f) for all x in the interval $[a, b]$ which satisfies

$$(-f)(x) = -f(x)$$

You are asked to prove that $F[a, b]$ is a vector space with respect to these definitions in Exercises 3.1.

Next we go on to prove some basic properties of general vector spaces, using the 10 axioms given in subsection 3.1.1.

3.1.3 Basic properties of general vector spaces

Proposition (3.1). Let V be a vector space and k be a real scalar. Then we have:

(a) For any real scalar k we have $k\mathbf{O} = \mathbf{O}$ where $\mathbf{O}$ is the zero vector.

(b) For the real number 0 and any vector $\mathbf{u}$ in V we have $0\mathbf{u} = \mathbf{O}$.

(c) For the real number -1 and any vector $\mathbf{u}$ in V we have $(-1)\mathbf{u} = -\mathbf{u}$.

(d) If $k\mathbf{u} = \mathbf{O}$, where k is a real scalar and $\mathbf{u}$ in V, then $k = 0$ or $\mathbf{u} = \mathbf{O}$.

We use the above 10 axioms given in subsection 3.1.1, to prove these properties.

(a) Proof.
Need to prove $k\mathbf{O} = \mathbf{O}$. By applying the above axiom 4, $\mathbf{u} + \mathbf{O} = \mathbf{u}$, we have

$$\begin{aligned} k\mathbf{O} &= k(\mathbf{O} + \mathbf{O}) \\ &= k\mathbf{O} + k\mathbf{O} \qquad \text{[applying axiom 8 which is } k(\mathbf{u} + \mathbf{v}) = k\mathbf{u} + k\mathbf{v}] \end{aligned}$$

Adding $-k\mathbf{O}$ to both sides of this, $k\mathbf{O} = k\mathbf{O} + k\mathbf{O}$, gives

$$\underbrace{k\mathbf{O} + (-k\mathbf{O})}_{=\mathbf{O}} = k\mathbf{O} + \underbrace{k\mathbf{O} + (-k\mathbf{O})}_{=\mathbf{O}} \qquad \text{[By axiom 5} \quad \mathbf{u} + (-\mathbf{u}) = \mathbf{O}]$$

$$\mathbf{O} = k\mathbf{O}$$

Hence we have our required result, $k\mathbf{O} = \mathbf{O}$.

■

(b) Proof.
Required to prove $0\mathbf{u} = \mathbf{O}$. For real numbers $0 = 0 + 0$ we have

$$\begin{aligned} 0\mathbf{u} &= (0 + 0)\mathbf{u} \\ &= 0\mathbf{u} + 0\mathbf{u} \qquad \text{[by axiom 9 } (k + c)\mathbf{u} = k\mathbf{u} + c\mathbf{u}] \end{aligned}$$

Adding $-0\mathbf{u}$ to both sides of this, $0\mathbf{u} = 0\mathbf{u} + 0\mathbf{u}$, gives

$$\underbrace{0\mathbf{u} + (-0\mathbf{u})}_{=\mathbf{O}} = 0\mathbf{u} + \underbrace{0\mathbf{u} + (-0\mathbf{u})}_{=\mathbf{O}}$$

$$\mathbf{O} = 0\mathbf{u}$$

We have result (b), that is $0\mathbf{u} = \mathbf{O}$.

■

(c) Proof.
Required to prove $(-1)\mathbf{u} = -\mathbf{u}$. Adding $\mathbf{u}$ to $(-1)\mathbf{u}$ gives

$$\begin{aligned} \mathbf{u} + (-1)\mathbf{u} &= 1\mathbf{u} + (-1)\mathbf{u} \\ &= (1 - 1)\mathbf{u} = 0\mathbf{u} \underbrace{=}_{\text{by part (b)}} \mathbf{O} \end{aligned}$$

By axiom 5 we have $\mathbf{u} + (-\mathbf{u}) = \mathbf{O}$. Equating this with the above, $\mathbf{u} + (-1)\mathbf{u} = \mathbf{O}$, we have

$$\mathbf{u} + (-1)\mathbf{u} = \mathbf{u} + (-\mathbf{u})$$

Adding $-\mathbf{u}$ to both sides of this gives:

$$\underbrace{-\mathbf{u} + \mathbf{u}}_{=\mathbf{O}} + (-1)\mathbf{u} = \underbrace{-\mathbf{u} + \mathbf{u}}_{=\mathbf{O}} + (-\mathbf{u})$$
$$\mathbf{O} + (-1)\mathbf{u} = \mathbf{O} + (-\mathbf{u})$$
$$(-1)\mathbf{u} = -\mathbf{u}$$

Hence we have our result, $(-1)\mathbf{u} = -\mathbf{u}$.

■

(d) Proof.
Need to prove that $k\mathbf{u} = \mathbf{O}$ implies $k = 0$ or $\mathbf{u} = \mathbf{O}$.
If $k \neq 0$ (non-zero) then

$$k\mathbf{u} = \mathbf{O}$$
$$\frac{1}{k}(k\mathbf{u}) = \underbrace{\frac{1}{k}\mathbf{O}}_{=\mathbf{O}} \qquad \left[\text{multiplying both sides by } \frac{1}{k}\right]$$
$$\left(\frac{1}{k}k\right)\mathbf{u} = \mathbf{O} \text{ gives } 1\mathbf{u} = \mathbf{O}$$

Since $1\mathbf{u} = \mathbf{u}$, therefore $\mathbf{u} = \mathbf{O}$ which is our required result.
If $k = 0$ then clearly the result holds because $0\mathbf{u} = \mathbf{O}$ by part (b).

■

Proposition (3.2). Let V be a vector space. The zero vector $\mathbf{O}$ (neutral element) of V is unique.

What does this proposition mean?
There is only one zero vector in the vector space V. Remember, by axiom 4 there is a vector called the zero vector denoted by $\mathbf{O}$ which satisfies

$$\mathbf{u} + \mathbf{O} = \mathbf{u} \text{ for every vector } \mathbf{u}$$

Proof.
Suppose there is another vector, say $\mathbf{w}$, which also satisfies

$$\mathbf{u} + \mathbf{w} = \mathbf{u}$$

What do we need to prove?
Required to prove that $\mathbf{w} = \mathbf{O}$. Equating these two, $\mathbf{u} + \mathbf{O} = \mathbf{u}$ and $\mathbf{u} + \mathbf{w} = \mathbf{u}$, gives

$$\mathbf{u} + \mathbf{O} = \mathbf{u} + \mathbf{w}$$

Adding $-\mathbf{u}$ to both sides of this yields:

$$-\mathbf{u}+(\mathbf{u}+\mathbf{O})=-\mathbf{u}+(\mathbf{u}+\mathbf{w})$$
$$\underbrace{(-\mathbf{u}+\mathbf{u})}_{=\mathbf{O}}+\mathbf{O}=\underbrace{(-\mathbf{u}+\mathbf{u})}_{=\mathbf{O}}+\mathbf{w}$$
$$\mathbf{O}+\mathbf{O}=\mathbf{O}+\mathbf{w} \text{ gives } \mathbf{O}=\mathbf{w}$$

Hence $\mathbf{w}=\mathbf{O}$, which means that the zero vector is *unique*.

■

Proposition (3.3). Let V be a vector space and k be a real scalar. The vector $-\mathbf{u}$ which satisfies axiom 5 is unique:

$$\mathbf{u}+(-\mathbf{u})=\mathbf{O}$$

Proof.
Suppose the vector $\mathbf{w}$ also satisfies this;

$$\mathbf{u}+\mathbf{w}=\mathbf{O}$$

What do we need to prove?
Required to prove that $\mathbf{w}=-\mathbf{u}$. Equating these two, $\mathbf{u}+(-\mathbf{u})=\mathbf{O}$ and $\mathbf{u}+\mathbf{w}=\mathbf{O}$, gives

$$\mathbf{u}+(-\mathbf{u})=\mathbf{u}+\mathbf{w}$$

Adding $-\mathbf{u}$ to both sides yields:

$$-\mathbf{u}+(\mathbf{u}+(-\mathbf{u}))=-\mathbf{u}+(\mathbf{u}+\mathbf{w})$$
$$\underbrace{(-\mathbf{u}+\mathbf{u})}_{=\mathbf{O}}+(-\mathbf{u})=\underbrace{(-\mathbf{u}+\mathbf{u})}_{=\mathbf{O}}+\mathbf{w}$$
$$\mathbf{O}+(-\mathbf{u})=\mathbf{O}+\mathbf{w} \quad \Rightarrow \quad -\mathbf{u}=\mathbf{w}$$

Hence the vector $-\mathbf{u}$ is unique.

■

Summary

A vector space is formed by a non-empty set within which the operation of vector addition and scalar multiplication satisfies the above 10 axioms.

Examples of vector spaces include the sets of matrices, polynomials and functions.

Some properties of vector spaces are:

(a) $k\mathbf{O}=\mathbf{O}$

(b) $0\mathbf{u}=\mathbf{O}$

(c) $(-1)\mathbf{u}=-\mathbf{u}$

(d) If $k\mathbf{u}=\mathbf{O}$ then $k=0$ or $\mathbf{u}=\mathbf{O}$

The zero vector $\mathbf{O}$ and the vector $-\mathbf{u}$ have the following properties

$$\mathbf{u} + (-\mathbf{u}) = \mathbf{O} \text{ and } \mathbf{u} + \mathbf{O} = \mathbf{u}$$

and are unique.

EXERCISES 3.1

(Brief solutions at end of book. Full solutions available at <http://www.oup.co.uk/companion/singh>.)

Throughout this exercise assume the usual operations of vector addition and scalar multiplication as defined in early chapters or in section 3.1.

1. Show that $\mathbb{R}^2$ is a vector space.
2. Show that $\mathbb{R}^3$ is a vector space.
3. Show that the set M_{22} of 2 by 2 matrices is a vector space.
4. Show that the set M_{23} of 2 by 3 matrices is a vector space.
5. Show that the set $F[a,\ b]$ which is the set of all functions defined on the interval $[a,\ b]$ is a vector space.
6. Show that the following set of matrices does not form a vector space:

$$\begin{pmatrix} 1 & 0 \\ 0 & a \end{pmatrix}$$

7. Show that matrices $\begin{pmatrix} a & 0 \\ 0 & b \end{pmatrix}$ form a vector space.
8. Show that the set of 2 by 2 non-invertible matrices do not form a vector space.
9. Let P_2 be the set of polynomials of degree 2 or less and

$$\mathbf{p}(x) = ax^2 + bx + c \text{ and } \mathbf{q}(x) = dx^2 + ex + f$$

be members of P_2.

Show that the set P_2 is a vector space with respect to the usual vector addition and scalar multiplication.

10. Consider the set $\mathbb{R}^2$ with the normal vector addition, but scalar multiplication defined by

$$k\begin{pmatrix} a \\ b \end{pmatrix} = \begin{pmatrix} 0 \\ kb \end{pmatrix}$$

Show that the set $\mathbb{R}^2$ with this scalar multiplication definition is not a vector space.

11. Let V be the set $\begin{pmatrix} a \\ b \end{pmatrix}$ in $\mathbb{R}^2$ where $a \geq 0$ and $b \geq 0$. Show that with respect to normal vector addition and scalar multiplication, the given set is not a vector space.
12. Let V be the set of rationals (fractions) $\mathbb{Q}$. Let vector addition be defined as the normal addition of rationals and scalar multiplication by the usual multiplication of rationals by a scalar, which is any real number.
    Show that this set is not a vector space with respect to this definition of vector addition and scalar multiplication.
13. Let V be the set of real numbers $\mathbb{R}$. Show that this set is a vector space with respect to the usual definition of vector addition and scalar multiplication.
14. Let V be the set of positive real numbers $\mathbb{R}^+$. Show that this set is not a vector space with respect to the usual definition of vector addition and scalar multiplication.
15. Explain why if the vector $\mathbf{u}$ is in the vector space V then the vector $-\mathbf{u}$ is also in V.
16. Let V be a vector space and $\mathbf{u}$ be a vector in V. Prove that
$$(-k)\mathbf{u} = -(k\mathbf{u}) = k(-\mathbf{u})$$
17. Let V be a vector space and $\mathbf{u}$ and $\mathbf{w}$ be vectors in V. Let $k \neq 0$ be a scalar. Prove that if $k\mathbf{u} = k\mathbf{w}$ then $\mathbf{u} = \mathbf{w}$.
18. Let V be a vector space with a non-zero vector $\mathbf{u}$ ($\neq \mathbf{O}$) and scalars k_1 and k_2. Prove that if $k_1\mathbf{u} = k_2\mathbf{u}$ then $k_1 = k_2$.
19. Let $\mathbf{u}$ and $\mathbf{w}$ be vectors in a vector space V. Prove that
$$-(\mathbf{u} + \mathbf{w}) = -\mathbf{u} - \mathbf{w}$$
20. Let $\mathbf{u}$ be a vector in a vector space V. Prove that
$$\underbrace{\mathbf{u} + \mathbf{u} + \mathbf{u} + \cdots + \mathbf{u}}_{n \text{ copies}} = n\mathbf{u}$$

SECTION 3.2 Subspace of a Vector Space

By the end of this section you will be able to

- prove some properties of subspaces
- understand what is meant by a linear combination and spanning set

In section 3.1 we discussed the whole vector space V. In this section we show that parts of the whole vector space also form a vector space in their own right. We will show that a non-empty set *within* V, which is *closed* under the basic operations of vector addition and scalar multiplication, is also a legitimate vector space.

3.2.1 Examples of vector subspaces

Let V be a vector space and S be a non-empty subset of V. If the set S satisfies all 10 axioms of a vector space with respect to the same vector addition and scalar multiplication as V then S is also a vector space. We say S is a subspace of V.

Definition (3.4). A non-empty subset S of a vector space V is called a **subspace** of V if it is also a vector space with respect to the same vector addition and scalar multiplication as V.

We illustrate this in Fig. 3.3.

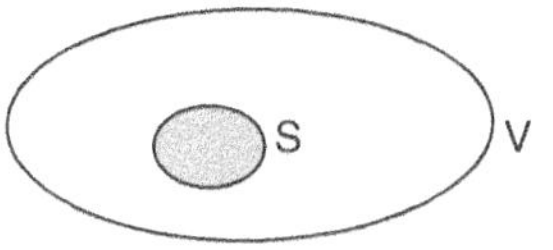

Figure 3.3

Note the difference between subspace and subset. A sub**set** is merely a specific set of elements chosen from V. A subset must also satisfy the 10 axioms of vector space to be called a sub**space**.

Example 3.4

Let V be the set $\mathbb{R}^2$ and vector addition and scalar multiplication be defined as normal. Let the set S be the single element $\mathbf{O}$ which in this case is the origin. Show that S is a subspace of V.

Solution

We know from the last section that $\mathbb{R}^2$ is a vector space and also that it contains the origin $\mathbf{O}$. This means that the given set S containing the single element $\mathbf{O}$ is a non-empty subset of V.

Vector addition and scalar multiplication in this set is defined as

$$\mathbf{O} + \mathbf{O} = \mathbf{O} \text{ and } k\mathbf{O} = \mathbf{O}, \text{ where } k \text{ is a scalar.}$$

By checking *all* 10 axioms of the last section, we can show that the set S is also a vector space. You are asked to do this in Exercises 3.2.

Hence by definition (3.4) we conclude that S is a subspace of $V = \mathbb{R}^2$.

We can represent this graphically in Fig. 3.4.

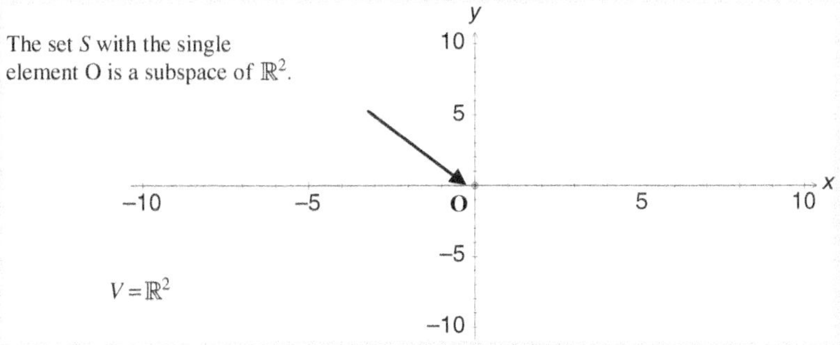

Figure 3.4

A set with a *single* element is called a **singleton set.**

More generally, we will use the following proposition to check if a given subset qualifies as a subspace.

(You will need to look again at the 10 axioms to understand the proof below.)

> Proposition (3.5). Let S be a non-empty subset of a vector space V. Then S is subspace of $V \Leftrightarrow$:
>
> (a) If $\mathbf{u}$ and $\mathbf{v}$ are vectors in the set S then the vector addition $\mathbf{u}+\mathbf{v}$ is also in S.
>
> (b) If $\mathbf{u}$ is a vector in S then for every scalar k we have, $k\mathbf{u}$ is also in S.

Note that this proposition means that we must have *closure* under both vector addition and scalar multiplication. This means that S is a subspace of $V \Leftrightarrow$ both the following are satisfied, as shown in Fig. 3.5.

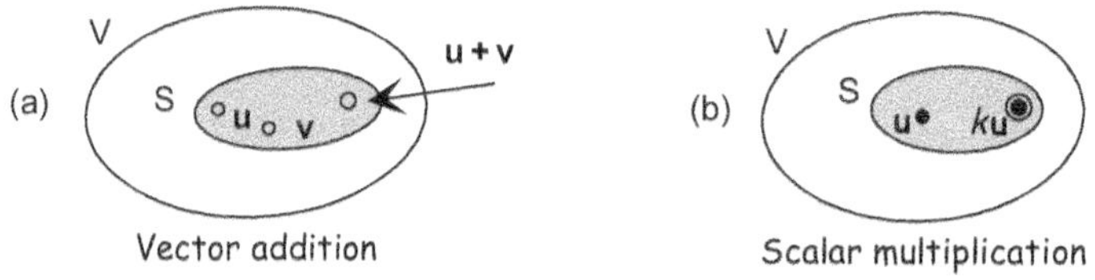

Figure 3.5

How do we prove this proposition?

($\Rightarrow$). We first assume that S is a subspace of the vector space V, and from this we deduce conditions (a) and (b) [show that we have closure under vector addition and scalar multiplication].

($\Leftarrow$). Then we assume conditions (a) and (b) are satisfied, and from this we deduce that the set S is a subspace of the vector space V.

Proof.

Let $\mathbf{u}$ and $\mathbf{v}$ be vectors in the set S.

($\Rightarrow$). Let S be a subspace of the vector space V. By the above definition (3.4) we have closure under vector addition and scalar multiplication because the set S is itself a vector space. [Remember, axioms 1 and 6 state that we have closure under vector addition and scalar multiplication]. Hence conditions (a) and (b) hold.

($\Leftarrow$). Assume conditions (a) and (b) are satisfied, that is we have closure under vector addition and scalar multiplication.

Required to prove that all 10 axioms of the last section are satisfied.

We have closure, therefore axioms 1 and 6 are satisfied. Axioms 2, 3, 7, 8, 9 and 10 are satisfied because these axioms are true for all vectors in the vector space V and vectors $\mathbf{u}$ and $\mathbf{v}$ are vectors in the vector space V.

For the set S we need to prove axioms 4 and 5 which are:

4. There is a zero vector $\mathbf{O}$ in V which satisfies

$$\mathbf{u}+\mathbf{O}=\mathbf{u} \qquad \text{for every vector } \mathbf{u} \text{ in } V$$

5. For every vector $\mathbf{u}$ there is a vector $-\mathbf{u}$ which satisfies the following:

$$\mathbf{u} + (-\mathbf{u}) = \mathbf{O}$$

We have to show (Fig. 3.6) that the zero vector, $\mathbf{O}$, and $-\mathbf{u}$ are also in S for any $\mathbf{u}$ in S.

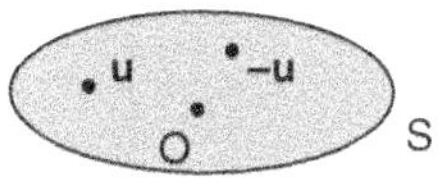

Figure 3.6

We are assuming closure under scalar multiplication (part (b)), which means that if $\mathbf{u}$ is in S then $k\mathbf{u}$ is also in S for any real scalar k. Substituting $k = 0$ gives

$$0\mathbf{u} = \mathbf{O} \text{ is also in } S$$

Substituting $k = -1$ gives

$$(-1)\mathbf{u} = -\mathbf{u} \text{ is also in } S$$

Hence the zero vector, $\mathbf{O}$, and $-\mathbf{u}$ are in the set S.

Now that we have shown all 10 axioms of a vector space are satisfied, we can conclude that S is a vector space, therefore it is a subspace of V.

■

What does Proposition (3.5) mean?

Proposition (3.5) makes life a lot easier because we only need to check closure under vector addition and scalar multiplication to show that a given set is a subspace. No need to check all 10 axioms.

In everyday English, closure refers to the act of closing or sealing, such as the lid for a container. It has the same meaning here: $\mathbf{u} + \mathbf{v}$ and $k\mathbf{u}$ are closed inside the subspace as shown in Fig. 3.7.

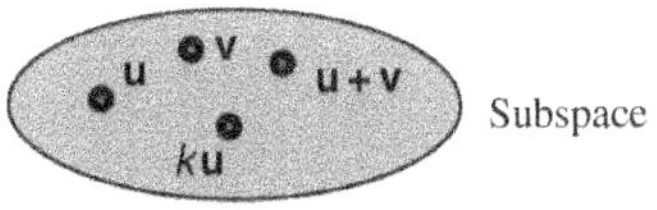

Figure 3.7

If vectors $\mathbf{u}$ and $\mathbf{v}$ are in subspace S then $\mathbf{u} + \mathbf{v}$ and $k\mathbf{u}$ cannot escape from S.

Example 3.5

Let V be the set $\mathbb{R}^2$ and vector addition and scalar multiplication be defined as normal. Let S be the set of vectors of the form $\begin{pmatrix} 0 \\ y \end{pmatrix}$. Show that S is a subspace of V.

(continued...)

Solution

We only need to check conditions (a) and (b) of Proposition (3.5). These are the closure conditions under vector addition and scalar multiplication. Since S is non-empty because $(0, 0)$ is in S so we need to show:

(a) If $\mathbf{u}$ and $\mathbf{v}$ are vectors in the set S then the vector addition $\mathbf{u}+\mathbf{v}$ is also in S.

(b) If $\mathbf{u}$ is a vector in S and k is any real scalar then $k\mathbf{u}$ is also in S.

Let $\mathbf{u} = \begin{pmatrix} 0 \\ a \end{pmatrix}$ and $\mathbf{v} = \begin{pmatrix} 0 \\ b \end{pmatrix}$ be vectors in S. Then

$$\mathbf{u}+\mathbf{v} = \begin{pmatrix} 0 \\ a \end{pmatrix} + \begin{pmatrix} 0 \\ b \end{pmatrix} = \begin{pmatrix} 0 \\ a+b \end{pmatrix} \qquad \left[\begin{array}{l}\text{closure under} \\ \text{vector addition}\end{array}\right]$$

which is in the set S and

$$k\mathbf{u} = k\begin{pmatrix} 0 \\ a \end{pmatrix} = \begin{pmatrix} 0 \\ ka \end{pmatrix} \qquad \left[\begin{array}{l}\text{closure under} \\ \text{scalar multiplication}\end{array}\right]$$

which is in S as well.

Conditions (a) and (b) are satisfied, therefore the given set S is a subspace of the vector space $\mathbb{R}^2$. Note that the set S is the y axis in the xy plane as shown in Fig. 3.8.

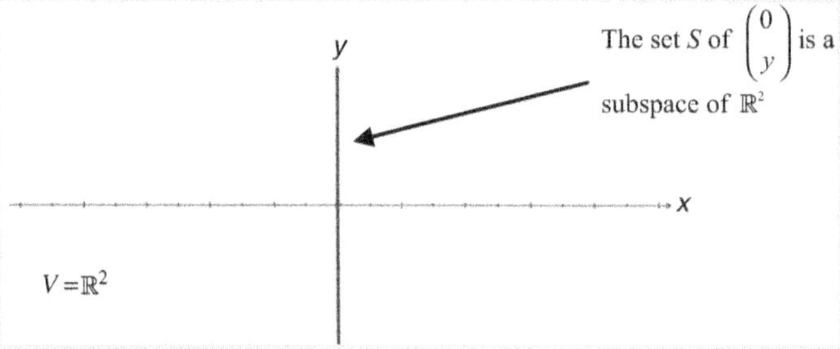

Figure 3.8

Example 3.6

Let S be the subset of vectors of the form $\begin{pmatrix} x \\ y \end{pmatrix}$ where $x \geq 0$ in the vector space $\mathbb{R}^2$. Show that S is not a subspace of $\mathbb{R}^2$.

Solution

How do we show that S is not a subspace of $\mathbb{R}^2$*?*

If we can show that we do not have closure under vector addition or scalar multiplication of vectors in S, then we can conclude that S is not a subspace. Consider the scalar $k = -1$ and the vector $\mathbf{u} = \begin{pmatrix} 1 \\ 2 \end{pmatrix}$; then clearly $\mathbf{u}$ is in the set S, but the scalar multiplication

$$k\mathbf{u} = (-1)\begin{pmatrix} 1 \\ 2 \end{pmatrix} = \begin{pmatrix} -1 \\ -2 \end{pmatrix}$$

is not in S.

Why not?

Because the first entry in the vector $\begin{pmatrix} -1 \\ -2 \end{pmatrix}$ is -1 which is less than 0 and the set S only contains vectors of the form $\begin{pmatrix} x \\ y \end{pmatrix}$ where the first entry $x \geq 0$ (greater than or equal to zero).

Hence S is *not* a subspace of $\mathbb{R}^2$ (Fig. 3.9).

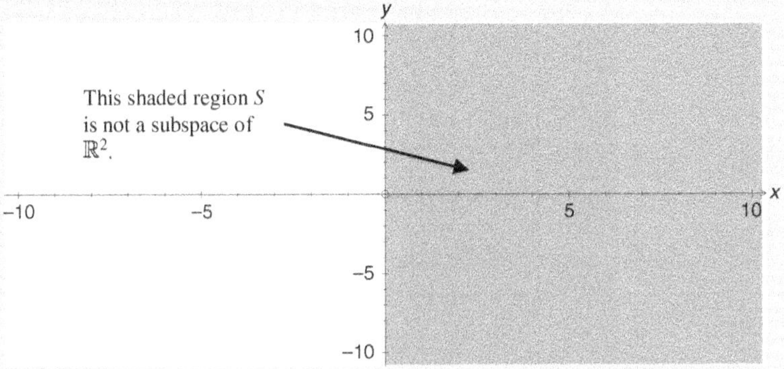

Figure 3.9

Clearly S is a subset but not a subspace of $\mathbb{R}^2$.

Note that by Example 3.5 the vertical axis $x = 0$ is a subspace of $\mathbb{R}^2$ but the whole region to the right of $x \geq 0$ is not.

Example 3.7

Let M_{22} be the set of 2 by 2 matrices. From the last section we know that this set is a vector space. Let S be a subset of M_{22}, containing matrices of the form $\begin{pmatrix} 1 & b \\ c & d \end{pmatrix}$.

Show that S is not a subspace of V.

Solution

We need to show that one of closure conditions (a) or (b) of Proposition (3.5) fails.

Let $\mathbf{u} = \begin{pmatrix} 1 & b \\ c & d \end{pmatrix}$ and $\mathbf{v} = \begin{pmatrix} 1 & 0 \\ 0 & 0 \end{pmatrix}$ be vectors in the subset S. Then

$$\mathbf{u} + \mathbf{v} = \begin{pmatrix} 1 & b \\ c & d \end{pmatrix} + \begin{pmatrix} 1 & 0 \\ 0 & 0 \end{pmatrix} = \begin{pmatrix} 1+1 & b+0 \\ c+0 & d+0 \end{pmatrix} = \begin{pmatrix} 2 & b \\ c & d \end{pmatrix}$$

Hence $\mathbf{u} + \mathbf{v}$ is not a member of the set S because the elements in S are of the form $\begin{pmatrix} 1 & b \\ c & d \end{pmatrix}$ but $\mathbf{u} + \mathbf{v} = \begin{pmatrix} 2 & b \\ c & d \end{pmatrix}$. The first entry in the matrix $\mathbf{u} + \mathbf{v}$ needs to be 1, not 2, to qualify as an element of the subset S. Therefore S is not a subspace of the vector space M_{22}.

3.2.2 Revision of linear combination

Linear combination combines the two fundamental operations of linear algebra – vector addition and scalar multiplication.

In the last chapter we introduced linear combination in $\mathbb{R}^n$. For example, we had

$$\mathbf{u} = k_1\mathbf{e}_1 + k_2\mathbf{e}_2 + \cdots + k_n\mathbf{e}_n \qquad (k\text{'s are scalars})$$

which is a **linear combination** of the standard unit vectors $\mathbf{e}_1,\ \mathbf{e}_2,\ \ldots$ and $\mathbf{e}_n$.

Similarly for general vector spaces we define linear combination as:

Definition (3.6). Let $\mathbf{v}_1,\ \mathbf{v}_2, \ldots$ and $\mathbf{v}_n$ be vectors in a vector space. If a vector $\mathbf{x}$ can be expressed as

$$\mathbf{x} = k_1\mathbf{v}_1 + k_2\mathbf{v}_2 + k_3\mathbf{v}_3 + \cdots + k_n\mathbf{v}_n \text{ (where } k\text{'s are scalars)}$$

then we say $\mathbf{x}$ is a **linear combination** of the vectors $\mathbf{v}_1,\ \mathbf{v}_2,\ \mathbf{v}_3,\ \ldots$ and $\mathbf{v}_n$.

Example 3.8

Let P_2 be the set of all polynomials of degree less than or equal to 2.

Let $\mathbf{v}_1 = t^2 - 1,\ \mathbf{v}_2 = t^2 + 3t - 5$ and $\mathbf{v}_3 = t$ be vectors in P_2.

Show that the quadratic polynomial

$$\mathbf{x} = 7t^2 - 15$$

is a linear combination of $\{\mathbf{v}_1,\ \mathbf{v}_2,\ \mathbf{v}_3\}$.

Solution

How do we show $\mathbf{x}$ is a linear combination of vectors $\mathbf{v}_1,\ \mathbf{v}_2$ and $\mathbf{v}_3$?

We need to find the values of the scalars $k_1,\ k_2$ and k_3 which satisfy

$$k_1\mathbf{v}_1 + k_2\mathbf{v}_2 + k_3\mathbf{v}_3 = \mathbf{x} \quad (*)$$

How can we determine these scalars?

By substituting $\mathbf{v}_1 = t^2 - 1,\ \mathbf{v}_2 = t^2 + 3t - 5,\ \mathbf{v}_3 = t$ and $\mathbf{x} = 7t^2 - 15$ into (*):

$$\begin{aligned} k_1\mathbf{v}_1 + k_2\mathbf{v}_2 + k_3\mathbf{v}_3 &= k_1(t^2 - 1) + k_2(t^2 + 3t - 5) + k_3t \\ &= k_1t^2 - k_1 + k_2t^2 + 3k_2t - 5k_2 + k_3t && \text{[expanding]} \\ &= (k_1 + k_2)t^2 + (3k_2 + k_3)t - (k_1 + 5k_2) && \text{[factorizing]} \\ &= 7t^2 - 15 && [\text{remember } \mathbf{x} = 7t^2 - 15] \end{aligned}$$

By equating coefficients of the last two lines

$$(k_1 + k_2)t^2 + (3k_2 + k_3)t - (k_1 + 5k_2) = 7t^2 - 15$$

gives

$$\begin{aligned} k_1 + k_2 &= 7 && [\text{equating } t^2] \\ 3k_2 + k_3 &= 0 && [\text{equating } t] \\ k_1 + 5k_2 &= 15 && [\text{equating constants}] \end{aligned}$$

Solving these equations gives the values of the scalars: $k_1 = 5$, $k_2 = 2$ and $k_3 = -6$.
Substituting these into $k_1\mathbf{v}_1 + k_2\mathbf{v}_2 + k_3\mathbf{v}_3 = \mathbf{x}$:

$$5\mathbf{v}_1 + 2\mathbf{v}_2 - 6\mathbf{v}_3 = \mathbf{x}$$

This means that adding 5 lots of $\mathbf{v}_1$, 2 lots of $\mathbf{v}_2$ and -6 lots of $\mathbf{v}_3$ gives the vector $\mathbf{x}$:

$$5(t^2 - 1) + 2(t^2 + 3t - 5) - 6t = 7t^2 - 15$$

You may like to check the algebra.
We conclude that $\mathbf{x}$ is a linear combination of $\{\mathbf{v}_1, \mathbf{v}_2, \mathbf{v}_3\}$.

The next proposition allows us to check that S is a subspace, by carrying out the test for scalar multiplication and vector addition in a single calculation.

Proposition (3.7). A non-empty subset S containing vectors $\mathbf{u}$ and $\mathbf{v}$ is a subspace of a vector space $V \Leftrightarrow$ any linear combination $k\mathbf{u} + c\mathbf{v}$ is also in S (k and c are scalars).

How do we prove this proposition?
Since it is a '$\Leftrightarrow$' statement we need to prove it both ways.

Proof.
($\Rightarrow$). Let S be a subspace of V, then S is a vector space. If $\mathbf{u}$ and $\mathbf{v}$ are in S then $k\mathbf{u} + c\mathbf{v}$ is also in S because the vector space S is closed under scalar multiplication and vector addition.
($\Leftarrow$). Assume $k\mathbf{u} + c\mathbf{v}$ is in S.

Substituting $k = c = 1$ into $k\mathbf{u} + c\mathbf{v}$ we have $\mathbf{u} + \mathbf{v}$ is also in S. Similarly for $c = 0$ we have $k\mathbf{u} + c\mathbf{v} = k\mathbf{u}$ is in S.

Hence we have closure under vector addition and scalar multiplication. By Proposition (3.5): S is subspace of $V \Leftrightarrow S$ is closed under vector addition and scalar multiplication.

We conclude that S is a subspace.

■

Proposition (3.7) is another test for a subspace. Hence a subspace of a vector space V is a non-empty subset S of V, such that for all vectors $\mathbf{u}$ and $\mathbf{v}$ in S and all scalars k and c we have $k\mathbf{u} + c\mathbf{v}$ is also in S.

Example 3.9

Let S be the subset of vectors of the form $(x \quad y \quad 0)^T$ in the vector space $\mathbb{R}^3$. Show that S is a subspace of $\mathbb{R}^3$.

(continued...)

Solution

How do we show that S is a subspace of $\mathbb{R}^3$?

We can use the above Proposition (3.7), which means we need to show that any linear combination $k\mathbf{u} + c\mathbf{v}$ is in S for any vectors $\mathbf{u}$ and $\mathbf{v}$ in S.

Clearly S is non-empty because the zero vector is in S. Let $\mathbf{u} = (a \quad b \quad 0)^T$ and $\mathbf{v} = (c \quad d \quad 0)^T$ be in S. Then for real scalars k_1 and k_2 we have

$$k_1\mathbf{u} + k_2\mathbf{v} = k_1\begin{pmatrix} a \\ b \\ 0 \end{pmatrix} + k_2\begin{pmatrix} c \\ d \\ 0 \end{pmatrix}$$

$$= \begin{pmatrix} k_1 a \\ k_1 b \\ 0 \end{pmatrix} + \begin{pmatrix} k_2 c \\ k_2 d \\ 0 \end{pmatrix} = \begin{pmatrix} k_1 a + k_2 c \\ k_1 b + k_2 d \\ 0 \end{pmatrix}$$

Hence $k_1\mathbf{u} + k_2\mathbf{v}$ is also in S.

By the above Proposition (3.7):

S is subspace of $V \Leftrightarrow$ any linear combination $k\mathbf{u} + c\mathbf{v}$ is also in S.

We conclude that the given set S is a subspace of the vector space $\mathbb{R}^3$.

You might find it easier to use this test (3.7) rather than (3.5) because you only need to recall that the linear combination is closed in S.

Note that the given set S in the above example describes the xy plane in three-dimensional space $\mathbb{R}^3$ as shown in Fig. 3.10:

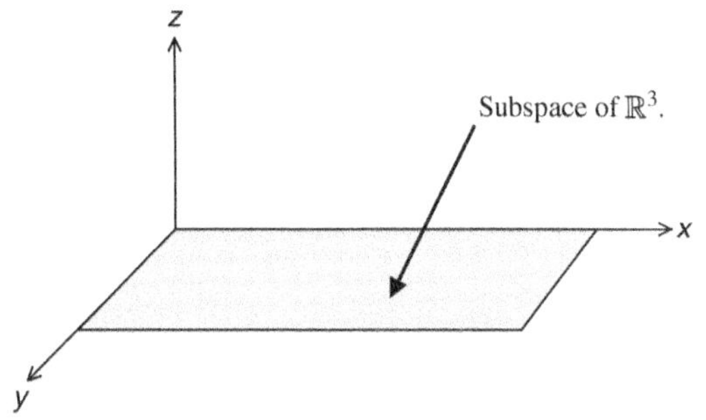

Figure 3.10

Sometimes we write $\begin{pmatrix} x \\ y \\ 0 \end{pmatrix}$ as the set $S = \left\{ \begin{pmatrix} x \\ y \\ z \end{pmatrix} \middle| \; z = 0 \right\}$ where the vertical line in the set means 'such that'. That is, S contains the set of vectors in $\mathbb{R}^3$ such that the last entry is zero.

3.2.3 Revision of spanning sets

Definition (3.8). If *every* vector in V can be produced by a linear combination of vectors $\mathbf{v}_1, \mathbf{v}_2, \mathbf{v}_3, \ldots$ and $\mathbf{v}_n$ then these vectors **span** or **generate** the vector space V. We write this as $span\{\mathbf{v}_1, \mathbf{v}_2, \mathbf{v}_3, \ldots, \mathbf{v}_n\}$.

For example, the standard unit vectors $\mathbf{e}_1 = \begin{pmatrix} 1 \\ 0 \end{pmatrix}$ and $\mathbf{e}_2 = \begin{pmatrix} 0 \\ 1 \end{pmatrix}$ span $\mathbb{R}^2$ because a linear combination of these vectors $\mathbf{e}_1$ and $\mathbf{e}_2$ can produce any vector in $\mathbb{R}^2$.

For example, the vector $\mathbf{u} = \begin{pmatrix} 1 \\ 1 \end{pmatrix}$ spans the line $y = x$ in $\mathbb{R}^2$ as Fig. 3.11(a) shows.

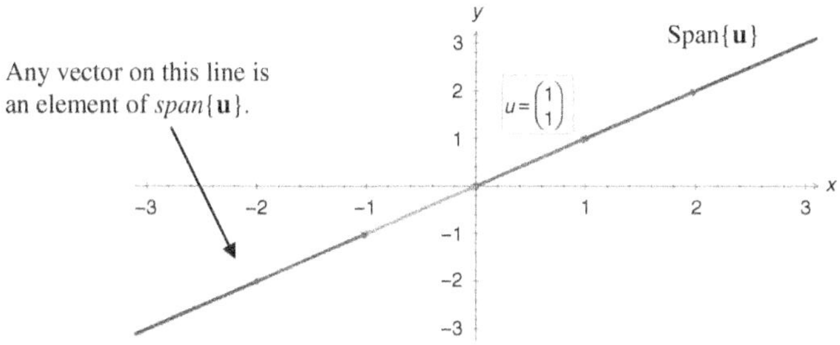

Figure 3.11(a)

Proposition (3.9). Let S be a non-empty subset of a vector space V. The set $span\{S\}$ is a subspace of the vector space V.

Proof – Exercises 3.2.

■

For example, in the above Fig. 3.11(a) *span* $\{\mathbf{u}\}$ is a subspace of $\mathbb{R}^2$. This is illustrated in Fig. 3.11(b).

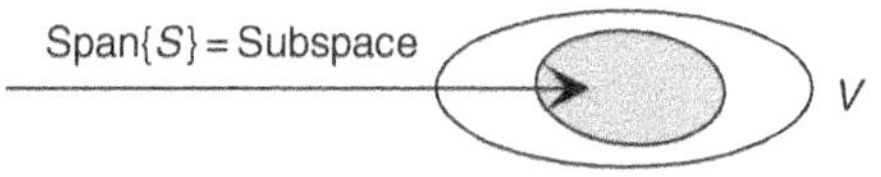

Figure 3.11(b)

Example 3.10

Let P_3 be the vector space containing the set of polynomials of degree 3 or less. Show that the set $\{1, t, t^2, t^3\}$ spans $\{P_3\}$.

Solution

Let $at^3 + bt^2 + ct + d$ be a general polynomial in the space P_3. Then we need to show that there are scalars k_1, k_2, k_3 and k_4 which satisfy the following:

$$k_1t^3 + k_2t^2 + k_3t + k_4 = at^3 + bt^2 + ct + d$$

Equating coefficients of t^3, t^2, t and constant gives $k_1 = a$, $k_2 = b$, $k_3 = c$ and $k_4 = d$ respectively. We have found scalars, $k_1 = a$, $k_2 = b$, $k_3 = c$ and $k_4 = d$, therefore the linear combination of $\{1, t, t^2, t^3\}$ can generate any vector in P_3 which means $\{1, t, t^2, t^3\}$ spans P_3.

Example 3.11

Let P_2 be the set of polynomials of degree 2 or less. Let the following vectors be in P_2:

$$\mathbf{v}_1 = t^2 - t - 1, \quad \mathbf{v}_2 = 6t^2 + 3t - 3 \text{ and } \mathbf{v}_3 = t^2 + 5t + 1$$

Determine whether the vector $\mathbf{u} = 3t^2 + 2t + 1$ is spanned by $\{\mathbf{v}_1, \mathbf{v}_2, \mathbf{v}_3\}$.

Solution

How do we check that the vector $\mathbf{u}$ *is spanned by* $\{\mathbf{v}_1, \mathbf{v}_2, \mathbf{v}_3\}$*?*

We need to see if scalars, k_1, k_2 and k_3 exist which satisfy the following linear combination:

$$k_1\mathbf{v}_1 + k_2\mathbf{v}_2 + k_3\mathbf{v}_3 = \mathbf{u}$$

Substituting $\mathbf{v}_1 = t^2 - t - 1$, $\mathbf{v}_2 = 6t^2 + 3t - 3$, $\mathbf{v}_3 = t^2 + 5t + 1$ and $\mathbf{u} = 3t^2 + 2t + 1$ into this:

$$\begin{aligned} k_1\mathbf{v}_1 + k_2\mathbf{v}_2 + k_3\mathbf{v}_3 &= k_1(t^2 - t - 1) + k_2(6t^2 + 3t - 3) + k_3(t^2 + 5t + 1) \\ &= (k_1 + 6k_2 + k_3)t^2 + (-k_1 + 3k_2 + 5k_3)t + (-k_1 - 3k_2 + k_3) \quad \text{[rearranging]} \\ &= 3t^2 + 2t + 1 \end{aligned}$$

Equating the coefficients of the last two lines gives:

$$\begin{aligned} k_1 + 6k_2 + k_3 &= 3 \quad &&\text{(equating } t^2\text{)} \\ -k_1 + 3k_2 + 5k_3 &= 2 \quad &&\text{(equating } t\text{)} \\ -k_1 - 3k_2 + k_3 &= 1 \quad &&\text{(equating constants)} \end{aligned}$$

What are we trying to find?

We need to find values of the scalars k_1, k_2 and k_3 which satisfy the above linear system. Writing the above system as an augmented matrix gives:

$$\begin{matrix} \text{R}_1 \\ \text{R}_2 \\ \text{R}_3 \end{matrix} \left(\begin{array}{ccc|c} 1 & 6 & 1 & 3 \\ -1 & 3 & 5 & 2 \\ -1 & -3 & 1 & 1 \end{array}\right)$$

Executing row operations, $\text{R}_2 + \text{R}_1$ and $\text{R}_3 + \text{R}_1$, gives:

$$\begin{matrix} \text{R}_1 \\ \text{R}_2^* = \text{R}_2 + \text{R}_1 \\ \text{R}_3^* = \text{R}_3 + \text{R}_1 \end{matrix} \left(\begin{array}{ccc|c} 1 & 6 & 1 & 3 \\ 0 & 9 & 6 & 5 \\ 0 & 3 & 2 & 4 \end{array}\right)$$

Carrying out the row operation $3\text{R}_3^* - \text{R}_2^*$:

$$\begin{matrix} \\ \text{R}_1 \\ \text{R}_2^* \\ 3\text{R}_3^* - \text{R}_2^* \end{matrix} \begin{matrix} k_1 \; k_2 \; k_3 \\ \left(\begin{array}{ccc|c} 1 & 6 & 1 & 3 \\ 0 & 9 & 6 & 5 \\ 0 & 0 & 0 & 7 \end{array}\right) \end{matrix}$$

From the bottom row we have $0k_1 + 0k_2 + 0k_3 = 0 = 7$, which means that the linear system is inconsistent, therefore we have no solution. There are no scalars (k's) which satisfy the above linear system. Hence the vector $\mathbf{u}$ does not exist in the *span*$\{\mathbf{v}_1, \mathbf{v}_2, \mathbf{v}_3\}$. This means that we cannot make $\mathbf{u}$ out of a linear combination of $\{\mathbf{v}_1, \mathbf{v}_2, \mathbf{v}_3\}$.

By Proposition (3.9) we have *span*$\{\mathbf{v}_1, \mathbf{v}_2, \mathbf{v}_3\}$ is a subspace of P_2 in the above example but $\mathbf{u} = 3t^2 + 2t + 1$ is not in the *span*$\{\mathbf{v}_1, \mathbf{v}_2, \mathbf{v}_3\}$. We can illustrate this as shown in Fig. 3.12.

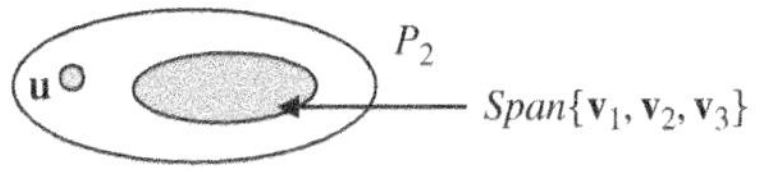

Figure 3.12

Example 3.12

Let $F[-\pi, \pi]$ be the vector space of continuous functions defined in the interval $[-\pi, \pi]$. Let the functions $\mathbf{f} = \cos^2(x)$ and $\mathbf{g} = \sin^2(x)$. Determine whether 2 is in *span*$\{\mathbf{f}, \mathbf{g}\}$.

Solution

We need to find scalars k and c such that

$$k\mathbf{f} + c\mathbf{g} = k\cos^2(x) + c\sin^2(x) = 2$$

From trigonometry we have the fundamental identity $\cos^2(x) + \sin^2(x) = 1$. Multiplying this by 2 gives us our result $2\cos^2(x) + 2\sin^2(x) = 2$.

Hence with $k = c = 2$ we have $k\mathbf{f} + c\mathbf{g} = 2$, therefore 2 is in the *span*$\{\mathbf{f}, \mathbf{g}\}$.

Example 3.13

Let M_{22} be the vector space of 2 by 2 matrices. Consider the matrices

$$\mathbf{A} = \begin{pmatrix} 1 & -1 \\ 1 & 2 \end{pmatrix}, \mathbf{B} = \begin{pmatrix} 0 & 2 \\ 0 & -1 \end{pmatrix} \text{ and } \mathbf{C} = \begin{pmatrix} 0 & 1 \\ 5 & 2 \end{pmatrix}$$

Determine whether the matrix $\mathbf{D} = \begin{pmatrix} 1 & 2 \\ -4 & -2 \end{pmatrix}$ is within the *span*$\{\mathbf{A}, \mathbf{B}, \mathbf{C}\}$.

Solution

We need to solve the linear combination

$$k_1\mathbf{A} + k_2\mathbf{B} + k_3\mathbf{C} = \mathbf{D}$$

for scalars k_1, k_2 and k_3:

$$\begin{aligned} k_1\mathbf{A} + k_2\mathbf{B} + k_3\mathbf{C} &= k_1\begin{pmatrix} 1 & -1 \\ 1 & 2 \end{pmatrix} + k_2\begin{pmatrix} 0 & 2 \\ 0 & -1 \end{pmatrix} + k_3\begin{pmatrix} 0 & 1 \\ 5 & 2 \end{pmatrix} \\ &= \begin{pmatrix} k_1 & -k_1 \\ k_1 & 2k_1 \end{pmatrix} + \begin{pmatrix} 0 & 2k_2 \\ 0 & -k_2 \end{pmatrix} + \begin{pmatrix} 0 & k_3 \\ 5k_3 & 2k_3 \end{pmatrix} \\ &= \begin{pmatrix} k_1 & -k_1 + 2k_2 + k_3 \\ k_1 + 5k_3 & 2k_1 - k_2 + 2k_3 \end{pmatrix} = \begin{pmatrix} 1 & 2 \\ -4 & -2 \end{pmatrix} \quad [\text{matrix } \mathbf{D}] \end{aligned}$$

Equating entries of the matrix gives

$$k_1 = 1, \quad -k_1 + 2k_2 + k_3 = 2, \quad k_1 + 5k_3 = -4 \text{ and } 2k_1 - k_2 + 2k_3 = -2$$

We have $k_1 = 1$, but how do we find k_2 and k_3?

Substituting $k_1 = 1$ into the third equation $k_1 + 5k_3 = -4$ yields

$$1 + 5k_3 = -4 \text{ gives } k_3 = -1$$

(continued...)

Substituting $k_1 = 1$ and $k_3 = -1$ into the last equation $2k_1 - k_2 + 2k_3 = -2$:

$$2(1) - k_2 + 2(-1) = -2$$
$$-k_2 = -2 \text{ which gives } k_2 = 2$$

Hence we have found scalars, $k_1 = 1, k_2 = 2$ and $k_3 = -1$, which satisfy

$$k_1\mathbf{A} + k_2\mathbf{B} + k_3\mathbf{C} = \mathbf{A} + 2\mathbf{B} - \mathbf{C} = \mathbf{D}$$

Therefore $\mathbf{D}$ belongs to *span*$\{\mathbf{A},\ \mathbf{B},\ \mathbf{C}\}$ because $\mathbf{D}$ can be made by a linear combination of $\mathbf{A}$, $\mathbf{B}$ and $\mathbf{C}$.

Summary

Proposition (3.5). The non-empty set S is a subspace of a vector space $\Leftrightarrow$:

(a) S is closed under vector addition.

(b) S is closed under scalar multiplication.

(3.6) If a vector $\mathbf{x}$ can be expressed as

$$\mathbf{x} = k_1\mathbf{v}_1 + k_2\mathbf{v}_2 + k_3\mathbf{v}_3 + \cdots + k_n\mathbf{v}_n \text{ (where } k\text{'s are scalars)}$$

then we say $\mathbf{x}$ is a *linear combination* of the vectors $\mathbf{v}_1, \mathbf{v}_2, \mathbf{v}_3, \ldots$ and $\mathbf{v}_n$.
(3.8) If *every* vector in V can be produced by a linear combination of vectors $\mathbf{v}_1, \mathbf{v}_2, \mathbf{v}_3, \ldots$ and $\mathbf{v}_n$ then we say these vectors *span* or *generate* the vector space V.

EXERCISES 3.2

(Brief solutions at end of book. Full solutions available at <http://www.oup.co.uk/companion/singh>.)
Throughout this exercise assume your scalars to be real numbers.

1. Consider Example 3.4. Show that S is a vector space with respect to the usual vector addition and scalar multiplication of $\mathbb{R}^2$.
2. Let S be the set of vectors $\begin{pmatrix} a \\ 0 \end{pmatrix}$ in the vector space $\mathbb{R}^2$. Show that S is a subspace of $\mathbb{R}^2$.
3. Let S be the set of vectors $\begin{pmatrix} a \\ a \end{pmatrix}$ in the vector space $\mathbb{R}^2$. Show that S is a subspace of $\mathbb{R}^2$.
4. Let S be the set of vectors $(0 \quad 0 \quad c)^T$ in the vector space $\mathbb{R}^3$. Show that S is a subspace of $\mathbb{R}^3$.

5. Let S be the set of vectors $\begin{pmatrix} 1 & b & c & d \end{pmatrix}^T$ in the vector space $\mathbb{R}^4$. Show that S is not a subspace of $\mathbb{R}^4$.

6. Let S be the set of vectors $\begin{pmatrix} a & b & c \end{pmatrix}^T$ where $a+b+c=0$ in the vector space $\mathbb{R}^3$. Determine whether S is a subspace of $\mathbb{R}^3$.

7. Let $S = \left\{ \begin{pmatrix} a \\ b \end{pmatrix} \,\middle|\, a \text{ and } b \text{ are integers} \right\}$ be a subset of $\mathbb{R}^2$. Show that S is not a subspace of $\mathbb{R}^2$.

8. Let $S = \left\{ \begin{pmatrix} a \\ b \\ c \end{pmatrix} \,\middle|\, a,\ b \text{ and } c \text{ are rational numbers} \right\}$ be a subset of $\mathbb{R}^3$. Show that S is not a subspace of $\mathbb{R}^3$.

9. Let M_{22} be the set of matrices of size 2 by 2. Let S be the subset of matrices of the form $\begin{pmatrix} a & 1 \\ c & d \end{pmatrix}$. Show that S is not a subspace of M_{22}.

10. Let M_{22} be the set of matrices of size 2 by 2. Let

$$S = \left\{ \begin{pmatrix} a & b \\ c & d \end{pmatrix} \,\middle|\, a,\ b,\ c,\ d \text{ are all integers} \right\}$$

Show that S is not a subspace of M_{22}.

11. Let S be the set of symmetric matrices (these are matrices $\mathbf{A}$ such that $\mathbf{A}^T = \mathbf{A}$, that is $\mathbf{A}$ transposed is equal to $\mathbf{A}$). Let V be the set of all matrices. Show that S is a subspace of V.

12. Let $\mathbf{v}_1 = t^2 - 1$, $\mathbf{v}_2 = t + 1$ and $\mathbf{v}_3 = 2t^2 + t - 1$ be vectors in P_2 where P_2 is the set of polynomials of degree 2 or less. Show that $\mathbf{x} = 7t^2 + 8t + 1$ is a linear combination of $\mathbf{v}_1$, $\mathbf{v}_2$ and $\mathbf{v}_3$.

13. Let $\mathbf{p}_1 = t^2 + 2t - 1$, $\mathbf{p}_2 = 2t + 1$ and $\mathbf{p}_3 = 5t^2 + 2t - 3$ be vectors in P_2. Show that the following are linear combinations of these vectors $\mathbf{p}_1$, $\mathbf{p}_2$ and $\mathbf{p}_3$:
(a) $\mathbf{x} = 4t^2 - 2t - 3$ (b) $\mathbf{x} = -2t^2 - 2$ (c) $\mathbf{x} = 6$

14. Let V be the set of real valued continuous functions. Then V is a vector space. Let $\mathbf{v}_1 = \sin^2(t)$ and $\mathbf{v}_2 = \cos^2(t)$. Show that the following are linear combinations of $\mathbf{v}_1$ and $\mathbf{v}_2$.
(a) $\mathbf{x} = 1$ (b) $\mathbf{x} = \pi$ (c) $\mathbf{x} = \cos(2t)$

15. Let S be a subspace of $\mathbb{R}^3$ given by $\begin{pmatrix} a \\ b \\ 0 \end{pmatrix}$. Let $\mathbf{u} = \begin{pmatrix} 1 \\ 2 \\ 0 \end{pmatrix}$ and $\mathbf{v} = \begin{pmatrix} -1 \\ 5 \\ 0 \end{pmatrix}$. Show that the vectors $\mathbf{u}$ and $\mathbf{v}$ span S.

16. Let P_2 be the set of polynomials of degree 2 or less. Let

$$\mathbf{p}_1 = t^2 + 3,\ \mathbf{p}_2 = 2t^2 + 5t + 6 \text{ and } \mathbf{p}_3 = 5t$$

Determine whether the vector $\mathbf{x} = t^2 + 3$ is in the span $\{\mathbf{p}_1,\ \mathbf{p}_2,\ \mathbf{p}_3\}$.

17. Let M_{22} be the set of 2 by 2 matrices. Consider the matrices

$$\mathbf{A} = \begin{pmatrix} 1 & 1 \\ 1 & 1 \end{pmatrix}, \ \mathbf{B} = \begin{pmatrix} -1 & 2 \\ 5 & 7 \end{pmatrix} \text{ and } \mathbf{C} = \begin{pmatrix} 2 & 6 \\ 8 & 0 \end{pmatrix}$$

Determine whether the matrix $\mathbf{D} = \begin{pmatrix} 7 & -3 \\ -14 & -26 \end{pmatrix}$ belongs to span $\{\mathbf{A}, \mathbf{B}, \mathbf{C}\}$.

18. Let $F[0, 2\pi]$ be the vector space of continuous functions. Let the functions $\mathbf{f} = \cos(2x)$ and $\mathbf{g} = \sin(2x)$. Determine whether the following are in span $\{\mathbf{f}, \mathbf{g}\}$:
(a) 0 (b) $\sin(2x)$ (c) $\cos^2(x) - \sin^2(x)$ (d) 1

19. Determine whether $x + 1$ and $(x + 1)^2$ span P_2.

20. Prove Proposition (3.9).

21. Let S be a subspace of a vector space V. Show that if T is a subset of S it may not be a subspace of V.

22. Let S and T be subspaces of a vectors space V. Prove or disprove that
(a) $S \cap T$ (b) $S \cup T$
are subspaces of V.
[$S \cap T$ is the **intersection** of S and T and is the set of elements which lie in both S and T. This means if $\mathbf{u} \in S \cap T$ then $\mathbf{u} \in S$ and $\mathbf{u} \in T$.
$S \cup T$ is the **union** of S and T and is the set all the elements in S or T.]

SECTION 3.3 Linear Independence and Basis

By the end of this section you will be able to

- test vectors for linear independence
- prove properties about linear independence
- determine whether given vectors are a basis for a vector space

We discussed the terms linear independence and basis during chapter 2, using examples in Euclidean space $\mathbb{R}^n$. In this section we expand these definitions to general vector spaces V.

From previous sections we know that the set of matrices, polynomials and functions form a vector space which means that they behave the same way although they are different objects.

A basis is a sort of coordinate system for vector space which we measure a vector against.

3.3.1 Linear independence

We extend the definition of linear independence to a general vector space V.

Definition (3.10). We say vectors $\mathbf{v}_1, \mathbf{v}_2, \mathbf{v}_3, \ldots$ and $\mathbf{v}_n$ in V are **linearly independent** $\Leftrightarrow$ the *only* real scalars $k_1, k_2, k_3, \ldots$ and k_n which satisfy:

$$k_1\mathbf{v}_1 + k_2\mathbf{v}_2 + k_3\mathbf{v}_3 + \cdots + k_n\mathbf{v}_n = \mathbf{O} \quad \text{are} \quad k_1 = k_2 = k_3 = \cdots = k_n = 0$$

What does this mean?

The only solution to the linear combination $k_1\mathbf{v}_1 + k_2\mathbf{v}_2 + k_3\mathbf{v}_3 + \cdots + k_n\mathbf{v}_n = \mathbf{O}$ requires all the scalars $k_1, k_2, k_3, \ldots$ and k_n to be equal to zero. In other words, you cannot make any one of the vectors $\mathbf{v}_j$ say, by a linear combination of the others.

Example 3.14

Let M_{22} be the vector space of size 2 by 2 matrices. Consider the matrices:

$$\mathbf{A} = \begin{pmatrix} 1 & 0 \\ 0 & 1 \end{pmatrix} \text{ and } \mathbf{B} = \begin{pmatrix} 0 & 2 \\ 2 & 0 \end{pmatrix}$$

Show that the matrices $\mathbf{A}$ and $\mathbf{B}$ are linearly independent.

Solution

*How do we show that matrices **A** and **B** are linearly independent?*

Required to show that $k\mathbf{A} + c\mathbf{B} = \mathbf{O}$ is true only when the scalars $k = c = 0$.

Let k and c be scalars, then by applying the above definition (3.10) to these matrices with the zero matrix $\mathbf{O} = \begin{pmatrix} 0 & 0 \\ 0 & 0 \end{pmatrix}$ we have

$$\begin{aligned} k\mathbf{A} + c\mathbf{B} &= k\begin{pmatrix} 1 & 0 \\ 0 & 1 \end{pmatrix} + c\begin{pmatrix} 0 & 2 \\ 2 & 0 \end{pmatrix} \\ &= \begin{pmatrix} k & 0 \\ 0 & k \end{pmatrix} + \begin{pmatrix} 0 & 2c \\ 2c & 0 \end{pmatrix} = \begin{pmatrix} k & 2c \\ 2c & k \end{pmatrix} = \begin{pmatrix} 0 & 0 \\ 0 & 0 \end{pmatrix} = \mathbf{O} \end{aligned}$$

Equating entries gives $k = 0$ and $2c = 0$. Since both (all) our scalars, $k = 0$ and $c = 0$, are zero, then by definition (3.10) we conclude that the given matrices $\mathbf{A}$ and $\mathbf{B}$ are linearly independent.

Linear independence means that we cannot make any one vector out of a combination of the others.

What does this mean in relation to Example 3.14?

It means that matrix $\mathbf{A}$ is not a multiple of matrix $\mathbf{B}$.

Example 3.15

Let V be the vector space of continuous functions defined on the real line.

Let $\mathbf{u} = \cos(x)$, $\mathbf{v} = \sin(x)$ and $\mathbf{w} = 2$ be vectors in V.

Show that the vectors $\mathbf{u}$, $\mathbf{v}$ and $\mathbf{w}$ are linearly independent.

(continued...)

Solution

By the above definition (3.10) we need to consider the linear combination

$$k_1\mathbf{u} + k_2\mathbf{v} + k_3\mathbf{w} = k_1 \cos(x) + k_2 \sin(x) + k_3(2) = 0 \qquad (*)$$

where k_1, k_2 and k_3 are scalars.

What do we need to prove?

We need to show that the given vectors $\mathbf{u}$, $\mathbf{v}$ and $\mathbf{w}$ are linearly independent.

How?

Required to show that all the scalars are equal to zero, that is $k_1 = k_2 = k_3 = 0$.

We need three equations, so we substitute three different values of x into (*). We choose values that help to simplify the equations.

Substituting $x = 0$ (to eliminate the sine term) into (*) gives

$$k_1 \underbrace{\cos(0)}_{=1} + k_2 \underbrace{\sin(0)}_{=0} + k_3(2) = k_1 + 2k_3 = 0 \qquad (1)$$

Next we substitute $x = \pi$ into (*):

$$k_1 \underbrace{\cos(\pi)}_{=-1} + k_2 \underbrace{\sin(\pi)}_{=0} + k_3(2) = -k_1 + 2k_3 = 0 \qquad (2)$$

Substituting $x = \pi/2$ into equation (*):

$$k_1 \underbrace{\cos\left(\frac{\pi}{2}\right)}_{=0} + k_2 \underbrace{\sin\left(\frac{\pi}{2}\right)}_{=1} + k_3(2) = k_2 + 2k_3 = 0 \qquad (3)$$

We can write these three equations - (1), (2) and (3) - in an augmented matrix and solve:

$$\begin{array}{c} \\ R_1 \\ R_2 \\ R_3 \end{array} \begin{array}{c} k_1\ k_2\ k_3 \\ \left(\begin{array}{ccc|c} 1 & 0 & 2 & 0 \\ -1 & 0 & 2 & 0 \\ 0 & 1 & 2 & 0 \end{array}\right) \end{array} \Rightarrow \begin{array}{c} \\ R_1 \\ R_2 + R_1 \\ R_3 \end{array} \begin{array}{c} k_1\ k_2\ k_3 \\ \left(\begin{array}{ccc|c} 1 & 0 & 2 & 0 \\ 0 & 0 & 4 & 0 \\ 0 & 1 & 2 & 0 \end{array}\right) \end{array}$$

By inspection we can see that $k_1 = k_2 = k_3 = 0$.

Since the only solution is $k_1 = k_2 = k_3 = 0$, the given vectors $\mathbf{u} = \cos(x)$, $\mathbf{v} = \sin(x)$ and $\mathbf{w} = 2$ are linearly independent.

In the above Example 3.15, how do we know which values of x to substitute?

Since the linear combination

$$k_1\mathbf{u} + k_2\mathbf{v} + k_3\mathbf{w} = k_1 \cos(x) + k_2 \sin(x) + k_3(2) = 0$$

is valid for any real value of x, we can try any real number. Normally we choose x values which simplify our arithmetic.

3.3.2 Linear dependence

What does linear dependence mean?

> Definition (3.11). The vectors $\mathbf{v}_1, \mathbf{v}_2, \mathbf{v}_3, \ldots$ and $\mathbf{v}_n$ in a vector space V are *linearly dependent* $\Leftrightarrow$ the scalars $k_1, k_2, k_3, \ldots$ and k_n are *not all zero* and satisfy
>
> $$k_1\mathbf{v}_1 + k_2\mathbf{v}_2 + k_3\mathbf{v}_3 + \cdots + k_n\mathbf{v}_n = \mathbf{O}$$

Example 3.16

Let V be the vector space of continuous functions defined on the real line.

Let $\mathbf{u} = \cos^2(x)$, $\mathbf{v} = \sin^2(x)$ and $\mathbf{w} = 2$ be vectors in V.

Show that the vectors $\mathbf{u}$, $\mathbf{v}$ and $\mathbf{w}$ are linearly dependent.

Solution

From our knowledge of trigonometry, we have the fundamental trigonometric identity

$$\cos^2(x) + \sin^2(x) = 1$$

Multiplying each side by 2 gives

$$2\cos^2(x) + 2\sin^2(x) = 2$$

$$2\mathbf{u} + 2\mathbf{v} = \mathbf{w} \qquad [\text{because } \mathbf{u} = \cos^2(x), \mathbf{v} = \sin^2(x) \text{ and } \mathbf{w} = 2]$$

$2\mathbf{u} + 2\mathbf{v} = \mathbf{w}$ implies that we can make vector $\mathbf{w}$ out of a linear combination of vectors $\mathbf{u}$ and $\mathbf{v}$, (2 lots of $\mathbf{u}$ plus 2 lots of $\mathbf{v}$), therefore vectors $\mathbf{u}$, $\mathbf{v}$ and $\mathbf{w}$ are linearly dependent.

Example 3.17

Let P_2 be the vector space of polynomials of degree 2 or less. Decide whether the following vectors in P_2 are linearly independent or dependent.

(a) $\mathbf{p} = 6t^2 + 8t + 2$ and $\mathbf{q} = 3t^2 + 4t + 1$

(b) $\mathbf{p} = 2t^2 + 3t + 2$ and $\mathbf{q} = t^2 + t + 1$

(c) $\mathbf{p} = t^2 + 3t - 1$, $\mathbf{q} = 2t^2 + 7t + 5$ and $\mathbf{r} = 7$

Solution

(a) *What do you notice about the first two given vectors,* $\mathbf{p} = 6t^2 + 8t + 2$ *and* $\mathbf{q} = 3t^2 + 4t + 1$*?*
The vector $\mathbf{p}$ is double vector $\mathbf{q}$, that is $\mathbf{p} = 2\mathbf{q}$ or $\mathbf{p} - 2\mathbf{q} = \mathbf{O}$.

What does this mean?

$\mathbf{p}$ and $\mathbf{q}$ are linearly dependent because we can make vector $\mathbf{p}$ by doubling $\mathbf{q}$.

Notice that the non-zero scalars are 1 and -2 because $\mathbf{p} - 2\mathbf{q} = (1)\mathbf{p} + (-2)\mathbf{q} = \mathbf{O}$.

(b) In this case the vector $\mathbf{p} = 2t^2 + 3t + 2$ is not a multiple of $\mathbf{q} = t^2 + t + 1$, which means that they are linearly independent.

(continued...)

(c) With three vectors it is much harder to spot any relationship between them, so we have to carry out the general procedure for solving simultaneous equations.

Let k_1, k_2 and k_3 be scalars. We need to find values for k_1, k_2 and k_3, so that they satisfy

$$k_1\mathbf{p} + k_2\mathbf{q} + k_3\mathbf{r} = \mathbf{O} \qquad (*)$$

Substituting our given vectors $\mathbf{p} = t^2 + 3t - 1$, $\mathbf{q} = 2t^2 + 7t + 5$, $\mathbf{r} = 7$ and $\mathbf{O} = 0$ into (*):

$$\begin{aligned} k_1\mathbf{p} + k_2\mathbf{q} + k_3\mathbf{r} &= k_1(t^2 + 3t - 1) + k_2(2t^2 + 7t + 5) + k_3(7) \\ &= k_1t^2 + 3k_1t - k_1 + 2k_2t^2 + 7k_2t + 5k_2 + 7k_3 \\ &\underset{\text{collecting like terms}}{=} (k_1 + 2k_2)t^2 + (3k_1 + 7k_2)t + (-k_1 + 5k_2 + 7k_3) \\ &= 0 \end{aligned}$$

Equating the coefficients of the last two lines gives

$$\begin{aligned} t^2 : &\quad k_1 + 2k_2 = 0 \\ t : &\quad 3k_1 + 7k_2 = 0 \\ \mathit{const} : &\quad -k_1 + 5k_2 + 7k_3 = 0 \end{aligned}$$

The only solution to these equations is $k_1 = k_2 = k_3 = 0$.
We conclude that the given vectors $\mathbf{p}$, $\mathbf{q}$ and $\mathbf{r}$ are linearly independent.

3.3.3 Properties of linear dependence and independence

Some of the properties associated with the linear independence and dependence of vectors in Euclidean space $\mathbb{R}^n$ were proved during chapter 2 in subsection 2.3.4. We can extend these results to the general vector space V. The proofs are very similar to the Euclidean space proofs that you were asked to show in chapter 2.

Proposition (3.12). Two vectors $\mathbf{u}$ and $\mathbf{v}$ in a vector space V are linearly **dependent** $\Leftrightarrow$ one of the vectors is a multiple of the other.

Proof.
($\Rightarrow$). Assume vectors $\mathbf{u}$ and $\mathbf{v}$ are linearly dependent.

What do we need to prove?
We need to prove one vector is a multiple of the other. The vectors $\mathbf{u}$ and $\mathbf{v}$ are linearly dependent, therefore by the above definition (3.11) there must be non-zero scalars k or c such that

$$k\mathbf{u} + c\mathbf{v} = \mathbf{O}$$

Suppose $k \neq 0$ then $k\mathbf{u} = -c\mathbf{v}$ which implies $\mathbf{u} = -\frac{c}{k}\mathbf{v}$. We have $\mathbf{u} = -\frac{c}{k}\mathbf{v}$, therefore the vector $\mathbf{u}$ is a multiple of the vector $\mathbf{v}$.

($\Leftarrow$). Now we prove it the other way. Assume that the vector $\mathbf{u}$ is a multiple of vector $\mathbf{v}$ which means that we can write this as

$$\mathbf{u} = m\mathbf{v} \qquad (m \text{ is a scalar})$$

Taking $-m\mathbf{v}$ from both sides, we have $\mathbf{u} - m\mathbf{v} = \mathbf{O}$. Since this linear combination $(1)\mathbf{u} - m\mathbf{v} = \mathbf{O}$ gives the zero vector with non-zero scalars, then by definition (3.11) we conclude that vectors $\mathbf{u}$ and $\mathbf{v}$ are linearly dependent.

■

Proposition (3.13). Let $S = \{\mathbf{v}_1, \mathbf{v}_2, \mathbf{v}_3, \ldots, \mathbf{v}_n\}$ be a set of linearly dependent vectors in a vector space V. Then the set of vectors $T = \{\mathbf{v}_1, \mathbf{v}_2, \ldots, \mathbf{v}_n, \mathbf{v}_{n+1}\}$ are also linearly dependent in V.

What does this proposition mean?

It means that if a set of n vectors are linearly dependent then the same n vectors plus another vector are also linearly dependent - the addition of more vectors does *not* change dependence.

Proof.

Since we are given that the vectors $S = \{\mathbf{v}_1, \mathbf{v}_2, \mathbf{v}_3, \ldots, \mathbf{v}_n\}$ are linearly dependent, we can write these as

$$k_1\mathbf{v}_1 + k_2\mathbf{v}_2 + k_3\mathbf{v}_3 + \cdots + k_n\mathbf{v}_n = \mathbf{O} \qquad (*)$$

where *all* the k's are not zero. Consider the linear combination

$$k_1\mathbf{v}_1 + k_2\mathbf{v}_2 + \cdots + k_n\mathbf{v}_n + \underbrace{k_{n+1}\mathbf{v}_{n+1}}_{\text{extra vector}}$$

If we take the case where $k_{n+1} = 0$ then we have

$$\underbrace{k_1\mathbf{v}_1 + k_2\mathbf{v}_2 + \cdots + k_n\mathbf{v}_n}_{=\mathbf{O} \text{ by } (*)} + \underbrace{k_{n+1}\mathbf{v}_{n+1}}_{\text{extra vector}} = \mathbf{O}$$

By (*) all the scalars are not zero in the last equation, therefore by definition (3.11) we conclude that the set of vectors $T = \{\mathbf{v}_1, \mathbf{v}_2, \ldots, \mathbf{v}_n, \mathbf{v}_{n+1}\}$ are linearly dependent.

■

Proposition (3.14). The vectors in the set $S = \{\mathbf{v}_1, \mathbf{v}_2, \mathbf{v}_3, \ldots, \mathbf{v}_n\}$ are linearly dependent $\Leftrightarrow$ one of these vectors, say $\mathbf{v}_k$, is a linear combination of the preceding vectors, that is

$$\mathbf{v}_k = c_1\mathbf{v}_1 + c_2\mathbf{v}_2 + c_3\mathbf{v}_3 + \cdots + c_{k-1}\mathbf{v}_{k-1}$$

What does this proposition mean?

This means that if the vectors $\mathbf{v}_1, \mathbf{v}_2, \mathbf{v}_3, \ldots$ and $\mathbf{v}_n$ are linearly dependent, then one of these vectors can be written in terms of the other vectors. This proposition also goes the other way, that is if one of the vectors can be written as a linear combination of the preceding vectors then these vectors are linearly dependent.

Proof.

($\Rightarrow$). Assume the vectors $S = \{\mathbf{v}_1, \mathbf{v}_2, \mathbf{v}_3, \ldots, \mathbf{v}_n\}$ are linearly dependent. We need to show that we can write the vector $\mathbf{v}_k$ as a linear combination of vectors $\mathbf{v}_1, \mathbf{v}_2, \ldots, \mathbf{v}_{k-1}$.

By the above definition (3.11):

Dependence: If $k_1, k_2, \ldots, k_n$ are not all zero then $k_1\mathbf{v}_1 + k_2\mathbf{v}_2 + \cdots + k_n\mathbf{v}_n = \mathbf{O}$

We can write the vectors in S as

$$c_1\mathbf{v}_1 + c_2\mathbf{v}_2 + c_3\mathbf{v}_3 + \cdots + c_n\mathbf{v}_n = \mathbf{O}$$

where *all* the scalars c's are *not* zero. Let k be the largest subscript for which $c_k \neq 0$ [not zero] then

$$c_1\mathbf{v}_1 + c_2\mathbf{v}_2 + c_3\mathbf{v}_3 + \cdots + c_{k-1}\mathbf{v}_{k-1} + c_k\mathbf{v}_k = \mathbf{O}$$

Carrying out vector algebra

$$\begin{aligned} c_k\mathbf{v}_k &= -c_1\mathbf{v}_1 - c_2\mathbf{v}_2 - c_3\mathbf{v}_3 - \cdots - c_{k-1}\mathbf{v}_{k-1} && \text{[transposing]} \\ \mathbf{v}_k &= -\frac{c_1}{c_k}\mathbf{v}_1 - \frac{c_2}{c_k}\mathbf{v}_2 - \frac{c_3}{c_k}\mathbf{v}_3 - \cdots - \frac{c_{k-1}}{c_k}\mathbf{v}_{k-1} && \text{[dividing by } c_k \neq 0\text{]} \\ &= d_1\mathbf{v}_1 + d_2\mathbf{v}_2 + d_3\mathbf{v}_3 + \cdots + d_{k-1}\mathbf{v}_{k-1} \text{ where } d_j = -\frac{c_j}{c_k} \end{aligned}$$

As can be seen, the vector $\mathbf{v}_k$ can be written as a linear combination of the preceding vectors.

($\Leftarrow$). Now we go the other way. Assume that we can write the vector $\mathbf{v}_k$ as a linear combination of the preceding vectors. This means that we have

$$\mathbf{v}_k = c_1\mathbf{v}_1 + c_2\mathbf{v}_2 + c_3\mathbf{v}_3 + \cdots + c_{k-1}\mathbf{v}_{k-1} \qquad \text{(c's are scalars)}$$

Rearranging this gives

$$c_1\mathbf{v}_1 + c_2\mathbf{v}_2 + c_3\mathbf{v}_3 + \cdots + c_{k-1}\mathbf{v}_{k-1} - \mathbf{v}_k = \mathbf{O}$$

Hence the vectors $\mathbf{v}_1, \mathbf{v}_2, \mathbf{v}_3, \ldots, \mathbf{v}_{k-1}$ and $\mathbf{v}_k$ are linearly dependent.

Why?

Because we can write the last line as

$$c_1\mathbf{v}_1 + c_2\mathbf{v}_2 + c_3\mathbf{v}_3 + \cdots + c_{k-1}\mathbf{v}_{k-1} + (-1)\mathbf{v}_k = \mathbf{O}$$

There is at least one non-zero scalar, -1.

By the previous Proposition (3.13):

If $S = \{\mathbf{v}_1, \ldots, \mathbf{v}_n\}$ are dependent then $T = \{\mathbf{v}_1, \ldots, \mathbf{v}_n, \mathbf{v}_{n+1}\}$ are also dependent.

We conclude that the vectors $\mathbf{v}_1,\ \mathbf{v}_2,\ \ldots, \mathbf{v}_k,\ \mathbf{v}_{k+1},\ \ldots$ and $\mathbf{v}_n$ (adding more vectors does not change dependency) are linearly dependent. This completes our proof.

■

Note that the above proposition also implies that if none of the vectors in a set can be written as a linear combination of the preceding vectors then the set is linearly independent. Hence to prove linear independence it is enough to show that none of the vectors is a linear combination of the preceding vectors.

Example 3.18

Let P_n be the vector space of polynomials of degree n or less.

Show that the set of vectors $\{1,\ t,\ t^2,\ t^3,\ \ldots,\ t^n\}$ are linearly independent.

Solution

Note that none of polynomials can be written as a linear combination of the preceding polynomials. For example, we cannot write the polynomial such as t^k as a linear combination of $\{1,\ t,\ t^2,\ t^3,\ \ldots,\ t^{k-1}\}$ with scalars c's:

$$c_0 1 + c_1 t + c_2 t^2 + c_3 t^3 + \cdots + c_{k-1} t^{k-1} \neq t^k \qquad \text{[not equal]}$$

Since none of the polynomials is a linear combination of the preceding polynomials we conclude that the given set $\{1,\ t,\ t^2,\ t^3,\ \ldots,\ t^n\}$ is linearly independent.

3.3.4 Basis vectors

This section extends the definition of basis vectors to the general vector space V.

During chapter 2 we covered basis in section 2.4.2, but only for the Euclidean space $\mathbb{R}^n$.

In this section we discuss the nature of a basis in a general vector space V.

You can think of a basis as the *axes* of a coordinate system that describes a vector space. This means that every vector in a vector space V can be written in terms of the coordinate system which is represented by the basis vectors. You can decompose each vector into its basis vectors.

Definition (3.15). Consider the n vectors $\mathbf{v}_1,\ \mathbf{v}_2,\ \mathbf{v}_3,\ \ldots$ and $\mathbf{v}_n$ in the vector space V. These vectors form a **basis** of $V \Leftrightarrow$

(i) $\mathbf{v}_1,\ \mathbf{v}_2,\ \mathbf{v}_3,\ \ldots$ and $\mathbf{v}_n$ span V

(ii) $\mathbf{v}_1,\ \mathbf{v}_2,\ \mathbf{v}_3,\ \ldots$ and $\mathbf{v}_n$ are linearly independent

We can write these n vectors $\mathbf{v}_1,\ \mathbf{v}_2,\ \mathbf{v}_3,\ \ldots$ and $\mathbf{v}_n$ as a set

$$B = \{\mathbf{v}_1,\ \mathbf{v}_2,\ \mathbf{v}_3,\ \ldots,\ \mathbf{v}_n\}$$

These are generally called the **basis vectors**.

What does the term '$\{\mathbf{v}_1, \mathbf{v}_2, \mathbf{v}_3, \ldots, \mathbf{v}_n\}$ spans V' mean?
If *every* vector in V can be produced by a linear combination of vectors $\{\mathbf{v}_1, \mathbf{v}_2, \mathbf{v}_3, \ldots, \mathbf{v}_n\}$ then we say that these vectors **span** or **generate** the vector space V. This is generally proved by considering an arbitrary vector in V and then showing that it is a linear combination of vectors $\mathbf{v}_1, \mathbf{v}_2, \mathbf{v}_3, \ldots$ and $\mathbf{v}_n$.

Next we show how to prove that a set of given vectors form a basis for a general vector space. The proofs are similar to the ones in the last chapter but we cannot visualize the vectors because we are dealing with general vector spaces.

Example 3.19

Let P_n be the vector space of polynomials of degree n or less.
Show that the set of vectors $S = \{1, t, t^2, \ldots, t^n\}$ is a basis for P_n.

Solution
How do we prove this result?
We need to show two things for S to qualify as a basis:

1. The set of vectors S must span P_n.
2. The set of vectors S must be linearly independent.

Span: Consider an arbitrary vector $c_0 + c_1 t + c_2 t^2 + c_3 t^3 + \ldots + c_n t^n$ in P_n. This can be written as a linear combination of $\{1, t, t^2, t^3, \ldots, t^n\}$ as follows:

$$k_0 + k_1 t + k_2 t^2 + k_3 t^3 + \cdots + k_n t^n = c_0 + c_1 t + c_2 t^2 + c_3 t^3 + \cdots + c_n t^n$$

where $k_j = c_j$ for $j = 1, 2, 3, \ldots$ and n. Hence the vectors $\{1, t, t^2, t^3, \ldots, t^n\}$ span P_n.

Linearly independent: We have already shown this in Example 3.18.

Both conditions (1 and 2) are satisfied, therefore the given vectors $S = \{1, t, t^2, \ldots, t^n\}$ form a basis for P_n.

These basis vectors $\{1, t, t^2, t^3, \ldots, t^n\}$ are generally known as the natural or standard basis for P_n. This is not the only basis for P_n; there are an infinite number of them. For example, you can multiply each element in this $\{1, t, t^2, t^3, \ldots, t^n\}$ by a non-zero number, and the set would then be another basis for P_n.

We can write any vector in terms of these basis vectors. This is similar to writing vectors in $\mathbb{R}^2$ in terms of $\mathbf{e}_1 = (1 \quad 0)^T$ representing the x axis and $\mathbf{e}_2 = (0 \quad 1)^T$ representing the y *axis*.

Example 3.20

Let M_{22} be the vector space containing matrices of size 2 by 2. Show that the following matrices are not a basis for M_{22}:

$$\mathbf{A} = \begin{pmatrix} 1 & 0 \\ 0 & 1 \end{pmatrix} \text{ and } \mathbf{B} = \begin{pmatrix} 0 & 1 \\ 1 & 0 \end{pmatrix}$$

Solution

How do we show this result?

We need to show either that the matrices $\mathbf{A}$ and $\mathbf{B}$ do *not* span M_{22} or that they are linearly *dependent.*

To show that matrices $\mathbf{A}$ and $\mathbf{B}$ do not span M_{22}, we only need to select a matrix in M_{22} and prove that a linear combination of $\mathbf{A}$ and $\mathbf{B}$ do not generate the selected matrix.

Consider the matrix $\mathbf{C} = \begin{pmatrix} 1 & 2 \\ 3 & 4 \end{pmatrix}$. Let k_1 and k_2 be scalars then

$$k_1\mathbf{A} + k_2\mathbf{B} = k_1\begin{pmatrix} 1 & 0 \\ 0 & 1 \end{pmatrix} + k_2\begin{pmatrix} 0 & 1 \\ 1 & 0 \end{pmatrix}$$

$$= \begin{pmatrix} k_1 & 0 \\ 0 & k_1 \end{pmatrix} + \begin{pmatrix} 0 & k_2 \\ k_2 & 0 \end{pmatrix} = \begin{pmatrix} k_1 & k_2 \\ k_2 & k_1 \end{pmatrix} = \begin{pmatrix} 1 & 2 \\ 3 & 4 \end{pmatrix}$$

By equating entries we have

$$\boxed{k_1 = 1,\ k_2 = 2,\ k_2 = 3 \text{ and } k_1 = 4}$$

This is inconsistent because we have two different values for k_1, 1 and 4, which means that we cannot obtain the matrix $\begin{pmatrix} 1 & 2 \\ 3 & 4 \end{pmatrix}$. Hence the given matrices $\mathbf{A} = \begin{pmatrix} 1 & 0 \\ 0 & 1 \end{pmatrix}$ and $\mathbf{B} = \begin{pmatrix} 0 & 1 \\ 1 & 0 \end{pmatrix}$ do not span M_{22} because we cannot generate the matrix $\mathbf{C}$, therefore these matrices cannot form a basis for M_{22}. The scalar k_2 also has two different values; 2 and 3.

[Note that matrices $\mathbf{A}$ and $\mathbf{B}$ are linearly independent but that is not enough to be a basis.]

In fact, the given matrices only produce matrices of the form $\begin{pmatrix} a & b \\ b & a \end{pmatrix}$, which means that we cannot obtain the matrix $\begin{pmatrix} a & b \\ c & d \end{pmatrix}$ where $a \neq d$ or $b \neq c$.

In the study of Fourier series, we discuss periodic (one which repeats itself) continuous functions. For example, the following is a periodic continuous function:

$$y = \frac{2}{\pi}\sin(x) + \frac{2}{3\pi}\sin(3x) + \frac{2}{5\pi}\sin(5x)$$

Figure 3.13 illustrates that this function repeats every 2π intervals.

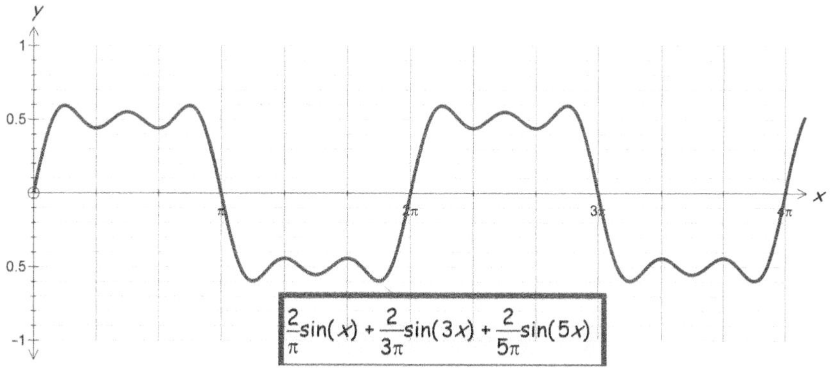

Figure 3.13

As you can see, this periodic function y is made up of a linear combination of $\sin(x)$, $\sin(3x)$ and $\sin(5x)$. Any function in the vector space of $F[0, 2\pi]$ of the form

$$y = k_1 \sin(x) + k_2 \sin(3x) + k_3 \sin(5x)$$

has a basis $\{\sin(x), \sin(3x), \sin(5x)\}$. Clearly these vectors span the space and you are asked to show linear independence of these in Exercises 3.3.

We can represent any periodic continuous function as a linear combination of sines and cosines. In general, a Fourier series will have the form

$$y = k_0 + k_1 \cos(x) + k_2 \cos(2x) + \cdots + c_1 \sin(x) + c_2 \sin(2x) + \cdots$$

where k's and c's are scalars. A basis for this general function is

$$\{1, \cos(x), \cos(2x), \cos(3x), \ldots, \sin(x), \sin(2x), \sin(3x), \ldots\}$$

We need 1 in the basis because this will create the constant term k_0 in the above. Hence we have an infinite number of basis vectors.

3.3.5 Uniqueness

Choosing a basis for a vector space is essentially choosing a coordinate system which allows each vector to be expressed uniquely.

Proposition (3.16). Let the set $B = \{\mathbf{v}_1, \mathbf{v}_2, \ldots, \mathbf{v}_n\}$ be a basis for a vector space V. Every vector in V can be expressed *uniquely* as a linear combination of the basis vectors.

What does this proposition mean?

Any vector in V written as the linear combination of basis vector has unique scalars.

Proof.

Consider an arbitrary vector $\mathbf{w}$ in V. Since $B = \{\mathbf{v}_1, \mathbf{v}_2, \mathbf{v}_3, \ldots, \mathbf{v}_n\}$ is a basis for the vector space V, there exist scalars $k_1, k_2, k_3, \ldots$ and k_n such that

$$k_1\mathbf{v}_1 + k_2\mathbf{v}_2 + k_3\mathbf{v}_3 + \cdots + k_n\mathbf{v}_n = \mathbf{w} \quad (\dagger)$$

Suppose we can write this vector $\mathbf{w}$ as another linear combination

$$c_1\mathbf{v}_1 + c_2\mathbf{v}_2 + c_3\mathbf{v}_3 + \cdots + c_n\mathbf{v}_n = \mathbf{w} \quad (\dagger\dagger)$$

where c's are scalars.

What do we need to prove?

Required to prove that the two sets of scalars are equal: $k_1 = c_1$, $k_2 = c_2, \ldots$ and $k_n = c_n$.

Subtracting the two **w** vectors, (†) – (††), gives

$$k_1\mathbf{v}_1 + k_2\mathbf{v}_2 + k_3\mathbf{v}_3 + \cdots + k_n\mathbf{v}_n - c_1\mathbf{v}_1 - c_2\mathbf{v}_2 - c_3\mathbf{v}_3 - \cdots - c_n\mathbf{v}_n = \mathbf{w} - \mathbf{w}$$
$$(k_1 - c_1)\mathbf{v}_1 + (k_2 - c_2)\mathbf{v}_2 + (k_3 - c_3)\mathbf{v}_3 + \cdots + (k_n - c_n)\mathbf{v}_n = \mathbf{O} \qquad (*)$$

Since $\mathbf{v}_1, \mathbf{v}_2, \mathbf{v}_3, \ldots$ and $\mathbf{v}_n$ are basis vectors, they are linearly independent, which means that all the bracketed terms (scalars) in (*) must be equal to zero (Definition (3.10)). We have

$$k_1 - c_1 = 0,\ k_2 - c_2 = 0,\ k_3 - c_3 = 0, \ldots \text{ and } k_n - c_n = 0$$
$$k_1 = c_1,\ k_2 = c_2,\ k_3 = c_3, \ldots \text{ and } k_n = c_n$$

Hence we have proved that the basis vector representation of any vector is unique.

■

Summary

Consider the linear combination:

$$k_1\mathbf{v}_1 + k_2\mathbf{v}_2 + k_3\mathbf{v}_3 + \cdots + k_n\mathbf{v}_n = \mathbf{O}$$

If the only solution is $k_1 = k_2 = k_3 = \cdots = k_n = 0$ then vectors $\mathbf{v}_1, \mathbf{v}_2, \cdots, \mathbf{v}_n$ are linearly independent, otherwise they are dependent.

Definition (3.15). Vectors $B = \{\mathbf{v}_1, \mathbf{v}_2, \mathbf{v}_3, \ldots, \mathbf{v}_n\}$ of the vector space V form a **basis** of $V \Leftrightarrow$

(i) $\{\mathbf{v}_1, \mathbf{v}_2, \mathbf{v}_3, \ldots, \mathbf{v}_n\}$ span V

(ii) $\{\mathbf{v}_1, \mathbf{v}_2, \mathbf{v}_3, \ldots, \mathbf{v}_n\}$ are linearly independent

EXERCISES 3.3

(Brief solutions at end of book. Full solutions available at <http://www.oup.co.uk/companion/singh>.)

1. Let M_{22} be the vector space of size 2 by 2 matrices. Decide whether the following matrices are linearly independent or dependent:

(a) $\mathbf{A} = \begin{pmatrix} 1 & 0 \\ 0 & 1 \end{pmatrix}$ and $\mathbf{B} = \begin{pmatrix} 0 & 1 \\ 1 & 0 \end{pmatrix}$ (b) $\mathbf{A} = \begin{pmatrix} 1 & 1 \\ 1 & 1 \end{pmatrix}$ and $\mathbf{B} = \begin{pmatrix} 2 & 2 \\ 2 & 2 \end{pmatrix}$

(c) $\mathbf{A} = \begin{pmatrix} 1 & 2 \\ 3 & 4 \end{pmatrix}$ and $\mathbf{B} = \begin{pmatrix} 2 & 2 \\ 2 & 2 \end{pmatrix}$ (d) $\mathbf{A} = \begin{pmatrix} 1 & 2 \\ 3 & 4 \end{pmatrix}$ and $\mathbf{B} = \begin{pmatrix} 2/5 & 4/5 \\ 6/5 & 8/5 \end{pmatrix}$

On a course in differential equations it is important that you can test the following functions in questions 2 and 3 for linear independence.

2. Let V be the vector space of continuous functions defined on the real line. Test the following vectors for linear independence:
(a) $\mathbf{f} = (x + 1)^2$ and $\mathbf{g} = x^2 + 2x + 1$ (b) $\mathbf{f} = 2$ and $\mathbf{g} = x^2$
(c) $\mathbf{f} = 1$ and $\mathbf{g} = e^x$ (d) $\mathbf{f} = \cos(x)$ and $\mathbf{g} = \sin(x)$
(e) $\mathbf{f} = \sin(x)$ and $\mathbf{g} = \sin(2x)$

3. Let V be the vector space of continuous functions defined on the real line. Test the following vectors for linear independence:
(a) $\mathbf{f} = \cos^2(x)$, $\mathbf{g} = \sin^2(x)$ and $\mathbf{h} = 5$
(b) $\mathbf{f} = \cos(2x)$, $\mathbf{g} = \sin^2(x)$ and $\mathbf{h} = \cos^2(x)$
(c) $\mathbf{f} = 1$, $\mathbf{g} = x$ and $\mathbf{h} = x^2$
(d) $\mathbf{f} = \sin(2x)$, $\mathbf{g} = \sin(x)\cos(x)$ and $\mathbf{h} = \cos(x)$
(e) $\mathbf{f} = e^x \sin(2x)$, $\mathbf{g} = e^x \sin(x)\cos(x)$ and $\mathbf{h} = e^x \cos(x)$
(f) $\mathbf{f} = 1$, $\mathbf{g} = e^x$ and $\mathbf{h} = e^{-x}$
(g) $\mathbf{f} = e^x$, $\mathbf{g} = e^{2x}$ and $\mathbf{h} = e^{3x}$

4. Let $\mathbf{f} = \sin(x)$, $\mathbf{g} = \sin(3x)$ and $\mathbf{h} = \sin(5x)$ be in the vector space of continuous functions $F[0,\ 2\pi]$. Show that these vectors are linearly independent.

5. Let M_{22} be the vector space of size 2 by 2 matrices. Show that the following matrices are a basis for M_{22}:

$$\mathbf{A} = \begin{pmatrix} 1 & 0 \\ 0 & 0 \end{pmatrix},\ \mathbf{B} = \begin{pmatrix} 0 & 1 \\ 0 & 0 \end{pmatrix},\ \mathbf{C} = \begin{pmatrix} 0 & 0 \\ 1 & 0 \end{pmatrix} \text{ and } \mathbf{D} = \begin{pmatrix} 0 & 0 \\ 0 & 1 \end{pmatrix}$$

[These are generally called the standard or natural basis for M_{22}]

We denote the vector space of polynomials of degree ≤ 2 by P_2 which is discussed in the next two questions.

6. Show that $\{1,\ t - 1,\ (t - 1)^2\}$ forms a basis for the vector space P_2. Write the vector $\mathbf{p} = t^2 + 1$ in terms of these basis vectors.

7. Show that $\{1,\ t^2 - 2t,\ 5(t - 1)^2\}$ does *not* form a basis for the vector space P_2.

8. Prove that if a set of vectors $\{\mathbf{v}_1,\ \mathbf{v}_2,\ \mathbf{v}_3, \ldots,\ \mathbf{v}_n\}$ is linearly independent then the set $\{k\mathbf{v}_1,\ k\mathbf{v}_2,\ k\mathbf{v}_3, \ldots,\ k\mathbf{v}_n\}$, where k is a non-zero scalar, is also linearly independent.

9. Prove that any non-zero vector $\mathbf{v}$ on its own is a linearly independent vector in a vector space V.

10. Prove that the zero vector, $\mathbf{O}$, on its own is a linearly dependent vector in a vector space V.

11. Consider the set of vectors $\{\mathbf{v}_1,\ \mathbf{v}_2,\ \mathbf{v}_3, \ldots,\ \mathbf{v}_n\}$. Prove that if any two vectors $\mathbf{v}_j = \mathbf{v}_m$ where $j \neq m$ in the set are equal then the set is linearly dependent.

12. Consider the set of linearly independent vectors $S = \{\mathbf{v}_1,\ \mathbf{v}_2,\ \mathbf{v}_3, \ldots,\ \mathbf{v}_n\}$. Prove that any non-empty subset of this is also linearly independent.

13. Consider the set of vectors $S = \{\mathbf{v}_1,\ \mathbf{v}_2,\ \mathbf{v}_3, \ldots,\ \mathbf{v}_n\}$ which spans a vector space V. Let $\mathbf{w}$ be a vector in V but not in the set S. Prove that $\{\mathbf{v}_1,\ \mathbf{v}_2,\ \mathbf{v}_3, \ldots,\ \mathbf{v}_n,\ \mathbf{w}\}$ spans V but is linearly dependent.

14. Consider the set of vectors $B = \{\mathbf{v}_1, \mathbf{v}_2, \mathbf{v}_3, \ldots, \mathbf{v}_n\}$ in a vector space V. Prove that if the set B is a basis for V and $S = \{\mathbf{v}_1, \mathbf{v}_2, \mathbf{v}_3, \ldots, \mathbf{v}_m\}$ is a set of linearly independent vectors in V then $m \leq n$.
15. *Prove that if $B_1 = \{\mathbf{v}_1, \mathbf{v}_2, \mathbf{v}_3, \ldots, \mathbf{v}_n\}$ and $B_2 = \{\mathbf{u}_1, \mathbf{u}_2, \mathbf{u}_3, \ldots, \mathbf{u}_m\}$ are bases for a vector space V then $n = m$. [That is every basis of a vector space has the same number of vectors.]
16. Prove that if the largest number of linearly independent vectors in a vector space V is n then any n linearly independent vectors form a basis for V.
17. *Let S be a subspace of a vector space V. Prove that if S and V have the same basis then $S = V$.

SECTION 3.4 Dimension

By the end of this section you will be able to

- understand what is meant by the dimension of a vector space
- determine the dimension of a vector space and subspace
- prove properties of finite dimensional vector spaces

We have used the word 'dimension' without really defining it. Our physical interpretation of a vector space of $\mathbb{R}^n$ is that it has n dimensions.

How can we find the dimensions of other vector spaces such as the set of matrices M_{mn} or polynomials P_n?

In this section we give the precise definition of dimension so that we can give a numerical value to the dimension of any vector space. This is a challenging section. You may need to look back at the definition of a basis of a vector space to understand the proofs later on in this section.

3.4.1 Introduction to dimension

What does the word 'dimension' mean in everyday language?

Dimension normally refers to the size of an object in a particular direction. For example, the height and diameter of a cylinder are the dimensions of the cylinder.

What does the term dimension of a vector space mean?

The definition of dimension is the number of basis vectors or axes needed to describe the given vector space V.

Definition (3.17). The number of vectors in a basis of a non-zero vector space V is the **dimension** of the space.

We will show later in this section that the number of vectors in a basis depends on two things: linear independence and spanning.

We will prove that if we have too many vectors in the basis set, they become linearly dependent and if there are too few vectors then they will not span the whole vector space.

Example 3.21

What is the dimension of $\mathbb{R}^2$ and $\mathbb{R}^3$?
Evaluate the dimension of $\mathbb{R}^n$.

Solution
From the previous chapter we know that $\mathbf{e}_1 = (1 \quad 0)^T$ and $\mathbf{e}_2 = (0 \quad 1)^T$ (unit vectors on the x and y axes respectively) form a basis for $\mathbb{R}^2$.
What is the dimension of $\mathbb{R}^2$?
Two vectors, $\mathbf{e}_1$ and $\mathbf{e}_2$, are needed for the basis of $\mathbb{R}^2$, therefore dimension of $\mathbb{R}^2$ is 2.
What is the dimension of $\mathbb{R}^3$?
Similarly, we have the unit vectors $\mathbf{e}_1 = (\ 1 \ \ 0 \ \ 0\ )^T$, $\mathbf{e}_2 = (\ 0 \ \ 1 \ \ 0\ )^T$ and $\mathbf{e}_3 = (\ 0 \ \ 0 \ \ 1\ )^T$ in the directions of x, y and z axes respectively, is the natural basis for $\mathbb{R}^3$, therefore the dimension of $\mathbb{R}^3$ is 3 because we have three vectors in the basis of $\mathbb{R}^3$.
What is the dimension of $\mathbb{R}^n$?
From the last section we have that the n vectors $\{\mathbf{e}_1, \mathbf{e}_2, \mathbf{e}_3, \ldots, \mathbf{e}_n\}$ form a natural basis for $\mathbb{R}^n$, so the dimension of $\mathbb{R}^n$ is n.

The number of vectors in the basis gives the dimension of the vector space V. This is often denoted by $\dim(V)$.

$$\dim(\mathbb{R}^2) = 2, \quad \dim(\mathbb{R}^3) = 3 \text{ and } \dim(\mathbb{R}^n) = n$$

*Why does the above definition (3.17) say **non-zero** space?*
Because the zero vector space $\{\mathbf{O}\}$ is linearly dependent, there are no vectors in the basis of $\{\mathbf{O}\}$, which means that it has dimension 0. (The zero vector does not need any axes.)

3.4.2 Finite dimensional vector space

What does the term finite dimensional vector space mean?
It is a vector space V which has a finite number of vectors in its basis.

Definition (3.18). In general, if a finite number of vectors form a basis for a vector space V then we say V is **finite dimensional**. Otherwise, the vector space V is known as infinite dimensional.

If the vector space V consists only of the zero vector then it is also finite dimensional.

Can you think of any finite dimensional vector spaces?
The Euclidean spaces - $\mathbb{R}^2, \mathbb{R}^3, \mathbb{R}^4, \ldots, \mathbb{R}^n$.

Are there any other examples of finite dimensional vector spaces?
The set P_2 of polynomials of degree 2 or less for example, or the set of all 2 by 2 matrices M_{22}. (These were covered in the previous section.)

Definition (3.19). In general, if n vectors $\{\mathbf{v}_1, \mathbf{v}_2, \mathbf{v}_3, \ldots, \mathbf{v}_n\}$ form a basis for a vector space V then we say that V is n-**dimensional**.

What is $\dim(M_{22})$ equal to?
The standard basis for M_{22} (matrices of size 2 by 2) from the Exercises 3.3 question 5 is

$$\mathbf{A} = \begin{pmatrix} 1 & 0 \\ 0 & 0 \end{pmatrix}, \mathbf{B} = \begin{pmatrix} 0 & 1 \\ 0 & 0 \end{pmatrix}, \mathbf{C} = \begin{pmatrix} 0 & 0 \\ 1 & 0 \end{pmatrix} \text{ and } \mathbf{D} = \begin{pmatrix} 0 & 0 \\ 0 & 1 \end{pmatrix}$$

Therefore $\dim(M_{22}) = 4$ because we have four matrices in $\{\mathbf{A}, \mathbf{B}, \mathbf{C}, \mathbf{D}\}$ which form a basis for M_{22}.

What is $\dim(P_2)$ equal to?
Remember that the standard basis for P_2 (the set of all polynomials of degree 2 or less) is the set $\{1, t, t^2\}$, which means $\dim(P_2) = 3$ since the basis consists of three vectors.

Table 3.1 shows some vector spaces and their dimensions.

Table 3.1 Some vector spaces and their dimensions

Vector space	Dimension
$\mathbb{R}^n$	n
P_n	$n+1$
M_{mn}	mn

The dimension does not need to be the number of vectors in the standard basis; it can be any basis for the given vector space because the number of vectors is the same, as the following theorem shows.

Theorem (3.20). Every basis for a finite dimensional vector space has the *same number* of vectors.

Proof.
We proved this in question 15 of Exercises 3.3.

■

Since every basis has the *same* number of vectors in a finite dimensional vector space, definition (3.17) means that:

$$\text{number of vectors in any basis} = \text{dimension of vector space}$$

Theorem (3.20) is important in linear algebra and definitely worth learning.

There are, of course, many infinite dimensional vector spaces, such as the set of *all* polynomials P (or P_∞). However, we confine ourselves to finite dimensional vector spaces for the remainder of this chapter.

3.4.3 Subspaces revisited

We discussed subspaces in section 3.2 of this chapter. A subspace is a non-empty subset S in a vector space V, which forms a vector space in its own right, with respect to the same vector addition and scalar multiplication as its parent set V. This is illustrated in Fig. 3.14.

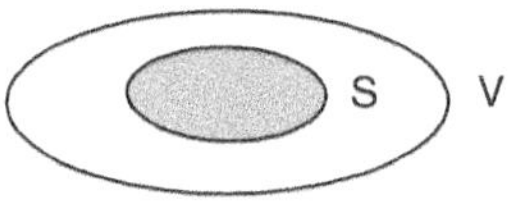

Figure 3.14

What is the dimension of a subspace of V?
That depends on the number of vectors in the basis of the subspace.

Example 3.22

Let $\mathbb{R}^2$ be a vector space and S be the set of vectors $\begin{pmatrix} 0 \\ a \end{pmatrix}$. In Example 3.5 we showed that this is a subspace of $\mathbb{R}^2$. Find a basis for the subspace S and determine $\dim(S)$.

Solution

Every vector of S is of the form $\begin{pmatrix} 0 \\ a \end{pmatrix}$, which we can express in terms of $\mathbf{e}_2 = \begin{pmatrix} 0 \\ 1 \end{pmatrix}$ as

$$\begin{pmatrix} 0 \\ a \end{pmatrix} = a\begin{pmatrix} 0 \\ 1 \end{pmatrix} = a\mathbf{e}_2$$

This vector $\mathbf{e}_2$ forms a basis for S, therefore $\dim(S) = 1$. This is shown graphically in Fig. 3.15.

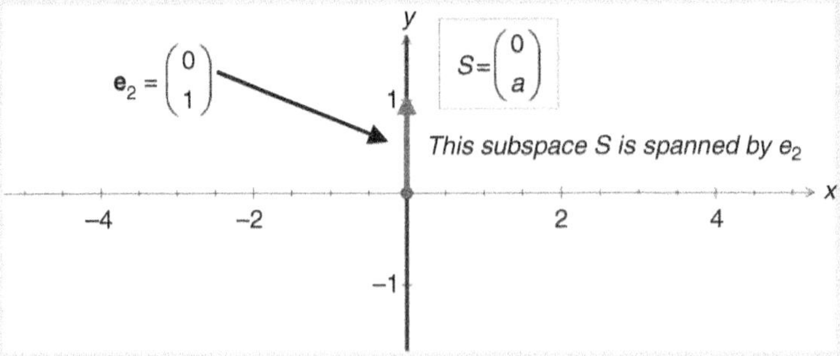

Figure 3.15

Figure 3.15 shows that the subspace S is the vertical axis (y axis) and $\mathbf{e}_2$ generates any vector in S, along this line.

Example 3.23

Let $\mathbf{u} = \begin{pmatrix} 2 \\ 0 \\ 0 \end{pmatrix}$, $\mathbf{v} = \begin{pmatrix} -1 \\ 0 \\ 0 \end{pmatrix}$ and $\mathbf{w} = \begin{pmatrix} 0 \\ 0 \\ 1 \end{pmatrix}$ span a subspace S of $\mathbb{R}^3$.

What is the dimension of S?

Solution

*What do you notice about the first two vectors **u** and **v**?*

They are linearly dependent because $\mathbf{u} = -2\mathbf{v}$. Note that the vector $\mathbf{w}$ is linearly independent of $\mathbf{u}$ and $\mathbf{v}$.

How many vectors are in the basis of the subspace S?

Two vectors $\mathbf{u}$ and $\mathbf{w}$ (or $\mathbf{v}$ and $\mathbf{w}$).

What is $\dim(S)$ *equal to?*

Two because we only have two vectors in the basis of S, that is $\dim(S) = 2$. This means that the given vectors $\mathbf{u}$, $\mathbf{v}$ and $\mathbf{w}$ span a subspace S which is a plane in $\mathbb{R}^3$ as demonstrated in Fig. 3.16.

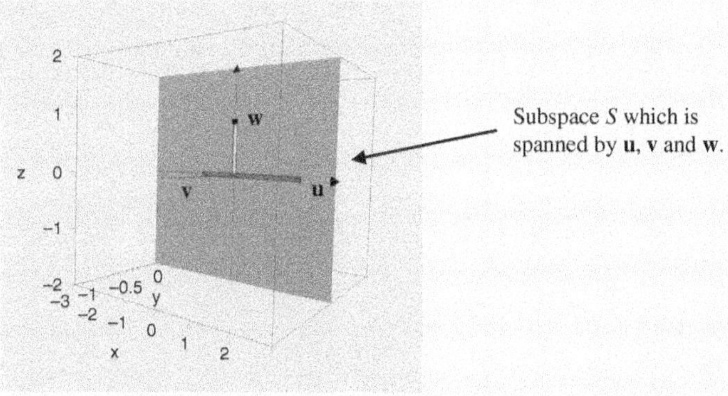

Figure 3.16

Note: a general misconception is that the dimension of a space is how many entries are required to specify a vector in that space. This is clearly wrong because in Example 3.23 above we require three entries to specify a vector in the subspace S but it has dimension 2. This means the space S in Fig. 3.16 can be spanned by two basis vectors or two axes rather than the three given vectors.

Example 3.24

Let S be a set of vectors in P_2 which are of the form

$$ax^2 + (a+b)x + (a+b) \text{ where } a \text{ and } b \text{ are scalars}$$

Then S is a subspace of P_2. Find a basis for S and determine $\dim(S)$.

Solution

We can write vectors in S as

$$\begin{aligned} ax^2 + (a+b)x + (a+b) &= ax^2 + ax + bx + (a+b) && \text{[expanding]} \\ &= a(x^2 + x + 1) + b(x+1) && \text{[factorizing]} \end{aligned}$$

(continued...)

Hence the vectors $x^2 + x + 1$ and $x + 1$ span S, because we have demonstrated that the original set can be generated by these two vectors

What else do we need to show for these vectors to be a basis for S?

We need to prove that they are linearly independent. Since $x^2 + x + 1$ and $x + 1$ are not multiples of each other, they are linearly independent.

Hence the two vectors $\{x^2 + x + 1, x + 1\}$ form a basis for S, therefore $\dim(S) = 2$.

3.4.4 Properties of finite dimensional vector spaces

In this section we show some important properties of bases and dimension. This is a demanding section because the proofs of propositions are lengthy.

Lemma (3.21). Let V be a finite n-dimensional vector space. We have the following:

(a) Let $\{\mathbf{v}_1, \mathbf{v}_2, \mathbf{v}_3, \ldots, \mathbf{v}_n\}$ be a set of linearly independent vectors. Then $\{\mathbf{v}_1, \mathbf{v}_2, \ldots, \mathbf{v}_m\}$ where $m > n$ (m is greater than n) is linearly dependent.

(b) If the n vectors $\{\mathbf{u}_1, \mathbf{u}_2, \mathbf{u}_3, \ldots, \mathbf{u}_n\}$ span V then $\{\mathbf{u}_1, \mathbf{u}_2, \mathbf{u}_3, \ldots, \mathbf{u}_m\}$ where $m < n$ does not span V.

What does part (a) mean?

In n-dimensional vector space, if you add additional vectors to n linearly independent vectors then the set becomes linearly dependent.

How do we prove this result?

By using proof by contradiction.

Proof of (a).

The number of basis vectors in V is n. Suppose that $\{\mathbf{v}_1, \mathbf{v}_2, \ldots, \mathbf{v}_m\}$ are linearly independent. Then we must have $m \le n$ (m is less than or equal to n).

Why?

Because by question 14 of Exercises 3.3 we know that the number of vectors in a linearly independent set must be less than or equal to n, $m \le n$ (the number of basis vectors).

However, we are given that $m > n$ (m is greater than n), therefore our supposition that $\{\mathbf{v}_1, \mathbf{v}_2, \ldots, \mathbf{v}_m\}$ is a linearly independent set of vectors must be wrong, so this set is linearly dependent.

■

What does part (b) mean?

If n vectors span V where V is an n-dimensional vector space, then less than n vectors cannot span V. Again, we use proof by contradiction.

Proof of (b).

Suppose $\{\mathbf{u}_1, \mathbf{u}_2, \mathbf{u}_3, \cdots, \mathbf{u}_m\}$ where $m < n$ span V, that is

$$span\{\mathbf{u}_1, \mathbf{u}_2, \mathbf{u}_3, \cdots, \mathbf{u}_m\} = V \qquad (*)$$

The dimension of the left hand side, $\dim(span\{\mathbf{u}_1, \mathbf{u}_2, \mathbf{u}_3, \ldots, \mathbf{u}_m\}) \leq m$ but the dimension of V is n. This is a contradiction because we are given $m < n$ and the dimension of both sides of (*) must be equal. Hence $\{\mathbf{u}_1, \mathbf{u}_2, \ldots, \mathbf{u}_m\}$ where $m < n$ cannot span V.

■

Theorem (3.22). Let V be a finite n-dimensional vector space. We have the following:

(a) Any linearly *independent* set of n vectors $\{\mathbf{v}_1, \mathbf{v}_2, \mathbf{v}_3, \ldots, \mathbf{v}_n\}$ form a basis for V.

(b) Any *spanning* set of n vectors $\{\mathbf{u}_1, \mathbf{u}_2, \mathbf{u}_3, \ldots, \mathbf{u}_n\}$ forms a basis for V.

How do we prove these results?

By using the definition of basis as described in the last section, that is:

Definition (3.15) A set of vectors is a basis for $V \Leftrightarrow$ it is linearly independent and spans V.

Proof of (a).

We are given that the vectors $\{\mathbf{v}_1, \mathbf{v}_2, \mathbf{v}_3, \ldots, \mathbf{v}_n\}$ are linearly independent so we only need to show that these vectors also span V. Suppose there is a vector $\mathbf{w}$ in V such that

$$\mathbf{w} = k_1\mathbf{v}_1 + k_2\mathbf{v}_2 + k_3\mathbf{v}_3 + \cdots + k_n\mathbf{v}_n + k_{n+1}\mathbf{v}_{n+1} \qquad (*)$$

where k's are scalars. [We are supposing that the given vectors do not span V that is why we have added an extra vector $\mathbf{v}_{n+1}$].

By the above Lemma (3.21) part (a):

In a n dimension space, vectors $\{\mathbf{v}_1, \mathbf{v}_2, \cdots, \mathbf{v}_m\}$ where $m > n$ are linearly dependent.

The set of vectors $\{\mathbf{v}_1, \mathbf{v}_2, \mathbf{v}_3, \ldots, \mathbf{v}_n, \mathbf{v}_{n+1}\}$ is linearly dependent so we can write the vector $\mathbf{v}_{n+1}$ in terms of its preceding vectors, that is

$$\mathbf{v}_{n+1} = c_1\mathbf{v}_1 + c_2\mathbf{v}_2 + c_3\mathbf{v}_3 + \cdots + c_n\mathbf{v}_n \text{ where } c\text{'s are scalars}$$

Substituting this into (*) gives

$$\begin{aligned}\mathbf{w} &= k_1\mathbf{v}_1 + k_2\mathbf{v}_2 + \cdots + k_n\mathbf{v}_n + k_{n+1}\mathbf{v}_{n+1} \\ &= k_1\mathbf{v}_1 + k_2\mathbf{v}_2 + \cdots + k_n\mathbf{v}_n + k_{n+1}(c_1\mathbf{v}_1 + c_2\mathbf{v}_2 + \cdots + c_n\mathbf{v}_n) \\ &= (k_1 + k_{n+1}c_1)\mathbf{v}_1 + (k_2 + k_{n+1}c_2)\mathbf{v}_2 + \cdots + (k_n + k_{n+1}c_n)\mathbf{v}_n\end{aligned}$$

Thus the vector $\mathbf{w}$ can be written as a linear combination of the given linearly independent vectors $\{\mathbf{v}_1, \mathbf{v}_2, \mathbf{v}_3, \ldots, \mathbf{v}_n\}$. Thus $\{\mathbf{v}_1, \mathbf{v}_2, \mathbf{v}_3, \cdots, \mathbf{v}_n\}$ spans V, therefore it forms a basis for V.

■

Proof of (b).

We are given that $\{\mathbf{u}_1, \mathbf{u}_2, \mathbf{u}_3, \ldots, \mathbf{u}_n\}$ spans the vector space V. Let $\mathbf{w}$ be an arbitrary vector of V, then we can write

$$\mathbf{w} = k_1\mathbf{u}_1 + k_2\mathbf{u}_2 + \cdots + k_{n-1}\mathbf{u}_{n-1} + k_n\mathbf{u}_n \qquad (\dagger)$$

where k's are scalars because $\{\mathbf{u}_1, \mathbf{u}_2, \mathbf{u}_3, \ldots, \mathbf{u}_n\}$ spans V.

What do we need to show for this set of vectors to be a basis for V?

Required to prove that the set of vectors under consideration $\{\mathbf{u}_1, \mathbf{u}_2, \mathbf{u}_3, \ldots, \mathbf{u}_n\}$ is linearly independent. Suppose $\{\mathbf{u}_1, \mathbf{u}_2, \mathbf{u}_3, \ldots, \mathbf{u}_n\}$ is linearly dependent then we can write the vector $\mathbf{u}_n$ in terms of its preceding vectors, that is

$$\mathbf{u}_n = c_1\mathbf{u}_1 + c_2\mathbf{u}_2 + c_3\mathbf{u}_3 + \cdots + c_{n-1}\mathbf{u}_{n-1} \text{ where } c\text{'s are scalars}$$

Substituting this into the above (†) gives

$$\begin{aligned} \mathbf{w} &= k_1\mathbf{u}_1 + k_2\mathbf{u}_2 + \cdots + k_{n-1}\mathbf{u}_{n-1} + k_n\mathbf{u}_n \\ &= k_1\mathbf{u}_1 + k_2\mathbf{u}_2 + \cdots + k_{n-1}\mathbf{u}_{n-1} + k_n(c_1\mathbf{u}_1 + c_2\mathbf{u}_2 + \cdots + c_{n-1}\mathbf{u}_{n-1}) \\ &= (k_1 + k_nc_1)\mathbf{u}_1 + (k_2 + k_nc_2)\mathbf{u}_2 + \cdots + (k_{n-1} + k_nc_{n-1})\mathbf{u}_{n-1} \end{aligned}$$

This shows that the above $n-1$ vectors $\{\mathbf{u}_1, \mathbf{u}_2, \mathbf{u}_3, \ldots, \mathbf{u}_{n-1}\}$ span V. This is impossible because, by the above Lemma (3.21), part (b):

If n vectors $\{\mathbf{u}_1, \mathbf{u}_2, \cdots, \mathbf{u}_n\}$ span V then $\{\mathbf{u}_1, \mathbf{u}_2, \cdots, \mathbf{u}_m\}$ where $m < n$ does *not* span V.

Fewer than n vectors *cannot span* V. Thus $\{\mathbf{u}_1, \mathbf{u}_2, \mathbf{u}_3, \ldots, \mathbf{u}_n\}$ is linearly independent which means it is a basis for the given vector space V.

■

Theorem (3.22) says two things:

1. The *maximum independent* set for an n-dimensional vector space is n vectors. If you add any more vectors then it becomes linearly dependent. The basis is a *maximum linearly independent* set.
2. The *minimum spanning* set for an n-dimensional vector space is n vectors. If you remove any of the vectors of the spanning set then it no longer spans V. The basis is a *minimum spanning* set.

Two Definitions of Basis

1. A basis for a vector space is the *largest independent* set of vectors.
2. A basis for a vector space is the *smallest spanning* set of vectors.

This means if you have n linearly independent vectors in an n-dimensional vector space V then it is a basis for V. You do not have to check that it spans V.

Also, if you have n spanning vectors in an n-dimensional vector space V then it is a basis for V. You do not have to check that the set is linearly independent.

A basis (axes) is the most efficient set of vectors used to describe a given vector space.

Let us apply the above theorem and see how it simplifies matters.

Example 3.25

Show that the following vectors form a basis for P_n:

$$S = \{1, t+1, (t+1)^2, (t+1)^3, \cdots, (t+1)^n\}$$

Solution

We have $\dim(P_n) = n + 1$, and there are $n + 1$ given vectors.

We need to show that the set of vectors in S are linearly independent or span the vector space P_n. Note that *none* of the polynomials in the set S can be written as a linear combination of the preceding polynomials in the set. This means that we *cannot* write $(t + 1)^m$ where $m \leq n$ in terms of the vectors before we get to $(t + 1)^m$.

What can you conclude about this set?

Since we cannot write any of the polynomials (vectors) in terms of the preceding polynomials (vectors), the set S must be linearly independent. Again, by the above Theorem (3.22) part (a), we conclude that the vectors in the set S are a basis for P_n.

Example 3.26

Let M_{22} be the vector space of all 2 by 2 matrices. Show that the following matrices do not form a basis for M_{22}:

$$\mathbf{A} = \begin{pmatrix} 1 & 0 \\ 0 & 3 \end{pmatrix}, \mathbf{B} = \begin{pmatrix} 0 & 1 \\ 2 & 0 \end{pmatrix} \text{ and } \mathbf{C} = \begin{pmatrix} 2 & 1 \\ 1 & 5 \end{pmatrix}$$

Solution

What is the dimension of M_{22}?

Recall from Table 3.1 in section 3.4.2 that $\dim(M_{22}) = 2 \times 2 = 4$. In the above list, we have three matrices so they cannot form a basis for M_{22}.

Why not?

Because $\dim(M_{22}) = 4$, and by Definition (3.17) we need exactly four matrices for a basis.

Example 3.27

Let $F[0,\ 2\pi]$ be a vector space of periodic continuous functions which is spanned by

$$S = \{\sin(x)\cos(x),\ \sin(2x),\ \cos(2x)\}$$

Show that these vectors do not form a basis for $F[0,\ 2\pi]$. Also find a basis for this space.

Solution

From trigonometry we have the identity

$$\sin(2x) = 2\sin(x)\cos(x)$$

Hence the first two vectors in S are dependent, therefore the set S cannot be a basis.

A basis can be established if we remove one of the vectors; $\sin(x)\cos(x)$ or $\sin(2x)$. A basis for the given space is $\{\sin(2x),\ \cos(2x)\}$.

Summary

(3.17). The number of vectors in a basis of a non-zero space V is the **dimension** of the space.
(3.20). Every basis for a finite dimensional vector space has the *same number* of vectors.
(3.22). Let V be a finite n-dimensional vector space. We have the following:

(a) Any linearly *independent* set of n vectors forms a basis for V.
(b) Any *spanning* set of n vectors forms a basis for V.

EXERCISES 3.4

(Brief solutions at end of book. Full solutions available at <http://www.oup.co.uk/companion/singh>.)

1. Find the dimension of the following vector spaces and exhibit a basis for each space:
(a) $\mathbb{R}^5$ (b) $\mathbb{R}^7$ (c) $\mathbb{R}^{11}$ (d) $\mathbb{R}^{13}$ (e) M_{33}
(f) M_{44} (g) M_{23} (h) P_3 (i) P_5 (j) $\mathbf{O}$

2. Let $\mathbb{R}^2$ be a vector space and S be the subspace given by the set of vectors $\begin{pmatrix} a \\ 0 \end{pmatrix}$. Find a basis for S and evaluate $\dim(S)$.

3. Let $\mathbf{u} = \begin{pmatrix} 0 & 0 & 3 \end{pmatrix}^T$, $\mathbf{v} = \begin{pmatrix} -5 & 0 & 0 \end{pmatrix}^T$ and $\mathbf{w} = \begin{pmatrix} 0 & 0 & 9 \end{pmatrix}^T$ span a subspace S of $\mathbb{R}^3$. Determine the dimension of S.

4. Let S be the set of vectors in P_2 which are of the form

$$at^2 + b$$

Then S is a subspace of V. Find a basis for S and determine $\dim(S)$.

5. Consider the subspace S which consists of matrices of the form $\begin{pmatrix} a & b \\ b & c \end{pmatrix}$ (this is the set of symmetric matrices) of the vector space M_{22}. Find a basis for S and evaluate the dimension of S.

6. Find the dimension of the following vector spaces and subspaces:
(a) M_{mn} (matrices of size m by n).
(b) The subspace S of P_3 given by the cubic polynomials $S = \{at^3 + bt^2 + c\}$.
(c) Let S be the set of vectors in P_3 which are of the form

$$S = \{at^3 + bt^2 + ct + d\}$$

7. Show that the following vectors are a basis for the corresponding vector spaces:

(a) $\mathbf{u} = \begin{pmatrix} 1 \\ 5 \end{pmatrix}$ and $\mathbf{v} = \begin{pmatrix} 2 \\ 1 \end{pmatrix}$ for $\mathbb{R}^2$.

(b) $\mathbf{u} = \begin{pmatrix} 0 \\ 3 \\ 4 \end{pmatrix}$, $\mathbf{v} = \begin{pmatrix} 1 \\ 1 \\ 1 \end{pmatrix}$ and $\mathbf{w} = \begin{pmatrix} 1 \\ 0 \\ 1 \end{pmatrix}$ for $\mathbb{R}^3$.

(c) $\{t^n, t^{n-1}, \ldots, t, 1\}$ for the vector space P_n.

(d) $\left\{\begin{pmatrix} 2 & 0 \\ 0 & 2 \end{pmatrix}, \begin{pmatrix} 0 & 3 \\ 0 & 2 \end{pmatrix}, \begin{pmatrix} 0 & 0 \\ 1 & 1 \end{pmatrix}, \begin{pmatrix} 0 & 0 \\ 0 & 5 \end{pmatrix}\right\}$ for M_{22}.

8. Explain why the following vectors are not a basis for the corresponding vector spaces:

(a) $\{1+t,\ 1+t^2,\ 1+2t+t^2,\ 1+2t\}$ for P_2.

(b) $\{1+t,\ 1+2t+t^2\}$ for P_2.

(c) $\{(1+t)^2,\ 1+t^2,\ 2+4t+2t^2\}$ for P_2.

(d) $\left\{\mathbf{A} = \begin{pmatrix} 1 & 0 \\ 0 & 1 \end{pmatrix}, \mathbf{B} = \begin{pmatrix} 2 & 0 \\ 0 & 2 \end{pmatrix}, \mathbf{C} = \begin{pmatrix} 0 & 0 \\ 1 & 1 \end{pmatrix}, \mathbf{D} = \begin{pmatrix} 0 & 0 \\ 0 & 5 \end{pmatrix}\right\}$ for M_{22}.

(e) $\left\{\mathbf{A} = \begin{pmatrix} 1 & 0 \\ 0 & 1 \end{pmatrix}, \mathbf{B} = \begin{pmatrix} 2 & 0 \\ 2 & 2 \end{pmatrix}, \mathbf{C} = \begin{pmatrix} 1 & 0 \\ 3 & 6 \end{pmatrix}, \mathbf{D} = \begin{pmatrix} 3 & 0 \\ 4 & 5 \end{pmatrix}\right\}$ for M_{22}.

9. Show that the following set of vectors is a basis for $\mathbb{R}^5$:

$$S = \left\{\begin{pmatrix} 1 \\ 2 \\ 1 \\ 0 \\ 0 \end{pmatrix}, \begin{pmatrix} -1 \\ -5 \\ 2 \\ 1 \\ 0 \end{pmatrix}, \begin{pmatrix} -3 \\ 5 \\ 2 \\ 0 \\ 0 \end{pmatrix}, \begin{pmatrix} 2 \\ 3 \\ 3 \\ 1 \\ 1 \end{pmatrix}, \begin{pmatrix} 1 \\ 0 \\ 0 \\ 0 \\ 0 \end{pmatrix}\right\}$$

10. Let S be a subspace of an n-dimensional vector space V. Prove that $\dim(S) \le n$.

11. Let S be a subspace of an n-dimensional vector space V. Prove that if $\dim(S) = n$ then $S = V$.

12. Let n be the dimension of a finite dimensional vector space. Prove that n is a positive integer or zero.

13. Let V be a one-dimensional vector space and vector $\mathbf{v} \neq \mathbf{O}$ be in V. Prove that $\mathbf{v}$ is a basis for V.

14. Let V be an n-dimensional vector space and $S = \{\mathbf{v}_1, \mathbf{v}_2, \mathbf{v}_3, \ldots, \mathbf{v}_n\}$ be a set of vectors in V such that **none** of the vectors is a linear combination of the preceding vectors. Prove that the set S forms a basis for V.

15. Let P be the vector space of all polynomials. Prove that P is infinite dimensional.

SECTION 3.5 Properties of a Matrix

By the end of this section you will be able to

- understand what is meant by the row and column space of a matrix
- determine a basis for the row space of a matrix
- find the rank of a matrix

In this section we examine the rows and columns of a matrix and define what is meant by rank. These are important concepts in solving linear equations which will be discussed in the next section.

You will need to know some of the work of chapter 1, such as elementary row operations and the definition of row echelon form. In this section we will not carry out the row operations but use MATLAB to place a given matrix into row echelon form. Of course, you don't have to use MATLAB; any appropriate software will do or you can use hand calculations.

3.5.1 Row and column vectors

What are the 'row vectors of a matrix'?

The row vectors of a matrix are the entries in the rows of a given matrix. For example, the row vectors of $\mathbf{A} = \begin{pmatrix} 1 & 2 & 3 \\ 4 & 5 & 6 \end{pmatrix}$ are $\begin{pmatrix} 1 \\ 2 \\ 3 \end{pmatrix}$ and $\begin{pmatrix} 4 \\ 5 \\ 6 \end{pmatrix}$ because the first row of matrix $\mathbf{A}$ has the entries 1, 2 and 3 and the second row has entries 4, 5 and 6.

What are the column vectors of $\mathbf{A} = \begin{pmatrix} 1 & 2 & 3 \\ 4 & 5 & 6 \end{pmatrix}$*?*

The matrix $\mathbf{A}$ has three columns vectors:

$$\begin{pmatrix} 1 \\ 4 \end{pmatrix}, \begin{pmatrix} 2 \\ 5 \end{pmatrix} \text{ and } \begin{pmatrix} 3 \\ 6 \end{pmatrix}$$

We generalize this by considering the m by n matrix:

$$\begin{array}{cc} & \begin{array}{cccc} \text{Col } 1 & \text{Col } 2 & \cdots & \text{Col } n \end{array} \\ \mathbf{A} = & \begin{pmatrix} a_{11} & a_{12} & \cdots & a_{1n} \\ a_{21} & a_{22} & \cdots & a_{2n} \\ \vdots & \vdots & \vdots & \vdots \\ a_{m1} & a_{m2} & \cdots & a_{mn} \end{pmatrix} \begin{array}{l} \text{Row } 1 \\ \text{Row } 2 \\ \vdots \\ \text{Row } m \end{array} \end{array}$$

What are the row vectors of this matrix $\mathbf{A}$*?*

The **row vectors** of $\mathbf{A}$ denoted $\mathbf{r}_1, \mathbf{r}_2, \ldots$ and $\mathbf{r}_m$ are given by:

$$\mathbf{r}_1 = \begin{pmatrix} a_{11} \\ a_{12} \\ \vdots \\ a_{1n} \end{pmatrix}, \quad \mathbf{r}_2 = \begin{pmatrix} a_{21} \\ a_{22} \\ \vdots \\ a_{2n} \end{pmatrix}, \cdots \text{ and } \mathbf{r}_m = \begin{pmatrix} a_{m1} \\ a_{m2} \\ \vdots \\ a_{mn} \end{pmatrix}$$

What are the column vectors of the matrix $\mathbf{A}$*?*

The **column vectors** of $\mathbf{A}$ denoted $\mathbf{c}_1, \mathbf{c}_2, \ldots$ and $\mathbf{c}_n$ are given by:

$$\mathbf{c}_1 = \begin{pmatrix} a_{11} \\ a_{21} \\ \vdots \\ a_{m1} \end{pmatrix}, \quad \mathbf{c}_2 = \begin{pmatrix} a_{12} \\ a_{22} \\ \vdots \\ a_{m2} \end{pmatrix}, \cdots \text{ and } \mathbf{c}_n = \begin{pmatrix} a_{1n} \\ a_{2n} \\ \vdots \\ a_{mn} \end{pmatrix}$$

For example, the row and column vectors of $\mathbf{B} = \begin{pmatrix} -3 & 6 \\ -5 & 2 \\ -2 & 7 \end{pmatrix}$ are

$\mathbf{r}_1 = \begin{pmatrix} -3 \\ 6 \end{pmatrix}$, $\mathbf{r}_2 = \begin{pmatrix} -5 \\ 2 \end{pmatrix}$, $\mathbf{r}_3 = \begin{pmatrix} -2 \\ 7 \end{pmatrix}$ and $\mathbf{c}_1 = \begin{pmatrix} -3 \\ -5 \\ -2 \end{pmatrix}$, $\mathbf{c}_2 = \begin{pmatrix} 6 \\ 2 \\ 7 \end{pmatrix}$ respectively.

3.5.2 Row and column space

The row space of a matrix **A** is the space spanned by the row vectors of **A**. Remember, the space spanned by vectors means the space containing all the linear combinations of these vectors.

*What is the row space of the above matrix **B**?*

It is the space, S, spanned by the vectors $\mathbf{r}_1 = \begin{pmatrix} -3 \\ 6 \end{pmatrix}$, $\mathbf{r}_2 = \begin{pmatrix} -5 \\ 2 \end{pmatrix}$ and $\mathbf{r}_3 = \begin{pmatrix} -2 \\ 7 \end{pmatrix}$.

Any linear combination of these vectors, $\mathbf{r}_1$, $\mathbf{r}_2$ and $\mathbf{r}_3$, is in the row space S. Hence

$$\text{Row Space } S = span\{\mathbf{r}_1, \mathbf{r}_2, \mathbf{r}_3\} = k_1 \begin{pmatrix} -3 \\ 6 \end{pmatrix} + k_2 \begin{pmatrix} -5 \\ 2 \end{pmatrix} + k_3 \begin{pmatrix} -2 \\ 7 \end{pmatrix}$$

where k_1, k_2 and k_3 are scalars. Each row vector has two entries of real numbers and we can show that S is a subspace of $\mathbb{R}^2$.

The row space S is the set of vectors $\mathbf{u}$ such that $\mathbf{u} = k_1\mathbf{r}_1 + k_2\mathbf{r}_2 + k_3\mathbf{r}_3$. This row space S, spanned by $\mathbf{r}_1$, $\mathbf{r}_2$ and $\mathbf{r}_3$, is the vector space given by

$$S = \{\mathbf{u} \mid \mathbf{u} = k_1\mathbf{r}_1 + k_2\mathbf{r}_2 + k_3\mathbf{r}_3\}$$

In this case, these vectors span the whole of $\mathbb{R}^2$ because no two vectors in S are multiplies of each other (they are linearly independent). The row space of the above matrix **B** occupies $\mathbb{R}^2$. Similarly the column space of a general matrix **A** is the space spanned by the column vectors of **A**. We formally define the row and column space as follows:

Definition (3.23). Let **A** be any matrix. Then

(a) The **row space** of the matrix **A** is the space spanned by the row vectors of matrix **A**.

(b) The **column space** of the matrix **A** is the space spanned by the column vectors of matrix **A**.

*What is the column space of the above matrix **B**?*

It is the space spanned by the vectors $\mathbf{c}_1 = \begin{pmatrix} -3 \\ -5 \\ -2 \end{pmatrix}$ and $\mathbf{c}_2 = \begin{pmatrix} 6 \\ 2 \\ 7 \end{pmatrix}$. Any linear combination of these vectors, $\mathbf{c}_1$ and $\mathbf{c}_2$, is in the column space of matrix **B**.

What sort of space is spanned by these vectors?

Each vector has three entries of real numbers and we can show that the column space is a subspace of $\mathbb{R}^3$. The vector space spanned by vectors $\mathbf{c}_1$ and $\mathbf{c}_2$ is actually a plane in $\mathbb{R}^3$ (Fig. 3.17).

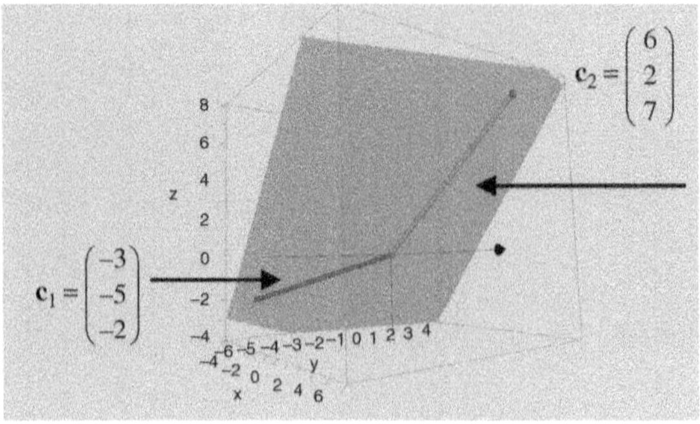

This dark shaded plane in $\mathbb{R}^3$ is the column space of matrix **B**. (Space spanned by column vectors $\mathbf{c}_1$ and $\mathbf{c}_2$.)

Figure 3.17

In general, we have the following proposition:

Proposition (3.24). Let **A** be a m by n matrix, that is we have

$$\mathbf{A} = \overbrace{\begin{pmatrix} a_{11} & \cdots & a_{1n} \\ \vdots & \ddots & \vdots \\ a_{m1} & \cdots & a_{mn} \end{pmatrix}}^{n \text{ columns}} \left.\vphantom{\begin{pmatrix} a \\ \vdots \\ a \end{pmatrix}}\right\} m \text{ rows}$$

(a) The **row space** of matrix **A** is a subspace of $\mathbb{R}^n$.

(b) The **column space** of matrix **A** is a subspace of $\mathbb{R}^m$.

Proof – See Exercises 3.5. ■

For example, the row and column space of matrix $\mathbf{A} = \begin{pmatrix} 1 & 4 \\ 5 & 9 \end{pmatrix}$ is a subspace of $\mathbb{R}^2$. The row space of matrix $\mathbf{B} = \begin{pmatrix} -1 & 3 & 9 \\ -5 & 2 & 6 \end{pmatrix}$ is a subspace of $\mathbb{R}^3$ but the column space is a subspace of $\mathbb{R}^2$.

3.5.3 Basis of a row space

Why do we want to find a basis for the row space of a matrix?

Recall that a basis is formed by the least number of vectors, or axes, required to describe the space. Our matrix may contain zero rows, or rows that are linearly dependent. By using row operations in the row space to reduce our matrix to its simplest form (or basis), we often end up with a much more efficient matrix, which is *row equivalent* to the original.

What does row equivalent matrix mean?
Two matrices are *row equivalent* if one is obtained from the other by elementary row operations. (This was discussed in chapter 1.) These operations are:

1. Multiplying a row by a non-zero constant.
2. Adding or subtracting a multiple of one row from another.
3. Interchanging rows.

Consider the matrix $\mathbf{A}$ with rows $\mathbf{a}_1$ and $\mathbf{a}_2$, converted using row operations, to the matrix $\mathbf{R}$:

$$\begin{matrix}\mathbf{a}_1\\ \mathbf{a}_2\end{matrix}\begin{pmatrix}1 & 2 & 3\\ 4 & 8 & 12\end{pmatrix} = \mathbf{A} \quad \Rightarrow \quad \begin{matrix}\mathbf{r}_1 = \mathbf{a}_1\\ \mathbf{r}_2 = \mathbf{a}_2 - 4\mathbf{a}_1\end{matrix}\begin{pmatrix}1 & 2 & 3\\ 0 & 0 & 0\end{pmatrix} = \mathbf{R}$$

Note that the rows of matrix $\mathbf{R}$, $\mathbf{r}_1$ and $\mathbf{r}_2$, are simply a linear combination of rows $\mathbf{a}_1$ and $\mathbf{a}_2$:

$$\mathbf{r}_1 = \mathbf{a}_1 \text{ and } \mathbf{r}_2 = \mathbf{a}_2 - 4\mathbf{a}_1$$

The rows of matrix $\mathbf{R}$ are a linear combination of the rows of matrix $\mathbf{A}$. This means that the row vectors of matrix $\mathbf{R}$ lie in the row space of matrix $\mathbf{A}$.

What can you predict about the row space of row equivalent matrices?
The space they occupy is equal.

For example, the row space S of the above matrix $\mathbf{A}$ is given by the vectors $\mathbf{v}$ in S such that:

$$\mathbf{v} = k_1\mathbf{a}_1 + k_2\mathbf{a}_2 = k_1\begin{pmatrix}1\\2\\3\end{pmatrix} + k_2\begin{pmatrix}4\\8\\12\end{pmatrix} = k_1\begin{pmatrix}1\\2\\3\end{pmatrix} + 4k_2\begin{pmatrix}1\\2\\3\end{pmatrix} = c\begin{pmatrix}1\\2\\3\end{pmatrix} = c\mathbf{r}_1$$

where $c = k_1 + 4k_2$

Hence all the vectors $\mathbf{v}$ are in the row space S of matrix $\mathbf{R}$. In fact, you can span the row space of matrix $\mathbf{A}$ with just one vector $\begin{pmatrix}1 & 2 & 3\end{pmatrix}^T$ which is the non-zero row in matrix $\mathbf{R}$. Actually the row space S created by matrix $\mathbf{A}$ is the same as the row space created by matrix $\mathbf{R}$. We have

$$\text{Row space of } \mathbf{A} = \text{Row space of } \mathbf{R}$$

Proposition (3.25). If matrices $\mathbf{A}$ and $\mathbf{R}$ are row equivalent then their *row spaces* are *equal*.

How do we prove this result?
By showing that

1. The row space of $\mathbf{A}$ is in row space of $\mathbf{R}$.
2. The row space of $\mathbf{R}$ is in row space of $\mathbf{A}$.

If both these conditions are satisfied then the row spaces of matrices $\mathbf{A}$ and $\mathbf{R}$ must be equal.

Proof.

Let **A** and **R** be row equivalent m by n matrices. Let the row vectors of **A** be $\mathbf{a}_1, \mathbf{a}_2, \mathbf{a}_3, \ldots, \mathbf{a}_m$ and the row vectors of **R** be $\mathbf{r}_1, \mathbf{r}_2, \mathbf{r}_3, \ldots, \mathbf{r}_m$. This means that we have

$$\mathbf{A} = \begin{pmatrix} \mathbf{a}_1 \\ \vdots \\ \mathbf{a}_m \end{pmatrix} \Rightarrow \mathbf{R} = \begin{pmatrix} \mathbf{r}_1 \\ \vdots \\ \mathbf{r}_m \end{pmatrix}$$

1) Matrices **A** and **R** are row equivalent, therefore the **r** row vectors are obtained from the **a** row vectors by elementary row operations. This means that every **r** row vector is a linear combination of the **a** row vectors. Therefore the row space of matrix **A** lies in the row space of matrix **R**.

2) Similarly, by considering the above argument the other way, we have that the row space of matrix **R** lies in the row space of matrix **A**.

Hence the row space of matrices **A** and **R** are equal.

■

Remember, we can use elementary row operations to put a matrix into row echelon form. Additionally, we can apply this to find a basis for the row space of a matrix as the following proposition states:

Proposition (3.26). If a matrix **R** is in row echelon form then its non-zero rows form a basis (set of axes) for the row space of matrix **R**.

Proof – See Exercises 3.5.

■

By combining the above two propositions (3.25) and (3.26) we have:

Proposition (3.27). If matrix **A** is row equivalent to matrix **R** where matrix **R** is in row echelon form then the non-zero rows of matrix **R** form a basis for the row space of matrix **A**.

Proof.

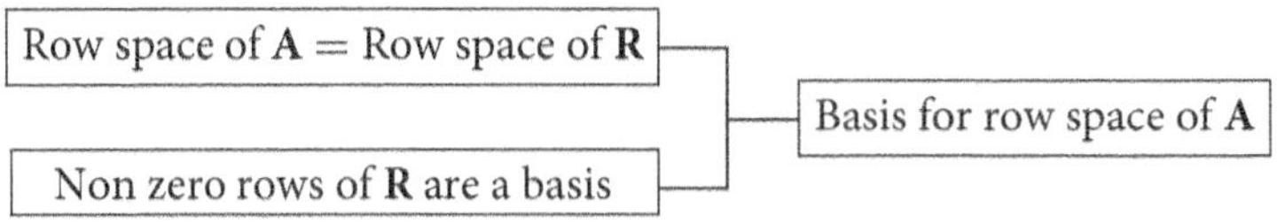

By the above Proposition (3.26) the basis of the row space of **R** is given by the non-zero rows of matrix **R**. By the above Proposition (3.25) we have

Row space of **A** = Row space of **R** [because **R** and **A** are row equivalent]

Thus the non-zero rows of matrix **R** form a basis for the row space of matrix **A**.

■

How can we use this proposition to find a basis for the row space of a given matrix?

By carrying out elementary row operations and placing the given matrix into row echelon form. We aim to achieve the following:

$$\mathbf{A} = \begin{pmatrix} \mathbf{a}_1 \\ \vdots \\ \mathbf{a}_m \\ \mathbf{a}_{m+1} \\ \vdots \end{pmatrix} \Longrightarrow \mathbf{R} = \begin{pmatrix} \mathbf{r}_1 \\ \vdots \\ \mathbf{r}_m \\ \mathbf{O} \\ \vdots \end{pmatrix}$$

These non-zero rows (vectors) form a basis for the row space of matrix **A**.

We can use MATLAB to place a matrix into row echelon form. Actually MATLAB and the command rref places a given matrix into *reduced* row echelon form. In the following examples we will use the reduced row echelon form but it is enough to find the row echelon form.

Example 3.28

Determine a basis for the row space of the following matrices:

(a) $\mathbf{A} = \begin{pmatrix} 1 & 2 \\ 2 & 4 \end{pmatrix}$ **(b)** $\mathbf{B} = \begin{pmatrix} -1 & 3 & 9 \\ -5 & 2 & 6 \end{pmatrix}$ **(c)** $\mathbf{C} = \begin{pmatrix} -3 & 6 \\ -5 & 2 \\ -2 & 7 \end{pmatrix}$

Solution

(a) Applying row operations we have

$$\mathbf{A} = \begin{pmatrix} 1 & 2 \\ 2 & 4 \end{pmatrix} \Longrightarrow \mathbf{R} = \begin{pmatrix} 1 & 2 \\ 0 & 0 \end{pmatrix} \leftarrow \text{Non-zero row}$$

The vector $\begin{pmatrix} 1 \\ 2 \end{pmatrix}$ is a basis for the row space of **A** (Fig. 3.18).

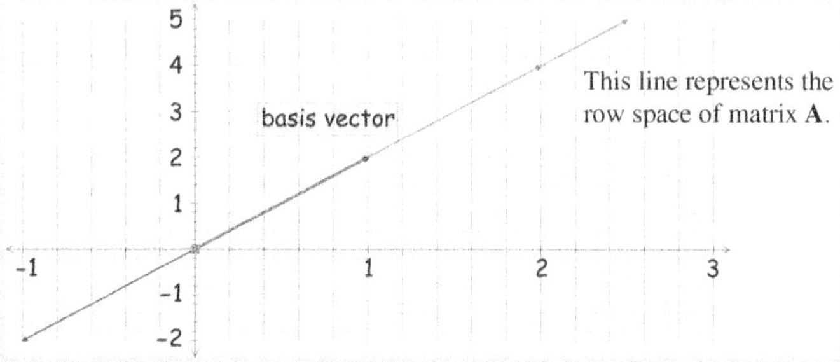

Figure 3.18

(b) Similarly, to find the reduced row echelon form of **B** we use MATLAB:

$$\mathbf{B} = \begin{pmatrix} -1 & 3 & 9 \\ -5 & 2 & 6 \end{pmatrix} \Longrightarrow \mathbf{R} = \begin{pmatrix} 1 & 0 & 0 \\ 0 & 1 & 3 \end{pmatrix}$$

(continued...)

What is a basis for the row space of matrix **B**?

The vectors $\left\{\begin{pmatrix}1\\0\\0\end{pmatrix}, \begin{pmatrix}0\\1\\3\end{pmatrix}\right\}$ form a basis for the row space of matrix **B**. Hence the row space of **B** is a subspace of $\mathbb{R}^3$.

(c) By using MATLAB, the reduced row echelon form of matrix **C** is given by:

$$\mathbf{C} = \begin{pmatrix}-3 & 6\\-5 & 2\\-2 & 7\end{pmatrix} \Longrightarrow \mathbf{R} = \begin{pmatrix}1 & 0\\0 & 1\\0 & 0\end{pmatrix}$$ Non-zero rows

Remember, it is the non-zero rows which form a basis for the row space. Thus the vectors $\left\{\begin{pmatrix}1\\0\end{pmatrix}, \begin{pmatrix}0\\1\end{pmatrix}\right\}$ form a basis for the row space of matrix **C**. We need two vectors $\left\{\begin{pmatrix}1\\0\end{pmatrix}, \begin{pmatrix}0\\1\end{pmatrix}\right\}$ to span the row space of **C** rather than the three given row vectors in matrix **C**. Remember, this basis span the whole of $\mathbb{R}^2$, so the row space of matrix **C** is $\mathbb{R}^2$.

The above example can easily be carried out by hand calculations because of the simple integer entries for the given matrices.

3.5.4 Basis of a spanned subspace of $\mathbb{R}^n$

Let $\mathbf{u} = \begin{pmatrix}1 & 2 & 3\end{pmatrix}^T$ and $\mathbf{v} = \begin{pmatrix}4 & 5 & 6\end{pmatrix}^T$ be vectors in $\mathbb{R}^3$. Let S be the space spanned by these vectors which is illustrated in Fig. 3.19.

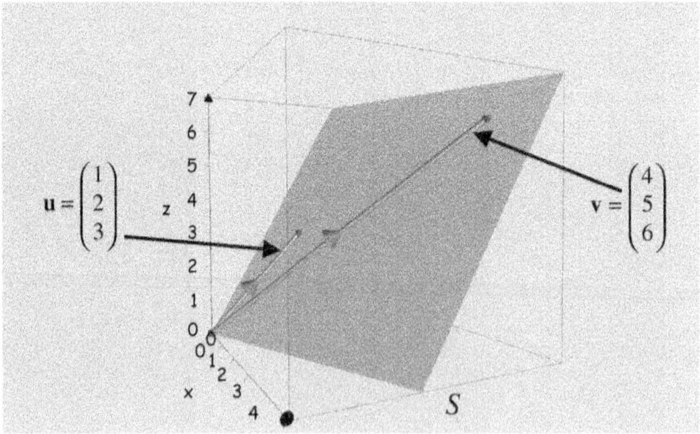

Vectors **u** and **v** span the dark shaded plane S in $\mathbb{R}^3$.

Figure 3.19

We are generally interested in finding a simple set of axes to describe this plane, or more formally, a basis for this space S.

How can we find such a basis?

Writing the vectors as rows of a matrix $\mathbf{A} = \begin{pmatrix} \mathbf{u} \\ \mathbf{v} \end{pmatrix}$ then the *row space* of $\mathbf{A}$ is the vector space spanned by vectors $\mathbf{u}$ and $\mathbf{v}$. By using row operations on $\mathbf{A}$, we can find a basis for the row space of a matrix $\mathbf{A}$ as we did in subsection 3.5.3 above.

The procedure of finding a basis for a subspace of $\mathbb{R}^n$ which is spanned by the vectors $\{\mathbf{r}_1, \mathbf{r}_2, \mathbf{r}_3, \ldots, \mathbf{r}_m\}$ is given by:

1. Form the matrix $\mathbf{A} = \begin{pmatrix} \mathbf{r}_1 \\ \vdots \\ \mathbf{r}_m \end{pmatrix}$. The row space of $\mathbf{A}$ is the space spanned by $\mathbf{r}_1, \mathbf{r}_2, \ldots, \mathbf{r}_m$.
2. Convert this matrix $\mathbf{A}$ into (reduced) row echelon form, $\mathbf{R}$ say.
3. The non-zero rows of matrix $\mathbf{R}$ form a basis for the vector space $span\{\mathbf{r}_1, \mathbf{r}_2, \ldots, \mathbf{r}_m\}$.

Example 3.29

Determine a basis for the vector space $S = span\{\mathbf{u}, \mathbf{v}, \mathbf{w}\}$ where

$$\mathbf{u} = \begin{pmatrix} 1 \\ 2 \\ 3 \end{pmatrix}, \mathbf{v} = \begin{pmatrix} 4 \\ 5 \\ 6 \end{pmatrix} \text{ and } \mathbf{w} = \begin{pmatrix} 7 \\ 8 \\ 9 \end{pmatrix}$$

Solution

Let $\mathbf{A}$ be the matrix given by $\mathbf{A} = \begin{pmatrix} \mathbf{u} \\ \mathbf{v} \\ \mathbf{w} \end{pmatrix}$, that is $\mathbf{A} = \begin{pmatrix} 1 & 2 & 3 \\ 4 & 5 & 6 \\ 7 & 8 & 9 \end{pmatrix}$. The row space of matrix $\mathbf{A}$ is the subspace spanned by vectors $\mathbf{u}$, $\mathbf{v}$ and $\mathbf{w}$. Thus

$$span\{\mathbf{u}, \mathbf{v}, \mathbf{w}\} = \text{Row space of } \mathbf{A}$$

What do we need to find?
The basis for the row space of the matrix $\mathbf{A}$.
How?
By finding the (reduced) row echelon form matrix $\mathbf{R}$ of $\mathbf{A}$ and then the non-zero rows of $\mathbf{R}$ form a basis (axes) for the row space of matrix $\mathbf{A}$. By using MATLAB we obtain:

$$\mathbf{A} = \begin{pmatrix} 1 & 2 & 3 \\ 4 & 5 & 6 \\ 7 & 8 & 9 \end{pmatrix} \Longrightarrow \begin{pmatrix} 1 & 0 & -1 \\ 0 & 1 & 2 \\ 0 & 0 & 0 \end{pmatrix} = \mathbf{R}$$

The non-zero rows of the above matrix $\mathbf{R}$ form a basis of the row space of $\mathbf{A}$:

$$\mathbf{r}_1 = \begin{pmatrix} 1 \\ 0 \\ -1 \end{pmatrix} \text{ and } \mathbf{r}_2 = \begin{pmatrix} 0 \\ 1 \\ 2 \end{pmatrix}$$

Thus a basis (axes) for the given vector space $S = span(\mathbf{u}, \mathbf{v}, \mathbf{w})$ is $B = \{\mathbf{r}_1, \mathbf{r}_2\}$ where $\mathbf{r}_1$ and $\mathbf{r}_2$ are vectors stated above. Note that S is a subspace of $\mathbb{R}^3$, but is actually a plane and can be spanned by two vectors, $\mathbf{r}_1$ and $\mathbf{r}_2$, rather than the three given vectors $\mathbf{u}$, $\mathbf{v}$ and $\mathbf{w}$.

Example 3.30

Determine a basis for the vector space $S = span\{\mathbf{u}, \mathbf{v}, \mathbf{w}\}$ where

$$\mathbf{u} = \begin{pmatrix} 1 \\ 3 \\ 5 \\ 6 \end{pmatrix}, \mathbf{v} = \begin{pmatrix} 1 \\ 7 \\ 8 \\ 9 \end{pmatrix} \text{ and } \mathbf{w} = \begin{pmatrix} 3 \\ 9 \\ 15 \\ 18 \end{pmatrix}$$

Solution

By applying the above procedure we have:

Step 1 and 2: Writing the matrix $\mathbf{A}$ and evaluating its reduced row echelon form:

$$\mathbf{A} = \begin{pmatrix} \mathbf{u} \\ \mathbf{v} \\ \mathbf{w} \end{pmatrix} = \begin{pmatrix} 1 & 3 & 5 & 6 \\ 1 & 7 & 8 & 9 \\ 3 & 9 & 15 & 18 \end{pmatrix} \Rightarrow \mathbf{R} = \begin{pmatrix} 1 & 0 & 2.75 & 3.75 \\ 0 & 1 & 0.75 & 0.75 \\ 0 & 0 & 0 & 0 \end{pmatrix}$$

$S = span\{\mathbf{u}, \mathbf{v}, \mathbf{w}\} =$ Row space of matrix $\mathbf{A}$.

Step 3:

What is a basis for $S = span\{\mathbf{u}, \mathbf{v}, \mathbf{w}\}$?

It is the non-zero rows of the reduced row echelon form matrix $\mathbf{R}$. We have

$$\mathbf{r}_1 = \begin{pmatrix} 1 \\ 0 \\ 2.75 \\ 3.75 \end{pmatrix} = \begin{pmatrix} 1 \\ 0 \\ 11/4 \\ 15/4 \end{pmatrix} = \frac{1}{4}\begin{pmatrix} 4 \\ 0 \\ 11 \\ 15 \end{pmatrix} \text{ and } \mathbf{r}_2 = \begin{pmatrix} 0 \\ 1 \\ 0.75 \\ 0.75 \end{pmatrix} = \begin{pmatrix} 0 \\ 1 \\ 3/4 \\ 3/4 \end{pmatrix} = \frac{1}{4}\begin{pmatrix} 0 \\ 4 \\ 3 \\ 3 \end{pmatrix}$$

Since the row space is the linear combination of vectors $\mathbf{r}_1$ and $\mathbf{r}_2$ so any non - zero scalar multiple of $\mathbf{r}_1$ and $\mathbf{r}_2$ is also a basis for the row space. (The basis gives the directions of the two axes for the space.) Hence, multiplying vectors $\mathbf{r}_1$ and $\mathbf{r}_2$ by 4 gives vectors:

$$\left\{(\,4 \;\; 0 \;\; 11 \;\; 15\,)^T, (\,0 \;\; 4 \;\; 3 \;\; 3\,)^T\right\}$$

which form a basis for $S = span\{\mathbf{u}, \mathbf{v}, \mathbf{w}\}$ and this is a subspace of $\mathbb{R}^4$. This S is a plane in $\mathbb{R}^4$ because we have two axes or two basis vectors for this space.

3.5.5 Rank of a matrix

Why is rank important in linear algebra?

Consider the linear system $\mathbf{Ax} = \mathbf{b}$. The augmented matrix $(\mathbf{A} \mid \mathbf{b})$ in row echelon form may produce zero rows, which means $0 = 0$, but these are not important in the solution of linear equations. It is the number (rank) of the non-zero rows in row echelon form which gives the solution of a linear system of equations. We will discover in the next section that the rank of matrix $\mathbf{A}$ and of the augmented matrix $(\mathbf{A} \mid \mathbf{b})$ tell us if there are no, a unique or an infinite number of solutions.

The rank of a matrix gives the *number* of linearly *independent* rows in a matrix which means that all the rows that are linearly *dependent* are counted as one. For example, the following matrix has a rank of 1:

$$\begin{matrix} R_1 \\ R_2 \end{matrix}\begin{pmatrix} 1 & 2 & 3 & 4 \\ 2 & 4 & 6 & 8 \end{pmatrix} \text{ can be transformed to } \begin{matrix} R_1 \\ R_2 - 2R_1 \end{matrix}\begin{pmatrix} 1 & 2 & 3 & 4 \\ 0 & 0 & 0 & 0 \end{pmatrix}$$

The second row is double the first, so carrying out row operations results in a single independent row. The rank of a matrix measures the amount of important information represented by the matrix.

An application of linear algebra is the transfer of digital data which is normally stored as a matrix. In these fields it is important that data is transferred as fast and efficiently as possible without losing any of it. The concept of a rank is critical here because a matrix with a lower rank takes up less memory and time to be transferred. Low rank matrices are much more efficient in the sense that they are much less computationally expensive to deal with.

Computer graphics rely on matrices to generate and manipulate images. The rank of the matrix tells you the dimension of the image. For example, the matrix $\mathbf{A} = \begin{pmatrix} 1 & 1 & 1 \\ 4 & 5 & 6 \\ 2 & 2 & 2 \end{pmatrix}$ transforms a vector in 3D onto a 2D plane because matrix $\mathbf{A}$ does not have 'full rank' (the top and bottom rows are linearly dependent) as shown in Fig. 3.20.

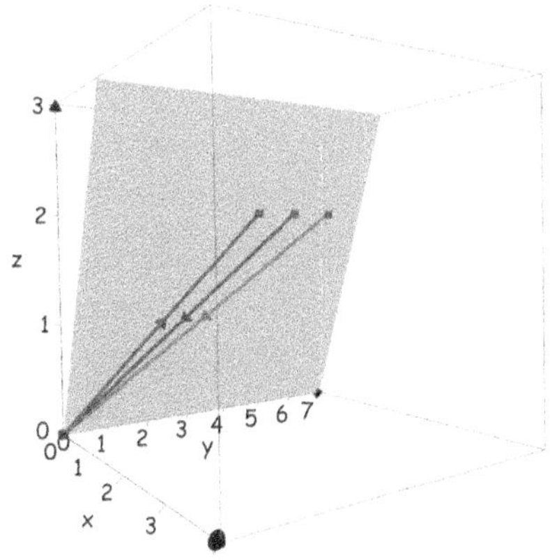

Matrix $\mathbf{A}$ will *only* transform vectors onto the plane shown and not anywhere else in $\mathbb{R}^3$.

Figure 3.20

We define the rank in terms of dimension.

Can you recall what the term dimension of a vector space means?

It is the least number of axes needed to describe the vector space, or in other words, the number of vectors in the basis of a vector space.

The *dimension* of the *row space* of a matrix is called *row rank* and the *dimension* of the *column space* is called the *column rank*. Note that the row rank of a given matrix $\mathbf{A}$ is the *number* of non-zero row vectors in row echelon form matrix $\mathbf{R}$ because the non-zero rows form a basis for the row space.

$$\mathrm{A} = \begin{pmatrix} \mathbf{a}_1 \\ \vdots \\ \mathbf{a}_m \\ \mathbf{a}_{m+1} \\ \vdots \end{pmatrix} \Rightarrow \mathrm{R} = \begin{pmatrix} \mathbf{r}_1 \\ \vdots \\ \mathbf{r}_m \\ \mathbf{O} \\ \vdots \end{pmatrix} \quad \left.\begin{matrix} \\ \\ \\ \end{matrix}\right\} m \text{ non-zero rows}$$

Row rank of matrix $\mathbf{A} = m$
The row rank of a matrix is called the **rank** of a matrix.

Definition (3.28). The **rank** of a matrix **A** is the row rank of **A**.

The rank of matrix **A** is denoted by $rank(\mathbf{A})$. Thus (3.28) says

$$rank(\mathbf{A}) = \text{row rank of } \mathbf{A}$$

What is the difference between rank and dimension?
Strictly speaking, the rank is an attribute of a matrix, while dimension is an attribute of a vector space.

Example 3.31

Determine the rank of the matrices **A**, **B** and **C** given in Example 3.28.

Solution
What is $rank(\mathbf{A})$ equal to?
From the solutions to Example 3.28 we have one vector in the basis of the row space of **A**, therefore the row rank is equal to 1 which means that $rank(\mathbf{A}) = 1$.
Similarly we have $rank(\mathbf{B}) = 2$ and $rank(\mathbf{C}) = 2$.

Example 3.32

Determine the rank of matrix $\mathbf{A} = \begin{pmatrix} 1 & 2 & 3 & 4 & 5 & 6 \\ 27 & 28 & 29 & 30 & 31 & 32 \\ 15 & 16 & 17 & 18 & 19 & 20 \\ 31 & 32 & 33 & 34 & 35 & 36 \\ 45 & 46 & 47 & 48 & 49 & 50 \end{pmatrix}$.

Solution
What do we need to do?
Place the given matrix **A** into row echelon form. By using MATLAB, we can find the reduced row echelon form matrix **R**:

$$\mathbf{R} = \begin{pmatrix} \boxed{1} & 0 & -1 & -2 & -3 & -4 \\ 0 & \boxed{1} & 2 & 3 & 4 & 5 \\ 0 & 0 & 0 & 0 & 0 & 0 \\ 0 & 0 & 0 & 0 & 0 & 0 \\ 0 & 0 & 0 & 0 & 0 & 0 \end{pmatrix} \begin{matrix} \leftarrow \text{Non-zero row} \\ \leftarrow \text{Non-zero row} \\ \\ \\ \\ \end{matrix}$$

*What is the rank of the given matrix **A** equal to?*
$rank(\mathbf{A}) = 2$ because we have two non-zero rows. In this case the rank tells you that two rows are important and three rows are redundant because the row space of matrix **A** can be spanned by the first two rows of matrix **R**.
If matrix **A** represents a digital image then we can compress this data to a matrix with rank 2, which is more computationally efficient than matrix **A**.

Proposition (3.29). Let $\mathbf{A}$ be any matrix. Then

$$\textit{Row rank of matrix } \mathbf{A} = \textit{Column rank of matrix } \mathbf{A}$$

This means

$$rank(\mathbf{A}) = \text{Row rank of } \mathbf{A} = \text{Column rank of } \mathbf{A}$$

Proof – Exercises 3.5.

■

The row and column rank of a matrix are equal.

We can also find a basis for the column space of a matrix by considering the columns of the reduced row echelon form matrix with leading 1's. In the above Example 3.32 the first two columns of matrix $\mathbf{R}$ have leading 1's. It can shown that the corresponding column vectors of matrix $\mathbf{A}$ form a basis for the column space of $\mathbf{A}$. Hence the following vectors are a basis:

$$(\ 1\ \ 27\ \ 15\ \ 31\ \ 45\ )^T \text{ and } (\ 2\ \ 28\ \ 16\ \ 32\ \ 46\ )^T$$

for the column space of matrix $\mathbf{A}$.

3.5.6 Rank and invertible matrices

In this subsection we discuss the relationships between rank and invertible matrices.

Proposition (3.30). Let $\mathbf{A}$ be an n by n matrix. The matrix $\mathbf{A}$ is invertible $\Leftrightarrow$ $rank\,(\mathbf{A}) = n$.

What does this mean?

Matrix is invertible $\Leftrightarrow$ it has *no* redundant rows. We say that the matrix $\mathbf{A}$ has **full rank**.

Proof.

($\Rightarrow$). We assume matrix $\mathbf{A}$ is invertible. By (1.35):

Theorem (1.35). Let $\mathbf{A}$ be a n by n matrix, then the following are equivalent:

(a) The matrix $\mathbf{A}$ is invertible.

(b) The reduced row echelon form of the matrix $\mathbf{A}$ is the identity matrix $\mathbf{I}$.

The reduced row echelon form of the matrix $\mathbf{A}$ is the identity n by n matrix $\mathbf{I}$. Thus there are n non-zero rows of $\mathbf{I}$, therefore $rank(\mathbf{A}) = n$.

($\Leftarrow$). In this case, we assume that $rank(\mathbf{A}) = n$ and we need to prove that the matrix $\mathbf{A}$ is invertible. Since $rank(\mathbf{A}) = n$, the reduced row echelon form of $\mathbf{A}$ has no zero rows. By (1.39);

Proposition (1.39). $\mathbf{R}$ has at least one row of zeros $\Leftrightarrow$ $\mathbf{A}$ is non-invertible (singular).

This means that matrix $\mathbf{A}$ must be invertible.

■

Hence if a square matrix is not of full rank then it is non-invertible.

Summary

(3.23) The *row space* of a matrix **A** is the vector space spanned by the row vectors of matrix **A**.
(3.25). If matrices **A** and **R** are row equivalent then their row spaces are equal.
(3.27). A basis for the row space of matrix **A** is the non-zero rows of the equivalent row echelon form matrix **R**.
The rank of a matrix is the *dimension* of the *row or column space* of the matrix.

EXERCISES 3.5

(Brief solutions at end of book. Full solutions available at <http://www.oup.co.uk/companion/singh>.)

1. For the following matrices find the row and column vectors:

(a) $\mathbf{A} = \begin{pmatrix} 1 & 2 \\ 3 & 4 \end{pmatrix}$ (b) $\mathbf{B} = \begin{pmatrix} 1 & 2 & 3 & 4 \\ 5 & 6 & 7 & 8 \\ 9 & 10 & 11 & 12 \end{pmatrix}$ (c) $\mathbf{C} = \begin{pmatrix} 1 & 2 \\ 3 & 4 \\ 5 & 6 \end{pmatrix}$

(d) $\mathbf{D} = \begin{pmatrix} 1 & 2 & 3 \\ 4 & 5 & 6 \end{pmatrix}$ (e) $\mathbf{E} = \begin{pmatrix} -1 & 2 & 5 \\ -3 & 7 & 0 \\ -8 & 1 & 3 \end{pmatrix}$ (f) $\mathbf{F} = \begin{pmatrix} -5 & 2 & 3 \\ 7 & 1 & 0 \\ -7 & 6 & 1 \\ -2 & 5 & 2 \end{pmatrix}$

2. Determine a basis for the row space of the matrices given in question 1. Also state the rank of the given matrix.

3. Determine a basis for the following subspace of $\mathbb{R}^n$ which are spanned by the vectors:

(a) $\mathbf{u} = \begin{pmatrix} 1 \\ 5 \end{pmatrix}, \mathbf{v} = \begin{pmatrix} 7 \\ 2 \end{pmatrix}$ (b) $\mathbf{u} = \begin{pmatrix} -1 \\ -3 \end{pmatrix}, \mathbf{v} = \begin{pmatrix} 4 \\ 12 \end{pmatrix}$

(c) $\mathbf{u} = \begin{pmatrix} 3 \\ 6 \\ 5 \end{pmatrix}, \mathbf{v} = \begin{pmatrix} 2 \\ 1 \\ 2 \end{pmatrix}, \mathbf{w} = \begin{pmatrix} 12 \\ 15 \\ 15 \end{pmatrix}$

(d) $\mathbf{u} = \begin{pmatrix} -1 \\ -2 \\ -5 \end{pmatrix}, \mathbf{v} = \begin{pmatrix} 0 \\ 1 \\ 5 \end{pmatrix}, \mathbf{w} = \begin{pmatrix} 2 \\ 3 \\ 2 \end{pmatrix}, \mathbf{x} = \begin{pmatrix} -4 \\ 1 \\ -7 \end{pmatrix}$

(e) $\mathbf{u} = \begin{pmatrix} 1 \\ 2 \\ 2 \\ 2 \end{pmatrix}, \mathbf{v} = \begin{pmatrix} -1 \\ 3 \\ 5 \\ 7 \end{pmatrix}, \mathbf{w} = \begin{pmatrix} 2 \\ -1 \\ -3 \\ -5 \end{pmatrix}, \mathbf{x} = \begin{pmatrix} 0 \\ 5 \\ 7 \\ 9 \end{pmatrix}$

4. Find a basis for the subspace of $\mathbb{R}^4$ which is given by $S = span(\mathbf{u}, \mathbf{v}, \mathbf{w}, \mathbf{x}, \mathbf{y})$ where

$$\mathbf{u} = \begin{pmatrix} -1 \\ 1 \\ -1 \\ 1 \end{pmatrix}, \mathbf{v} = \begin{pmatrix} -5 \\ -9 \\ -7 \\ -1 \end{pmatrix}, \mathbf{w} = \begin{pmatrix} 0 \\ 7 \\ 1 \\ 3 \end{pmatrix}, \mathbf{x} = \begin{pmatrix} -6 \\ -1 \\ -7 \\ 3 \end{pmatrix}, \mathbf{y} = \begin{pmatrix} 2 \\ 5 \\ 3 \\ 1 \end{pmatrix}.$$

[You can convert the decimal entries of the reduced row echelon form matrix into fractional entries by using the command rats(rref(A)) in MATLAB. In MATLAB, the rats() command converts the argument into a rational number.]

5. Determine a basis for the column space of the matrices given in question 1 by taking the transpose of the given matrices. Also state the rank of the given matrix.
6. Let $\mathbf{A}$ be any matrix. Prove that $rank(\mathbf{A}) = rank(\mathbf{A}^T)$.
7. Let $\mathbf{A}$ be an invertible matrix. Prove that $rank(\mathbf{A}) = rank(\mathbf{A}^{-1})$.
8. Let $\mathbf{A}$ be a square n by n matrix. Prove the following:

   (a) $\mathbf{A}$ has rank $n \Leftrightarrow$ the linear system $\mathbf{Ax} = \mathbf{O}$ has the trivial solution $\mathbf{x} = \mathbf{O}$.

   (b) $\mathbf{A}$ has rank $n \Leftrightarrow$ the linear system $\mathbf{Ax} = \mathbf{b}$ has a unique solution.

9. Let $\mathbf{A}$ be a square n by n matrix whose row vectors are given by the set of vectors, $S = \{\mathbf{r}_1, \mathbf{r}_2, \mathbf{r}_3, \ldots, \mathbf{r}_n\}$:

$$\mathrm{A} = \begin{pmatrix} \mathbf{r}_1 \\ \vdots \\ \mathbf{r}_n \end{pmatrix}$$

   Prove that matrix $\mathbf{A}$ is invertible $\Leftrightarrow S$ is a set of linearly independent vectors.

10. Let $\mathbf{A}$ be a square n by n matrix whose column vectors are given by the set of vectors; $S = \{\mathbf{c}_1, \mathbf{c}_2, \mathbf{c}_3, \ldots, \mathbf{c}_n\}$:

$$\mathbf{A} = (\mathbf{c}_1 \ \cdots \ \mathbf{c}_n)$$

   Prove that matrix $\mathbf{A}$ is invertible $\Leftrightarrow S$ is a set of linearly independent vectors.

11. Prove that the row space of a matrix $\mathbf{A}$ is identical to the row space of $k\,\mathbf{A}$, where k is a non-zero scalar.
12. Let $\mathbf{A}$ be any matrix and k be a non-zero scalar. Show that $rank(k\mathbf{A}) = rank(\mathbf{A})$.
13. Prove Proposition (3.24).
14. Prove that if $\mathbf{R}$ is the reduced row echelon matrix then the non-zero rows of $\mathbf{R}$ are linearly independent.
15. Prove that if $\mathbf{A}$ is a matrix and the reduced row echelon form matrix $\mathbf{R}$ of $\mathbf{A}$ contains zero rows then the rows of $\mathbf{A}$ are linearly dependent.
16. Prove Proposition (3.26).
17. Let $\mathbf{A}$ be any matrix whose rows are given by the set of linear independent vectors $S = \{\mathbf{r}_1, \mathbf{r}_2, \mathbf{r}_3, \ldots, \mathbf{r}_n\}$. Prove that $rank(\mathbf{A}) = n$.
18. For the following you may assume:

    The columns of matrix $\mathbf{A}$ with the corresponding leading 1's in reduced row echelon form matrix $\mathbf{R}$ are a basis for the column space. Prove Proposition (3.29).

SECTION 3.6 Linear Systems Revisited

By the end of this section you will be able to

- understand and determine the null space and nullity of a matrix
- prove some properties of null space, nullity and rank of a matrix
- determine solutions to a non-homogeneous system of linear equations

In the last section we concentrated on the row space of a matrix, but in this section we examine the part played by the *column space* of a matrix in the solution of the linear system $\mathbf{Ax} = \mathbf{b}$.

In this section we answer the critical question of linear algebra:

What conditions provide infinite, unique or no solutions to a linear system?

You will need to ensure you are familiar with the concepts of row space, column space and the rank of a matrix to analyse solutions of linear systems.

3.6.1 Null space

We consider the homogeneous system first, which is a linear system of equations written in matrix form as $\mathbf{Ax} = \mathbf{O}$. By using elementary row operations we solve the equivalent system $\mathbf{Rx} = \mathbf{O}$ where matrix $\mathbf{R}$ is row equivalent to matrix $\mathbf{A}$.

Example 3.33

Solve the homogeneous system of linear equations:

$$\begin{aligned} x + 2y + 3z &= 0 \\ 4x + 5y + 6z &= 0 \\ 7x + 8y + 9z &= 0 \end{aligned}$$

Solution

We have the linear system $\mathbf{Ax} = \mathbf{O}$ where $\mathbf{A}$ is the coefficient matrix, $\mathbf{x}$ is the vector of unknowns and $\mathbf{O}$ is the zero vector.

How can we find the unknowns x, y and z?

We can write out the augmented matrix and then convert this into reduced row echelon form by using hand calculations or MATLAB with command rref(A):

$$(\mathbf{A} \mid \mathbf{O}) = \left(\begin{array}{ccc|c} 1 & 2 & 3 & 0 \\ 4 & 5 & 6 & 0 \\ 7 & 8 & 9 & 0 \end{array}\right) \Longrightarrow \begin{array}{c} \begin{array}{ccc} x & y & z \end{array} \\ \left(\begin{array}{ccc|c} 1 & 0 & -1 & 0 \\ 0 & 1 & 2 & 0 \\ 0 & 0 & 0 & 0 \end{array}\right) \end{array}$$

From the middle row of the matrix on the right hand side we have

$$y + 2z = 0 \text{ which gives } y = -2z$$

None of the rows start with z so z is a free variable.

Let $z = s$ where s is any real number then $y = -2s$ and from the top row we have

$$x - z = 0 \text{ which gives } x = z = s$$

Our solution set is $x = s,\ y = -2s$ and $z = s$ which we can write in vector form as

$$\mathbf{x} = \begin{pmatrix} x \\ y \\ z \end{pmatrix} = \begin{pmatrix} s \\ -2s \\ s \end{pmatrix} = s\begin{pmatrix} 1 \\ -2 \\ 1 \end{pmatrix} \text{ where } s \text{ is a } \textit{parameter}$$

This *free variable* s is any real number and the solution is the set of *all* points on the axis (thick line) as shown in Fig. 3.21:

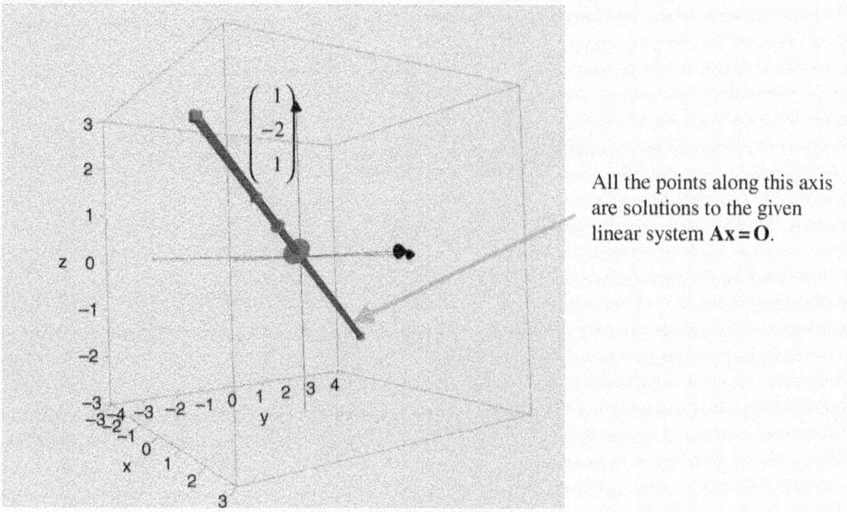
All the points along this axis are solutions to the given linear system Ax = O.
z
x
y

Figure 3.21

The solution $\mathbf{x} = s\begin{pmatrix} 1 \\ -2 \\ 1 \end{pmatrix}$ is a scalar multiple of the vector $\mathbf{u} = \begin{pmatrix} 1 \\ -2 \\ 1 \end{pmatrix}$, so this vector $\mathbf{u}$ is a basis vector on the axis shown for the solution space. Note that this solution space is a subspace of $\mathbb{R}^3$.

The vector or solution space, call it N, of a homogeneous system $\mathbf{Ax} = \mathbf{O}$ is called the **null space** of matrix $\mathbf{A}$. In general, we have:

Definition (3.31). Let $\mathbf{A}$ be any matrix. The *set of vectors* $\mathbf{x}$ in N (or $N(\mathbf{A})$) which satisfy the homogeneous system $\mathbf{Ax} = \mathbf{O}$ is called the **null space** of matrix $\mathbf{A}$. The dimension of this null space is called the **nullity** of matrix $\mathbf{A}$ and is denoted by $nullity(\mathbf{A})$.

For the above Example 3.33 the null space is the set of vectors $N = \{s\mathbf{u} \mid s \in \mathbb{R}\}$ which is shown as an axis in Fig. 3.21. In this case, $nullity(\mathbf{A}) = 1$ because we have one axis or a single vector $\mathbf{u}$, which is a basis for N. Hence the null space of matrix $\mathbf{A}$ is of dimension 1.

How can we write the general solution to $\mathbf{Ax} = \mathbf{O}$?

$$\begin{pmatrix} a_{11} & \cdots & a_{1n} \\ \vdots & \ddots & \vdots \\ a_{m1} & \cdots & a_{mn} \end{pmatrix} \begin{pmatrix} x_1 \\ \vdots \\ x_n \end{pmatrix} = \begin{pmatrix} 0 \\ \vdots \\ 0 \end{pmatrix}$$

Multiplying these matrices we have:

$$\begin{array}{ccccccc} a_{11}x_1 & + & \cdots & + & a_{1n}x_n & = & 0 \\ \vdots & & & & \vdots & \vdots & \vdots \\ a_{m1}x_1 & + & \cdots & + & a_{mn}x_n & = & 0 \end{array}$$

We can write this in terms of the column vectors $\mathbf{c}_1, \mathbf{c}_2, \ldots$ and $\mathbf{c}_n$ of matrix $\mathbf{A}$:

$$\overset{\mathbf{c}_1}{\begin{pmatrix} a_{11} \\ \vdots \\ a_{m1} \end{pmatrix}} x_1 + \cdots + \overset{\mathbf{c}_n}{\begin{pmatrix} a_{1n} \\ \vdots \\ a_{mn} \end{pmatrix}} x_n = \begin{pmatrix} 0 \\ \vdots \\ 0 \end{pmatrix} \qquad (*)$$

Note that the left hand side is a linear combination of the column vectors $\mathbf{c}_1, \mathbf{c}_2, \ldots$ and $\mathbf{c}_n$ of matrix $\mathbf{A}$. Recall that this linear combination is the *column space* of the matrix $\mathbf{A}$. The null space consists of vectors $\mathbf{x} = (\, x_1 \;\; x_2 \;\; \cdots \;\; x_n \,)^T$ such that they satisfy the linear combination (*).

The null space is a non-empty set.

How do we know it is non-empty?
Because the homogeneous system $\mathbf{Ax} = \mathbf{O}$ always has the trivial solution $\mathbf{x} = \mathbf{O}$ ($x_1 = \cdots = x_n = 0$) so we know the null space of matrix $\mathbf{A}$ is *not* empty.

Proposition (3.32). If $\mathbf{A}$ is a matrix with n columns then the null space $N(\mathbf{A})$ is a subspace of $\mathbb{R}^n$.

Proof – Exercises 3.6.

■

The null space of a matrix is a vector space.

Example 3.34

Determine the null space, nullity and rank of the matrix $\mathbf{B} = \begin{pmatrix} 1 & 2 & 3 & 4 \\ 5 & 6 & 7 & 8 \end{pmatrix}$.

Solution

How do we find the null space of the given matrix $\mathbf{B}$*?*
The null space is the set of vectors $\mathbf{x}$ which satisfies the homogeneous system $\mathbf{Bx} = \mathbf{O}$:

$$(\mathbf{B} \mid \mathbf{O}) = \left(\begin{array}{cccc|c} 1 & 2 & 3 & 4 & 0 \\ 5 & 6 & 7 & 8 & 0 \end{array}\right) \Longrightarrow \begin{array}{c} \begin{array}{cccc} x & y & z & w \end{array} \\ \left(\begin{array}{cccc|c} 1 & 0 & -1 & -2 & 0 \\ 0 & 1 & 2 & 3 & 0 \end{array}\right) \end{array} = (\mathbf{R} \mid \mathbf{O})$$

The matrix on the right hand side **R** is actually the reduced row echelon form of matrix **B**.
Expanding out the bottom row gives:

$$y + 2z + 3w = 0 \text{ yields } y = -2z - 3w$$

None of the rows begin with z and w, so these are our free variables. Let $z = s$ and $w = t$ where t and s are any real numbers. We have $y = -2z - 3w = -2s - 3t$.
Expanding the top row we have

$$x - z - 2w = 0 \quad \text{gives} \quad x = z + 2w$$

Substituting $z = s$ and $w = t$ into $x = z + 2w$ gives $x = s + 2t$. The solution of $\mathbf{Bx} = \mathbf{O}$ is the null space of matrix **B** given by the set of vectors $\mathbf{x} = (\, x \;\; y \;\; z \;\; w \,)^T$ which has the entries $x = s + 2t$, $y = -2s - 3t$, $z = s$ and $w = t$:

$$\mathbf{x} = \begin{pmatrix} x \\ y \\ z \\ w \end{pmatrix} = \begin{pmatrix} s + 2t \\ -2s - 3t \\ s \\ t \end{pmatrix} = s\begin{pmatrix} 1 \\ -2 \\ 1 \\ 0 \end{pmatrix} + t\begin{pmatrix} 2 \\ -3 \\ 0 \\ 1 \end{pmatrix}$$

Let $\mathbf{u} = (\, 1 \;\; -2 \;\; 1 \;\; 0 \,)^T$ and $\mathbf{v} = (\, 2 \;\; -3 \;\; 0 \;\; 1 \,)^T$ then we can write the null space N as

$$N = \{s\mathbf{u} + t\mathbf{v} \;\mid\; t \in \mathbb{R},\; s \in \mathbb{R}\}$$

Substituting $s = t = 0$ into $s\mathbf{u} + t\mathbf{v}$ gives the zero vector **O**, which of course is in the null space N. We can substitute any real numbers for s and t to obtain an infinite number of vectors in N.
Note that N is a subspace of $\mathbb{R}^4$ because vectors **u** and **v** have four real entries.
What can we say about the vectors ***u*** *and* ***v****?*
The vectors **u** and **v** span the null space N and they are also linearly independent.
How do we know these vectors are linearly independent?
Because **u** and **v** are *not* multiples of each other. This means that vectors **u** and **v** form a basis for the null space N. (Actually the null space N is a plane in $\mathbb{R}^4$ because we have two axes or basis vectors, **u** and **v**.)
What is the nullity of matrix ***B*** *equal to?*

$$nullity(\mathbf{B}) = 2$$

This means the set of vectors **x** which satisfy $\mathbf{Bx} = \mathbf{O}$ is of dimension 2. Hence we require two axes or basis vectors to describe the null space of matrix **B**.
What is the rank of the given matrix ***B*** *equal to?*
The above reduced row echelon form matrix **R** has two non-zero rows, therefore

$$rank(\mathbf{B}) = 2$$

Using the procedure outlined in this Example 3.34 gives us a basis for the null space.

Example 3.35

Determine the null space, nullity and rank of the matrix $\mathbf{C} = \begin{pmatrix} 1 & 3 \\ 5 & 15 \end{pmatrix}$.

(continued...)

Solution

The null space N is the set of vectors $\mathbf{x}$ which satisfy $\mathbf{Cx} = \mathbf{O}$.

How do we find this vector space?

By applying elementary row operations to convert into reduced row echelon form:

$$(\mathbf{C} \mid \mathbf{O}) = \left(\begin{array}{cc|c} 1 & 3 & 0 \\ 5 & 15 & 0 \end{array}\right) \Longrightarrow (\mathbf{R} \mid \mathbf{O}) = \left(\begin{array}{cc|c} \overset{x}{1} & \overset{y}{3} & 0 \\ 0 & 0 & 0 \end{array}\right) \text{ non-zero row}$$

We only have *one* non-zero row in the reduced row echelon form matrix **R**, therefore

$$rank(\mathbf{C}) = 1$$

Considering the equivalent homogeneous system $\mathbf{Rx} = \mathbf{O}$ we have

$$x + 3y = 0 \text{ implies } x = -3y$$

Let $y = s$ then $x = -3y = -3s$ where s is any real number. Hence

$$\mathbf{x} = \begin{pmatrix} x \\ y \end{pmatrix} = \begin{pmatrix} -3s \\ s \end{pmatrix} = s\begin{pmatrix} -3 \\ 1 \end{pmatrix}$$

Let $\mathbf{u} = \begin{pmatrix} -3 \\ 1 \end{pmatrix}$ then the null space $N = \{s\mathbf{u} \mid s \in \mathbb{R}\}$ and the graph of this space is:

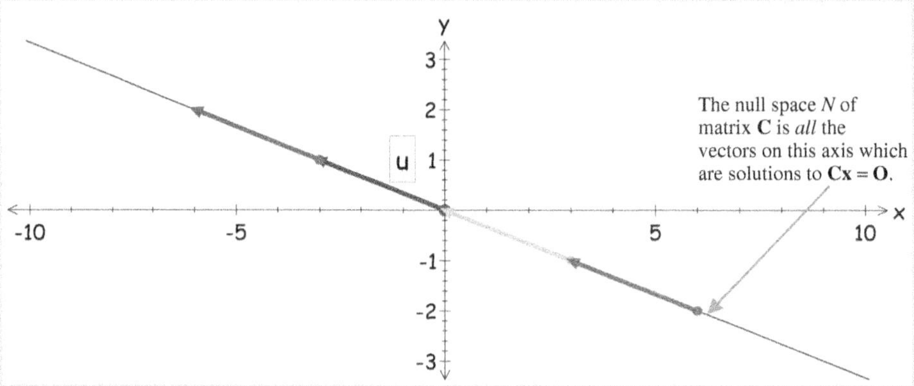
y
x
3
2
1
-1
-2
-3
-10
-5
5
10
u
The null space N of matrix C is all the vectors on this axis which are solutions to Cx = O.

Figure 3.22

Note that the vector **u** spans the null space and is linearly independent so it is a basis for N.

*What is the nullity of matrix **C** equal to?*

It is the dimension of the null space N, which is 1, because it has only one basis vector **u** (or one free variable - s). We have $nullity(\mathbf{C}) = 1$.

We can also evaluate the nullity of a matrix by finding the number of free variables in the system.

Proposition (3.33). The number of free variables in the system $\mathbf{Ax} = \mathbf{O}$ is equal to the nullity of the matrix **A**.

Proof – Exercises 3.6.

■

3.6.2 Properties of rank and nullity

In the above examples 3.33, 3.34 and 3.35 we had

$$\begin{aligned} nullity(\mathbf{A}) + rank(\mathbf{A}) &= 1 + 2 = 3 \\ nullity(\mathbf{B}) + rank(\mathbf{B}) &= 2 + 2 = 4 \\ nullity(\mathbf{C}) + rank(\mathbf{C}) &= 1 + 1 = 2 \end{aligned}$$

Can you see any relationship between the nullity, rank and the number of unknowns in vector $\mathbf{x}$*?*

Nullity + Rank = Number of unknowns

In general, if we have a m by n matrix $\mathbf{A}$:

$$\underbrace{\begin{pmatrix} a_{11} & \cdots & a_{1n} \\ \vdots & \ddots & \vdots \\ a_{m1} & \cdots & a_{mn} \end{pmatrix}}_{n \text{ columns}} \begin{pmatrix} x_1 \\ \vdots \\ x_n \end{pmatrix} = \begin{pmatrix} 0 \\ \vdots \\ 0 \end{pmatrix} \qquad [\mathbf{Ax} = \mathbf{O}]$$

Then

$$nullity(\mathbf{A}) + rank(\mathbf{A}) = n$$

Note that n is the number of columns of matrix $\mathbf{A}$ which is the total number of unknowns in the homogeneous system $\mathbf{Ax} = \mathbf{O}$. This result normally has the grand title of 'The Dimension Theorem of Matrices'.

Theorem (3.34). The dimension theorem of matrices (rank-nullity theorem):
If $\mathbf{A}$ is a matrix with n columns (number of unknowns) then

$$nullity(\mathbf{A}) + rank(\mathbf{A}) = n$$

Proof.
Let $\mathbf{R}$ be the reduced row echelon form of matrix $\mathbf{A}$ and let it have rank r, which means that it has r non-zero rows located at the top of the matrix. By the definition of reduced row echelon form, we know that each of these r rows has a leading 1.

$$\mathbf{R} = \overbrace{\begin{pmatrix} \boxed{1} & & & & & \\ 0 & \ddots & & & & \\ \vdots & & \boxed{1} & & \cdots & \\ 0 & 0 & 0 & \cdots & \cdots & 0 \\ 0 & 0 & 0 & \cdots & \cdots & 0 \end{pmatrix}}^{n \text{ unknows}} \begin{matrix} \leftarrow \\ \vdots \\ \leftarrow \\ \\ \\ \end{matrix} \left.\begin{matrix} \\ \\ \\ \end{matrix}\right\} r \text{ non-zero rows}$$

There are n unknowns in total and r of these can be expressed in terms of the remaining $n - r$ unknowns. This means there are $n - r$ free variables in the system. By the above Proposition (3.33):

The number of free variables in the system $\mathbf{Ax} = \mathbf{O}$ is equal to the nullity of the matrix $\mathbf{A}$. We have $nullity(\mathbf{A}) = n - r$ and transposing this gives us our result

$$nullity(\mathbf{A}) + r = n \qquad \text{which is} \qquad nullity(\mathbf{A}) + rank(\mathbf{A}) = n$$

■

What is the nullity of a matrix **A** *which is of full rank?*

Full rank means $rank(\mathbf{A}) = n$, therefore

$$nullity(\mathbf{A}) = n - rank(\mathbf{A}) = n - n = 0$$

Hence the dimension of the null space is 0, so null space only contains the trivial solution - zero vector, $\mathbf{x} = \mathbf{O}$. In case of full rank, $\mathbf{Ax} = \mathbf{O} \Rightarrow \mathbf{x} = \mathbf{O}$ is the only solution.

Example 3.36

Determine the nullity and rank of the matrix:

$$\mathbf{A} = \begin{pmatrix} 1 & 2 & 3 & 4 & 5 & 6 & 7 \\ 8 & 9 & 10 & 11 & 12 & 13 & 14 \\ 15 & 16 & 17 & 18 & 19 & 20 & 21 \\ 22 & 23 & 24 & 25 & 26 & 27 & 28 \\ 29 & 30 & 31 & 32 & 33 & 34 & 35 \end{pmatrix}$$

Solution

By using MATLAB, the reduced row echelon form matrix $\mathbf{R}$ of $\mathbf{A}$ is given by

$$\mathbf{R} = \begin{pmatrix} 1 & 0 & -1 & -2 & -3 & -4 & -5 \\ 0 & 1 & 2 & 3 & 4 & 5 & 6 \\ 0 & 0 & 0 & 0 & 0 & 0 & 0 \\ 0 & 0 & 0 & 0 & 0 & 0 & 0 \\ 0 & 0 & 0 & 0 & 0 & 0 & 0 \end{pmatrix} \begin{matrix} \leftarrow \text{non-zero row} \\ \leftarrow \text{non-zero row} \\ \\ \\ \\ \end{matrix}$$

What is rank $(\mathbf{A})$ *equal to?*

Since there are two non-zero rows in matrix $\mathbf{R}$, therefore $rank(\mathbf{A}) = 2$. Note that the matrix has five rows but there are only two linearly independent rows.

What is $nullity(\mathbf{A})$ *equal to?*

$nullity(\mathbf{A})$ is the dimension of the null space and can be evaluated by:

Theorem (3.34) $nullity(\mathbf{A}) + rank(\mathbf{A}) = n$

Substituting $rank(\mathbf{A}) = 2$ and $n = 7$ (the number of columns of matrix $\mathbf{A}$) we have

$$nullity(\mathbf{A}) + 2 = 7 \text{ gives } nullity(\mathbf{A}) = 5$$

This means that the set of vectors $\mathbf{x}$ which satisfy $\mathbf{Ax} = \mathbf{O}$ is of dimension five. We need five axes or five basis vectors to describe the vectors $\mathbf{x}$ in the null space of matrix $\mathbf{A}$.

Sometimes it is more convenient to write $nullity(\mathbf{A}) + rank(\mathbf{A}) = n$ as

$$nullity(\mathbf{A}) = n - rank(\mathbf{A})$$

3.6.3 Non-homogeneous linear systems

Now we consider linear equations of the form $\mathbf{Ax} = \mathbf{b}$ where $\mathbf{b} \neq \mathbf{O}$ (not zero). Equations of the form $\mathbf{Ax} = \mathbf{b}$, where $\mathbf{b} \neq \mathbf{O}$, are called **non-homogeneous** linear equations. Throughout the remaining part of this section we assume $\mathbf{b} \neq \mathbf{O}$.

Example 3.37

Solve the above Example 3.35 with $\mathbf{Cx} = \mathbf{b}$ where $\mathbf{b} = (7 \quad 35)^T$:

$$\begin{aligned} x + 3y &= 7 \\ 5x + 15y &= 35 \end{aligned}$$

Solution

We first write out the augmented matrix and then evaluate the reduced row echelon form:

$$\left(\begin{array}{cc|c} 1 & 3 & 7 \\ 5 & 15 & 35 \end{array}\right) \Longrightarrow \begin{array}{c} \begin{matrix} x & y \end{matrix} \\ \left(\begin{array}{cc|c} 1 & 3 & 7 \\ 0 & 0 & 0 \end{array}\right) \end{array}$$

Expanding the top row of this right hand matrix we have

$$x + 3y = 7 \text{ which gives } x = 7 - 3y$$

Let $y = s$, then we have $x = 7 - 3y = 7 - 3s$.

We have $x = 7 - 3s$ and $y = s$. The general solution $\mathbf{x}$ is

$$\mathbf{x} = \begin{pmatrix} x \\ y \end{pmatrix} = \begin{pmatrix} 7 - 3s \\ s \end{pmatrix} = s\begin{pmatrix} -3 \\ 1 \end{pmatrix} + \begin{pmatrix} 7 \\ 0 \end{pmatrix}$$

The homogeneous solution $\mathbf{x}_H$ to $\mathbf{Cx} = \mathbf{O}$ in Example 3.35 was $\mathbf{x}_H = s(-3 \quad 1)^T$.

What do you notice?

Solving this non-homogeneous system $\mathbf{Cx} = \mathbf{b}$ where $\mathbf{b} = (7 \quad 35)^T$ gives a two-part solution: the homogeneous solution; $\mathbf{x}_H = s(-3 \quad 1)^T$ plus an extra term, $(7 \quad 0)^T$, called the particular solution, which we denote by $\mathbf{x}_P$.

$\mathbf{x}_H = s\begin{pmatrix} -3 \\ 1 \end{pmatrix}$ gives us the slope $-\frac{1}{3}$, and $\begin{pmatrix} 7 \\ 0 \end{pmatrix}$ moves the line horizontally by seven units to the solution $\mathbf{x}$. We demonstrate this in Fig. 3.23.

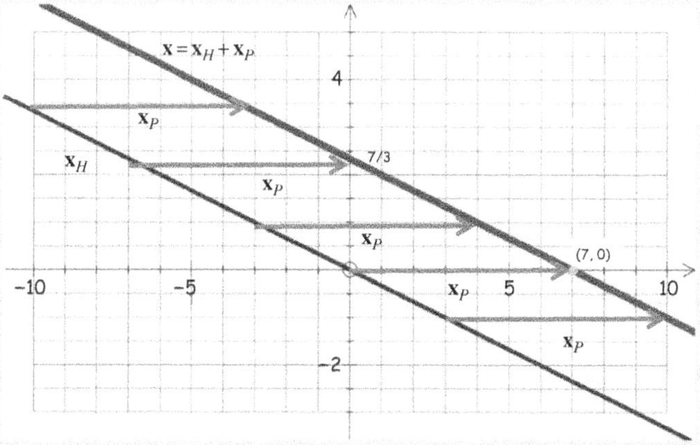

Figure 3.23

(continued...)

Hence our general solution to $\mathbf{Cx} = \mathbf{b}$ is $\mathbf{x} = \mathbf{x}_H + \mathbf{x}_P$ where

$$\mathbf{x}_H = s(-3\ \ 1)^T \qquad \text{and} \qquad \mathbf{x}_P = (7\ \ 0)^T$$

Notice that in the above example we have

$$\mathbf{Cx}_H = \mathbf{O} \text{ and } \mathbf{Cx}_P = \mathbf{b}$$

because $\mathbf{x}_H$ is the homogeneous solution and $\mathbf{x}_P$ is the particular solution. Combining these to solve the non-homogeneous system $\mathbf{Cx} = \mathbf{b}$ we have

$$\begin{aligned} \mathbf{Cx}_H + \mathbf{Cx}_P &= \mathbf{O} + \mathbf{b} \\ \mathbf{C}(\mathbf{x}_H + \mathbf{x}_P) &= \mathbf{b} \qquad \Rightarrow \qquad \mathbf{x} = \mathbf{x}_H + \mathbf{x}_P \end{aligned}$$

Hence the general solution of $\mathbf{Cx} = \mathbf{b}$ is given by:

$$\mathbf{x} = (\text{homogeneous solution}) + (\text{particular solution})$$

In general, for non-homogeneous systems we have the following:

Proposition (3.35). Let $\mathbf{x}_P$ be the particular solution of $\mathbf{Ax} = \mathbf{b}$ and $\mathbf{x}_H$ be the solution to the homogeneous system $\mathbf{Ax} = \mathbf{O}$. All the solutions of $\mathbf{Ax} = \mathbf{b}$ are of the form $\mathbf{x}_P + \mathbf{x}_H$.

What does this proposition mean?
The solution of $\mathbf{Ax} = \mathbf{b}$ consists of two parts:

$$(\text{homogeneous solution}) + (\text{particular solution})$$

Remember, the homogeneous solution $\mathbf{x}_H$ (vector) belongs to the null space of matrix $\mathbf{A}$.

Proof.
Let $\mathbf{x}$ be the solution of $\mathbf{Ax} = \mathbf{b}$ then

$$\begin{aligned} \mathbf{A}(\mathbf{x} - \mathbf{x}_P) &= \mathbf{Ax} - \mathbf{Ax}_P \\ &= \mathbf{b} - \mathbf{b} = \mathbf{O} \end{aligned}$$

Since we have $\mathbf{A}(\mathbf{x} - \mathbf{x}_P) = \mathbf{O}$, therefore $\mathbf{x} - \mathbf{x}_P$ is the homogeneous solution, that is $\mathbf{x}_H = \mathbf{x} - \mathbf{x}_P$. Hence we have our result $\mathbf{x} = \mathbf{x}_P + \mathbf{x}_H$.

We need to show *all* the solutions are of this format $\mathbf{x}_P + \mathbf{x}_H$.

Let $\mathbf{x}'$ be a solution of $\mathbf{Ax} = \mathbf{O}$, then

$$\mathbf{A}(\mathbf{x} + \mathbf{x}') = \mathbf{Ax} + \mathbf{Ax}' = \mathbf{Ax} + \mathbf{O} = \mathbf{b} + \mathbf{O} = \mathbf{b}$$

Hence $\mathbf{x} + \mathbf{x}'$ is a solution of $\mathbf{Ax} = \mathbf{b}$.

We conclude that *all* the solutions are of this form $\mathbf{x} = \mathbf{x}_P + \mathbf{x}_H$.

■

Example 3.38

Solve the non-homogeneous linear system:

$$\begin{aligned} x - 2y + 2z + 0.03t &= 1.7 \\ -y + z - 0.02t &= -1.6 \\ x - y \quad\quad\quad 0.01t &= 0 \end{aligned}$$

Solution

Writing the augmented matrix and evaluating the reduced row echelon form gives

$$\left(\begin{array}{cccc|c} 1 & -2 & 2 & 0.03 & 1.7 \\ 0 & -1 & 1 & -0.02 & -1.6 \\ 1 & -1 & 0 & 0.01 & 0 \end{array}\right) \Longrightarrow \begin{array}{c} \begin{array}{cccc} x & y & z & t \end{array} \\ \left(\begin{array}{cccc|c} 1 & 0 & 0 & 0.07 & 4.9 \\ 0 & 1 & 0 & 0.06 & 4.9 \\ 0 & 0 & 1 & 0.04 & 3.3 \end{array}\right) \end{array}$$

The general solution is

$$x + 0.07t = 4.9,\ y + 0.06t = 4.9 \text{ and } z + 0.04t = 3.3$$

$$x = 4.9 - 0.07t,\quad y = 4.9 - 0.06t \text{ and } z = 3.3 - 0.04t$$

where t is our free variable. In vector form we have

$$\mathbf{x} = \begin{pmatrix} x \\ y \\ z \\ t \end{pmatrix} = \begin{pmatrix} 4.9 - 0.07t \\ 4.9 - 0.06t \\ 3.3 - 0.04t \\ t \end{pmatrix} = \underbrace{\begin{pmatrix} 4.9 \\ 4.9 \\ 3.3 \\ 0 \end{pmatrix}}_{\text{particular sol'n}=\mathbf{x}_P} + \underbrace{\begin{pmatrix} -0.07 \\ -0.06 \\ -0.04 \\ 1 \end{pmatrix} t}_{\text{homogeneous sol'n}=\mathbf{x}_H}$$

This is an example of an **underdetermined** system: a system with *more unknowns* than equations, which gives infinitely many solutions provided it is consistent. The row echelon form matrix **R** of an underdetermined system has the following shape:

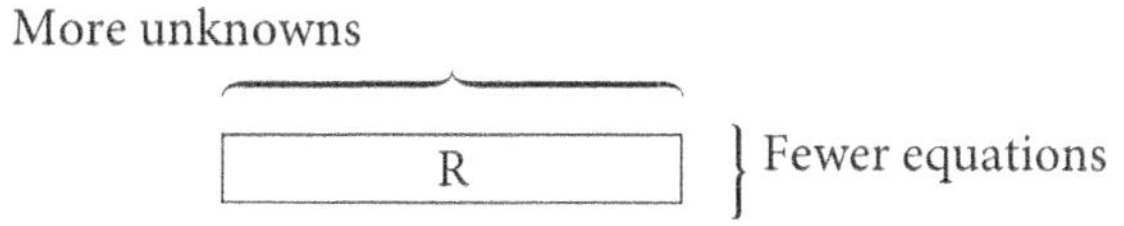

An **overdetermined** system is a system with *more equations* than unknowns. The row echelon form matrix **R** has the following shape:

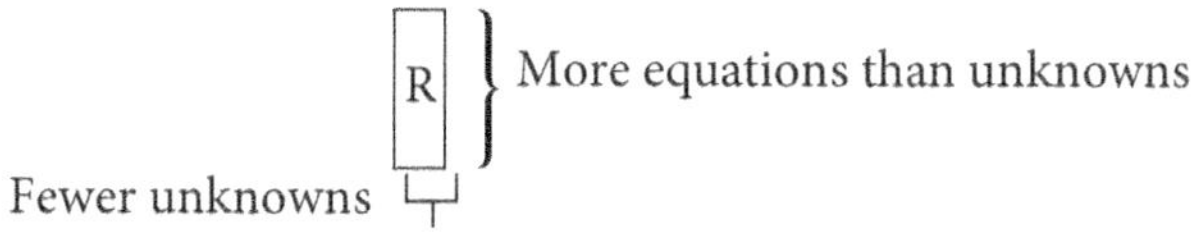

3.6.4 Properties of non-homogeneous linear systems

Proposition (3.36). The linear system $\mathbf{Ax} = \mathbf{b}$ has a solution $\Leftrightarrow$ $\mathbf{b}$ can be generated by the column space of matrix $\mathbf{A}$. (Fig. 3.24)

Figure 3.24

Proof – Exercises 3.6.

■

For the homogeneous system, $\mathbf{Ax} = \mathbf{O}$, we have the zero vector $\mathbf{O}$ in the column space of matrix $\mathbf{A}$. However, in the non-homogeneous case, $\mathbf{Ax} = \mathbf{b}$, we cannot guarantee that the vector $\mathbf{b}$ is in the column space of matrix $\mathbf{A}$. The vector $\mathbf{b}$ must be in the *column space* in order for $\mathbf{Ax} = \mathbf{b}$ to have a solution.

Proposition (3.37). Let $\mathbf{A}$ be a m by n matrix and $\mathbf{b} \neq \mathbf{O}$ then the linear system $\mathbf{Ax} = \mathbf{b}$ has a solution $\Leftrightarrow$ $rank(\mathbf{A} \mid \mathbf{b}) = rank(\mathbf{A})$.

What does this proposition mean in simple English?
The ranks of augmented matrix $(\mathbf{A} \mid \mathbf{b})$ and $\mathbf{A}$ must be equal for $\mathbf{Ax} = \mathbf{b}$ to have a solution. Also, the other way, if $\mathbf{Ax} = \mathbf{b}$ has a solution then the ranks of $\mathbf{A}$ and $(\mathbf{A} \mid \mathbf{b})$ are equal.

Proof.
$(\Rightarrow)$. Let $\mathbf{Ax} = \mathbf{b}$ have a solution.

How do we prove $rank(\mathbf{A} \mid \mathbf{b}) = rank(\mathbf{A})$?
We show that the following are impossible:
(a) $rank(\mathbf{A} \mid \mathbf{b}) > rank(\mathbf{A})$ (b) $rank(\mathbf{A} \mid \mathbf{b}) < rank(\mathbf{A})$

(a) Suppose $rank(\mathbf{A} \mid \mathbf{b}) > rank(\mathbf{A})$ where $rank(\mathbf{A}) = p$ say. Then we have p non-zero rows in the row echelon form $\mathbf{R}$ of matrix $\mathbf{A}$ but *more than* p non-zero rows in the augmented matrix $(\mathbf{R} \mid \mathbf{b}')$:

$$\begin{array}{c} \\ p \text{ non-zero rows} \left\{ \begin{array}{c} \\ \\ \\ \end{array} \right. \\ \\ \\ \end{array} \begin{array}{c} \begin{array}{ccccccc} x_1 & x_2 & \cdots & & & x_n & \end{array} \\ \left(\begin{array}{cccccc|c} r_{11} & & & & & & b'_1 \\ 0 & \ddots & & & & & \vdots \\ \vdots & & r_{pj} & & \cdots & & b' \\ \hline 0 & 0 & 0 & \cdots & \cdots & 0 & b'_{p+1} \\ \hline \vdots & \vdots & \vdots & \cdots & \cdots & 0 & \vdots \end{array} \right) \end{array} \begin{array}{l} \\ \\ \\ \\ \leftarrow (p+1)\text{ th non-zero row} \\ \\ \end{array}$$

Expanding along the $(p + 1)$th row we have

$$0x_1 + 0x_2 + \cdots + 0x_{n-1} + 0x_n = b'_{p+1} \quad (*)$$

However, b'_{p+1} *cannot* equal zero.

Why not?
Because by our supposition $rank(\mathbf{A} \mid \mathbf{b}) > rank(\mathbf{A}) = p$ which means $(p + 1)th$ row must be non-zero, therefore $b'_{p+1} \neq 0$.

It is impossible to have the solution (*) because the left hand side is zero but the right hand is non-zero. We have a contradiction, which means that our supposition $rank(\mathbf{A} \mid \mathbf{b}) > rank(\mathbf{A})$ must be wrong.

(b) We *cannot* have $rank(\mathbf{A} \mid \mathbf{b}) < rank(\mathbf{A})$ or $rank(\mathbf{A}) > rank(\mathbf{A} \mid \mathbf{b})$.

Why not?

Because this would mean we have:

Same matrix as **R** on the left.

$$\mathrm{R} = \begin{pmatrix} r_{11} & & r_{1n} \\ & & \\ \hline 0\ldots & & 0 \end{pmatrix} \Big\}\ \text{Non-zero rows} \qquad (\mathrm{R}|\mathrm{b}') = \left(\begin{array}{ccc|c} r_{11} & & r_{1n} & b_1' \\ \hline 0\ldots & & 0 & 0 \\ 0\ldots & & 0 & 0 \end{array}\right) \Big\}\ \text{Non-zero rows}$$

The number of non-zero rows in matrix **R** is *greater than* the number of non-zero rows in the augmented matrix $(\mathbf{R} \mid \mathbf{b}')$. This is impossible because the same matrix **R** *cannot* have more non-zero rows in matrix **R** than $(\mathbf{R} \mid \mathbf{b}')$.

Since both $rank(\mathbf{A} \mid \mathbf{b}) > rank(\mathbf{A})$ and $rank(\mathbf{A} \mid \mathbf{b}) < rank(\mathbf{A})$ are false, we must have $rank(\mathbf{A}) = rank(\mathbf{A} \mid \mathbf{b})$

($\Leftarrow$). We assume $rank(\mathbf{A}) = rank(\mathbf{A} \mid \mathbf{b})$. Writing this out we have

$$rank\begin{pmatrix} r_{11} & \cdots & & r_{1n} \\ \hline 0 & 0 & \cdots & 0 \\ \vdots & & & \vdots \end{pmatrix} = rank\left(\begin{array}{cccc|c} \mathbf{c}_1 & \mathbf{c}_2 & \cdots & \mathbf{c}_n & \\ r_{11} & \cdots & & r_{1n} & b_1 \\ & & & & \vdots \\ \hline 0 & 0 & \cdots & 0 & 0 \\ \vdots & & & \vdots & \vdots \end{array}\right)$$

Expanding out the augmented matrix in terms of equations, we have **b** is in the column space of matrix **A**. By the above Proposition (3.36) the linear system $\mathbf{Ax} = \mathbf{b}$ has a solution

■

The above Proposition (3.37) can be used as a test to see if there are any solutions to the given linear system. Note that the above proposition is also saying:

$$rank(\mathbf{A} \mid \mathbf{b}) \neq rank(\mathbf{A}) \Leftrightarrow \text{No solution}$$

But how do we test for a unique solution or an infinite number of solutions?

The next proposition gives the number of solutions to a non-homogeneous linear system.

Proposition (3.38). Consider the linear system $\mathbf{Ax} = \mathbf{b}$ where **A** has n columns and $\mathbf{b} \neq \mathbf{O}$. (**A** has n columns means that there are n unknowns in the system.)

(a) $rank(\mathbf{A}) = rank(\mathbf{A} \mid \mathbf{b}) = n$ (Full rank) $\Leftrightarrow$ the linear system has a *unique* solution.

(b) $rank(\mathbf{A}) = rank(\mathbf{A} \mid \mathbf{b}) < n \Leftrightarrow$ the linear system has an *infinite* number of solutions.

(c) $rank(\mathbf{A}) \neq rank(\mathbf{A} \mid \mathbf{b}) \Leftrightarrow$ the linear system has *no* solution.

Proof – Exercises 3.6. ■

Example 3.39

Determine whether the following systems have infinite, unique or no solutions (you do not need to find them):

(a) $\begin{aligned} x - y - 2z - 3w &= 5 \\ -4x + 4y + 8z + 12w &= 2 \end{aligned}$ **(b)** $\begin{aligned} x + 2y + 3z &= 1 \\ 4x + 5y + 6z &= 2 \\ 7x + 8y + 8z &= 3 \end{aligned}$

Solution

(a) Writing out the coefficient matrix, $\mathbf{A}$, augmented matrix, $(\mathbf{A} \mid \mathbf{b})$, and placing the given matrices into reduced row echelon form we have:

$$\mathbf{A} = \begin{pmatrix} 1 & -1 & -2 & -3 \\ -4 & 4 & 8 & 12 \end{pmatrix} \Longrightarrow \begin{pmatrix} 1 & -1 & -2 & -3 \\ 0 & 0 & 0 & 0 \end{pmatrix} \begin{matrix} \leftarrow \text{non-zero row} \\ \end{matrix}$$

We have $rank(\mathbf{A}) = 1$.

$$(\mathbf{A} \mid \mathbf{b}) = \left(\begin{array}{cccc|c} 1 & -1 & -2 & -3 & 5 \\ -4 & 4 & 8 & 12 & 2 \end{array}\right) \Longrightarrow \left(\begin{array}{cccc|c} 1 & -1 & -2 & -3 & 0 \\ 0 & 0 & 0 & 0 & 1 \end{array}\right) \begin{matrix} \leftarrow \text{non-zero row} \\ \leftarrow \text{non-zero row} \end{matrix}$$

Hence $rank(\mathbf{A} \mid \mathbf{b}) = 2$.

We have $rank(\mathbf{A}) = 1$ but $rank(\mathbf{A} \mid \mathbf{b}) = 2$, so $rank(\mathbf{A})$ does not equal $rank(\mathbf{A} \mid \mathbf{b})$. Therefore the given system has *no* solution.

(b) Similarly we have

$$(\mathbf{A} \mid \mathbf{b}) = \left(\begin{array}{ccc|c} 1 & 2 & 3 & 1 \\ 4 & 5 & 6 & 2 \\ 7 & 8 & 8 & 3 \end{array}\right) \Longrightarrow \left(\begin{array}{ccc|c} 1 & 0 & -1 & -1/3 \\ 0 & 1 & 2 & 2/3 \\ 0 & 0 & 1 & 0 \end{array}\right) \begin{matrix} \leftarrow \text{non-zero row} \\ \leftarrow \text{non-zero row} \\ \leftarrow \text{non-zero row} \end{matrix}$$

What is the rank of $\mathbf{A}$ *and* $(\mathbf{A} \mid \mathbf{b})$ *equal to?*

In both cases, we have $rank(\mathbf{A}) = rank(\mathbf{A} \mid \mathbf{b}) = 3$ which means it is of full rank because the matrix $\mathbf{A}$ has three columns, so we have a unique solution.

The matrix $\mathbf{A}$ in part (b) is of full rank so by proposition (3.30):

Let $\mathbf{A}$ be a n by n matrix. The matrix $\mathbf{A}$ is invertible $\Leftrightarrow rank(\mathbf{A}) = n$.

The matrix $\mathbf{A}$ is invertible.

We can also deduce the following results:

Proposition (3.39). Let $\mathbf{A}$ be an n by n matrix then the following are equivalent:

(a) Matrix $\mathbf{A}$ is invertible.

(b) $rank(\mathbf{A}) = n$ (full rank)

(c) Null space of matrix $\mathbf{A}$ only contains the zero vector.

(d) $nullity(\mathbf{A}) = 0$

(e) Rows of matrix $\mathbf{A}$ are linearly independent. (See question 9 of Exercises 3.5.)

(f) Columns of matrix $\mathbf{A}$ are linearly independent. (See question 10 of Exercises 3.5.)

It can be shown that a basis for the column space of a matrix is given by the columns containing the leading 1's in reduced row echelon form:

Proposition (3.40). The columns which have leading 1's of a matrix **A** form a basis for the column space of **A**.

Proof – Website.

■

Summary

(3.31) The set of vectors **x** which satisfy the homogeneous system $\mathbf{Ax} = \mathbf{O}$ is called the null space of **A**, and the dimension of this space is called the nullity of matrix **A**.

(3.33) If the matrix **A** has n columns then $nullity(\mathbf{A}) + rank(\mathbf{A}) = n$.

A non-homogeneous system $\mathbf{Ax} = \mathbf{b}$ has the general two part solution:

$$\mathbf{x} = \text{(homogeneous solution)} + \text{(particular solution)}$$

EXERCISES 3.6

(Brief solutions at end of book. Full solutions available at <http://www.oup.co.uk/companion/singh>.)

1. Determine and sketch the null space of the following matrices:

(a) $\begin{pmatrix} 1 & 0 \\ 0 & 1 \end{pmatrix}$ (b) $\begin{pmatrix} 1 & 2 \\ 2 & 4 \end{pmatrix}$ (c) $\begin{pmatrix} 1 & 0 \\ 1 & 2 \\ 6 & 10 \end{pmatrix}$ (d) $\begin{pmatrix} 1 & 2 \\ 3 & 4 \\ 5 & 6 \end{pmatrix}$

2. Determine the solution of the following homogeneous systems:

(a) $\begin{aligned} x - 2y - 3z &= 0 \\ 4x - 5y - 6z &= 0 \\ 7x - 8y - 9z &= 0 \end{aligned}$ (b) $\begin{aligned} 2x - 2y - 2z &= 0 \\ 4x - 4y - 4z &= 0 \\ 8x - 8y - 8z &= 0 \end{aligned}$

(c) $\begin{aligned} 2x + 9y - 3z &= 0 \\ 5x + 6y - z &= 0 \\ 9x + 8y - 9z &= 0 \end{aligned}$ (d) $\begin{aligned} -3x + y - z &= 0 \\ 2x + 5y - 7z &= 0 \\ 4x + 8y - 4z &= 0 \end{aligned}$

3. Determine the null space, rank and nullity of the following matrices:

(a) $\mathbf{A} = \begin{pmatrix} 1 & -2 & -3 \\ 4 & -5 & -6 \\ 7 & -8 & -9 \end{pmatrix}$ (b) $\mathbf{B} = \begin{pmatrix} 2 & -2 & -2 \\ 4 & -4 & -4 \\ 8 & -8 & -8 \end{pmatrix}$

(c) $\mathbf{C} = \begin{pmatrix} 2 & 9 & -3 \\ 5 & 6 & -1 \\ 9 & 8 & -9 \end{pmatrix}$ (d) $\mathbf{D} = \begin{pmatrix} -3 & 1 & -1 \\ 2 & 5 & -7 \\ 4 & 8 & -4 \end{pmatrix}$

4. Determine the null space of the matrix:

$$\mathbf{A} = \begin{pmatrix} 1 & 2 & 3 & 4 & 5 & 6 & 7 \\ 8 & 9 & 10 & 11 & 12 & 13 & 14 \\ 15 & 16 & 17 & 18 & 19 & 20 & 21 \\ 22 & 23 & 24 & 25 & 26 & 27 & 28 \\ 29 & 30 & 31 & 32 & 33 & 34 & 35 \end{pmatrix}$$

5. Determine bases for row, column and null space of the following matrices. Also state their rank and nullity.

(a) $\mathbf{A} = \begin{pmatrix} 1 & -4 & -9 \\ 2 & 5 & -7 \end{pmatrix}$ (b) $\mathbf{B} = \begin{pmatrix} 1 & 3 \\ 2 & 5 \\ -14 & -37 \end{pmatrix}$ (c) $\mathbf{C} = \begin{pmatrix} 1 & 3 & -9 & 5 \\ 2 & 6 & 7 & 1 \\ 1 & 3 & -8 & 1 \end{pmatrix}$

6. Solve the following non-homogeneous system of linear equations:

(a) $\begin{aligned} 2x + 5y + 7z + 10w &= 3 \\ x + y + 2z + 5w &= 6 \end{aligned}$ (b) $\begin{aligned} 2x - y - 4z &= 13 \\ 3x + 3y - 5z &= 13 \\ 3x - 4y + 10z &= -10 \end{aligned}$

7. Determine whether the following systems have infinite, unique or no solutions:

(a) $\begin{aligned} 2x + 8y &= 12 \\ 7x + 28y &= 42 \end{aligned}$ (b) $\begin{aligned} 2x - 3y - 6z + 12w &= 2 \\ 3x - 5y - 7z + 16w &= 5 \end{aligned}$

(c) $\begin{aligned} x + 2y + 3z &= 1 \\ 4x + 5y + 6z &= 2 \\ 7x + 8y + 9z &= 4 \end{aligned}$ (d) $\begin{aligned} 2x + 5y - 3z - 7w &= 0 \\ x + y - 4z - 8w &= 9 \\ 3x + 4y + w &= 6 \\ 5x + 21y - z + 3w &= 2 \end{aligned}$

8. Check to see if the following vectors **u** are in the null space of corresponding matrices:

(a) $\mathbf{u} = \begin{pmatrix} 1 \\ 2 \\ 3 \end{pmatrix}$, $\mathbf{A} = \begin{pmatrix} 1 & 1 & -1 \\ 2 & -1 & 0 \\ 5 & 2 & -3 \end{pmatrix}$ (b) $\mathbf{u} = \begin{pmatrix} 1 \\ 1 \\ 1 \end{pmatrix}$, $\mathbf{B} = \begin{pmatrix} 1 & 4 & -5 \\ -7 & 5 & 2 \end{pmatrix}$

(c) $\mathbf{u} = \begin{pmatrix} 2 \\ 1 \\ 5 \end{pmatrix}$, $\mathbf{C} = \begin{pmatrix} 1 & 2 & 3 & 4 \\ 5 & 6 & 7 & 8 \end{pmatrix}$ (d) $\mathbf{u} = \begin{pmatrix} 1 \\ 3 \\ -4 \\ 7 \end{pmatrix}$, $\mathbf{D} = \begin{pmatrix} 1 & -2 & 6 & 1 \\ 3 & -6 & 7 & 8 \\ 5 & 2 & 1 & 7 \\ 1 & 6 & 3 & 2 \end{pmatrix}$

9. Prove that a n by n matrix **A** is invertible $\Leftrightarrow$ *nullity* $(\mathbf{A}) = 0$.

10. Prove that elementary row operations do *not* change the null space of a matrix.

11. Let **A** be an m by n matrix. Prove that every vector in the null space of matrix **A** is orthogonal to every vector in the row space of matrix **A**.

12. Prove Proposition (3.32).

13. Prove Proposition (3.36).

14. Prove Proposition (3.38).

MISCELLANEOUS EXERCISES 3

In this exercise you may check your numerical answers using MATLAB.

3.1. Find a basis for the null space of **A**, where

$$\mathbf{A} = \begin{bmatrix} 1 & -2 & 1 & 1 \\ -1 & 2 & 0 & 1 \\ 2 & -4 & 1 & 0 \end{bmatrix}$$

University of Western Ontario, Canada

3.2. Let **A** be a 7×5 matrix, and suppose that the homogeneous linear system $\mathbf{AX} = \mathbf{O}$ is uniquely solvable. Answer the following questions:

(i) What is the rank of **A**?

(ii) If the linear system $\mathbf{AX} = \mathbf{B}$ is solvable, is it then uniquely solvable?

A. $rank(\mathbf{A}) = 7$; no.
B. $rank(\mathbf{A}) = 5$; yes.
C. $rank(\mathbf{A}) = 7$; yes.
D. $rank(\mathbf{A}) = 2$; yes.
E. $rank(\mathbf{A}) = 5$; no.
F. $rank(\mathbf{A}) = 2$; no.

University of Ottawa, Canada

3.3. If we denote by $C_1,\ C_2,\ C_3,\ C_4,\ C_5$ the columns of the matrix

$$\mathbf{A} = \begin{bmatrix} 1 & 3 & -2 & 0 & 2 \\ 2 & 6 & -5 & -2 & 4 \\ 0 & 0 & 5 & 10 & 0 \\ 2 & 6 & 0 & 8 & 4 \end{bmatrix}$$

then a basis of the column space of **A** is
A. $C_1,\ C_2,\ C_3$
B. $C_1,\ C_3,\ C_4$
C. $C_1,\ C_3$
D. $C_1,\ C_2$
E. $C_1,\ C_3,\ C_5$
F. $C_1,\ C_4$

University of Ottawa, Canada

3.4. Let **A** be an $m \times n$ matrix, let **b** be a vector in $\mathbb{R}^m$, and suppose that **v** is a solution of $\mathbf{Ax} = \mathbf{b}$.

(a) Prove that if **w** is a solution of $\mathbf{Ax} = \mathbf{O}$, then $\mathbf{v} + \mathbf{w}$ is a solution of $\mathbf{Ax} = \mathbf{b}$.

(b) Prove that for any solution **u** to $\mathbf{Ax} = \mathbf{b}$, there is a solution to $\mathbf{Ax} = \mathbf{O}$.

Illinois State University, USA

3.5. Let **A** be an $n \times n$ matrix such that $\mathbf{Ax} = \mathbf{b}$ has exactly one solution for each $\mathbf{b}$ in $\mathbb{R}^n$. What three other things (or facts) can you say about **A**.

Illinois State University, USA

3.6. For each of the following situations, provide an example of the item requested, or explain why such an example could not exist.

(a) A homogeneous linear system of equations with no solutions.

(b) Two matrices, **A** and **B**, that are not the same size but so that $\mathbf{AB} = \mathbf{BA}$.

(c) A set of four vectors that spans $\mathbb{R}^4$ but is not a basis for $\mathbb{R}^4$.

(d) A three-dimensional subspace of the vector space formed by the set of 2×2 matrices.

Saint Michael's College, Vermont, USA
(part question)

3.7. Which of the following are true?

(a) If V is a vector space of dimension n, then every set of n linearly independent vectors in V is a spanning set of V.

(b) P_2 contains a basis of polynomials p satisfying $p(0) = 2$.

(c) $\{1, \sin^2(x), \cos^2(x)\}$ are linearly independent subset of $F[0, 2\pi]$.

University of Ottawa, Canada

3.8. Let V be a vector space. Which of the following statements are true?

(a) If $\{\mathbf{u}, \mathbf{v}, \mathbf{w}\}$ is a linearly independent subset of V, then also $\{\mathbf{u}, \mathbf{v}\}$ is linearly independent.

(b) Every spanning set of V contains a basis of V.

(c) If $\dim(V) = n$ then every set of n linearly independent vectors of V is a basis.

University of Ottawa, Canada

3.9. It is given that $\mathbf{A} = \begin{bmatrix} 1 & 0 & -2 & 1 & 3 \\ -1 & 1 & 5 & -1 & -3 \\ 0 & 2 & 6 & 0 & 1 \\ 1 & 1 & 1 & 1 & 4 \end{bmatrix}$ and its reduced row echelon form is given by $\mathbf{B} = \begin{bmatrix} 1 & 0 & -2 & 1 & 0 \\ 0 & 1 & 3 & 0 & 0 \\ 0 & 0 & 0 & 0 & 1 \\ 0 & 0 & 0 & 0 & 0 \end{bmatrix}$.

(a) Find the rank of **A**.

(b) Find the nullity of **A**.

(c) Find a basis for the column space of **A**.

(d) Find a basis for the row space of **A**.

(e) Find a basis for the null space of **A**.

Purdue University, USA

3.10. (a) Let $\mathbf{u}_1, \ldots, \mathbf{u}_k$ be vectors in a subspace U of $\mathbb{R}^n$. Define what it means

(i) for $\mathbf{u}_1, \ldots, \mathbf{u}_k$ to be linearly independent

(ii) for $\mathbf{u}_1, \ldots, \mathbf{u}_k$ to span U

(b) Determine the null space of the matrix

$$\mathbf{A} = \begin{pmatrix} 1 & 3 & -4 \\ 2 & -1 & -1 \\ -2 & -6 & 8 \end{pmatrix}$$

(c) Determine whether the vectors

$$\mathbf{v}_1 = \begin{bmatrix} -5 \\ 4 \\ -6 \end{bmatrix}, \mathbf{v}_2 = \begin{bmatrix} -1 \\ 2 \\ -1 \end{bmatrix}, \mathbf{v}_3 = \begin{bmatrix} 0 \\ -2 \\ 3 \end{bmatrix}$$

in $\mathbb{R}^3$ are linearly independent or linearly dependent. If they are linearly dependent, express one of the vectors as a linear combination of the other two.

University of Southampton, UK
(part question)

3.11. (a) When is a subset V of $\mathbb{R}^n$ a *subspace*? Give two distinct examples of subspaces of $\mathbb{R}^n$.

(b) What is meant by saying that the vectors $\mathbf{v}_1, \mathbf{v}_2, \ldots, \mathbf{v}_s$ in $\mathbb{R}^n$ are linearly independent?

How is the linear span of the vectors $\mathbf{v}_1, \mathbf{v}_2, \ldots, \mathbf{v}_s$ defined?

(c) Show that the following vectors in $\mathbb{R}^3$ are linearly dependent:

$$\mathbf{v}_1 = (\ 1,\ 1,\ 1\ ),\ \mathbf{v}_2 = (\ 2,\ 3,\ -2\ ),\ \mathbf{v}_3 = (\ -2\ \ -5\ \ 10\ )$$

What is the dimension of the linear span of $\mathbf{v}_1, \mathbf{v}_2, \mathbf{v}_3$?

University of Sussex, UK

3.12. Consider the following matrix $\mathbf{A}$ and vector $\vec{b}$:

$$\mathbf{A} = \begin{bmatrix} 2 & 6 & -4 \\ -2 & -5 & 4 \\ 4 & 11 & -8 \end{bmatrix}, \quad \vec{b} = \begin{bmatrix} 0 \\ 1 \\ -1 \end{bmatrix}$$

(a) By using row reduction, reduce the augmented matrix $\begin{bmatrix} \mathbf{A} & \vec{b} \end{bmatrix}$.

(b) Using part (a) of this question, explain why $\mathbf{A}$ and $\begin{bmatrix} \mathbf{A} & \vec{b} \end{bmatrix}$ have the same rank.

(c) Find in vector form the general (or complete) solution to $\mathbf{A}\vec{x} = \vec{b}$.
[Note that $\vec{b}$ is the vector $\mathbf{b}$].

University of New Brunswick, Canada

3.13. Compute the reduced row echelon form of the matrix

$$\mathbf{A} = \begin{pmatrix} 1 & 2 & 3 & 1 & 2 & 3 \\ 2 & 4 & 6 & 1 & 2 & 5 \\ 3 & 6 & 9 & 4 & 8 & 3 \end{pmatrix} \in M_{3\times 6}(\mathbb{R})$$

Determine a basis of the null space $N(A)$.

RWTH Aachen University, Germany

3.14. (a) Define the terms *null- space* $N(A)$ *of* $\mathbf{A}$, *column space* $C(A)$ *of* $\mathbf{A}$, *rank* $\mathbf{A}$, *nullity* $\mathbf{A}$, where $\mathbf{A} \in M_{m\times n}(\mathbb{R})$.

(b) For the following matrix $\mathbf{A}$, find the reduced row echelon form of $\mathbf{A}$, a basis for $N(A)$ and rank $\mathbf{A}$ and nullity $\mathbf{A}$:

$$\mathbf{A} = \begin{bmatrix} 1 & -1 & 2 & 0 \\ 2 & -2 & 1 & 3 \\ -2 & 2 & -7 & 3 \end{bmatrix}$$

Also explain why $(1, 5, 1)^T$ and $(0, 3, 3)^T$ form a basis for $C(A)$.

University of Queensland, Australia

3.15. (a) Determine whether the following subsets are subspaces (giving reasons for your answers).

(i) $U = \left\{ \begin{pmatrix} a & b \\ c & d \end{pmatrix} \in M(2, 2) \;\middle|\; a^2 = d^2 \right\}$ in $M(2, 2)$
[$M(2, 2)$ is the vector space M_{22}].

(ii) $V = \{p(x) \in P_n \mid p(3) = 0\}$ in P_n

(iii) $W = \{(x, y, z, t) \in \mathbb{R}^4 \mid y = z + t\}$ in $\mathbb{R}^4$

(b) Find a basis for the real vector space $\mathbb{R}^3$ containing the vector $(3, 5, -4)$.

(c) Do the following sets form a basis for V? If not, determine whether they are linearly independent, a spanning set for V, or neither.

(i) $\{(1, 0, 1), (1, 1, 0), (0, 1, 1), (1\ 1\ 1)\}$ for $V = \mathbb{R}^3$.

(ii) $\{5, 2 + x - 3x^2, 4x - 1\}$ for $V = P_2$

City University, London, UK

3.16. Let V and W be two subspaces of $\mathbb{R}^n$. Define the set

$$V + W = \{\mathbf{u} \in \mathbb{R}^n \mid \mathbf{u} = \mathbf{v} + \mathbf{w} \text{ for some } \mathbf{v} \in V \text{ and some } \mathbf{w} \in W\}.$$

Prove that $V + W$ is a subspace of $\mathbb{R}^n$.

Illinois State University, USA
(part question)

3.17. For the given set S and vector $\mathbf{v}$, determine if $\mathbf{v}$ is in the span of S.

(a) $S = \left\{ \begin{bmatrix} 1 & 0 \\ -1 & 0 \end{bmatrix}, \begin{bmatrix} 0 & 1 \\ 0 & 1 \end{bmatrix}, \begin{bmatrix} 1 & 1 \\ 0 & 0 \end{bmatrix}, \begin{bmatrix} 0 & 0 \\ 1 & 1 \end{bmatrix} \right\}, \mathbf{v} = \begin{bmatrix} 3 & 1 \\ -1 & 3 \end{bmatrix}$

(b) $S = \{1 + x,\ x + x^2,\ x + x^3,\ 1 + x + x^2 + x^3\}$, $\mathbf{v} = 2 - 3x + 4x^2 + x^3$

Illinois State University, USA

3.18. Let $\mathbf{B} = \begin{bmatrix} 1 & 2 & 2 & 8 \\ 2 & 4 & 4 & 13 \\ 1 & 1 & 1 & 5 \end{bmatrix}$.

Reduce **B** to reduced row echelon form by performing appropriate elementary row operations. Hence, or otherwise, find the following.

(a) A basis for $NULL(\mathbf{B})$

(b) A basis for $Col(\mathbf{B})$

(c) The nullity of **B**.

(d) The rank of **B**.

University of Sydney, Australia

3.19. A square matrix **A** is called *skew-symmetric* if $\mathbf{A}^T = -\mathbf{A}$. Prove the set of n-by-n skew-symmetric matrices is a subspace of all n-by-n matrices.

Harvey Mudd College, California, USA

3.20. Consider the vectors:

$$\begin{aligned} \mathbf{v}_1 &= (\ 2\ \ 4\ \ 1\ \ 1\ ) \\ \mathbf{v}_2 &= (\ 4\ \ 7\ \ 2\ \ 2\ ) \\ \mathbf{v}_3 &= (\ 6\ \ 8\ \ 7\ \ 5\ ) \end{aligned}$$

(a) Are these vectors linearly independent? Justify.

(b) Let these vectors span a vector space V. Find the dimension of V and a set of basis vectors for V.

(c) Does the vector $\mathbf{v} = [\ 2\ \ 1\ \ 1\ \ 1\ ]$ belong to the vector space V. Justify.

(d) Let $\mathbf{v}_1$, $\mathbf{v}_2$ and $\mathbf{v}_3$ be arranged as the first, second and third rows of a 3×4 matrix **A**. Let the null space of **A** be the vector space W. Determine the dimension of W.

(e) Find a set of basis vectors for W.

National University of Singapore

3.21. If **A** is a 5×3 matrix, show that the rows of **A** are linearly dependent.

Clark University, USA

3.22. Can a 4×10 matrix **A** have a null space of dimension 2? Why or why not?

University of South Carolina, USA
(part question)

3.23. Start with this 2 by 4 matrix

$$\mathbf{A} = \begin{bmatrix} 2 & 3 & 1 & -1 \\ 6 & 9 & 3 & -2 \end{bmatrix}$$

(a) Find all special solutions $\mathbf{Ax} = \mathbf{O}$ and **describe the null space** of **A**.

(b) Find the complete solutions – meaning all solutions (x_1, x_2, x_3, x_4) – to

$$\mathbf{Ax} = \begin{bmatrix} 2x_1 + 3x_2 + x_3 - x_4 \\ 6x_1 + 9x_2 + 3x_3 - 2x_4 \end{bmatrix} = \begin{bmatrix} 1 \\ 2 \end{bmatrix}$$

(c) When an m by n matrix $\mathbf{A}$ has rank $r = m$, the system $\mathbf{Ax} = \mathbf{b}$ can be solved for which b(best answer)? How many special solutions to $\mathbf{Ax} = \mathbf{O}$?

Massachusetts Institute of Technology, USA

3.24. Find a real number c such that the vectors $\begin{pmatrix} 1 \\ 2 \\ 4 \end{pmatrix}, \begin{pmatrix} 2 \\ 3 \\ 5 \end{pmatrix}, \begin{pmatrix} 2 \\ 7 \\ c \end{pmatrix}$ do not form a basis of $\mathbb{R}^3$.

University of California, Berkeley, USA
(part question)

3.25. Find a 3×3 matrix whose null space (kernel) is the span of the vector $(1, 2, 3)$, and whose column space (image) is the span of the vectors $(4, 5, 6)$ and $(7, 8, 9)$.

University of California, Berkeley, USA

3.26. (a) Show that $\{\cos(x), \cos(2x), \cos(3x)\}$ is linearly independent.

(b) Show that $(1+x)^3, (1-x)^3, (1+2x)^3, (1-2x)^3, (1+3x)^3$ are linearly dependent functions. State clearly any general fact about vector spaces that you use to justify your assertion.

Hint: First show that these functions are in the subspace spanned by $1, x, x^2, x^3$.

McGill University, Canada

3.27. (a) Prove or disprove the following statement:

$Span((1, 2, -1, -2), (2, 1, 2, -1)) = Span((-1, 4, -7, -4), (8, 7, 4, -7))$

(b) If $\mathbf{u}, \mathbf{v}, \mathbf{w}$ are linearly independent vectors in $\mathbb{R}^n$ for which values of k are the vectors $k\mathbf{u} + \mathbf{v}, \mathbf{v} + k\mathbf{w}, \mathbf{w} + k\mathbf{u}$ are linearly independent?

McGill University, Canada

3.28. True or False? Give a complete justification for your answers.

(a) Let $\mathbf{A}$ be an $n \times n$ matrix. If there is a vector $\mathbf{y} \in \mathbb{R}^n$ so that the equation $\mathbf{Ax} = \mathbf{y}$ has more than one solution, then the columns of $\mathbf{A}$ span $\mathbb{R}^n$.

(b) The subset $H = \{(x, y) \in \mathbb{R}^2 \mid y = x^2\}$ of $\mathbb{R}^2$ is a vector subspace of $\mathbb{R}^2$.

(c) Let $\mathbf{A}$ be a 5×6 matrix. Then the null space of $\mathbf{A}$ is a vector subspace of $\mathbb{R}^5$.

University of Maryland, Baltimore County, USA

3.29. Suppose that $\mathbf{v}_1, \ldots, \mathbf{v}_n$ are vectors in $\mathbb{R}^n$ and that $\mathbf{A}$ is an $n \times n$ matrix. If $\mathbf{Av}_1, \ldots, \mathbf{Av}_n$ form a basis of $\mathbb{R}^n$, show that $\mathbf{v}_1, \ldots, \mathbf{v}_n$ form a basis of $\mathbb{R}^n$ and that $\mathbf{A}$ is invertible.

University of California, Berkeley, USA

Janet Drew

is Professor of Astrophysics and Assistant Dean of School (Research) at the University of Hertfordshire, UK.

Tell us about yourself and your work.

I am a research astrophysicist and university professor who qualified via a physics undergraduate degree and astrophysics PhD. Over the past 30 years I have worked in the UK successively at University College London, Cambridge, Oxford and Imperial College – moving to Hertfordshire a few years ago. For two-thirds of my research career I have operated as a theoretical astronomer specialising in the modelling of ionised outflows from hot stars and accretion disks in various settings. More recently the emphasis has shifted to observational astronomy through leading the international consortium of scientists carrying out digital optical broadband surveys, with narrowband H-alpha, of the Milky Way (see e.g. http://www.iphas.org).

How do you use linear algebra in your job?

It was more important to me as a theoretical astrophysicist, because then the modelling involved the numerical solution of mildly non-linear systems of simultaneous equations, generally rendered in linear form to make them tractable. Although less important now, it hasn't gone away because it turns out that one of the best methods of establishing a global uniform calibration to wide-field astronomical survey data is to set up and solve a large system of simultaneous equations (of dimension several thousand).

How important is linear algebra?

Very. The need to obtain reliable solutions to systems of linear equations is extremely common in the world of physical science. It's everywhere, frankly. As I set out on my research career, after obtaining my PhD, one of the first book purchases I needed to make was a textbook in numerical analysis (the real-world computational application of linear algebra, essentially) to sit by my side as I found ways to model radiation transport effects in the statistical equilibrium of many-level hydrogen atoms. Hydrogen, incidentally, is far and away the most abundant and therefore most important element in the Universe – it has to be described well in understanding its gas content both in stars and in the space in between them.

What are the challenges connected with the subject?

If I may subtly redefine the subject as numerical analysis, one of the challenges is to find ways to test that you really are solving the system(s) of equations you want to solve in non-trivial cases. The issue of convergence (and efficiency) arises too, for schemes that aim to deal with non-linearity through iteration. It doesn't usually take too long to set up an algorithm, but it takes a whole heap longer to validate them.

Have you any particular messages that you would like to give to students starting off studying linear algebra?

. . . Just that it really has more applications than you can even begin to imagine when you first meet the nuts-and-bolts mathematics at school and undergraduate level. Matrices are your friends: treat them nicely and they will do so much for you.

4 Inner Product Spaces

SECTION 4.1 Introduction to Inner Product Spaces

By the end of this section you will be able to

- understand what is meant by an inner product space
- prove some properties of inner product spaces
- define and prove properties of the norm of a vector

In chapter 2, we introduced the idea of a dot product for a Euclidean n-space. In this chapter, we extend the concept of a dot product to general vector spaces. The general operation that will replace the dot product in general vector spaces is called an *inner product*. In fact, the dot product was an example of a specific operation more commonly referred to as an inner product, in a Euclidean n-space.

General vector spaces such as the set of polynomials, matrices and continuous functions are fundamentally identical in structure to the Euclidean n-space. Because of this, all of the proofs we introduced for the dot product in Euclidean n-space hold for *any* inner product space. An inner product space is a vector space with additional structure which caters for distance, angle and projection of vectors. We need inner products to define these three notions for general vector spaces.

So far we have only looked at the fundamental linear algebra operations of scalar multiplication and vector addition. We have not mentioned anything about an inner product that involves multiplication of vectors with each other in the general vector space, or inner space.

4.1.1 Definition of inner product

How did we define the inner or dot product in chapter 2?

Let $\mathbf{u} = \begin{pmatrix} u_1 \\ \vdots \\ u_n \end{pmatrix}$ and $\mathbf{v} = \begin{pmatrix} v_1 \\ \vdots \\ v_n \end{pmatrix}$ be vectors in $\mathbb{R}^n$ then the inner product of $\mathbf{u}$ and $\mathbf{v}$ denoted by $\mathbf{u} \cdot \mathbf{v}$ is

(2.4) $$\mathbf{u} \cdot \mathbf{v} = \mathbf{u}^T\mathbf{v} = u_1v_1 + u_2v_2 + u_3v_3 + \cdots + u_nv_n$$

Remember, the answer was a scalar *not* a vector. This inner product was named the dot product (also called the scalar product) in $\mathbb{R}^n$. This is the usual (or standard) inner product in $\mathbb{R}^n$ but there are many other types of inner products in $\mathbb{R}^n$.

For the general vector space, the inner product is denoted by $\langle \mathbf{u},\ \mathbf{v} \rangle$ rather than $\mathbf{u} \cdot \mathbf{v}$. For the general vector space, the definition of inner product is based on Proposition (2.6) of chapter 2 and is given by:

Definition (4.1). An inner product on a real vector space V is an operation which assigns to each pair of vectors, $\mathbf{u}$ and $\mathbf{v}$, a *unique* real number $\langle \mathbf{u},\ \mathbf{v} \rangle$ which satisfies the following axioms for *all* vectors $\mathbf{u}$, $\mathbf{v}$ and $\mathbf{w}$ in V and *all* scalars k.

(i) $\langle \mathbf{u},\ \mathbf{v} \rangle = \langle \mathbf{v},\ \mathbf{u} \rangle$ [commutative law]

(ii) $\langle \mathbf{u} + \mathbf{v},\ \mathbf{w} \rangle = \langle \mathbf{u},\ \mathbf{w} \rangle + \langle \mathbf{v},\ \mathbf{w} \rangle$ [distributive law]

(iii) $\langle k\mathbf{u},\ \mathbf{v} \rangle = k\langle \mathbf{u},\ \mathbf{v} \rangle$ [taking out the scalar k]

(iv) $\langle \mathbf{u},\ \mathbf{u} \rangle \geq 0$ and we have $\langle \mathbf{u},\ \mathbf{u} \rangle = 0 \Leftrightarrow \mathbf{u} = \mathbf{O}$ [Means the inner product between the same vectors is zero or positive.]

A real vector space which satisfies these axioms is called a **real inner product space**. Note that evaluating $\langle\ ,\ \rangle$ gives a real number (scalar) *not* a vector. Next we give some examples of inner product spaces.

4.1.2 Examples of inner products

Example 4.1

Show that the Euclidean n space, $\mathbb{R}^n$, with the dot product as defined in (2.4) on the previous page is indeed an example of an inner product space.

Solution

See Proposition (2.6) of chapter 2.

The dot product is just one example of an inner product. In fact, there are an infinite number of ways we can define an inner product. The next example demonstrates another inner product on $\mathbb{R}^2$ defined by a matrix sandwiched between two vectors.

Example 4.2

Let the Euclidean 2 space, $\mathbb{R}^2$, be a vector space with

$$\langle \mathbf{u},\ \mathbf{w} \rangle = \mathbf{u}^T \mathbf{A} \mathbf{w} \text{ where } \mathbf{A} = \begin{pmatrix} 1 & 2 \\ 2 & 5 \end{pmatrix}$$

Show that this is an inner product for $\mathbb{R}^2$. [Similar to the dot product which is $\mathbf{u}^T\mathbf{w}$, but this time we are required to multiply the vector $\mathbf{w}$ by the matrix $\mathbf{A}$ and then carry out the dot product.]

Solution

Exercises 4.1.

Example 4.3

Let P_2 be the vector space of polynomials of degree 2 or less. Let

$$\mathbf{p} = c_0 + c_1x + c_2x^2 \text{ and } \mathbf{q} = d_0 + d_1x + d_2x^2$$

be polynomials in P_2. Show that the multiplication of coefficients:

$$\langle \mathbf{p}, \mathbf{q} \rangle = c_0d_0 + c_1d_1 + c_2d_2$$

defines an inner product on P_2.
[For example, if $\mathbf{p} = 2 + 3x + 5x^2$ and $\mathbf{q} = 4 - x + 7x^2$ then

$$\langle \mathbf{p}, \mathbf{q} \rangle = (2 \times 4) + (3 \times (-1)) + (5 \times 7) = 40]$$

Solution

How do we show $\langle \mathbf{p}, \mathbf{q} \rangle = c_0d_0 + c_1d_1 + c_2d_2$ is an inner product on P_2?
Checking *all* four axioms of Definition (4.1).

(i) We need to check $\langle \mathbf{p}, \mathbf{q} \rangle = \langle \mathbf{q}, \mathbf{p} \rangle$ holds for the multiplication of coefficients:

$$\begin{aligned}\langle \mathbf{p}, \mathbf{q} \rangle &= c_0d_0 + c_1d_1 + c_2d_2 \\ &= d_0c_0 + d_1c_1 + d_2c_2 \qquad \left[\begin{array}{l}\text{remember, the order of multiplication}\\ \text{does not matter}\end{array}\right] \\ &= \langle \mathbf{q}, \mathbf{p} \rangle\end{aligned}$$

(ii) We need to check $\langle \mathbf{p} + \mathbf{q}, \mathbf{r} \rangle = \langle \mathbf{p}, \mathbf{r} \rangle + \langle \mathbf{q}, \mathbf{r} \rangle$:
Adding the two quadratics $\mathbf{p} = c_0 + c_1x + c_2x^2$ and $\mathbf{q} = d_0 + d_1x + d_2x^2$ gives:

$$\mathbf{p} + \mathbf{q} = (c_0 + d_0) + (c_1 + d_1)x + (c_2 + d_2)x^2$$

Let $\mathbf{r} = e_0 + e_1x + e_2x^2$. Finding the inner product of $\mathbf{p} + \mathbf{q}$ and $\mathbf{r}$:

$$\begin{aligned}\langle \mathbf{p} + \mathbf{q}, \mathbf{r} \rangle &= \left\langle (c_0 + d_0) + (c_1 + d_1)x + (c_2 + d_2)x^2,\ e_0 + e_1x + e_2x^2 \right\rangle \\ &= (c_0 + d_0)e_0 + (c_1 + d_1)e_1 + (c_2 + d_2)e_2 \qquad [\text{multiplying coefficients}] \\ &= c_0e_0 + d_0e_0 + c_1e_1 + d_1e_1 + c_2e_2 + d_2e_2\end{aligned}$$

Evaluating the addition of the other inner products:

$$\begin{aligned}\langle \mathbf{p}, \mathbf{r} \rangle + \langle \mathbf{q}, \mathbf{r} \rangle &= \underbrace{c_0e_0 + c_1e_1 + c_2e_2}_{=\langle \mathbf{p}, \mathbf{r} \rangle} + \underbrace{d_0e_0 + d_1e_1 + d_2e_2}_{=\langle \mathbf{q}, \mathbf{r} \rangle} \\ &= c_0e_0 + d_0e_0 + c_1e_1 + d_1e_1 + c_2e_2 + d_2e_2 \qquad [\text{rearranging}] \\ &= \langle \mathbf{p} + \mathbf{q}, \mathbf{r} \rangle \qquad [\text{from above}]\end{aligned}$$

Hence part (ii) is satisfied – $\langle \mathbf{p} + \mathbf{q}, \mathbf{r} \rangle = \langle \mathbf{p}, \mathbf{r} \rangle + \langle \mathbf{q}, \mathbf{r} \rangle$.

(continued...)

(iii) We need to check $\langle k\mathbf{p}, \mathbf{q}\rangle = k\langle \mathbf{p}, \mathbf{q}\rangle$ [taking out the scalar k]:

$$\begin{aligned}\langle k\mathbf{p}, \mathbf{q}\rangle &= \left\langle k\left(c_0 + c_1x + c_2x^2\right), d_0 + d_1x + d_2x^2\right\rangle \\ &= \left\langle kc_0 + kc_1x + kc_2x^2, d_0 + d_1x + d_2x^2\right\rangle && \text{[opening brackets]} \\ &= kc_0d_0 + kc_1d_1 + kc_2d_2 && \text{[multiplying coefficients]} \\ &= k\left(c_0d_0 + c_1d_1 + c_2d_2\right) && \text{[taking out a common factor]} \\ &= k\langle \mathbf{p}, \mathbf{q}\rangle\end{aligned}$$

Hence part (iii) is satisfied.

(iv) We need to show $\langle \mathbf{p}, \mathbf{p}\rangle \geq 0$ and that we have $\langle \mathbf{p}, \mathbf{p}\rangle = 0 \Leftrightarrow \mathbf{p} = \mathbf{O}$:

$$\begin{aligned}\langle \mathbf{p}, \mathbf{p}\rangle &= \left\langle c_0 + c_1x + c_2x^2, \ c_0 + c_1x + c_2x^2\right\rangle \\ &= c_0c_0 + c_1c_1 + c_2c_2 = (c_0)^2 + (c_1)^2 + (c_2)^2 \geq 0\end{aligned}$$

Also

$$\langle \mathbf{p}, \mathbf{p}\rangle = (c_0)^2 + (c_1)^2 + (c_2)^2 = 0 \ \Leftrightarrow \ c_0 = c_1 = c_2 = 0$$

If $c_0 = c_1 = c_2 = 0$ then $\mathbf{p} = \mathbf{O}$.
All four axioms are satisfied, therefore multiplication of coefficients:

$$\langle \mathbf{p}, \mathbf{q}\rangle = c_0d_0 + c_1d_1 + c_2d_2$$

forms an inner product on P_2.

In the above example, we defined the inner product as multiplication of each coefficient of the quadratic polynomial:

$$\langle \mathbf{p}, \mathbf{q}\rangle = c_0d_0 + c_1d_1 + c_2d_2$$

However, if we define this as the *addition* of the coefficients:

$$\langle \mathbf{p}, \mathbf{q}\rangle = \left(c_0 + d_0\right) + \left(c_1 + d_1\right) + \left(c_2 + d_2\right)$$

then $\langle \mathbf{p}, \mathbf{q}\rangle$ is *not* an inner product.

Why not?
Because axiom (iv) of Definition (4.1) fails, that is $\langle \mathbf{p}, \mathbf{p}\rangle \geq 0$ is false.

For example, if $\mathbf{p} = -1 - x - x^2$ then

$$\begin{aligned}\langle \mathbf{p}, \mathbf{p}\rangle &= \left\langle -1 - x - x^2, -1 - x - x^2\right\rangle \\ &= (-1 - 1) + (-1 - 1) + (-1 - 1) = -3 < 0\end{aligned}$$

Hence $\langle \mathbf{p}, \mathbf{p} \rangle \not\geq 0$ [not greater than or equal to zero].

Next we define an inner product on the vector space of matrices, M_{22}. For this we need to define what is meant by the trace of a matrix:

The **trace** of a matrix is the sum of its leading diagonal elements, that is

$$trace \begin{pmatrix} \boxed{a} & b \\ c & \boxed{d} \end{pmatrix} = a + d$$

One example of an inner product on the vector space of matrices is the following:

Let M_{22} be the vector space of 2 by 2 matrices and inner product on M_{22} be defined by

$$\langle \mathbf{A}, \mathbf{B} \rangle = tr\left(\mathbf{B}^T \mathbf{A}\right)$$

where tr is the trace. This is an inner product on M_{22} – you may like to check all 4 axioms.

If we define the inner product as our normal matrix multiplication $\langle \mathbf{A}, \mathbf{B} \rangle = \mathbf{AB}$ on M_{22} then this is *not* an inner product.

Why not?

Because the commutative law does not hold, that is

$$\langle \mathbf{A}, \mathbf{B} \rangle = \mathbf{AB} \neq \mathbf{BA} = \langle \mathbf{B}, \mathbf{A} \rangle \text{ [not equal]}$$

There are many other examples of inner product spaces which you are asked to verify in Exercises 4.1. Next we move onto proving properties of general inner products.

4.1.3 Properties of inner products

Proposition (4.2). Let $\mathbf{u}$, $\mathbf{v}$ and $\mathbf{w}$ be vectors in a real inner product space V and k be any real scalar. We have the following properties of inner products:

(i) $\langle \mathbf{u}, \mathbf{O} \rangle = \langle \mathbf{O}, \mathbf{v} \rangle = 0$ (ii) $\langle \mathbf{u}, k\mathbf{v} \rangle = k\langle \mathbf{u}, \mathbf{v} \rangle$

(iii) $\langle \mathbf{u}, \mathbf{v} + \mathbf{w} \rangle = \langle \mathbf{u}, \mathbf{v} \rangle + \langle \mathbf{u}, \mathbf{w} \rangle$

How do we prove these properties?

We use the axioms of inner products stated in Definition (4.1).

Proof of (i).

We can write the zero vector as $0(\mathbf{O})$ because $0(\mathbf{O}) = \mathbf{O}$. Using the axioms of definition (4.1) we have

$$\begin{aligned} \langle \mathbf{u}, \mathbf{O} \rangle &= \langle \mathbf{u}, 0(\mathbf{O}) \rangle \\ &= \langle 0(\mathbf{O}), \mathbf{u} \rangle && \left[\text{by part (i) of (4.1) which is } \langle \mathbf{u}, \mathbf{v} \rangle = \langle \mathbf{v}, \mathbf{u} \rangle\right] \\ &= 0\langle \mathbf{O}, \mathbf{u} \rangle && \left[\text{by part (iii) of (4.1) which is } \langle k\mathbf{u}, \mathbf{v} \rangle = k\langle \mathbf{u}, \mathbf{v} \rangle\right] \\ &= 0 \end{aligned}$$

Similarly $\langle \mathbf{O}, \mathbf{v} \rangle = 0$.

■

Proof of (ii).
The inner product is commutative, $\langle \mathbf{u}, \mathbf{v} \rangle = \langle \mathbf{v}, \mathbf{u} \rangle$, which means we can switch the vectors around. We have

$$\begin{aligned} \langle \mathbf{u}, k\mathbf{v} \rangle &= \langle k\mathbf{v}, \mathbf{u} \rangle && [\text{switching vectors } \langle \mathbf{u}, \mathbf{v} \rangle = \langle \mathbf{v}, \mathbf{u} \rangle] \\ &= k\langle \mathbf{v}, \mathbf{u} \rangle && [\text{by part (iii) of (4.1) which is } \langle k\mathbf{u}, \mathbf{v} \rangle = k\langle \mathbf{u}, \mathbf{v} \rangle] \\ &= k\langle \mathbf{u}, \mathbf{v} \rangle && [\text{switching vectors } \langle \mathbf{u}, \mathbf{v} \rangle = \langle \mathbf{v}, \mathbf{u} \rangle] \end{aligned}$$

■

Proof of (iii).
We have

$$\begin{aligned} \langle \mathbf{u}, \mathbf{v} + \mathbf{w} \rangle &= \langle \mathbf{v} + \mathbf{w}, \mathbf{u} \rangle && [\text{switching vectors } \langle \mathbf{u}, \mathbf{v} \rangle = \langle \mathbf{v}, \mathbf{u} \rangle] \\ &= \langle \mathbf{v}, \mathbf{u} \rangle + \langle \mathbf{w}, \mathbf{u} \rangle && \left[\begin{array}{l} \text{by part (ii) of (4.1) which is} \\ \langle \mathbf{v} + \mathbf{w}, \mathbf{u} \rangle = \langle \mathbf{v}, \mathbf{u} \rangle + \langle \mathbf{w}, \mathbf{u} \rangle \end{array}\right] \\ &= \langle \mathbf{u}, \mathbf{v} \rangle + \langle \mathbf{u}, \mathbf{w} \rangle && [\text{switching vectors } \langle \mathbf{u}, \mathbf{v} \rangle = \langle \mathbf{v}, \mathbf{u} \rangle] \end{aligned}$$

■

4.1.4 The norm or length of a vector

From chapter 2 we know that we can find the length (norm) $\|\mathbf{u}\|$ of a vector $\mathbf{u}$.

Do you remember how the (Euclidean) norm was defined?
The norm or length of a vector $\mathbf{u}$ in $\mathbb{R}^n$ was defined by Pythagoras' Theorem: $\|\mathbf{u}\| = \sqrt{\mathbf{u} \cdot \mathbf{u}}$.

The norm is defined in the same manner for the general vector space V. Let $\mathbf{u}$ be a vector in V then the norm denoted by $\|\mathbf{u}\|$ is defined as

(4.3) $$\|\mathbf{u}\| = \sqrt{\langle \mathbf{u}, \mathbf{u} \rangle} \ [\text{positive root}]$$

Note that for the general vector space we *cannot* use the definition for the dot product because that is only defined for Euclidean space, $\mathbb{R}^n$, and in this chapter we are examining inner products on general vector spaces. The norm of a vector $\mathbf{u}$ is a real number which gives the size of the vector $\mathbf{u}$.

Generally to find the norm $\|\mathbf{u}\|$ we find $\|\mathbf{u}\|^2 = \langle \mathbf{u}, \mathbf{u} \rangle$ and then take the square root of the result.

How do we find the norm of a function or a matrix?
The norm of a matrix $\mathbf{A}$ in the vector space M_{mn} is given by the above (4.3):
$\|\mathbf{A}\| = \sqrt{\langle \mathbf{A}, \mathbf{A} \rangle}$ where $\langle \mathbf{A}, \mathbf{A} \rangle$ is the inner product of $\mathbf{A}$ and $\mathbf{A}$.

Norm measures the magnitude of things. For example, if take our earlier inner product defined for matrices, $\langle \mathbf{A}, \mathbf{B} \rangle = tr\left(\mathbf{B}^T\mathbf{A}\right)$, then we can evaluate the magnitude of $\|\mathbf{A}\|$ and $\|\mathbf{B}\|$ by calculating the norms.

Let $\mathbf{A} = \begin{pmatrix} 1 & 2 \\ 3 & 4 \end{pmatrix}$ and $\mathbf{B} = \begin{pmatrix} 10 & 20 \\ 30 & 40 \end{pmatrix}$, find $\|\mathbf{A}\|$ and $\|\mathbf{B}\|$.

We have

$$\|\mathbf{A}\|^2 = \langle \mathbf{A}, \mathbf{A} \rangle = tr\left[\begin{pmatrix} 1 & 2 \\ 3 & 4 \end{pmatrix}^T \begin{pmatrix} 1 & 2 \\ 3 & 4 \end{pmatrix}\right]$$

$$= tr\left[\begin{pmatrix} 1 & 3 \\ 2 & 4 \end{pmatrix}\begin{pmatrix} 1 & 2 \\ 3 & 4 \end{pmatrix}\right] = tr\left[\begin{pmatrix} 10 & * \\ * & 20 \end{pmatrix}\right] = 10 + 20 = 30$$

Hence $\|\mathbf{A}\| = \sqrt{30}$. Since the inner product is the trace of the matrix, we don't need to worry what the entries on the other diagonal are; that is why we have placed $*$ in those positions.

Similarly

$$\|\mathbf{B}\| = \sqrt{3000} = 10\sqrt{30} = 10\|\mathbf{A}\| \quad [\text{because } \|\mathbf{A}\| = \sqrt{30}]$$

Notice that the entries in matrix **B** are 10 times the entries in matrix **A**. The norm of matrix **B** is 10 times the norm of matrix **A**.

Why are we interested in the magnitude of a matrix?

Remember, matrices are functions applied to vectors such as $\mathbf{x}$, say. We generally want to compare the size of vectors $\mathbf{Ax}$ with $\mathbf{Bx}$, which is given by their norms $\|\mathbf{Ax}\| = \|\mathbf{A}\|\,\|\mathbf{x}\|$ and $\|\mathbf{Bx}\| = \|\mathbf{B}\|\,\|\mathbf{x}\|$. In the above example, the vector $\mathbf{Bx}$ is 10 times larger in magnitude than vector $\mathbf{Ax}$.

In the following example we apply the norm to vector spaces of continuous functions.

Example 4.4

Let $C[0, 1]$ be the vector space of continuous functions on the closed interval $[0, 1]$. Let $f(x) = x$ and $g(x) = x^2 - 1$ be functions in $C[0, 1]$ and their inner product be given by

$$\langle \mathbf{f}, \mathbf{g} \rangle = \int_0^1 f(x)\, g(x)\, dx$$

This inner product requires us to multiply two vectors and then sum them over $[0, 1]$. [Similar to the dot product but in place of a finite sum we have an integral because we are dealing with continuous functions.] Determine:

(i) $\langle \mathbf{f}, \mathbf{g} \rangle$ **(ii)** $\|\mathbf{f}\|$ **(iii)** $\|\mathbf{g}\|$ **(iv)** $\|\mathbf{f} - \mathbf{g}\|$ **(v)** $\|\mathbf{f} - \mathbf{h}\|$ where $h(x) = x^3$

Solution

(i) The inner product $\langle \mathbf{f}, \mathbf{g} \rangle = \int_0^1 f(x)\, g(x)\, dx$ with $f(x) = x$ and $g(x) = x^2 - 1$ gives

(continued...)

$$\langle \mathbf{f}, \mathbf{g} \rangle = \int_0^1 f(x)\,g(x)\,dx$$

$$= \int_0^1 x\left(x^2 - 1\right) dx \quad \left[\text{substituting } f(x) = x \text{ and } g(x) = x^2 - 1\right]$$

$$= \int_0^1 \left(x^3 - x\right) dx = \left[\frac{x^4}{4} - \frac{x^2}{2}\right]_0^1 = \left[\frac{1}{4} - \frac{1}{2}\right] = -\frac{1}{4}$$

(ii) As stated above, to find $\|\mathbf{f}\|$ we determine $\|\mathbf{f}\|^2 = \langle \mathbf{f}, \mathbf{f} \rangle$ and then take the square root of the result:

$$\|\mathbf{f}\|^2 = \langle \mathbf{f}, \mathbf{f} \rangle$$

$$= \int_0^1 f(x)\,f(x)\,dx \underset{\text{substituting } f(x)=x}{=} \int_0^1 xx\,dx = \int_0^1 x^2\,dx = \left[\frac{x^3}{3}\right]_0^1 = \frac{1}{3} \quad \left[\text{because } xx = x^2\right]$$

What is $\|\mathbf{f}\|$ equal to?
Square root of $\|\mathbf{f}\|^2 = \frac{1}{3}$ which is $\|\mathbf{f}\| = \frac{1}{\sqrt{3}}$.
What does this $\|\mathbf{f}\| = \frac{1}{\sqrt{3}}$ signify?
The norm $\|\mathbf{f}\| = \left(\int_0^1 [f(x)]^2\right)^{1/2} dx$ involves integrating (summing) the function squared, and then taking the square root which measures an average value of the function, called the *root mean square – rms*.

$\|\mathbf{f}\| = 1/\sqrt{3} = 0.577$ (3dp) is the rms value of the function $f(x) = x$ between 0 and 1.

(iii) Similarly we have

$$\|\mathbf{g}\|^2 = \langle \mathbf{g}, \mathbf{g} \rangle = \int_0^1 g(x)\,g(x)\,dx$$

$$= \int_0^1 \left(x^2 - 1\right)\left(x^2 - 1\right) dx \quad \left[\text{substituting } g(x) = x^2 - 1\right]$$

$$= \int_0^1 \left(x^4 - 2x^2 + 1\right) dx = \left[\frac{x^5}{5} - 2\frac{x^3}{3} + x\right]_0^1 = \frac{1}{5} - \frac{2}{3} + 1 = \frac{8}{15} \quad [\text{integrating}]$$

What is $\|\mathbf{g}\|$ equal to?
We need to take the square root of $\frac{8}{15}$ to find $\|\mathbf{g}\|$. Hence $\|\mathbf{g}\| = \sqrt{\frac{8}{15}}$ signifies the rms value, which is an average value of the function $g(x) = x^2 - 1$ between 0 and 1. Note that this $\|\mathbf{g}\|$ is always a positive value, despite $g(x) = x^2 - 1$ being negative between 0 and 1, because we have squared the function and then taken the positive square root.

(iv) Similarly to find $\|\mathbf{f}-\mathbf{g}\|$ we first determine $\|\mathbf{f}-\mathbf{g}\|^2$ and then take the square root:

$$\begin{aligned}
\|\mathbf{f}-\mathbf{g}\|^2 &= \langle \mathbf{f}-\mathbf{g},\ \mathbf{f}-\mathbf{g}\rangle \\
&= \int_0^1 [f(x)-g(x)]\,[f(x)-g(x)]\,dx \\
&= \int_0^1 \left[x-x^2-1\right]\left[x-x^2-1\right]dx && \text{[substituting]} \\
&= \int_0^1 \left(x^2-x^3-x-x^3+x^4+x^2-x+x^2+1\right)dx && \text{[expanding]} \\
&= \int_0^1 \left(x^4-2x^3+3x^2-2x+1\right)dx && \text{[simplifying]} \\
&= \left[\frac{x^5}{5}-2\frac{x^4}{4}+3\frac{x^3}{3}-2\frac{x^2}{2}+x\right]_0^1 && \text{[integrating]} \\
&= \left[\frac{x^5}{5}-\frac{x^4}{2}+x^3-x^2+x\right]_0^1 = \left[\frac{1}{5}-\frac{1}{2}+1-1+1\right] = \frac{7}{10}
\end{aligned}$$

Hence $\|\mathbf{f}-\mathbf{g}\| = \sqrt{7/10}$.

(v) Similarly $\|\mathbf{f}-\mathbf{h}\| = \sqrt{8/105}$.

$\|\mathbf{f}-\mathbf{g}\|$ measures the distance between functions **f** and **g** and is critical in the study of mathematical and numerical analysis. A measurement of distance between functions tells us how close together these functions are.

The **distance** or **metric** between two vectors **u** and **v** is denoted by $d(\mathbf{u},\mathbf{v})$ and is defined as

$$d(\mathbf{u},\mathbf{v}) = \|\mathbf{u}-\mathbf{v}\| \tag{4.4}$$

In the above Example 4.4, the distance between **f** and **g** was given by $\|\mathbf{f}-\mathbf{g}\| = \sqrt{7/10}$ and the distance between **f** and **h** was $\|\mathbf{f}-\mathbf{h}\| = \sqrt{8/105}$. Clearly **f** is closer to **h** than **g** because $\sqrt{8/105} = 0.276$ is smaller than $\sqrt{7/10} = 0.837$.

Remember, we have introduced **f** and **g** as vectors in a vector space of $C[0, 1]$. We can view this as shown in Fig. 4.1.

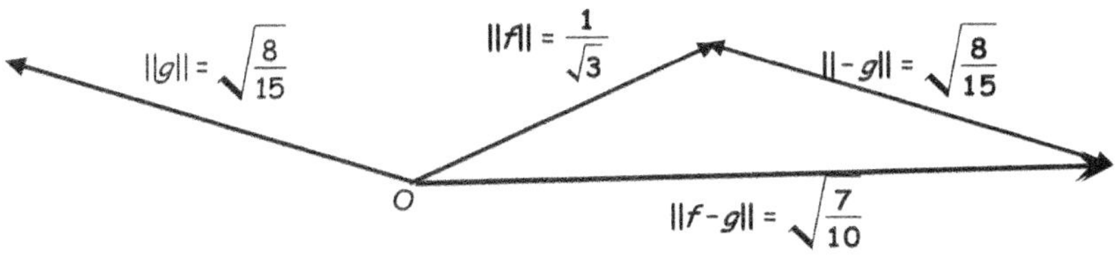

Figure 4.1

We use inner products to measure how close one function is to another. This is important in the study of numerical analysis.

What does $\|\mathbf{f}-\mathbf{g}\| = 0$ *mean?*
The functions $\mathbf{f}$ and $\mathbf{g}$ are equal because

$$\|\mathbf{f}-\mathbf{g}\| = 0 \;\Rightarrow\; \langle \mathbf{f}-\mathbf{g}, \mathbf{f}-\mathbf{g}\rangle = 0 \;\Rightarrow\; \mathbf{f} = \mathbf{g}$$

Next we look at a norm on the set of polynomials P_n.

Example 4.5

Determine the norm $\|\,\|$ of the quadratic (vector) in P_2, where the inner product is given by

$$\langle \mathbf{p}, \mathbf{q}\rangle = c_0d_0 + c_1d_1 + c_2d_2 \qquad \text{[multiplication of coefficients]}$$

for **(i)** $\mathbf{p} = x^2 + x + 1$ **(ii)** $\mathbf{q} = 2x^2 + 2x + 2$ **(iii)** $\mathbf{r} = c_2x^2 + c_1x + c_0$

Solution

We can plot these vectors in 3d space with axes labelled x^2, x and constant. These axes are *not* perpendicular to each other as we will show in section 4.3.

Since $\mathbf{p} = x^2 + x + 1$, the x^2, x and constant coefficients for $\mathbf{p}$ are 1, 1 and 1 respectively. Similarly we plot the other vectors (Fig. 4.2).

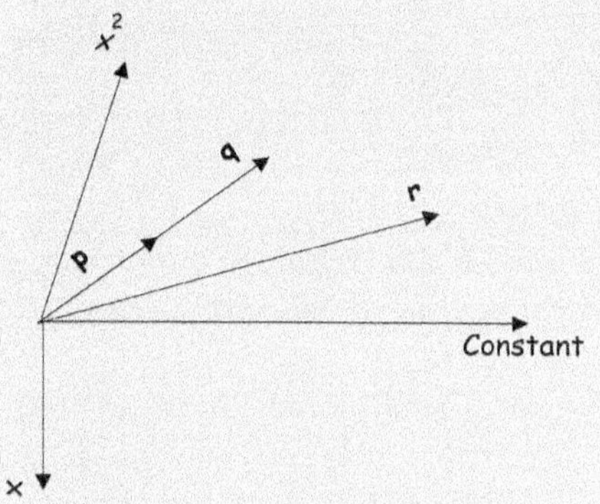

Figure 4.2

We will show in section 4.3 that the constant and x axes are, in fact, perpendicular to each other. However, the x^2 axis is not perpendicular to these.

(i) By definition (4.3) above we have $\|\mathbf{p}\| = \sqrt{\langle \mathbf{p}, \mathbf{p}\rangle}$ so first we find $\|\mathbf{p}\|^2 = \langle \mathbf{p}, \mathbf{p}\rangle$ and then take the square root:

$$\langle \mathbf{p}, \mathbf{p}\rangle = \left\langle x^2 + x + 1,\; x^2 + x + 1\right\rangle = 1^2 + 1^2 + 1^2 = 3$$

Taking the square root yields $\|\mathbf{p}\| = \sqrt{3}$.

(ii) Similarly we have

$$\langle \mathbf{q}, \mathbf{q}\rangle = \left\langle 2x^2 + 2x + 2,\; 2x^2 + 2x + 2\right\rangle = 2^2 + 2^2 + 2^2 = 12$$

Hence $\|\mathbf{q}\| = \sqrt{12} = \sqrt{4 \times 3} = 2\sqrt{3}$.

(iii) Again, repeating this process for $\mathbf{r} = c_2x^2 + c_1x + c_0$:

$$\langle \mathbf{r}, \mathbf{r}\rangle = c_0c_0 + c_1c_1 + c_2c_2 = (c_0)^2 + (c_1)^2 + (c_2)^2$$

Therefore $\|\mathbf{r}\| = \sqrt{\langle \mathbf{r}, \mathbf{r}\rangle} = \sqrt{(c_0)^2 + (c_1)^2 + (c_2)^2}$.

This $\|\mathbf{r}\|$ gives the size of the quadratic by applying Pythagoras to each of the coefficients (legs), c_0, c_1 and c_2.

Note that, for this Example 4.5, we had $\|\mathbf{p}\| = \sqrt{3}$ and $\|\mathbf{q}\| = 2\sqrt{3}$, which means that the quadratic $\mathbf{q} = 2x^2 + 2x + 2$ is twice the size of $\mathbf{p} = x^2 + x + 1$, and note that $\mathbf{q} = 2\mathbf{p}$.

Some norms are *not* defined in terms of the inner product.

Consider the Euclidean space $\mathbb{R}^n$ and let $\mathbf{x} = \begin{pmatrix} x_1 & x_2 & \cdots & x_n \end{pmatrix}^T$ be a vector in $\mathbb{R}^n$, then the following are important norms in this space:

$\|\mathbf{x}\|_1 = |x_1| + |x_2| + |x_3| + \cdots + |x_n|$ [This is called the **one norm**]

$\|\mathbf{x}\|_2 = \sqrt{|x_1|^2 + |x_2|^2 + |x_3|^2 + \cdots + |x_n|^2}$ [This is called the **two norm**]

$\|\mathbf{x}\|_\infty = \max\left(|x_j|\right)$ where $j = 1, 2, \ldots, n$ [This is called the **infinity norm**]

Note that $\|\mathbf{x}\|_\infty = \max\left(|x_j|\right)$ where $j = 1, 2, \ldots, n$ means that we select the maximum absolute value out of $x_1, x_2, x_3, \ldots$ and x_n. This norm measures the maximum absolute value. For example, if we have the vector $\mathbf{x} = \begin{pmatrix} -1 & 4 & -9 & 7 \end{pmatrix}^T$ in $\mathbb{R}^4$ then

$$\|\mathbf{x}\|_1 = |-1| + |4| + |-9| + |7| = 21$$

$$\|\mathbf{x}\|_2 = \sqrt{|-1|^2 + |4|^2 + |-9|^2 + |7|^2} = \sqrt{147} = 12.12 \ (2\text{ dp})$$

$$\|\mathbf{x}\|_\infty = \max\left(|-1|, |4|, |-9|, |7|\right) = \max(1, 4, 9, 7) = 9$$

Which norm should we use?

In a physical situation, you must decide which of these norms suits your problem. For example, suppose we want to measure the noise on a signal, which means that we need a way to measure the variation from the main signal. For the above defined norms, $\|\mathbf{x}\|_\infty$ would measure the largest difference, $\|\mathbf{x}\|_2$ would measure the added average variation and $\|\mathbf{x}\|_1$ would measure the total difference from the main signal. Once you have decided which measure is most appropriate then you try to minimize this.

4.1.5 Properties of the norm of a vector

Next we state certain properties of the norm of a vector.

Proposition (4.5). Let V be an inner product space and $\mathbf{u}$ and $\mathbf{v}$ be vectors in V. If k is any real scalar then we have the following properties of norms:

(i) $\|\mathbf{u}\| \geq 0$ [non-negative] (ii) $\|\mathbf{u}\| = 0 \Leftrightarrow \mathbf{u} = \mathbf{O}$ (iii) $\|k\,\mathbf{u}\| = |k|\,\|\mathbf{u}\|$

[Note that for a real scalar k we have $\sqrt{k^2} = |k|$ where $|k|$ is the modulus of k.]

How do we prove these results?

We use the definition of the norm given above:

(4.3) $$\|\mathbf{u}\| = \sqrt{\langle \mathbf{u}, \mathbf{u} \rangle}$$

Proof of part (i).
By the definition of a norm (4.3) $\|\mathbf{u}\| = \sqrt{\langle \mathbf{u}, \mathbf{u} \rangle}$ we have

$$\|\mathbf{u}\| = \sqrt{\langle \mathbf{u}, \mathbf{u} \rangle} \geq 0$$

■

Proof of part (ii) which is $\|\mathbf{u}\| = 0 \Leftrightarrow \mathbf{u} = \mathbf{O}$.
Again, by the norm definition (4.3) we have

$$\|\mathbf{u}\| = \sqrt{\langle \mathbf{u}, \mathbf{u} \rangle} = 0 \quad \Rightarrow \quad \mathbf{u} = \mathbf{O}$$

and

$$\mathbf{u} = \mathbf{O} \quad \Rightarrow \quad \|\mathbf{u}\| = \sqrt{\langle \mathbf{O}, \mathbf{O} \rangle} = 0$$

■

Proof of part (iii), which is $\|k\mathbf{u}\| = |k| \, \|\mathbf{u}\|$.
Applying the norm definition (4.3), we first find $\|k\mathbf{u}\|^2$ and then we take the square root:

$$\begin{aligned} \|k\mathbf{u}\|^2 &= \langle k\mathbf{u}, k\mathbf{u} \rangle \\ &= k\langle \mathbf{u}, k\mathbf{u} \rangle \\ &= kk\langle \mathbf{u}, \mathbf{u} \rangle = k^2 \langle \mathbf{u}, \mathbf{u} \rangle \end{aligned}$$

Taking the square root gives

$$\|k\mathbf{u}\| = \sqrt{k^2 \langle \mathbf{u}, \mathbf{u} \rangle} = \sqrt{k^2}\sqrt{\langle \mathbf{u}, \mathbf{u} \rangle} = |k| \, \|\mathbf{u}\|$$

This is our required result.

■

Summary

(4.1). An inner product is a function on a vector space which satisfies the four axioms.
The norm or length of a vector denoted $\|\mathbf{u}\|$, is defined as $\|\mathbf{u}\| = \sqrt{\langle \mathbf{u}, \mathbf{u} \rangle}$.

EXERCISES 4.1

(Brief solutions at end of book. Full solutions available at <http://www.oup.co.uk/companion/singh>.)

1. Consider the inner product defined on $C[0, 1]$ defined in Example 4.4.
   If $f(x) = x$ and $g(x) = x^2$ then determine the following:
   (a) $\langle \mathbf{f}, \mathbf{g} \rangle$ (b) $\langle \mathbf{g}, \mathbf{f} \rangle$ (c) $\langle 3\mathbf{f}, \mathbf{g} \rangle$ (d) $\langle \mathbf{f}, \mathbf{f} \rangle$
   (e) $\|\mathbf{f}\|$ (f) $\langle \mathbf{g}, \mathbf{g} \rangle$ (g) $\|\mathbf{g}\|$

2. Let $C[-1, 1]$ be the vector space of all the continuous functions on the interval $[-1, 1]$. The following is an inner product on $C[-1, 1]$:

$$\langle \mathbf{f}, \mathbf{g} \rangle = \int_{-1}^{1} f(x)g(x)\,dx$$

If $f(x) = x$ and $g(x) = x^3$ then determine the following:

(a) $\langle \mathbf{f}, \mathbf{g} \rangle$ (b) $\langle \mathbf{g}, \mathbf{f} \rangle$ (c) $\langle 3\mathbf{f}, \mathbf{g} \rangle$ (d) $\langle \mathbf{f}, \mathbf{f} \rangle$

(e) $\|\mathbf{f}\|$ (f) $\langle \mathbf{g}, \mathbf{g} \rangle$ (g) $\|\mathbf{g}\|$

3. Let $P_2(x)$ be the vector space of polynomials of degree 2 or less. Consider the inner product on this space given in Example 4.3.

If $\mathbf{p} = 2 - 3x + 5x^2$ and $\mathbf{q} = 7 + 5x - 4x^2$ then determine the following:

(a) $\langle \mathbf{p}, \mathbf{q} \rangle$ (b) $\langle \mathbf{q}, \mathbf{p} \rangle$ (c) $\langle \mathbf{p}, -3\mathbf{q} \rangle$ (d) $\langle \mathbf{p}, \mathbf{p} \rangle$

(e) $\|\mathbf{p}\|$ (f) $\langle \mathbf{q}, \mathbf{q} \rangle$ (g) $\|\mathbf{q}\|$

4. Show that $\langle \mathbf{u}, \mathbf{w} \rangle = \mathbf{u}^T \mathbf{A}\mathbf{w}$, as defined in Example 4.2, is indeed an inner product.

5. Let M_{22} be the vector space of 2 by 2 matrices and inner product on M_{22} be defined by

$$\langle \mathbf{A}, \mathbf{B} \rangle = tr\left(\mathbf{B}^T \mathbf{A}\right)$$

where tr is the trace. For $\mathbf{A} = \begin{pmatrix} 1 & 2 \\ 3 & 4 \end{pmatrix}$, $\mathbf{B} = \begin{pmatrix} 5 & 6 \\ 7 & 8 \end{pmatrix}$ and $\mathbf{C} = \begin{pmatrix} -1 & 1 \\ 2 & 5 \end{pmatrix}$ determine

(a) $\langle \mathbf{A}, \mathbf{B} \rangle$ (b) $\langle 5\mathbf{A}, \mathbf{B} \rangle$ (c) $\langle -\mathbf{A}, -\mathbf{B} \rangle$ (d) $\|\mathbf{A}\|$ (e) $\|\mathbf{B}\|$

(f) $\langle \mathbf{A}, \mathbf{C} \rangle$ (g) $\langle \mathbf{B}, \mathbf{C} \rangle$ (h) $\langle \mathbf{A} + \mathbf{B}, \mathbf{C} \rangle$ (i) $\langle \mathbf{A}, \mathbf{C} + \mathbf{B} \rangle$

6. Let V be an inner product space. Prove the following results in the inner product space:

(a) $\langle \mathbf{u}, \mathbf{v} - \mathbf{w} \rangle = \langle \mathbf{u}, \mathbf{v} \rangle - \langle \mathbf{u}, \mathbf{w} \rangle$

(b) $\langle \mathbf{u} - \mathbf{v}, \mathbf{w} \rangle = \langle \mathbf{u}, \mathbf{w} \rangle - \langle \mathbf{v}, \mathbf{w} \rangle$

(c) $\langle k_1\mathbf{u}, k_2\mathbf{v} \rangle = k_1k_2 \langle \mathbf{u}, \mathbf{v} \rangle$

(d) $\langle k_1\mathbf{u} + k_2\mathbf{v}, k_3\mathbf{w} + k_4\mathbf{x} \rangle = k_1k_3 \langle \mathbf{u}, \mathbf{w} \rangle + k_1k_4 \langle \mathbf{u}, \mathbf{x} \rangle + k_2k_3 \langle \mathbf{v}, \mathbf{w} \rangle + k_2k_4 \langle \mathbf{v}, \mathbf{x} \rangle$

(e) $\langle k_1\mathbf{u}_1 + k_2\mathbf{u}_2 + \cdots + k_n\mathbf{u}_n, \mathbf{w} \rangle = k_1 \langle \mathbf{u}_1, \mathbf{w} \rangle + k_2 \langle \mathbf{u}_2, \mathbf{w} \rangle + \cdots + k_n \langle \mathbf{u}_n, \mathbf{w} \rangle$

7. Let $C[0, 1]$ be the vector space of all the continuous functions on the interval $[0, 1]$. Consider the normal inner product of this space given in Example 4.4.

If $f(x) = x + 1$, $g(x) = x^2$ and $h(x) = x - 1$ then determine the following:

(a) $\langle \mathbf{f}, \mathbf{g} \rangle$ (b) $\langle \mathbf{g}, \mathbf{h} \rangle$ (c) $\langle \mathbf{f}, \mathbf{h} \rangle$ (d) $\langle \mathbf{f} + \mathbf{g}, \mathbf{h} \rangle$

(e) $\langle \mathbf{f}, \mathbf{h} + \mathbf{g} \rangle$ (f) $\langle \mathbf{f} + 3\mathbf{g}, \mathbf{h} \rangle$ (g) $\langle \mathbf{f} - 3\mathbf{g}, \mathbf{h} \rangle$ (h) $\langle \mathbf{f} - \mathbf{g}, \mathbf{h} \rangle$

(i) $\langle 2\mathbf{f} + 5\mathbf{g}, 6\mathbf{h} \rangle$ (j) $\langle -10\mathbf{f}, 2\mathbf{h} + 5\mathbf{g} \rangle$

8. Let $C[0, 1]$ be the vector space of all the continuous functions on the interval $[0, 1]$. Explain why the following is not an inner product on $C[0, 1]$:

$$\langle \mathbf{f}, \mathbf{g} \rangle = \int_0^1 [f(x) - g(x)]\, dx$$

9. Show that $\langle \mathbf{u}, \mathbf{v} \rangle = \mathbf{u}^T \mathbf{A} \mathbf{v}$, where $\mathbf{A} = \begin{pmatrix} 1 & 2 \\ 3 & 4 \end{pmatrix}$ is *not* an inner product on $\mathbb{R}^2$.

10. Show that $\langle \mathbf{u}, \mathbf{v} \rangle = u_1 v_1 + u_2 v_2 + u_3 v_3 - u_4 v_4$, where $\mathbf{u} = \begin{pmatrix} u_1 & u_2 & u_3 & u_4 \end{pmatrix}^T$ and $\mathbf{v} = \begin{pmatrix} v_1 & v_2 & v_3 & v_4 \end{pmatrix}^T$ are vectors in $\mathbb{R}^4$, is *not* an inner product on $\mathbb{R}^4$.

11. Let $\mathbf{u}$, $\mathbf{v}$ and $\mathbf{w}$ be vectors in an inner product space V. If

$$\langle \mathbf{u}, \mathbf{v} \rangle = -1, \ \langle \mathbf{u}, \mathbf{w} \rangle = 5, \ \langle \mathbf{v}, \mathbf{w} \rangle = 3, \ \|\mathbf{v}\| = 2, \ \|\mathbf{u}\| = 4 \text{ and } \|\mathbf{w}\| = 7,$$

then determine the following:

(a) $\langle \mathbf{u}, \mathbf{v} + \mathbf{w} \rangle$ (b) $\langle 2\mathbf{u} + 3\mathbf{v}, 5\mathbf{v} - 2\mathbf{w} \rangle$ (c) $\|\mathbf{u} - \mathbf{v}\|$ (d) $\|\mathbf{u} - \mathbf{v} - \mathbf{w}\|$

12. In (4.4) we defined the distance function as

$$d(\mathbf{u}, \mathbf{v}) = \|\mathbf{u} - \mathbf{v}\|$$

where $\mathbf{u}$ and $\mathbf{v}$ are vectors in V. Prove the following properties of this distance function:

(a) $d(\mathbf{u}, \mathbf{v}) = d(\mathbf{v}, \mathbf{u})$ (b) $d(\mathbf{u}, \mathbf{v}) \geq 0$ (c) $d(\mathbf{u}, \mathbf{v}) = 0 \Leftrightarrow \mathbf{u} = \mathbf{v}$

(d) $d(k\mathbf{u}, k\mathbf{v}) = |k|\, d(\mathbf{u}, \mathbf{v})$

SECTION 4.2 Inequalities and Orthogonality

By the end of this section you will be able to

- state and prove Cauchy-Schwarz and Minkowski inequalities
- understand what is meant by orthogonal, normalized and orthonormal vectors

A common application of the inner product is to establish when two or more vectors are orthogonal, or perpendicular to each other.

Why is orthogonality important?

For general vector spaces such as polynomials, matrices and functions, a simple basis is one where the set of vectors are mutually orthogonal (perpendicular) and have unit length. We will discuss the concept of a basis in the next section.

First, we do a numerical example involving the norm and the inner product in $\mathbb{R}^2$. In this example, the inner product is the usual dot product for $\mathbb{R}^2$.

Example 4.6

Let $\mathbf{u} = \begin{pmatrix} 1 \\ 2 \end{pmatrix}$ and $\mathbf{v} = \begin{pmatrix} 3 \\ 4 \end{pmatrix}$ be in $\mathbb{R}^2$. Determine the following with respect to the dot product:

(i) $|\langle \mathbf{u}, \mathbf{v} \rangle|$ **(ii)** $\|\mathbf{u}\| \, \|\mathbf{v}\|$ **(iii)** $\|\mathbf{u}\| + \|\mathbf{v}\|$ **(iv)** $\|\mathbf{u} + \mathbf{v}\|$

Solution

(i) We have $\langle \mathbf{u}, \mathbf{v} \rangle = \mathbf{u} \cdot \mathbf{v} = \begin{pmatrix} 1 \\ 2 \end{pmatrix} \cdot \begin{pmatrix} 3 \\ 4 \end{pmatrix} = (1 \times 3) + (2 \times 4) = 11$. Hence $|\langle \mathbf{u}, \mathbf{v} \rangle| = |11| = 11$.

(ii) Remember, the symbol $\| \|$ means the length of the vector. By Pythagoras we have

$$\|\mathbf{u}\| = \left\| \begin{pmatrix} 1 \\ 2 \end{pmatrix} \right\| = \sqrt{1^2 + 2^2} = \sqrt{5} \quad \text{and} \quad \|\mathbf{v}\| = \left\| \begin{pmatrix} 3 \\ 4 \end{pmatrix} \right\| = \sqrt{3^2 + 4^2} = 5$$

Multiplying these we have $\|\mathbf{u}\| \, \|\mathbf{v}\| = \sqrt{5} \times 5 = 11.1803$ (4dp).

(iii) Adding both the above results in part (ii) we have

$$\|\mathbf{u}\| + \|\mathbf{v}\| = \sqrt{5} + 5 = 7.2361 \text{ (4dp)}$$

(iv) Similarly by using Pythagoras we have

$$\|\mathbf{u} + \mathbf{v}\| = \left\| \begin{pmatrix} 1 \\ 2 \end{pmatrix} + \begin{pmatrix} 3 \\ 4 \end{pmatrix} \right\| = \left\| \begin{pmatrix} 4 \\ 6 \end{pmatrix} \right\| = \sqrt{4^2 + 6^2} = \sqrt{52} = 7.\,2111 \text{ (4 dp)}$$

We can illustrate these in $\mathbb{R}^2$ (Fig. 4.3):

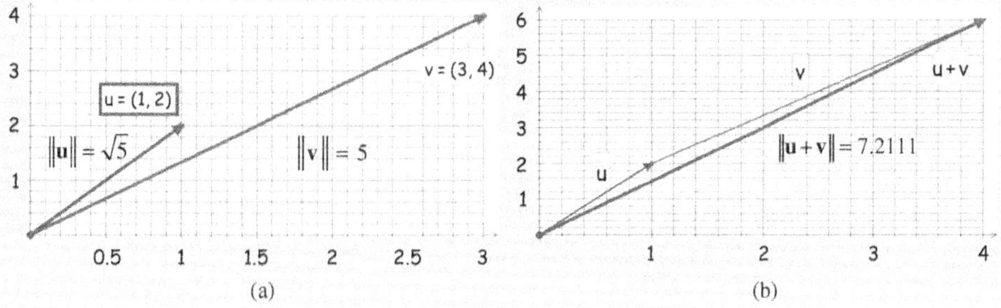

Figure 4.3

4.2.1 Inequalities

In this subsection we prove two inequalities:

$$|\langle \mathbf{u}, \mathbf{v} \rangle| \leq \|\mathbf{u}\| \, \|\mathbf{v}\| \qquad \text{Cauchy–Schwarz inequality}$$
$$\|\mathbf{u} + \mathbf{v}\| \leq \|\mathbf{u}\| + \|\mathbf{v}\| \qquad \text{Minkowski inequality}$$

First we state and prove the Cauchy–Schwarz inequality. It is an important inequality which is used in many applications and fields of mathematics. It is a slightly more involved proof because we apply one particular result of quadratic equations in order to prove the Cauchy–Schwarz inequality.

Under what conditions is the general quadratic $ax^2 + bx + c \geq 0$ (greater than or equal to zero)?
Well, the graph of such a quadratic lies above (or on) the x axis, as shown in Fig. 4.4:

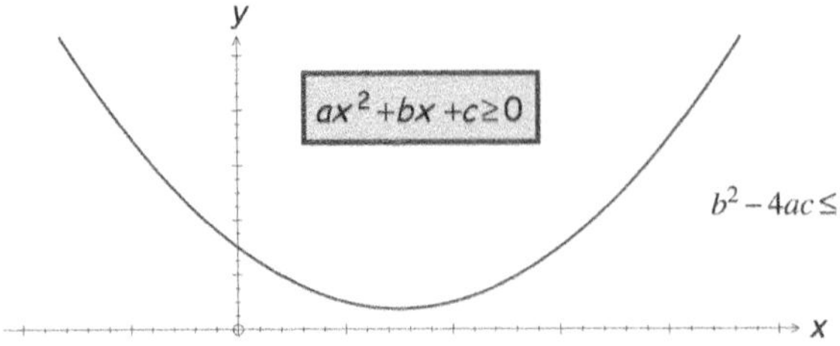

Figure 4.4

For the quadratic, $ax^2 + bx + c$, to be greater than or equal to 0 for all x means that either the graph does not cross the x axis or it touches it at just one point. This is satisfied when

$$b^2 - 4ac \leq 0 \qquad (\dagger)$$

This is where the discriminant, $b^2 - 4ac$, is less than or equal to zero. [Remember, the discriminant in this case means that it discriminates between real and complex roots of a quadratic equation.]

The Cauchy–Schwarz inequality for Euclidean space $\mathbb{R}^n$ was covered in section 2.2.3.

It demonstrates a universal inequality between inner products and norms, and was given by:

$$|\mathbf{u} \cdot \mathbf{v}| \leq \|\mathbf{u}\| \, \|\mathbf{v}\|$$

Here we expressed the inequality in terms of the dot product, but we can extend this definition to the general inner product:

$$|\langle \mathbf{u}, \mathbf{v} \rangle| \leq \|\mathbf{u}\| \, \|\mathbf{v}\|$$

The Cauchy–Schwarz inequality (4.6). Let $\mathbf{u}$ and $\mathbf{v}$ be vectors in an inner product space, then

$$|\langle \mathbf{u}, \mathbf{v} \rangle| \leq \|\mathbf{u}\| \, \|\mathbf{v}\|$$

We can see this inequality for the above Example 4.6:

$$|\langle \mathbf{u}, \mathbf{v} \rangle| \underbrace{=}_{\text{by part (i)}} 11 \leq 11.1803 \underbrace{=}_{\text{by part (ii)}} \|\mathbf{u}\| \, \|\mathbf{v}\|$$

Note that the Cauchy–Schwarz inequality connects the notion of an inner product with the notion of length.

[Hint: Consider the inner product $\langle k\mathbf{u} + \mathbf{v}, \; k\mathbf{u} + \mathbf{v} \rangle$ where k is a scalar.]

Proof.

How do we prove this result?

By the hint, expanding the given inner product $\langle k\mathbf{u}+\mathbf{v},\ k\mathbf{u}+\mathbf{v}\rangle$ and by using the rules on inner products established in the last section:

$$\begin{aligned}\langle k\mathbf{u}+\mathbf{v},\ k\mathbf{u}+\mathbf{v}\rangle &= \langle k\mathbf{u},\ k\mathbf{u}\rangle + \langle k\mathbf{u},\ \mathbf{v}\rangle + \langle \mathbf{v},\ k\mathbf{u}\rangle + \langle \mathbf{v},\ \mathbf{v}\rangle \\ &= k^2\langle \mathbf{u},\ \mathbf{u}\rangle + \underbrace{k\langle \mathbf{u},\ \mathbf{v}\rangle + k\langle \mathbf{v},\ \mathbf{u}\rangle}_{=2k\langle \mathbf{u},\ \mathbf{v}\rangle \text{ because } \langle \mathbf{u},\ \mathbf{v}\rangle=\langle \mathbf{v},\ \mathbf{u}\rangle} + \langle \mathbf{v},\ \mathbf{v}\rangle \\ &= k^2\|\mathbf{u}\|^2 + 2k\langle \mathbf{u},\ \mathbf{v}\rangle + \|\mathbf{v}\|^2 \quad \left[\text{remember } \langle \mathbf{u},\ \mathbf{u}\rangle = \|\mathbf{u}\|^2\right]\end{aligned}$$

By Definition (4.1) Axiom (iv), we have $\langle \mathbf{w},\ \mathbf{w}\rangle \geq 0$, which means that the inner product between the same vectors is ≥ 0:

$$\langle k\mathbf{u}+\mathbf{v},\ k\mathbf{u}+\mathbf{v}\rangle \geq 0 \qquad [\text{remember } k\mathbf{u}+\mathbf{v} \text{ is a vector}]$$

Substituting this into the above expansion we have

$$k^2\|\mathbf{u}\|^2 + 2k\langle \mathbf{u},\ \mathbf{v}\rangle + \|\mathbf{v}\|^2 \geq 0 \qquad (*)$$

We consider this as a quadratic equation by taking k as our variable. We can write this as

$$ak^2 + bk + c \geq 0 \text{ where } a = \|\mathbf{u}\|^2,\ b = 2\langle \mathbf{u},\ \mathbf{v}\rangle \text{ and } c = \|\mathbf{v}\|^2$$

The quadratic is ≥ 0, therefore we can substitute these $a = \|\mathbf{u}\|^2$, $b = 2\langle \mathbf{u},\ \mathbf{v}\rangle$ and $c = \|\mathbf{v}\|^2$ into (†), which is given above by; $b^2 - 4ac \leq 0$:

$$\begin{aligned}b^2 - 4ac = (2\langle \mathbf{u},\ \mathbf{v}\rangle)^2 - 4\|\mathbf{u}\|^2\|\mathbf{v}\|^2 &\leq 0 \\ (2\langle \mathbf{u},\ \mathbf{v}\rangle)^2 &\leq 4\|\mathbf{u}\|^2\|\mathbf{v}\|^2 \\ 4(\langle \mathbf{u},\ \mathbf{v}\rangle)^2 &\leq 4\|\mathbf{u}\|^2\|\mathbf{v}\|^2 \\ (\langle \mathbf{u},\ \mathbf{v}\rangle)^2 &\leq \|\mathbf{u}\|^2\|\mathbf{v}\|^2 \quad [\text{cancelling 4's}] \\ \sqrt{(\langle \mathbf{u},\ \mathbf{v}\rangle)^2} &\leq \sqrt{\|\mathbf{u}\|^2\|\mathbf{v}\|^2} \quad [\text{square root}] \\ |\langle \mathbf{u},\ \mathbf{v}\rangle| &\leq \|\mathbf{u}\|\,\|\mathbf{v}\| \quad \left[\text{remember, } \sqrt{x^2} = |x|\right]\end{aligned}$$

This is our required result, the Cauchy–Schwarz inequality.

■

An important inequality for *addition* of norms of vectors is the following:

The Minkowski (pronounced 'Minkofski') or triangular inequality (4.7). Let V be an inner product space. For *all* vectors $\mathbf{u}$ and $\mathbf{v}$ we have

$$\|\mathbf{u}+\mathbf{v}\| \leq \|\mathbf{u}\| + \|\mathbf{v}\|$$

What does this mean?
In $\mathbb{R}^2$ we have (Fig. 4.5):

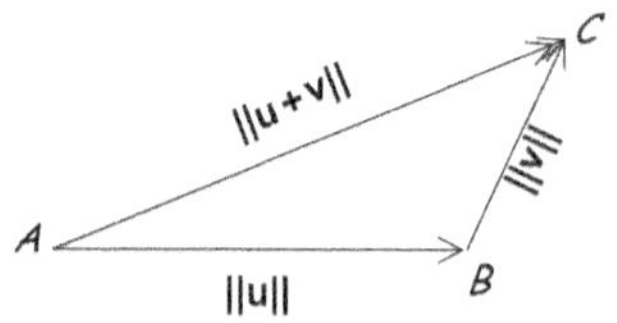

Figure 4.5

[Remember, this is the donkey theorem: if a donkey is at station A and the food is located at station C then the donkey would travel the shortest distance along AC not AB and then BC.]

In the above Example 4.6 (see Fig. 4.3(b)) we had the inequality:

$$\|\mathbf{u}+\mathbf{v}\| \underbrace{=}_{\text{by part (iv)}} 7.2111 \le 7.2361 \underbrace{=}_{\text{by part (iii)}} \|\mathbf{u}\| + \|\mathbf{v}\|$$

Proof – Exercises 4.2.

■

Note that the Cauchy–Schwarz inequality establishes an inequality connecting the *multiplication* of vectors:

$$|\langle \mathbf{u}, \mathbf{v}\rangle| \le \|\mathbf{u}\| \, \|\mathbf{v}\|$$

while the Minkowski inequality is an inequality relating *addition* of vectors:

$$\|\mathbf{u}+\mathbf{v}\| \le \|\mathbf{u}\| + \|\mathbf{v}\|$$

4.2.2 Orthogonal vectors

Can you remember what orthogonal vectors meant in the Euclidean space $\mathbb{R}^n$?
Two vectors $\mathbf{u}$ and $\mathbf{v}$ in $\mathbb{R}^n$ are said to be **orthogonal** or **perpendicular** $\Leftrightarrow$

(2.5) $$\mathbf{u}\cdot\mathbf{v} = 0$$

Hence for the general vector space V with an inner product we have:

Definition (4.8). Two vectors $\mathbf{u}$ and $\mathbf{v}$ in the vector space V are said to be **orthogonal** $\Leftrightarrow \langle \mathbf{u}, \mathbf{v}\rangle = 0$

This is a fundamental and very useful result in linear algebra.

If vectors $\mathbf{u}$ and $\mathbf{v}$ are orthogonal, we say that $\mathbf{u}$ is orthogonal to $\mathbf{v}$, or vice versa that is $\mathbf{v}$ is orthogonal to $\mathbf{u}$ because $\langle \mathbf{u}, \mathbf{v}\rangle = \langle \mathbf{v}, \mathbf{u}\rangle = 0$.

Consider the vectors $\mathbf{u}$, $\mathbf{v}$, $\mathbf{w}$ and $\mathbf{x}$ in $\mathbb{R}^2$ (Fig. 4.6):

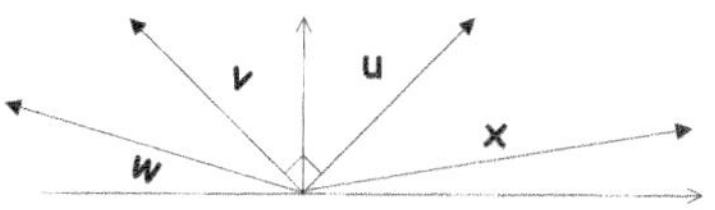

$\langle \mathbf{u}, \mathbf{x} \rangle$ is positive
$\langle \mathbf{u}, \mathbf{w} \rangle$ is negative
$\langle \mathbf{u}, \mathbf{v} \rangle = 0$

Figure 4.6

You may recall from chapter 2 that vectors acting in the same direction have a positive dot product and vectors in opposite directions have a negative dot product. If they are perpendicular (**u** and **v**) then the dot product is zero.

Applications such as signal processing, communication and radar systems rely on inner products such as $S = \left| \left\langle \frac{\mathbf{x}}{\|\mathbf{x}\|}, \frac{\mathbf{y}}{\|\mathbf{y}\|} \right\rangle \right|$ where **x** and **y** are non-zero signals (vectors) in an inner product space. In the exercises, you are asked to show that this expression S lies between 0 and 1. S measures the degree to which the two signals are alike. A value of S close to 1 means that the signals are similar in nature. A value of S close to 0 means that the signals are very different but not necessarily orthogonal.

This expression S is useful in a matched filter detector. The detector uses S to compare a set of signals against a target signal to decide which signal is most like the target signal.

Another application of orthogonality is in information retrieval, because vector spaces can be used to represent documents. A document and a query term can be represented by vectors **d** and **q** respectively. If $\langle \mathbf{q}, \mathbf{d} \rangle = 0$ then vectors **q** and **d** are orthogonal, which means that the query term does not exist in the document.

To know that two functions or a set of functions are orthogonal is of incredibly powerful mathematical value, because you can use the theorems of linear algebra to develop approximations that are as close as you like to difficult functions that may be impossible to calculate in any other way. This is particularly true in the fields of mathematical physics such as Fourier analysis, polynomial series approximations and Legendre polynomials.

Orthogonality signifies a certain kind of 'independence', or a complete absence of interference. If the inner product is zero then one vector has *no* interference on the other.

Example 4.7

Show that the vectors $\mathbf{p} = x^2 + x + 1$ and $\mathbf{q} = -x^2 - x + 2$ are orthogonal in the vector space of polynomials of degree ≤ 2, P_2, where an inner product is given by

$$\langle \mathbf{p}, \mathbf{q} \rangle = c_2 d_2 + c_1 d_1 + c_0 d_0 \quad \text{[multiplying coefficients]}$$

($\mathbf{p} = c_2 x^2 + c_1 x + c_0$ and $\mathbf{q} = d_2 x^2 + d_1 x + d_0$).

Solution

For $\mathbf{p} = x^2 + x + 1$ and $\mathbf{q} = -x^2 - x + 2$ we have

$$\begin{aligned} \langle \mathbf{p}, \mathbf{q} \rangle &= \langle x^2 + x + 1, -x^2 - x + 2 \rangle \\ &= \langle (1)x^2 + (1)x + 1, (-1)x^2 + (-1)x + 2 \rangle \\ &= 1(-1) + 1(-1) + 1(2) = 0 \end{aligned}$$

(continued...)

We conclude that the given vectors $\mathbf{p}$ and $\mathbf{q}$ are orthogonal because $\langle \mathbf{p}, \mathbf{q} \rangle = 0$.
We can illustrate this as shown in Fig. 4.7.

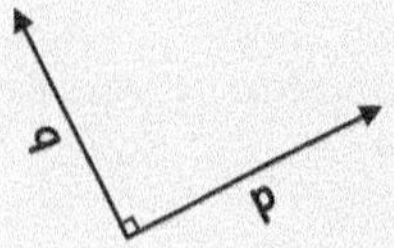

Figure 4.7

Example 4.8

Let the vector space of matrices of size 2 by 2, M_{22}, have an inner product defined by

$$\langle \mathbf{A}, \mathbf{B} \rangle = tr\left(\mathbf{B}^T \mathbf{A}\right) \text{ where } tr \text{ is trace}$$

Show that the matrices $\mathbf{A} = \begin{pmatrix} -1 & -1 \\ 1 & 5 \end{pmatrix}$ and $\mathbf{B} = \begin{pmatrix} 1 & 2 \\ 3 & 0 \end{pmatrix}$ are orthogonal with respect to each other.

Remember, the trace of a matrix is the addition of leading diagonal entries.

Solution

$$\langle \mathbf{A}, \mathbf{B} \rangle = tr\left(\mathbf{B}^T \mathbf{A}\right) = tr\left[\begin{pmatrix} 1 & 2 \\ 3 & 0 \end{pmatrix}^T \begin{pmatrix} -1 & -1 \\ 1 & 5 \end{pmatrix}\right]$$

$$\underset{\text{transposing}}{=} tr\left[\begin{pmatrix} 1 & 3 \\ 2 & 0 \end{pmatrix} \begin{pmatrix} -1 & -1 \\ 1 & 5 \end{pmatrix}\right] \underset{\text{multiplying}}{=} tr\begin{pmatrix} 2 & * \\ * & -2 \end{pmatrix} = 2 - 2 = 0$$

Hence matrices $\mathbf{A}$ and $\mathbf{B}$ are orthogonal, because $\langle \mathbf{A}, \mathbf{B} \rangle = 0$.

The structure of Euclidean vector spaces is exactly the same structure we find in other vector spaces. In the next example you will see this structure extended to accommodate the integrals of sine and cosine products.

Example 4.9

(Fourier series). Show that the vectors $f(t) = \cos(t)$ and $g(t) = \sin(t)$ are orthogonal in the set of continuous functions $C[0, \pi]$ with respect to the inner product given by

$$\langle \mathbf{f}, \mathbf{g} \rangle = \int_0^{\pi} f(t)\, g(t)\, dt$$

Solution

We have

$$\langle \mathbf{f}, \mathbf{g} \rangle = \langle \cos(t), \sin(t) \rangle = \int_0^{\pi} \cos(t)\sin(t)\,dt \qquad (*)$$

How do we integrate $\cos(t)\sin(t)$*?*

Use the following trigonometric identity:

$$\cos(t)\sin(t) = \frac{1}{2}\sin(2t) \qquad (\dagger)$$

Substituting this into the above we have

$$\begin{aligned}
\int_0^{\pi} \cos(t)\sin(t)\,dt &= \frac{1}{2}\int_0^{\pi} \sin(2t)\,dt && \left[\text{applying the identity } (\dagger)\right] \\
&= \frac{1}{2}\left[-\frac{\cos(2t)}{2}\right]_0^{\pi} && \left[\text{because } \int \sin(kt)\,dt = -\frac{\cos(kt)}{k}\right] \\
&= -\frac{1}{4}\left[\cos(2\pi) - \cos(0)\right] && \left[\begin{array}{l}\text{substituting limits and} \\ \text{taking out } -1/2\end{array}\right] \\
&= -\frac{1}{4}\left[1 - 1\right] = 0 && \left[\text{because } \cos(2\pi) = \cos(0) = 1\right]
\end{aligned}$$

By (*) we conclude that $\langle \mathbf{f}, \mathbf{g} \rangle = 0$, which means that $f(t) = \cos(t)$ and $g(t) = \sin(t)$ are orthogonal in $C[0, \pi]$. Hence sine and cosine functions are orthogonal in $C[0, \pi]$. This is an important result in the study of Fourier series.

Next we prove Pythagoras' Theorem which is applicable to vectors in general vector space with an inner product.

Pythagoras' Theorem (4.9). Let the vectors $\mathbf{u}$ and $\mathbf{v}$ be *orthogonal* in an inner product space V then

$$\|\mathbf{u} + \mathbf{v}\|^2 = \|\mathbf{u}\|^2 + \|\mathbf{v}\|^2$$

Proof.

Expanding the left hand side we have

$$\begin{aligned}
\|\mathbf{u} + \mathbf{v}\|^2 &= \langle \mathbf{u} + \mathbf{v}, \mathbf{u} + \mathbf{v} \rangle && \left[\text{by definition of norm, } \|\mathbf{w}\|^2 = \langle \mathbf{w}, \mathbf{w} \rangle\right] \\
&= \langle \mathbf{u}, \mathbf{u} \rangle + \underbrace{\langle \mathbf{u}, \mathbf{v} \rangle}_{=0} + \underbrace{\langle \mathbf{v}, \mathbf{u} \rangle}_{=0} + \langle \mathbf{v}, \mathbf{v} \rangle && \left[\mathbf{u}, \mathbf{v} \text{ are orthogonal so } \langle \mathbf{u}, \mathbf{v} \rangle = \langle \mathbf{v}, \mathbf{u} \rangle = 0\right] \\
&= \|\mathbf{u}\|^2 + \|\mathbf{v}\|^2 && \left[\text{remember, } \langle \mathbf{w}, \mathbf{w} \rangle = \|\mathbf{w}\|^2\right]
\end{aligned}$$

■

Proposition (4.10). Every vector in an inner product space V is orthogonal to the zero vector, $\mathbf{O}$.

Proof.
Let $\mathbf{v}$ be an arbitrary vector in V. Then

$$\langle \mathbf{v}, \mathbf{O} \rangle = 0 \text{ [because by (4.2) part (i) we have } \langle \mathbf{v}, \mathbf{O} \rangle = 0]$$

Since $\mathbf{v}$ was arbitrary, every vector in V is orthogonal to the zero vector, $\mathbf{O}$.

■

4.2.3 Normalizing vectors

What is a unit vector?
A unit vector $\mathbf{u}$ is a vector of length 1 or a norm of 1, that is $\|\mathbf{u}\| = 1$.

The process of converting a given vector into a unit vector is called **normalizing**.

Proposition (4.11). Every non-zero vector $\mathbf{w}$ in an inner product space V can be normalized by setting $\mathbf{u} = \frac{\mathbf{w}}{\|\mathbf{w}\|}$.

Proof – Exercises 4.2.

■

We write the normalized vector $\mathbf{w}$ as $\widehat{\mathbf{w}}$, which is pronounced as 'w hat'. We have $\mathbf{u} = \widehat{\mathbf{w}}$.

Example 4.10

Normalize the vector $\mathbf{p} = 5x^2 - 2x + 1$ in P_2 with respect to the usual inner product of multiplying coefficients.

Solution
The normalized vector $\widehat{\mathbf{p}}$ is given by $\widehat{\mathbf{p}} = \mathbf{p}/\|\mathbf{p}\|$, where $\mathbf{p} = 5x^2 - 2x + 1$.
What is $\|\mathbf{p}\|$ equal to?
By the definition of the norm (4.3) we have $\|\mathbf{p}\|^2 = \langle \mathbf{p}, \mathbf{p} \rangle$, which we can evaluate by using the given inner product:

$$\|\mathbf{p}\|^2 = \langle \mathbf{p}, \mathbf{p} \rangle = \left\langle 5x^2 - 2x + 1,\ 5x^2 - 2x + 1 \right\rangle$$

$$= (5 \times 5) + (-2 \times (-2)) + (1 \times 1) = 30 \quad \text{[multiplying coefficients]}$$

Taking the square root of $\|\mathbf{p}\|^2 = 30$ gives $\|\mathbf{p}\| = \sqrt{30}$. Hence the normalized (unit) vector is

$$\widehat{\mathbf{p}} = \frac{\mathbf{p}}{\|\mathbf{p}\|} = \frac{1}{\|\mathbf{p}\|}\mathbf{p} = \frac{1}{\sqrt{30}}\underbrace{\left(5x^2 - 2x + 1\right)}_{=\mathbf{p}}$$

4.2.4 Orthonormal set

A set of vectors which are orthogonal (perpendicular) to each other is called an **orthogonal set**.

A set of vectors in which *all* the vectors have a norm or length of 1 is called a **normalized set**.

A set of perpendicular unit vectors is called an **orthonormal** set. This is a set of vectors which are *both* orthogonal and normalized.

For example, the set of vectors $\mathbf{e}_1 = \begin{pmatrix} 1 & 0 \end{pmatrix}^T$ and $\mathbf{e}_2 = \begin{pmatrix} 0 & 1 \end{pmatrix}^T$ in $\mathbb{R}^2$ are orthonormal (perpendicular unit) vectors with the inner product as the dot product:

$$\langle \mathbf{e}_1, \mathbf{e}_2 \rangle = \begin{pmatrix} 1 \\ 0 \end{pmatrix} \cdot \begin{pmatrix} 0 \\ 1 \end{pmatrix} = (1 \times 0) + (0 \times 1) = 0$$

The vectors $\mathbf{e}_1$ and $\mathbf{e}_2$ are orthogonal (perpendicular). The norms or lengths of these vectors:

$$\|\mathbf{e}_1\| = \|\mathbf{e}_2\| = 1$$

Thus the vectors $\mathbf{e}_1$ and $\mathbf{e}_2$ are orthonormal (perpendicular unit vectors) because they are *both* orthogonal and normalized vectors.

Examples of orthonormal sets are shown in Fig. 4.8 for $\mathbb{R}^2$ and $\mathbb{R}^3$:

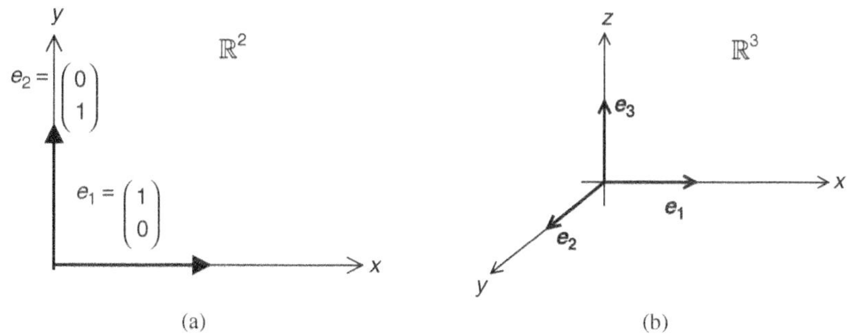

Figure 4.8

Remember that these are the standard basis vectors and they are also an orthonormal (perpendicular unit) basis for $\mathbb{R}^2$ and $\mathbb{R}^3$. These perpendicular unit vectors make a convenient basis as we will discuss in the next section.

Definition (4.12). Let V be a finite-dimensional vector space with an inner product. A set of vectors

$$B = \{\mathbf{u}_1, \mathbf{u}_2, \mathbf{u}_3, \ldots, \mathbf{u}_n\}$$

for V is called an **orthonormal set** if they are

(i) orthogonal, that is $\langle \mathbf{u}_i, \mathbf{u}_j \rangle = 0$ for $i \neq j$

(ii) normalized, that is $\|\mathbf{u}_j\| = 1$ for $j = 1, 2, 3, \ldots, n$

Normalized vectors have a norm (or length) of 1.
Orthogonal (different) vectors are orthogonal (perpendicular) to each other.

Example 4.11

Show that $B = \{\mathbf{e}_1, \mathbf{e}_2, \mathbf{e}_3\}$ forms an orthonormal (perpendicular unit) set of vectors for $\mathbb{R}^3$ with the inner product given by the dot product. (See Fig. 4.8(b).)

Solution

Remember, $\mathbf{e}_1 = (1\ 0\ 0)^T$, $\mathbf{e}_2 = (0\ 1\ 0)^T$ and $\mathbf{e}_3 = (0\ 0\ 1)^T$. We first check for orthogonality:

$$\langle \mathbf{e}_1, \mathbf{e}_2 \rangle = \mathbf{e}_1 \cdot \mathbf{e}_2 = \begin{pmatrix} 1 \\ 0 \\ 0 \end{pmatrix} \cdot \begin{pmatrix} 0 \\ 1 \\ 0 \end{pmatrix} = (1 \times 0) + (0 \times 1) + (0 \times 0) = 0$$

Similarly $\langle \mathbf{e}_1, \mathbf{e}_3 \rangle = \langle \mathbf{e}_2, \mathbf{e}_3 \rangle = 0$.

We have $\langle \mathbf{e}_1, \mathbf{e}_2 \rangle = \langle \mathbf{e}_1, \mathbf{e}_3 \rangle = \langle \mathbf{e}_2, \mathbf{e}_3 \rangle = 0$, therefore the set $B = \{\mathbf{e}_1, \mathbf{e}_2, \mathbf{e}_3\}$ is orthogonal, which means that each of the vectors $\{\mathbf{e}_1, \mathbf{e}_2, \mathbf{e}_3\}$ is perpendicular to the others.

What else do we need to prove?

The norm (or length) of each vector is 1, that is we need to show

$$\|\mathbf{e}_1\| = \|\mathbf{e}_2\| = \|\mathbf{e}_3\| = 1$$

Verify that the length of the vectors $\|\mathbf{e}_1\| = \|\mathbf{e}_2\| = \|\mathbf{e}_3\| = 1$.

Thus $B = \{\mathbf{e}_1, \mathbf{e}_2, \mathbf{e}_3\}$ forms an orthonormal (perpendicular unit) set of vectors in $\mathbb{R}^3$, because each vector is orthogonal to the others and all the vectors have a norm of 1.

This $B = \{\mathbf{e}_1, \mathbf{e}_2, \mathbf{e}_3\}$ is *not* only an orthonormal set in $\mathbb{R}^3$, but actually forms an orthonormal *basis* for $\mathbb{R}^3$. The most useful type of basis is an orthonormal basis.

An orthonormal basis turns out to be a great tool used in the fields of differential equations and quantum mechanics.

Let's examine an orthonormal set for the vector space of polynomials.

Example 4.12

Let P_3 be the inner product space of polynomials of degree 3 or less with the usual inner product generated by multiplying coefficients. Let $\{\mathbf{p}, \mathbf{q}\}$ be an orthonormal set of vectors in P_3. Find $\|\mathbf{p} + \mathbf{q}\|$.

Solution

Since $\{\mathbf{p}, \mathbf{q}\}$ is an orthonormal set, by definition the vectors are orthogonal and normalized (Fig. 4.9).

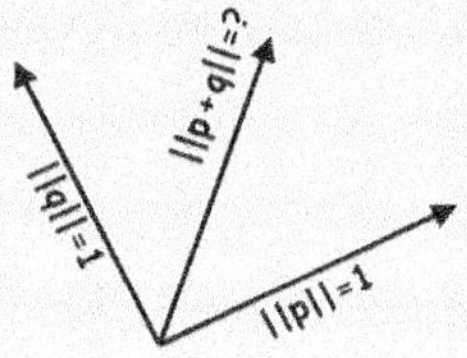

Figure 4.9

Applying Pythagoras (4.9), we have

$$\|\mathbf{p}+\mathbf{q}\|^2 = \|\mathbf{p}\|^2 + \|\mathbf{q}\|^2 = 1 + 1 = 2$$

Taking the square root gives $\|\mathbf{p}+\mathbf{q}\| = \sqrt{2}$.

The next example is more demanding because it relies on integration, and there are an infinite number of vectors – we have an infinite dimensional vector space. You will study this vector space in a subject called *Fourier series*. In Fourier series, a periodic function (a function which repeats itself in a given period) can be written as a linear combination of sines and cosines.

In the context of periodic functions, the inner product can be used to generate the Fourier coefficients.

You might be able to evaluate most of the integrals, but since this is not a course on calculus, the integrals are calculated for you.

Example 4.13

Let $C[0, 2\pi]$ be the vector space of continuous functions on the interval $[0, 2\pi]$ and the inner product on this vector space be defined by

$$\langle \mathbf{f}_n, \mathbf{g}_n \rangle = \int_0^{2\pi} f_n(t)\, g_n(t)\, dt$$

Let S be the set $\{1, \cos(t), \cos(2t), \cos(3t), \ldots, \sin(t), \sin(2t), \sin(3t), \ldots\}$. By letting $\mathbf{f}_0 = \cos(0) = 1$, $\mathbf{f}_n = \cos(nt)$ and $\mathbf{g}_n = \sin(nt)$ show that the set S is orthogonal and normalize these vectors to create an orthonormal set.

[Results: for any positive integers n and m we have

(i) $\int_0^{2\pi} \cos(nt)\sin(mt)\, dt = 0$,

(ii) $\int_0^{2\pi} \cos(nt)\, dt = 0$,

(iii) $\int_0^{2\pi} \sin(nt)\, dt = 0$,

(iv) $\int_0^{2\pi} \cos(nt)\cos(mt)\, dt = 0$ provided $n \neq m$,

(v) $\int_0^{2\pi} \sin(nt)\sin(mt)\, dt = 0$ provided $n \neq m$ and

(vi) $\int_0^{2\pi} \cos^2(nt)\, dt = \int_0^{2\pi} \sin^2(nt)\, dt = \pi$.]

(continued...)

Solution

How do we show that the given set of vectors in S are orthogonal?

By using the integration results in the brackets above. First we test $\mathbf{f}_0$ with the other vectors:

$$\langle \mathbf{f}_0, \mathbf{f}_n \rangle = \int_0^{2\pi} [1 \times \cos(nt)]\, dt = \int_0^{2\pi} \cos(nt)\, dt \underbrace{=}_{\text{by result (ii)}} 0$$

$$\langle \mathbf{f}_0, \mathbf{g}_n \rangle = \int_0^{2\pi} [1 \times \sin(nt)]\, dt = \int_0^{2\pi} \sin(nt)\, dt \underbrace{=}_{\text{by result (iii)}} 0$$

Next we test $\mathbf{f}_n$ with $\mathbf{g}_m$:

$$\langle \mathbf{f}_n, \mathbf{g}_m \rangle = \int_0^{2\pi} \cos(nt)\sin(mt)\, dt \underbrace{=}_{\text{by result (i)}} 0$$

Lastly we test $\mathbf{f}_n$ with $\mathbf{f}_m$ and $\mathbf{g}_n$ with $\mathbf{g}_m$ where $m \neq n$:

$$\langle \mathbf{f}_n, \mathbf{f}_m \rangle = \int_0^{2\pi} \cos(nt)\cos(mt)\, dt \underbrace{=}_{\text{by result (iv)}} 0 \text{ provided } m \neq n$$

$$\langle \mathbf{g}_n, \mathbf{g}_m \rangle = \int_0^{2\pi} \sin(nt)\sin(mt)\, dt \underbrace{=}_{\text{by result (v)}} 0 \text{ provided } m \neq n$$

Hence the inner product of *all distinct* vectors is zero, which means that the set $\{1, \cos(t), \cos(2t), \cos(3t), \ldots, \sin(t), \sin(2t), \sin(3t), \ldots\}$ is orthogonal.

Normalizing:

First we determine $\|\mathbf{f}_0\|^2$, $\|\mathbf{f}_n\|^2$ and $\|\mathbf{g}_n\|^2$ and then we take the square root:

$$\|\mathbf{f}_0\|^2 = \langle \mathbf{f}_0, \mathbf{f}_0 \rangle$$
$$= \int_0^{2\pi} (1 \times 1)\, dt$$
$$\underbrace{=}_{\text{integrating}} [t]_0^{2\pi} = 2\pi - 0 = 2\pi$$

$$\|\mathbf{f}_n\|^2 = \langle \mathbf{f}_n, \mathbf{f}_n \rangle$$
$$= \int_0^{2\pi} \cos(nt) \times \cos(nt)\, dt = \int_0^{2\pi} \cos^2(nt)\, dt \underbrace{=}_{\text{by result (vi)}} \pi$$

$$\|\mathbf{g}_n\|^2 = \langle \mathbf{g}_n, \mathbf{g}_n \rangle$$
$$= \int_0^{2\pi} \sin(nt) \times \sin(nt)\, dt = \int_0^{2\pi} \sin^2(nt)\, dt \underbrace{=}_{\text{by result (vi)}} \pi$$

Taking the square root of these results $\|\mathbf{f}_0\|^2 = 2\pi$, $\|\mathbf{f}_n\|^2 = \pi$ and $\|\mathbf{g}_n\|^2 = \pi$ gives $\|\mathbf{f}_0\| = \sqrt{2\pi}$, $\|\mathbf{f}_n\| = \sqrt{\pi}$ and $\|\mathbf{g}_n\| = \sqrt{\pi}$ respectively.

How can we normalize these (convert into unit vectors) $\mathbf{f}_0$, $\mathbf{f}_n$ and $\mathbf{g}_n$?

By using Proposition (4.11) $\mathbf{u} = \frac{\mathbf{w}}{\|\mathbf{w}\|}$ which means that we divide each vector by its norm. For all $n \geq 1$ we have

$$\frac{\mathbf{f}_0}{\|\mathbf{f}_0\|} = \frac{1}{\sqrt{2\pi}}, \quad \frac{\mathbf{f}_n}{\|\mathbf{f}_n\|} = \frac{\cos(nt)}{\sqrt{\pi}} \text{ and } \frac{\mathbf{g}_n}{\|\mathbf{g}_n\|} = \frac{\sin(nt)}{\sqrt{\pi}}$$

Hence the set $\left\{\frac{1}{\sqrt{2\pi}}, \frac{\cos(t)}{\sqrt{\pi}}, \frac{\cos(2t)}{\sqrt{\pi}}, \ldots, \frac{\sin(t)}{\sqrt{\pi}}, \frac{\sin(2t)}{\sqrt{\pi}}, \ldots\right\}$ is an orthonormal set of functions.

This orthonormal set is critical in the study of Fourier series. Note that the sines and cosines are orthogonal to each other in the vector space $C[0, 2\pi]$.

Summary

(4.8) Vectors $\mathbf{u}$ and $\mathbf{v}$ are orthogonal $\Leftrightarrow$ $\langle \mathbf{u}, \mathbf{v} \rangle = 0$.

(4.12) A set of perpendicular unit vectors

$$B = \{\mathbf{u}_1, \mathbf{u}_2, \mathbf{u}_3, \ldots, \mathbf{u}_n\}$$

for vector space V is called an orthonormal set.

EXERCISES 4.2

(Brief solutions at end of book. Full solutions available at <http://www.oup.co.uk/companion/singh>.)

1. Show that the following vectors in $\mathbb{R}^2$ are orthogonal with respect to the dot product. Also plot them on $\mathbb{R}^2$.

   (a) $\mathbf{u} = \begin{pmatrix} 1 \\ 1 \end{pmatrix}$ and $\mathbf{v} = \begin{pmatrix} -1 \\ 1 \end{pmatrix}$ (b) $\mathbf{u} = \begin{pmatrix} -2 \\ -3 \end{pmatrix}$ and $\mathbf{v} = \begin{pmatrix} 3 \\ -2 \end{pmatrix}$

   (c) $\mathbf{u} = \begin{pmatrix} -4 \\ 5 \end{pmatrix}$ and $\mathbf{v} = \begin{pmatrix} 5 \\ 4 \end{pmatrix}$ (d) $\mathbf{u} = \begin{pmatrix} 2 \\ 7 \end{pmatrix}$ and $\mathbf{v} = \begin{pmatrix} -7 \\ 2 \end{pmatrix}$

2. Show that the vectors $\mathbf{u} = \begin{pmatrix} a \\ b \end{pmatrix}$ and $\mathbf{v} = \begin{pmatrix} -b \\ a \end{pmatrix}$ are orthogonal in $\mathbb{R}^2$ with respect to the dot product.

3. Show that the following vectors in $\mathbb{R}^3$ form an orthogonal set with respect to the dot product.

(a) $\mathbf{u} = \begin{pmatrix} 5 \\ 0 \\ 0 \end{pmatrix}$, $\mathbf{v} = \begin{pmatrix} 0 \\ 1 \\ 0 \end{pmatrix}$ and $\mathbf{w} = \begin{pmatrix} 0 \\ 0 \\ 10 \end{pmatrix}$

(b) $\mathbf{u} = \begin{pmatrix} 0 \\ 0 \\ 0 \end{pmatrix}$, $\mathbf{v} = \begin{pmatrix} 1 \\ 1 \\ -1 \end{pmatrix}$ and $\mathbf{w} = \begin{pmatrix} -3 \\ 10 \\ 7 \end{pmatrix}$

4. Let the vector space of size 2 by 2 matrices M_{22} have an inner product defined by $\langle \mathbf{A}, \mathbf{B} \rangle = tr\left(\mathbf{B}^T\mathbf{A}\right)$ where tr is the trace of the matrix.

   (a) Show that the matrices $\mathbf{A} = \begin{pmatrix} 3 & 7 \\ 5 & 4 \end{pmatrix}$ and $\mathbf{B} = \begin{pmatrix} 2 & 1 \\ 7 & -12 \end{pmatrix}$ are orthogonal.

   (b) Determine $\|\mathbf{A} + \mathbf{B}\|$.

5. Determine the values of k so that the following vectors are orthogonal in $\mathbb{R}^4$ with respect to the dot product.
   (a) $\mathbf{u} = (1\ 2\ 3\ 4)^T$, $\mathbf{v} = \left(-2\ 3\ k\ 5\right)^T$ (b) $\mathbf{u} = \left(k\ -1\ k\ 1\right)^T$, $\mathbf{v} = \left(2\ 4\ k\ 5\right)^T$

6. Let $C[0, 1]$ be the vector space of continuous functions on the interval $[0, 1]$ with an inner product defined by

$$\langle \mathbf{f}, \mathbf{g} \rangle = \int_0^1 f(x)g(x)\, dx$$

   Verify the Cauchy–Schwarz inequality for
   (a) $f(x) = x$ and $g(x) = x - 1$ (b) $f(x) = 1$ and $g(x) = e^x$

7. Show that any non-zero vectors $\mathbf{x}$ and $\mathbf{y}$ in an inner product space satisfy:

$$0 \le \left| \left\langle \frac{\mathbf{x}}{\|\mathbf{x}\|}, \frac{\mathbf{y}}{\|\mathbf{y}\|} \right\rangle \right| \le 1$$

8. Let $C[-\pi, \pi]$ be the vector space of continuous functions with an inner product defined by

$$\langle \mathbf{f}, \mathbf{g} \rangle = \int_{-\pi}^{\pi} f(x)g(x)\, dx$$

   (a) Show that $\mathbf{f}$ and $\mathbf{g}$ are orthogonal for $f(x) = \cos(x)$ and $g(x) = \sin(x)$.
   (b) Verify the Cauchy–Schwarz inequality for $\mathbf{f}$ and $\mathbf{g}$ given in part (a).
   (c) Verify the Minkowski (triangular) inequality for $\mathbf{f}$ and $\mathbf{g}$ given in part (a).
   (d) Normalize these vectors $\mathbf{f}$ and $\mathbf{g}$.

   [You may use the following result: $\int_{-\pi}^{\pi} \sin^2(x)\, dx = \int_{-\pi}^{\pi} \cos^2(x)\, dx = \pi$]

9. Let $C[0, \pi]$ be the vector space of continuous functions on the interval $[0, \pi]$ with an inner product defined by

$$\langle \mathbf{f}, \mathbf{g}_n \rangle = \int_0^{\pi} f(t) g_n(t)\, dt$$

Let $f(t) = 1$ and $g_n(t) = \sin(2nt)$ where $n = 1, 2, 3, \ldots$

Show that the set $\{1, \sin(2t), \sin(4t), \sin(6t), \ldots\}$ is orthogonal. Normalize this set of vectors and write down the orthonormal set.

[You may use: $\int_0^{\pi} \sin(nt)\sin(mt) dt = 0$ provided $n \neq m$ and $\int_0^{\pi} \sin^2(nt)\, dt = \frac{\pi}{2}$]

10. Show that the following set S is an orthonormal set of 2 by 2 matrices M_{22} with an inner product defined by

$$\langle \mathbf{A}, \mathbf{B} \rangle = tr\left(\mathbf{B}^T \mathbf{A}\right)$$

$S = \{\mathbf{A}, \mathbf{B}, \mathbf{C}, \mathbf{D}\}$ where $\mathbf{A} = \begin{pmatrix} 1 & 0 \\ 0 & 0 \end{pmatrix}$, $\mathbf{B} = \begin{pmatrix} 0 & 1 \\ 0 & 0 \end{pmatrix}$, $\mathbf{C} = \begin{pmatrix} 0 & 0 \\ 1 & 0 \end{pmatrix}$ and $\mathbf{D} = \begin{pmatrix} 0 & 0 \\ 0 & 1 \end{pmatrix}$

11. Suppose a guitar string is modelled between 0 and 1 as shown in Fig. 4.10.

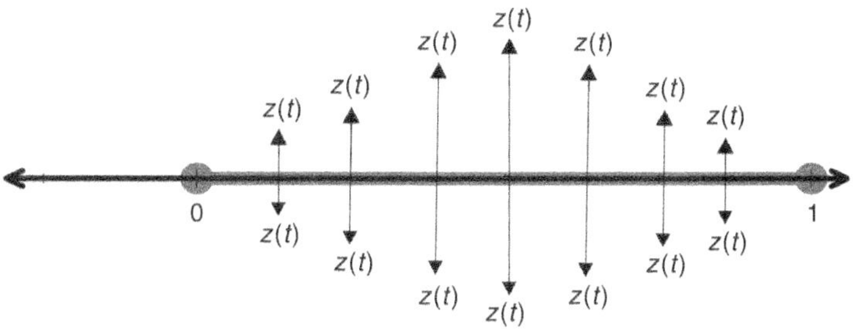

Figure 4.10

Consider a vertical force $f(t)$ at each point of the string and a vertical displacement of $z(t)$ at that point. The amount of energy released is given by the inner product:

$$\langle f(t), z(t) \rangle = \int_0^1 f(t)\, z(t)\, dt$$

Find the energy released if $f(t) = 100\sin(100t)$ and $z(t) = \cos(10t)$.

[Hint: $2\sin(A)\cos(B) = \sin(A + B) + \sin(A - B)$]

12. Prove Proposition (4.11).

13. Let $\mathbf{u}$ and $\mathbf{v}$ be orthogonal vectors in an inner product space. Prove that

$$\|\mathbf{u} - \mathbf{v}\|^2 = \|\mathbf{u}\|^2 + \|\mathbf{v}\|^2$$

14. Let $\mathbf{u}$ and $\mathbf{v}$ be orthonormal vectors in an inner product space V. Determine
(a) $\|\mathbf{u}+\mathbf{v}\|$ (b) $\|\mathbf{u}-\mathbf{v}\|$

15. (i) Let $\{\mathbf{u}_1, \mathbf{u}_2, \mathbf{u}_3, \ldots, \mathbf{u}_n\}$ be an orthogonal set of vectors in an inner product space. Prove Pythagoras' theorem for these vectors, that is

$$\|\mathbf{u}_1+\mathbf{u}_2+\mathbf{u}_3+\cdots+\mathbf{u}_n\|^2 = \|\mathbf{u}_1\|^2+\|\mathbf{u}_2\|^2+\|\mathbf{u}_3\|^2+\cdots+\|\mathbf{u}_n\|^2$$

(ii) Let $\{\mathbf{f}_1, \mathbf{f}_2, \mathbf{f}_3, \ldots, \mathbf{f}_n\}$ be an orthonormal set of vectors in inner product space of $C[0, \pi]$. Find the length $\|\mathbf{f}_1+\mathbf{f}_2+\mathbf{f}_3+\cdots+\mathbf{f}_n\|$.

16. Prove that the vectors $\{\mathbf{u}_1, \mathbf{u}_2, \ldots, \mathbf{u}_n\}$ are orthogonal $\Leftrightarrow$ $\{k_1\mathbf{u}_1, k_2\mathbf{u}_2, \ldots, k_n\mathbf{u}_n\}$ are orthogonal, where the k's are non-zero scalars. Is the result still valid if any of k's are zero?

17. Prove Minkowski inequality (4.7).

SECTION 4.3 Orthonormal Bases

By the end of this section you will be able to

- understand what is meant by an orthonormal basis
- find an orthonormal basis by using the Gram-Schmidt process

4.3.1 Introduction to an orthonormal bases

Why is an orthonormal basis important?

Generally, it is easier to work with an orthonormal basis (axes) rather than any other basis. For example, in $\mathbb{R}^2$, try working with a basis of $\mathbf{u} = \begin{pmatrix} 1 \\ 0 \end{pmatrix}$ and $\mathbf{v} = \begin{pmatrix} 0.9 \\ 0.1 \end{pmatrix}$. Writing $\mathbf{w} = \begin{pmatrix} 1 \\ 1 \end{pmatrix}$ in terms of these basis (axes) vectors $\mathbf{u}$ and $\mathbf{v}$ we have (Fig. 4.11).

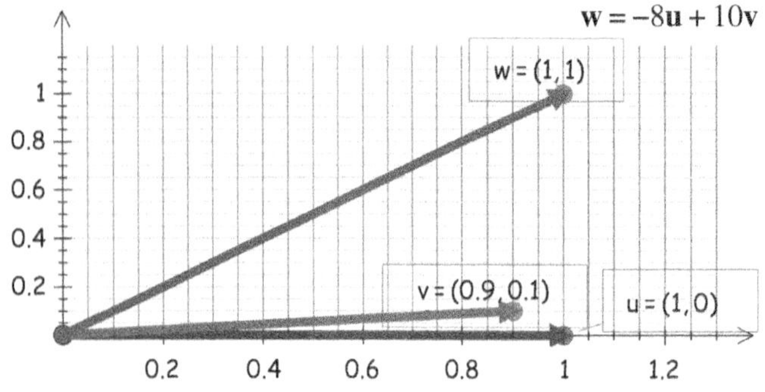

Figure 4.11

$$\mathbf{w} = \begin{pmatrix} 1 \\ 1 \end{pmatrix} = -8\begin{pmatrix} 1 \\ 0 \end{pmatrix} + 10\begin{pmatrix} 0.9 \\ 0.1 \end{pmatrix}$$
$$= -8\mathbf{u} + 10\mathbf{v}$$

Writing $\mathbf{w} = -8\mathbf{u} + 10\mathbf{v}$ involves a lot more arithmetic than expressing this vector $\mathbf{w}$ in our usual orthonormal basis $\mathbf{e}_1$ and $\mathbf{e}_2$ as $\mathbf{w} = \mathbf{e}_1 + \mathbf{e}_2$ because $\mathbf{e}_1 = \begin{pmatrix} 1 & 0 \end{pmatrix}^T$ and $\mathbf{e}_2 = \begin{pmatrix} 0 & 1 \end{pmatrix}^T$ are the unit vectors in the x and y directions respectively.

In an n-dimensional vector space there are n orthogonal (perpendicular) axes or basis vectors. We will show that an orthogonal set of n vectors is automatically linearly independent, and therefore forms a legitimate basis (normalizing is just a matter of scale). Generally, it is easier to show that vectors are orthogonal rather than linearly independent.

For Fourier series (which is used in signal processing), an example of an orthogonal basis is

$$\{1, \ \sin(nx), \ \cos(nx)\} \text{ where } n \text{ is a positive integer}$$

What do you think the term orthonormal basis means?

An orthonormal basis is a set of vectors which are normalized *and* are orthogonal to each other. They form a basis (axes) for the vector space.

Examples of orthonormal (perpendicular unit) basis are shown in Fig. 4.12 for $\mathbb{R}^2$ and $\mathbb{R}^3$:

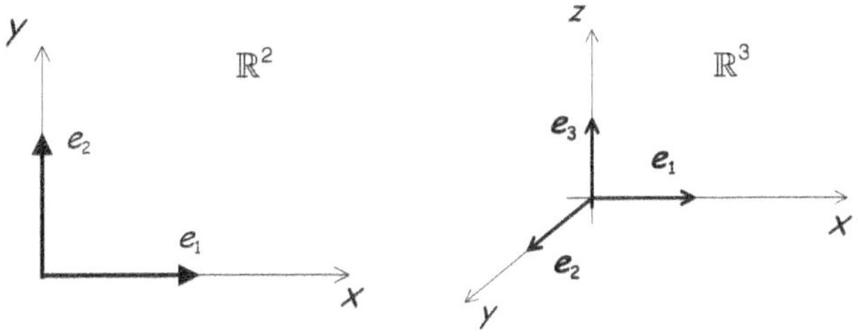

Figure 4.12

Note that our usual x, y and z axes are orthogonal to each other. The vectors $\{\mathbf{e}_1, \ \mathbf{e}_2\}$ form an orthonormal (perpendicular unit) basis for $\mathbb{R}^2$ and the set $\{\mathbf{e}_1, \ \mathbf{e}_2, \ \mathbf{e}_3\}$ forms an orthonormal (perpendicular unit) basis for $\mathbb{R}^3$.

In general, the set $B = \{\mathbf{e}_1, \ \mathbf{e}_2, \ \mathbf{e}_3, \ldots, \ \mathbf{e}_n\}$ forms an orthonormal (perpendicular unit) basis for $\mathbb{R}^n$ with respect to the dot product. Remember, $\mathbf{e}_k = \begin{pmatrix} 0 & \cdots & 1 & 0 & \cdots \end{pmatrix}^T$ (1 in the kth position and zeros everywhere else.)

Definition (4.13). Let V be a finite dimensional vector space with an inner product. A set of basis vectors

$$B = \{\mathbf{u}_1, \ \mathbf{u}_2, \ \mathbf{u}_3, \ldots, \ \mathbf{u}_n\}$$

for V is called an **orthonormal basis** if they are

(i) Orthogonal, that is $\langle \mathbf{u}_i, \mathbf{u}_j \rangle = 0$ for $i \neq j$

(ii) Normalized, that is $\|\mathbf{u}_j\| = 1$ for $j = 1, 2, 3, \ldots, n$

What do you notice about this definition?

It's the same as the definition of an orthonormal set (4.12) given in the last section, but this time the set of vectors are the *basis* vectors of the vector space.

Can you remember what the term basis means?

It's a set of vectors which represent the axes of the vector space and satisfies two conditions:

1) Span the space 2) Linearly independent

Basis is like a coordinate system of a vector space – you can write any vector in terms of basis vectors.

For example, the normalized sine and cosine functions form an orthonormal basis for the vector space of periodic continuous functions, $C[0, 2\pi]$. (See Example 4.13 of section 4.2.4.)

To check that a given set of vectors form a basis for a vector space V can be tedious.

Why?

Because normally we need to check two things about a basis:

1) the vectors span V 2) the vectors are linearly independent.

From the last chapter, we know that it is enough to show that for any n-dimensional vector space, n linearly *independent* vectors form a *basis* for this vector space. We only need to check condition (2). Next we show that an *orthogonal* set is linearly *independent*, that is

$$\text{orthogonality} \Rightarrow \text{linear independence}$$

4.3.2 Properties of an orthonormal basis

In this subsection, we prove that if we have an orthogonal set of vectors then they are linearly independent. We use this to prove that in an n-dimensional vector space, any set of n orthogonal non-zero vectors forms a basis for that vector space. (Generally, checking orthogonality is easier than checking linear independence.)

Proposition (4.14). If $\{\mathbf{v}_1, \mathbf{v}_2, \mathbf{v}_3, \ldots, \mathbf{v}_n\}$ is an *orthogonal* set of non-zero vectors in an inner product space then the elements of this set are *linearly independent*.

How do we prove this proposition?

We want to prove that the vectors $\{\mathbf{v}_1, \mathbf{v}_2, \mathbf{v}_3, \ldots, \mathbf{v}_n\}$ are linearly independent, so we consider the linear combination:

$$k_1\mathbf{v}_1 + k_2\mathbf{v}_2 + \cdots + k_n\mathbf{v}_n = \mathbf{O}$$

and show that *all* the scalars are zero: $k_1 = 0,\ k_2 = 0,\ k_3 = 0, \ldots$ and $k_n = 0$.

Why?

Because from chapter 3 we have:

(3.10). Vectors $\mathbf{v}_1, \mathbf{v}_2, \ldots, \mathbf{v}_n$ are **linearly independent** $\Leftrightarrow$ the only solution to $k_1\mathbf{v}_1 + k_2\mathbf{v}_2 + k_3\mathbf{v}_3 + \cdots + k_n\mathbf{v}_n = \mathbf{O}$ is $k_1 = k_2 = k_3 = \cdots = k_n = 0$

Proof.

Consider the linear combination of the vectors $\{\mathbf{v}_1, \mathbf{v}_2, \mathbf{v}_3, \ldots, \mathbf{v}_n\}$ and equate them to the zero vector, $\mathbf{O}$:

$$k_1\mathbf{v}_1 + k_2\mathbf{v}_2 + k_3\mathbf{v}_3 + \cdots + k_n\mathbf{v}_n = \mathbf{O} \qquad (\dagger)$$

Consider the inner product of an arbitrary vector $\mathbf{v}_j$ in the set $\{\mathbf{v}_1, \mathbf{v}_2, \mathbf{v}_3, \ldots, \mathbf{v}_n\}$ with the zero vector given in $(\dagger)$.

What is the inner product $\langle k_1\mathbf{v}_1 + k_2\mathbf{v}_2 + k_3\mathbf{v}_3 + \cdots + k_n\mathbf{v}_n, \mathbf{v}_j\rangle$ equal to?

0, because

$$\langle k_1\mathbf{v}_1 + k_2\mathbf{v}_2 + k_3\mathbf{v}_3 + \cdots + k_n\mathbf{v}_n, \mathbf{v}_j\rangle = \langle \mathbf{O}, \mathbf{v}_j\rangle = 0 \qquad (*)$$

We are given that $\{\mathbf{v}_1, \mathbf{v}_2, \mathbf{v}_3, \ldots, \mathbf{v}_n\}$ is orthogonal, this means that the inner product between two *different* vectors is *zero*:

$$\langle \mathbf{v}_i, \mathbf{v}_j\rangle = 0 \text{ if } i \neq j$$

Expanding (*) and using $\langle \mathbf{v}_i, \mathbf{v}_j\rangle = 0$ if $i \neq j$ we have

$$\begin{aligned}
\langle k_1\mathbf{v}_1 + \cdots + k_n\mathbf{v}_n, \mathbf{v}_j\rangle &= \langle k_1\mathbf{v}_1, \mathbf{v}_j\rangle + \cdots + \langle k_j\mathbf{v}_j, \mathbf{v}_j\rangle + \cdots + \langle k_n\mathbf{v}_n, \mathbf{v}_j\rangle \\
&\underbrace{=}_{\text{taking out scalars}} k_1\langle \mathbf{v}_1, \mathbf{v}_j\rangle + \cdots + k_j\langle \mathbf{v}_j, \mathbf{v}_j\rangle + \cdots + k_n\langle \mathbf{v}_n, \mathbf{v}_j\rangle \\
&= k_1\underbrace{\langle \mathbf{v}_1, \mathbf{v}_j\rangle}_{=0}\underbrace{+\cdots+}_{=0} k_j\langle \mathbf{v}_j, \mathbf{v}_j\rangle \underbrace{+\cdots+}_{=0} k_n\underbrace{\langle \mathbf{v}_n, \mathbf{v}_j\rangle}_{=0} \\
&= k_1(0) + 0\cdots 0 + k_j\|\mathbf{v}_j\|^2 + 0\cdots 0 + k_n(0) = k_j\|\mathbf{v}_j\|^2
\end{aligned}$$

The last line follows from the definition of the norm $\langle \mathbf{u}, \mathbf{u}\rangle = \|\mathbf{u}\|^2$. The *only* non-zero contribution is the inner product of the vectors $\mathbf{v}_j$ and $\mathbf{v}_j$. By (*) we know that all this is equal to zero, which means that we have

$$k_j\|\mathbf{v}_j\|^2 = 0$$

$\mathbf{v}_j$ is one of the *non-zero* vectors among the set $\{\mathbf{v}_1, \mathbf{v}_2, \mathbf{v}_3, \ldots, \mathbf{v}_n\}$, therefore $\|\mathbf{v}_j\|^2 \neq 0$ [not equal to 0]. Thus to satisfy the above $k_j\|\mathbf{v}_j\|^2 = 0$ we must have $k_j = 0$. Since $\mathbf{v}_j$ was an arbitrary vector, we conclude that *all* the scalars k's in $(\dagger)$ must be zero, which proves that an *orthogonal* set of vectors is linearly independent.

■

What does this proposition mean?

It means that *orthogonality* implies linear *independence*. For example, the vectors $\{\mathbf{e}_1, \mathbf{e}_2, \mathbf{e}_3\}$ are orthogonal (perpendicular), therefore linearly independent.

We can go further, as the next proposition states.

Corollary (4.15). In an n-dimensional inner product space V, any set of n orthogonal non-zero vectors forms a basis (or axes) for V.

Which tools do we use to prove this result?

Theorem (3.22) (a) of the last chapter which says:

Any linearly *independent* set of n vectors $\{\mathbf{v}_1, \mathbf{v}_2, \mathbf{v}_3, \ldots, \mathbf{v}_n\}$ forms a basis for V.

Proof.

By the previous Proposition (4.14), we know that the set of n orthogonal vectors are linearly independent.

By Theorem (3.22) we conclude that n orthogonal vectors form a basis for V. ■

What does this corollary mean?

If we have an n-dimensional vector space with an inner product then *any* n orthogonal (perpendicular) non-zero vectors form a set of basis (axes) vectors for that vector space.

However, vectors which are linearly independent may *not* be orthogonal. For example, the vectors $\mathbf{u} = (3\ 1)^T$ and $\mathbf{v} = (1\ 2)^T$ in $\mathbb{R}^2$ are linearly independent (we cannot make $\mathbf{v}$ from a scalar multiple of $\mathbf{u}$, and vice versa) but *not* orthogonal as you can see in Fig. 4.13.

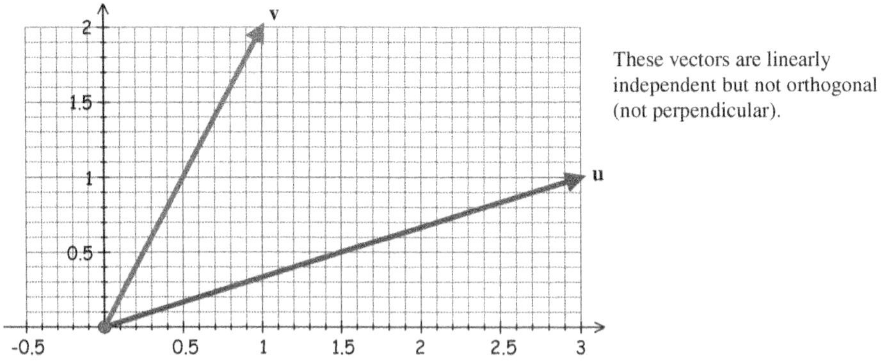

Figure 4.13

Hence orthogonal vectors are independent, but independent vectors are *not* necessarily orthogonal. We have

$$\text{orthogonality} \Rightarrow \text{independence but independence} \not\Rightarrow \text{orthogonality}$$

Example 4.14

Show that $\mathbf{u} = \begin{pmatrix} -1 \\ 1 \end{pmatrix}$ and $\mathbf{v} = \begin{pmatrix} -1 \\ -1 \end{pmatrix}$ are orthogonal in $\mathbb{R}^2$ with respect to the dot product and write down an orthonormal (perpendicular unit) basis (axes) for $\mathbb{R}^2$.

Solution

How do we show that the given vectors **u** *and* **v** *are orthogonal (perpendicular)?*

By showing the inner (dot) product of these vectors is zero:

$$\langle \mathbf{u}, \mathbf{v} \rangle = \mathbf{u} \cdot \mathbf{v} = \begin{pmatrix} -1 \\ 1 \end{pmatrix} \cdot \begin{pmatrix} -1 \\ -1 \end{pmatrix} = (-1 \times (-1)) + (1 \times (-1)) = 0$$

Hence vectors **u** and **v** are orthogonal (perpendicular).

How do we normalize these vectors?

By using Proposition (4.11) $\frac{\mathbf{w}}{\|\mathbf{w}\|}$, which means dividing a vector by its length.

What is length $\|\mathbf{u}\|$ *equal to?*

$$\|\mathbf{u}\|^2 = \langle \mathbf{u}, \mathbf{u} \rangle = \mathbf{u} \cdot \mathbf{u} = \begin{pmatrix} -1 \\ 1 \end{pmatrix} \cdot \begin{pmatrix} -1 \\ 1 \end{pmatrix} = (-1)^2 + 1^2 = 2$$

Taking the square root gives $\|\mathbf{u}\| = \sqrt{2}$. Similarly we have $\|\mathbf{v}\| = \sqrt{2}$.

Thus the normalized (unit) vectors are

$$\frac{\mathbf{u}}{\|\mathbf{u}\|} = \frac{1}{\sqrt{2}} \begin{pmatrix} -1 \\ 1 \end{pmatrix} \text{ and } \frac{\mathbf{v}}{\|\mathbf{v}\|} = \frac{1}{\sqrt{2}} \begin{pmatrix} -1 \\ -1 \end{pmatrix}$$

An orthonormal (perpendicular unit) basis (axes) for $\mathbb{R}^2$ is $\left\{ \frac{1}{\sqrt{2}} \begin{pmatrix} -1 \\ 1 \end{pmatrix}, \frac{1}{\sqrt{2}} \begin{pmatrix} -1 \\ -1 \end{pmatrix} \right\}$.

We can plot this orthonormal basis (vectors representing new axes) for $\mathbb{R}^2$ (Fig. 4.14).

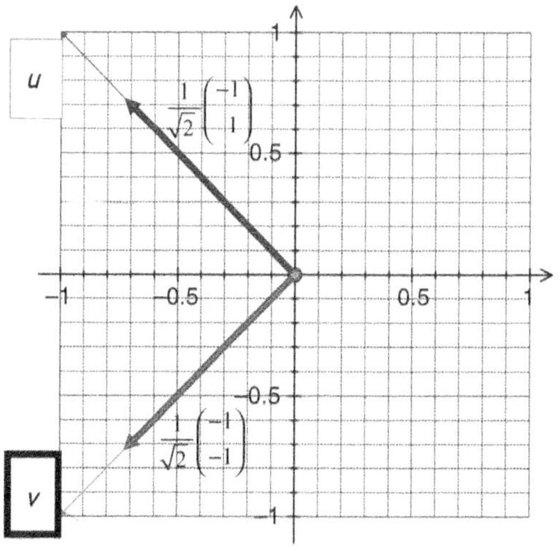

Figure 4.14

4.3.3 Generating orthogonal vectors

Consider two given vectors **u** and **v** in $\mathbb{R}^2$. We can project the vector **v** onto vector **u**.

What does projection mean?

Projection is the procedure, for example, of showing a film on a screen; we say the film has been projected onto a screen. Similarly we project the vector **v** onto **u** as shown in Fig. 4.15.

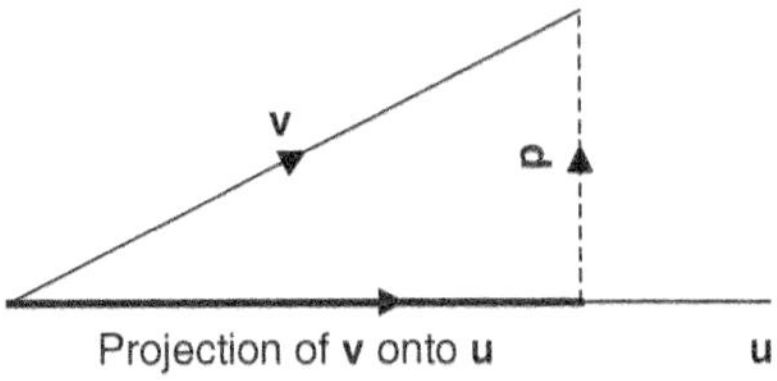

Figure 4.15

What is the projection of **v** *onto* **u** *equal to?*

Since it is in the direction of vector **u**, it is a scalar multiple of **u**. Let's nominate this scalar by the letter k, so we have

$$\text{Projection of } \mathbf{v} \text{ onto } \mathbf{u} = k\mathbf{u}$$

Let **p** be the perpendicular vector shown in Fig. 4.15. Adding the vectors gives

$$\mathbf{v} = k\mathbf{u} + \mathbf{p}$$

The vector **p** is orthogonal (perpendicular) to the vector **u**, therefore $\langle \mathbf{u}, \mathbf{p} \rangle = 0$. This means that the projection of the perpendicular vector **p** onto vector **u** is zero.

Taking the inner product of this, $\mathbf{v} = k\mathbf{u} + \mathbf{p}$, with the vector **u**, we have

$$\begin{aligned} \langle \mathbf{u}, \mathbf{v} \rangle &= \langle \mathbf{u}, (k\mathbf{u} + \mathbf{p}) \rangle \\ &= k\langle \mathbf{u}, \mathbf{u} \rangle + \langle \mathbf{u}, \mathbf{p} \rangle \\ &= k\|\mathbf{u}\|^2 + 0 && \left[\begin{array}{l}\text{because } \mathbf{p} \text{ and } \mathbf{u} \text{ are perpendicular,} \\ \text{therefore } \langle \mathbf{u}, \mathbf{p} \rangle = 0 \text{ and } \langle \mathbf{u}, \mathbf{u} \rangle = \|\mathbf{u}\|^2\end{array}\right] \\ &= k\|\mathbf{u}\|^2 \end{aligned}$$

Rearranging this $\langle \mathbf{u}, \mathbf{v} \rangle = k\|\mathbf{u}\|^2$ to make k the subject of the formula gives

$$k = \frac{\langle \mathbf{u}, \mathbf{v} \rangle}{\|\mathbf{u}\|^2} \qquad (\dagger)$$

Rearranging the above $\mathbf{v} = k\mathbf{u} + \mathbf{p}$ to make **p** the subject:

$$\begin{aligned} \text{orthogonal vector } \mathbf{p} &= \mathbf{v} - k\mathbf{u} && [\mathbf{p} = \mathbf{v} - [\text{projection of } \mathbf{v} \text{ onto } \mathbf{u}]] \\ &= \mathbf{v} - \frac{\langle \mathbf{u}, \mathbf{v} \rangle}{\|\mathbf{u}\|^2}\mathbf{u} && [\text{by } (\dagger)] \end{aligned}$$

We have perpendicular vector **p** (Fig. 4.16):

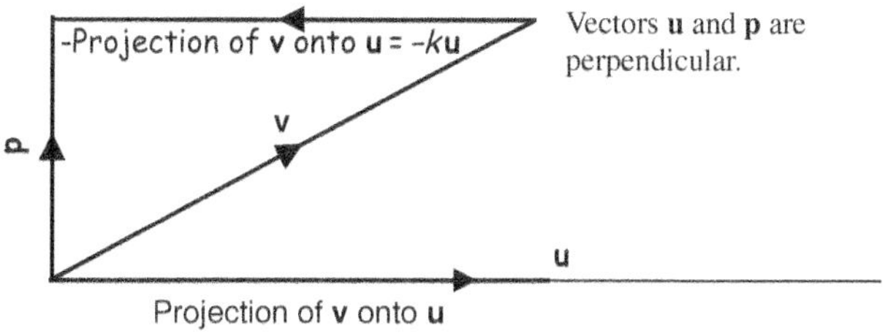

Vectors **u** and **p** are perpendicular.

Figure 4.16

Hence we have created orthogonal vectors **p** and **u** out of the given non-orthogonal vectors **u** and **v**.

It is important to note that the projection of orthogonal vectors is zero. This is what we were hinting at in the last section; *orthogonality signifies a certain kind of independence or a complete absence of interference.*

Example 4.15

Let $\mathbf{v}_1 = \begin{pmatrix} 3 \\ 0 \end{pmatrix}$ and $\mathbf{v}_2 = \begin{pmatrix} 1 \\ 2 \end{pmatrix}$ be vectors in $\mathbb{R}^2$. Construct a pair of orthogonal (perpendicular) vectors $\{\mathbf{p}_1, \mathbf{p}_2\}$ from this non-orthogonal set $\{\mathbf{v}_1, \mathbf{v}_2\}$.

Solution

We start with one of the given vectors, $\mathbf{v}_1$ say, and call this vector $\mathbf{p}_1$. Hence $\mathbf{p}_1 = \mathbf{v}_1$. We construct a vector which is orthogonal to $\mathbf{p}_1 = \mathbf{v}_1$.

By the above formula, orthogonal vector $\mathbf{p}_2 = \mathbf{v} - \dfrac{\langle \mathbf{u}, \mathbf{v} \rangle}{\|\mathbf{u}\|^2}\mathbf{u}$ with $\mathbf{u} = \mathbf{v}_1 = \mathbf{p}_1$ and $\mathbf{v} = \mathbf{v}_2$:

$$\text{orthogonal vector } \mathbf{p}_2 = \mathbf{v}_2 - \frac{\langle \mathbf{p}_1, \mathbf{v}_2 \rangle}{\|\mathbf{p}_1\|^2}\mathbf{p}_1$$

Substituting $\mathbf{p}_1 = \mathbf{v}_1 = \begin{pmatrix} 3 \\ 0 \end{pmatrix}$ and $\mathbf{v}_2 = \begin{pmatrix} 1 \\ 2 \end{pmatrix}$ into this formula gives:

$$\text{orthogonal vector } \mathbf{p}_2 = \begin{pmatrix} 1 \\ 2 \end{pmatrix} - \frac{\left\langle \begin{pmatrix} 3 \\ 0 \end{pmatrix}, \begin{pmatrix} 1 \\ 2 \end{pmatrix} \right\rangle}{\left\| \begin{pmatrix} 3 \\ 0 \end{pmatrix} \right\|^2} \begin{pmatrix} 3 \\ 0 \end{pmatrix} \qquad (*)$$

Evaluating each component of (*):

$$\left\langle \begin{pmatrix} 3 \\ 0 \end{pmatrix}, \begin{pmatrix} 1 \\ 2 \end{pmatrix} \right\rangle = \begin{pmatrix} 3 \\ 0 \end{pmatrix} \cdot \begin{pmatrix} 1 \\ 2 \end{pmatrix} = (3 \times 1) + (0 \times 2) = 3, \quad \left\| \begin{pmatrix} 3 \\ 0 \end{pmatrix} \right\|^2 = \begin{pmatrix} 3 \\ 0 \end{pmatrix} \cdot \begin{pmatrix} 3 \\ 0 \end{pmatrix} = 3^2 + 0^2 = 9$$

Putting these values into (*) yields

$$\text{orthogonal vector } \mathbf{p}_2 = \begin{pmatrix} 1 \\ 2 \end{pmatrix} - \frac{3}{9}\begin{pmatrix} 3 \\ 0 \end{pmatrix} = \begin{pmatrix} 1 \\ 2 \end{pmatrix} - \frac{1}{3}\begin{pmatrix} 3 \\ 0 \end{pmatrix} = \begin{pmatrix} 1-1 \\ 2-0 \end{pmatrix} = \begin{pmatrix} 0 \\ 2 \end{pmatrix}$$

Illustrating the orthogonal vector $\mathbf{p}_2$ and the given vectors $\mathbf{v}_1$ and $\mathbf{v}_2$ as shown in Fig. 4.17.

(continued...)

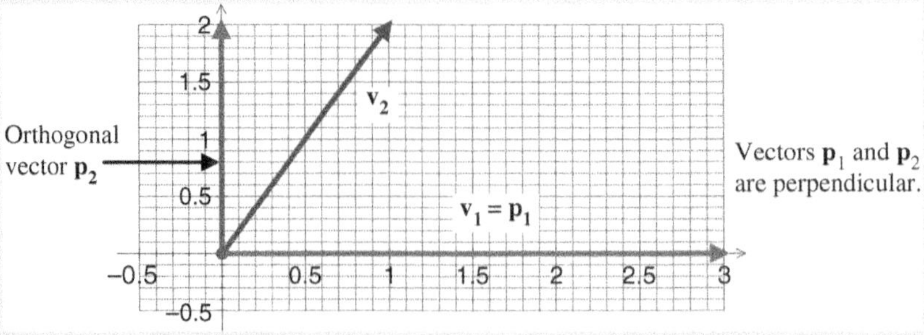

Figure 4.17

Note that we start with one of the given vectors, $\mathbf{v}_1$ say, then we create a vector orthogonal (perpendicular) to it by using the other given vector $\mathbf{v}_2$. We have achieved the following:

$$\left\{\mathbf{v}_1 = \begin{pmatrix} 3 \\ 0 \end{pmatrix}, \mathbf{v}_2 = \begin{pmatrix} 1 \\ 2 \end{pmatrix}\right\} \Rightarrow \left\{\mathbf{p}_1 = \begin{pmatrix} 3 \\ 0 \end{pmatrix}, \mathbf{p}_2 = \begin{pmatrix} 0 \\ 2 \end{pmatrix}\right\}$$

We can extend this procedure to any finite dimensional vector space and create an **orthogonal basis** for the vector space. Suppose we have a basis $\{\mathbf{v}_1, \mathbf{v}_2, \mathbf{v}_3, \ldots, \mathbf{v}_n\}$, and from this we want create an orthogonal basis $\{\mathbf{p}_1, \mathbf{p}_2, \mathbf{p}_3, \ldots, \mathbf{p}_n\}$. The procedure is:

1. Let $\mathbf{p}_1 = \mathbf{v}_1$, that is $\mathbf{p}_1$ equals one of the given vectors.
2. We create vector $\mathbf{p}_2$, which is orthogonal to $\mathbf{p}_1 = \mathbf{v}_1$, by using the second given vector $\mathbf{v}_2$. Hence we apply the above stated formula $\mathbf{p}_2 = \mathbf{v}_2 - \dfrac{\langle \mathbf{v}_2, \mathbf{p}_1 \rangle}{\|\mathbf{p}_1\|^2}\mathbf{p}_1$.
3. We create vector $\mathbf{p}_3$, which is orthogonal to both $\mathbf{p}_1 = \mathbf{v}_1$ and $\mathbf{p}_2$, by using the third given vector $\mathbf{v}_3$. The formula for this is similarly produced:

$$\mathbf{p}_3 = \mathbf{v}_3 - \frac{\langle \mathbf{v}_3, \mathbf{p}_1 \rangle}{\|\mathbf{p}_1\|^2}\mathbf{p}_1 - \frac{\langle \mathbf{v}_3, \mathbf{p}_2 \rangle}{\|\mathbf{p}_2\|^2}\mathbf{p}_2$$

4. We carry on producing vectors which are orthogonal (perpendicular) to the previous created vectors $\mathbf{p}_1, \mathbf{p}_2, \ldots, \mathbf{p}_k$ by using the next given vector $\mathbf{v}_{k+1}$.

These steps are known as the Gram–Schmidt process.

4.3.4 The Gram–Schmidt process

Given any arbitrary basis $\{\mathbf{v}_1, \mathbf{v}_2, \mathbf{v}_3, \ldots, \mathbf{v}_n\}$ for a finite dimensional inner product space, we can find an orthogonal basis $\{\mathbf{p}_1, \mathbf{p}_2, \mathbf{p}_3, \ldots, \mathbf{p}_n\}$ by the Gram–Schmidt process which is described next:

Gram–Schmidt process (4.16).

Let $\mathbf{p}_1 = \mathbf{v}_1$

$$\mathbf{p}_2 = \mathbf{v}_2 - \frac{\langle \mathbf{v}_2, \mathbf{p}_1 \rangle}{\|\mathbf{p}_1\|^2}\mathbf{p}_1$$

$$\mathbf{p}_3 = \mathbf{v}_3 - \frac{\langle \mathbf{v}_3, \mathbf{p}_1 \rangle}{\|\mathbf{p}_1\|^2}\mathbf{p}_1 - \frac{\langle \mathbf{v}_3, \mathbf{p}_2 \rangle}{\|\mathbf{p}_2\|^2}\mathbf{p}_2$$

$$\mathbf{p}_4 = \mathbf{v}_4 - \frac{\langle \mathbf{v}_4, \mathbf{p}_1 \rangle}{\|\mathbf{p}_1\|^2}\mathbf{p}_1 - \frac{\langle \mathbf{v}_4, \mathbf{p}_2 \rangle}{\|\mathbf{p}_2\|^2}\mathbf{p}_2 - \frac{\langle \mathbf{v}_4, \mathbf{p}_3 \rangle}{\|\mathbf{p}_3\|^2}\mathbf{p}_3$$

$$\vdots \quad \vdots \qquad \vdots \qquad\qquad \vdots \qquad\qquad \vdots$$

$$\mathbf{p}_n = \mathbf{v}_n - \frac{\langle \mathbf{v}_n, \mathbf{p}_1 \rangle}{\|\mathbf{p}_1\|^2}\mathbf{p}_1 - \frac{\langle \mathbf{v}_n, \mathbf{p}_2 \rangle}{\|\mathbf{p}_2\|^2}\mathbf{p}_2 - \frac{\langle \mathbf{v}_n, \mathbf{p}_3 \rangle}{\|\mathbf{p}_3\|^2}\mathbf{p}_3 - \cdots - \frac{\langle \mathbf{v}_n, \mathbf{p}_{n-1} \rangle}{\|\mathbf{p}_{n-1}\|^2}\mathbf{p}_{n-1}$$

These vectors $\mathbf{p}$ are orthogonal, therefore a basis.

What else do we need to do to form an orthonormal basis?

Normalize $\{\mathbf{p}_1, \mathbf{p}_2, \mathbf{p}_3, \ldots, \mathbf{p}_n\}$ by using Proposition (4.11) which is $\frac{\mathbf{p}_j}{\|\mathbf{p}_j\|}$. [Dividing by their norms (lengths).]

The proof of Gram–Schmidt process (4.16) for an inner product space is given on the website <http://www.oup.co.uk/companion/singh>.

Example 4.16

Transform the basis $\mathbf{v}_1 = \begin{pmatrix} 2 \\ 1 \end{pmatrix}$ and $\mathbf{v}_2 = \begin{pmatrix} 1 \\ 1 \end{pmatrix}$ in $\mathbb{R}^2$ to an orthonormal (perpendicular unit) basis (axes) for $\mathbb{R}^2$ with respect to the dot product. Also plot this orthonormal basis and the given basis on two different graphs.

Solution

Using the above Gram–Schmidt process (4.16) means that we need to find vectors $\mathbf{p}_1$ and $\mathbf{p}_2$ which are orthogonal (perpendicular) to each other.

We have $\mathbf{p}_1 = \mathbf{v}_1 = \begin{pmatrix} 2 \\ 1 \end{pmatrix}$ which is one of the given vectors. By the above Gram–Schmidt process (4.16) we have

$$\mathbf{p}_2 = \mathbf{v}_2 - \frac{\langle \mathbf{v}_2, \mathbf{p}_1 \rangle}{\|\mathbf{p}_1\|^2}\mathbf{p}_1 \qquad (*)$$

We need to evaluate each component of (*).

What is the inner product $\langle \mathbf{v}_2, \mathbf{p}_1 \rangle$ equal to?

$$\langle \mathbf{v}_2, \mathbf{p}_1 \rangle = \mathbf{v}_2 \cdot \mathbf{p}_1 = \begin{pmatrix} 1 \\ 1 \end{pmatrix} \cdot \begin{pmatrix} 2 \\ 1 \end{pmatrix} = (1 \times 2) + (1 \times 1) = 3$$

(continued...)

What is $\|\mathbf{p}_1\|^2$ equal to?

$$\|\mathbf{p}_1\|^2 = \langle \mathbf{p}_1, \mathbf{p}_1 \rangle = \mathbf{p}_1 \cdot \mathbf{p}_1$$

$$= \begin{pmatrix} 2 \\ 1 \end{pmatrix} \cdot \begin{pmatrix} 2 \\ 1 \end{pmatrix} = 2^2 + 1^2 = 5 \qquad \text{[Pythagoras]}$$

Substituting all these, $\mathbf{v}_2 = \begin{pmatrix} 1 \\ 1 \end{pmatrix}$, $\langle \mathbf{v}_2, \mathbf{p}_1 \rangle = 3$, $\|\mathbf{p}_1\|^2 = 5$ and $\mathbf{p}_1 = \begin{pmatrix} 2 \\ 1 \end{pmatrix}$, into (*) gives

$$\mathbf{p}_2 = \mathbf{v}_2 - \frac{\langle \mathbf{v}_2, \mathbf{p}_1 \rangle}{\|\mathbf{p}_1\|^2} \mathbf{p}_1$$

$$= \begin{pmatrix} 1 \\ 1 \end{pmatrix} - \frac{3}{5} \begin{pmatrix} 2 \\ 1 \end{pmatrix}$$

$$= \begin{pmatrix} 1 - 6/5 \\ 1 - 3/5 \end{pmatrix} = \begin{pmatrix} -1/5 \\ 2/5 \end{pmatrix} = \frac{1}{5} \begin{pmatrix} -1 \\ 2 \end{pmatrix}$$

Our vectors $\mathbf{p}_1 = \begin{pmatrix} 2 \\ 1 \end{pmatrix}$ and $\mathbf{p}_2 = \frac{1}{5} \begin{pmatrix} -1 \\ 2 \end{pmatrix}$ are an orthogonal (perpendicular) basis.

But how do we make these into an orthonormal (perpendicular unit) basis?

By normalizing these vectors (converting into unit vectors), which is achieved by dividing the vectors by their norms (lengths). Call these new unit vectors $\mathbf{u}_1$ and $\mathbf{u}_2$, and they are

$$\mathbf{u}_1 = \frac{\mathbf{p}_1}{\|\mathbf{p}_1\|} \quad \text{and} \quad \mathbf{u}_2 = \frac{\mathbf{p}_2}{\|\mathbf{p}_2\|}$$

From above $\|\mathbf{p}_1\|^2 = 5$ and taking square root gives $\|\mathbf{p}_1\| = \sqrt{5}$. We need to normalize the other perpendicular vector $\mathbf{p}_2$, which means we need to convert it to a unit vector:

$$\|\mathbf{p}_2\|^2 = \langle \mathbf{p}_2, \mathbf{p}_2 \rangle = \frac{1}{5} \begin{pmatrix} -1 \\ 2 \end{pmatrix} \cdot \frac{1}{5} \begin{pmatrix} -1 \\ 2 \end{pmatrix}$$

$$= \frac{1}{25} \left[(-1)^2 + 2^2 \right] = \frac{5}{25} = \frac{1}{5}$$

Thus $\|\mathbf{p}_2\| = \frac{1}{\sqrt{5}}$. Our normalized (unit) vectors are

$$\mathbf{u}_1 = \frac{\mathbf{p}_1}{\|\mathbf{p}_1\|} = \frac{1}{\|\mathbf{p}_1\|} \mathbf{p}_1 = \frac{1}{\sqrt{5}} \begin{pmatrix} 2 \\ 1 \end{pmatrix}$$

$$\mathbf{u}_2 = \frac{1}{\|\mathbf{p}_2\|} \mathbf{p}_2 = \frac{1}{\frac{1}{\sqrt{5}}} \frac{1}{5} \begin{pmatrix} -1 \\ 2 \end{pmatrix} = \frac{1}{\sqrt{5}} \begin{pmatrix} -1 \\ 2 \end{pmatrix} \quad \left[\text{By Law of Indices } \frac{1}{\sqrt{5}} 5 = \sqrt{5} \right]$$

Our orthonormal (perpendicular unit) basis is the set $\{\mathbf{u}_1, \mathbf{u}_2\}$ where these are given above:

$$\mathbf{u}_1 = \frac{1}{\sqrt{5}} \begin{pmatrix} 2 \\ 1 \end{pmatrix} \quad \text{and} \quad \mathbf{u}_2 = \frac{1}{\sqrt{5}} \begin{pmatrix} -1 \\ 2 \end{pmatrix}$$

Plotting these orthonormal (perpendicular unit) basis and our given basis in $\mathbb{R}^2$ (Fig. 4.18):

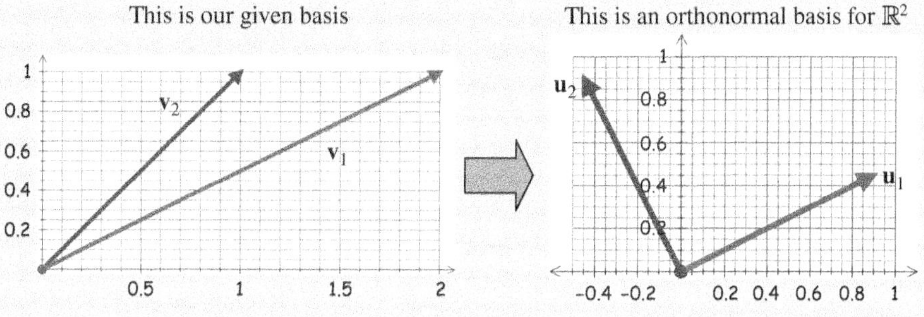

Figure 4.18

From Fig. 4.18 you should observe that $\mathbf{u}_1$ is in the same direction as the given vector $\mathbf{v}_1$ but has a length of 1. Remember, we start with $\mathbf{p}_1 = \mathbf{v}_1$, and $\mathbf{u}_1$ is a unit vector in that direction.

We can generally ignore the fraction (scalars) because vectors are orthogonal independent of scalars, as was established in question 16 of Exercises 4.2 which says:

Vectors $\{\mathbf{v}_1, \mathbf{v}_2, \ldots, \mathbf{v}_n\}$ are orthogonal $\Leftrightarrow$ $\{k_1\mathbf{v}_1, k_2\mathbf{v}_2, \ldots, k_n\mathbf{v}_n\}$ are orthogonal.

For example, in the above case we have $\mathbf{p}_2 = \frac{1}{5}\begin{pmatrix}-1\\2\end{pmatrix}$. Ignore the scalar (fraction) $\frac{1}{5}$ and normalize (convert to unit) the vector $\begin{pmatrix}-1\\2\end{pmatrix}$:

$$\left\|\begin{pmatrix}-1\\2\end{pmatrix}\right\|^2 = \begin{pmatrix}-1\\2\end{pmatrix}\cdot\begin{pmatrix}-1\\2\end{pmatrix} = (-1)^2 + 2^2 = 5 \text{ implies } \left\|\begin{pmatrix}-1\\2\end{pmatrix}\right\| = \sqrt{5}$$

Normalizing this vector gives $\mathbf{u}_2 = \frac{1}{\sqrt{5}}\begin{pmatrix}-1\\2\end{pmatrix}$, which is the same unit vector $\mathbf{u}_2$ as in the above Example 4.16. It makes the arithmetic a lot easier if we ignore fractions or any other scalars.

The orthonormal basis is *not* unique. For instance, if you do *not* remove the scalar (fraction), you could have a *different* orthonormal basis. Also, if we apply the Gram–Schmidt process in reverse order, that is $\{\mathbf{v}_2, \mathbf{v}_1\}$, then the result may be a *different* orthonormal basis. In question 2 of Exercises 4.3, we have switched vectors $\mathbf{v}_1$ and $\mathbf{v}_2$ of Example 4.16. See if you obtain the same or different set of vectors for the orthonormal basis.

We can also apply the Gram–Schmidt process to other vector spaces. Next we apply this to vector space of polynomials.

Example 4.17

Let P_2 be the vector space of polynomials of degree 2 or less with inner product given by:

$$\langle \mathbf{q}, \mathbf{r} \rangle = \int_{-1}^{1} q(x)\, r(x)\, dx$$

Transform the standard basis $\{1, x, x^2\}$, which is *not* orthogonal, to an orthogonal basis.

Solution

Let $\mathbf{v}_1 = 1$, $\mathbf{v}_2 = x$, and $\mathbf{v}_3 = x^2$.

The first vector is given $\mathbf{p}_1 = \mathbf{v}_1 = 1$. Next we find a vector which is orthogonal to this by using (4.16) which is:

$$\mathbf{p}_2 = \mathbf{v}_2 - \frac{\langle \mathbf{v}_2, \mathbf{p}_1 \rangle}{\|\mathbf{p}_1\|^2}\mathbf{p}_1 \qquad (*)$$

We need to find each of these components.

What is the inner product $\langle \mathbf{v}_2, \mathbf{p}_1 \rangle$ equal to?

$$\langle \mathbf{v}_2, \mathbf{p}_1 \rangle = \int_{-1}^{1} x(1)\, dx = \int_{-1}^{1} x\, dx = \left[\frac{x^2}{2}\right]_{-1}^{1} = \frac{1}{2}\left[1^2 - (-1)^2\right] = 0$$

Since the inner product is zero, so the vectors $\mathbf{v}_2$ and $\mathbf{p}_1$ are already orthogonal. Putting this into (*) yields

$$\mathbf{p}_2 = \mathbf{v}_2 - (0)\,\mathbf{p}_1 = \mathbf{v}_2 = x \quad [\text{because we are given } \mathbf{v}_2 = x]$$

What else do we need to find?

The perpendicular vector $\mathbf{p}_3$ by applying Gram-Schmidt process (4.16) which is

$$\mathbf{p}_3 = \mathbf{v}_3 - \frac{\langle \mathbf{v}_3, \mathbf{p}_1 \rangle}{\|\mathbf{p}_1\|^2}\mathbf{p}_1 - \frac{\langle \mathbf{v}_3, \mathbf{p}_2 \rangle}{\|\mathbf{p}_2\|^2}\mathbf{p}_2 \qquad (**)$$

Evaluating each of the components gives

$$\langle \mathbf{v}_3, \mathbf{p}_1 \rangle = \int_{-1}^{1} x^2(1)\, dx = \int_{-1}^{1} x^2\, dx = \left[\frac{x^3}{3}\right]_{-1}^{1} = \frac{1}{3}\left[1^3 - (-1)^3\right] = \frac{2}{3}$$

$$\langle \mathbf{v}_3, \mathbf{p}_2 \rangle = \int_{-1}^{1} x^2(x)\, dx = \int_{-1}^{1} x^3\, dx = \left[\frac{x^4}{4}\right]_{-1}^{1} = \frac{1}{4}\left[1^4 - (-1)^4\right] = 0$$

$$\|\mathbf{p}_1\|^2 = \langle \mathbf{p}_1, \mathbf{p}_1 \rangle = \int_{-1}^{1} 1(1)\, dx = [x]_{-1}^{1} = [1 - (-1)] = 2$$

$$\|\mathbf{p}_2\|^2 = \langle \mathbf{p}_2, \mathbf{p}_2 \rangle = \int_{-1}^{1} x(x)\, dx = \int_{-1}^{1} x^2\, dx = \frac{2}{3}$$

Substituting all these, $\langle \mathbf{v}_3, \mathbf{p}_1 \rangle = \frac{2}{3}$, $\|\mathbf{p}_1\|^2 = 2$, $\langle \mathbf{v}_3, \mathbf{p}_2 \rangle = 0$, and $\|\mathbf{p}_2\|^2 = \frac{2}{3}$ into (**) yields

$$\begin{aligned}\mathbf{p}_3 &= \mathbf{v}_3 - \frac{\langle \mathbf{v}_3, \mathbf{p}_1 \rangle}{\|\mathbf{p}_1\|^2}\mathbf{p}_1 - \frac{\langle \mathbf{v}_3, \mathbf{p}_2 \rangle}{\|\mathbf{p}_2\|^2}\mathbf{p}_2 \\ &= x^2 - \frac{2/3}{2}(1) - \frac{0}{2/3}x = x^2 - \frac{1}{3}\end{aligned}$$

Our three orthogonal polynomials (vectors) are

$$\mathbf{p}_1 = 1, \quad \mathbf{p}_2 = x \quad \text{and} \quad \mathbf{p}_3 = x^2 - \frac{1}{3}$$

In this example, we have transferred the standard basis to an orthogonal basis:

$$\{1,\ x,\ x^2\} \quad \Rightarrow \quad \{1,\ x,\ x^2 - 1/3\}$$

If we scale the above obtained orthogonal polynomials

$$\mathbf{p}_1 = 1, \quad \mathbf{p}_2 = x \quad \text{and} \quad \mathbf{p}_3 = x^2 - 1/3$$

so that at $x = 1$ they have a value of 1, we get

$$\mathbf{p}_1 = 1, \quad \mathbf{p}_2 = x \quad \text{and} \quad \mathbf{p}_3' = \frac{1}{2}\left(3x^2 - 1\right)$$

These three orthogonal polynomials $\mathbf{p}_1$, $\mathbf{p}_2$ and $\mathbf{p}_3'$ are the first three **Legendre polynomials**. An important property of Legendre polynomials is that they must be orthogonal with respect to the inner product given in Example 4.17. Legendre polynomials crop up in differential equations.

There are also other orthogonal polynomials such as **Chebyshev** (pronounced 'chebishef') **polynomials** which are important in approximation theory. In approximation theory we try to provide the 'best' polynomial approximation $p(x)$ to a function $f(x)$ with a series expansion of orthogonal polynomials. By 'best' we mean that we construct a polynomial $p(x)$ so that the distance function or the error $\|f(x) - p(x)\|$ is as small as possible for all x in $[a,\ b]$, where $[a,\ b]$ is the interval our function is defined on.

This is different from a Taylor expansion, which gives the best approximation at a point.

Notice that the structure you see with dot product in Euclidean space is exactly the same structure you have with these polynomials.

It is interesting to know: if we have two given vector spaces U and V, are they essentially identical? We can think of U and V as identical if they have the same structure but only the nature of their elements or points differ. We say that vector spaces U and V are **isomorphic**.

For example, P_2 and $\mathbb{R}^3$ are isomorphic because they have the same structure but the elements such as $(1\ 2\ 3)^T$ in $\mathbb{R}^3$ differ from $1 + 2x + 3x^2$ in P_2. We can plot these in three-dimensional space with our normal orthogonal x, y and z axes in $\mathbb{R}^3$ and constant, x and $x^2 - 1/3$ orthogonal axes in P_2.

In the next chapter we will prove a very important result in linear algebra:

Every n-dimensional real vector space is isomorphic to $\mathbb{R}^n$.

This means that all the non-Euclidean vector spaces we have looked at in this chapter, such as the set of polynomials P_n, matrices M_{mn} and continuous functions on $C[a,\ b]$, are identical in structure to $\mathbb{R}^n$, provided they have the same dimension.

Proposition (4.17). Let $\{\mathbf{v}_1, \mathbf{v}_2, \mathbf{v}_3, \ldots, \mathbf{v}_n\}$ be an *orthonormal* set of vectors in an inner product space V of dimension n. Let $\mathbf{u}$ be a vector in V then

$$\mathbf{u} = \langle \mathbf{u},\ \mathbf{v}_1\rangle \mathbf{v}_1 + \langle \mathbf{u},\ \mathbf{v}_2\rangle \mathbf{v}_2 + \langle \mathbf{u},\ \mathbf{v}_3\rangle \mathbf{v}_3 + \cdots + \langle \mathbf{u},\ \mathbf{v}_n\rangle \mathbf{v}_n$$

Proof – Exercises 4.3.

Note, the scalars are given by inner products.

■

Summary

(4.13). Let V be a finite dimensional vector space with an inner product. A set of basis vectors

$$B = \{\mathbf{u}_1, \mathbf{u}_2, \mathbf{u}_3, \ldots, \mathbf{u}_n\}$$

for V is called an **orthonormal basis** if they are both orthogonal and normalized.

The Gram–Schmidt process (4.16) produces an orthogonal basis for a finite dimensional vector space.

EXERCISES 4.3

(Brief solutions at end of book. Full solutions available at <http://www.oup.co.uk/companion/singh>.)

1. Transform the following basis vectors in $\mathbb{R}^2$ to an orthonormal basis for $\mathbb{R}^2$ with respect to the dot product. Also plot the orthonormal basis.

   (a) $\mathbf{v}_1 = \begin{pmatrix} 1 \\ 0 \end{pmatrix}$ and $\mathbf{v}_2 = \begin{pmatrix} 2 \\ 1 \end{pmatrix}$ (b) $\mathbf{v}_1 = \begin{pmatrix} 1 \\ 3 \end{pmatrix}$ and $\mathbf{v}_2 = \begin{pmatrix} 2 \\ -1 \end{pmatrix}$

   (c) $\mathbf{v}_1 = \begin{pmatrix} 2 \\ 3 \end{pmatrix}$ and $\mathbf{v}_2 = \begin{pmatrix} 4 \\ 5 \end{pmatrix}$ (d) $\mathbf{v}_1 = \begin{pmatrix} -2 \\ -5 \end{pmatrix}$ and $\mathbf{v}_2 = \begin{pmatrix} -3 \\ -1 \end{pmatrix}$

2. Consider Example 4.16 with the vectors $\mathbf{v}_1$ and $\mathbf{v}_2$ the other way around, that is

$$\mathbf{v}_1 = \begin{pmatrix} 1 \\ 1 \end{pmatrix} \text{ and } \mathbf{v}_2 = \begin{pmatrix} 2 \\ 1 \end{pmatrix}$$

   Transform these basis vectors in $\mathbb{R}^2$ to an orthonormal basis for $\mathbb{R}^2$ with respect to the dot product.

   What do you notice about your answer in relation to Example 4.16?

3. Transform the basis $\mathbf{v}_1 = \begin{pmatrix} 1 & 1 & 1 \end{pmatrix}^T$, $\mathbf{v}_2 = \begin{pmatrix} 1 & 1 & 0 \end{pmatrix}^T$ and $\mathbf{v}_3 = \begin{pmatrix} 2 & 0 & 0 \end{pmatrix}^T$ in $\mathbb{R}^3$ to an orthonormal (perpendicular unit) basis for $\mathbb{R}^3$ with respect to the dot product.

4. Transform the following basis vectors in $\mathbb{R}^3$ to an orthonormal basis for $\mathbb{R}^3$ with respect to the dot product.

   (a) $\mathbf{v}_1 = \begin{pmatrix} 1 & 0 & 1 \end{pmatrix}^T$, $\mathbf{v}_2 = \begin{pmatrix} 3 & 1 & 1 \end{pmatrix}^T$ and $\mathbf{v}_3 = \begin{pmatrix} -1 & -1 & -1 \end{pmatrix}^T$

   (b) $\mathbf{v}_1 = \begin{pmatrix} 2 & 2 & 2 \end{pmatrix}^T$, $\mathbf{v}_2 = \begin{pmatrix} -1 & 0 & -1 \end{pmatrix}^T$ and $\mathbf{v}_3 = \begin{pmatrix} -1 & 2 & 3 \end{pmatrix}^T$

   (c) $\mathbf{v}_1 = \begin{pmatrix} 1 & 2 & 0 \end{pmatrix}^T$, $\mathbf{v}_2 = \begin{pmatrix} 2 & 0 & 2 \end{pmatrix}^T$ and $\mathbf{v}_3 = \begin{pmatrix} 1 & 0 & 3 \end{pmatrix}^T$

5. Transform the following vectors which span a subspace of $\mathbb{R}^4$ to an orthonormal basis for this subspace with respect to the dot product.

   (a) $\mathbf{v}_1 = \begin{pmatrix} 1 & 0 & 3 & 0 \end{pmatrix}^T, \mathbf{v}_2 = \begin{pmatrix} 1 & 2 & 1 & 0 \end{pmatrix}^T$ and $\mathbf{v}_3 = \begin{pmatrix} 2 & 1 & 0 & 0 \end{pmatrix}^T$

   (b) $\mathbf{v}_1 = \begin{pmatrix} 1 & 1 & 5 & 2 \end{pmatrix}^T, \mathbf{v}_2 = \begin{pmatrix} -3 & 3 & 4 & -2 \end{pmatrix}^T$ and $\mathbf{v}_3 = \begin{pmatrix} -1 & -2 & 2 & 5 \end{pmatrix}^T$

   (c) $\mathbf{v}_1 = \begin{pmatrix} 1 & 2 & 3 & 4 \end{pmatrix}^T, \mathbf{v}_2 = \begin{pmatrix} 2 & 1 & 1 & 0 \end{pmatrix}^T$ and $\mathbf{v}_3 = \begin{pmatrix} 3 & 0 & -1 & 3 \end{pmatrix}^T$

6. Normalize the orthogonal polynomials found in Example 4.17.
7. Transform the basis $\mathbf{v}_1 = x^2$, $\mathbf{v}_2 = x$, $\mathbf{v}_3 = 1$ to an orthogonal basis for P_2 with respect to the inner product given in Example 4.17. Compare this with the orthogonal basis obtained in Example 4.17.
8. Write the following polynomials in P_2 as linear combinations of the Legendre polynomials $\mathbf{p}_1 = 1$, $\mathbf{p}_2 = x$ and $\mathbf{p}_3 = \frac{1}{2}\left(3x^2 - 1\right)$:

   (a) $x^2 + x + 1$ (b) $2x^2 - 1$ (c) 3 (d) x^2 (e) $5x + 2$

9. Prove Proposition (4.17).

SECTION 4.4 Orthogonal Matrices

By the end of this section you will be able to

- understand what is meant by an orthogonal matrix
- factorize a given matrix

4.4.1 Orthogonal matrices

One of the most tedious problems in linear algebra is to find the inverse of a matrix. For a 3 by 3 or larger matrix, it becomes a monotonous task to determine the inverse by hand calculations. However, there is one set of matrices called *orthogonal matrices* where the inverse can be obtained by transposition of the matrix. It is straightforward to find the transpose of a matrix.

Orthogonal matrices arise naturally when working with orthonormal bases. Working with orthonormal bases is very handy because it allows you to use a formula like Pythagoras or it allows you to work in the field of Fourier series.

Orthogonal matrices are important in subjects such as numerical analysis because these matrices have good numerical stability.

Definition (4.18). A square matrix $\mathbf{Q} = \begin{pmatrix} \mathbf{v}_1 & \mathbf{v}_2 & \mathbf{v}_3 & \cdots & \mathbf{v}_n \end{pmatrix}$, whose columns $\mathbf{v}_1, \mathbf{v}_2, \mathbf{v}_3, \ldots, \mathbf{v}_n$ are orthonormal (perpendicular unit) vectors, is called an **orthogonal matrix**.

An example of an orthogonal matrix is the identity matrix.

Example 4.18

Let $\mathbf{Q} = \begin{pmatrix} \mathbf{u}_1 & \mathbf{u}_2 & \mathbf{u}_3 \end{pmatrix}$ where $\mathbf{u}_1 = \frac{1}{\sqrt{3}}\begin{pmatrix} 1 \\ 1 \\ 1 \end{pmatrix}$, $\mathbf{u}_2 = \frac{1}{\sqrt{6}}\begin{pmatrix} 1 \\ 1 \\ -2 \end{pmatrix}$ and $\mathbf{u}_3 = \frac{1}{\sqrt{2}}\begin{pmatrix} 1 \\ -1 \\ 0 \end{pmatrix}$ which is an orthonormal basis for $\mathbb{R}^3$. Determine $\mathbf{Q}^T\mathbf{Q}$.

Solution

What does the notation $\mathbf{Q}^T$ mean?

The transpose of the matrix $\mathbf{Q}$. We have

$$\mathbf{Q}^T\mathbf{Q} = \begin{pmatrix} \mathbf{u}_1 & \mathbf{u}_2 & \mathbf{u}_3 \end{pmatrix}^T \begin{pmatrix} \mathbf{u}_1 & \mathbf{u}_2 & \mathbf{u}_3 \end{pmatrix}$$

$$= \begin{pmatrix} \mathbf{u}_1^T \\ \mathbf{u}_2^T \\ \mathbf{u}_3^T \end{pmatrix} \begin{pmatrix} \mathbf{u}_1 & \mathbf{u}_2 & \mathbf{u}_3 \end{pmatrix} \quad \left[\begin{array}{l}\text{When transposing the column}\\ \text{vector } \mathbf{u}_1 \text{ becomes the row}\\ \text{vector } \mathbf{u}_1^T.\end{array}\right]$$

$$= \begin{pmatrix} 1/\sqrt{3} & 1/\sqrt{3} & 1/\sqrt{3} \\ 1/\sqrt{6} & 1/\sqrt{6} & -2/\sqrt{6} \\ 1/\sqrt{2} & -1/\sqrt{2} & 0 \end{pmatrix} \begin{pmatrix} 1/\sqrt{3} & 1/\sqrt{6} & 1/\sqrt{2} \\ 1/\sqrt{3} & 1/\sqrt{6} & -1/\sqrt{2} \\ 1/\sqrt{3} & -2/\sqrt{6} & 0 \end{pmatrix}$$

To evaluate this by hand is rather tedious, unless we can factor out some of the square roots:

$$\frac{1}{\sqrt{3}} = \frac{\sqrt{2}}{\sqrt{2}\sqrt{3}} = \frac{\sqrt{2}}{\sqrt{6}} \quad \left[\begin{array}{l}\text{multiplying numerator}\\ \text{and denominator by} \sqrt{2}\end{array}\right]$$

$$\frac{1}{\sqrt{2}} = \frac{\sqrt{3}}{\sqrt{3}\sqrt{2}} = \frac{\sqrt{3}}{\sqrt{6}} \quad \left[\begin{array}{l}\text{multiplying numerator}\\ \text{and denominator by} \sqrt{3}\end{array}\right]$$

Replacing $1/\sqrt{3}$ and $1/\sqrt{2}$ in terms of $1/\sqrt{6}$ in the above:

$$\mathbf{Q}^T\mathbf{Q} = \begin{pmatrix} \sqrt{2}/\sqrt{6} & \sqrt{2}/\sqrt{6} & \sqrt{2}/\sqrt{6} \\ 1/\sqrt{6} & 1/\sqrt{6} & -2/\sqrt{6} \\ \sqrt{3}/\sqrt{6} & -\sqrt{3}/\sqrt{6} & 0 \end{pmatrix} \begin{pmatrix} \sqrt{2}/\sqrt{6} & 1/\sqrt{6} & \sqrt{3}/\sqrt{6} \\ \sqrt{2}/\sqrt{6} & 1/\sqrt{6} & -\sqrt{3}/\sqrt{6} \\ \sqrt{2}/\sqrt{6} & -2/\sqrt{6} & 0 \end{pmatrix}$$

$$= \frac{1}{\sqrt{6}}\begin{pmatrix} \sqrt{2} & \sqrt{2} & \sqrt{2} \\ 1 & 1 & -2 \\ \sqrt{3} & -\sqrt{3} & 0 \end{pmatrix} \frac{1}{\sqrt{6}}\begin{pmatrix} \sqrt{2} & 1 & \sqrt{3} \\ \sqrt{2} & 1 & -\sqrt{3} \\ \sqrt{2} & -2 & 0 \end{pmatrix} \quad \left[\begin{array}{l}\text{taking out a}\\ \text{scalar multiple}\\ 1/\sqrt{6}\end{array}\right]$$

$$= \frac{1}{\sqrt{6}}\frac{1}{\sqrt{6}}\begin{pmatrix} 6 & 0 & 0 \\ 0 & 6 & 0 \\ 0 & 0 & 6 \end{pmatrix} = \frac{1}{6}\begin{pmatrix} 6 & 0 & 0 \\ 0 & 6 & 0 \\ 0 & 0 & 6 \end{pmatrix} = \begin{pmatrix} 1 & 0 & 0 \\ 0 & 1 & 0 \\ 0 & 0 & 1 \end{pmatrix}$$

What do you notice about your final result in the above example?

We end up with the identity matrix $\mathbf{I}$, that is $\mathbf{Q}^T\mathbf{Q} = \mathbf{I}$. This is *no coincidence*. There is a general result which says that if matrix $\mathbf{Q}$ is orthogonal then $\mathbf{Q}^T\mathbf{Q} = \mathbf{I}$. We can also go the other way, that is if $\mathbf{Q}^T\mathbf{Q} = \mathbf{I}$ then matrix $\mathbf{Q}$ is orthogonal.

Proposition (4.19). Let $\mathbf{Q} = \begin{pmatrix} \mathbf{v}_1 & \mathbf{v}_2 & \mathbf{v}_3 & \cdots & \mathbf{v}_n \end{pmatrix}$ be a square matrix. Then $\mathbf{Q}$ is an orthogonal matrix $\Leftrightarrow$ $\mathbf{Q}^T\mathbf{Q} = \mathbf{I}$.

How do we prove this result?
We have $\Leftrightarrow$ in the statement, so we need to prove it both ways, $\Rightarrow$ and $\Leftarrow$.

Proof.
($\Rightarrow$). We assume that $\mathbf{Q} = \begin{pmatrix} \mathbf{v}_1 & \mathbf{v}_2 & \mathbf{v}_3 & \cdots & \mathbf{v}_n \end{pmatrix}$ is an orthogonal matrix, which means that $\mathbf{v}_1, \mathbf{v}_2, \ldots, \mathbf{v}_n$ is an orthonormal (perpendicular unit) set of vectors in $\mathbb{R}^n$. Required to prove $\mathbf{Q}^T\mathbf{Q} = \mathbf{I}$. We carry out the matrix multiplication:

$$\mathbf{Q}^T\mathbf{Q} = \begin{pmatrix} \mathbf{v}_1 & \mathbf{v}_2 & \mathbf{v}_3 & \cdots & \mathbf{v}_n \end{pmatrix}^T \begin{pmatrix} \mathbf{v}_1 & \mathbf{v}_2 & \mathbf{v}_3 & \cdots & \mathbf{v}_n \end{pmatrix}$$

$$= \begin{pmatrix} \mathbf{v}_1^T \\ \mathbf{v}_2^T \\ \vdots \\ \mathbf{v}_n^T \end{pmatrix} \begin{pmatrix} \mathbf{v}_1 & \mathbf{v}_2 & \mathbf{v}_3 & \cdots & \mathbf{v}_n \end{pmatrix} \quad \begin{bmatrix} \text{transposing to convert columns} \\ \text{to rows} \end{bmatrix}$$

$$= \begin{pmatrix} \mathbf{v}_1^T\mathbf{v}_1 & \mathbf{v}_1^T\mathbf{v}_2 & \cdots & \mathbf{v}_1^T\mathbf{v}_n \\ \mathbf{v}_2^T\mathbf{v}_1 & \mathbf{v}_2^T\mathbf{v}_2 & \cdots & \mathbf{v}_2^T\mathbf{v}_n \\ \vdots & \vdots & & \vdots \\ \mathbf{v}_n^T\mathbf{v}_1 & \mathbf{v}_n^T\mathbf{v}_2 & \cdots & \mathbf{v}_n^T\mathbf{v}_n \end{pmatrix} \quad \begin{bmatrix} \text{carrying out matrix} \\ \text{multiplication – row by} \\ \text{column} \end{bmatrix}$$

Remember, our destination is to show that the final matrix in the above is the identity.

What is the first entry $\mathbf{v}_1^T\mathbf{v}_1$ in the above matrix equal to?
This is the dot product because we defined this in chapter 2:

$$\mathbf{u} \cdot \mathbf{v} = \mathbf{u}^T\mathbf{v} \tag{2.4}$$

Applying this to $\mathbf{v}_1^T\mathbf{v}_1$ we have

$$\mathbf{v}_1^T\mathbf{v}_1 = \mathbf{v}_1 \cdot \mathbf{v}_1$$

What is $\mathbf{v}_1 \cdot \mathbf{v}_1$ equal to?
By the definition of the norm (length) of chapter 2: (2.8) $\|\mathbf{u}\| = \sqrt{\mathbf{u} \cdot \mathbf{u}}$

Hence $\mathbf{v}_1 \cdot \mathbf{v}_1 = \|\mathbf{v}_1\|^2$. Remember, $\mathbf{v}_1$ is an orthonormal (perpendicular unit) vector which means it has a length of 1. Therefore $\mathbf{v}_1 \cdot \mathbf{v}_1 = \|\mathbf{v}_1\|^2 = 1$. This means that the first entry of the matrix $\mathbf{Q}^T\mathbf{Q}$ is $\mathbf{v}_1^T\mathbf{v}_1 = 1$. Similarly all along the leading diagonal of $\mathbf{Q}^T\mathbf{Q}$, the entries are 1 because

$$\mathbf{v}_2^T\mathbf{v}_2 = \mathbf{v}_3^T\mathbf{v}_3 = \mathbf{v}_4^T\mathbf{v}_4 = \cdots = \mathbf{v}_n^T\mathbf{v}_n = 1$$

What does the second entry $\mathbf{v}_1^T\mathbf{v}_2$ in the matrix $\mathbf{Q}^T\mathbf{Q}$ equal to?
Again, by (2.4) we have $\mathbf{v}_1^T\mathbf{v}_2 = \mathbf{v}_1 \cdot \mathbf{v}_2$.

What is $\mathbf{v}_1 \cdot \mathbf{v}_2$ equal to?

Remember, we are assuming vectors $\mathbf{v}_1$ and $\mathbf{v}_2$ are perpendicular unit vectors because they belong to an orthonormal set of vectors. Vectors $\mathbf{v}_1$ and $\mathbf{v}_2$ are perpendicular so $\mathbf{v}_1 \cdot \mathbf{v}_2 = 0$. Similarly the dot product of two *different* vectors $\mathbf{v}_i$ and $\mathbf{v}_j$ is zero, that is

$$\mathbf{v}_i^T\mathbf{v}_j = \mathbf{v}_i \cdot \mathbf{v}_j = 0 \qquad \text{provided } i \text{ does } not \text{ equal } j$$

This means that *all* the remaining entries are equal to zero. Hence the matrix $\mathbf{Q}^T\mathbf{Q}$ has 1's along the leading diagonal and 0's everywhere else, which means that it is the identity matrix $\mathbf{I}$.

We have proved our required result $\mathbf{Q}^T\mathbf{Q} = \mathbf{I}$.

($\Leftarrow$). See Exercises 4.4.

■

From this $\mathbf{Q}^T\mathbf{Q} = \mathbf{I}$, can you deduce any other result?

Since $\mathbf{Q}^T\mathbf{Q} = \mathbf{I}$, we conclude that the *inverse* of the matrix $\mathbf{Q}$ is $\mathbf{Q}^T$.

Proposition (4.20). $\mathbf{Q}$ is an orthogonal matrix $\Leftrightarrow$ $\mathbf{Q}^{-1} = \mathbf{Q}^T$.

This is a powerful result because the inverse of any orthogonal matrix is the transpose of the matrix. It is a lot easier to transpose a matrix, rather than converting to reduced row echelon form.

Proof.

($\Rightarrow$). From the previous proposition, we have $\mathbf{Q}^T\mathbf{Q} = \mathbf{I}$ for an orthogonal matrix $\mathbf{Q}$. Remember, $\mathbf{Q}$ is a square matrix, so it has both left and right inverse and they are equal to each other.

By the definition of the inverse matrix of chapter 1:

(1.16). A square matrix $\mathbf{A}$ is said to be *invertible* if there is a matrix $\mathbf{B}$ such that $\mathbf{AB} = \mathbf{BA} = \mathbf{I}$ where $\mathbf{B}$ is denoted by $\mathbf{A}^{-1}$

Hence $\mathbf{Q}^T$ is the inverse of matrix $\mathbf{Q}$, which means that $\mathbf{Q}^{-1} = \mathbf{Q}^T$.

($\Leftarrow$). Required to prove that $\mathbf{Q}^{-1} = \mathbf{Q}^T$ implies that $\mathbf{Q}$ is orthogonal. Right multiplying $\mathbf{Q}^{-1} = \mathbf{Q}^T$ by $\mathbf{Q}$ gives $\mathbf{Q}^{-1}\mathbf{Q} = \mathbf{I} = \mathbf{Q}^T\mathbf{Q}$. By the above Proposition (4.19):

$$\mathbf{Q} \text{ is an orthogonal matrix } \Leftrightarrow \mathbf{Q}^T\mathbf{Q} = \mathbf{I}$$

We already have $\mathbf{Q}^T\mathbf{Q} = \mathbf{I}$, so $\mathbf{Q}$ is orthogonal.

■

4.4.2 Properties of orthogonal matrices

Example 4.19

Let $\mathbf{Q} = \frac{1}{\sqrt{2}}\begin{pmatrix} 1 & 1 \\ 1 & -1 \end{pmatrix}$, $\mathbf{u} = \begin{pmatrix} 1 \\ 2 \end{pmatrix}$ and $\mathbf{w} = \begin{pmatrix} 3 \\ 1 \end{pmatrix}$. Determine the dot products:

(i) $\mathbf{Qu} \cdot \mathbf{Qw}$ **(ii)** $\mathbf{u} \cdot \mathbf{w}$ **(iii)** *What do you notice about your results?*

Solution

(i) Evaluating the dot products:

$$\mathbf{Qu} = \frac{1}{\sqrt{2}}\begin{pmatrix} 1 & 1 \\ 1 & -1 \end{pmatrix}\begin{pmatrix} 1 \\ 2 \end{pmatrix} = \frac{1}{\sqrt{2}}\begin{pmatrix} 3 \\ -1 \end{pmatrix}$$

$$\mathbf{Qw} = \frac{1}{\sqrt{2}}\begin{pmatrix} 1 & 1 \\ 1 & -1 \end{pmatrix}\begin{pmatrix} 3 \\ 1 \end{pmatrix} = \frac{1}{\sqrt{2}}\begin{pmatrix} 4 \\ 2 \end{pmatrix}$$

$$\mathbf{QuQw} = \frac{1}{\sqrt{2}}\begin{pmatrix} 3 \\ -1 \end{pmatrix} \cdot \frac{1}{\sqrt{2}}\begin{pmatrix} 4 \\ 2 \end{pmatrix} = \frac{1}{2}[(3 \times 4) + (-1 \times 2)] = 5$$

(ii) Similarly, we have

$$\mathbf{u} \cdot \mathbf{w} = \begin{pmatrix} 1 \\ 2 \end{pmatrix} \cdot \begin{pmatrix} 3 \\ 1 \end{pmatrix} = (1 \times 3) + (2 \times 1) = 5$$

(iii) Note that $\mathbf{Qu} \cdot \mathbf{Qw} = \mathbf{u} \cdot \mathbf{w}$.

This result $\mathbf{Qu} \cdot \mathbf{Qw} = \mathbf{u} \cdot \mathbf{w}$ for the last example is no coincidence. This is a general result that if $\mathbf{Q}$ is an orthogonal matrix then the dot product of $\mathbf{Qu}$ and $\mathbf{Qw}$ is exactly $\mathbf{u} \cdot \mathbf{w}$. Actually:

Proposition (4.21). Let $\mathbf{Q}$ be an n by n matrix and $\mathbf{u}$ and $\mathbf{w}$ be any vectors in $\mathbb{R}^n$. Then

$$\mathbf{Q} \text{ is an orthogonal matrix} \Leftrightarrow \mathbf{Qu} \cdot \mathbf{Qw} = \mathbf{u} \cdot \mathbf{w}$$

What does this proposition mean?

This means that an orthogonal matrix preserves the dot product, and if the dot product is preserved then $\mathbf{Q}$ is an orthogonal matrix.

Proof.

($\Rightarrow$). Assume matrix $\mathbf{Q}$ is orthogonal. Required to prove $\mathbf{Qu} \cdot \mathbf{Qw} = \mathbf{u} \cdot \mathbf{w}$. Let us examine the left hand side of this $\mathbf{Qu} \cdot \mathbf{Qw} = \mathbf{u} \cdot \mathbf{w}$. We use

(2.4) $\mathbf{u} \cdot \mathbf{w} = \mathbf{u}^T\mathbf{w}$

Applying this formula (2.4) to $\mathbf{Qu} \cdot \mathbf{Qw}$ we have

$$\begin{aligned}
\mathbf{Qu} \cdot \mathbf{Qw} &= (\mathbf{Qu})^T \mathbf{Qw} \\
&= \left(\mathbf{u}^T\mathbf{Q}^T\right)\mathbf{Qw} && \left[\text{by (1.19) (d) } (\mathbf{AB})^T = \mathbf{B}^T\mathbf{A}^T\right] \\
&= \mathbf{u}^T\left(\mathbf{Q}^T\mathbf{Q}\right)\mathbf{w} && \left[\text{by (1.16)(a) } (\mathbf{AB})\,\mathbf{C} = \mathbf{A}\,(\mathbf{BC})\right] \\
&= \mathbf{u}^T\,(\mathbf{I})\,\mathbf{w} && \left[\text{because } \mathbf{Q} \text{ is orthogonal so } \mathbf{Q}^T\mathbf{Q} = \mathbf{I}\right] \\
&= \mathbf{u}^T\mathbf{w} \underbrace{=}_{\text{By (2.4)}} \mathbf{u} \cdot \mathbf{w}
\end{aligned}$$

This proves that the dot product is preserved under matrix multiplication by an orthogonal matrix.
($\Leftarrow$). Now we go the other way. Assume $\mathbf{Qu} \cdot \mathbf{Qw} = \mathbf{u} \cdot \mathbf{w}$ for all vectors $\mathbf{u}$ and $\mathbf{w}$, and prove that $\mathbf{Q}$ is an orthogonal matrix. We have:

$$\begin{aligned}\mathbf{Qu} \cdot \mathbf{Qw} &= (\mathbf{Qu})^T \mathbf{Qw} \\ &= \left(\mathbf{u}^T\mathbf{Q}^T\right)\mathbf{Qw} \\ &= \mathbf{u}^T\left(\mathbf{Q}^T\mathbf{Q}\right)\mathbf{w} = \mathbf{u} \cdot \mathbf{w} \quad \text{[by assumption]}\end{aligned}$$

Using (2.4) $\mathbf{u} \cdot \mathbf{w} = \mathbf{u}^T\mathbf{w}$ on the right hand side of the last line, we have

$$\mathbf{u}^T\left(\mathbf{Q}^T\mathbf{Q}\right)\mathbf{w} = \mathbf{u}^T\mathbf{w} \text{ implies that } \mathbf{Q}^T\mathbf{Q} = \mathbf{I}$$

By the above Proposition (4.19): $\mathbf{Q}$ is an orthogonal matrix $\Leftrightarrow \mathbf{Q}^T\mathbf{Q} = \mathbf{I}$
We conclude that $\mathbf{Q}$ is an orthogonal matrix because $\mathbf{Q}^T\mathbf{Q} = \mathbf{I}$, which is our required result.

■

Example 4.20

Let $\mathbf{Q} = \frac{1}{\sqrt{2}}\begin{pmatrix} 1 & 1 \\ 1 & -1 \end{pmatrix}$ and $\mathbf{u} = \begin{pmatrix} 1 \\ 2 \end{pmatrix}$. Determine the lengths:

(i) $\|\mathbf{u}\|$ **(ii)** $\|\mathbf{Qu}\|$ **(iii)** Plot the vectors $\mathbf{u}$ and $\mathbf{Qu}$ on $\mathbb{R}^2$.

Solution

(i) Using Pythagoras, we have

$$\|\mathbf{u}\|^2 = \mathbf{u} \cdot \mathbf{u} = \begin{pmatrix} 1 \\ 2 \end{pmatrix} \cdot \begin{pmatrix} 1 \\ 2 \end{pmatrix} = 1^2 + 2^2 = 5 \quad \Rightarrow \|\mathbf{u}\| = \sqrt{5}$$

(ii) Similarly we have

$$\|\mathbf{Qu}\|^2 = \mathbf{Qu} \cdot \mathbf{Qu} \underset{\text{by Example 4.19}}{=} \frac{1}{\sqrt{2}}\begin{pmatrix} 3 \\ -1 \end{pmatrix} \cdot \frac{1}{\sqrt{2}}\begin{pmatrix} 3 \\ -1 \end{pmatrix} = \frac{1}{2}\left[3^2 + (-1)^2\right] = 5$$

Hence $\|\mathbf{Qu}\| = \sqrt{5}$.
(iii) A plot of $\mathbf{u}$ and $\mathbf{Qu}$ is shown in Fig. 4.19.
Note that since lengths $\|\mathbf{u}\| = \|\mathbf{Qu}\| = \sqrt{5}$, the vectors $\mathbf{u}$ and $\mathbf{Qu}$ lie on the circle which has a radius of $\sqrt{5}$ from the origin.

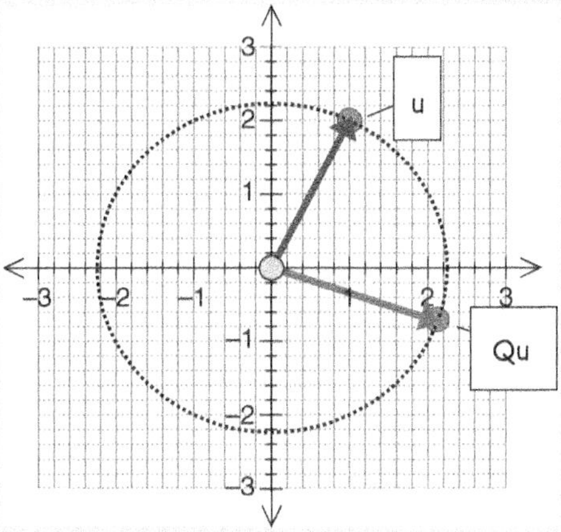

Figure 4.19

For an orthogonal matrix **Q** the length of vectors **Qu** and **u** are equal.

Proposition (4.22). Let **Q** be an n by n matrix and **u** be a vector in $\mathbb{R}^n$. We have

$$\mathbf{Q} \text{ is an orthogonal matrix} \Leftrightarrow \|\mathbf{Qu}\| = \|\mathbf{u}\|$$

What does this proposition mean?

Applying an orthogonal matrix **Q** to a vector **u** may change the direction but *not* the length. An orthogonal matrix **Q** in **Qu** will just rotate the vector **u** (Fig. 4.20).

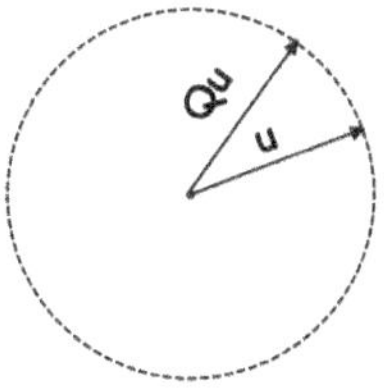

Figure 4.20

An orthogonal matrix preserves the length of a vector and if a transformation matrix preserves length then it is an orthogonal matrix.

Proof – Exercises 4.4.

■

4.4.3 Triangular matrices

What are triangular matrices?

Definition (4.23). A triangular matrix is an n by n matrix where all entries to one side of the leading diagonal are zero.

For example, the following are triangular matrices:

$$\text{(a)}\begin{pmatrix} 1 & 2 & 3 \\ 0 & 4 & 5 \\ 0 & 0 & 6 \end{pmatrix} \text{ and (b) } \begin{pmatrix} 1 & 0 & 0 \\ 2 & 3 & 0 \\ 4 & 5 & 6 \end{pmatrix}$$

leading diagonal

(a) is an example of an upper triangular matrix.

(b) is an example of a lower triangular matrix.

Another type of matrix is a diagonal matrix.

Do you know what is meant by a diagonal matrix?

Definition (4.24). A diagonal matrix is an n by n matrix where *all* entries to both sides of the leading diagonal are zero.

Can you think of an example of a diagonal matrix?

The identity matrix $\mathbf{I} = \begin{pmatrix} 1 & 0 & 0 \\ 0 & 1 & 0 \\ 0 & 0 & 1 \end{pmatrix}$. Another example is $\begin{pmatrix} 1 & 0 & 0 \\ 0 & 2 & 0 \\ 0 & 0 & 3 \end{pmatrix}$.

A diagonal matrix is both an upper and lower triangular matrix.

4.4.4 QR factorization

Let $\mathbf{A}$ be an m by n matrix with linearly independent columns. We can factorize matrix $\mathbf{A}$ into

$$\mathbf{A} = \mathbf{QR} \tag{4.25}$$

where $\mathbf{Q}$ is an orthogonal matrix and $\mathbf{R}$ is an upper triangular matrix.

Why factorize the matrix $\mathbf{A}$ into an orthogonal matrix $\mathbf{Q}$ and upper triangular matrix $\mathbf{R}$?

1. So that we can efficiently solve linear systems such as $\mathbf{Ax} = \mathbf{b}$. This linear system can be written as

$$(\mathbf{QR})\,\mathbf{x} = \mathbf{Q}\,(\mathbf{Rx}) = \mathbf{b} \qquad (*)$$

Left multiplying this (*) by the inverse of $\mathbf{Q}$ which is $\mathbf{Q}^{-1} = \mathbf{Q}^T$ gives

$$\underbrace{\mathbf{Q}^T\mathbf{Q}}_{=\mathbf{I}}(\mathbf{Rx}) = \mathbf{Rx} = \mathbf{Q}^T\mathbf{b}$$

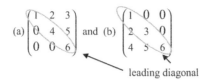

$\mathbf{Q}^T\mathbf{b}$ is a vector $\mathbf{c}$ say, that is $\mathbf{Q}^T\mathbf{b} = \mathbf{c}$. The given linear system $\mathbf{Ax} = \mathbf{b}$ has been reduced to solving $\mathbf{Rx} = \mathbf{c}$, where $\mathbf{R}$ is an upper triangular matrix. We can solve this system $\mathbf{Rx} = \mathbf{c}$ by back substitution because $\mathbf{R}$ is an upper triangular matrix:

$$\mathbf{Rx} = \mathbf{c} \text{ in matrix form is } \begin{pmatrix} r_{11} & & r_{1n} \\ & \ddots & \\ \bigcirc & & r_{nn} \end{pmatrix} \begin{pmatrix} x_1 \\ \vdots \\ x_n \end{pmatrix} = \begin{pmatrix} c_1 \\ \vdots \\ c_n \end{pmatrix}$$

The advantage of this method is that once we have factorized matrix $\mathbf{A}$ then we can solve $\mathbf{Ax} = \mathbf{b}$ for various different vectors $\mathbf{b}$ in one go.

2. Factorizing a matrix is also applied in solving least squares approximation which is:

   Find the vector $\mathbf{x}$ such that $\|\mathbf{Ax} - \mathbf{b}\|$ is a minimum

3. Additionally QR factorization is used to find the eigenvalues and eigenvectors (to be discussed in chapter 7) of a matrix by numerical means.

   We need to find an orthogonal matrix $\mathbf{Q}$ and an upper triangular matrix $\mathbf{R}$ such that $\mathbf{A} = \mathbf{QR}$.

How can we find such matrices?

Since $\mathbf{Q}$ is an orthogonal matrix, its columns are orthonormal (perpendicular unit) vectors:

$$\mathbf{Q} = \begin{pmatrix} \mathbf{q}_1 & \mathbf{q}_2 & \mathbf{q}_3 & \cdots & \mathbf{q}_n \end{pmatrix}$$

But how do we get these orthonormal vectors from the matrix A?

Well, let matrix $\mathbf{A} = (\,\mathbf{a}_1\ \mathbf{a}_2\ \mathbf{a}_3\ \cdots\ \mathbf{a}_n\,)$ where the vectors $\mathbf{a}_1,\ \mathbf{a}_2, \ldots,\ \mathbf{a}_n$ are linearly independent. Then by applying the Gram–Schmidt process we can transform these $\mathbf{a}$ vectors into $\mathbf{q}$ vectors because Gram–Schmidt produces orthonormal vectors.

Applying the Gram–Schmidt process (4.16) and normalization from the last section we have:

Let $\mathbf{a}_1 = \frac{\mathbf{q}_1}{\|\mathbf{q}_1\|}$. Rearranging this gives $\mathbf{a}_1 = r_{11}\mathbf{q}_1$ where $r_{11} = \frac{1}{\|\mathbf{q}_1\|}$.

Also $\mathbf{q}_2 = \frac{1}{\|\mathbf{q}_2\|}\left[\mathbf{a}_2 - \frac{\langle \mathbf{v}_2, \mathbf{q}_1\rangle}{\|\mathbf{q}_1\|^2}\mathbf{q}_1\right]$. Transposing this gives

$$\mathbf{a}_2 = \frac{\langle \mathbf{v}_2, \mathbf{q}_1\rangle}{\|\mathbf{q}_1\|^2}\mathbf{q}_1 + \|\mathbf{q}_2\|\,\mathbf{q}_2$$

We can write this as $\mathbf{a}_2 = r_{12}\mathbf{q}_1 + r_{22}\mathbf{q}_2$ where $r_{12} = \frac{\langle \mathbf{v}_2, \mathbf{q}_1\rangle}{\|\mathbf{q}_1\|^2}$ and $r_{22} = \|\mathbf{q}_2\|$.
Expanding by applying the Gram–Schmidt process we have

$$\begin{aligned}
\mathbf{a}_1 &= r_{11}\mathbf{q}_1 \\
\mathbf{a}_2 &= r_{12}\mathbf{q}_1 + r_{22}\mathbf{q}_2 \\
\mathbf{a}_3 &= r_{13}\mathbf{q}_1 + r_{23}\mathbf{q}_2 + r_{33}\mathbf{q}_3 \\
&\ \ \vdots \qquad\quad \vdots \qquad\quad \vdots \\
\mathbf{a}_n &= r_{1n}\mathbf{q}_1 + r_{2n}\mathbf{q}_2 + r_{3n}\mathbf{q}_3 + \cdots + r_{nn}\mathbf{q}_n
\end{aligned}$$

We can write this in matrix form as

$$\begin{aligned}\mathbf{A} &= (\,\mathbf{a}_1\ \mathbf{a}_2\ \mathbf{a}_3\ \cdots\ \mathbf{a}_n\,)\\ &= (\,r_{11}\mathbf{q}_1\quad r_{12}\mathbf{q}_1 + r_{22}\mathbf{q}_2\quad r_{13}\mathbf{q}_1 + r_{23}\mathbf{q}_2 + r_{33}\mathbf{q}_3\quad \cdots\quad r_{1n}\mathbf{q}_1 + r_{2n}\mathbf{q}_2 + r_{3n}\mathbf{q}_3 + \cdots + r_{nn}\mathbf{q}_n\,)\\ &= \underbrace{(\mathbf{q}_1\ \ \mathbf{q}_2\ \ \mathbf{q}_3\ \ \cdots\ \ \mathbf{q}_n)}_{=\mathbf{Q}}\underbrace{\begin{pmatrix} r_{11} & r_{12} & \cdots & r_{1n}\\ 0 & r_{22} & \cdots & r_{2n}\\ 0 & 0 & \ddots & \vdots\\ 0 & \vdots & \cdots & r_{nn}\end{pmatrix}}_{=\mathbf{R}}\\ &= \mathbf{QR}\end{aligned}$$

The problem is how do we find the upper triangular matrix **R**? Since **Q** is orthogonal we know the inverse of matrix **Q** is $\mathbf{Q}^T$. Left multiplying both sides of $\mathbf{A} = \mathbf{QR}$ by $\mathbf{Q}^T$ gives

$$\mathbf{Q}^T\mathbf{A} = \underbrace{\mathbf{Q}^T\mathbf{Q}}_{=\mathbf{I}}\mathbf{R} = \mathbf{IR} = \mathbf{R}$$

Hence the upper triangular matrix **R** is given by

$$\mathbf{R} = \mathbf{Q}^T\mathbf{A} \tag{4.26}$$

Let us apply this **QR** factorization to a particular matrix **A**.

Example 4.21

Find the **QR** factorization of $\mathbf{A} = \begin{pmatrix} 1 & 1 & 2\\ 1 & 1 & 0\\ 1 & 0 & 0\end{pmatrix}$.

Solution

The column vectors of **A** are $\mathbf{a}_1 = (\,1\ \ 1\ \ 1\,)^T$, $\mathbf{a}_2 = (\,1\ \ 1\ \ 0\,)^T$ and $\mathbf{a}_3 = (\,2\ \ 0\ \ 0\,)^T$. These are the same vectors given in question 3 of Exercises 4.3. By applying the Gram-Schmidt process, we found the following orthonormal vectors:

$$\mathbf{Q} = (\,\mathbf{q}_1\ \ \mathbf{q}_2\ \ \mathbf{q}_3\,) \text{ where } \mathbf{q}_1 = \frac{1}{\sqrt{3}}\begin{pmatrix}1\\1\\1\end{pmatrix}, \mathbf{q}_2 = \frac{1}{\sqrt{6}}\begin{pmatrix}1\\1\\-2\end{pmatrix} \text{ and } \mathbf{q}_3 = \frac{1}{\sqrt{2}}\begin{pmatrix}1\\-1\\0\end{pmatrix}$$

The matrix **Q** and its transpose were simplified in Example 4.18:

$$\mathbf{Q} = \frac{1}{\sqrt{6}}\begin{pmatrix}\sqrt{2} & 1 & \sqrt{3}\\ \sqrt{2} & 1 & -\sqrt{3}\\ \sqrt{2} & -2 & 0\end{pmatrix} \text{ and } \mathbf{Q}^T = \frac{1}{\sqrt{6}}\begin{pmatrix}\sqrt{2} & \sqrt{2} & \sqrt{2}\\ 1 & 1 & -2\\ \sqrt{3} & -\sqrt{3} & 0\end{pmatrix}$$

Why do we need to find the transpose of the matrix $\mathbf{Q}$*?*
Because the upper triangular matrix $\mathbf{R} = \mathbf{Q}^T\mathbf{A}$:

$$\mathbf{R} = \mathbf{Q}^T\mathbf{A} = \frac{1}{\sqrt{6}}\begin{pmatrix} \sqrt{2} & \sqrt{2} & \sqrt{2} \\ 1 & 1 & -2 \\ \sqrt{3} & -\sqrt{3} & 0 \end{pmatrix}\begin{pmatrix} 1 & 1 & 2 \\ 1 & 1 & 0 \\ 1 & 0 & 0 \end{pmatrix}$$

$$= \frac{1}{\sqrt{6}}\begin{pmatrix} 3\sqrt{2} & 2\sqrt{2} & 2\sqrt{2} \\ 0 & 2 & 2 \\ 0 & 0 & 2\sqrt{3} \end{pmatrix}$$

You may like to check the following by carrying out the matrix multiplication:

$$\mathbf{A} = \begin{pmatrix} 1 & 1 & 2 \\ 1 & 1 & 0 \\ 1 & 0 & 0 \end{pmatrix} = \mathbf{QR} = \frac{1}{\sqrt{6}}\begin{pmatrix} \sqrt{2} & 1 & \sqrt{3} \\ \sqrt{2} & 1 & -\sqrt{3} \\ \sqrt{2} & -2 & 0 \end{pmatrix}\frac{1}{\sqrt{6}}\begin{pmatrix} 3\sqrt{2} & 2\sqrt{2} & 2\sqrt{2} \\ 0 & 2 & 2 \\ 0 & 0 & 2\sqrt{3} \end{pmatrix}$$

Summary

Orthogonal matrices are square matrices whose columns are orthonormal (perpendicular unit) vectors.

We can factorize an m by n matrix $\mathbf{A}$ whose column vectors are linearly independent into $\mathbf{A} = \mathbf{QR}$ where $\mathbf{Q}$ is an orthogonal matrix and $\mathbf{R}$ is an upper triangular matrix.

EXERCISES 4.4

(Brief solutions at end of book. Full solutions available at <http://www.oup.co.uk/companion/singh>.)

1. In each of the following cases, show that $\mathbf{Q}^T\mathbf{Q} = \mathbf{I}$:

   (a) $\mathbf{Q} = \frac{1}{\sqrt{2}}\begin{pmatrix} 1 & 1 \\ 1 & -1 \end{pmatrix}$ (b) $\mathbf{Q} = \frac{1}{5}\begin{pmatrix} 3 & 4 \\ 4 & -3 \end{pmatrix}$

   (c) $\mathbf{Q} = \begin{pmatrix} \cos(\theta) & \sin(\theta) \\ \sin(\theta) & -\cos(\theta) \end{pmatrix}$

2. Determine $\mathbf{Q}^{-1}$ for each of the matrices in question 1.
3. Determine which of the following matrices are orthogonal. If they are orthogonal find their inverse.

(a) $\mathbf{A} = \begin{pmatrix} 1 & 0 & 0 \\ 0 & 2 & 0 \\ 0 & 0 & 3 \end{pmatrix}$ (b) $\mathbf{B} = \frac{1}{2}\begin{pmatrix} 1 & 1 \\ 1 & 1 \\ 1 & -1 \\ 1 & -1 \end{pmatrix}$

(c) $\mathbf{C} = \begin{pmatrix} 1/\sqrt{2} & 1/\sqrt{3} & 1/\sqrt{6} \\ -1/\sqrt{2} & 1/\sqrt{3} & 1/\sqrt{6} \\ 0 & 1/\sqrt{3} & -2/\sqrt{6} \end{pmatrix}$

4. Let $\mathbf{A} = \begin{pmatrix} a & b \\ c & d \end{pmatrix}$. What can you say about the size and angle between the column vectors of $\mathbf{A}$, where $\mathbf{A}$ is an orthogonal matrix?

5. Find the third column vector of an orthogonal matrix whose first two columns are:

$$\begin{pmatrix} 1/\sqrt{3} \\ 1/\sqrt{3} \\ 1/\sqrt{3} \end{pmatrix} \text{ and } \begin{pmatrix} 1/\sqrt{2} \\ 0 \\ -1/\sqrt{2} \end{pmatrix}$$

6. Determine $\mathbf{Qu}$ where $\mathbf{Q} = \begin{pmatrix} \cos(\theta) & \sin(\theta) \\ \sin(\theta) & -\cos(\theta) \end{pmatrix}$ and $\mathbf{u} = \begin{pmatrix} 1 \\ 0 \end{pmatrix}$. Plot your result on $\mathbb{R}^2$. What effect does the orthogonal matrix $\mathbf{Q}$ have on the vector $\mathbf{u}$?

7. Let $\mathbf{A} = \begin{pmatrix} 2 & -1 & -1 \\ 2 & 0 & 2 \\ 2 & -1 & 3 \end{pmatrix}$. The column vectors of $\mathbf{A}$ were the vectors given in Exercises 4.3 question 4(b).

   (a) Determine the $\mathbf{QR}$ factorization of matrix $\mathbf{A}$.

   (b) Solve the linear system $\mathbf{Ax} = \mathbf{b}$ where $\mathbf{b} = \begin{pmatrix} -3 & 8 & 9 \end{pmatrix}^T$.

   (c) Solve the linear system $\mathbf{Ax} = \mathbf{b}$ where $\mathbf{b} = \begin{pmatrix} -5 & 12 & 11 \end{pmatrix}^T$.

8. Prove the converse of Proposition (4.19).

9. Prove Proposition (4.22).

10. Prove that if $\mathbf{A}$ is a square matrix with the property $\|\mathbf{Au}\| = 1$ for all unit vectors $\mathbf{u}$ then $\mathbf{A}$ is an orthogonal matrix.

11. If $\mathbf{Q}$ is orthogonal prove that $\mathbf{Q}^T$ is also orthogonal.

12. Prove that the columns and rows of an orthogonal matrix are linearly independent.

MISCELLANEOUS EXERCISES 4

(Brief solutions at end of book. Full solutions available at <http://www.oup.co.uk/companion/singh>.)

4.1. Decide whether the following statement is *True* or *False*. Justify your answer.

Every linearly independent set in $\mathbb{R}^n$ is an orthogonal set.

Washington State University, USA
(part question)

4.2. The vectors

$$\mathbf{X}_1 = \begin{bmatrix} 1, & 1, & 1, & 1 \end{bmatrix}, \mathbf{X}_2 = \begin{bmatrix} 1, & 0, & 0, & 1 \end{bmatrix}, \mathbf{X}_3 = \begin{bmatrix} 0, & 2, & 1, & -1 \end{bmatrix}$$

are a basis of a subspace U of $\mathbb{R}^4$. Find an orthogonal basis of U.

University of Ottawa, Canada

4.3. Let $S = Span\left\{\begin{bmatrix} 1 \\ 0 \\ 1 \\ 0 \end{bmatrix}, \begin{bmatrix} 0 \\ 1 \\ 1 \\ 1 \end{bmatrix}, \begin{bmatrix} 1 \\ 1 \\ 2 \\ 2 \end{bmatrix}\right\}$.

(a) Find a basis for S.

(b) Find an orthonormal basis for S.

Purdue University, USA

4.4. (a) Which bases of a Euclidean space V are called orthogonal? Orthonormal?

(b) Show that $\mathbf{f}_1 = \begin{pmatrix} -1 \\ 0 \\ 2 \end{pmatrix}$, $\mathbf{f}_2 = \begin{pmatrix} 0 \\ -2 \\ 3 \end{pmatrix}$ and $\mathbf{f}_3 = \begin{pmatrix} 1 \\ 1 \\ 4 \end{pmatrix}$ form a basis of $\mathbb{R}^3$.

(c) Find the orthogonal basis of $\mathbb{R}^3$ which is the output of the Gram–Schmidt orthogonalization applied to the basis from part (b). (The inner product on $\mathbb{R}^3$ is the standard one.)

University of Dublin, Ireland

4.5. (a) Prove that an orthogonal set of non-zero vectors in $\mathbb{R}^n$ is linearly independent.
(b) Let $B = \{\mathbf{v}_1, \ldots, \mathbf{v}_n\}$ be an orthonormal basis of $\mathbb{R}^n$. Show that for $\mathbf{w} \in \mathbb{R}^n$,

$$\mathbf{w} = (\mathbf{v}_1 \cdot \mathbf{w})\,\mathbf{v}_1 + (\mathbf{v}_2 \cdot \mathbf{w})\,\mathbf{v}_2 + \cdots + (\mathbf{v}_n \cdot \mathbf{w})\,\mathbf{v}_n$$

University of Manchester, UK
(part question)

4.6. (a) Define what it means for vectors $\mathbf{w}_1, \mathbf{w}_2, \mathbf{w}_3$ in $\mathbb{R}^3$ to be orthonormal.
(b) Apply the Gram–Schmidt process to the vectors

$$\mathbf{v}_1 = \begin{bmatrix} 0 \\ 1 \\ 1 \end{bmatrix}, \mathbf{v}_2 = \begin{bmatrix} 1 \\ 1 \\ 0 \end{bmatrix}, \mathbf{v}_3 = \begin{bmatrix} 5 \\ 4 \\ 6 \end{bmatrix}$$

in $\mathbb{R}^3$ to find an orthonormal set in $\mathbb{R}^3$.

University of Southampton, UK
(part question)

4.7. Consider the real vector space $\mathbb{R}^4$ with real inner product given by

$$\langle \mathbf{x}, \mathbf{y} \rangle = x_1y_1 + x_2y_2 + x_3y_3 + x_4y_4$$

for $\mathbf{x} = (\, x_1, \; x_2, \; x_3, \; x_4 \,), \mathbf{y} = (\, y_1, \; y_2, \; y_3, \; y_4 \,) \in \mathbb{R}^4$.

(a) Define the norm of a vector $\mathbf{x} = (\, x_1, \; x_2, \; x_3, \; x_4 \,) \in \mathbb{R}^4$ with respect to the above inner product. What is the norm of $(\, 2, \; -3, \; 5, \; 1 \,)$?

(b) When do we say that two vectors $\mathbf{x}, \mathbf{y} \in \mathbb{R}^4$ are orthogonal? Are $(\, 0, \; 1, \; -1, \; 1 \,)$ and $(\, 6, \; 3, \; 3, \; 27 \,)$ orthogonal?

(c) What is an orthonormal set of vectors in $\mathbb{R}^4$?
Is $\left\{ (\, 0, \; 0, \; 1, \; 0 \,), \left(\frac{1}{\sqrt{6}}, \; \frac{-2}{\sqrt{6}}, \; 0, \; \frac{1}{\sqrt{6}} \right), \left(\frac{2}{\sqrt{5}}, \; \frac{1}{\sqrt{5}}, \; 0, \; 0 \right) \right\}$ an orthonormal set? Justify your answer.

(d) Use the Gram–Schmidt process to construct an orthonormal basis for $\mathbb{R}^4$ starting from the basis

$$\{(\, 1, \; 0, \; 0, \; 0 \,), (\, 1, \; 1, \; 0, \; 1 \,), (\, 0, \; 1, \; 1, \; 1 \,), (\, 0, \; 1, \; -1, \; 0 \,)\}$$

City University, London, UK

4.8. Consider the real vector space $M\,(2, \; 2)$ with real inner product given by

$$\langle \mathbf{A}, \mathbf{B} \rangle = tr\left(\mathbf{B}^T \mathbf{A} \right)$$

for all $\mathbf{A}, \; \mathbf{B} \in M\,(2, \; 2)$.

[$M\,(2, \; 2)$ is M_{22} which is the set of 2×2 matrices and tr is the trace of the matrix.]

(a) Define the norm of a matrix $\mathbf{A} \in M\,(2, \; 2)$ with respect to the above inner product. What is the norm of $\begin{pmatrix} 2 & -3 \\ 5 & 1 \end{pmatrix}$?

(b) When do we say that two matrices $\mathbf{A}, \; \mathbf{B} \in M\,(2, \; 2)$ are orthogonal (with respect to the above inner product)? Are $\begin{pmatrix} 0 & 1 \\ -1 & 1 \end{pmatrix}$ and $\begin{pmatrix} 6 & 3 \\ 3 & 27 \end{pmatrix}$ orthogonal?

(c) What is an orthonormal set of matrices in $M\,(2, \; 2)$ (with respect to the above inner product)? Is

$$\left\{ \mathbf{A}_1 = \begin{pmatrix} 1 & 0 \\ 0 & 0 \end{pmatrix}, \; \mathbf{A}_2 = \begin{pmatrix} 0 & 1/\sqrt{2} \\ 0 & 1/\sqrt{2} \end{pmatrix}, \; \mathbf{A}_3 = \begin{pmatrix} 0 & 0 \\ 1 & 0 \end{pmatrix} \right\}$$

an orthonormal set? Justify your answer.

(d) Let $\mathbf{B} = \begin{pmatrix} 0 & 1 \\ -1 & 0 \end{pmatrix}$. Using (c) and the fact that $\{\mathbf{A}_1, \; \mathbf{A}_2, \; \mathbf{A}_3, \; \mathbf{B}\}$ is a basis for $M\,(2, \; 2)$, find a matrix $\mathbf{A}_4$ such that $\{\mathbf{A}_1, \; \mathbf{A}_2, \; \mathbf{A}_3, \; \mathbf{A}_4\}$ is an orthonormal basis for $M\,(2, \; 2)$.

(e) Find $\lambda_1,\ \lambda_2,\ \lambda_3,\ \lambda_4 \in \mathbb{R}$ such that

$$\begin{pmatrix} 1 & 0 \\ 0 & 1 \end{pmatrix} = \lambda_1\mathbf{A}_1 + \lambda_2\mathbf{A}_2 + \lambda_3\mathbf{A}_3 + \lambda_4\mathbf{A}_4$$

City University, London, UK

4.9. Suppose $\mathbf{q}_1,\ \mathbf{q}_2,\ \mathbf{q}_3$ are orthonormal vectors in $\mathbb{R}^4$ (not $\mathbb{R}^3$).

(a) What is the length of the vector $\mathbf{v} = 2\mathbf{q}_1 - 3\mathbf{q}_2 + 2\mathbf{q}_3$?

(b) What four vectors does Gram–Schmidt produce when it orthonormalizes the vectors $\mathbf{q}_1,\ \mathbf{q}_2,\ \mathbf{q}_3,\ \mathbf{u}$?

(c) If $\mathbf{u}$ in part (b) is the vector $\mathbf{v}$ in part (a), why does the Gram–Schmidt break down?

Massachusetts Institute of Technology, USA (part question)

4.10. Consider the vectors

$$\mathbf{v}_1 = [0,\ 1,\ 0,\ 1,\ 0],\ \mathbf{v}_2 = [0,\ 1,\ 1,\ 0,\ 0],\ \mathbf{v}_3 = [0,\ 1,\ 0,\ 1,\ 1]$$

in $\mathbb{R}^5$. Find $\mathbf{w}_1,\ \mathbf{w}_2,\ \mathbf{w}_3$ in $\mathbb{R}^5$ such that $\mathbf{w}_i \cdot \mathbf{w}_j = 0$ for $i \neq j$ (i and j between 1 and 3), and such that $Span\left(\{\mathbf{w}_1,\ \mathbf{w}_2,\ \mathbf{w}_3\}\right) = Span\left(\{\mathbf{v}_1,\ \mathbf{v}_2,\ \mathbf{v}_3\}\right)$ for $i = 1,\ 2,\ 3$.

University of California, Berkeley, USA

4.11. (a) Can you find vectors $\mathbf{v} = \left[\ v_1,\ v_2\ \right]$ and $\mathbf{w} = \left[\ w_1,\ w_2\ \right]$ in $\mathbb{R}^2$ which are orthogonal with respect to the inner product $\langle \mathbf{v},\ \mathbf{w}\rangle = 3v_1w_1 + 2v_2w_2$, but not orthonormal with respect to the usual dot product?

(b) If your answer to (a) is 'yes', find $\mathbf{v}$ and $\mathbf{w}$. If your answer to (a) is 'no', give a reason why you cannot find $\mathbf{v}$ and $\mathbf{w}$.

University of Toronto, Canada

4.12. [Edited version of question.] Let P_3 be the set of all polynomials with degree less than or equal to 3. We will think of them as a subspace of $C[-1,\ 1]$, the continuous functions from the interval $[-1,\ 1]$ to the reals. P_3 has an inner product given by $\langle \mathbf{f},\ \mathbf{g}\rangle = \int_{-1}^{1} f(x)\,g(x)\ dx$. P_3 has what we call a *standard* basis $\left\{1,\ \mathbf{x},\ \mathbf{x}^2,\ \mathbf{x}^3\right\}$. Let V be the space spanned by $\{\mathbf{1},\ \mathbf{x}\}$. Find an orthonormal basis for V.

Johns Hopkins University, USA

4.13. Define, for $\mathbf{A}$ and $\mathbf{B}$ in $M_{2,\,2}(\mathbb{R})$

$$\langle \mathbf{A},\ \mathbf{B}\rangle = tr\left(\mathbf{A}^T\mathbf{B}\right)$$

(Recall, if $\mathbf{M} = \left[m_{ij}\right] \in M_{n,\,n}(F),\ tr(\mathbf{M}) = \sum_{i=1}^{n} m_{ii}$.)

(a) Show that $\langle \mathbf{A},\ \mathbf{B}\rangle$ is an inner product on $M_{2,\,2}(\mathbb{R})$.

(b) Find the distance between $\mathbf{A} = \begin{bmatrix} 1 & 2 \\ 1 & 0 \end{bmatrix}$ and $\mathbf{B} = \begin{bmatrix} 3 & 3 \\ 1 & 2 \end{bmatrix}$ in this inner product space.

(c) Find the angle between $\mathbf{A} = \begin{bmatrix} 1 & 2 \\ 1 & 0 \end{bmatrix}$ and $\mathbf{B} = \begin{bmatrix} 3 & 3 \\ 1 & 2 \end{bmatrix}$ in this inner product space.

University of Toronto, Canada

4.14. Consider the linear space V of polynomials $\mathbf{f}(t) = c_0 + c_1 t + c_2 t^2$ on $0 \le t \le 1$ with inner product

$$\langle \mathbf{f}, \mathbf{g} \rangle = \int_0^1 \mathbf{f}(t)\ \mathbf{g}(t)\ dt$$

Find a basis for the subspace S of all $\mathbf{f}$ in V orthogonal to $1 + t$ satisfying the additional restriction equation $\mathbf{f}(\frac{1}{2}) = 0$.

University of Utah, USA

4.15. Define the inner product of two polynomials $\mathbf{f}$ and $\mathbf{g}$ by the rule

$$\langle \mathbf{f}, \mathbf{g} \rangle = \int_0^1 \mathbf{f}(x)\ \mathbf{g}(1 - x) dx$$

Using this definition of the inner product, find an orthogonal basis for the vector space of all polynomials of degree ≤ 2.

Columbia University, New York, USA

Sample questions

4.16. Explain why $\langle \mathbf{A}, \mathbf{B} \rangle = tr\,(\mathbf{AB})$ where $\mathbf{A}, \mathbf{B} \in M_{22}$ is *not* an inner product.

4.17. Prove Proposition (4.17).

4.18. Let P_3 be the vector space of cubic polynomials with the inner product given by

$$\langle \mathbf{f}, \mathbf{g} \rangle = \int_{-1}^1 f(x)\ g(x) dx$$

Convert the standard basis $\{\mathbf{1}, \mathbf{x}, \mathbf{x}^2, \mathbf{x}^3\}$ to an orthonormal basis for P_3.
[The resulting polynomials are called the **Legendre** (normalized) polynomials.]

4.19. Find an orthogonal basis for the column space $C\,(\mathbf{A})$ of the matrix

$$\mathbf{A} = \begin{bmatrix} 1 & 0 & 2 & 1 \\ 0 & 1 & 3 & 3 \\ 2 & 5 & 1 & 0 \\ 0 & 4 & 0 & 0 \end{bmatrix}$$

4.20. Let V be an inner product space. Prove that if the vector $\mathbf{u}$ is orthogonal for every $\mathbf{v} \in V$ then $\mathbf{u} = \mathbf{O}$.

4.21. Let V be an inner product space. Prove that if the vector $\mathbf{u}$ is orthogonal to $\mathbf{v}$ then every scalar multiple of $\mathbf{u}$ is also orthogonal to $\mathbf{v}$.

4.22. Let $\{\mathbf{v}_1,\ \mathbf{v}_2,\ \mathbf{v}_3, \ldots,\ \mathbf{v}_n\}$ be a set of orthonormal vectors in an inner product space V. Prove that

$$\langle x_1\mathbf{v}_1 + x_2\mathbf{v}_2 + \cdots + x_n\mathbf{v}_n,\ y_1\mathbf{v}_1 + y_2\mathbf{v}_2 + \cdots + y_n\mathbf{v}_n\rangle = x_1y_1 + x_2y_2 + \cdots + x_ny_n$$

where x_i and y_i are scalars.

4.23. The distance function between two vectors $\mathbf{u}$ and $\mathbf{v}$ in an inner product space V is denoted as $d(\mathbf{u},\ \mathbf{v})$ and defined in (4.4) as

$$d(\mathbf{u},\ \mathbf{v}) = \|\mathbf{u} - \mathbf{v}\|$$

Prove the following results:

(a) If $\mathbf{u} \neq \mathbf{v}$ then $d(\mathbf{u},\ \mathbf{v}) > 0$.

(b) $d(\mathbf{u},\ \mathbf{v}) \leq d(\mathbf{u},\ \mathbf{w}) + d(\mathbf{w},\ \mathbf{v})$ where $\mathbf{w} \in V$.

Anshul Gupta

is a Research Staff Member in the HPC Group, Business Analytics and Mathematical Sciences, at the IBM Thomas J. Watson Research Center in New York.

Tell us about yourself and your work.

Although my background is in computer science, my work is at the confluence of computer science and numerical linear algebra. It often involves applying computer science techniques to develop scalable parallel algorithms and software for numerical methods used in scientific computing and optimization.

How do you use linear algebra in your job, how important is linear algebra and what are the challenges connected with the subject?

I both use and develop linear algebra algorithms and software. The biggest challenge that I face is that there is no software for many numerical linear algebra algorithms that are published in linear algebra journals, such as SIMAX, SISC, etc. The corresponding papers typically present limited experimental evidence of the effectiveness of the proposed techniques and a reliable comparison with competing techniques is even harder to find.

What are the key issues in your area of linear algebra research?

There are two important issues:

1) lack of resilience of linear algebra algorithms and software to soft errors on massively parallel machines,
2) scalability limiting effects of global communication and synchronization on linear algebra software on massively parallel machines.

Addressing both these issues requires the developing of novel algorithmic techniques.

Have you any particular messages that you would like to give to students starting off studying linear algebra?

Always pay attention to how a linear algebra algorithm will perform on real problems on real computers. Asymptotic bounds based on operation counts are almost obsolete when the operations are virtually free and almost all the cost is in data transfer. Different algorithms with similar complexities can have different spatial and temporal locality properties, which have a huge impact on the behaviour of the algorithm in a realistic setting.

5 Linear Transformations

SECTION 5.1 Introduction to Linear Transformations

By the end of this section you will be able to

- determine what is meant by a transformation
- prove some properties of linear transformations (mappings)

In this chapter we look at functions defined on vector spaces called transformations. Most of the theorems in linear algebra can be formulated in terms of linear transformations which we will discuss below. Transformations also give matrices a geometric flavour. For example, rotations, reflections and projections of vectors can be described by writing a transformation as a matrix.

Transformations are particularly useful in computer graphics and games. Examples of transformations which can be used to create an animation are shown in Fig. 5.1.

Sometimes linear algebra is described as the study of linear transformations on vector spaces.

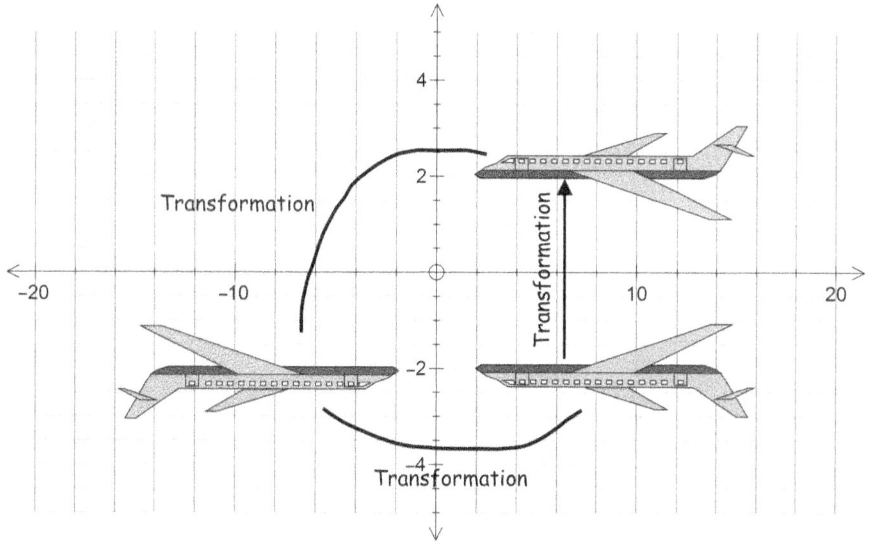

Figure 5.1

5.1.1 Functions

What does the term function mean?

In everyday language it means the purpose of something or someone. In mathematics, a function is a process that takes an input value, and generates a *unique* output value.

More formally:

Definition (5.1). A function operating from set A to set B is a rule which assigns *every element* of A to one and *only one* element in the set B.

A function f from A to B is denoted by $f : A \to B$. The set where the function starts from (A) is called the **domain** and where it arrives (B) is called the **codomain**. The function f takes a particular value, such as x, and converts it to a value y. This is denoted by

$$f(x) = y$$

The actual values that the function f takes in the set B is called the **range or image** (Fig. 5.2).

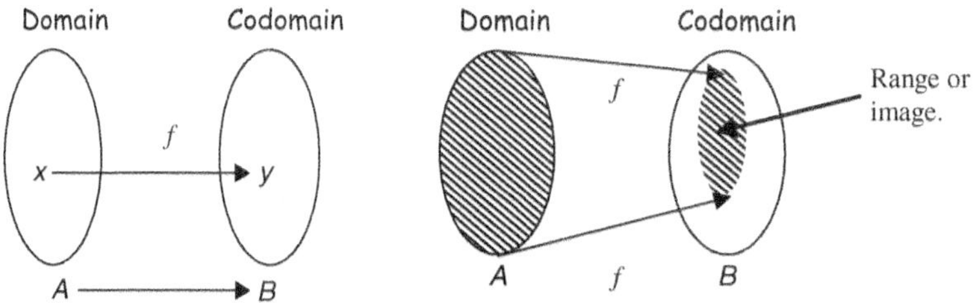

Figure 5.2

In this chapter we discuss a function between two vector spaces such as V and W. For example, $f : \mathbb{R}^2 \to \mathbb{R}^3$ is a function between two vector spaces (Euclidean spaces) where $\mathbb{R}^2$ is the domain and $\mathbb{R}^3$ the codomain.

In general, if for a function f we have $f : V \to W$ (a function from V to W), where V and W are vector spaces, then the function f is called a **transformation** or **map** from V to W. Transformation is just another word for function and works on vectors as well as numbers. Also we tend to use the letter T for transformation, that is $T : V \to W$.

In some mathematical literature you will see the term *map* used to mean transformation. They are *both* the same thing. We will use transformation.

Example 5.1

Consider $T : \mathbb{R}^2 \to \mathbb{R}^2$ (transformation from $\mathbb{R}^2$ to $\mathbb{R}^2$) defined by $T\left(\begin{bmatrix} x \\ y \end{bmatrix}\right) = \begin{pmatrix} -y \\ x \end{pmatrix}$.

Determine $T\left(\begin{bmatrix} 1 \\ 2 \end{bmatrix}\right)$ and plot both $\begin{bmatrix} 1 \\ 2 \end{bmatrix}$ and $T\left(\begin{bmatrix} 1 \\ 2 \end{bmatrix}\right)$ on $\mathbb{R}^2$.

Solution

How do we find $T\left(\begin{bmatrix}1\\2\end{bmatrix}\right)$*?*

By substituting $x = 1$ and $y = 2$ into $T\left(\begin{bmatrix}x\\y\end{bmatrix}\right) = \begin{pmatrix}-y\\x\end{pmatrix}$:

$$T\left(\begin{bmatrix}1\\2\end{bmatrix}\right) = \begin{pmatrix}-2\\1\end{pmatrix}$$

Plotting these on $\mathbb{R}^2$ we have (Fig. 5.3).

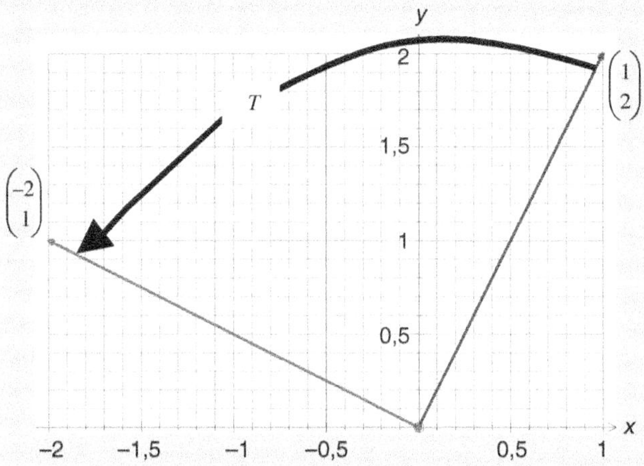

Figure 5.3

The transformation T rotates the vector $\begin{pmatrix}1\\2\end{pmatrix}$ by 90° in an anti-clockwise direction with centre at the origin.

Example 5.2

Consider the transformation $T : \mathbb{R}^3 \to \mathbb{R}^2$ defined by $T\left(\begin{bmatrix}x\\y\\z\end{bmatrix}\right) = \begin{pmatrix}x-y\\x-z\end{pmatrix}$. Determine

(i) $T\left(\begin{bmatrix}1\\2\\3\end{bmatrix}\right)$ **(ii)** $T\left(\begin{bmatrix}-1\\-2\\-3\end{bmatrix}\right)$

Solution

This type of transformation is common in computer graphics because it represents something in three dimensions $(\mathbb{R}^3)$ onto a two-dimensional screen $(\mathbb{R}^2)$.

(continued...)

(i) For $T\left(\begin{bmatrix}1\\2\\3\end{bmatrix}\right)$ we substitute $x=1, y=2$ and $z=3$ into $T\left(\begin{bmatrix}x\\y\\z\end{bmatrix}\right)=\begin{pmatrix}x-y\\x-z\end{pmatrix}$:

$$T\left(\begin{bmatrix}1\\2\\3\end{bmatrix}\right)=\begin{pmatrix}1-2\\1-3\end{pmatrix}=\begin{pmatrix}-1\\-2\end{pmatrix}$$

(ii) Similarly we have

$$T\left(\begin{bmatrix}-1\\-2\\-3\end{bmatrix}\right)=\begin{pmatrix}-1-(-2)\\-1-(-3)\end{pmatrix}=\begin{pmatrix}-1+2\\-1+3\end{pmatrix}=\begin{pmatrix}1\\2\end{pmatrix}$$

Note that the transformation T maps a vector from $\mathbb{R}^3$ (3-space) to a vector in $\mathbb{R}^2$ (2-space). (Transforms a vector in 3d onto a plane.)

5.1.2 Linear transformations (linear mappings)

What do you think the term 'linear transformation' means?

A linear transformation between vector spaces is a special transformation (mapping) which preserves the fundamental linear algebra operations - scalar multiplication and vector addition.

The formal definition is:

Definition (5.2). A transformation $T : V \to W$ is called a **linear transformation** $\Leftrightarrow$ for *all* vectors $\mathbf{u}$ and $\mathbf{v}$ in the vector space V and for any scalar k we have

(a) $T(\mathbf{u}+\mathbf{v}) = T(\mathbf{u}) + T(\mathbf{v})$ [T preserves vector addition]

(b) $T(k\mathbf{u}) = kT(\mathbf{u})$ [T preserves scalar multiplication]

For example, the transformations given in the above examples 5.1 and 5.2 are linear. For T to be a linear transformation or a linear map it needs to satisfy *both* conditions, (a) and (b).

What do these conditions mean?

They both mean that for a linear transformation it does *not* matter whether you carry out the vector operation first and then transform, or vice versa.

A transformation is a function with an input and an output (Fig. 5.4).

Figure 5.4

A linear transformation is one where the output changes proportionately to the input. A real-life example is cooking recipes – double your ingredients and you double your output.

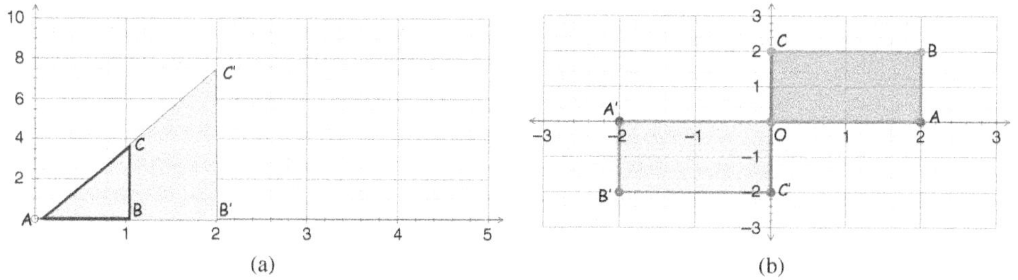

Figure 5.5

We can treat matrices as linear transformations which act on points or vectors (Fig. 5.5).

Fig. 5.5(a) shows an example of a transformation which enlarges an object. (Triangle ABC is enlarged to $AB'C'$ – doubles each side.) The matrix transformation in this case is $\begin{pmatrix} 2 & 0 \\ 0 & 2 \end{pmatrix}$.

Fig. 5.5(b) shows an example of a transformation which rotates the points of the rectangle anti-clockwise by 180° about the origin. The matrix transformation is $\begin{pmatrix} -1 & 0 \\ 0 & -1 \end{pmatrix}$.

Example 5.3

Show that $T : \mathbb{R}^2 \to \mathbb{R}^2$ defined by $T(\mathbf{x}) = \mathbf{A}\mathbf{x}$ where $\mathbf{A} = \begin{pmatrix} 1 & 0 \\ 0 & -1 \end{pmatrix}$ is a linear transformation.

Solution

What does this transformation do?

Let $\mathbf{x} = \begin{bmatrix} x \\ y \end{bmatrix}$ then $T\left(\begin{bmatrix} x \\ y \end{bmatrix}\right) = \begin{pmatrix} 1 & 0 \\ 0 & -1 \end{pmatrix}\begin{pmatrix} x \\ y \end{pmatrix} = \begin{pmatrix} x \\ -y \end{pmatrix}$.

The transformation T places a negative in front of the y. The result is shown in Fig. 5.6.

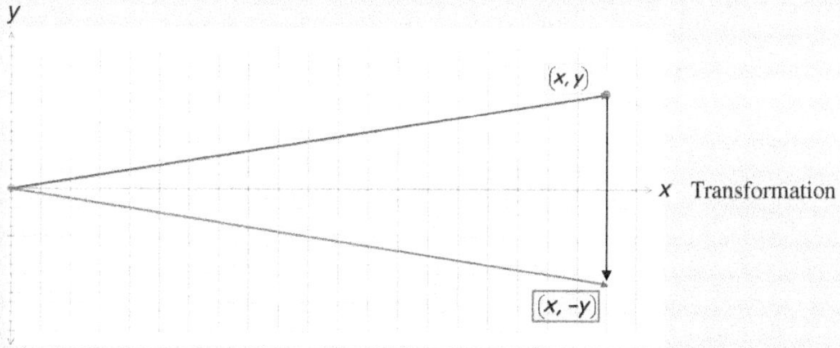

Figure 5.6

(continued...)

The given transformation *reflects* the vector in the horizontal axis. We need to show that this is a linear transformation.

Let $\mathbf{u} = \begin{bmatrix} x \\ y \end{bmatrix}$ and $\mathbf{v} = \begin{bmatrix} s \\ t \end{bmatrix}$ be our vectors in $\mathbb{R}^2$.

How do we show that the given T is linear?

Show that T preserves both vector addition and scalar multiplication. This means that we need to check that both conditions (a) and (b) of the above Definition (5.2) hold.

Checking condition (a):

$$T(\mathbf{u}+\mathbf{v}) = T\left(\begin{bmatrix} x \\ y \end{bmatrix} + \begin{bmatrix} s \\ t \end{bmatrix}\right)$$

$$= T\left(\begin{bmatrix} x+s \\ y+t \end{bmatrix}\right) \underset{\substack{\text{applying the given} \\ \text{transformation}}}{=} \begin{pmatrix} x+s \\ -(y+t) \end{pmatrix} = \begin{pmatrix} x+s \\ -y-t \end{pmatrix}$$

Now we need to show that $T(\mathbf{u}) + T(\mathbf{v})$ is equal to this vector on the right, $\begin{pmatrix} x+s \\ -y-t \end{pmatrix}$:

$$T(\mathbf{u}) + T(\mathbf{v}) = \begin{pmatrix} x \\ -y \end{pmatrix} + \begin{pmatrix} s \\ -t \end{pmatrix} = \begin{pmatrix} x+s \\ -y-t \end{pmatrix}$$

Thus we have shown condition (a), that is $T(\mathbf{u}+\mathbf{v}) = T(\mathbf{u}) + T(\mathbf{v})$.

In this example, linear transformation means that we can add the vectors $\mathbf{u}$ and $\mathbf{v}$ and then reflect in the horizontal axis, or we can reflect first and then add. In either case, we end up with the same vector because $T(\mathbf{u}+\mathbf{v}) = T(\mathbf{u}) + T(\mathbf{v})$.

Checking condition (b):

We need to demonstrate that T preserves scalar multiplication, that is $T(k\,\mathbf{u}) = k\,T(\mathbf{u})$, where k is any scalar:

$$T(k\mathbf{u}) = T\left(k\begin{bmatrix} x \\ y \end{bmatrix}\right) = T\left(\begin{bmatrix} kx \\ ky \end{bmatrix}\right) = \begin{pmatrix} kx \\ -ky \end{pmatrix} \quad \left[\text{applying } T\left(\begin{bmatrix} x \\ y \end{bmatrix}\right) = \begin{pmatrix} x \\ -y \end{pmatrix}\right]$$

What do we do next?

Show that $k\,T(\mathbf{u})$ is equal to the above vector on the right, $\begin{pmatrix} kx \\ -ky \end{pmatrix}$:

$$k\,T(\mathbf{u}) = kT\left(\begin{bmatrix} x \\ y \end{bmatrix}\right) = k\begin{pmatrix} x \\ -y \end{pmatrix} = \begin{pmatrix} kx \\ -ky \end{pmatrix} = T(k\,\mathbf{u}) \quad \text{[from above]}$$

Thus we have our result. Again, we can change the order, that is first apply scalar multiplication and then transform, or vice versa. We end up with the same vector, $T(k\,\mathbf{u}) = k\,T(\mathbf{u})$.

This means the given transformation T satisfies *both* conditions (a) and (b) of definition:

(5.2) (a) $T(\mathbf{u}+\mathbf{v}) = T(\mathbf{u}) + T(\mathbf{v})$ and (b) $T(k\,\mathbf{u}) = k\,T(\mathbf{u})$.

Therefore T is a linear transformation.

Let V be a vector space. Then a *linear mapping or transformation* $T : V \to V$ (the domain and codomain are the *same* vector space) is called a **linear operator**. Linear operators are important in subjects such as functional analysis and quantum mechanics because they preserve vector addition and scalar multiplication.

Is the transformation T given in the above Example 5.3 a linear operator?

Yes, because the transform goes from $\mathbb{R}^2$ to $\mathbb{R}^2$ (same vector spaces), that is $T : \mathbb{R}^2 \to \mathbb{R}^2$.

Example 5.4

Show that the transformation $T : \mathbb{R}^2 \to \mathbb{R}^3$ defined by $T\left(\begin{bmatrix} x \\ y \end{bmatrix}\right) = \begin{pmatrix} x+y \\ x-y \\ y \end{pmatrix}$ is linear. This is not a linear *operator* because T goes from $\mathbb{R}^2$ to $\mathbb{R}^3$ (different vector spaces). The transformation T in this case transforms a vector in the plane onto 3d space (Fig. 5.7).

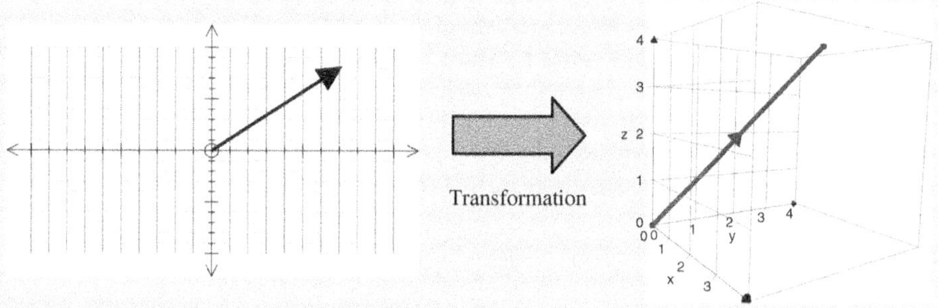

Figure 5.7

Solution

Let $\mathbf{u} = \begin{pmatrix} x \\ y \end{pmatrix}$ and $\mathbf{v} = \begin{pmatrix} s \\ t \end{pmatrix}$. As for the above Example 5.3 we have:

$$T(\mathbf{u}+\mathbf{v}) = T\left(\begin{bmatrix} x \\ y \end{bmatrix} + \begin{bmatrix} s \\ t \end{bmatrix}\right)$$

$$= T\left(\begin{bmatrix} x+s \\ y+t \end{bmatrix}\right) \underset{\substack{\text{applying the given} \\ \text{transformation}}}{=} \begin{pmatrix} (x+s)+(y+t) \\ (x+s)-(y+t) \\ y+t \end{pmatrix} = \begin{pmatrix} x+s+y+t \\ x-y+s-t \\ y+t \end{pmatrix}$$

We need to show that $T(\mathbf{u}) + T(\mathbf{v})$ is equal to the vector on the right hand side.

$$T(\mathbf{u}) + T(\mathbf{v}) = T\left(\begin{bmatrix} x \\ y \end{bmatrix}\right) + T\left(\begin{bmatrix} s \\ t \end{bmatrix}\right)$$

$$\underset{\substack{\text{applying the given} \\ \text{transformation}}}{=} \begin{pmatrix} x+y \\ x-y \\ y \end{pmatrix} + \begin{pmatrix} s+t \\ s-t \\ t \end{pmatrix} = \begin{pmatrix} x+y+s+t \\ x-y+s-t \\ y+t \end{pmatrix}$$

This is identical to $T(\mathbf{u}+\mathbf{v})$, therefore $T(\mathbf{u}+\mathbf{v}) = T(\mathbf{u}) + T(\mathbf{v})$, which means that T preserves vector addition. Similarly we can check $T(k\,\mathbf{u}) = k\,T(\mathbf{u})$ for any scalar k. Therefore by

(5.2) (a) $T(\mathbf{u}+\mathbf{v}) = T(\mathbf{u}) + T(\mathbf{v})$ and (b) $T(k\,\mathbf{u}) = k\,T(\mathbf{u})$

the given transformation T is a linear transformation.

A square function such as $f(x) = x^2$ is also a transformation. In this case, we have $f : \mathbb{R} \to \mathbb{R}$ given by $f(x) = x^2$.

Is f a linear transformation?

We need to check both conditions (a) and (b) of definition (5.2):

$$f(x+y) = (x+y)^2 = x^2 + 2xy + y^2$$
$$f(x) + f(y) = x^2 + y^2$$

Since $x^2 + 2xy + y^2 \neq x^2 + y^2$ [not equal], therefore

$$f(x+y) \neq f(x) + f(y)$$

What does this mean?

It means that f is *not* a linear function or linear transformation because condition (a) of (5.2) fails. [The output does *not* change in proportion; we have an extra $2xy$ with $f(x+y)$.]

Example 5.5

Let V be an inner product space. Show that $T : V \to \mathbb{R}$ defined by $T(\mathbf{u}) = \|\mathbf{u}\|$ is *not* a linear transformation. [This transformation measures the norm or length of a vector.]

Solution

How can we show that the given transformation is not linear?

If the transformation T *fails* one (or both) of the conditions of (5.2), that is:

(a) $T(\mathbf{u} + \mathbf{v}) = T(\mathbf{u}) + T(\mathbf{v})$ (b) $T(k\,\mathbf{u}) = k\,T(\mathbf{u})$

then T is *not* a linear transformation. Checking condition (a):

$$T(\mathbf{u} + \mathbf{v}) = \|\mathbf{u} + \mathbf{v}\|$$
$$T(\mathbf{u}) = \|\mathbf{u}\| \text{ and } T(\mathbf{v}) = \|\mathbf{v}\|$$

However, by Minkowski's inequality of chapter 4:

(4.7) $\|\mathbf{u} + \mathbf{v}\| \leq \|\mathbf{u}\| + \|\mathbf{v}\|$ [less than or equal to],

we have

$$\begin{aligned} T(\mathbf{u} + \mathbf{v}) &= \|\mathbf{u} + \mathbf{v}\| \\ &\leq \|\mathbf{u}\| + \|\mathbf{v}\| = T(\mathbf{u}) + T(\mathbf{v}) \end{aligned}$$

Hence $T(\mathbf{u} + \mathbf{v}) \leq T(\mathbf{u}) + T(\mathbf{v})$, therefore condition (a) fails because we don't have equality and so the given transformation is *not* linear. In this case, T is *not* a linear transformation because the length of the vector $\mathbf{u} + \mathbf{v}$ is *not* equal to the length of $\mathbf{u}$ plus the length of $\mathbf{v}$.

Example 5.6

Let V be a vector space. Show that the identity transformation $T : V \to V$ defined by $T(\mathbf{u}) = \mathbf{u}$ is a linear operator.

Solution

This is the identity transformation because T maps a given vector $\mathbf{u}$ back to $\mathbf{u}$ (Fig. 5.8).

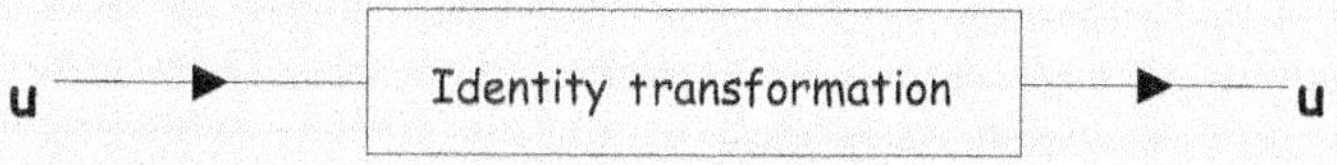

Figure 5.8

How do we show that this transformation is linear?

Clearly, the output is proportional to the input but intuition does *not* constitute a rigorous test. We use our definition of linear transformation. Check the two conditions of (5.2):

(a) $T(\mathbf{u}+\mathbf{v}) = T(\mathbf{u}) + T(\mathbf{v})$ and (b) $T(k\mathbf{u}) = kT(\mathbf{u})$

Let $\mathbf{u}$ and $\mathbf{v}$ be arbitrary vectors in the vector space V. We have

$$\begin{aligned} T(\mathbf{u}+\mathbf{v}) &= \mathbf{u}+\mathbf{v} \quad \text{[as transformation has the same input and output]} \\ &= T(\mathbf{u}) + T(\mathbf{v}) \end{aligned}$$

Hence condition (a) is satisfied. Let k be any scalar, then

$$T(k\mathbf{u}) = k\mathbf{u} = kT(\mathbf{u})$$

Thus condition (b) is satisfied. Therefore by the above Definition (5.2), we conclude that the given identity transformation, T, is a linear operator.

5.1.3 Properties of linear transformations

In this subsection, we highlight some of the properties of linear transformation.

Proposition (5.3). Let V and W be vector spaces and $\mathbf{u}$ and $\mathbf{v}$ be vectors in V. Let $T: V \to W$ be a linear transformation, then we have the following:

(a) $T(\mathbf{O}) = \mathbf{O}$ where $\mathbf{O}$ is the zero vector.

(b) $T(-\mathbf{u}) = -T(\mathbf{u})$

(c) $T(\mathbf{u}-\mathbf{v}) = T(\mathbf{u}) - T(\mathbf{v})$

Proof of (a).

We can write the zero vector $\mathbf{O}$ as $0\mathbf{u}$:

$$T(\mathbf{O}) = T(0\mathbf{u}) = \underbrace{0T(\mathbf{u})}_{\text{because } T \text{ is linear}} = \mathbf{O}$$

Thus $T(\mathbf{O}) = \mathbf{O}$.

■

Proof of (b).
We write $-\mathbf{u}$ as $(-1)\,\mathbf{u}$. We have

$$T(-\mathbf{u}) = T((-1)\,\mathbf{u}) \underbrace{=}_{\text{because } T \text{ is linear}} (-1)\,T(\mathbf{u}) = -T(\mathbf{u})$$

■

Proof of (c).
We write $\mathbf{u} - \mathbf{v}$ as $\mathbf{u} + (-\mathbf{v})$:

$$\begin{aligned} T(\mathbf{u} - \mathbf{v}) &= T(\mathbf{u} + (-\mathbf{v})) \\ &= T(\mathbf{u}) + T(-\mathbf{v}) && [\text{because } T \text{ is linear}] \\ &= T(\mathbf{u}) - T(\mathbf{v}) && [\text{by part (b) } T(-\mathbf{v}) = -T(\mathbf{v})] \end{aligned}$$

■

Another important property of linear transformation is the following:

Proposition (5.4). Let V and W be vector spaces. Let $T: V \to W$ be a linear transformation and $\{\mathbf{u}_1, \mathbf{u}_2, \mathbf{u}_3, \ldots, \mathbf{u}_n\}$ be vectors in the vector space V such that

$$\mathbf{v} = k_1\mathbf{u}_1 + k_2\mathbf{u}_2 + \cdots + k_n\mathbf{u}_n$$

where the k's are scalars. This means that the vector $\mathbf{v}$ is a linear combination of the $\mathbf{u}_{\text{subscript}}$ vectors. Then

$$T(\mathbf{v}) = k_1T(\mathbf{u}_1) + k_2T(\mathbf{u}_2) + \cdots + k_nT(\mathbf{u}_n)$$

How do we prove this proposition?
By using mathematical induction. The three steps of proof by mathematical induction are:

1. Prove the given result for some base case $n = 1$ or $n = m_0$.
2. Assume that the result is true for $n = m$.
3. Prove the result for $n = m + 1$.

Proof.

1. We can take out the scalar $T(k_1\mathbf{u}_1) = k_1T(\mathbf{u}_1)$ because T is linear. The result holds for $n = 1$.
2. We assume that the result is true for $n = m$, that is

$$T(k_1\mathbf{u}_1 + k_2\mathbf{u}_2 + \cdots + k_m\mathbf{u}_m) = k_1T(\mathbf{u}_1) + \cdots + k_mT(\mathbf{u}_m) \qquad (*)$$

3. Required to prove the result for $n = m + 1$:

$$\begin{aligned}
&T\left(k_1\mathbf{u}_1 + \cdots + k_m\mathbf{u}_m + k_{m+1}\mathbf{u}_{m+1}\right) \\
&= T\left(\left[k_1\mathbf{u}_1 + \cdots + k_m\mathbf{u}_m\right] + k_{m+1}\mathbf{u}_{m+1}\right) \\
&= T\left(\left[k_1\mathbf{u}_1 + \cdots + k_m\mathbf{u}_m\right]\right) + T\left(k_{m+1}\mathbf{u}_{m+1}\right) \quad [\text{because } T \text{ is linear}] \\
&= \underbrace{k_1 T\left(\mathbf{u}_1\right) + \cdots + k_m T\left(\mathbf{u}_m\right)}_{\text{By } (*)} + k_{m+1} T\left(\mathbf{u}_{m+1}\right)
\end{aligned}$$

Hence the result is true for $n = m + 1$.
By mathematical induction we have our required result.

■

We can use both these propositions (5.3) and (5.4) to determine whether a given transformation is linear or not. If any one of these properties (5.3) (a), (b), (c) or (5.4) is *not* satisfied for a given transformation then we conclude that T is *not* a linear transformation.

In the exercises, you are asked to show that if $T(\mathbf{O}) \neq \mathbf{O}$ [not equal] then T is *not linear*.

Example 5.7

Let P_2 be the vector space of polynomials of degree 2 or less. The polynomials are of the form $c_2x^2 + c_1x + c_0$ where the c's are the coefficients.

Show that the transformation $T : P_2 \to P_2$ defined by

$$T\left(c_2x^2 + c_1x + c_0\right) = c_2x^2 + c_1x + \left(c_0 + 1\right)$$

is *not* a linear transformation.

Solution

We can try testing (5.3) (a) which says that $T(\mathbf{O}) = \mathbf{O}$.

What is the zero vector in this case?

The zero vector is the zero polynomial which means that *all* the coefficients, c's, are equal to zero.

$$\begin{aligned}
T(\mathbf{O}) &= T\left(0x^2 + 0x + 0\right) \\
&= 0x^2 + 0x + (0 + 1) = 1 \quad [\text{applying transformation}]
\end{aligned}$$

Since 1 is *not* the zero vector, therefore T is *not* a linear transformation because we have $T(\mathbf{O}) \neq \mathbf{O}$ [not equal].

Example 5.8

Consider the linear transformation $T : \mathbb{R}^2 \to \mathbb{R}^2$ defined by

$$T\left(\mathbf{u}\right) = \mathbf{A}\mathbf{u} \text{ where } \mathbf{A} = \begin{pmatrix} 1 & 1 \\ -1 & 1 \end{pmatrix}$$

(continued...)

(i) The five corners of a house are given by the column vectors in the matrix:

$$\mathbf{H} = \begin{pmatrix} 0 & 0 & 2 & 4 & 4 \\ 0 & 3 & 5 & 3 & 0 \end{pmatrix}$$

Transform each of the column vectors of $\mathbf{H}$ under T.

(ii) Describe the effect of the transformation T on the column vectors of $\mathbf{H}$.

Solution

(i) Remember, the matrix multiplication $\mathbf{AH}$ means matrix $\mathbf{A}$ acts on each column vector of $\mathbf{H}$:

$$\mathbf{AH} = \begin{pmatrix} 1 & 1 \\ -1 & 1 \end{pmatrix}\begin{pmatrix} 0 & 0 & 2 & 4 & 4 \\ 0 & 3 & 5 & 3 & 0 \end{pmatrix} = \begin{pmatrix} 0 & 3 & 7 & 7 & 4 \\ 0 & 3 & 3 & -1 & -4 \end{pmatrix}$$

(ii) Plotting our results (Fig. 5.9):

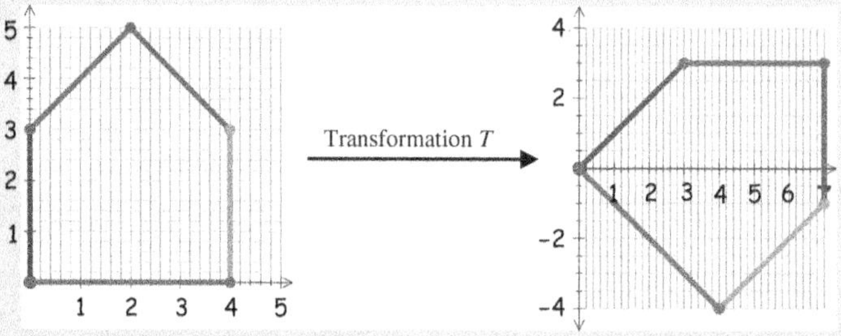

Figure 5.9

The transformation T rotates the given vectors (corners) by 45° in a clockwise direction.

Summary

A transformation or mapping is a function which assigns *every* element in the domain (start) to a *unique* element in the codomain (arrival).

A transformation between two vector spaces V and W denoted $T : V \to W$ is linear $\Leftrightarrow$

(a) $T(\mathbf{u} + \mathbf{v}) = T(\mathbf{u}) + T(\mathbf{v})$ for all vectors $\mathbf{u}$ and $\mathbf{v}$

(b) $T(k\mathbf{u}) = kT(\mathbf{u})$ for any scalar k

EXERCISES 5.1

(Brief solutions at end of book. Full solutions available at <http://www.oup.co.uk/companion/singh>.)

1. Consider the transformation $T : \mathbb{R}^2 \to \mathbb{R}^2$ given by $T(\mathbf{u}) = \mathbf{Au}$ where $\mathbf{A} = \begin{pmatrix} 0 & 1 \\ 1 & 0 \end{pmatrix}$.

Determine the vector $T(\mathbf{u})$ where $\mathbf{u}$ is given by:

(a) $\mathbf{u}=\begin{pmatrix}1\\3\end{pmatrix}$ (b) $\mathbf{u}=\begin{pmatrix}-1\\5\end{pmatrix}$ (c) $\mathbf{u}=\begin{pmatrix}\sqrt{2}\\1\end{pmatrix}$ (d) $\mathbf{u}=\begin{pmatrix}-2\\-3\end{pmatrix}$

For each of these vectors $\mathbf{u}$ and $T(\mathbf{u})$ plot them on the same plane $\mathbb{R}^2$.

2. Consider the transformation $T:\mathbb{R}^3\to\mathbb{R}^2$ given by $T\left(\begin{bmatrix}x\\y\\z\end{bmatrix}\right)=\begin{pmatrix}xy\\xz\end{pmatrix}$. Determine the vector $T(\mathbf{u})$ where $\mathbf{u}$ is given by:

(a) $\mathbf{u}=\begin{pmatrix}1\\3\\5\end{pmatrix}$ (b) $\mathbf{u}=\begin{pmatrix}-1\\-4\\4\end{pmatrix}$ (c) $\mathbf{u}=\begin{pmatrix}\sqrt{2}\\\sqrt{8}\\\sqrt{18}\end{pmatrix}$ (d) $\mathbf{u}=\begin{pmatrix}-2\\-3\\-4\end{pmatrix}$

3. Let $T:\mathbb{R}^3\to\mathbb{R}^3$ be given by $T\left(\begin{bmatrix}x\\y\\z\end{bmatrix}\right)=\begin{pmatrix}x+y\\y+z\\z+x\end{pmatrix}$. Find $T(\mathbf{u})$ where

(a) $\mathbf{u}=\begin{pmatrix}2\\4\\7\end{pmatrix}$ (b) $\mathbf{u}=\begin{pmatrix}-3\\8\\-6\end{pmatrix}$ (c) $\mathbf{u}=\begin{pmatrix}\pi\\2\pi\\5\pi\end{pmatrix}$ (d) $\mathbf{u}=\begin{pmatrix}1/2\\2/3\\3/4\end{pmatrix}$

4. Explain why $T:U\to V$ given by $T(\mathbf{u})=\pm\sqrt{\mathbf{u}}$ is not a transformation.

5. Determine whether the following transformations (mappings) are linear:

(a) $T:\mathbb{R}^2\to\mathbb{R}^2$ given by (i) $T\left(\begin{bmatrix}x\\y\end{bmatrix}\right)=\begin{pmatrix}y\\x\end{pmatrix}$ (ii) $T\left(\begin{bmatrix}x\\y\end{bmatrix}\right)=\begin{pmatrix}x^2\\y^2\end{pmatrix}$

(b) $T:\mathbb{R}^3\to\mathbb{R}^2$ given by $T\left(\begin{bmatrix}x\\y\\z\end{bmatrix}\right)=\begin{pmatrix}xy\\xz\end{pmatrix}$

(c) $T:\mathbb{R}^3\to\mathbb{R}^3$ given by (i) $T\left(\begin{bmatrix}x\\y\\z\end{bmatrix}\right)=\begin{pmatrix}x+y\\y+z\\z+x\end{pmatrix}$ (ii) $T\left(\begin{bmatrix}x\\y\\z\end{bmatrix}\right)=\begin{pmatrix}\sqrt{x}\\\sqrt{y}\\\sqrt{z}\end{pmatrix}$

6. Let P_2 be the vector space of polynomials of degree 2 or less. Decide whether the following transformations are linear:

(a) $T:P_2\to P_2$ given by $T\left(c_2x^2+c_1x+c_0\right)=c_0x^2+c_1x+c_2$
(b) $T:P_2\to P_2$ given by $T\left(c_2x^2+c_1x+c_0\right)=c_2^2x^2+c_1^2x+c_0^2$

7. Let M_{nn} be the vector space of size n by n matrices. Decide whether the following transformations (mappings) are linear:

(a) $T:M_{22}\to M_{22}$ given by $T(\mathbf{A})=\mathbf{A}^T$ where $\mathbf{A}^T$ is the transpose of the matrix $\mathbf{A}$.
(b) $T:M_{22}\to\mathbb{R}$ given by $T(\mathbf{A})=tr(\mathbf{A})$ where $tr(\mathbf{A})$ is the trace of the matrix $\mathbf{A}$. [Remember, the trace of a matrix is the addition of all the leading diagonal elements of the matrix.]

(c) $T : M_{nn} \to \mathbb{R}$ given by $T(\mathbf{A}) = a_{11}a_{22}a_{33}\cdots a_{nn}$ where a_{jj} are the entries along the leading diagonal of a n by n matrix.

8. Show that the transpose of any square matrix is a linear transformation (mapping).

9. Let $T : C[0,1] \to \mathbb{R}$ be a transformation (mapping) given by $T(f) = \int_0^1 f(x)dx$ where f is a continuous function defined in the interval $[0,1]$. Show that T is a linear transformation.

10. Let $T : V \to W$ be a transform such that $T(\mathbf{O}) \neq \mathbf{O}$. Prove that T is *not* a linear transform.

11. Let $T : V \to W$ be the zero transform, that is $T(\mathbf{v}) = \mathbf{O}$ for all vectors $\mathbf{v} \in V$. Show that T is a linear transformation (mapping).

12. Let $T : V \to W$ be a linear transformation of an n-dimensional vector space into a vector space W. Let $\{\mathbf{v}_1, \mathbf{v}_2, \mathbf{v}_3, \ldots, \mathbf{v}_n\}$ be a basis for V. Prove that if $\mathbf{u}$ is any vector in V then we can write $T(\mathbf{u})$ as a linear combination of

$$\{T(\mathbf{v}_1), T(\mathbf{v}_2), T(\mathbf{v}_3), \ldots, T(\mathbf{v}_n)\}$$

SECTION 5.2 Kernel and Range of a Linear Transformation

By the end of this section you will be able to

- understand what is meant by the kernel and range of a linear transform
- derive some properties of the kernel and range

This is a challenging section which will use some of the results from earlier chapters. In particular we will need to apply some propositions from chapter 3 to prove some of the properties of the kernel and range. Try to do the proofs on your own first, and if you get stuck then have a look at the text. This is a good way of understanding a proof.

The next two sections have similarities with subsection 3.6.1, where we defined: null space, nullity and rank of a matrix. In this section we describe the kernel (null space), nullity and range of a linear transformation.

Why is the kernel and range important?

Generally the kernel and range tell you which information has been carried over in the transform and what has been lost.

How do we know this?

By one of the most important results in linear algebra which is:

Let $T : V \to W$ be a linear transform where V is of finite dimension (dim). Then

$$\dim(\text{kernel}(T)) + \dim(\mathit{range}(T)) = \dim(V)$$

This result says that all the information is contained in these two sets – kernel and range. This is equivalent to the dimension theorem (3.34) that we mentioned in chapter 3.

How is the kernel and range connected to a linear system of equations?

A matrix, say $\mathbf{A}$, can be used to represent a linear transformation, so finding the kernel means finding all the vectors $\mathbf{x}$ such that $\mathbf{Ax} = \mathbf{O}$. The linear system $\mathbf{Ax} = \mathbf{b}$ has a solution if and only if vector $\mathbf{b}$ is in the range (image) of the linear transformation.

5.2.1 Kernel of a linear transformation (mapping)

The notation $\mathbf{Ax} = \mathbf{b}$ means that the matrix $\mathbf{A}$ transforms (maps) the vector $\mathbf{x}$ to the vector $\mathbf{b}$. Matrix $\mathbf{A}$ is a function with input $\mathbf{x}$ and output $\mathbf{b}$ (Fig. 5.10).

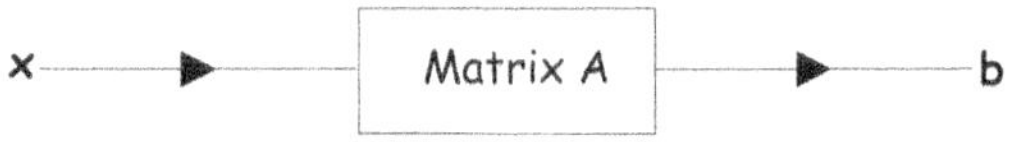

Figure 5.10

Generally a function f in mathematics acts on numbers, but the matrix $\mathbf{A}$ acts on vectors.

Example 5.9

Find $\mathbf{v}$ such that $T(\mathbf{v}) = \mathbf{O}$ for the linear transformation $T : \mathbb{R}^2 \to \mathbb{R}^2$ defined by

$$T(\mathbf{v}) = \mathbf{Av} \text{ where } \mathbf{A} = \begin{pmatrix} 1 & 1 \\ -1 & 1 \end{pmatrix}$$

Solution

Let $\mathbf{v} = (x \;\; y)^T$ and $\mathbf{A}$ be the given matrix, then

$$T\left(\begin{pmatrix} x \\ y \end{pmatrix}\right) = \begin{pmatrix} 1 & 1 \\ -1 & 1 \end{pmatrix}\begin{pmatrix} x \\ y \end{pmatrix} = \begin{pmatrix} 0 \\ 0 \end{pmatrix} \quad \left[\text{Using } T(\mathbf{v}) = \mathbf{Av} = \mathbf{O} = \begin{pmatrix} 0 \\ 0 \end{pmatrix}\right]$$

Opening out the matrix gives the simultaneous equations:

$$\left.\begin{aligned} x + y &= 0 \\ -x + y &= 0 \end{aligned}\right\} \Rightarrow x = 0 \quad \text{and} \quad y = 0$$

Hence $\mathbf{v} = (x \;\; y)^T = (0 \;\; 0)^T = \mathbf{O}$.

The set of vectors in the starting vector space which are transformed to the zero vector is called the **kernel** (pronounced 'kur-nl') of the transformation and is denoted by $\ker(T)$.

For the above Example 5.9 we have $\ker(T) = \{\mathbf{O}\}$, that is the kernel of T is *only* the zero vector. There are *no* other vectors in $\mathbb{R}^2$ which get transformed to the zero vector under T. (This $\ker(T) = \{\mathbf{O}\}$ is the **null space** of the matrix $\mathbf{A}$.) Note that $T(\mathbf{v}) = \mathbf{O}$ is equivalent to solving $\mathbf{Av} = \mathbf{O}$, because matrix $\mathbf{A}$ represents the transform T.

We will show later that if the kernel is just the zero vector, $\mathbf{O}$, then we can move things back, that is the linear transform has an inverse. It means that all the information was carried over. In general, there may exist other vectors besides the zero vector which are also transformed to the zero vector as Example 5.10 below demonstrates.

For a general linear transform, $T: V \to W$, the set of vectors in the domain V which arrive at the zero vector $\mathbf{O}$ in W is called the kernel of T and is illustrated in Fig. 5.11.

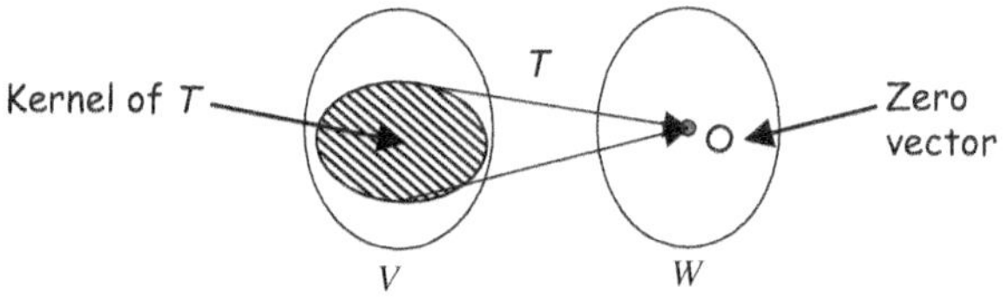

Figure 5.11

In everyday English language, kernel means softer part inside a shell. In linear algebra, kernel represents those parts which get transformed to the zero vector.

Definition (5.5). Let $T: V \to W$ be a linear transform (map). The set of *all* vectors $\mathbf{v}$ in V that are transformed to the zero vector in W is called the **kernel** of T, denoted by $\ker(T)$. It is the set of vectors $\mathbf{v}$ in V such that $T(\mathbf{v}) = \mathbf{O}$.

Example 5.10

Consider the zero linear transformation $T: V \to W$ such that $T(\mathbf{v}) = \mathbf{O}$ for all vectors $\mathbf{v}$ in V. Find $\ker(T)$.

Solution

All the vectors in V arrive at the zero vector under T, that is $T(\mathbf{v}) = \mathbf{O}$, therefore $\ker(T) = V$. This means *all* of the input set V is the kernel of T (Fig. 5.12).

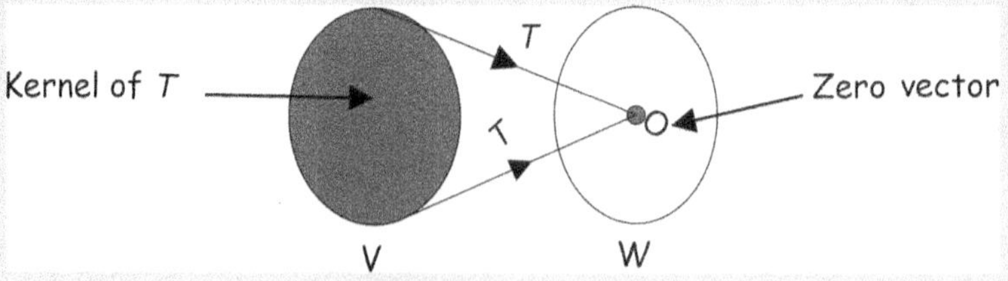

Figure 5.12

Example 5.11

(i) Consider the linear transformation $T : \mathbb{R}^3 \to \mathbb{R}^3$ given by $T\left(\begin{pmatrix} x \\ y \\ z \end{pmatrix}\right) = \begin{pmatrix} x \\ y \\ 0 \end{pmatrix}$. Find $\ker(T)$.

(ii) Show that $\mathbf{u} - \mathbf{v}$ is in $\ker(T)$ where $\mathbf{u} = (1 \ \ 2 \ \ 3)^T$ and $\mathbf{v} = (1 \ \ 2 \ \ 5)^T$.

Solution

(i) We need to find the vectors in $\mathbb{R}^3$ which arrive at the zero vector under T.
Which vectors are transformed (mapped) to the zero vector?

Inspecting the given transformation $T\left(\begin{pmatrix} x \\ y \\ z \end{pmatrix}\right) = \begin{pmatrix} x \\ y \\ 0 \end{pmatrix}$, we arrive at the zero vector if $x = 0, y = 0$ and $z = r$, where r is any real number. Therefore

$$T\left(\begin{pmatrix} 0 \\ 0 \\ r \end{pmatrix}\right) = \begin{pmatrix} 0 \\ 0 \\ 0 \end{pmatrix} = \mathbf{O}$$

Thus the kernel of T in this case is $\left\{ \begin{pmatrix} 0 \\ 0 \\ r \end{pmatrix} \middle| \ r \text{ is a real number} \right\}$. The kernel of the given transformation T is the z axis in $\mathbb{R}^3$ as illustrated in Fig. 5.13.

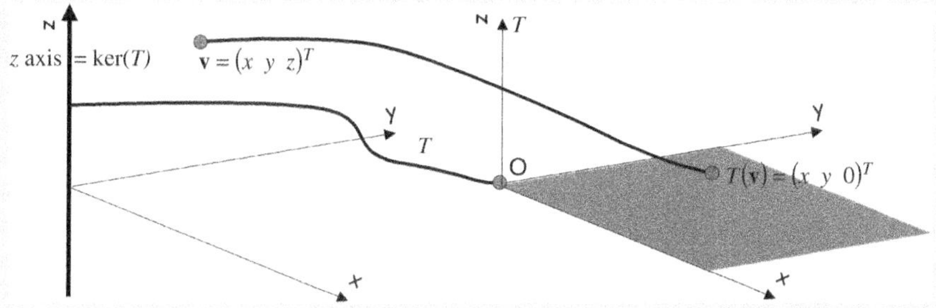

Figure 5.13

The given T transforms any point in $\mathbb{R}^3$ to the shaded plane shown in Fig. 5.13.

(ii) We have

$$\mathbf{u} - \mathbf{v} = (1 \ \ 2 \ \ 3)^T - (1 \ \ 2 \ \ 5)^T = (0 \ \ 0 \ \ -2)^T$$

Transforming this vector $\mathbf{u} - \mathbf{v}$:

$$T(\mathbf{u} - \mathbf{v}) = T\left((0 \ \ 0 \ \ -2)^T\right) = (0 \ \ 0 \ \ 0)^T = \mathbf{O}$$

Since $T(\mathbf{u} - \mathbf{v}) = \mathbf{O}$ the vector $\mathbf{u} - \mathbf{v}$ is in $\ker(T)$. Using T is linear on $T(\mathbf{u} - \mathbf{v}) = \mathbf{O}$ gives

$$T(\mathbf{u} - \mathbf{v}) = T(\mathbf{u}) - T(\mathbf{v}) = \mathbf{O} \text{ implies } T(\mathbf{u}) = T(\mathbf{v})$$

(continued...)

$\mathbf{u} - \mathbf{v}$ is in $\ker(T)$ means that vectors $\mathbf{u}$ and $\mathbf{v}$ arrive at the same point under T (Fig. 5.14).

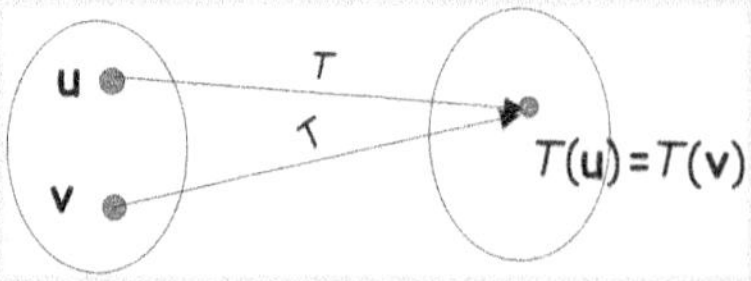

Figure 5.14

Hence the kernel detects when two points are transformed to the same point.

The linear transformation in Example 5.11 projects vectors in $\mathbb{R}^3$ onto a plane in $\mathbb{R}^3$. This type of linear transformation crops up in computer graphics. In computer graphics the concept of two vectors being transformed to the same point means that one object is 'blocking' the view of the other on the screen.

If the vector $\mathbf{u} - \mathbf{v}$ is in $\ker(T)$ for a linear transformation T, which projects vectors onto a screen, then $T(\mathbf{u})$ is blocking $T(\mathbf{v})$ or vice versa on the screen.

5.2.2 Properties of the kernel

Proposition (5.6). Let $T : V \to W$ be a linear transformation (mapping) between the vector spaces V and W. Then the kernel of T is a subspace of the vector space V.

What does this mean?

The set of vectors $\mathbf{v}$ in V such that $T(\mathbf{v}) = \mathbf{O}$ is a subspace of V which can be illustrated as shown in Fig. 5.15.

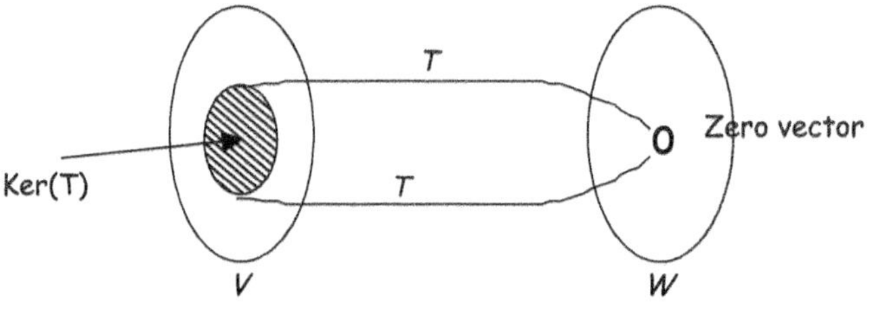

Figure 5.15

How do we prove this result?

By using Proposition (3.7) of chapter 3 which states:

A non-empty subset S with vectors $\mathbf{u}$ and $\mathbf{v}$ is a subspace $\Leftrightarrow$ $k\mathbf{u} + c\mathbf{v}$ is also in S.

In this case, the set S in question is $\ker(T)$.

Proof.

How do we know that $\ker(T)$ *is non-empty?*

Because by Proposition (5.3) (a) of the last section we have $T(\mathbf{O}) = \mathbf{O}$ for a linear transform T. Hence the zero vector $\mathbf{O}$ is in $\ker(T)$ so it *cannot* be empty.

What else do we need to show?

(3.7) says that if $\mathbf{u}$ and $\mathbf{v}$ are vectors in $\ker(T)$ then any linear combination $k\mathbf{u} + c\mathbf{v}$ (k and c are scalars) is also in $\ker(T)$.

Let $\mathbf{u}$ and $\mathbf{v}$ be vectors in $\ker(T)$, consider the vector $k\mathbf{u} + c\mathbf{v}$:

$$\begin{aligned} T(k\mathbf{u} + c\mathbf{v}) &= kT(\mathbf{u}) + cT(\mathbf{v}) && [\text{because } T \text{ is linear}] \\ &= k\mathbf{O} + c\mathbf{O} && \left[\begin{array}{l} T(\mathbf{u}) = T(\mathbf{v}) = \mathbf{O} \text{ because} \\ \mathbf{u} \text{ and } \mathbf{v} \text{ are in } \ker(T) \end{array}\right] \\ &= \mathbf{O} \end{aligned}$$

We have $T(k\mathbf{u} + c\mathbf{v}) = \mathbf{O}$, therefore $k\mathbf{u} + c\mathbf{v}$ is in $\ker(T)$.

Any linear combination of vectors $\mathbf{u}$ and $\mathbf{v}$ are in $\ker(T)$, therefore $\ker(T)$ is a subspace of V.

■

Since $\ker(T)$ is a subspace of the vector space V we have the following definition:

Definition (5.7). Let $T : V \to W$ be a linear transform. Then $\ker(T)$ is also called the **null space** of T and the dimension of $\ker(T)$ is called the **nullity** of T which is denoted by $\text{nullity}(T)$.

This definition is identical to the definition of null space of a matrix given in chapter 3.

Let $T(\mathbf{u}) = \mathbf{A}\mathbf{u}$ be a linear transformation then $\ker(T)$, or the null space of T, is the general solution $\mathbf{u}$ such that $\mathbf{A}\mathbf{u} = \mathbf{O}$. [Definition (3.31).]

This means that $\ker(T)$ is the set of input vectors $\mathbf{u}$ that arrive at the zero vector by applying matrix $\mathbf{A}$. Remember, this is the same as the null space of matrix $\mathbf{A}$. The null space definition given above is more generic because it applies to any linear transformation as you will see in later examples.

Example 5.12

Let $T : \mathbb{R}^3 \to \mathbb{R}^2$ be given by $T(\mathbf{x}) = \mathbf{A}\mathbf{x}$ where $\mathbf{x}$ is in $\mathbb{R}^3$ and

$$\mathbf{A} = \begin{pmatrix} 1 & 2 & 3 \\ 4 & 5 & 6 \end{pmatrix}$$

Find **(i)** $\ker(T)$ **(ii)** null space of T **(iii)** $\text{nullity}(T)$

Solution

(i) $\ker(T)$ is the solution space of

$$T(\mathbf{x}) = \mathbf{A}\mathbf{x} = \mathbf{O}$$

Hence it is the null space of matrix $\mathbf{A}$. We found null spaces of matrices in section 3.6.

(continued...)

How?
By placing the matrix $\mathbf{A}$ into (reduced) row echelon form $\mathbf{R}$ and then solving $\mathbf{Rx} = \mathbf{O}$:

$$\mathbf{A} = \begin{pmatrix} 1 & 2 & 3 \\ 4 & 5 & 6 \end{pmatrix} \Longrightarrow \mathbf{R} = \begin{pmatrix} 1 & 0 & -1 \\ 0 & 1 & 2 \end{pmatrix}$$

Letting $\mathbf{x} = \begin{pmatrix} x & y & z \end{pmatrix}^T$ then $\mathbf{Rx} = \mathbf{O}$ gives:

$$\begin{aligned} x - z &= 0 \\ y + 2z &= 0 \end{aligned}$$

From the first equation, we have $x = z$ and the bottom equation gives $y = -2z$. Let $z = r$, where r is any real number, then $x = r$, $y = -2r$ and $z = r$. This means that $\ker(T)$ is the solution:

$$\mathbf{x} = \begin{pmatrix} x \\ y \\ z \end{pmatrix} = \begin{pmatrix} r \\ -2r \\ r \end{pmatrix} = r\begin{pmatrix} 1 \\ -2 \\ 1 \end{pmatrix}$$

We have $\ker(T) = \left\{ r\mathbf{u} \mid \mathbf{u} = \begin{pmatrix} 1 & -2 & 1 \end{pmatrix}^T \text{ and } r \text{ is any real number} \right\}$.

(ii) The null space of T and matrix $\mathbf{A}$ is $\ker(T) = \left\{ r\mathbf{u} \mid \mathbf{u} = \begin{pmatrix} 1 & -2 & 1 \end{pmatrix}^T \right\}$.

(iii) *What does nullity(T) mean?*
It is the dimension of the null space or $\ker(T)$. The dimension is the number of vectors in the basis of $\ker(T)$. There is only one vector in the basis of $\ker(T)$ which is $\mathbf{u} = \begin{pmatrix} 1 & -2 & 1 \end{pmatrix}^T$, therefore $\text{nullity}(T) = 1$.

The challenge is finding $\ker(T)$, the kernel of the given linear transformation (mapping). Once we have $\ker(T)$ then the dimension of this subspace, given by $\text{nullity}(T)$ is normally straightforward, we only need to find the number of basis vectors for $\ker(T)$.

However, finding a basis can be more challenging.

Example 5.13

Let P_2 and P_3 be the vector space of polynomials of degrees less than or equal to 2 and 3 respectively. Consider the linear transformation $T : P_3 \to P_2$ given by

$$T(\mathbf{p}) = p'(x)$$

where $p(x)$ is a cubic polynomial and $p'(x)$ represents the first derivative of $p(x)$.

Determine $\text{nullity}(T)$.

Solution

How do we find nullity(T)?
First we find $\ker(T)$.

What is $\ker(T)$ *equal to?*
$\ker(T)$ is the set of polynomials that arrive at the zero vector under the given linear transform T. This transform T differentiates a given polynomial $p(x)$. We have

$$T(\mathbf{p}) = p'(x) = 0$$

Which polynomials gives zero after differentiation?
The constant polynomials, that is

$$T(\mathbf{c}) = c' = 0 \text{ where } c \text{ is a constant}$$

Thus $\ker(T) = \{c \mid c \text{ is any real number}\}$.
What is the dimension of this?
A basis for $\ker(T)$ is $\{1\}$, therefore dimension of $\ker(T)$ is 1 because there is only *one* vector in the basis. We have

$$\text{nullity}(T) = 1$$

5.2.3 The range (image) of a linear transformation

We briefly defined what is meant by the range (image) of a function in section 5.1.

What is the range of a linear transformation $T : V \to W$*?*
Figure 5.16 illustrates the meaning of the range or image of a linear transform.

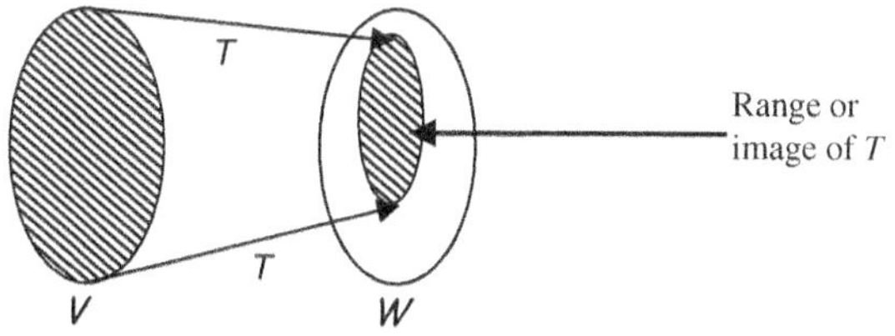

Figure 5.16

The range of a linear transformation T is the set of vectors we arrive at after applying T.

Definition (5.8). Let $T : V \to W$ be a linear transform. The **range** (image) of the linear transform is the set of *all* the output vectors $\mathbf{w}$ in W for which there are input vectors $\mathbf{v}$ in V such that

$$T(\mathbf{v}) = \mathbf{w}$$

The range is the set of output vectors in W that are images of the input vectors in V under the transform T. We can write the range(T) of $T : V \to W$ in set theory notation as follows:

$$\text{range}(T) = \{T(\mathbf{v}) \mid \mathbf{v} \text{ in } V\} \quad (5.9)$$

Note that the range(T) may *not* include *all* of the arrival vector space W as you can see in the above Fig. 5.16.

Example 5.14

Consider the zero linear transformation $T : V \to W$ such that $T(\mathbf{v}) = \mathbf{O}$ for all vectors $\mathbf{v}$ in V. Find range(T) or in other words the image of T.

Solution

All vectors arrive at a single destination – the zero vector under T, therefore:

$$\text{range}(T) = \mathbf{O} \quad [\text{zero vector}]$$

Thus the range or image is a single element $\{\mathbf{O}\}$. This is illustrated in Fig. 5.17.

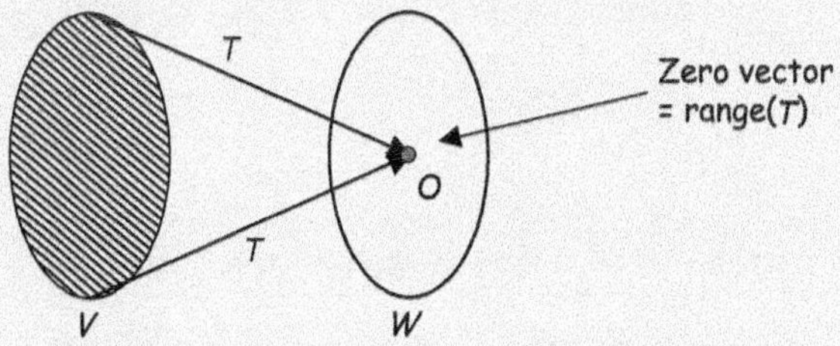

Figure 5.17

Note that the ker(T) is the whole of the *start* vector space V, ker(T) $= V$, but range of T is just the zero vector, $\mathbf{O}$.

It is important to note the difference between kernel and range. The kernel of a transformation lies in the domain (start) and range lies in the codomain (arrival).

A common term for *range* is *image*. We will use both.

Example 5.15

Consider the linear transformation $T : P_2 \to \mathbb{R}$ such that $T(\mathbf{p}) = \int_0^1 p(x)dx$ where P_2 is the vector space of polynomials of degree 2 or less. Determine range (T). [You can check that T is indeed a linear transformation.]

Solution

Let $p(x) = ax^2 + bx + c$ because $p(x)$ is a polynomial of degree 2 or less, then

$$T(p(x)) = \int_0^1 (ax^2 + bx + c)dx$$

$$\underset{\text{integrating}}{=} \left[\frac{ax^3}{3} + \frac{bx^2}{2} + cx\right]_0^1 = \frac{a}{3} + \frac{b}{2} + c \quad \begin{bmatrix}\text{substituting limits} \\ x = 1 \text{ and } x = 0\end{bmatrix}$$

We have a, b and c are any real numbers, therefore $T(\mathbf{p}) = \frac{a}{3} + \frac{b}{2} + c$ can be any real number.

What is range(T) equal to?

The set of all the real numbers, that is range$(T) = \mathbb{R}$. We have all of P_2 being transformed to the set of real numbers $\mathbb{R}$ (Fig. 5.18).

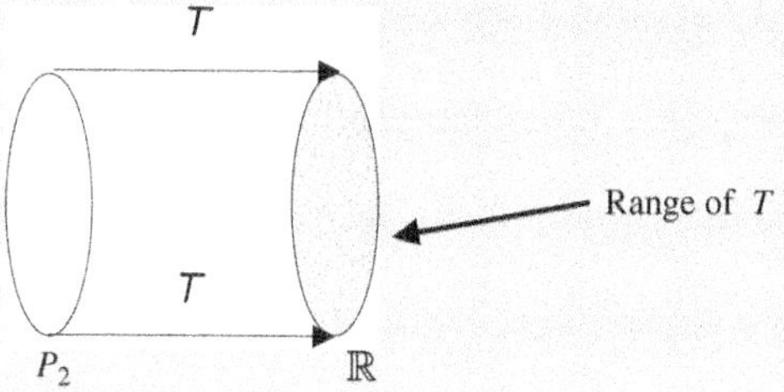

Figure 5.18

5.2.4 Properties of the range of a linear transformation

Note that the range of a linear transform $T : V \to W$ is composed of vectors in the arrival vector space W. (The vectors we arrive at in W after applying T to vectors in V.) Next we show that the range of T is a subspace of the arrival vector space W.

Proposition (5.10). Let V and W be vector spaces and $T : V \to W$ be a linear transformation. The range or image of the transformation T is a subspace of the arrival vector space W.

What does this mean?

The shaded part in W in Fig. 5.16 is a subspace of W.

How do we prove this proposition?

Again, by using Proposition (3.7) of chapter 3:

A non-empty subset S with vectors $\mathbf{u}$ and $\mathbf{v}$ is a subspace $\Leftrightarrow k\mathbf{u} + c\mathbf{v}$ is also in S.

Proof.

For a linear transform we have $T(\mathbf{O}) = \mathbf{O}$, therefore $\mathbf{O}$ is in the set of range(T), which means the range is non-empty.

How do we prove that range(T) is a subspace of the arrival space W?

By showing any linear combination is also in the range. Let $\mathbf{u}$ and $\mathbf{w}$ be vectors in range(T) then we need to show that

$$k\mathbf{u} + c\mathbf{w} \text{ is also in range}(T) \quad [k \text{ and } c \text{ are scalars}]$$

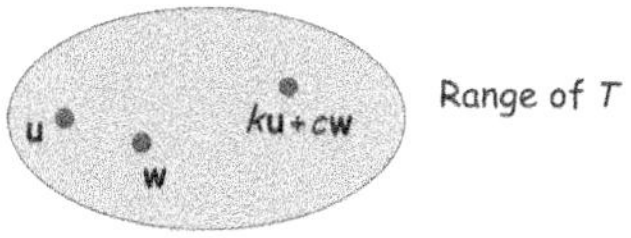

Figure 5.19

Required to prove that the set range(T) is *closed*, which means that $k\mathbf{u} + c\mathbf{w}$ cannot escape from range(T).

Since $\mathbf{u}$ and $\mathbf{w}$ are in range(T), there must exist input vectors $\mathbf{x}$ in V and $\mathbf{y}$ in V such that

$$T(x) = \mathbf{u} \text{ and } T(\mathbf{y}) = \mathbf{w}$$

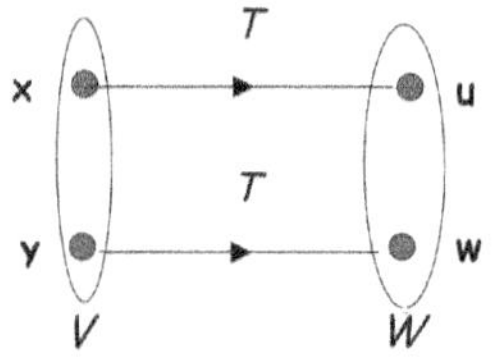

Figure 5.20

Consider the vector $k\mathbf{u} + c\mathbf{w}$:

$$\begin{aligned} k\mathbf{u} + c\mathbf{w} &= k\,T(\mathbf{x}) + c\,T(\mathbf{y}) \\ &= T(k\mathbf{x}) + T(c\mathbf{y}) && [\text{because } T \text{ is linear}] \\ &= T(k\mathbf{x} + c\mathbf{y}) && [\text{because } T \text{ is linear}] \end{aligned}$$

V is a vector space, therefore the linear combination $k\mathbf{x} + c\mathbf{y}$ is in V. We have

$$T(k\mathbf{x} + c\mathbf{y}) = k\mathbf{u} + c\mathbf{w}$$

therefore $k\mathbf{u} + c\mathbf{w}$ is in range(T). Hence by (3.7), we conclude that range(T) is a subspace of the vector space W.

■

Summary

Let $T : V \to W$ be a linear transformation. We have the following:

(5.5) The set of *all* the vectors $\mathbf{v}$ in V such that $T(\mathbf{v}) = \mathbf{O}$ is called the **kernel** of T.

(5.9) The **range** (or **image**) of T in W is defined by range(T) $= \left\{ T(\mathbf{v}) \,\middle|\, \mathbf{v} \text{ in } V \right\}$.

EXERCISES 5.2

(Brief solutions at end of book. Full solutions available at <http://www.oup.co.uk/companion/singh>.)

1. Find the kernel of the following linear transformations:

(a) $T : \mathbb{R}^2 \to \mathbb{R}^2$ given by $T(\mathbf{v}) = \mathbf{A}\mathbf{v}$ where $\mathbf{A} = \begin{pmatrix} 1 & 0 \\ 0 & 1 \end{pmatrix}$.

(b) $T : \mathbb{R}^2 \to \mathbb{R}^2$ given by $T(\mathbf{v}) = \mathbf{A}\mathbf{v}$ where $\mathbf{A} = \begin{pmatrix} 1 & 1 \\ 1 & 1 \end{pmatrix}$.

(c) $T : \mathbb{R}^2 \to \mathbb{R}^2$ given by $T\left(\begin{pmatrix} x \\ y \end{pmatrix}\right) = \begin{pmatrix} 0 \\ 0 \end{pmatrix}$.

(d) $T : \mathbb{R}^3 \to \mathbb{R}^3$ given by $T\left(\begin{pmatrix} x \\ y \\ z \end{pmatrix}\right) = \begin{pmatrix} 0 \\ 0 \\ x \end{pmatrix}$.

(e) $T : \mathbb{R}^3 \to \mathbb{R}^2$ given by $T\left(\begin{pmatrix} x \\ y \\ z \end{pmatrix}\right) = \begin{pmatrix} y - z \\ x - z \end{pmatrix}$.

2. Let $T : M_{nn} \to M_{nn}$ be the linear transformation given by the transpose, that is

$$T(\mathbf{A}) = \mathbf{A}^T$$

Determine $\ker(T)$ and $\text{range}(T)$. [M_{nn} is an n by n square matrix].

3. Let $T : P_2 \to P_1$ be the linear transformation of differentiation

$$T(\mathbf{p}) = p'(x)$$

Determine $\ker(T)$ and $\text{range}(T)$.

4. Let $T : V \to W$ be a linear transformation. Show that if $\mathbf{u} \in \ker(T)$ and $\mathbf{v} \in \ker(T)$ then for any scalars k and c the vector $(k\,\mathbf{u} + c\,\mathbf{v}) \in \ker(T)$.

5. Let $T : V \to W$ be a linear transformation. Prove that if $\mathbf{v} \in \ker(T)$ then $-\mathbf{v} \in \ker(T)$.

6. Consider the zero linear transformation $T : V \to W$ defined by $T(\mathbf{v}) = \mathbf{O}$ for all $\mathbf{v}$ in the domain V. Prove that $\ker(T) = V$.

7. Consider the identity linear transformation $T : V \to W$ defined by $T(\mathbf{v}) = \mathbf{v}$. Prove the following results:
(a) $\ker(T) = \{\mathbf{O}\}$ (b) $\text{range}(T) = V$

8. Let $T : V \to W$ be a linear transformation and vectors $\mathbf{u}, \mathbf{v}$ be in V.
Prove that if $T(\mathbf{u}) = \mathbf{x}$ and $T(\mathbf{v}) = \mathbf{x}$ then $(\mathbf{u} - \mathbf{v}) \in \ker(T)$.

9. Let $T : V \to W$ be a linear transformation. Prove that if $S_1 = \{\mathbf{v}_1, \mathbf{v}_2, \ldots, \mathbf{v}_n\}$ spans the domain V then $S_2 = \{T(\mathbf{v}_1), T(\mathbf{v}_2), \ldots, T(\mathbf{v}_n)\}$ spans range (T).

10. Let $T : V \to W$ be a linear transformation. Prove that if $B = \{\mathbf{v}_1, \mathbf{v}_2, \ldots, \mathbf{v}_n\}$ is a basis for the domain V then $S = \{T(\mathbf{v}_1), T(\mathbf{v}_2), \ldots, T(\mathbf{v}_n)\}$ is a basis for $\text{range}(T)$.

11. Let $T : V \to W$ be a linear transformation. Let S be a subspace of V and $T(S) = \left\{T(s) \middle| s \in S\right\}$. Prove that $T(S)$ is a subspace of $\text{range}(T)$.

SECTION 5.3 Rank and Nullity

By the end of this section you will be able to

- find the rank and nullity of a linear transformation

5.3.1 Rank of a linear transformation (mapping)

In chapter 3 we discussed what is meant by the rank of a matrix and why it is important. The rank of a matrix is the maximum number of linearly independent rows in the matrix, or the dimension of the row space of the matrix. Remember, in some matrices many of the rows vanish because they are linear combinations of the other rows (linearly dependent). The rank is important in transferring data because a transformation with a lower rank takes up less memory and time to be transferred. Low rank transformations are much more computationally efficient.

The rank of a linear transformation tells us how much information has been transformed over and is measured as the dimension of the range. The rank also tells us whether information has been lost by the linear transform.

Definition (5.11). Let $T : V \to W$ be a linear transform (map) and range(T) be the range. Then the dimension of range(T) is called the **rank** of T denoted rank(T) (Fig. 5.21).

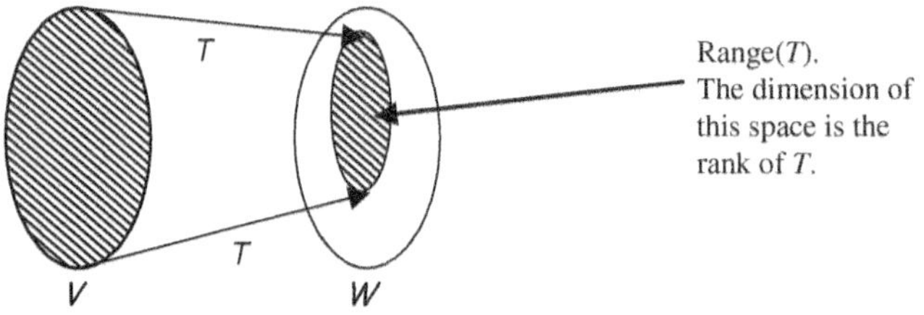
T
T
Range(T).
The dimension of
this space is the
rank of T.
V
W

Figure 5.21

Next we state one of the most important results of linear algebra. The proof of this theorem is given towards the end of the section because it is long and requires you to recall some definitions given in previous chapters.

Dimension theorem (also called the rank-nullity theorem) (5.12). Let $T : V \to W$ be a linear transformation from an n-dimensional vector space V to a vector space W. Then

$$\text{rank}(T) + \text{nullity}(T) = n$$

What does this formula mean?

It means

$$\dim\left(\text{range}(T)\right) + \dim\left(\ker(T)\right) = n$$

where dim represents dimension. This is same as the dimension theorem (3.34) of chapter 3. This suggests that adding the dimension of the range and kernel gives the dimension of the start vector space V. As stated earlier, this result says that all the information is contained in these two sets – kernel and range.

Why is the rank of a linear transformation important?

The rank gives us how much information has been carried over by the transform. If the rank of the linear transform $T : V \to W$ is equal to the dimension of the start vector space V then *all* the information has been moved over and we can go back; that is, the linear transform has an inverse. We have:

1. If $\text{rank}(T) = \dim(V)$ then all the information has been carried over by T.
2. If $\text{rank}(T) < \dim(V)$ then some information has been lost by T.
3. If $\text{rank}(T) = 0 < \dim(V)$ then virtually all the information has been lost by T.

Example 5.16

Determine the rank and nullity of the linear transformation (mapping) $T : P_2 \to \mathbb{R}$:

$$T(\mathbf{p}) = \int_0^1 p(x)dx$$

where P_2 is the vector space of polynomials of degree 2 or less.

Solution

What do you notice about the given linear transform?

It's the same transform as in Example 5.15 of the last section. We have already established that $\text{range}(T) = \mathbb{R}$ in Example 5.15.

What is the rank of T?

By the above definition (5.11) we have

$$rank(T) = \dim(\text{range}(T)) = \dim(\mathbb{R})$$

$\dim(\mathbb{R}) = 1$ because of the general rule $\dim(\mathbb{R}^m) = m$. Thus $rank(T) = 1$.

We also need to find the nullity of T which is the dimension of the kernel.

How?

By using the above formula (5.12):

$$\dim(\text{range}(T)) + \dim(\ker(T)) = n$$

What is the value of n, the dimension of P_2 equal to?

From chapter 3, Table 3.1 states that $\dim(P_k) = k + 1$ so $\dim(P_2) = 3$.

Substituting $n = \dim(P_2) = 3$ and $\dim(\text{range}(T)) = 1$ into

$$\dim(\text{range}(T)) + \dim(\ker(T)) = n$$

gives

$$1 + \dim(\ker(T)) = 3 \quad \text{which yields} \quad \dim(\ker(T)) = 2$$

(continued...)

Hence the nullity of T is 2 and the rank of T is 1.

Since $rank(T) = 1$ but dimension of the start vector space P_2 is 3, some of the information has been lost by the linear transform T.

We are given that $T : P_2 \to \mathbb{R}$ and we know that $\dim(P_2) = 3$. However, the dimension of the arrival vector space $\mathbb{R}$ is 1. Clearly the given transform T has lost some information by going from dimension 3 to 1.

Example 5.17

Find the rank and nullity of the linear transform, $T : \mathbb{R}^2 \to \mathbb{R}^2$ defined by

$$T\left(\begin{pmatrix} x \\ y \end{pmatrix}\right) = \begin{pmatrix} 2x - y \\ 6x - 3y \end{pmatrix}$$

Solution

Since T is linear we can find the rank and nullity of T.

How do we find these?

Apply the dimension theorem formula (5.12):

$$\dim(\text{range}(T)) + \dim(\ker(T)) = n$$

What is n equal to?

Remember, n is the dimension of the domain (start vector space) which is $\mathbb{R}^2$ because we are given $T : \mathbb{R}^2 \to \mathbb{R}^2$. Hence $n = \dim(\mathbb{R}^2) = 2$.

We need to find the dimension of the kernel of T.

What is $\ker(T)$ equal to?

Remember, $\ker(T)$ represents those vectors in $\mathbb{R}^2$ which arrive at the zero vector under the transformation T:

$$T\left(\begin{pmatrix} x \\ y \end{pmatrix}\right) = \begin{pmatrix} 2x - y \\ 6x - 3y \end{pmatrix} = \begin{pmatrix} 0 \\ 0 \end{pmatrix}$$

We have the simultaneous equations:

$$\left.\begin{aligned} 2x - y &= 0 \\ 6x - 3y &= 0 \end{aligned}\right\} \text{ gives } y = 2x$$

Let $x = s$ where s is any real number, then $y = 2s$. Thus

$$\ker(T) = \left\{ s\begin{pmatrix} 1 \\ 2 \end{pmatrix} \,\middle|\, s \text{ is a real number} \right\}$$

What is the dimension of $\ker(T)$?

A basis (vector representing axis) for $\ker(T)$ is $\{(1\ \ 2)^T\}$, therefore $\dim(\ker(T)) = 1$ because we only have *one* vector in the basis. Hence nullity $(T) = 1$.

To find the rank which is the dimension of the range, we substitute $n = 2$ and $\dim(\ker(T)) = 1$ into the above formula (5.12):

$$\dim(\text{range}(T)) + \dim(\ker(T)) = n$$

to get $\dim(\text{range}(T)) + 1 = 2$. This gives $\dim(\text{range}(T)) = 1$. Therefore rank$(T) = 1$.

Hence $rank(T) = nullity(T) = 1$.

$rank(T) = 1$ means that the actual arrival space is of dimension 1. However, the dimension of the start vector space $\mathbb{R}^2$ is 2, so some information has been lost by T because T transforms vectors from dimension 2 to 1.

5.3.2 Kernel and rank of the linear transformation $T(\mathbf{x}) = \mathbf{A}\mathbf{x}$

Proposition (5.13). Let $T : \mathbb{R}^n \to \mathbb{R}^m$ be a linear transformation given by $T(\mathbf{x}) = \mathbf{A}\mathbf{x}$. Then $range(T)$ is the column space of $\mathbf{A}$.

Proof.
See Exercises 5.3.

■

In the next example, we convert the given matrix into its (reduced) row echelon form. The method of conversion has been covered so many times in previous chapters that we will not detail the steps, but simply write down the final (reduced) row echelon form.

Example 5.18

Let $T : \mathbb{R}^4 \to \mathbb{R}^2$ be given by $T(\mathbf{x}) = \mathbf{A}\mathbf{x}$ where $\mathbf{x}$ is in $\mathbb{R}^4$ and

$$\mathbf{A} = \begin{pmatrix} 1 & 3 & 4 & 5 \\ 2 & 6 & -8 & -6 \end{pmatrix}$$

Find a basis for **(i)** $\ker(T)$ **(ii)** $range(T)$

Solution

(i) The kernel is a subspace of $\mathbb{R}^4$ because $\mathbb{R}^4$ is the start vector space of T.

What is $\ker(T)$ *equal to?*

It is the null space $\mathbf{x}$ of matrix $\mathbf{A}$.

How do we find a basis for $\ker(T)$*?*

We need to place the above matrix $\mathbf{A}$ into (reduced) row echelon form $\mathbf{R}$ and then solve $\mathbf{R}\mathbf{x} = \mathbf{O}$. The general solution $\mathbf{x}$ gives a basis (axes) for the kernel. We can use MATLAB or hand calculations to obtain the reduced row echelon form:

$$\mathbf{A} = \begin{pmatrix} 1 & 3 & 4 & 5 \\ 2 & 6 & -8 & -6 \end{pmatrix} \Longrightarrow \begin{pmatrix} 1 & 3 & 0 & 1 \\ 0 & 0 & 1 & 1 \end{pmatrix} = \mathbf{R}$$

Let $\mathbf{x} = (x \;\; y \;\; z \;\; w)^T$. Expanding the rows of $\mathbf{R}\mathbf{x} = \mathbf{O}$ gives the simultaneous equations:

$$\begin{aligned} x + 3y \qquad + w &= 0 \qquad (1) \\ z + w &= 0 \qquad (2) \end{aligned}$$

(continued...)

From the bottom equation (2) we have $z = -w$. Let $w = s$ where s is any real number. We have $z = -w = -s$. Substituting $w = s$ into the top equation (1) gives

$$x + 3y \quad + s = 0 \quad \text{implies} \quad x = -3y - s$$

Let $y = t$ where t is any real number. We have $x = -3y - s = -3t - s$. Hence $x = -3t - s$, $y = t$, $z = -s$ and $w = s$:

$$\mathbf{x} = \begin{pmatrix} x \\ y \\ z \\ w \end{pmatrix} = \begin{pmatrix} -3t - s \\ t \\ -s \\ s \end{pmatrix} = s \begin{pmatrix} -1 \\ 0 \\ -1 \\ 1 \end{pmatrix} + t \begin{pmatrix} -3 \\ 1 \\ 0 \\ 0 \end{pmatrix}$$

Therefore a set B of basis (axes) vectors for the kernel of T is given by

$$B = \left\{ \begin{pmatrix} -1 & 0 & -1 & 1 \end{pmatrix}^T, \begin{pmatrix} -3 & 1 & 0 & 0 \end{pmatrix}^T \right\}$$

Hence $nullity(T) = 2$ because we have two basis (axes) vectors for $\ker(T)$.

(ii) *What is $range(T)$ equal to?*

By the above Proposition (5.13):

$range(T)$ is the column space of $\mathbf{A}$.

The column space of $\mathbf{A}$ gives $range(T)$. We convert the columns of matrix $\mathbf{A}$ into rows by taking the transpose of $\mathbf{A}$.

A set B' of basis (axes) vectors for the range of T can be found by determining the reduced row echelon form of $\mathbf{A}^T$ (matrix $\mathbf{A}$ transposed). The non-zero rows of this reduced row echelon form give a basis for the range of T. We have

$$\mathbf{A}^T = \begin{pmatrix} 1 & 2 \\ 3 & 6 \\ 4 & -8 \\ 5 & -6 \end{pmatrix} \Longrightarrow \begin{pmatrix} 1 & 0 \\ 0 & 1 \\ 0 & 0 \\ 0 & 0 \end{pmatrix} = \mathbf{R}' \quad \begin{matrix} \longleftarrow \text{non-zero row} \\ \longleftarrow \text{non-zero row} \\ \\ \\ \end{matrix}$$

The first two rows of the reduced row echelon form matrix $\mathbf{R}'$ is a basis for the range of T:

$$B' = \left\{ \begin{pmatrix} 1 & 0 \end{pmatrix}^T, \begin{pmatrix} 0 & 1 \end{pmatrix}^T \right\}$$

Since the range of T has two basis (axes) vectors so $rank(T) = 2$.

Remember, adding the rank and nullity of T gives the dimension of the start vector space:

$$\text{rank}(T) + \text{nullity}(T) = 2 + 2 = 4 = \dim\left(\mathbb{R}^4\right)$$

5.3.3 Proof of the dimension theorem

Now we prove the dimension theorem which was:

(5.12). Let $T : V \to W$ be a linear transformation from an n dimensional vector space V to a vector space W. Then $\text{rank}(T) + \text{nullity}(T) = n$.

Proof.
Let $\dim(\ker(T)) = k$ and $\{\mathbf{v}_1, \mathbf{v}_2, \mathbf{v}_3, \ldots, \mathbf{v}_k\}$ be a set of basis (axes) vectors for $\ker(T)$. Since $\ker(T)$ is a subspace of the n-dimensional space V, therefore $k \leq n$. We consider the two cases, $k = n$ and $k < n$:

Case 1: $k = n$.

What does $k = n$ mean?
It means that *all* the vectors in V are transformed to the zero vector $\mathbf{O}$, that is

$$T(\mathbf{v}) = \mathbf{O} \text{ for } all\ \mathbf{v} \text{ in } V$$

What is the range of T?
All the vectors arrive at the zero vector $\mathbf{O}$ so $range(T) = \{\mathbf{O}\}$.

What is dimension of $range(T) = \{\mathbf{O}\}$?
By section 3.4 of chapter 3 we have $\dim(\{\mathbf{O}\}) = 0$.

Hence we have $\dim(\ker(T)) = n$ and $\dim(\text{range}(T)) = 0$:

$$\dim(\text{range}(T)) + \dim(\ker(T)) = 0 + n = n$$

This is our required result.

Case 2: Consider $k < n$. Let $\{\mathbf{v}_1, \mathbf{v}_2, \ldots, \mathbf{v}_k, \mathbf{v}_{k+1}, \ldots, \mathbf{v}_n\}$ be a set of basis (axes) vectors for the vector space V where $\{\mathbf{v}_1, \mathbf{v}_2, \mathbf{v}_3, \ldots, \mathbf{v}_k\}$ is a basis (axes) for $\ker(T)$ as stated above. This is possible because we are given that the dimension of the vector space V is n and $\dim(\ker(T)) = k < n$.

Let $\mathbf{v}$ be an arbitrary vector in the start vector space V, then we can write this vector $\mathbf{v}$ as a linear combination of the basis (axes) vectors, that is

$$\mathbf{v} = c_1\mathbf{v}_1 + c_2\mathbf{v}_2 + \cdots + c_k\mathbf{v}_k + \cdots + c_n\mathbf{v}_n$$

where the c_j's are scalars.

Thus the vector $T(\mathbf{v})$ is in $\text{range}(T)$ and we have

$$\begin{aligned} T(\mathbf{v}) &= T(c_1\mathbf{v}_1 + c_2\mathbf{v}_2 + \cdots + c_k\mathbf{v}_k + c_{k+1}\mathbf{v}_{k+1} + \cdots + c_n\mathbf{v}_n) \\ &= \underbrace{T(c_1\mathbf{v}_1 + c_2\mathbf{v}_2 + \cdots + c_k\mathbf{v}_k)}_{\text{is in } \ker(T)} + T(c_{k+1}\mathbf{v}_{k+1} + \cdots + c_n\mathbf{v}_n) \qquad [\text{because } T \text{ is linear}] \\ &= \mathbf{O} + T(c_{k+1}\mathbf{v}_{k+1} + \cdots + c_n\mathbf{v}_n) \\ &= c_{k+1}T(\mathbf{v}_{k+1}) + \cdots + c_nT(\mathbf{v}_n) \qquad [\text{because } T \text{ is linear}] \end{aligned}$$

We have

$$T(\mathbf{v}) \quad = \quad c_{k+1}T(\mathbf{v}_{k+1}) \quad + \quad \cdots \quad + \quad c_nT(\mathbf{v}_n) \qquad (*)$$

where $\mathbf{v}$ is an arbitrary vector in the start vector space V, and $T(\mathbf{v})$ is in $\text{range}(T)$. Thus the vectors on the right hand side of (*), that is $S = \{T(\mathbf{v}_{k+1}), T(\mathbf{v}_{k+2}), \ldots, T(\mathbf{v}_n)\}$ span (or generate) $\text{range}(T)$. We need to show that these vectors in the set S form a basis for $\text{range}(T)$.

We have already established that these vectors span range(T), but what do we need to show for these vectors in S to form a basis for range(T)?
Show that they are linearly independent.

How?
Consider the linear combination

$$d_{k+1}T(\mathbf{v}_{k+1}) + d_{k+2}T(\mathbf{v}_{k+2}) + \cdots + d_nT(\mathbf{v}_n) = \mathbf{O}$$

where the d_j's are scalars.

How do we show linear independence?
We need to prove that *all* the scalars d_j's are zero.

Since T is linear we have

$$\begin{aligned} d_{k+1}T(\mathbf{v}_{k+1}) + d_{k+2}T(\mathbf{v}_{k+2}) + \cdots + d_nT(\mathbf{v}_n) &= \mathbf{O} \\ T(d_{k+1}\mathbf{v}_{k+1} + d_{k+2}\mathbf{v}_{k+2} + \cdots + d_n\mathbf{v}_n) &= \mathbf{O} \end{aligned}$$

This means that the vector $d_{k+1}\mathbf{v}_{k+1} + d_{k+2}\mathbf{v}_{k+2} + \cdots + d_n\mathbf{v}_n$ is in $\ker(T)$. Since $\{\mathbf{v}_1, \mathbf{v}_2, \mathbf{v}_3, \cdots, \mathbf{v}_k\}$ is a basis for $\ker(T)$, we have

$$\begin{aligned} c_1\mathbf{v}_1 + c_2\mathbf{v}_2 + c_3\mathbf{v}_3 + \cdots + c_k\mathbf{v}_k &= d_{k+1}\mathbf{v}_{k+1} + d_{k+2}\mathbf{v}_{k+2} + \cdots + d_n\mathbf{v}_n \\ c_1\mathbf{v}_1 + c_2\mathbf{v}_2 + c_3\mathbf{v}_3 + \cdots + c_k\mathbf{v}_k - d_{k+1}\mathbf{v}_{k+1} - d_{k+2}\mathbf{v}_{k+2} - \cdots - d_n\mathbf{v}_n &= \mathbf{O} \end{aligned}$$

Remember, at the start of the proof we had $\{\mathbf{v}_1, \mathbf{v}_2, \ldots, \mathbf{v}_k, \ldots, \mathbf{v}_n\}$ as a set of basis (axes) vectors for the vector space V, therefore they are linearly independent so we have

$$c_1 = c_2 = \cdots = c_k = d_{k+1} = d_{k+2} = \cdots = d_n = 0$$

All the d_j's are zero, therefore $S = \{T(\mathbf{v}_{k+1}), T(\mathbf{v}_{k+2}), \ldots, T(\mathbf{v}_n)\}$ is a linearly independent set of vectors. Thus $S = \{T(\mathbf{v}_{k+1}), T(\mathbf{v}_{k+2}), \ldots, T(\mathbf{v}_n)\}$ forms a set of basis (axes) vectors for range(T).

How many vectors are there in this basis?
There are $n - k$ vectors in this basis. Hence the dimension of this space is given by $\dim(\text{range}(T)) = n - k$. We have our result

$$\dim(\text{range}(T)) + \dim(\ker(T)) = (n - k) + k = n$$

This completes our proof.

■

We can apply the dimension theorem to check whether a given transformation is linear or not.

Example 5.19

Consider the transform $T : \mathbb{R}^3 \to \mathbb{R}^2$ given by $T\left[\begin{pmatrix} x \\ y \\ z \end{pmatrix}\right] = \begin{pmatrix} xz \\ xy \end{pmatrix}$. Test this transform for linearity.

Solution

The value of n in the dimension theorem (5.12) is dimension of the start vector space:

$$n = \dim\left(\mathbb{R}^3\right) = 3$$

What is the kernel equal to in this case?

It is the set of vectors in the start vector space which are transformed to the zero vector:

$$T\left[\begin{pmatrix} x \\ y \\ z \end{pmatrix}\right] = \begin{pmatrix} xz \\ xy \end{pmatrix} = \begin{pmatrix} 0 \\ 0 \end{pmatrix} \quad \Rightarrow \quad x = 0, y = r \text{ and } z = s \text{ where } r, s \text{ are any real numbers}$$

The entries of vector $\mathbf{x}$ such that $T(\mathbf{x}) = \mathbf{O}$ gives $x = 0, y = r$ and $z = s$:

$$\mathbf{x} = \begin{pmatrix} x \\ y \\ z \end{pmatrix} = \begin{pmatrix} 0 \\ r \\ s \end{pmatrix} = r\begin{pmatrix} 0 \\ 1 \\ 0 \end{pmatrix} + s\begin{pmatrix} 0 \\ 0 \\ 1 \end{pmatrix}$$

A set of basis (axes) vectors for the kernel of T is $\left\{(0 \ \ 1 \ \ 0)^T, (0 \ \ 0 \ \ 1)^T\right\}$. This means $nullity(T) = 2$.

The range of T is given by $\begin{pmatrix} xz \\ xy \end{pmatrix}$ which is $\mathbb{R}^2$. Hence $\dim(\mathbb{R}^2) = rank(T) = 2$.

We have

$$rank(T) + nullity(T) = 2 + 2 = 4 \neq 3 = n$$

Since the dimension theorem fails, the given transformation cannot be linear.

Summary

(5.11) The dimension of range(T) is called the rank of T.

(5.12) Dimension or rank-nullity theorem is

$$rank(T) + nullity(T) = n$$

where n is the dimension of the domain V.

EXERCISES 5.3

(Brief solutions at end of book. Full solutions available at <http://www.oup.co.uk/companion/singh>.)

1. Determine
   (i) ker(T) (ii) nullity(T) (iii) range(T) (iv) rank(T)

for the following linear transformations without using the dimension theorem:

(a) $T : \mathbb{R}^2 \to \mathbb{R}^2$ given by $T(\mathbf{v}) = \mathbf{Av}$ where $\mathbf{A} = \begin{pmatrix} 1 & 2 \\ 1 & 2 \end{pmatrix}$.

(b) $T : \mathbb{R}^3 \to \mathbb{R}^3$ given by $T(\mathbf{v}) = \mathbf{Av}$ where $\mathbf{A} = \begin{pmatrix} 1 & 0 & 0 \\ 0 & 1 & 0 \\ 0 & 0 & 1 \end{pmatrix}$.

(c) $T : \mathbb{R}^3 \to \mathbb{R}^3$ given by $T(\mathbf{v}) = \mathbf{Av}$ where $\mathbf{A} = \begin{pmatrix} 1 & 3 & 5 \\ 2 & 6 & 10 \\ 4 & 12 & 20 \end{pmatrix}$.

(d) $T : P_3 \to P_3$ given by $T(\mathbf{p}) = x\mathbf{p}'$.

(e) $T : P_3 \to P_2$ given by $T(\mathbf{p}) = \mathbf{p}'$.

(f) $T : P_3 \to \mathbb{R}$ given by $T(\mathbf{p}) = \int_0^1 p(x)\, dx$.

(g) $T : M_{22} \to P_1$ given by $T\left(\begin{pmatrix} a & b \\ c & d \end{pmatrix}\right) = (a+c)\,x + (b+d)$

2. For each of the examples in parts (a) to (g) in question 1 verify the dimension theorem (5.12) which says $\text{nullity}(T) + \text{rank}(T) = n$ where n is the dimension of the domain of the linear transformation T.

3. Let $T : \mathbb{R}^5 \to \mathbb{R}^3$ be given by $T(\mathbf{u}) = \mathbf{Au}$ where $\mathbf{u}$ is in $\mathbb{R}^5$ and

$$\mathbf{A} = \begin{pmatrix} 1 & 3 & 4 & 2 & 1 \\ 2 & 6 & -7 & -2 & 5 \\ 4 & 12 & 6 & 4 & 6 \end{pmatrix}$$

Find a basis for (i) $\ker(T)$ (ii) *range*(T)

4. Prove Proposition (5.13).

SECTION 5.4 Inverse Linear Transformations

By the end of this section you will be able to

- prove properties of a one-to-one and onto transformations
- test which transforms have an inverse
- understand what is meant by isomorphism

In this section we state the conditions that allow an inverse transformation to exist. Inverse transformations are important because we often want to be able to undo a transformation. We need a way back from our destination to our starting point and this can only be achieved by the inverse transformation.

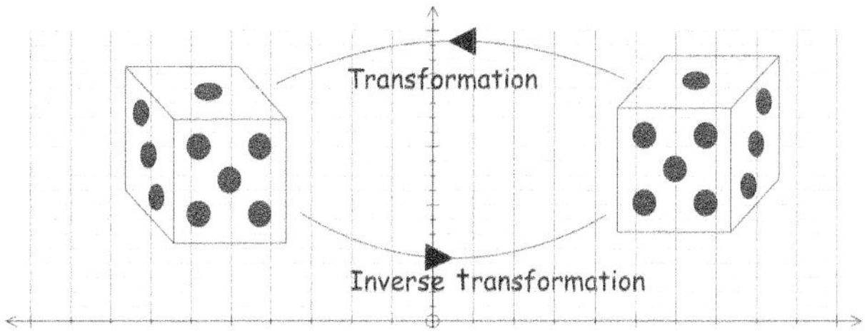

Figure 5.22

Figure 5.22 illustrates what is meant by an inverse transform.

Transformations are used in cryptography, where a message is encoded by a linear transform T. In order to decode this message we use the inverse transform of T.

Before we examine inverse transformations we need to discuss what is meant by **one-to-one** and **onto** because they are closely related to the inverse. Having established these definitions we prove some properties of these concepts.

This is a challenging section because you are required to prove a number of results. However, if you learn the definitions thoroughly and can apply these with confidence then the proofs should be straightforward.

5.4.1 One-to-one (injective) transformations

What do you think a one-to-one linear transformation is?

It's a linear transformation T in which every vector in the domain (start) arrives at a *different* vector in the range under T (Fig. 5.23).

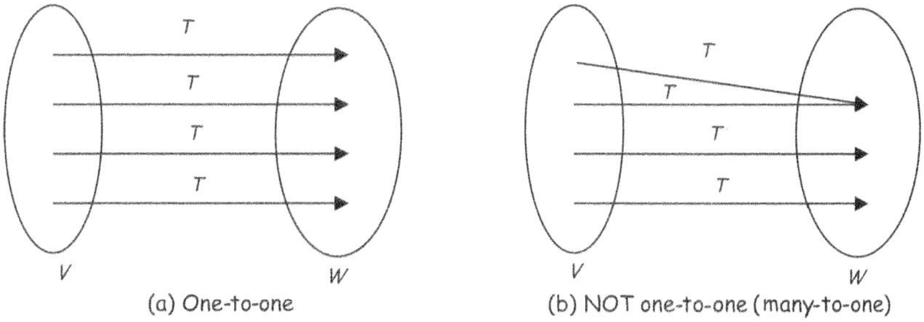

Figure 5.23

Fig. 5.23(a) shows a *one-to-one* transformation while Fig. 5.23(b) is *not* one-to-one (it is actually a many-to-one) transformation. Next, we give a formal definition of one-to-one.

Definition (5.14). Let $T: V \to W$ be a linear transform, $\mathbf{u}$ and $\mathbf{v}$ be in the domain V. The transform T is **one-to-one** $\Leftrightarrow$

$$\mathbf{u} \neq \mathbf{v} \quad \text{implies} \quad T(\mathbf{u}) \neq T(\mathbf{v})$$

What does this definition mean?

Different vectors such as $\mathbf{u}$ and $\mathbf{v}$ where $\mathbf{u} \neq \mathbf{v}$ have *distinct* images $T(\mathbf{u}) \neq T(\mathbf{v})$ under the linear transformation T. Two *different* start vectors arrive at *different* destinations.

In other literature on linear algebra, terms such as **1–1** or **injective** are used for one-to-one. We will stick with one-to-one.

Example 5.20

Consider the linear identity transformation $T : V \to W$ defined by $T(\mathbf{v}) = \mathbf{v}$ for *all* $\mathbf{v}$ in V. Show that this transformation T is one-to-one (injective).

Solution

Let $\mathbf{u}$ and $\mathbf{v}$ be different vectors in V, that is $\mathbf{u} \neq \mathbf{v}$. Applying the given identity transformation we have

$$T(\mathbf{u}) = \mathbf{u} \text{ and } T(\mathbf{v}) = \mathbf{v}$$

We have $\mathbf{u} \neq \mathbf{v}$, therefore $T(\mathbf{u}) = \mathbf{u} \neq \mathbf{v} = T(\mathbf{v})$. Hence $T(\mathbf{u}) \neq T(\mathbf{v})$, so by the above Definition (5.14):

The transform T is one-to-one $\Leftrightarrow$ $\mathbf{u} \neq \mathbf{v}$ implies $T(\mathbf{u}) \neq T(\mathbf{v})$

The given transformation T is one-to-one because different vectors, $\mathbf{u} \neq \mathbf{v}$, arrive at different destinations, $T(\mathbf{u}) \neq T(\mathbf{v})$.

Example 5.21

Let $T : V \to W$ be the zero linear transformation defined by $T(\mathbf{v}) = \mathbf{O}$ for *all* $\mathbf{v}$ in V. Show that this transformation is *not* one-to-one (*not* injective).

Solution

Let $\mathbf{u}$ and $\mathbf{v}$ be different vectors in V, that is $\mathbf{u} \neq \mathbf{v}$. Then applying the given transformation:

$$T(\mathbf{u}) = \mathbf{O} \text{ and } T(\mathbf{v}) = \mathbf{O}$$

This gives $T(\mathbf{u}) = T(\mathbf{v}) = \mathbf{O}$. We have different vectors, $\mathbf{u} \neq \mathbf{v}$, arriving at the same destination, $T(\mathbf{u}) = T(\mathbf{v})$, therefore by (5.14) we conclude that the given zero linear transformation is *not* one-to-one.

Another test for one-to-one is:
T is *one-to-one* $\Leftrightarrow$

(5.15) $$T(\mathbf{u}) = T(\mathbf{v}) \text{ implies } \mathbf{u} = \mathbf{v}$$

[If we have arrived at the same destination $T(\mathbf{u}) = T(\mathbf{v})$ then we have started with the same vectors $\mathbf{u} = \mathbf{v}$.]

Example 5.22

Let $T : \mathbb{R}^2 \to \mathbb{R}^2$ be the linear transformation defined by

$$T\left(\begin{bmatrix} x \\ y \end{bmatrix}\right) = \begin{pmatrix} x - y \\ x + y \end{pmatrix}$$

Show that this transformation is one-to-one (injective).

Solution

This time we use result (5.15) for the given T.

Let $\mathbf{u} = \begin{bmatrix} a \\ b \end{bmatrix}$ and $\mathbf{v} = \begin{bmatrix} c \\ d \end{bmatrix}$. Applying the given transformation to vectors $\mathbf{u}$ and $\mathbf{v}$ we have

$$T(\mathbf{u}) = T\left(\begin{bmatrix} a \\ b \end{bmatrix}\right) = \begin{pmatrix} a - b \\ a + b \end{pmatrix} \text{ and } T(\mathbf{v}) = T\left(\begin{bmatrix} c \\ d \end{bmatrix}\right) = \begin{pmatrix} c - d \\ c + d \end{pmatrix}$$

If $T(\mathbf{u}) = T(\mathbf{v})$ then we have $\begin{pmatrix} a - b \\ a + b \end{pmatrix} = \begin{pmatrix} c - d \\ c + d \end{pmatrix}$ and equating corresponding entries gives:

$$\left.\begin{aligned} a - b &= c - d \\ a + b &= c + d \end{aligned}\right\} \quad \text{implies } a = c \text{ and } b = d$$

Thus the solution of the above simultaneous equations is $a = c$ and $b = d$ which gives

$$\mathbf{u} = \begin{bmatrix} a \\ b \end{bmatrix} = \begin{bmatrix} c \\ d \end{bmatrix} = \mathbf{v}$$

We have $T(\mathbf{u}) = T(\mathbf{v})$ implies $\mathbf{u} = \mathbf{v}$.

This means that the same arrival vectors $T(\mathbf{u}) = T(\mathbf{v})$ implies that we started with the same vectors $\mathbf{u} = \mathbf{v}$ so by the above result (5.15) the given transformation T is one-to-one.

5.4.2 Properties of one-to-one transformations

An easier check for one-to-one transformation is the following:

Proposition (5.16). Let $T : V \to W$ be a linear transformation between the vector spaces V and W. Then T is one-to-one $\Leftrightarrow$ $\ker(T) = \{\mathbf{O}\}$.

Why do we need another test for one-to-one?

This is a much simpler test for one-to-one. The proposition means that $\ker(T) = \{\mathbf{O}\}$ is equivalent to T is one-to-one. The kernel of T measures one-to-one.

How do we prove this result?

By going both ($\Rightarrow$ and $\Leftarrow$) ways. ($\Rightarrow$) First assume T is one-to-one and derive $\ker(T) = \{\mathbf{O}\}$ and then ($\Leftarrow$) assume $\ker(T) = \{\mathbf{O}\}$ and derive T is one-to-one.

Proof.

($\Rightarrow$) Assume T is one-to-one. Since T is linear, by result (5.3) part (a):

If T is linear then $T(\mathbf{O}) = \mathbf{O}$

We have $T(\mathbf{O}) = \mathbf{O}$. Suppose there is a vector $\mathbf{u}$ in V such that $T(\mathbf{u}) = \mathbf{O}$ then

$$T(\mathbf{u}) = \mathbf{O} = T(\mathbf{O}) \text{ implies } T(\mathbf{u}) = T(\mathbf{O})$$

Since we are assuming T is one-to-one, by
(5.15) $T(\mathbf{u}) = T(\mathbf{v})$ implies $\mathbf{u} = \mathbf{v}$
we have

$$T(\mathbf{u}) = T(\mathbf{O}) \text{ implies } \mathbf{u} = \mathbf{O}$$

Thus the *only* vector transformed under T to the zero vector is $\mathbf{O}$, which gives

$$\ker(T) = \{\mathbf{O}\}$$

($\Leftarrow$) Now going the other way, we assume $\ker(T) = \{\mathbf{O}\}$ and we need to prove T is one-to-one. Let $\mathbf{u}$ and $\mathbf{v}$ be vectors in V which arrive at the same destination, $T(\mathbf{u}) = T(\mathbf{v})$. We have

$$\begin{aligned} T(\mathbf{u}) &= T(\mathbf{v}) \\ T(\mathbf{u}) - T(\mathbf{v}) &= \mathbf{O} \\ T(\mathbf{u} - \mathbf{v}) &= \mathbf{O} \qquad \left[\text{because } T \text{ is Linear}\right] \end{aligned}$$

This means that $(\mathbf{u} - \mathbf{v})$ is in $\ker(T) = \{\mathbf{O}\}$. Thus

$$\mathbf{u} - \mathbf{v} = \mathbf{O} \Rightarrow \mathbf{u} = \mathbf{v}$$

By statement (5.15):
T is one-to-one $\Leftrightarrow$ $T(\mathbf{u}) = T(\mathbf{v})$ implies $\mathbf{u} = \mathbf{v}$
We have $T(\mathbf{u}) = T(\mathbf{v})$ implies $\mathbf{u} = \mathbf{v}$, therefore T is a one-to-one transformation.
We have proved our result both ways, therefore T is one-to-one $\Leftrightarrow$ $\ker(T) = \{\mathbf{O}\}$.

■

Example 5.23

Show that the linear transformation T given in Example 5.22 is one-to-one by using proposition (5.16).

Solution

What do we need to show?

Required to prove that the kernel of T of Example 5.22 is the zero vector.

How?

Applying the given transformation and equating to zero (because we are trying to find the kernel) yields

$$T\left(\begin{pmatrix} x \\ y \end{pmatrix}\right) = \begin{pmatrix} x - y \\ x + y \end{pmatrix} = \begin{pmatrix} 0 \\ 0 \end{pmatrix} \text{ implies } \left.\begin{aligned} x - y = 0 \\ x + y = 0 \end{aligned}\right\} \text{ gives } x = 0, y = 0$$

Thus the kernel of T is $(0 \quad 0)^T = \mathbf{O}$. Therefore $\ker(T) = \{\mathbf{O}\}$ (is the zero vector) which means that T is one-to-one.

There are many examples where the kernel of a transformation T is *not* equal to the zero vector. For example, if we consider the derivative transformation S of polynomials then the kernel of S is the set of all constants which are *not* the zero vector.

Another test for one-to-one is the following:

Proposition (5.17). Let $T : V \to W$ be a linear transformation. T is one-to-one $\Leftrightarrow$ $nullity(T) = 0$.

(Remember, the nullity of the transformation T is the dimension of the kernel of T.)

Proof – Exercises 5.4.

■

5.4.3 Onto (surjective) linear transformations

What do you think the term onto transformation means?

An illustration of an onto transformation is shown in Fig. 5.24.

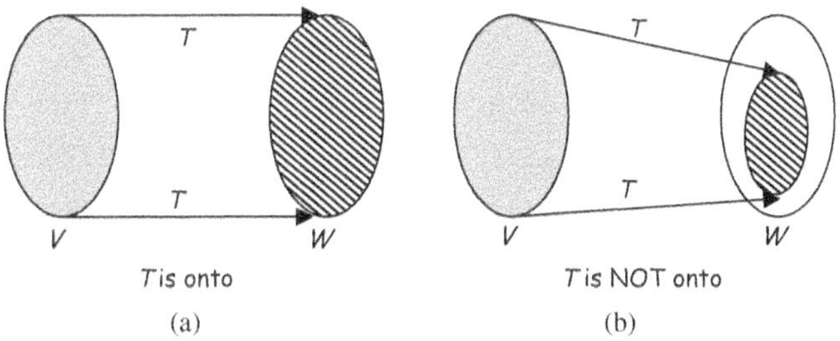

Figure 5.24

An *onto* transformation is when all the information carried over by T fills the whole arrival vector space W.

How can we write this in mathematical terms?

Definition (5.18). Let $T : V \to W$ be a linear transform. The transform T is **onto** $\Leftrightarrow$ for every $\mathbf{w}$ in the arrival vector space W there exists at least one $\mathbf{v}$ in the start vector space V such that

$$\mathbf{w} = T(\mathbf{v})$$

In other words $T : V \to W$ is an *onto* transformation $\Leftrightarrow$ $range(T) = W$. This means the arriving vectors of T fill *all* of W. We can write this as a proposition:

Proposition (5.19). A linear transformation $T : V \to W$ is *onto* $\Leftrightarrow$ $range(T) = W$.

Proof – Exercises 5.4.

■

In other mathematical literature, or your lecture notes, you might find the term **surjective** to mean onto. We will use onto.

Example 5.24

Show that the linear transformation $T : \mathbb{R}^2 \to \mathbb{R}^2$ given by

$$T(\mathbf{v}) = \mathbf{Av} \text{ where } \mathbf{A} = \begin{pmatrix} 1 & 0 \\ 0 & 1 \end{pmatrix}$$

is an onto (surjective) transformation.

Solution

How do we show that the given transformation is onto?

Show that the range of T is $\mathbb{R}^2$.

Let $\mathbf{v} = \begin{pmatrix} x \\ y \end{pmatrix}$ then applying the given transformation to this vector yields:

$$T\left(\begin{pmatrix} x \\ y \end{pmatrix}\right) = \begin{pmatrix} 1 & 0 \\ 0 & 1 \end{pmatrix}\begin{pmatrix} x \\ y \end{pmatrix} = \begin{pmatrix} x \\ y \end{pmatrix}$$

What is the range of T?

Since $T\left(\begin{pmatrix} x \\ y \end{pmatrix}\right) = \begin{pmatrix} x \\ y \end{pmatrix}$ we have $range(T) = \mathbb{R}^2$, which means that the range of T fills the whole arrival vector space $\mathbb{R}^2$. Hence the given transformation T is *onto.*

Example 5.25

Show that the linear differentiation transformation $T : P_3 \to P_3$ given by

$$T(\mathbf{p}) = \mathbf{p}'$$

is *not* onto (*not* surjective).

Solution

Let $\mathbf{p} = ax^3 + bx^2 + cx + d$ because $\mathbf{p}$ is a member of P_3 which is the vector space of polynomials of degree 3 or less. Applying the given transformation to $\mathbf{p} = ax^3 + bx^2 + cx + d$ yields:

$$\begin{aligned} T(ax^3 + bx^2 + cx + d) &= (ax^3 + bx^2 + cx + d)' \\ &= 3ax^2 + 2bx + c \qquad [\text{differentiating}] \end{aligned}$$

We have $T(\mathbf{p}) = 3ax^2 + 2bx + c$ and this is a quadratic (degree 2) polynomial (*not* a cubic, degree 3, polynomial) which means that it is a member of P_2. (Remember, P_2 is the vector space of polynomials of degree 2 or less.) Thus the range of T is P_2 and because $P_2 \neq P_3$ (*not* equal) the given transformation is *not* onto (or *not* surjective) (Fig. 5.25).

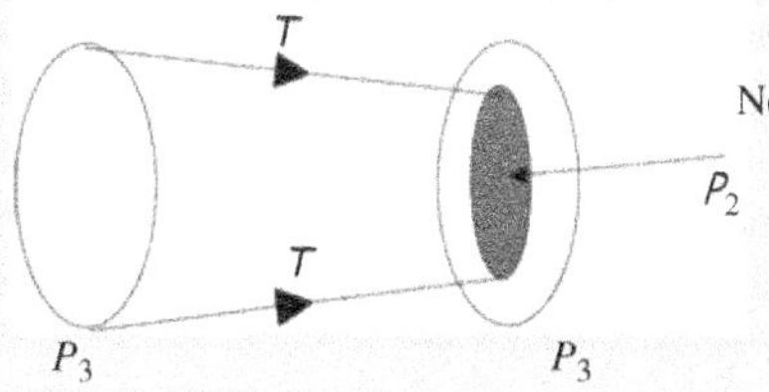

Figure 5.25

The given transform T does *not* fill the whole of the arrival vector space P_3.

5.4.4 Properties of onto (surjective) linear transformations

Proposition (5.20). Let $T : V \to W$ be a linear transformation. Then T is onto $\Leftrightarrow$

$$rank(T) = \dim(W)$$

This proposition means that if the dimension of the range is equal to the dimension of the arrival vector space then the transformation T is onto. Also, if T is onto then the dimensions of the range and codomain are equal. (The range = codomain.)

Proof – Exercises 5.4.

■

Proposition (5.21). If $T : V \to W$ is a linear transformation and $\dim(V) = \dim(W)$ then T is a one-to-one transformation $\Leftrightarrow$ T is onto.

This means that if the start and arrival vector spaces are of equal dimension then a transformation which is *one-to-one* is also *onto* and an *onto* transformation is also *one-to-one.* We get both (one-to-one and onto) or neither provided the start and arrival sets have the same dimension (Fig. 5.26).

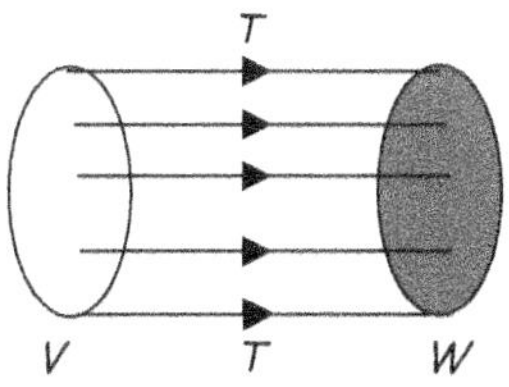

Figure 5.26

How do we prove the given result?
($\Rightarrow$) First we assume that T is one-to-one and deduce T is onto and then ($\Leftarrow$) we assume T is onto and deduce that T is one-to-one.

Proof.
Let the start and arrival vector spaces have the same dimension, $\dim(V) = \dim(W) = n$.
($\Rightarrow$) Assume T is one-to-one. By Proposition (5.17):
T is one-to-one $\Leftrightarrow$ $nullity(T) = 0$
We have $nullity(T) = 0$. By the dimension theorem (5.12):
$rank(T) + nullity(T) = n$ (where n is the dimension of V)
We have $rank(T) = n$ because $nullity(T) = 0$. Since $\dim(W) = n$ so

$$rank(T) = \dim(W) = n$$

By the above Proposition (5.20):

$$T : V \to W \text{ is onto } \Leftrightarrow rank(T) = \dim(W)$$

We conclude that T is onto.
($\Leftarrow$) Conversely we assume that T is onto and prove it is one-to-one.

How?
We prove $nullity(T) = 0$ then by Proposition (5.17) we have one-to-one:
T is one-to-one $\Leftrightarrow$ $nullity(T) = 0$

We are assuming T is onto, therefore $W = range(T)$. We have

$$rank(T) = \dim(W) = n \quad \text{[remember, at the start we let } \dim(W) = n]$$

We also have $\dim(V) = n$, therefore substituting this and $rank(T) = n$ into the dimension theorem

$$rank(T) + nullity(T) = \dim(V)$$

gives

$$n + nullity(T) = n \quad \Rightarrow \quad nullity(T) = 0$$

Since the $nullity(T) = 0$ so by (5.17) we conclude that T is one-to-one.

■

Proposition (5.22). If $T : V \to W$ is a linear transformation and $\dim(V) = \dim(W)$ then T is *both* one-to-one and onto $\Leftrightarrow$ $\ker(T) = \{\mathbf{O}\}$.

Proof – Exercises 5.4.

■

Proposition (5.22) makes life easier.

Why?
It means that to prove a linear transformation $T : V \to W$ with $\dim(V) = \dim(W)$ is *both* one-to-one and onto, we only need to show $\ker(T) = \{\mathbf{O}\}$.

Why is this important?

If the kernel is the zero vector, $\mathbf{O}$, then you know that everything was carried over by the transform, and you can move things back (i.e. your linear transformation has an inverse).

Example 5.26

Let $T : \mathbb{R}^2 \to \mathbb{R}^2$ be a linear transformation given by $T\left(\begin{pmatrix} x \\ y \end{pmatrix}\right) = \begin{pmatrix} x - y \\ x + y \end{pmatrix}$.

Show that T is *both* onto and one-to-one.

Solution

What is the dimension of $\mathbb{R}^2$?

2. Both our start and arrival vector space have the same dimension:

$$\dim\left(\mathbb{R}^2\right) = \dim\left(\mathbb{R}^2\right) = 2$$

We can use the above Proposition (5.22) to show that kernel of T is the zero vector, $\ker(T) = \{\mathbf{O}\}$. We have

$$T\left(\begin{pmatrix} x \\ y \end{pmatrix}\right) = \begin{pmatrix} x - y \\ x + y \end{pmatrix} = \begin{pmatrix} 0 \\ 0 \end{pmatrix}$$

Solving these simultaneous equations:

$$\left.\begin{matrix} x - y = 0 \\ x + y = 0 \end{matrix}\right\} \text{ gives } x = 0 \text{ and } y = 0 \text{ implies } \ker(T) = \begin{pmatrix} 0 \\ 0 \end{pmatrix} = \mathbf{O}$$

Since $\ker(T) = \{\mathbf{O}\}$, by (5.22) the given transformation is both *one-to-one* and *onto*.

The above example shows a simple test for a transformation to be one-to-one and onto. Just check $\ker(T) = \{\mathbf{O}\}$ provided the dimension of start and arrival vector spaces are equal.

A transformation which is *both* one-to-one and onto is called a **bijection**, or we say the transform is **bijective**.

5.4.5 Inverse linear transformations

For inverse transformations to exist we need the given transformation to be **bijective** (one-to-one and onto). This is why the work on these topics preceded this subsection.

What does the term 'inverse linear transform' mean?

In everyday language, inverse means opposite or in reverse. An inverse linear transform undoes the linear transform. Figure 5.27 overleaf illustrates this.

It shows a linear transform $T : V \to W$ in which T takes the vector $\mathbf{u}$ to $T(\mathbf{u})$ and *inverse* T takes $T(\mathbf{u})$ back to $\mathbf{u}$.

How do we denote inverse T?

T^{-1} denotes inverse T (recall T^{-1} does *not* equal $1/T$.)

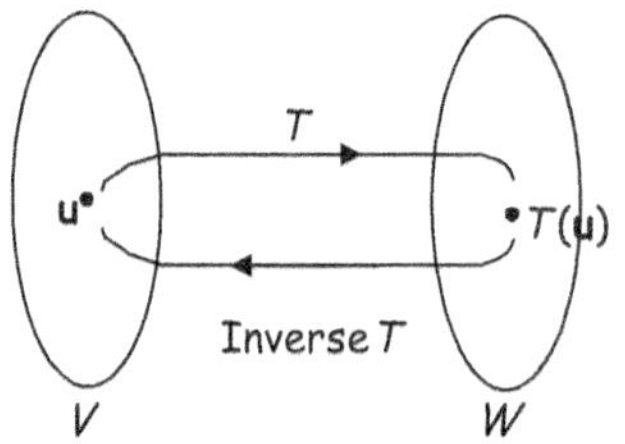

Figure 5.27

Next we give the formal definition of T^{-1}.

Definition (5.23). Let $T : V \to W$ be a *bijective* linear transform. The inverse transformation $T^{-1} : W \to V$ is defined as

$$\mathbf{v} = T^{-1}(\mathbf{w}) \Leftrightarrow T(\mathbf{v}) = \mathbf{w}$$

If T has an inverse transform, we refer to T as being **invertible**. To be invertible, T must be *both* one-to-one *and* onto (bijective) as the above proposition states. This means that if a given linear transform is bijective (one-to-one and onto) then it has an inverse transform.

Example 5.27

Let $T : \mathbb{R}^2 \to \mathbb{R}^2$ be a linear transformation given by

$$T\left(\begin{pmatrix} x \\ y \end{pmatrix}\right) = \begin{pmatrix} x - y \\ x + y \end{pmatrix}$$

Find the inverse transformation T^{-1}.

Solution

We can only find T^{-1} if T is *both* one-to-one and onto.

How do we show this?

We have already established that the given T is one-to-one and onto in Example 5.26, because we have the same transformation T.

How do we determine T^{-1}?

Let our arrival points be $a = x - y$ and $b = x + y$, then express our starting points x and y in terms of a and b.

Why?

Because we have the situation shown in Fig. 5.28.

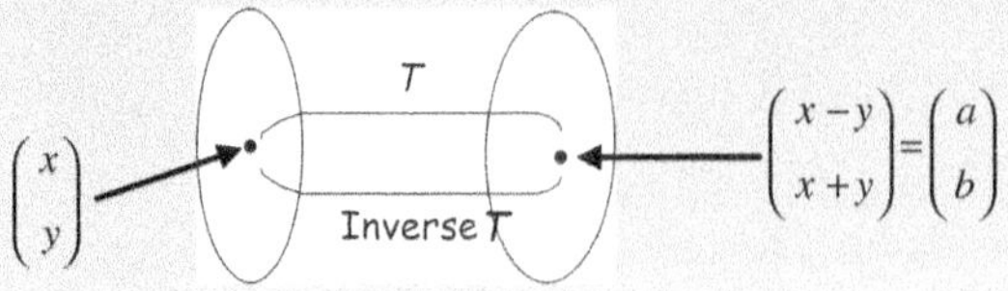

Figure 5.28

Adding these equations:

$$\begin{array}{rl} & x - y = a \\ + & \underline{x + y = b} \\ & 2x + 0 = a + b \text{ which gives } x = \dfrac{a+b}{2} = \dfrac{1}{2}(a+b) \end{array}$$

Subtracting these equations:

$$\begin{array}{rl} & x - y = a \\ - & \underline{(x + y = b)} \\ & 0 - 2y = a - b \text{ which gives } y = \dfrac{b-a}{2} = \dfrac{1}{2}(b-a) \end{array}$$

Thus the inverse transformation $T^{-1} : \mathbb{R}^2 \to \mathbb{R}^2$ is given by

$$T^{-1}\left(\begin{pmatrix} a \\ b \end{pmatrix}\right) = \begin{pmatrix} x \\ y \end{pmatrix} = \begin{pmatrix} \frac{1}{2}(a+b) \\ \frac{1}{2}(b-a) \end{pmatrix} = \frac{1}{2}\begin{pmatrix} a+b \\ b-a \end{pmatrix}$$

Hence we have $T^{-1}\left(\begin{pmatrix} a \\ b \end{pmatrix}\right) = \frac{1}{2}\begin{pmatrix} a+b \\ b-a \end{pmatrix}$.

Example 5.28

Let $T : \mathbb{R}^2 \to \mathbb{R}^2$ be a linear transformation given by $T(\mathbf{x}) = \mathbf{A}\mathbf{x}$ where $\mathbf{A} = \begin{pmatrix} 1 & 3 \\ 2 & 6 \end{pmatrix}$.

Find the inverse transformation T^{-1}.

Solution

Since the start and arrival vector spaces are of *same* dimension $\mathbb{R}^2$, there is a chance that the given transform is *invertible*.

If we can show that the kernel of T is the zero vector, $\mathbf{O}$, then by (5.22) we can say T is a bijection and so it is invertible.

How can we find the kernel of T?

Let vector $\mathbf{x} = \begin{pmatrix} x \\ y \end{pmatrix}$ be in the start vector space, which arrives at the zero vector under T. We need to find $\mathbf{x}$ such that $T(\mathbf{x}) = \mathbf{A}\mathbf{x} = \mathbf{O}$. The augmented matrix is given by:

$$\left(\begin{array}{cc|c} 1 & 3 & 0 \\ 2 & 6 & 0 \end{array}\right) \Longrightarrow \begin{array}{c} \begin{array}{cc} x & y \end{array} \\ \left(\begin{array}{cc|c} 1 & 3 & 0 \\ 0 & 0 & 0 \end{array}\right) \end{array}$$

We have $x + 3y = 0$ implies $x = -3y$. Let $y = r$, where r is any real number, then $x = -3r$:

$$\ker(T) = \begin{pmatrix} x \\ y \end{pmatrix} = \begin{pmatrix} -3r \\ r \end{pmatrix} = r\begin{pmatrix} -3 \\ 1 \end{pmatrix} \neq \mathbf{O}$$

Hence $\ker(T) \neq \mathbf{O}$ [not zero] so the given transform does *not* have an inverse.

Proposition (5.24). Let $T : V \to W$ be a linear transform which is *both* one-to-one and onto. Then the inverse transform $T^{-1} : W \to V$ is also linear.

Proof – Exercises 5.4.

■

5.4.6 Isomorphism

At the end of section 4.3.4 of chapter 4 we described isomorphic vector spaces, but next we give the formal definition:

Definition (5.25). If the linear transformation $T : V \to W$ is *invertible* then we say that the vector spaces V and W are **isomorphic**. Such a transformation T is called an **isomorphism**.

What does isomorphism say about the vector spaces?
An isomorphism between vector spaces means these spaces are identical from a mathematical viewpoint, even though they are different spaces.

Well, what does this mean?
It means that *isomorphic* vector spaces have an *identical* structure. They have similar properties with respect to the fundamental linear algebra operations of vector addition and scalar multiplication.

We can make an analogy with music. Consider a low C note on a piano or a guitar, as an example. Although the piano and guitar have different sounds, both the low C notes on the piano and on the guitar will be in tune and is a representation of the same vibration. Both instruments will produce a sound vibration with the same characteristics and properties. We can think of the piano and guitar as being isomorphic on a particular range of notes.

Why is isomorphism useful?
If we didn't look for isomorphism then we would be reinventing the wheel each time by studying the same things over and over again for different vector spaces. An isomorphism between two vector spaces preserves the structure.

Example 5.29

Show that vector spaces P_n and $\mathbb{R}^{n+1}$ are isomorphic.

Solution
Remember, P_n is the vector space of polynomials of degree n or less and $\mathbb{R}^{n+1}$ is the $n+1$ Euclidean space. We need to create a linear transformation between P_n and $\mathbb{R}^{n+1}$.
How?
We can write every polynomial $\mathbf{p}$ in P_n as

$$\mathbf{p} = p(x) = c_0 + c_1x + c_2x^2 + \cdots + c_nx^n$$

The set of standard basis (axes) vectors for P_n is $\{1, x, x^2, \ldots, x^n\}$. We can describe this vector $\mathbf{p}$ by writing it in terms of the coordinates $(c_0, c_1, c_2, \ldots, c_n)$. This coordinate can also correspond to our Euclidean space $\mathbb{R}^{n+1}$. Hence, we write the transformation as $T : P_n \to \mathbb{R}^{n+1}$ given by

$$T(c_0 + c_1x + c_2x^2 + \cdots + c_nx^n) = \begin{pmatrix} c_0 \\ c_1 \\ \vdots \\ c_n \end{pmatrix}$$

Verify that this transformation is indeed linear.

In order to show P_n and $\mathbb{R}^{n+1}$ are isomorphic we need to prove T is invertible.

How?

Since the dimensions of the two vector spaces are equal, $\dim(P_n) = \dim(\mathbb{R}^{n+1}) = n + 1$, we only need to prove that kernel of T is the zero vector, that is $\ker(T) = \{\mathbf{O}\}$.

Remember, the kernel of T are those vectors in P_n which arrive at the zero vector under T. The zero vector in the arrival vector space $\mathbb{R}^{n+1}$ is given by

$$\begin{pmatrix} c_0 & c_1 & \cdots & c_n \end{pmatrix}^T = \begin{pmatrix} 0 & 0 & \cdots & 0 \end{pmatrix}^T = \mathbf{O}$$

Hence all the coefficient c's are equal to zero:

$$T\left(c_0 + c_1x + c_2x^2 + \cdots + c_nx^n\right) = \begin{pmatrix} c_0 \\ c_1 \\ \vdots \\ c_n \end{pmatrix} = \begin{pmatrix} 0 \\ 0 \\ \vdots \\ 0 \end{pmatrix} \Longrightarrow \begin{matrix} c_0 & = & 0 \\ c_1 & = & 0 \\ \vdots & \vdots & \vdots \\ c_n & = & 0 \end{matrix}$$

This gives $\mathbf{p} = 0 + 0x + 0x^2 + \cdots + 0x^n = \mathbf{O}$. Hence $\ker(T) = \{\mathbf{O}\}$, so T is one-to-one and onto which means it is invertible.

Since T is invertible, the vector spaces connecting T are isomorphic. Hence P_n and $\mathbb{R}^{n+1}$ are isomorphic.

Lemma (5.26). Let V and W be finite-dimensional real vector spaces of equal dimension and $T : V \to W$ be an isomorphism. If $\{\mathbf{v}_1, \mathbf{v}_2, \ldots, \mathbf{v}_n\}$ is a basis (axes) for V then

$$\{T(\mathbf{v}_1), T(\mathbf{v}_2), \ldots, T(\mathbf{v}_n)\}$$

is a basis (axes) for W.

Proof – Exercises 5.4.

■

Next we prove a powerful result.

Theorem (5.27). Let V and W be finite-dimensional real vector spaces. Then V is isomorphic to $W \Leftrightarrow$

$$\dim(V) = \dim(W)$$

Why is this result important?

Because it says that any two vector spaces of the same dimension are identical in structure. Given two vector spaces V and W then are they essentially the same. We can think of V and W as identical if they have the same structure and only the nature of their elements differs.

Proof.

If V and W are of dimension 0 then clearly the result holds. Assume the dimensions of V and W are not zero.

($\Leftarrow$). Let $T : V \to W$ be a transformation. Required to prove V and W are isomorphic which means that we have to show that T is invertible.

How do we show that T is invertible?

Prove that $\ker(T) = \{\mathbf{O}\}$.

Since both the vector spaces are of same dimension, n say, they have the same number of vectors in a basis (axes). Let $\{\mathbf{v}_1, \mathbf{v}_2, \mathbf{v}_3, \ldots, \mathbf{v}_n\}$ be a basis for V. By the above Lemma (5.27) we have $\{T(\mathbf{v}_1), T(\mathbf{v}_2), T(\mathbf{v}_3), \ldots, T(\mathbf{v}_n)\}$ is a basis (axes) for W.

Let $T : V \to W$ be the transformation:

$$T(k_1\mathbf{v}_1 + k_2\mathbf{v}_2 + k_3\mathbf{v}_3 + \cdots + k_n\mathbf{v}_n) = k_1T(\mathbf{v}_1) + k_2T(\mathbf{v}_2) + \cdots + k_nT(\mathbf{v}_n)$$

where k's are scalars.

Verify that this transformation is linear.

Let $\mathbf{u}$ be a vector in vector space V which is transformed to the zero vector under T, that is $T(\mathbf{u}) = \mathbf{O}$. Required to prove $\mathbf{u} = \mathbf{O}$ because we need $\ker(T) = \{\mathbf{O}\}$.

We can write the vector $\mathbf{u}$ as a linear combination of the basis (axes) vectors of V:

$$\mathbf{u} = k_1\mathbf{v}_1 + k_2\mathbf{v}_2 + k_3\mathbf{v}_3 + \cdots + k_n\mathbf{v}_n \qquad (*)$$

We have

$$\begin{aligned} T(\mathbf{u}) &= T\left(k_1\mathbf{v}_1 + k_2\mathbf{v}_2 + k_3\mathbf{v}_3 + \cdots + k_n\mathbf{v}_n\right) \\ &= k_1T(\mathbf{v}_1) + k_2T(\mathbf{v}_2) + \cdots + k_nT(\mathbf{v}_n) \\ &= \mathbf{O} \qquad \left[\text{because } T(\mathbf{u}) = \mathbf{O}\right] \end{aligned}$$

Since $\{T(\mathbf{v}_1), T(\mathbf{v}_2), T(\mathbf{v}_3), \ldots, T(\mathbf{v}_n)\}$ is a basis for W, these vectors are linearly independent, which means that all the scalars are equal to zero:

$$k_1 = k_2 = k_3 = \cdots = k_n = 0$$

Substituting these scalars into (*) gives

$$\mathbf{u} = k_1\mathbf{v}_1 + k_2\mathbf{v}_2 + k_3\mathbf{v}_3 + \cdots + k_n\mathbf{v}_n = \mathbf{O}$$

Hence $\ker(T) = \{\mathbf{O}\}$, so the transformation is invertible, which implies that start V and arrival W vector spaces are isomorphic.

($\Rightarrow$). Exercises 5.4 question 19(c).

■

By the above Theorem (5.27) we have:

Proposition (5.28). Every n-dimensional real vector space is isomorphic to $\mathbb{R}^n$.

This means that all the non-Euclidean vector spaces we have looked at in previous chapters, such as the set of polynomials P_n, matrices M_{mn} and continuous functions $C[a, b]$, are identical in structure to $\mathbb{R}^n$, provided they have the same dimension. This is the most useful isomorphism in linear algebra, because any n-dimensional vector space can be recast in terms of the familiar Euclidean space $\mathbb{R}^n$.

For example, the vector space M_{22} is isomorphic to $\mathbb{R}^4$. However, M_{23} is *not* isomorphic to $\mathbb{R}^5$ because M_{23} is of dimension 6 but $\mathbb{R}^5$ is of dimension 5.

Summary

Let $T: V \to W$.

The transform T is one-to-one $\Leftrightarrow$ $\ker(T) = \{\mathbf{O}\}$.

The transform T is onto $\Leftrightarrow$ $range(T) = W$.

(5.23) If T is bijective then the inverse transformation $T^{-1}: W \to V$ is defined as

$$\mathbf{v} = T^{-1}(\mathbf{w}) \Leftrightarrow T(\mathbf{v}) = \mathbf{w}$$

Vector spaces V and W are isomorphic if there is an invertible linear transformation between V and W.

EXERCISES 5.4

(Brief solutions at end of book. Full solutions available at <http://www.oup.co.uk/companion/singh>.)

1. Show that the following linear transformations are one-to-one by using Propositions (5.14) or (5.15):

   (a) $T: \mathbb{R}^2 \to \mathbb{R}^2$ given by $T(\mathbf{v}) = \mathbf{Iv}$ where $\mathbf{I} = \begin{pmatrix} 1 & 0 \\ 0 & 1 \end{pmatrix}$.

   (b) $T: \mathbb{R}^2 \to \mathbb{R}^2$ given by $T\left(\begin{pmatrix} x \\ y \end{pmatrix}\right) = \begin{pmatrix} y \\ x \end{pmatrix}$.

   (c) $T: \mathbb{R}^2 \to \mathbb{R}^2$ given by $T\left(\begin{pmatrix} x \\ y \end{pmatrix}\right) = \begin{pmatrix} x+y \\ x-y \end{pmatrix}$.

2. Let M_{mn} be the vector space of m by n matrices. Consider the linear transformation $T: M_{mn} \to M_{nm}$ defined by $T(\mathbf{A}) = \mathbf{A}^T$. Show that T is one-to-one. (Remember, $\mathbf{A}^T$ is the transpose of the matrix $\mathbf{A}$).

3. Determine whether the following linear transformations are one-to-one and/or onto:

(a) $T:\mathbb{R}^2 \to \mathbb{R}^2$ given by $T\left(\begin{pmatrix} x \\ y \end{pmatrix}\right) = \begin{pmatrix} x \\ x \end{pmatrix}$.

(b) $T:\mathbb{R}^3 \to \mathbb{R}^3$ given by $T\left(\begin{pmatrix} x \\ y \\ z \end{pmatrix}\right) = \begin{pmatrix} 0 \\ y \\ z \end{pmatrix}$.

(c) $T:\mathbb{R}^3 \to \mathbb{R}^3$ given by $T\left(\begin{pmatrix} x \\ y \\ z \end{pmatrix}\right) = \begin{pmatrix} z \\ y \\ x \end{pmatrix}$.

(d) $T:\mathbb{R}^2 \to \mathbb{R}^3$ given by $T\left(\begin{pmatrix} x \\ y \end{pmatrix}\right) = \begin{pmatrix} 2x+3y \\ x+y \\ 0 \end{pmatrix}$.

4. Let $T:\mathbb{R}^3 \to \mathbb{R}^3$ be the linear transformation defined by

$$T(\mathbf{u}) = \mathbf{A}\mathbf{u} \text{ where } \mathbf{A} = \begin{pmatrix} 1 & 0 & 0 \\ 0 & 1 & 0 \\ 0 & 0 & 1 \end{pmatrix}$$

Show that T is one-to-one and onto.

5. Show that the linear transformation $T: P_3 \to P_2$ given by $T(\mathbf{p}) = \mathbf{p}'$ where $\mathbf{p}'$ is the derivative of $\mathbf{p}$, is *not* one-to-one but is onto.

6. Show that the linear transformation $T: P_n \to P_n$ given by $T(\mathbf{p}) = \mathbf{p}''$ where $\mathbf{p}''$ is the second derivative of $\mathbf{p}$ is *not* one-to-one nor onto.

7. Let $T:\mathbb{R}^3 \to \mathbb{R}^2$ be the linear transformation defined by $T\left(\begin{pmatrix} x \\ y \\ z \end{pmatrix}\right) = \begin{pmatrix} y \\ z \end{pmatrix}$.

Show that T is *not* one-to-one but is onto.

8. Show that the linear transformation $T: P_2 \to P_3$ given by

$$T(ax^2 + bx + c) = ax + (b + c)$$

is neither one-to-one nor onto.

9. Show that the linear transformation $T: P_n \to P_n$ given by

$$T(\mathbf{p}) = \mathbf{p}'$$

is neither one-to-one nor onto.

10. Prove Proposition (5.17).

11. Prove the following:

If $T: V \to W$ is a linear one-to-one transformation then for every vector $\mathbf{w}$ in *range*(T) there exists a *unique* vector $\mathbf{v}$ in V such that $T(\mathbf{v}) = \mathbf{w}$.

12. Prove Proposition (5.19).

13. Prove Proposition (5.20).
14. Prove Proposition (5.22).
15. Prove Proposition (5.24).
16. Prove Lemma (5.26).
17. Let $T : \mathbb{R}^2 \to \mathbb{R}^2$ be a linear transformation given by

$$T\left(\begin{pmatrix} x \\ y \end{pmatrix}\right) = \begin{pmatrix} 2x + y \\ x - y \end{pmatrix}$$

Find the inverse linear transformation.

18. Show that $T : M_{22} \to \mathbb{R}^4$, given by

$$T\left(\begin{bmatrix} a & b \\ c & d \end{bmatrix}\right) = (a \;\; b \;\; c \;\; d)^T$$

is an isomorphism.

19. Prove the following properties of isomorphism.
    (a) A linear transform $T : V \to W$ is an isomorphism $\Leftrightarrow \ker(T) = \{\mathbf{O}\}$.
    (b) If $T : V \to W$ is an isomorphism then $T^{-1} : W \to V$ is also an isomorphism.
    (c) If vector spaces V and W are isomorphic then $\dim(V) = \dim(W)$.

SECTION 5.5 The Matrix of a Linear Transformation

By the end of this section you will be able to

- find the standard matrix of a linear transformation
- determine the transformation matrix for non-standard bases
- apply the transformation matrix

Two of the most powerful tools in mathematics are linear algebra and calculus. In this section we establish a link between them by representing a derivative in terms of a matrix.

We start by showing that any linear transformation can be written in matrix form. This simply requires us to write our linear transform equations in the form $\mathbf{Au}$, where the variables $x, y, z \ldots$ are stored in the vector $\mathbf{u}$, leaving behind their coefficients in the matrix $\mathbf{A}$. Writing a linear transformation in this way has untold benefits. With the numerical components of the transformation held in isolation, we can use some conventional arithmetic to manipulate them.

For example, if the matrix of coefficients performs a rotation and a number of rotations need to be performed in succession to create an animation, then you can just multiply by the matrix as many times as required. Each application of the matrix performs another rotation.

Using the transformation equations or using the matrix alternative are equivalent. You won't lose information if you use the matrix, and in many cases, using the matrix is simply much more efficient

In the next section we cover composition of transformations. We find that if transformations are written as matrices, then this radically reduces the work involved in the composition of linear transformations.

5.5.1 The standard matrix for a linear transformation (mapping)

We can describe a linear transformation $T:\mathbb{R}^n \to \mathbb{R}^m$ by a matrix as the following example demonstrates.

Example 5.30

Let $T:\mathbb{R}^2 \to \mathbb{R}^2$ be the linear operator (start and arrival spaces are identical) given by

$$T\left(\begin{pmatrix} x \\ y \end{pmatrix}\right) = \begin{pmatrix} 2x - y \\ x + 3y \end{pmatrix}$$

Write this transformation as $T(\mathbf{u}) = \mathbf{Au}$ where $\mathbf{u} = (x\ y)^T$.

Solution

Writing $\begin{pmatrix} 2x - y \\ x + 3y \end{pmatrix}$ in the form $\mathbf{Au}$ where $\mathbf{u} = \begin{pmatrix} x \\ y \end{pmatrix}$ is given by

$$\begin{pmatrix} 2x - y \\ x + 3y \end{pmatrix} = \begin{pmatrix} 2 & -1 \\ 1 & 3 \end{pmatrix}\begin{pmatrix} x \\ y \end{pmatrix} = \mathbf{Au} \text{ where } \mathbf{A} = \begin{pmatrix} 2 & -1 \\ 1 & 3 \end{pmatrix}$$

Thus $T(\mathbf{u}) = \mathbf{Au}$. Since $T:\mathbb{R}^2 \to \mathbb{R}^2$ so the matrix $\mathbf{A}$ is of size 2 by 2.

In this section we write a general linear transformation $T:\mathbb{R}^n \to \mathbb{R}^m$ as $T(\mathbf{u}) = \mathbf{Au}$. We say that $\mathbf{A}$ is the *matrix representation* of the linear transformation T. The transformation T acts on the vector $\mathbf{u}$. Similarly the matrix $\mathbf{A}$ acts on the vector $\mathbf{u}$ to give $\mathbf{Au}$.

Is it always going to be possible to write any linear transformation in matrix representation?
Yes, provided we have finite-dimensional vector spaces.

How are we going to show this?
Consider a general linear transformation $T:V \to W$. We examine the basis (axes) vectors in V and look at the images of these under the given transformation T. To find $T(\mathbf{v})$ for any $\mathbf{v}$ in V we only need to know how T behaves towards these basis (axes) vectors because $\mathbf{v}$ can be written as a linear combination of the basis (axes) vectors.

First we consider a linear transformation with respect to the standard basis for $\mathbb{R}^n$; $\{\mathbf{e}_1, \mathbf{e}_2, \mathbf{e}_3, \ldots, \mathbf{e}_n\}$ which represents our normal x, y, z, ... axes in $\mathbb{R}^n$. Remember, $\mathbf{e}_1$ is the unit vector in the x direction, $\mathbf{e}_2$ is the unit vector in the y direction....

Proposition (5.29). Let $T : \mathbb{R}^n \to \mathbb{R}^m$ be a linear transformation and $\{\mathbf{e}_1, \mathbf{e}_2, \mathbf{e}_3, \ldots, \mathbf{e}_n\}$ be the standard basis for $\mathbb{R}^n$. If

$$T(\mathbf{e}_1) = \begin{pmatrix} c_{11} \\ c_{21} \\ \vdots \\ c_{m1} \end{pmatrix}, \quad T(\mathbf{e}_2) = \begin{pmatrix} c_{12} \\ c_{22} \\ \vdots \\ c_{m2} \end{pmatrix}, \cdots \text{ and } T(\mathbf{e}_n) = \begin{pmatrix} c_{1n} \\ c_{2n} \\ \vdots \\ c_{mn} \end{pmatrix}$$

then we can write the transformation $T(\mathbf{u})$ as

$$T(\mathbf{u}) = \mathbf{A}\mathbf{u}$$

where $\mathbf{A} = \begin{pmatrix} T(\mathbf{e}_1) & T(\mathbf{e}_2) & \cdots & T(\mathbf{e}_n) \end{pmatrix} = \begin{pmatrix} c_{11} & c_{12} & & c_{1n} \\ c_{21} & c_{22} & & c_{2n} \\ \vdots & \vdots & \cdots & \vdots \\ c_{m1} & c_{m2} & & c_{mn} \end{pmatrix}$ and $\mathbf{u}$ is in $\mathbb{R}^n$.

A is called the **standard matrix** and is of size m by n.

Note that the first column of matrix **A** is the column vector $T(\mathbf{e}_1)$, the second column of matrix **A** is the column vector $T(\mathbf{e}_2)$, ... and the last column is the column vector $T(\mathbf{e}_n)$.

Proof.

Let $\mathbf{u}$ be an arbitrary vector in $\mathbb{R}^n$ and $\{\mathbf{e}_1, \mathbf{e}_2, \ldots, \mathbf{e}_n\}$ be the standard basis (axes) vectors of the n-dimensional vector space $\mathbb{R}^n$. By the definition of standard basis we can write

$$\mathbf{u} = \begin{pmatrix} u_1 \\ u_2 \\ \vdots \\ u_n \end{pmatrix} = u_1 \underbrace{\begin{pmatrix} 1 \\ 0 \\ \vdots \\ 0 \end{pmatrix}}_{=\mathbf{e}_1} + u_2 \underbrace{\begin{pmatrix} 0 \\ 1 \\ 0 \\ \vdots \end{pmatrix}}_{=\mathbf{e}_2} + \cdots + u_n \underbrace{\begin{pmatrix} 0 \\ 0 \\ \vdots \\ 1 \end{pmatrix}}_{=\mathbf{e}_n} = u_1\mathbf{e}_1 + u_2\mathbf{e}_2 + \cdots + u_n\mathbf{e}_n$$

where u_j's are scalars and are the coordinates of $\mathbf{u}$ with respect to the standard basis (axes) $\{\mathbf{e}_1, \mathbf{e}_2, \mathbf{e}_3, \ldots, \mathbf{e}_n\}$. Applying the linear transformation T we have

$$\begin{aligned} T(\mathbf{u}) &= T(u_1\mathbf{e}_1 + u_2\mathbf{e}_2 + \cdots + u_n\mathbf{e}_n) && [\text{because } \mathbf{u} = u_1\mathbf{e}_1 + u_2\mathbf{e}_2 + \cdots + u_n\mathbf{e}_n] \\ &= u_1 T(\mathbf{e}_1) + u_2 T(\mathbf{e}_2) + \cdots + u_n T(\mathbf{e}_n) && [\text{because } T \text{ is linear}] \\ &\underset{\text{Given in Proposition}}{=} u_1 \underbrace{\begin{pmatrix} c_{11} \\ c_{21} \\ \vdots \\ c_{m1} \end{pmatrix}}_{=T(\mathbf{e}_1)} + u_2 \underbrace{\begin{pmatrix} c_{12} \\ c_{22} \\ \vdots \\ c_{m2} \end{pmatrix}}_{=T(\mathbf{e}_2)} + \cdots + u_n \underbrace{\begin{pmatrix} c_{1n} \\ c_{2n} \\ \vdots \\ c_{mn} \end{pmatrix}}_{=T(\mathbf{e}_n)} && (1) \end{aligned}$$

Let the matrix $\mathbf{A} = \begin{pmatrix} T(\mathbf{e}_1) & T(\mathbf{e}_2) & \cdots & T(\mathbf{e}_n) \end{pmatrix} = \begin{pmatrix} c_{11} & c_{12} & & c_{1n} \\ c_{21} & c_{22} & \cdots & c_{2n} \\ \vdots & \vdots & & \vdots \\ c_{m1} & c_{m2} & & c_{mn} \end{pmatrix}$.

Applying the matrix **A** to the vector **u** gives **Au**, which is the linear combination of the column vectors of **A**:

$$\mathbf{Au} = \begin{pmatrix} c_{11} & c_{12} & & c_{1n} \\ c_{21} & c_{22} & \cdots & c_{2n} \\ \vdots & \vdots & & \vdots \\ c_{m1} & c_{m2} & & c_{mn} \end{pmatrix} \begin{pmatrix} u_1 \\ u_2 \\ \vdots \\ u_n \end{pmatrix} = u_1 \begin{pmatrix} c_{11} \\ c_{21} \\ \vdots \\ c_{m1} \end{pmatrix} + u_2 \begin{pmatrix} c_{12} \\ c_{22} \\ \vdots \\ c_{m2} \end{pmatrix} + \cdots + u_n \begin{pmatrix} c_{1n} \\ c_{2n} \\ \vdots \\ c_{mn} \end{pmatrix} \quad (2)$$

The right hand sides of (1) and (2) are identical, so we have our result $T(\mathbf{u}) = \mathbf{Au}$.

■

Example 5.31

Let $T : \mathbb{R}^2 \to \mathbb{R}^2$ be the linear operator given by

$$T\left(\begin{bmatrix} x \\ y \end{bmatrix}\right) = \begin{pmatrix} 2x - y \\ x + 3y \end{pmatrix}$$

Determine the standard matrix **A** which represents T.

Solution

What is the standard basis for $\mathbb{R}^2$*?*

$\mathbf{e}_1 = (1\ 0)^T$ is the unit vector in the x direction and $\mathbf{e}_2 = (0\ 1)^T$ is the unit vector in the y direction. Applying the given linear operator to these basis (axes) vectors $\{\mathbf{e}_1, \mathbf{e}_2\}$ yields:

$$T(\mathbf{e}_1) = T\left(\begin{bmatrix} 1 \\ 0 \end{bmatrix}\right) = \begin{pmatrix} 2(1) - 0 \\ 1 + 3(0) \end{pmatrix} = \begin{pmatrix} 2 \\ 1 \end{pmatrix} \quad \left[\text{using } T\left(\begin{bmatrix} x \\ y \end{bmatrix}\right) = \begin{pmatrix} 2x - y \\ x + 3y \end{pmatrix}\right]$$

$$T(\mathbf{e}_2) = T\left(\begin{bmatrix} 0 \\ 1 \end{bmatrix}\right) = \begin{pmatrix} 2(0) - 1 \\ 0 + 3(1) \end{pmatrix} = \begin{pmatrix} -1 \\ 3 \end{pmatrix} \quad \left[\text{using } T\left(\begin{bmatrix} x \\ y \end{bmatrix}\right) = \begin{pmatrix} 2x - y \\ x + 3y \end{pmatrix}\right]$$

Note, $T(\mathbf{e}_1)$ gives the x coefficients and $T(\mathbf{e}_2)$ gives the y coefficients, because $\mathbf{e}_1$ and $\mathbf{e}_2$ are the unit vectors in these directions respectively.

What is the standard matrix **A** *equal to in this case?*

By the above Proposition (5.29) we have $\mathbf{A} = \begin{pmatrix} T(\mathbf{e}_1) & T(\mathbf{e}_2) \end{pmatrix} = \begin{pmatrix} 2 & -1 \\ 1 & 3 \end{pmatrix}$.

Note that we have the same matrix as in Example 5.30, which represents the given transformation T. We have $T(\mathbf{u}) = \mathbf{Au}$ where $\mathbf{u} = (x\ y)^T$.

Example 5.32

Let $T : \mathbb{R}^3 \to \mathbb{R}^2$ be the linear transformation given by

$$T\left(\begin{bmatrix} x \\ y \\ z \end{bmatrix}\right) = \begin{pmatrix} 3x + y + z \\ x - 3y - z \end{pmatrix}$$

Determine the standard matrix $\mathbf{A}$ which represents T.

Solution

What is the standard basis for $\mathbb{R}^3$*?*

$$\mathbf{e}_1 = \begin{bmatrix} 1 & 0 & 0 \end{bmatrix}^T,\ \mathbf{e}_2 = \begin{bmatrix} 0 & 1 & 0 \end{bmatrix}^T \text{ and } \mathbf{e}_3 = \begin{bmatrix} 0 & 0 & 1 \end{bmatrix}^T$$

where $\mathbf{e}_1$, $\mathbf{e}_2$ and $\mathbf{e}_3$ are unit vectors in the x, y and z directions respectively.

We first examine how T acts on these basis (axes) vectors. Applying the given linear transformation to the basis (axis) vector $\mathbf{e}_1$ yields:

$$T(\mathbf{e}_1) = T\left(\begin{bmatrix} 1 \\ 0 \\ 0 \end{bmatrix}\right) = \begin{pmatrix} 3(1) + 0 + 0 \\ 1 - 3(0) - 0 \end{pmatrix} = \begin{pmatrix} 3 \\ 1 \end{pmatrix} \quad \left[\text{because } T\left(\begin{bmatrix} x \\ y \\ z \end{bmatrix}\right) = \begin{pmatrix} 3x + y + z \\ x - 3y - z \end{pmatrix}\right]$$

Similarly, applying T to $\mathbf{e}_2$ and $\mathbf{e}_3$ gives $T(\mathbf{e}_2) = \begin{pmatrix} 1 \\ -3 \end{pmatrix}$ and $T(\mathbf{e}_3) = \begin{pmatrix} 1 \\ -1 \end{pmatrix}$ respectively.

What is the standard matrix $\mathbf{A}$ *equal to?*

By (5.29) which says $\mathbf{A} = (T(\mathbf{e}_1) \ \ T(\mathbf{e}_2) \ \ \cdots \ \ T(\mathbf{e}_n))$, we write these vectors as the first, second and last columns of matrix $\mathbf{A}$:

$$\mathbf{A} = (T(\mathbf{e}_1) \ \ T(\mathbf{e}_2) \ \ T(\mathbf{e}_3)) = \begin{pmatrix} 3 & 1 & 1 \\ 1 & -3 & -1 \end{pmatrix}$$

We can check this result by evaluating $\mathbf{Au}$, where $\mathbf{u} = (x \ \ y \ \ z)^T$:

$$\mathbf{Au} = \begin{pmatrix} 3 & 1 & 1 \\ 1 & -3 & -1 \end{pmatrix}\begin{pmatrix} x \\ y \\ z \end{pmatrix} = \begin{pmatrix} 3x + y + z \\ x - 3y - z \end{pmatrix}$$

This is the given transformation, therefore matrix $\mathbf{A}$ represents transformation T.

Normally the standard matrix can be found by inspection.

How?

From Example 5.32, we have the transformation and standard matrix as

$$T\left(\begin{bmatrix} x \\ y \\ z \end{bmatrix}\right) = \begin{pmatrix} 3x + y + z \\ x - 3y - z \end{pmatrix} \text{ and standard matrix is } \mathbf{A} = \begin{matrix} \ \ x & \ \ y & \ \ z \\ \end{matrix} \atop \begin{pmatrix} 3 & 1 & 1 \\ 1 & -3 & -1 \end{pmatrix}$$

$\mathbf{A}$ is the standard matrix because we use the standard basis to write $\mathbf{A}$.
In the above example $T : \mathbb{R}^3 \to \mathbb{R}^2$ so the matrix $\mathbf{A}$ is of size 2 by 3.

5.5.2 Transformation matrix for non-standard bases

Why use non-standard bases?

Examining a vector in a different basis (axes) may bring out structure related to that basis, which is hidden in the standard representation. It may be a relevant and useful structure. For example, we used to measure the motion of the planets in a basis (axes) with the earth at the centre. Then we discovered that putting the sun at the centre made life simpler - orbits were measured against a basis with the sun at the focus.

For some motions, such as projectiles, our standard basis (xy axes) may be the most suitable, but for studying other kinds of motions, such as orbits, a polar basis (r, θ) may work better.

If we use latitudes and longitudes to work out a map then we have been effectively using spherical polar coordinates (r, θ, φ) rather than our standard xyz axes.

Another example is trying to find the forces on an aeroplane as shown in Fig. 5.29. The components parallel and perpendicular to the aeroplane are a lot more useful than the horizontal and vertical components.

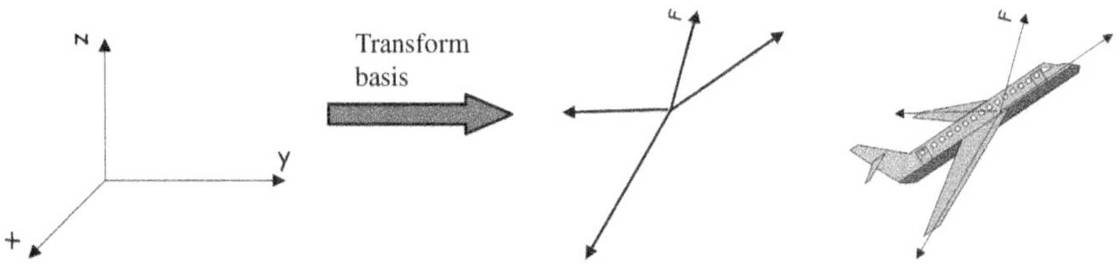

Figure 5.29

In computer games and 3D design software we often want to rotate our xyz axes (basis) to obtain new axes (basis) which are a lot more useful. (See question 7 of Exercises 5.5.)

In crystal structures, we need to use a basis which gives a cleaner set of coordinates called Miller indices. The Miller indices are coordinates used to specify direction and planes in a crystal or lattice. A vector from the origin to the lattice point is normally written in appropriate basis (axes) vectors and then the coordinates are given by the Miller indices.

Many problems in physics can be simplified due to their symmetrical properties if the right basis (axes) is chosen. Choosing a basis (axes) wisely can greatly reduce the amount of arithmetic you have to do.

For the remaining part of this section we will need to use the notation $[\mathbf{u}]_B$, which means the coordinates of the vector $\mathbf{u}$ with respect to a basis (axes) B.

Example 5.33

Write the vector $\mathbf{u} = \begin{pmatrix} 3 \\ 9 \end{pmatrix}$ of $\mathbb{R}^2$ in terms of new basis (axes) vectors $B = \left\{ \mathbf{b}_1 = \begin{pmatrix} 1 \\ 1 \end{pmatrix}, \mathbf{b}_2 = \begin{pmatrix} 1 \\ 4 \end{pmatrix} \right\}$.

Solution

This means find $[\mathbf{u}]_B$. Let k and c be scalars, then we write the linear combination

$$k\begin{pmatrix}1\\1\end{pmatrix}+c\begin{pmatrix}1\\4\end{pmatrix}=\begin{pmatrix}3\\9\end{pmatrix}$$

Solving the simultaneous equations

$$\left.\begin{aligned}k+c&=3\\k+4c&=9\end{aligned}\right\}\quad \text{gives } k=1, c=2$$

We have $\mathbf{u}=\begin{pmatrix}3\\9\end{pmatrix}=1\mathbf{b}_1+2\mathbf{b}_2=1\begin{pmatrix}1\\1\end{pmatrix}+2\begin{pmatrix}1\\4\end{pmatrix}$ or coordinates with respect to basis B is $\begin{bmatrix}3\\9\end{bmatrix}_B=\begin{pmatrix}1\\2\end{pmatrix}$

The coordinates of the vector $\begin{pmatrix}3\\9\end{pmatrix}$ is $\begin{pmatrix}1\\2\end{pmatrix}$ with respect to the basis $B=\left\{\mathbf{b}_1=\begin{pmatrix}1\\1\end{pmatrix}, \mathbf{b}_2=\begin{pmatrix}1\\4\end{pmatrix}\right\}$ (Fig. 5.30).

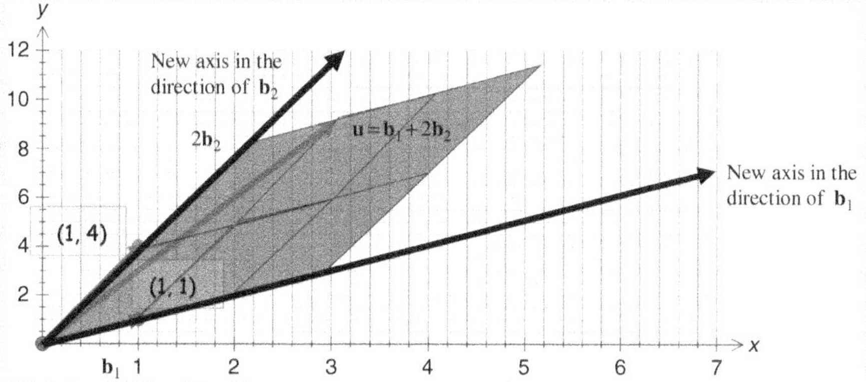

Figure 5.30

Note that $\mathbf{u}=(3\ \ 9)^T$ coordinates are calculated with respect to a standard basis $\{\mathbf{e}_1,\mathbf{e}_2\}$ or our normal xy axes, but the coordinates with respect to the new basis $B=\{\mathbf{b}_1,\mathbf{b}_2\}$ is $(1\ \ 2)^T$.

We assume that if we have the vector $k_1\,\mathbf{u}_1+k_2\,\mathbf{u}_2+\cdots+k_n\,\mathbf{u}_n$ where k's are scalars, then

(5.30) $$\left[k_1\,\mathbf{u}_1+k_2\,\mathbf{u}_2+\cdots+k_n\,\mathbf{u}_n\right]_B=k_1\,[\mathbf{u}_1]_B+k_2\,[\mathbf{u}_2]_B+\cdots+k_n\,[\mathbf{u}_n]_B$$

Proposition (5.31). Let $T: V\to W$ be a linear transformation and V and W be finite-dimensional vector spaces with basis $B=\{\mathbf{v}_1,\mathbf{v}_2,\ldots,\mathbf{v}_n\}$ and $C=\{\mathbf{w}_1,\mathbf{w}_2,\ldots,\mathbf{w}_m\}$ respectively. Then we can write the transformation as

$$[T(\mathbf{u})]_C=\mathbf{A}\,[\mathbf{u}]_B$$

where $[T(\mathbf{u})]_C$ are the coordinates of $T(\mathbf{u})$ with respect to the basis (axes) C, $[\mathbf{u}]_B$ are the coordinates of $\mathbf{u}$ with respect to the basis (axes) B and $\mathbf{A}$ is a matrix given by

$$\mathbf{A} = \big([T(\mathbf{v}_1)]_C \;\; [T(\mathbf{v}_2)]_C \;\; \cdots \;\; [T(\mathbf{v}_n)]_C\big)$$

What does this notation mean?

Figure 5.31 illustrates this:

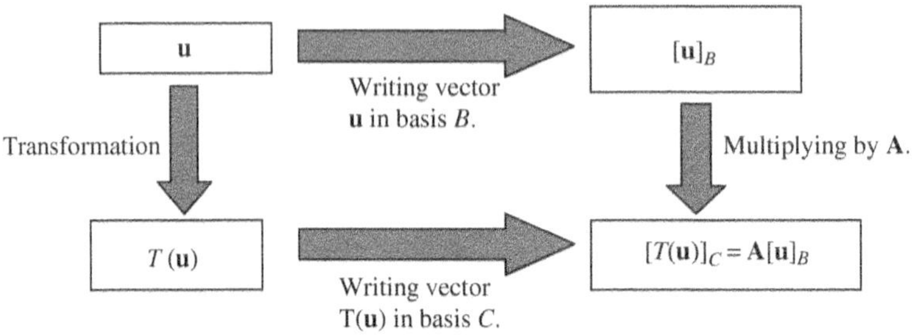

Figure 5.31

Proof.

Let $\mathbf{u}$ be an arbitrary vector in the n-dimensional vector space V with basis $\{\mathbf{v}_1, \mathbf{v}_2, \ldots, \mathbf{v}_n\}$. By the definition of basis we can write $\mathbf{u}$ as a linear combination of basis (axes) vectors:

$$\mathbf{u} = k_1\mathbf{v}_1 + k_2\mathbf{v}_2 + \cdots + k_n\mathbf{v}_n$$

where k's are scalars.

The coordinates of vector $\mathbf{u}$ with respect to basis (axes) $B = \{\mathbf{v}_1, \ldots, \mathbf{v}_n\}$ is $[\mathbf{u}]_B = \begin{pmatrix} k_1 \\ \vdots \\ k_n \end{pmatrix}$.

Applying the rules of linear transformation we have

$$T(\mathbf{u}) = T\left(k_1\mathbf{v}_1 + k_2\mathbf{v}_2 + \cdots + k_n\mathbf{v}_n\right) \quad \left[\text{because } \mathbf{u} = k_1\mathbf{v}_1 + k_2\mathbf{v}_2 + \cdots + k_n\mathbf{v}_n\right]$$

$$\underbrace{=}_{\text{because } T \text{ is linear}} k_1T(\mathbf{v}_1) + k_2T(\mathbf{v}_2) + \cdots + k_nT(\mathbf{v}_n) \qquad (^*)$$

Note that $T(\mathbf{v}_1)$, $T(\mathbf{v}_2), \ldots,$ and $T(\mathbf{v}_n)$ are all in the arrival vector space W. Let $C = \{\mathbf{w}_1, \mathbf{w}_2, \ldots, \mathbf{w}_m\}$ be a set of basis (axes) vectors for W then we can write

$$\begin{aligned}
T(\mathbf{v}_1) &= c_{11}\mathbf{w}_1 + c_{21}\mathbf{w}_2 + \cdots + c_{m1}\mathbf{w}_m \\
T(\mathbf{v}_2) &= c_{12}\mathbf{w}_1 + c_{22}\mathbf{w}_2 + \cdots + c_{m2}\mathbf{w}_m \\
\vdots \quad & \quad\;\; \vdots \quad\;\; \vdots \quad\;\; \vdots \qquad \cdots \qquad \vdots \\
T(\mathbf{v}_n) &= c_{1n}\mathbf{w}_1 + c_{2n}\mathbf{w}_2 + \cdots + c_{mn}\mathbf{w}_m
\end{aligned}$$

where the c's are scalars. We can write these as coordinates with respect to the basis (axes) C:

$$[T(\mathbf{v}_1)]_C = \begin{pmatrix} c_{11} \\ c_{21} \\ \vdots \\ c_{m1} \end{pmatrix}, \ [T(\mathbf{v}_2)]_C = \begin{pmatrix} c_{12} \\ c_{22} \\ \vdots \\ c_{m2} \end{pmatrix}, \ \ldots \text{ and } [T(\mathbf{v}_n)]_C = \begin{pmatrix} c_{1n} \\ c_{2n} \\ \vdots \\ c_{mn} \end{pmatrix} \qquad (\dagger)$$

Let $\mathbf{A} = \begin{pmatrix} c_{11} & c_{12} & & c_{1n} \\ c_{21} & c_{22} & \cdots & c_{2n} \\ \vdots & \vdots & & \vdots \\ c_{m1} & c_{m2} & & c_{mn} \end{pmatrix}$ and from above $[\mathbf{u}]_B = \begin{pmatrix} k_1 \\ k_2 \\ \vdots \\ k_n \end{pmatrix}$ then $\mathbf{A}[\mathbf{u}]_B$ is given by

$$\mathbf{A}[\mathbf{u}]_B = \begin{pmatrix} c_{11} & c_{12} & & c_{1n} \\ c_{21} & c_{22} & \cdots & c_{2n} \\ \vdots & \vdots & & \vdots \\ c_{m1} & c_{m2} & & c_{mn} \end{pmatrix} \begin{pmatrix} k_1 \\ k_2 \\ \vdots \\ k_n \end{pmatrix} \qquad (^{**})$$

Examining $[T(\mathbf{u})]_C$ we have

$$\begin{aligned}
[T(\mathbf{u})]_C &= \left[k_1 T(\mathbf{v}_1) + k_2 T(\mathbf{v}_2) + \cdots + k_n T(\mathbf{v}_n)\right]_C \qquad \left[\text{from } (^*)\right] \\
&\underbrace{=}_{\text{by (5.30)}} k_1 [T(\mathbf{v}_1)]_C + k_2 [T(\mathbf{v}_2)]_C + \cdots + k_n [T(\mathbf{v}_n)]_C \\
&= k_1 \begin{pmatrix} c_{11} \\ c_{21} \\ \vdots \\ c_{m1} \end{pmatrix} + k_2 \begin{pmatrix} c_{12} \\ c_{22} \\ \vdots \\ c_{m2} \end{pmatrix} + \cdots + k_n \begin{pmatrix} c_{1n} \\ c_{2n} \\ \vdots \\ c_{mn} \end{pmatrix} \qquad \left[\text{by } (\dagger)\right] \\
&= \begin{pmatrix} c_{11} & c_{12} & & c_{1n} \\ c_{21} & c_{22} & \cdots & c_{2n} \\ \vdots & \vdots & & \vdots \\ c_{m1} & c_{m2} & & c_{mn} \end{pmatrix} \begin{pmatrix} k_1 \\ k_2 \\ \vdots \\ k_n \end{pmatrix} \underbrace{=}_{\text{by } (**)} \mathbf{A}[\mathbf{u}]_B
\end{aligned}$$

Hence we have our result $[T(\mathbf{u})]_C = \mathbf{A}[\mathbf{u}]_B$.

■

Example 5.34

Let $T : \mathbb{R}^3 \to \mathbb{R}^2$ be the linear transformation defined by

$$T\left(\begin{pmatrix} x \\ y \\ z \end{pmatrix}\right) = \begin{pmatrix} 3x + y + z \\ x - 3y - z \end{pmatrix}$$

(continued...)

Find the transformation matrix $\mathbf{A}$ with respect to the bases for $\mathbb{R}^3$ and $\mathbb{R}^2$ given by:

$$B = \left\{ \mathbf{v}_1 = \begin{pmatrix} 1 \\ 1 \\ 1 \end{pmatrix}, \mathbf{v}_2 = \begin{pmatrix} -1 \\ 0 \\ 1 \end{pmatrix}, \mathbf{v}_3 = \begin{pmatrix} 0 \\ 0 \\ 1 \end{pmatrix} \right\} \text{ and } C = \left\{ \mathbf{w}_1 = \begin{pmatrix} 1 \\ 2 \end{pmatrix}, \mathbf{w}_2 = \begin{pmatrix} -1 \\ 1 \end{pmatrix} \right\} \text{ respectively}$$

Note that B is a set of basis (axes) vectors for the start vector space $\mathbb{R}^3$ while C is a set of basis (axes) vectors for the arrival vector space $\mathbb{R}^2$.

Solution

Applying the given linear transformation T to each of the basis (axes) vectors $B = \{\mathbf{v}_1, \mathbf{v}_2, \mathbf{v}_3\}$:

$$T(\mathbf{v}_1) = T\left(\begin{pmatrix} 1 \\ 1 \\ 1 \end{pmatrix}\right) = \begin{pmatrix} 3(1)+1+1 \\ 1-3(1)-1 \end{pmatrix} = \begin{pmatrix} 5 \\ -3 \end{pmatrix} \quad \left[\text{because } T\left(\begin{pmatrix} x \\ y \\ z \end{pmatrix}\right) = \begin{pmatrix} 3x+y+z \\ x-3y-z \end{pmatrix}\right]$$

Similarly (verify this) we have

$$T(\mathbf{v}_2) = \begin{pmatrix} -2 \\ -2 \end{pmatrix} \text{ and } T(\mathbf{v}_3) = \begin{pmatrix} 1 \\ -1 \end{pmatrix}$$

What else do we need to find?

We need to write each of these arriving vectors, $T(\mathbf{v}_1), T(\mathbf{v}_2)$ and $T(\mathbf{v}_3)$, as the coordinates of the basis (axes) $C = \left\{ \mathbf{w}_1 = \begin{pmatrix} 1 \\ 2 \end{pmatrix}, \mathbf{w}_2 = \begin{pmatrix} -1 \\ 1 \end{pmatrix} \right\}$:

$$T(\mathbf{v}_1) = \begin{pmatrix} 5 \\ -3 \end{pmatrix} = a\mathbf{w}_1 + b\mathbf{w}_2 = a\begin{pmatrix} 1 \\ 2 \end{pmatrix} + b\begin{pmatrix} -1 \\ 1 \end{pmatrix} \quad (1)$$

$$T(\mathbf{v}_2) = \begin{pmatrix} -2 \\ -2 \end{pmatrix} = c\mathbf{w}_1 + d\mathbf{w}_2 = c\begin{pmatrix} 1 \\ 2 \end{pmatrix} + d\begin{pmatrix} -1 \\ 1 \end{pmatrix} \quad (2)$$

$$T(\mathbf{v}_3) = \begin{pmatrix} 1 \\ -1 \end{pmatrix} = e\mathbf{w}_1 + f\mathbf{w}_2 = e\begin{pmatrix} 1 \\ 2 \end{pmatrix} + f\begin{pmatrix} -1 \\ 1 \end{pmatrix} \quad (3)$$

How can we find the matrix A?

By the above Proposition (5.31) we have $\mathbf{A} = \left([T(\mathbf{v}_1)]_C \;\; [T(\mathbf{v}_2)]_C \;\; [T(\mathbf{v}_3)]_C\right)$ which in this case is

$$\mathbf{A} = \begin{pmatrix} a & c & e \\ b & d & f \end{pmatrix}.$$

Why?

Because by (1), (2) and (3) we have

$$[T(\mathbf{v}_1)]_C = \begin{pmatrix} a \\ b \end{pmatrix}, [T(\mathbf{v}_2)]_C = \begin{pmatrix} c \\ d \end{pmatrix} \text{ and } [T(\mathbf{v}_3)]_C = \begin{pmatrix} e \\ f \end{pmatrix}$$

Remember, the scalars in the linear combination of $\mathbf{w}_1$ and $\mathbf{w}_2$ give the coordinates of vectors $T(\mathbf{v}_1), T(\mathbf{v}_2)$ and $T(\mathbf{v}_3)$ with respect to basis (axes) C.

How can we determine the unknowns $a, b, c, \ldots, f$?
We need to solve the three pairs of simultaneous equations (1), (2) and (3) in the above. Consider the first pair (1):

$$\begin{pmatrix} 5 \\ -3 \end{pmatrix} = a\begin{pmatrix} 1 \\ 2 \end{pmatrix} + b\begin{pmatrix} -1 \\ 1 \end{pmatrix} = \begin{pmatrix} a \\ 2a \end{pmatrix} + \begin{pmatrix} -b \\ b \end{pmatrix} = \begin{pmatrix} a-b \\ 2a+b \end{pmatrix}$$

What is the solution to this pair of simultaneous equations?

$$\left.\begin{aligned} a-b &= 5 \\ 2a+b &= -3 \end{aligned}\right\} \text{ gives } a = \frac{2}{3} \text{ and } b = -\frac{13}{3}$$

Similarly, we can find the solutions of the other two simultaneous equations. They are

(2) $$\left.\begin{aligned} c-d &= -2 \\ 2c+d &= -2 \end{aligned}\right\} \text{ gives } c = -\frac{4}{3} \text{ and } d = \frac{2}{3}$$

(3) $$\left.\begin{aligned} e-f &= 1 \\ 2e+f &= -1 \end{aligned}\right\} \text{ gives } e = 0 \text{ and } f = -1$$

What is the matrix $\mathbf{A}$ equal to?

$$\mathbf{A} = \begin{pmatrix} a & c & e \\ b & d & f \end{pmatrix} = \begin{pmatrix} 2/3 & -4/3 & 0 \\ -13/3 & 2/3 & -1 \end{pmatrix}$$

This means that $[T(\mathbf{u})]_C = \mathbf{A}[\mathbf{u}]_B$ which is

$$[T(\mathbf{u})]_C = \begin{pmatrix} 2/3 & -4/3 & 0 \\ -13/3 & 2/3 & -1 \end{pmatrix}[\mathbf{u}]_B$$

To write the vector $T(\mathbf{u})$ in terms of basis (axes) C, denoted $[T(\mathbf{u})]_C$, we multiply the vector $\mathbf{u}$ written in basis (axes) B, denoted $[\mathbf{u}]_B$, by the matrix $\mathbf{A}$.

Example 5.35

For the linear transformation given in the above Example 5.34, use the matrix $\mathbf{A}$ to find $T(\mathbf{u})$ where $\mathbf{u} = (1 \;\; 0 \;\; 2)^T$ with respect to the same bases

$$B = \left\{ \mathbf{v}_1 = \begin{pmatrix} 1 \\ 1 \\ 1 \end{pmatrix}, \mathbf{v}_2 = \begin{pmatrix} -1 \\ 0 \\ 1 \end{pmatrix}, \mathbf{v}_3 = \begin{pmatrix} 0 \\ 0 \\ 1 \end{pmatrix} \right\} \text{ and } C = \left\{ \mathbf{w}_1 = \begin{pmatrix} 1 \\ 2 \end{pmatrix}, \mathbf{w}_2 = \begin{pmatrix} -1 \\ 1 \end{pmatrix} \right\}$$

Solution

We need to find $[T(\mathbf{u})]_C = \mathbf{A}[\mathbf{u}]_B$. First we determine $[\mathbf{u}]_B$.
What does $[\mathbf{u}]_B$ represent?
Remember, the entries in the given vector $\mathbf{u} = (1 \;\; 0 \;\; 2)^T$ are the coordinates with respect to the standard basis (xyz axes). However, we need to write $\mathbf{u}$ with respect to the given basis B because $[\mathbf{u}]_B$ is the coordinates of the vector $\mathbf{u}$ with respect to the basis (axes) vectors B.

(continued...)

$$\mathbf{u} = \begin{pmatrix} 1 \\ 0 \\ 2 \end{pmatrix} = a\mathbf{v}_1 + b\mathbf{v}_2 + c\mathbf{v}_3 = a\begin{pmatrix} 1 \\ 1 \\ 1 \end{pmatrix} + b\begin{pmatrix} -1 \\ 0 \\ 1 \end{pmatrix} + c\begin{pmatrix} 0 \\ 0 \\ 1 \end{pmatrix} \quad (*)$$

These a, b and c are *not* the same values as for Example 5.34.
How can we find these?
Solve the simultaneous equations by expanding (*):

$$\left.\begin{aligned} a - b &= 1 \\ a &= 0 \\ a + b + c &= 2 \end{aligned}\right\} \text{ gives } a = 0,\ b = -1 \text{ and } c = 3$$

What are the coordinates of $\mathbf{u}$ with respect to the basis (axes) B, $[\mathbf{u}]_B$, equal to?

$$[\mathbf{u}]_B = \begin{pmatrix} a \\ b \\ c \end{pmatrix} = \begin{pmatrix} 0 \\ -1 \\ 3 \end{pmatrix}$$

Note that $\mathbf{u} = (1 \ \ 0 \ \ 2)^T$ gives the coordinates of $\mathbf{u}$ with respect to our standard basis (*xyz* axes) but $[\mathbf{u}]_B = (0 \ \ -1 \ \ 3)^T$ gives the coordinates of $\mathbf{u}$ with respect to the given basis $B = \{\mathbf{v}_1, \mathbf{v}_2, \mathbf{v}_3\}$.
From the above Proposition (5.31): $[T(\mathbf{u})]_C = \mathbf{A}[\mathbf{u}]_B$
We found matrix $\mathbf{A}$ in the previous example, so using this $\mathbf{A}$ we have

$$[T(\mathbf{u})]_C = \mathbf{A}[\mathbf{u}]_B = \begin{pmatrix} 2/3 & -4/3 & 0 \\ -13/3 & 2/3 & -1 \end{pmatrix}[\mathbf{u}]_B$$

and evaluating this right hand side gives

$$[T(\mathbf{u})]_C = \begin{pmatrix} 2/3 & -4/3 & 0 \\ -13/3 & 2/3 & -1 \end{pmatrix}\begin{pmatrix} 0 \\ -1 \\ 3 \end{pmatrix} = \begin{pmatrix} 4/3 \\ -11/3 \end{pmatrix}$$

We can illustrate the above process as shown in Fig. 5.32.

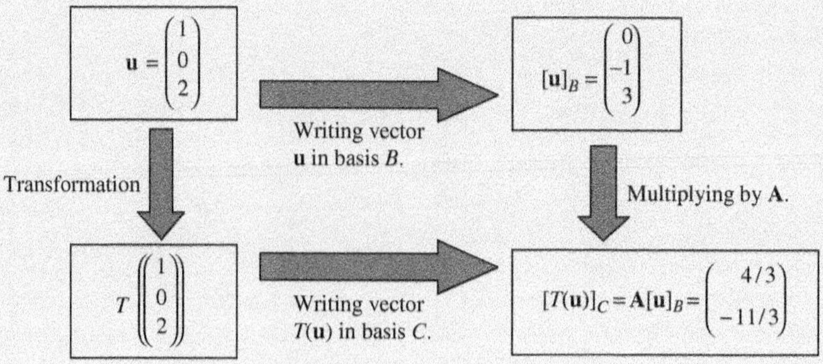

Figure 5.32

The coordinates of $T(\mathbf{u})$ where $\mathbf{u} = (1 \ \ 0 \ \ 2)^T$ with respect to the given basis C are $\begin{pmatrix} 4/3 \\ -11/3 \end{pmatrix}$.

We have measured the vector $T(\mathbf{u})$ against the new axes represented by $\mathbf{w}_1$ and $\mathbf{w}_2$:

$$T(\mathbf{u}) = \frac{4}{3}\mathbf{w}_1 - \frac{11}{3}\mathbf{w}_2$$

This is illustrated in Fig. 5.33.

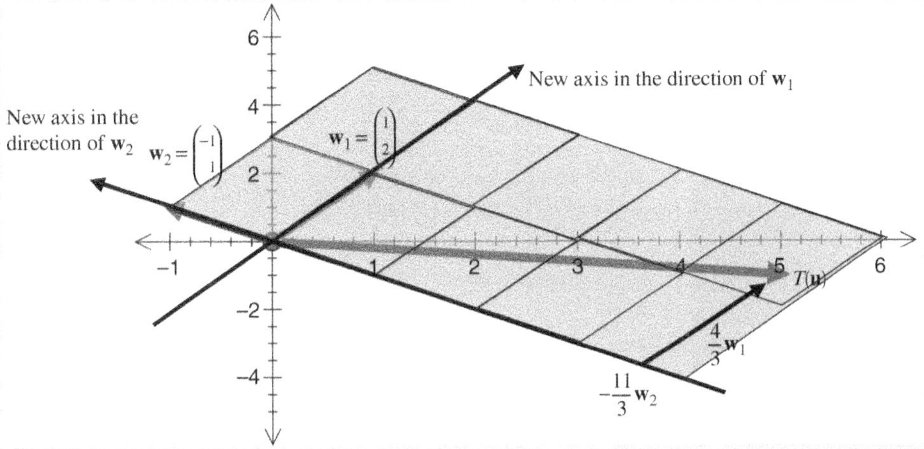

Figure 5.33

However, the coordinates of $T(\mathbf{u})$ with respect to our standard basis (*xy* axes) are given by:

$$T\left(\begin{pmatrix} 1 \\ 0 \\ 2 \end{pmatrix}\right) = \frac{4}{3}\mathbf{w}_1 - \frac{11}{3}\mathbf{w}_2 = \frac{4}{3}\begin{pmatrix} 1 \\ 2 \end{pmatrix} - \frac{11}{3}\begin{pmatrix} -1 \\ 1 \end{pmatrix} \quad \left[\begin{array}{l}\text{because } \mathbf{w}_1 \text{ and } \mathbf{w}_2 \\ \text{are vectors in basis } C\end{array}\right]$$

$$= \begin{pmatrix} 4/3 \\ 8/3 \end{pmatrix} - \begin{pmatrix} -11/3 \\ 11/3 \end{pmatrix} = \begin{pmatrix} 5 \\ -1 \end{pmatrix}$$

You can check this $T(\mathbf{u}) = \begin{pmatrix} 5 \\ -1 \end{pmatrix}$ by applying the given transformation directly with

$$\mathbf{u} = (1 \ \ 0 \ \ 2)^T.$$

5.5.3 A matrix for the derivative

Example 5.36

Let the linear transformation $T : P_2 \to P_1$ be defined by

$$T(\mathbf{p}) = \mathbf{p}'$$

(continued...)

where $\mathbf{p}'$ is the derivative of $\mathbf{p}$. Determine the matrix $\mathbf{A}$ which represents the given transformation with respect to the ordered bases

$$B = [1, x, x^2] \text{ and } C = [1, x]$$

for P_2 and P_1 respectively. Find $T\left(2x^2 + 3x + 1\right)$ by using this matrix $\mathbf{A}$.

Solution

First we look at how T behaves towards the basis (axes) vectors. Applying the given linear transformation to the vectors in basis (axes) $B = \left[1, x, x^2\right]$ we have

$$\begin{aligned} T(1) &= 1' = 0 && \text{[differentiating]} \\ T(x) &= x' = 1 && \text{[differentiating]} \\ T(x^2) &= (x^2)' = 2x && \text{[differentiating]} \end{aligned}$$

We need to write each of these as the coordinates of the basis (axes) $C = [1, x]$:

$$\begin{aligned} T(1) &= 0 = a(1) + b(x) && \text{gives } a = 0 \text{ and } b = 0 \\ T(x) &= 1 = c(1) + d(x) && \text{gives } c = 1 \text{ and } d = 0 \\ T(x^2) &= 2x = e(1) + f(x) && \text{gives } e = 0 \text{ and } f = 2 \end{aligned}$$

What is our matrix $\mathbf{A}$ equal to?

$$\mathbf{A} = \left([T(1)]_C \;\; [T(x)]_C \;\; [T(x^2)]_C\right) = \begin{pmatrix} a & c & e \\ b & d & f \end{pmatrix} = \begin{pmatrix} 0 & 1 & 0 \\ 0 & 0 & 2 \end{pmatrix}$$

For $T\left(2x^2 + 3x + 1\right)$ we have $\mathbf{p} = 2x^2 + 3x + 1$ so $[\mathbf{p}]_B = \begin{pmatrix} 1 \\ 3 \\ 2 \end{pmatrix}$ (coefficients of 1, x and x^2, and it must be in this order, because the given basis is ordered, $B = \left[1, x, x^2\right]$. Remember, a basis is a set of vectors which represent axes, so the order of the basis vectors matters.):

$$[T(\mathbf{p})]_C = \mathbf{A}[\mathbf{p}]_B = \begin{pmatrix} 0 & 1 & 0 \\ 0 & 0 & 2 \end{pmatrix} \begin{pmatrix} 1 \\ 3 \\ 2 \end{pmatrix} = \begin{pmatrix} 3 \\ 4 \end{pmatrix}$$

The entries 3 and 4 in the right hand column vector are the coefficients of the basis $C = [1, x]$, which means that we have

$$T\left(2x^2 + 3x + 1\right) = 3(1) + 4(x) = 3 + 4x$$

Checking this result by differentiating the quadratic:

$$T\left(2x^2 + 3x + 1\right) = \left(2x^2 + 3x + 1\right)' = 4x + 3.$$

Thus the matrix $\mathbf{A}$ does give the derivative of the quadratic polynomial $2x^2 + 3x + 1$.

We can find the derivative of any quadratic polynomial by using the matrix **A**. For example, to find $\left(3x^2+5x+7\right)'$ we use the above matrix **A** as follows:

$$\begin{pmatrix} 0 & 1 & 0 \\ 0 & 0 & 2 \end{pmatrix}\begin{pmatrix} 7 \\ 5 \\ 3 \end{pmatrix}=\begin{pmatrix} 5 \\ 6 \end{pmatrix}$$

Hence $\left(3x^2+5x+7\right)'=5+6x$.

If we consider the derivative of the general quadratic $\left(ax^2+bx+c\right)'$ then

$$\begin{pmatrix} 0 & 1 & 0 \\ 0 & 0 & 2 \end{pmatrix}\begin{pmatrix} c \\ b \\ a \end{pmatrix}=\begin{pmatrix} b \\ 2a \end{pmatrix} \text{ implies } \left(ax^2+bx+c\right)'=b+2ax$$

There is another advantage of using matrix form: we can write a matrix for a different non-standard basis such as $B=\left[1, 1-x, (1-x)^2\right]$. You are asked to find such a matrix in Exercises 5.5. It is not clear what happens using calculus, whereas the matrix form makes it transparent.

Note that a problem in differentiation can be converted to a problem in arithmetic of matrices. This is always going to be the case as long as we have a linear transformation T between two vector spaces U and V because by Proposition (5.31) we have

$$[T(\mathbf{u})]_C=\mathbf{A}[\mathbf{u}]_B$$

where B and C are bases for U and V respectively.

Summary

(5.31) Let $T: V \to W$ be a linear transformation and V and W be finite-dimensional vector spaces with bases $B=\{\mathbf{v}_1, \mathbf{v}_2, \mathbf{v}_3, \ldots, \mathbf{v}_n\}$ and $C=\{\mathbf{w}_1, \mathbf{w}_2, \mathbf{w}_3, \ldots, \mathbf{w}_m\}$ respectively. Then

$$[T(\mathbf{u})]_C=\mathbf{A}[\mathbf{u}]_B$$

where $\mathbf{u}$ is a vector in V and $\mathbf{A}=\left([T(\mathbf{v}_1)]_C \;\; [T(\mathbf{v}_2)]_C \;\; \cdots \;\; [T(\mathbf{v}_n)]_C\right)$.

EXERCISES 5.5

(Brief solutions at end of book. Full solutions available at <http://www.oup.co.uk/companion/singh>.)

1. Determine the standard matrix for the following linear transformations:

(a) $T\left(\begin{pmatrix} x \\ y \end{pmatrix}\right)=\begin{pmatrix} x+y \\ 2x+2y \end{pmatrix}$ (b) $T\left(\begin{pmatrix} x \\ y \end{pmatrix}\right)=\begin{pmatrix} x+y \\ -x-y \end{pmatrix}$

(c) $T\left(\begin{pmatrix} x \\ y \end{pmatrix}\right) = \begin{pmatrix} 2x + 3y \\ x - 5y \end{pmatrix}$

(d) $T\left((x \ y \ z)^T\right) = (x + y + z \quad x - y - z \quad 2x + y - z)^T$

(e) $T\left((x \ y \ z)^T\right) = (2x - y \quad 4x - y + 3z \quad 7x - y - z)^T$

(f) $T\left((x \ y \ z \ w)^T\right) = (x - y + z - 3w \quad -x + 3y - 7z - w \quad 9x + 5y + 6z + 12w \quad x)^T$

(g) $T\left((x \ y \ z \ w)^T\right) = \mathbf{O}$

2. Determine the standard matrix of the following linear transformations *without* reading off the coefficients of x, y, z.

(a) $T\left(\begin{pmatrix} x \\ y \end{pmatrix}\right) = \begin{pmatrix} x - y \\ x + 2y \end{pmatrix}$ (b) $T\left(\begin{pmatrix} x \\ y \end{pmatrix}\right) = \begin{pmatrix} 3x - 2y \\ 5y - x \end{pmatrix}$

(c) $T\left(\begin{pmatrix} x \\ y \\ z \end{pmatrix}\right) = \begin{pmatrix} x - y - z \\ x + y + z \end{pmatrix}$ (d) $T\left(\begin{pmatrix} x \\ y \\ z \end{pmatrix}\right) = \begin{pmatrix} 0 \\ 0 \\ 0 \end{pmatrix}$

(e) $T\left(\begin{pmatrix} x \\ y \\ z \end{pmatrix}\right) = \begin{pmatrix} -3x - 5y - 6z \\ -2x + 7y + 5z \\ 0 \end{pmatrix}$

3. The linear transform $T : \mathbb{R}^2 \to \mathbb{R}^3$ satisfies the following:

$$T\left([1 \quad 0]^T\right) = (1 \ \ 2 \ \ 3)^T \text{ and } T\left([0 \quad 1]^T\right) = (4 \ \ 5 \ \ 6)^T$$

Determine the matrix $\mathbf{A}$ which represents this transformation T.

4. For the linear transformation $T : P_2 \to P_1$ given by

$$T(\mathbf{p}) = \mathbf{p}'$$

where $\mathbf{p}'$ is the derivative of $\mathbf{p}$, determine the matrix $\mathbf{A}$ which represents the given transformation with respect to the ordered bases

$$B = \left[1, 1 - x, (1 - x)^2\right] \text{ and } C = [1, \ x]$$

for P_2 and P_1 respectively. Find $T\left(2x^2 + 3x + 1\right)$ by using this matrix $\mathbf{A}$.

5. For the linear transformation $T : P_3 \to P_2$ given by

$$T(\mathbf{p}) = \mathbf{p}'$$

where $\mathbf{p}'$ is the derivative of $\mathbf{p}$. Determine the matrix $\mathbf{A}$ which represents the given transformation with respect to the ordered bases in each of the following cases:

(i) $B = \left[1, x, x^2, x^3\right]$ and $C = \left[1, x, x^2\right]$ for P_3 and P_2 respectively.

(ii) $B = \left[1, x, x^2, x^3\right]$ and $C = \left[x^2, x, 1\right]$ for P_3 and P_2 respectively.

(iii) $B = \left[x^3, x^2, x, 1\right]$ and $C = \left[1, x, x^2\right]$ for P_3 and P_2 respectively.

What do you notice about the matrix $\mathbf{A}$ for parts (i), (ii) and (iii)?
In each case find $T\left(-1+3x-7x^2-2x^3\right)$ by using this matrix $\mathbf{A}$.

6. Let $T:\mathbb{R}^2\to\mathbb{R}^2$ be the linear operator defined by $T\left(\begin{pmatrix}x\\y\end{pmatrix}\right)=\begin{pmatrix}x+y\\x-y\end{pmatrix}$.
   Find the transformation matrix $\mathbf{A}$ with respect to the basis

$$B=\left\{\mathbf{v}_1=\begin{pmatrix}1\\2\end{pmatrix},\mathbf{v}_2=\begin{pmatrix}0\\-1\end{pmatrix}\right\}$$

   Find $T\left(\begin{pmatrix}-3\\1\end{pmatrix}\right)$ by using this matrix $\mathbf{A}$ and directly by using the given transformation.

7. In a computer game we want to change our standard basis $\{\mathbf{e}_1,\mathbf{e}_2\}$ by a rotation of 45° anti-clockwise which is given by the matrix $\mathbf{A}=\begin{pmatrix}\cos(45^\circ) & -\sin(45^\circ)\\ \sin(45^\circ) & \cos(45^\circ)\end{pmatrix}$.
   Determine the new basis and write the coordinates of the vector $\mathbf{v}=(2\ \ 1)^T$ in terms of the new basis. Sketch the new basis and the vector $\mathbf{v}$.

8. Let $T:\mathbb{R}^2\to\mathbb{R}^3$ be defined by

$$T\left(\begin{bmatrix}x & y\end{bmatrix}^T\right)=(-x\quad -y\quad x+3y)^T$$

   Determine the transformation matrix $\mathbf{A}$ with respect to the bases

$$B=\left\{\mathbf{v}_1=(1\ \ 2)^T,\ \mathbf{v}_2=(1\ \ 1)^T\right\}\ \text{and}$$
$$C=\left\{\mathbf{w}_1=(1\ \ 0\ \ 1)^T,\ \mathbf{w}_2=(1\ \ 2\ \ 0)^T,\ \mathbf{w}_3=(0\ \ 1\ \ 1)^T\right\}$$

   for $\mathbb{R}^2$ and $\mathbb{R}^3$ respectively. Find $T\left(\begin{bmatrix}2 & 1\end{bmatrix}^T\right)$ by using this matrix $\mathbf{A}$.

9. Let $T:M_{22}\to M_{22}$ be defined by

$$T(\mathbf{X})=\mathbf{X}^T\quad(\text{where }\mathbf{X}^T\text{ is the transpose of the matrix }\mathbf{X})$$

   Determine the transformation matrix $\mathbf{A}$ with respect to the bases

$$B=C=\left\{\mathbf{m}_1=\begin{pmatrix}1&0\\0&0\end{pmatrix},\mathbf{m}_2=\begin{pmatrix}0&1\\0&0\end{pmatrix},\mathbf{m}_3=\begin{pmatrix}0&0\\1&0\end{pmatrix},\mathbf{m}_4=\begin{pmatrix}0&0\\0&1\end{pmatrix}\right\}$$

   for M_{22} where M_{22} is the 2 by 2 matrices. Find $T\left(\begin{pmatrix}1&2\\3&4\end{pmatrix}\right)$ by using this matrix $\mathbf{A}$.

10. Let V be vector space spanned by the set $\{\sin(x),\cos(x)\}$. Let T be the differential linear operator $T:V\to V$ given by

$$T(\mathbf{f})=\mathbf{f}'\text{ where }\mathbf{f}'\text{ is the derivative of }\mathbf{f}$$

Determine the transformation matrix **A** with respect to the ordered bases

$$B = \left[\sin(x), \cos(x)\right]$$

By using this transformation matrix **A** find $T(\mathbf{g})$ where
(i) $\mathbf{g} = 5\sin(x) + 2\cos(x)$ (ii) $\mathbf{g} = m\sin(x) + n\cos(x)$

11. Let $T : P_2 \to P_2$ be the linear operator given by

$$T(\mathbf{p}) = \mathbf{p}\,(x+3)$$

This means that $T\left(ax^2 + bx + c\right) = a\,(x+3)^2 + b\,(x+3) + c$. Determine the transformation matrix **A** with respect to the ordered basis $B = \left[1, x, x^2\right]$.

Find $T\left(q + nx + mx^2\right)$ by using this matrix **A** and directly by applying the given transformation.

12. Let V be a vector space spanned by the set $\left\{\sin(x), \cos(x), e^x\right\}$. Let T be the differential linear operator $T : V \to V$ given by

$$T(\mathbf{f}) = \mathbf{f}' \text{ where } \mathbf{f}' \text{ is the derivative of } \mathbf{f}$$

Determine the transformation matrix **A** with respect to the ordered basis:

$$B = \left[\sin(x), \quad \cos(x), \quad e^x\right]$$

By using this transformation matrix **A**, find $T(\mathbf{g})$ where
(i) $\mathbf{g} = -\sin(x) + 4\cos(x) - 2e^x$ (ii) $\mathbf{g} = m\sin(x) + n\cos(x) + pe^x$

13. Let V be a vector space spanned by the set $\left\{e^{2x}, xe^{2x}, x^2e^{2x}\right\}$. Let T be the differential linear operator $T : V \to V$ given by

$$T\,(\mathbf{f}) = \mathbf{f}' \text{ where } \mathbf{f}' \text{ is the derivative of } \mathbf{f}$$

Determine the transformation matrix **A** with respect to the ordered basis:

$$B = \left[e^{2x}, xe^{2x}, x^2e^{2x}\right]$$

By using this matrix find $T\left(ae^{2x} + bxe^{2x} + cx^2e^{2x}\right)$.

14. Let $T : V \to V$ be the identity linear operator on an n-dimensional vector space V which is given by

$$T(\mathbf{v}) = \mathbf{v} \quad \text{for } all \text{ vectors } \mathbf{v} \text{ in } V$$

Prove that the matrix **A** for the linear transformation T with respect to any basis B is the identity matrix $\mathbf{I}_n$.

SECTION 5.6 Composition and Inverse Linear Transformations

By the end of this section you will be able to

- understand what is meant by composition of linear transforms
- prove some properties of composition of transformations
- prove that the inverse matrix represents the inverse transform T^{-1}

5.6.1 Composition of linear transformations (mappings)

Remember that linear transforms are functions and you should be familiar with the concept of a function.

What does composition mean?
Composition means making something by combining parts.

What do you think composition of linear transformation means?
It is the linear transformation created by putting together two or more linear transformations.

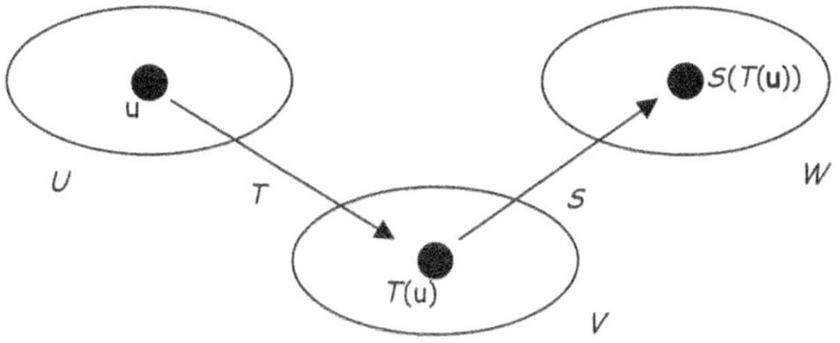

Figure 5.34

Fig. 5.34 shows a linear transformation $T : U \to V$ which takes the vector $\mathbf{u}$ to $T(\mathbf{u})$ and then the linear transformation $S : V \to W$ takes $T(\mathbf{u})$ to $S(T(\mathbf{u}))$. This composition of T and S is denoted by $S \circ T$. The formal definition is the following:

Definition (5.32). Let $T : U \to V$ and $S : V \to W$ be linear transforms and $\mathbf{u}$ and $T(\mathbf{u})$ be in the domain (start vector spaces) of U and V respectively. Then the **composition** of these two transforms, S and T, denoted by $S \circ T$ is defined by $S \circ T : U \to W$ (from U to W) and

$$(S \circ T)(\mathbf{u}) = S(T(\mathbf{u}))$$

What does $S(T(\mathbf{u}))$ mean?
Means *first* apply the transform T to the vector $\mathbf{u}$ to get $T(\mathbf{u})$ and *then* apply the transform S to this vector to give $S(T(\mathbf{u}))$.

Example 5.37

Let $T : \mathbb{R}^2 \to \mathbb{R}^2$ and $S : \mathbb{R}^2 \to \mathbb{R}^2$ be linear operators given by

$$T\left(\begin{bmatrix} x \\ y \end{bmatrix}\right) = \begin{pmatrix} x - y \\ x + y \end{pmatrix} \text{ and } S\left(\begin{bmatrix} x \\ y \end{bmatrix}\right) = \begin{pmatrix} 2x + 3y \\ x - 5y \end{pmatrix}$$

Determine $(S \circ T)\left(\begin{bmatrix} 2 \\ 3 \end{bmatrix}\right)$ and $(T \circ S)\left(\begin{bmatrix} 2 \\ 3 \end{bmatrix}\right)$.

Solution

$(S \circ T)\left(\begin{bmatrix} 2 \\ 3 \end{bmatrix}\right)$ means apply the transform T to the vector $\begin{bmatrix} 2 \\ 3 \end{bmatrix}$ and then apply transform S to this result. By using the above Definition (5.32) we have

$$\begin{aligned}
(S \circ T)\left(\begin{bmatrix} 2 \\ 3 \end{bmatrix}\right) &= S\left(T\left(\begin{bmatrix} 2 \\ 3 \end{bmatrix}\right)\right) \\
&= S\left(\begin{bmatrix} 2 - 3 \\ 2 + 3 \end{bmatrix}\right) && \left[\text{applying } T\left(\begin{bmatrix} x \\ y \end{bmatrix}\right) = \begin{pmatrix} x - y \\ x + y \end{pmatrix}\right] \\
&= S\left(\begin{bmatrix} -1 \\ 5 \end{bmatrix}\right) = \begin{pmatrix} 2(-1) + 3(5) \\ -1 - 5(5) \end{pmatrix} && \left[\text{applying } S\left(\begin{bmatrix} x \\ y \end{bmatrix}\right) = \begin{pmatrix} 2x + 3y \\ x - 5y \end{pmatrix}\right] \\
&= \begin{pmatrix} 13 \\ -26 \end{pmatrix}
\end{aligned}$$

Similarly, carrying out the transformations the other way, first S then T, we have [verify this]:

$$(T \circ S)\left(\begin{bmatrix} 2 \\ 3 \end{bmatrix}\right) = T\left(S\left(\begin{bmatrix} 2 \\ 3 \end{bmatrix}\right)\right) = \begin{pmatrix} 26 \\ 0 \end{pmatrix}$$

These vectors are illustrated in Fig. 5.35.

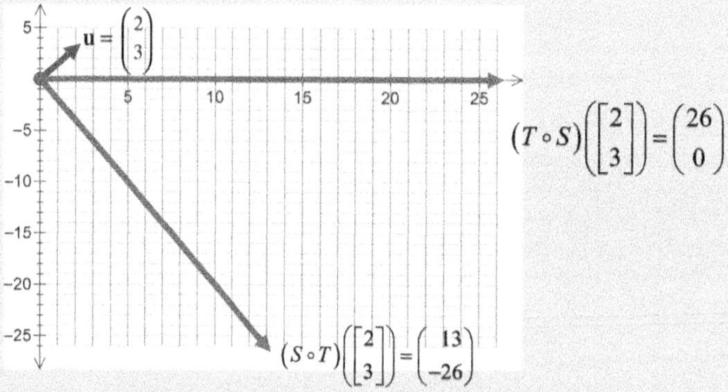

Figure 5.35

Note that $(S \circ T)\left(\begin{bmatrix} 2 \\ 3 \end{bmatrix}\right) = \begin{pmatrix} 13 \\ -26 \end{pmatrix}$ does *not equal* $(T \circ S)\left(\begin{bmatrix} 2 \\ 3 \end{bmatrix}\right) = \begin{pmatrix} 26 \\ 0 \end{pmatrix}$.

In general, $(S \circ T) \neq (T \circ S)$ [not equal]. This means that *changing* the order of transformation gives a *different* result.

Next, we prove that if S and T are linear transformations then the composite of these $S \circ T$ is also a linear transformation.

Proposition (5.33). Let $T : U \to V$ and $S : V \to W$ be linear transforms. Then the *composition transformation* $S \circ T : U \to W$ is also linear.

Remember, linear transformations preserve the fundamental linear algebra operations of vector addition and scalar multiplication. This proposition means that combining two linear transformations preserves scalar multiplication and vector addition. This is a useful property to have because we don't have to worry about the order of doing things.

How do we prove this result?

To prove that $S \circ T$ is a *linear* transformation we need to show two things:

1. $(S \circ T)(\mathbf{u} + \mathbf{v}) = (S \circ T)(\mathbf{u}) + (S \circ T)(\mathbf{v})$ [vector addition]
2. $(S \circ T)(k\mathbf{u}) = k(S \circ T)(\mathbf{u})$ [scalar multiplication]

where $\mathbf{u}$ and $\mathbf{v}$ are in the start vector space U of transform T and k is a scalar.

Proof.

1) Let $\mathbf{u}$ and $\mathbf{v}$ be vectors in U. Then by the above definition of composition (5.32) we have

$$\begin{aligned}(S \circ T)(\mathbf{u}+\mathbf{v}) &= S[T(\mathbf{u}+\mathbf{v})] \\ &= S[T(\mathbf{u}) + T(\mathbf{v})] && [\text{because } T \text{ is linear}] \\ &= S[T(\mathbf{u})] + S[T(\mathbf{v})] && [\text{because } S \text{ is linear}] \\ &= (S \circ T)(\mathbf{u}) + (S \circ T)(\mathbf{v}) && [\text{because } S(T(\mathbf{u})) = (S \circ T)(\mathbf{u})]\end{aligned}$$

2) Similarly we have

$$\begin{aligned}(S \circ T)(k\mathbf{u}) &= S(T(k\mathbf{u})) \\ &= S(k\,T(\mathbf{u})) && [\text{because } T \text{ is linear}] \\ &= k\,S(T(\mathbf{u})) && [\text{because } S \text{ is linear}] \\ &= k(S \circ T)(\mathbf{u}) && [\text{because } S(T(\mathbf{u})) = (S \circ T)(\mathbf{u})]\end{aligned}$$

Since both the above conditions, 1 and 2, are satisfied, $S \circ T$ is a linear transformation.

■

5.6.2 Matrix of composition linear transformations

In the last section we showed that we can write a linear transformation as a matrix. In this section we prove that the matrix of the *composite* transformation is the *multiplication* of the matrices that represent each transformation. Multiplying matrices is painless compared to combining linear transformations by using the given formula.

The next proposition is complicated because of the different bases used. If we used standard bases then the result would be a lot simpler – $(S \circ T)(\mathbf{u}) = \mathbf{BA}(\mathbf{u})$. However, we want to have the flexibility of using any bases.

Proposition (5.34). Let $T : U \to V$ and $S : V \to W$ be linear transforms and U, V and W be finite-dimensional vector spaces with bases B, C and D respectively. If

$$[T(\mathbf{u})]_C = \mathbf{A}[\mathbf{u}]_B \text{ and } [S(\mathbf{v})]_D = \mathbf{B}[\mathbf{v}]_C$$

where $\mathbf{u}$ is a vector in U and $\mathbf{v}$ is a vector in V then

$$[(S \circ T)(\mathbf{u})]_D = \mathbf{BA}[\mathbf{u}]_B$$

Here the bold $\mathbf{B}$ is a matrix and italic B is a basis. We could remove all the subscripts if we used the standard bases and the proposition would be $(S \circ T)(\mathbf{u}) = \mathbf{BA}(\mathbf{u})$.

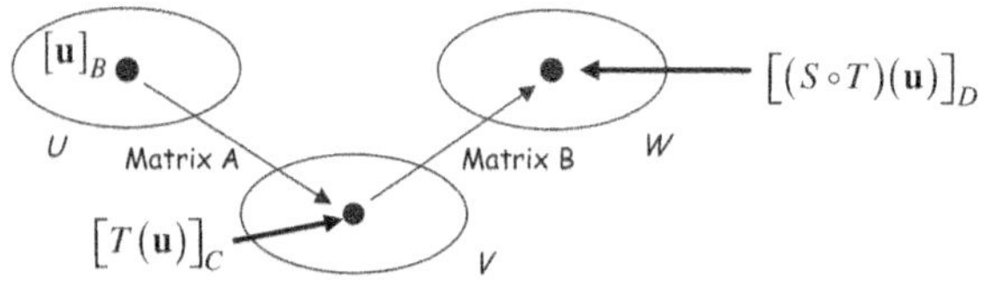

Figure 5.36

This means that the matrix multiplication $\mathbf{BA}$ represents the composite transform $S \circ T$.

Proof.

Let $\mathbf{u}$ be an arbitrary vector in the start vector space U. Then by assumption we have

$$[T(\mathbf{u})]_C = \mathbf{A}[\mathbf{u}]_B \qquad (*)$$

Let the vector $\mathbf{v} = T(\mathbf{u})$ be in the vector space V. Then by the other assumption we have

$$[S(T(\mathbf{u}))]_D = \mathbf{B}[T(\mathbf{u})]_C \qquad (**)$$

Note that $S(T(\mathbf{u})) = (S \circ T)(\mathbf{u})$ by the definition of composition. We have

$$\begin{aligned} [(S \circ T)(\mathbf{u})]_D &= [S(T(\mathbf{u}))]_D \\ &= \mathbf{B}[T(\mathbf{u})]_C && [\text{by } (**)] \\ &= \mathbf{B}(\mathbf{A}[\mathbf{u}]_B) = \mathbf{BA}[\mathbf{u}]_B && [\text{by } (*)] \end{aligned}$$

Hence this proves our required result that the composite transformation $S \circ T$ can be represented by the matrix multiplication $\mathbf{BA}$.

■

Example 5.38

Let $T : \mathbb{R}^2 \to \mathbb{R}^2$ and $S : \mathbb{R}^2 \to \mathbb{R}^2$ be linear operators given by:

$$T\left(\begin{bmatrix} x \\ y \end{bmatrix}\right) = \begin{pmatrix} x - y \\ x + y \end{pmatrix} \text{ and } S\left(\begin{bmatrix} x \\ y \end{bmatrix}\right) = \begin{pmatrix} 2x + 3y \\ x - 5y \end{pmatrix}$$

Determine the standard matrices for the compositions $(S \circ T)$ and $(T \circ S)$.

Also, find how the composite transforms act on vector $\mathbf{u} = \begin{pmatrix} 2 \\ 3 \end{pmatrix}$, $(S \circ T)(\mathbf{u})$ and $(T \circ S)(\mathbf{u})$.

Solution

How do we find the standard matrix for the given transformation T?

Read off the coefficients of x and y. Let $\mathbf{A}$ be this matrix then

$$\mathbf{A} = \begin{pmatrix} 1 & -1 \\ 1 & 1 \end{pmatrix} \quad \left[\text{because } T\left(\begin{bmatrix} x \\ y \end{bmatrix}\right) = \begin{pmatrix} x - y \\ x + y \end{pmatrix}\right]$$

Let $\mathbf{B}$ be the standard matrix for the other given linear transformation S:

$$\mathbf{B} = \begin{pmatrix} 2 & 3 \\ 1 & -5 \end{pmatrix} \quad \left[\text{because } S\left(\begin{bmatrix} x \\ y \end{bmatrix}\right) = \begin{pmatrix} 2x + 3y \\ x - 5y \end{pmatrix}\right]$$

By the above Proposition (5.34) and using the standard basis we have

$$\begin{aligned} (S \circ T)(\mathbf{u}) &= \mathbf{BA}[\mathbf{u}] \\ &= \begin{pmatrix} 2 & 3 \\ 1 & -5 \end{pmatrix}\begin{pmatrix} 1 & -1 \\ 1 & 1 \end{pmatrix}[\mathbf{u}] = \begin{pmatrix} 5 & 1 \\ -4 & -6 \end{pmatrix}[\mathbf{u}] \end{aligned}$$

The composite transform $S \circ T$ on vector $\mathbf{u} = \begin{bmatrix} 2 \\ 3 \end{bmatrix}$ is given by $(S \circ T)\left(\begin{bmatrix} 2 \\ 3 \end{bmatrix}\right)$:

$$(S \circ T)\left(\begin{bmatrix} 2 \\ 3 \end{bmatrix}\right) = \begin{pmatrix} 5 & 1 \\ -4 & -6 \end{pmatrix}\begin{pmatrix} 2 \\ 3 \end{pmatrix} = \begin{pmatrix} 13 \\ -26 \end{pmatrix}$$

Similarly, applying the composite transform $T \circ S$ on vector $\mathbf{u}$ gives

$$\begin{aligned} (T \circ S)(\mathbf{u}) &= \mathbf{AB}[\mathbf{u}] \\ &= \begin{pmatrix} 1 & -1 \\ 1 & 1 \end{pmatrix}\begin{pmatrix} 2 & 3 \\ 1 & -5 \end{pmatrix}[\mathbf{u}] = \begin{pmatrix} 1 & 8 \\ 3 & -2 \end{pmatrix}[\mathbf{u}] \end{aligned}$$

Also in the above manner we have

$$(T \circ S)\left(\begin{bmatrix} 2 \\ 3 \end{bmatrix}\right) = \begin{pmatrix} 1 & 8 \\ 3 & -2 \end{pmatrix}\begin{pmatrix} 2 \\ 3 \end{pmatrix} = \begin{pmatrix} 26 \\ 0 \end{pmatrix}$$

Note that these answers agree with the previous Example 5.37 because we have the same transformations.

The calculations in this example are easier than Example 5.37 because we have reduced the combining of linear transformations to a matrix arithmetic problem.

If a linear transformation represented a rotation, then a number of rotations can be evaluated by multiplying matrices. This is a much smoother task than applying the formula to the composition of many linear transformations.

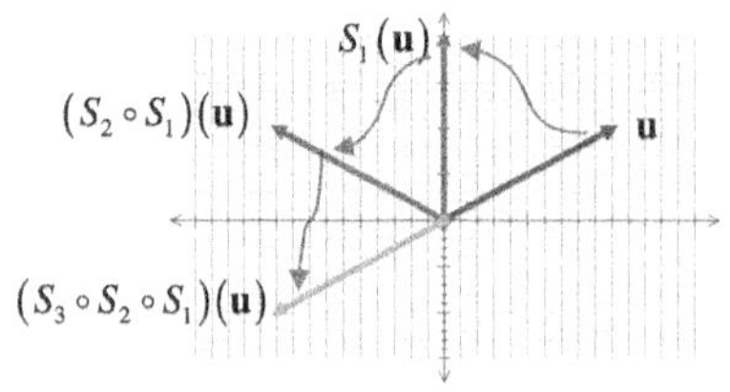

Figure 5.37

For example, consider the rotation transformations S_1, S_2 and S_3 on vector $\mathbf{u}$ (Fig. 5.37). If the transformation S_j is represented by the matrix $\mathbf{A}_j$ then

$$(S_3 \circ S_2 \circ S_1)(\mathbf{u}) = \underbrace{\mathbf{A}_3\mathbf{A}_2\mathbf{A}_1}_{\text{multiplication of matrices}} (\mathbf{u})$$

Multiplying the matrices on the right hand side is a lot simpler than evaluation of $S_3 \circ S_2 \circ S_1$.

Since matrix multiplication is *not* commutative so the composition of transforms is *not* commutative:

$$\mathbf{BA} \neq \mathbf{AB} \text{ [not equal], therefore } (T \circ S) \neq (S \circ T)$$

Changing the order of the transformations changes the resulting vector as you should have noticed for the above examples.

Example 5.39

Let V be the space spanned by $B = \{e^x, \quad xe^x, \quad x^2e^x\}$. Let $T : V \to V$ and $S : V \to V$ be defined by

$$T(\mathbf{p}) = \mathbf{p}' \text{ (differential operator)} \quad \text{and} \quad S(\mathbf{p}) = \mathbf{p} \text{ (identity operator)}$$

where $\mathbf{p} = ae^x + bxe^x + cx^2e^x$. Determine the matrix for the composite transform $S \circ T$ and find $(S \circ T)\left(e^x + 2xe^x + 3x^2e^x\right)$.

Solution

We need to find the matrix which represents the given differential operator T.

How?

By finding the transformation of each vector in $\{e^x, xe^x, x^2e^x\}$:

$$T\left(e^x\right) = \left(e^x\right)' = e^x \quad \text{[differentiating]}$$

$$T\left(xe^x\right) = \left(xe^x\right)' = e^x + xe^x \quad \text{[differentiating by product rule]}$$

$$T(x^2e^x) = \left(x^2e^x\right)' = 2xe^x + x^2e^x \quad \text{[differentiating by product rule]}$$

We need to write these in terms of the vectors $\{e^x, xe^x, x^2e^x\}$:

$$T(e^x) = e^x = 1(e^x) + 0(xe^x) + 0(x^2e^x)$$
$$T(xe^x) = e^x + xe^x = 1(e^x) + 1(xe^x) + 0(x^2e^x)$$
$$T(x^2e^x) = 2xe^x + x^2e^x = 0(e^x) + 2(xe^x) + 1(x^2e^x)$$

What is the matrix that represents the transform T equal to?
We use (5.31) of the last section: $\mathbf{A} = \left([T(\mathbf{v}_1)]_C \ \ [T(\mathbf{v}_2)]_C \ \ \cdots \ \ [T(\mathbf{v}_n)]_C\right)$

$$\mathbf{A} = \left(T(e^x) \ \ T(xe^x) \ \ T(x^2e^x)\right) = \begin{pmatrix} 1 & 1 & 0 \\ 0 & 1 & 2 \\ 0 & 0 & 1 \end{pmatrix}$$

What is the matrix for the identity transformation $S(\mathbf{p}) = \mathbf{p}$?
Let $\mathbf{B}$ be the matrix for this transformation S. By question 14 of Exercises 5.5:
If S is an identity linear operator then the matrix for S is the identity matrix $\mathbf{I}_n$.

$$\mathbf{B} = \mathbf{I} \qquad \text{[Identity matrix]}$$

Thus the matrix for composite transform $S \circ T$ is $\mathbf{BA}$ which is

$$\mathbf{BA} = \mathbf{IA} = \mathbf{A}$$

We need to evaluate $(S \circ T)\left(e^x + 2xe^x + 3x^2e^x\right)$. Rewriting

$$e^x + 2xe^x + 3x^2e^x = 1(e^x) + 2(xe^x) + 3(x^2e^x) = \mathbf{p}$$

Hence $[\mathbf{p}]_B = (1 \ \ 2 \ \ 3)^T$:

$$(S \circ T)\left(1(e^x) + 2(xe^x) + 3(x^2e^x)\right) = \mathbf{BA}[\mathbf{p}]_B \underset{\text{because } \mathbf{BA}=\mathbf{A}}{=} \mathbf{A}[\mathbf{p}]_B = \begin{pmatrix} 1 & 1 & 0 \\ 0 & 1 & 2 \\ 0 & 0 & 1 \end{pmatrix}\begin{pmatrix} 1 \\ 2 \\ 3 \end{pmatrix} = \begin{pmatrix} 3 \\ 8 \\ 3 \end{pmatrix}$$

Hence using these values gives

$$(S \circ T)\left(e^x + 2xe^x + 3x^2e^x\right) = 3e^x + 8xe^x + 3x^2e^x$$

Verify this result by applying the differential operator T to the vector $e^x + 2xe^x + 3x^2e^x$ and then the identity, and you should end up with the above result $3e^x + 8xe^x + 3x^2e^x$.

5.6.3 Invertible linear transformations

What is an invertible linear transformation?

From section 5.4.5 we have

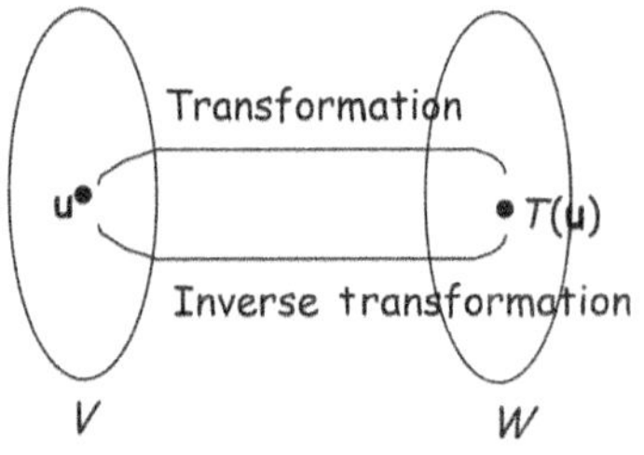

Figure 5.38

Fig. 5.38 shows a linear transformation $T : V \to W$ (from V to W) which takes $\mathbf{u}$ to $T(\mathbf{u})$. The inverse transformation denoted by $T^{-1} : W \to V$ (from W to V) takes $T(\mathbf{u})$ back to $\mathbf{u}$.

Definition (5.35). Let $T : V \to W$ be a linear transform. Then T is invertible $\Leftrightarrow$ there exists a transformation T^{-1} such that

(a) $(T^{-1} \circ T) = I_V$ where I_V is the identity transformation on the start vector space V

(b) $(T \circ T^{-1}) = I_W$ where I_W is the identity transformation on the arrival vector space W

If a linear transform T is invertible we say it is an **invertible linear transform**.

Proposition (5.36). Let $T : V \to W$ be a linear transform of finite-dimensional spaces V and W. Then T is **invertible** $\Leftrightarrow$ the matrix $\mathbf{A}$ which represents the transformation T is invertible.

Proof – Exercises 5.6.

■

We found an inverse transformation in section 5.4.5. We use the same technique established in that section to find the matrix for T^{-1} in the following example.

Example 5.40

Let $T : \mathbb{R}^2 \to \mathbb{R}^2$ be the linear operator given by $T\left(\begin{bmatrix} x \\ y \end{bmatrix}\right) = \begin{pmatrix} x - y \\ x + y \end{pmatrix}$.

(i) Determine the matrix for the linear operator $T^{-1} : \mathbb{R}^2 \to \mathbb{R}^2$.

(ii) Show that the identity matrix $\mathbf{I}$ represents the composite function $T \circ T^{-1}$.

T is the same transform given in Example 5.27 of section 5.4.5. *(continued...)*

Solution

(i) By Example 5.27 we have the inverse transformation T^{-1}:

$$T^{-1}\left(\begin{bmatrix} x \\ y \end{bmatrix}\right) = \frac{1}{2}\begin{pmatrix} x+y \\ -x+y \end{pmatrix}$$

What is the standard matrix, call it $\mathbf{A}$, which represents T^{-1}?

It is the coefficients of x and y, which means the matrix is $\mathbf{A} = \frac{1}{2}\begin{pmatrix} 1 & 1 \\ -1 & 1 \end{pmatrix}$.

What is the matrix, call it $\mathbf{B}$, which represents the given transformation $T\left(\begin{bmatrix} x \\ y \end{bmatrix}\right) = \begin{pmatrix} x-y \\ x+y \end{pmatrix}$?

Again, this is the coefficients of x and y as column vectors. We have $\mathbf{B} = \begin{pmatrix} 1 & -1 \\ 1 & 1 \end{pmatrix}$.

(ii) *How do we show that the identity matrix $\mathbf{I}$ represents the composite transform $T \circ T^{-1}$?*

By showing $\mathbf{BA} = \mathbf{I}$ because $\mathbf{BA}$ represents the composite transform $T \circ T^{-1}$. We have

$$\mathbf{BA} = \frac{1}{2}\begin{pmatrix} 1 & -1 \\ 1 & 1 \end{pmatrix}\begin{pmatrix} 1 & 1 \\ -1 & 1 \end{pmatrix} = \frac{1}{2}\begin{pmatrix} 2 & 0 \\ 0 & 2 \end{pmatrix} = \begin{pmatrix} 1 & 0 \\ 0 & 1 \end{pmatrix} = \mathbf{I}$$

In the next proposition, we state and prove that we can find the *inverse* transformation of a given transform by determining the *inverse* matrix $\mathbf{A}^{-1}$.

Proposition (5.37). Let $T : V \to V$ be an invertible linear transform (actually a linear operator) of finite-dimensional space V. Let $\mathbf{A}$ be a matrix representing the transform T with respect to a basis for V. Then $\mathbf{A}^{-1}$ is the matrix representing the inverse transform T^{-1} with respect to the same basis.

How do we prove this result?

We use the above Proposition (5.36) which says:

T is *invertible* $\Leftrightarrow$ the matrix $\mathbf{A}$ which represents the transformation T is *invertible.*

Proof.

By (5.36) we know that the matrix $\mathbf{A}$ is invertible. Since T is invertible we have

$$T^{-1} \circ T = I_V$$

This means that $T^{-1} \circ T$ is the identity transformation. Let $\mathbf{B}$ be the matrix representing T^{-1} then we need to show $\mathbf{B} = \mathbf{A}^{-1}$.

By Proposition (5.35) we have that the matrix multiplication $\mathbf{BA}$ represents the composite transform $T^{-1} \circ T$ which is the identity. Thus by the result of question 14 of the last Exercises 5.5 which says:

If T is an identity linear operator then the matrix for T is the identity matrix $\mathbf{I}_n$

We have $\mathbf{BA} = \mathbf{I}$, because $T^{-1} \circ T$ is the identity transformation.

What is the matrix **B** *equal to?*
Because $\mathbf{BA} = \mathbf{I}$, therefore $\mathbf{B} = \mathbf{A}^{-1}$ which is our required result.

■

Example 5.41

Let $T : \mathbb{R}^3 \to \mathbb{R}^3$ be the linear operator given by $T\left(\begin{bmatrix} x \\ y \\ z \end{bmatrix}\right) = \begin{pmatrix} x + 5y + 3z \\ 2x + 3y + z \\ 3x + 4y + z \end{pmatrix}$.

Determine the inverse transformation $T^{-1} : \mathbb{R}^3 \to \mathbb{R}^3$.

Solution

How do we find the inverse transformation T^{-1}?

First we write out the standard matrix for T and then we find the inverse of this matrix. The standard matrix for T is given by reading the coefficients of x, y and z:

$$\mathbf{A} = \begin{pmatrix} 1 & 5 & 3 \\ 2 & 3 & 1 \\ 3 & 4 & 1 \end{pmatrix} \quad \left[\text{Because } T\left(\begin{bmatrix} x \\ y \\ z \end{bmatrix}\right) = \begin{pmatrix} x + 5y + 3z \\ 2x + 3y + z \\ 3x + 4y + z \end{pmatrix}\right]$$

For the inverse transformation T^{-1} we need to find the inverse matrix $\mathbf{A}^{-1}$.

How?

We can use row operations or apply MATLAB. You can verify (in your own time) that the following is the inverse matrix $\mathbf{A}^{-1}$:

$$\mathbf{A}^{-1} = \begin{pmatrix} -1 & 7 & -4 \\ 1 & -8 & 5 \\ -1 & 11 & -7 \end{pmatrix}$$

What is the inverse transformation T^{-1} equal to?

The entries of the first column of this matrix $\mathbf{A}^{-1}$ are the x coefficients, the entries in the second column are the y coefficients and the entries in the last column are the z coefficients:

$$T^{-1}\left(\begin{bmatrix} x \\ y \\ z \end{bmatrix}\right) = \begin{pmatrix} -x + 7y - 4z \\ x - 8y + 5z \\ -x + 11y - 7z \end{pmatrix} \quad \left[\text{because } \mathbf{A}^{-1} = \begin{pmatrix} -1 & 7 & -4 \\ 1 & -8 & 5 \\ -1 & 11 & -7 \end{pmatrix}\right]$$

Finding the inverse matrix is *no* easy task but it is still simpler than trying to find the inverse transformation by transposing formulae.

Example 5.42

Let $T : P_2 \to P_2$ be the linear operator given by

$$T(\mathbf{p}) = (\mathbf{p}x)' \text{ where } \mathbf{p} = ax^2 + bx + c$$

and $B = \left[x^2, x, 1\right]$ be an ordered basis (axes) for P_2. Determine an expression for the inverse transformation $T^{-1}(\mathbf{p})$ with respect to the same basis B.

Solution

What do we need to find first?

The matrix which represents the given transformation T.

How can we locate this matrix?

By finding the transformation $T(\mathbf{p}) = (\mathbf{p}x)'$ of each basis (axes) vector in $B = [x^2, x, 1]$:

$$\begin{aligned} T(x^2) &= (x^2 x)' = (x^3)' = 3x^2 \\ T(x) &= (xx)' = (x^2)' = 2x \\ T(1) &= (1x)' = (x)' = 1 \end{aligned}$$

We need to write these in terms of the basis vectors $B = [x^2, x, 1]$:

$$\begin{aligned} T(x^2) &= 3x^2 = 3(x^2) + 0(x) + 0\,(1) \\ T(x) &= 2x = 0(x^2) + 2(x) + 0\,(1) \\ T(1) &= 1 = 0(x^2) + 0(x) + 1\,(1) \end{aligned}$$

What is the matrix $\mathbf{A}$ *which represents the given transformation T equal to?*

$$\mathbf{A} = \begin{pmatrix} 3 & 0 & 0 \\ 0 & 2 & 0 \\ 0 & 0 & 1 \end{pmatrix}$$

To find the inverse transformation T^{-1} we need to find $\mathbf{A}^{-1}$. By using row operations we have:

$$\mathbf{A}^{-1} = \begin{pmatrix} 1/3 & 0 & 0 \\ 0 & 1/2 & 0 \\ 0 & 0 & 1 \end{pmatrix}$$

With $\mathbf{p} = ax^2 + bx + c$ we can evaluate the coefficients of x^2, x and constants of the inverse transformation T^{-1} as follows:

$$\begin{pmatrix} 1/3 & 0 & 0 \\ 0 & 1/2 & 0 \\ 0 & 0 & 1 \end{pmatrix} \begin{pmatrix} a \\ b \\ c \end{pmatrix} = \begin{pmatrix} a/3 \\ b/2 \\ c \end{pmatrix}$$

The entries in the right hand vector are the coefficients of x^2, x and constants respectively. This means that the inverse transformation is given by

$$T^{-1}(\mathbf{p}) = T^{-1}(ax^2 + bx + c) = \frac{a}{3}x^2 + \frac{b}{2}x + c$$

Note that in the above Example 5.42 we were given the transformation

$$\begin{aligned} T(\mathbf{p}) = (\mathbf{p}x)' &= \left(\left[ax^2 + bx + c\right]x\right)' \\ &= \left(ax^3 + bx^2 + cx\right)' = 3ax^2 + 2bx + c \end{aligned}$$

In this transformation we have multiplied a general quadratic $ax^2 + bx + c$ by x and then found the derivative of this result. The inverse transformation is

$$T^{-1}(\mathbf{p}) = T^{-1}\left(ax^2 + bx + c\right) = \frac{a}{3}x^2 + \frac{b}{2}x + c$$

This is the same as integrating a general quadratic $ax^2 + bx + c$ which gives $\frac{a}{3}x^3 + \frac{b}{2}x^2 + cx$ and then dividing this result by x which yields $\frac{a}{3}x^2 + \frac{b}{2}x + c = T^{-1}(\mathbf{p})$. Note that this is the reverse process of the given transformation T. We can illustrate this as shown in Fig. 5.39.

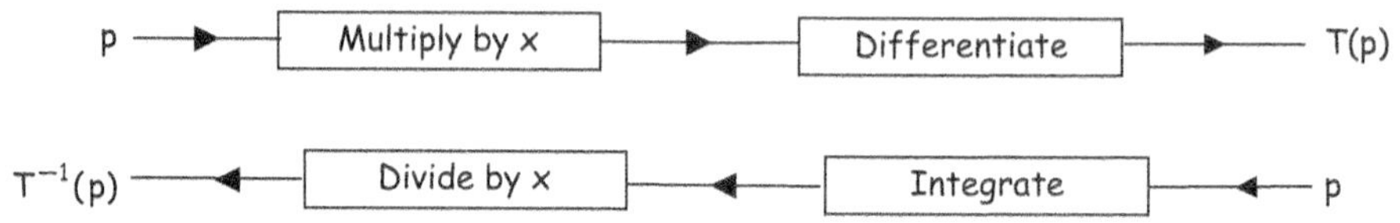

Figure 5.39

5.6.4 Operations of invertible linear operators

Proposition (5.38). Let $T : V \to V$ and $S : V \to V$ be invertible linear operators of finite-dimensional space V. Then $S \circ T$ is invertible and

$$(S \circ T)^{-1} = T^{-1} \circ S^{-1}$$

Remember, inverses undo the last operation first as illustrated in Fig. 5.39 above.

How do we prove this result?

We use the above propositions and the matrix result of chapter 1:

(1.27) $$(\mathbf{BA})^{-1} = \mathbf{A}^{-1}\mathbf{B}^{-1}$$

Proof.

Let the matrices **A** and **B** represent the given linear transformations T and S respectively. By Proposition (5.34)

$$(S \circ T)(\mathbf{u}) = \mathbf{BA}(\mathbf{u})$$

We know that the matrix multiplication **BA** represents the composite transform $S \circ T$. By applying the above Proposition (5.37):

If matrix **A** represents the transform T then $\mathbf{A}^{-1}$ represents the inverse T^{-1}

We have that the matrix representation of $(S \circ T)^{-1}$ is $(\mathbf{BA})^{-1}$. By the above (1.27) we have

$$(\mathbf{BA})^{-1} = \mathbf{A}^{-1}\mathbf{B}^{-1}$$

The matrix $\mathbf{A}^{-1}$ represents the transform T^{-1} and $\mathbf{B}^{-1}$ represents the transform S^{-1}. Thus $\mathbf{A}^{-1}\mathbf{B}^{-1}$ represents the composite transform $T^{-1} \circ S^{-1}$. This means that we have our required result

$$(S \circ T)^{-1} = T^{-1} \circ S^{-1}$$

■

Summary

The composite transform denoted $S \circ T$ is defined as

(5.32) $$(S \circ T)(\mathbf{u}) = S(T(\mathbf{u}))$$

(5.37) Let $\mathbf{A}$ be a matrix representing the transform T then $\mathbf{A}^{-1}$ is the matrix representing the inverse transform T^{-1}.

EXERCISES 5.6

(Brief solutions at end of book. Full solutions available at <http://www.oup.co.uk/companion/singh>.)

1. Consider the linear operators $T : \mathbb{R}^2 \to \mathbb{R}^2$ and $S : \mathbb{R}^2 \to \mathbb{R}^2$ given by

$$T\left(\begin{bmatrix} x \\ y \end{bmatrix}\right) = \begin{pmatrix} y \\ x \end{pmatrix} \text{ and } S\left(\begin{bmatrix} x \\ y \end{bmatrix}\right) = \begin{pmatrix} y \\ 0 \end{pmatrix}$$

   (i) Determine the vector $(S \circ T)\left(\begin{bmatrix} 1 \\ 2 \end{bmatrix}\right)$ and $(T \circ S)\left(\begin{bmatrix} 1 \\ 2 \end{bmatrix}\right)$.

   (ii) Determine the vector $(S \circ T)\left(\begin{bmatrix} 1 \\ 1 \end{bmatrix}\right)$ and $(T \circ S)\left(\begin{bmatrix} 1 \\ 1 \end{bmatrix}\right)$.

2. Let $T : P_2 \to P_1$ and $S : P_2 \to P_2$ be defined by

$$T(\mathbf{p}) = \mathbf{p}' \text{ and } S(\mathbf{p}) = ax^2 + bx \quad \text{where} \quad \mathbf{p} = ax^2 + bx + c$$

   Determine (a) $(T \circ S)(\mathbf{p})$ (b) $(S \circ T)(\mathbf{p})$ (c) $(T \circ T)(\mathbf{p})$ (d) $(S \circ S)(\mathbf{p})$

3. Let the linear operators $T : \mathbb{R}^2 \to \mathbb{R}^2$ and $S : \mathbb{R}^2 \to \mathbb{R}^2$ be given by

$$T\left(\begin{bmatrix} x \\ y \end{bmatrix}\right) = \begin{pmatrix} x + 2y \\ x - 3y \end{pmatrix} \text{ and } S\left(\begin{bmatrix} x \\ y \end{bmatrix}\right) = \begin{pmatrix} 2x + y \\ -x - y \end{pmatrix}$$

Determine the standard matrices for the compositions $S \circ T$ and $T \circ S$. Also find

(i) $(S \circ T)\left(\begin{bmatrix} 1 \\ 5 \end{bmatrix}\right)$ (ii) $(T \circ S)\left(\begin{bmatrix} 1 \\ 5 \end{bmatrix}\right)$ (iii) $(T \circ T)\left(\begin{bmatrix} 1 \\ 5 \end{bmatrix}\right)$ (iv) $(S \circ S)\left(\begin{bmatrix} 1 \\ 5 \end{bmatrix}\right)$

4. Let the linear transformations $T : \mathbb{R}^3 \to \mathbb{R}^3$ and $S : \mathbb{R}^3 \to \mathbb{R}^3$ be given by

$$T\left(\begin{bmatrix} x \\ y \\ z \end{bmatrix}\right) = \begin{pmatrix} 2x - y + 10z \\ 3x - 5y + 6z \\ x + 3y - 9z \end{pmatrix} \text{ and } S\left(\begin{bmatrix} x \\ y \\ z \end{bmatrix}\right) = \begin{pmatrix} -x - y - z \\ x + y + 2z \\ x + 2y + 3z \end{pmatrix}$$

Determine the standard matrices for the compositions $S \circ T$ and $T \circ S$. Also find

(i) $(S \circ T)\left(\begin{bmatrix} 1 \\ 2 \\ 3 \end{bmatrix}\right)$ (ii) $(T \circ S)\left(\begin{bmatrix} 1 \\ 2 \\ 3 \end{bmatrix}\right)$ (iii) $(T \circ T)\left(\begin{bmatrix} 1 \\ 2 \\ 3 \end{bmatrix}\right)$

(iv) $(S \circ S)\left(\begin{bmatrix} 1 \\ 2 \\ 3 \end{bmatrix}\right)$

5. (a) Let V be the space spanned by $\{e^x,\ xe^x,\ x^2e^x\}$. Let $T : V \to V$ and $S : V \to V$ be linear operators defined by

$$T(\mathbf{p}) = \mathbf{p}' \text{ and } S(\mathbf{p}) = ce^x + bxe^x + ax^2e^x \text{ where } \mathbf{p} = ae^x + bxe^x + cx^2e^x$$

Determine the matrix for

(i) $S \circ T$ (ii) $T \circ S$ (iii) $T \circ T$ (iv) $S \circ S$

What do you notice about the transformation $S \circ S$? Find $S^{-1}(\mathbf{p})$.

(b) Show that the linear operator T is invertible and find T^{-1}. Use T^{-1} to determine

$$\int x^2e^x dx$$

6. Let $T : P_2 \to P_2$ and $S : P_2 \to P_2$ be linear operators defined by

$$T\left(ax^2 + bx + c\right) = -ax^2 - bx - c \text{ and } S\left(ax^2 + bx + c\right) = -bx - c$$

and $B = \left[1,\ x,\ x^2\right]$ be an ordered basis for P_2.

Determine the matrix which represents the following transformations.

(i) $S \circ T$ (ii) $T \circ S$ (iii) $T \circ T$ (iv) $S \circ S$

Also find $(S \circ T)\left(1 + 2x + 3x^2\right)$, $(T \circ S)\left(1 + 2x + 3x^2\right)$, $(T \circ T)\left(1 + 2x + 3x^2\right)$ and $(S \circ S)\left(1 + 2x + 3x^2\right)$.

7. Let $T : \mathbb{R}^2 \to \mathbb{R}^2$ be the linear operator given by $T\left(\begin{bmatrix} x \\ y \end{bmatrix}\right) = \begin{pmatrix} x - y \\ x + y \end{pmatrix}$.

Determine the standard matrices for transformations T and $T^{-1} : \mathbb{R}^2 \to \mathbb{R}^2$.

8. Let $T : \mathbb{R}^3 \to \mathbb{R}^3$ be the linear operator given by $T\left(\begin{bmatrix} x \\ y \\ z \end{bmatrix}\right) = \begin{pmatrix} x + y + z \\ x + y - z \\ x - y - z \end{pmatrix}$. Find the inverse transformation $T^{-1} : \mathbb{R}^3 \to \mathbb{R}^3$. [Use MATLAB to find the inverse.]

9. Let $T : P_3 \to P_3$ be the linear operator defined by

$$T(\mathbf{p}) = (\mathbf{p}x)' \text{ where } \mathbf{p} = ax^3 + bx^2 + cx + d$$

and $B = \left[x^3, x^2, x, 1\right]$ be an ordered basis for P_3. Determine an expression for $T^{-1}(\mathbf{p})$ with respect to the same basis B.

10. Let $T : \mathbb{R}^3 \to \mathbb{R}^3$ be the linear operator given by

$$T\left(\begin{bmatrix} x \\ y \\ z \end{bmatrix}\right) = \begin{pmatrix} x + y + z \\ 2x + 2y + z \\ y + z \end{pmatrix}$$

Find $T^{-1} : \mathbb{R}^3 \to \mathbb{R}^3$.

11. Determine which of the following transformations are invertible. If they are invertible then find the inverse transformation.

(a) $T : \mathbb{R}^3 \to \mathbb{R}^2$ given by $T\left(\begin{bmatrix} x \\ y \\ z \end{bmatrix}\right) = \begin{pmatrix} x - z \\ x + y + z \end{pmatrix}$

(b) $T : \mathbb{R}^3 \to \mathbb{R}^3$ given by $T\left(\begin{bmatrix} x \\ y \\ z \end{bmatrix}\right) = \begin{pmatrix} 0 \\ x + y + z \\ x - y - z \end{pmatrix}$

(c) $T : P_3 \to P_3$ given by $T(\mathbf{p}) = \mathbf{p}$ where $\mathbf{p} = ax^3 + bx^2 + cx + d$.

In computer graphics, each pixel on a screen can be viewed as a point or a vector. The following matrices carry out the corresponding operations on the vector or point:

$\mathbf{R}_\theta = \begin{pmatrix} \cos(\theta) & -\sin(\theta) \\ \sin(\theta) & \cos(\theta) \end{pmatrix}$ *rotates a vector by an angle of θ anti-clockwise about the origin.*

$\mathbf{RF}_x = \begin{pmatrix} 1 & 0 \\ 0 & -1 \end{pmatrix}$ *reflects a vector in the x axis.*

12. Let $T, Q : \mathbb{R}^2 \to \mathbb{R}^2$ be the linear operators acting on $\mathbf{x} = (x \;\; y)^T$ be given by

$$T(\mathbf{x}) = \mathbf{R}_\theta \mathbf{x} \qquad \text{and} \qquad Q(\mathbf{x}) = \mathbf{RF}_x \mathbf{x}$$

(i) What is the net effect of $(Q \circ Q)(\mathbf{x})$?

(ii) Show that if we apply the linear operator Q an *odd* number of times on the vector $\mathbf{x}$ then the net effect is reflection in the x axis.

(ii) Show that the composite transform $T \circ T$ rotates the vector $\mathbf{x}$ by an angle of 2θ.

13. The probabilities of injury (I) and death (D) on urban roads (U) and motorways (M) are given by the matrix $\mathbf{A}$:

$$\mathbf{A} = \begin{matrix} & \begin{matrix} \text{I} & \text{D} \end{matrix} & \\ & \begin{pmatrix} 1/3 & 2/3 \\ 2/3 & 1/3 \end{pmatrix} & \begin{matrix} \text{M} \\ \text{U} \end{matrix} \end{matrix}$$

A linear operator $T : \mathbb{R}^2 \to \mathbb{R}^2$ is given by $T(\mathbf{x}) = \mathbf{A}\mathbf{x}$. The nth vector $\mathbf{x}_n$ is defined as

$$\mathbf{x}_n = \underbrace{T \circ T \circ T \circ \cdots \circ T}_{n \text{ copies}}(\mathbf{x}_0) \text{ where } \mathbf{x}_0 = (1 \quad 0)^T$$

(i) Use MATLAB or some other software to find x_1, x_2, x_3, x_4 and x_5 correct to 4dp.

(ii) By using MATLAB, or otherwise, predict a value for x_n for large n.

14. Let $T : V \to W$ be a linear transformation. Let I_V and I_W be the identity linear transformations on V and W respectively. Prove that
(i) $T \circ I_V = T$ (ii) $I_W \circ T = T$

15. Let $T : U \to V$ and $S : V \to W$ be linear transformations and U, $V$and W be finite-dimensional vector spaces. Prove that $(S \circ T) \neq (T \circ S)$.

16. Let $T : U \to V$ and $S : V \to W$ be linear transformations, and U, V and W be finite-dimensional vector spaces. Prove that

$$k\,(S \circ T) = (kS) \circ T = S \circ (kT) \text{ where } k \text{ is any scalar.}$$

17. Let $T : V \to V$ be an invertible linear operator. Prove that T^{-1} is linear.

18. Let $T : V \to V$ be an invertible linear operator. Prove that $\left(T^{-1}\right)^{-1} = T$.

19. Let $T : V \to W$ be an invertible linear transformation. Prove that the inverse transformation $T^{-1} : W \to V$ is unique.

20. Prove Proposition (5.36).

MISCELLANEOUS EXERCISES 5

(Brief solutions at end of book. Full solutions available at <http://www.oup.co.uk/companion/singh>.)

[In chapter 5 we used the term transformation to mean map.]

5.1. Show that the function $T : \mathbb{R}^2 \to \mathbb{R}^3$ is a linear transformation.

$$T\begin{bmatrix} x_1 \\ x_2 \end{bmatrix} = \begin{bmatrix} x_1 + 5x_2 \\ 0 \\ 2x_1 - 3x_2 \end{bmatrix}$$

University of Puget Sound, USA

5.2. Let $T : \mathbb{R}^3 \to \mathbb{R}^2$ be the map defined by

$$T[\,x_1,\, x_2,\, x_3\,]^T = [\,2x_1 - 3x_2 + 4x_3,\, -x_1 + x_2\,]^T$$

Show that T is a linear map and find its standard matrix.

University of Ottawa, Canada

5.3. Give an example or state no such example exists of:
A function $f : \mathbb{R}^2 \to \mathbb{R}^2$ that is not a linear transformation.

Illinois State University, USA
(part question)

5.4. Let $T(x_1,\ x_2,\ x_3) = (3x_2 + 2x_3,\ 3x_1 - 4x_2)$.

(a) Find the standard matrix for T.

(b) Is T one-to-one?

(c) Is T onto?

University of South Carolina, USA

5.5. (a) Let S denote the subset of **all** vectors $\vec{x}$ in $\mathbb{R}^2$ for which $\vec{x} \cdot \vec{u} = 0$ for some fixed $\vec{u} \in \mathbb{R}^2$. Is S a subspace of $\mathbb{R}^2$ or not? Give reasons for your answer.

(b) Let $\vec{u} = \begin{bmatrix} 1 \\ -1 \end{bmatrix}$ and $\vec{v} = \begin{bmatrix} 0 \\ -1 \end{bmatrix}$ be two vectors of $\mathbb{R}^2$, and let $T : \mathbb{R}^2 \to \mathbb{R}^2$ be a linear transformation for which $T\left(\vec{u}\right) = \begin{bmatrix} 2 \\ 1 \end{bmatrix}$ and $T\left(\vec{v}\right) = \begin{bmatrix} 1 \\ 3 \end{bmatrix}$.

(i) Find $T\left(\vec{e_1}\right)$ and $T\left(\vec{e_2}\right)$.

(ii) Find the matrix representation of T with respect to the standard basis $\left\{\vec{e_1}, \vec{e_2}\right\}$.

(iii) For what vector $\vec{w}$ of $\mathbb{R}^2$ is $T\left(\vec{w}\right) = \begin{bmatrix} 1 \\ 8 \end{bmatrix}$?

University of New Brunswick, Canada

5.6. Let the operator $S : \mathbb{R}^3 \to \mathbb{R}^4$ be defined by

$$S(x_1, x_2, x_3) = (x_1 - 4x_2 + 2x_3,\ 2x_1 + 7x_2 - x_3,\ -x_1 - 8x_2 + 2x_3,\ 2x_1 + x_2 + x_3)$$

(a) Find the standard matrix of S.

(b) Find a basis for the range of S.

University of Western Ontario, Canada

5.7. (a) Given the vector spaces, V, W, over the field K, state what is meant by a linear transformation $T : V \to W$.

(b) What is meant by the *kernel* of T, ker T?

(c) Show that T is injective if and only if ker $T = \{\mathbf{O}\}$.

(d) If $T : \mathbb{R}^3 \to \mathbb{R}^2$ is given by the formula,

$$T(x,\ y,\ z) = (x + y,\ x + y + z),$$

find ker T.

University of Sussex, UK

5.8. Let $T : \mathbb{R}^4 \to \mathbb{R}^3$ be a linear transformation defined by

$$T\left(\begin{bmatrix} x_1 \\ x_2 \\ x_3 \\ x_4 \end{bmatrix}\right) = \begin{pmatrix} 1 & 1 & 1 & 0 \\ 1 & 1 & -1 & 1 \\ 0 & 0 & 2 & 1 \end{pmatrix} \begin{pmatrix} x_1 \\ x_2 \\ x_3 \\ x_4 \end{pmatrix}$$

(a) Find a basis for the set of vectors in $\mathbb{R}^3$ that are in the image of T. Do the columns of the matrix associated with T span $\mathbb{R}^3$?

(b) Is T onto? Explain clearly why or why not.

(c) Find a basis for the null space of T.

(d) Is T one-to-one? Explain why or why not.

Mount Holyoke College, Massachusetts, USA

5.9. Suppose $T : \mathbb{R}^2 \to \mathbb{R}^3$ is a linear transformation such that $T\begin{bmatrix} 3 \\ -5 \end{bmatrix} = \begin{bmatrix} 1 \\ -1 \\ 2 \end{bmatrix}$ and $T\begin{bmatrix} -1 \\ 2 \end{bmatrix} = \begin{bmatrix} 3 \\ 0 \\ -2 \end{bmatrix}$. Determine $T\begin{bmatrix} 7 \\ -11 \end{bmatrix}$.

University of Puget Sound, USA

5.10. Let $L : \mathbb{R}^3 \to \mathbb{R}^3$ be a linear transformation such that

$$L\left(\begin{bmatrix} -2 \\ 1 \\ -2 \end{bmatrix}\right) = \begin{bmatrix} -3 \\ 1 \\ 2 \end{bmatrix}, \quad L\left(\begin{bmatrix} 3 \\ 2 \\ -1 \end{bmatrix}\right) = \begin{bmatrix} 2 \\ -2 \\ 1 \end{bmatrix}, \quad L\left(\begin{bmatrix} -1 \\ -1 \\ 1 \end{bmatrix}\right) = \begin{bmatrix} -1 \\ 2 \\ 4 \end{bmatrix}$$

Find $L\left(\begin{bmatrix} 1 \\ 1 \\ -1 \end{bmatrix}\right)$.

Purdue University, USA

5.11. Let $\mathbf{A}$ be a 6×5 matrix such that the nullity of $\mathbf{A}$ is 0. Let $T : \mathbb{R}^n \to \mathbb{R}^m$ be the linear transformation defined by $T(\mathbf{x}) = \mathbf{A}\mathbf{x}$. Answer each of the following questions.
Be sure to justify your answers.

(a) What is the value of m? What is the value of n?

(b) What is the maximum number of linearly independent vectors in the range of T?

(c) Is T onto?

(d) Is T one-to-one?

Illinois State University, USA

5.12. The images of the unit vectors in $\mathbb{R}^2$ under the linear transformation $T : \mathbb{R}^2 \to \mathbb{R}^3$ are given as $T(\mathbf{e}_1) = \begin{bmatrix} 2 \\ 1 \\ h \end{bmatrix}$, and $T(\mathbf{e}_2) = \begin{bmatrix} 3 \\ k \\ 0 \end{bmatrix}$. Determine all the values of the parameters h and k for which T is one-to-one.

Washington State University, USA

5.13. Let $C^1[0, 1]$ represent the vector space of continuously differentiable functions defined on the interval $[0, 1]$ and let $C[0, 1]$ be the vector space of continuous functions on $[0, 1]$. Define the transformation $T : C^1[0, 1] \to C[0, 1]$ by

$$f \to Tf \text{ where } (Tf)(x) = \frac{df}{dx}(x) - 2f(x) \text{ for } 0 \le x \le 1$$

Thus, $T(e^x) = \frac{d(e^x)}{dx} - 2e^x = -e^x$, i.e, $e^x \xrightarrow{T} -e^x$.

(a) What is the zero vector in the space $C[0, 1]$?

(b) Show that T is a linear transformation.

(c) Determine the kernel of T. What is dim ker T?

Harvey Mudd College, California, USA

5.14. If $T : V \to W$ is a linear mapping, show that

(a) $\ker(T)$ is a subspace of V.

(b) $\text{im}(T)$ is a subspace of W.

(c) For the linear transformation T given below, find the dimensions of the kernel and the image respectively.

$$T : \mathbb{R}^3 \to \mathbb{R}^3 \text{ given by } T(x,\ y,\ z) = (x + 2y - z,\ \ y + z,\ \ x + y - 2z);$$

University of Sussex, UK

5.15. Let $T : V \to W$ be a linear transformation from the vector spaces V to the vector space W. Let $S = \{\mathbf{v}_1, \mathbf{v}_2, \ldots, \mathbf{v}_n\}$ be a set of vectors in V. Suppose that the set of vectors $\{T(\mathbf{v}_1), T(\mathbf{v}_2), \ldots, T(\mathbf{v}_n)\}$ is a *linearly independent* set of vectors in W. Prove that S must be a linearly independent set in V. Produce a counter-example to show that the converse is generally false.

Harvey Mudd College, California, USA

5.16. Let $T : P_2 \to P_3$ be defined by

$$T\left(a_0 + a_1 t + a_2 t^2\right) := 3a_1 + 2a_2 t + a_0 t^2 + (a_1 + a_2)\, t^3$$

(a) This is a linear mapping. What are the two conditions that one has to check in order to prove this? Check one of them.

(b) Equip these two spaces P_2 and P_3 with the bases $B = \left\{1, t, t^2\right\}$ and $C = \left\{1, t, t^2, t^3\right\}$, respectively. Compute the matrix $\mathbf{A}$ that represents T with respect to these two bases.

(c) Demonstrate how to work with this matrix by computing $T\left(2 + 5t - t^2\right)$, using $\mathbf{A}$.

(d) Is it possible to do the same with $T(p) := t^3 + p(t)$? (If yes, don't do it!)

University of New York, USA

5.17. (a) Give the definition of a linear transformation T from $\mathbb{R}^n \to \mathbb{R}^k$.

(b) (i) Show that $T : \mathbb{R}^3 \to \mathbb{R}^4$ given by

$$T\left(\left[x, y, z\right]\right) = \left[x + 2y, 2y - 3z, z + x, x\right]$$

is a linear transformation.

(ii) Find the standard matrix representation of T.

(c) Let $T : \mathbb{R}^n \to \mathbb{R}^k$ be a linear transformation and let $\beta = \left\{\mathbf{b}_1, \mathbf{b}_2, \ldots, \mathbf{b}_n\right\}$ be a basis of $\mathbb{R}^n$. Show that, for any $\mathbf{v} \in \mathbb{R}^n$, the vector $T(\mathbf{v})$ is uniquely determined by the vectors $T(\mathbf{b}_1), T(\mathbf{b}_2), \ldots, T(\mathbf{b}_n)$.

University of Toronto, Canada

5.18. Consider the subspace V of $C(\mathbb{R})$ spanned by the set

$$\beta = \left\{1 + x, 1 - x, \quad e^x, \quad (1 + x)\, e^x\right\}$$

(a) Show that β is a basis for V (by showing that it is linearly independent).

(b) Find the coordinates $[\mathbf{u}]^\beta$ in the ordered basis β, where

$$\mathbf{u} = 1 - xe^x$$

(c) Let $D : V \to V$ be the differentiation operator $D : f \mapsto \frac{\mathrm{d}f}{\mathrm{d}x}$.

Find the matrix representation $[D]_\beta^\beta$.

(d) Verify that

$$[D\mathbf{u}]^\beta = [D]_\beta^\beta\, [\mathbf{u}]^\beta$$

Loughborough University, UK

5.19. (a) Determine whether the following maps are linear or not. Justify your answers.

(i) $f : \mathbb{R}^2 \to P_5 : (a, \ b) \to (a + b)x^5$.

(ii) $f : M(2, 2) \to \mathbb{R}^2 : \begin{pmatrix} a & b \\ c & d \end{pmatrix} \to (ad, \ bc)$.

(b) Let V, W be two real finite-dimensional vector spaces and let $f : V \to W$ be a linear map. Define what is meant by the image, the kernel, the rank and the nullity of f and state the rank-nullity theorem.

(c) Let $f : P_2 \to \mathbb{R}^3$ be a linear map with nullity $f = 1$. Determine whether the map f is injective, surjective, both or neither. Justify your answer.

(d) Consider the map $f : \mathbb{R}^4 \to \mathbb{R}^3$ given by

$$f(x, \ y, \ z, \ t) = (x + y, \ 0, \ z + t)$$

for all $(x, \ y, \ z, \ t) \in \mathbb{R}^4$. Determine whether f is injective, surjective, both or neither and find a basis for the kernel of f and a basis for the image of f.

City University, London, UK

5.20. Let $T_1\left(\begin{bmatrix} x \\ y \end{bmatrix}\right) = \begin{bmatrix} 2x+y \\ -x+y \end{bmatrix}$ and $T_2\left(\begin{bmatrix} x \\ y \end{bmatrix}\right) = \begin{bmatrix} x \\ x+2y \end{bmatrix}$.

Which of the following matrices represents the composition $T_1 \circ T_2$ with respect to the standard basis for $\mathbb{R}^2$?

A. $\begin{bmatrix} 3 & 1 \\ 0 & 3 \end{bmatrix}$ B. $\begin{bmatrix} 3 & 4 \\ 0 & 1 \end{bmatrix}$ C. $\begin{bmatrix} 3 & 2 \\ 0 & 2 \end{bmatrix}$ D. $\begin{bmatrix} 2 & 1 \\ 0 & 3 \end{bmatrix}$ E. $\begin{bmatrix} 2 & 0 \\ 1 & 3 \end{bmatrix}$

University of Toronto, Canada

5.21. Suppose S and T are the linear transformations given by

$$S\left((x_1,\ x_2,\ x_3)\right) = (3x_1 + 5x_2 - x_3,\ 4x_2 + 3x_3,\ x_1 - x_2 + 4x_3)$$

and $T(\mathbf{x}) = \mathbf{A}\mathbf{x}$, where

$$\mathbf{A} = \begin{pmatrix} 1 & 0 & 5 \\ 2 & -1 & 0 \\ 4 & 1 & 0 \end{pmatrix}$$

Find the matrix $\mathbf{C}$ so that $(S \circ T)(\mathbf{x}) = S\left(T(\mathbf{x})\right) = \mathbf{C}\mathbf{x}$.

University of Maryland, Baltimore County, USA

5.22. Let $\phi : \mathbb{R}^4 \to \mathbb{R}^4$ be the mapping given by $\phi(\mathbf{x}) = \mathbf{A}\mathbf{x}$ with

$$\mathbf{A} = \begin{pmatrix} 1 & 0 & -1 & 1 \\ 1 & 1 & 1 & 1 \\ 1 & 2 & 3 & 4 \\ -1 & 5 & 2 & 2 \end{pmatrix}$$

(a) Is ϕ a linear mapping? Justify your answer. Let $\mathbf{e}_1 = (1,\ 0,\ 0,\ 0)^T$. What is $\phi(\mathbf{e}_1)$?

(b) Compute the determinant of $\mathbf{A}$. *[Determinants are covered in the next chapter.]*

(c) Find a basis for the subspaces $\ker(\phi)$ and $\text{im}(\phi)$. Is the mapping ϕ invertible?

Jacobs University, Germany

5.23. (a) Let V be a vector space of dimension n with a basis

$$B = \left\{\vec{u_1}, \ldots, \vec{u_n}\right\}$$

(i) Give definition of a linear transformation on V.

(ii) If T is a linear transformation on V, define what is meant by the matrix of T with respect to the basis B.

(b) Let V be the vector space of all functions of the form $f(x) = a\sin x + b\cos x$ for arbitrary real constants a and b. Select in V the basis $B = \{\sin x, \cos x\}$.

Find the matrix $[T]_B$ of the linear transformation

$$T\left(f(x)\right) = f'(x) + f''(x)$$

with respect to the basis B. Here, $f'(x)$ and $f''(x)$ denote the first and the second derivative of the function $f(x)$,

$$f'(x) = \frac{d}{dx}f(x), \qquad f''(x) = \frac{d}{dx}f'(x) = \frac{d^2}{dx^2}f(x)$$

(c) Is the operator T from part (b) invertible? If so, find the matrix of the inverse operator T^{-1} in the basis B.

(d) Using the operator T^{-1}, find a function $f(x)$ in V such that

$$f''(x) + f'(x) = 2\sin x + 3\cos x$$

University of Manchester, UK

5.24. Let V be a vector space over $\mathbb{R}$, and let $T : V \to \mathbb{R}^n$ be a one-to-one linear transformation. Show that for $\mathbf{u}, \mathbf{v} \in V$ the formula

$$\langle \mathbf{u}, \mathbf{v} \rangle_V := \langle T(\mathbf{u}), T(\mathbf{v}) \rangle$$

defines an inner product on V. (Here $\langle -, - \rangle$ is the standard inner product on $\mathbb{R}^n$.)

University of California, Berkeley, USA

5.25. Let $\mathbf{v}_1$ and $\mathbf{v}_2$ be independent vectors in V and let $T : V \to W$ be a one-to-one linear transformation of V into W. Prove that $T(\mathbf{v}_1)$ and $T(\mathbf{v}_2)$ are independent vectors in W.

University of Toronto, Canada (part question)

5.26. True or false:

If V is a vector space and $T : V \to V$ is an injective linear transformation, then T is surjective. (Be careful.)

University of California, Berkeley, USA (part question)

5.27. Suppose A is a linear operator on a 10-dimensional space V, such that $A^2 = 0$.

(a) Show that $\mathrm{Im}(A) \subset Ker(A)$.

(b) Show that the rank of A is at most 5. (Hint: (a) and the rank-nullity theorem might help.)

Stanford University, USA (part question)

Sample questions

5.28. Let M_{nn} be the vector space of size n by n matrices. Let $\mathbf{B}$ be a matrix of M_{nn} and $T : M_{nn} \to M_{nn}$ be a transform such that $T(\mathbf{A}) = \mathbf{AB} + \mathbf{BA}$ where $\mathbf{A} \in M_{nn}$

Show that T is a linear transform.

5.29. Let $T : \mathbb{R}^2 \to \mathbb{R}^2$ be given by $T\left(\begin{pmatrix} x \\ y \end{pmatrix}\right) = \begin{pmatrix} e^x \\ e^y \end{pmatrix}$. Show that T is **not** linear.

Petros Drineas

is Associate Professor of Computer Science at the Rensselaer Polytechnic Institute in New York.

Tell us about yourself and your work.

Linear algebra is a fundamental tool in the design and analysis of algorithms for data mining and statistical data analysis applications: numerous modern, massive datasets are modelled as matrices. Naturally, a solid background in linear algebra is a must if your work will eventually involve processing large-scale data, and many highly rewarding jobs in the future will involve the analysis of Big Data. Think, for example, of the following data matrix: the rows of the matrix correspond to customers of your favourite online store, and the columns correspond to the products available at the store. Let the (i,j)-th entry of the matrix be the 'utility' of product j for customer i. Then, a recommender system like the ones used by many such online stores, seeks to recommend high utility products to customers, with the obvious hope that the customer will purchase the product. Since it is not possible to ask every customer to reveal his or her preference for every product, a recommender system tries to *infer* products of high utility for a customer from a small sample of known entries of the matrix. A lot of my work focuses on problems of this type: how can we reconstruct missing entries of a matrix from a small sample of rows and/or columns and/or elements of the matrix? Solving this problem better than the current state of the art would have made you a millionaire in the Netflix challenge a couple of years ago!

A related question focuses on identifying influential rows and/or columns and/or elements of a matrix. Imagine, for example, that you are interested in approximating the product of two matrices by selecting a few rows of the first matrix and a few columns of the second matrix. Which rows and or columns will return the best approximation? Such problems are of particular interest when dealing with massive matrices that cannot be easily stored in random access memory (RAM). Similar sampling-based algorithms are very useful in approximating singular values and singular vectors of matrices. These operations are of particular importance in data mining, since they lie at the heart of principal components analysis (PCA), a fundamental dimensionality reduction technique that helps visualize, denoise and interpret high-dimensional datasets.

The challenges in my line of work lie at the intersection of linear algebra and probability theory. This 'marriage' of the two domains as well as the formal study of the effects of sampling rows/columns/elements of matrices is relatively recent, and you can trace it to a few influential papers in the late 1990s. On the one hand, probability theory provides the fundamental tools that help quantify the 'distance' between the original matrix and the sampled one. On the other hand, linear algebraic tools (for example, matrix perturbation theory), allow us to quantify how characteristic properties (e.g. singular values, singular vectors, etc.) of the original matrix can be approximated using the sampled matrix.

Have you any particular messages that you would like to give to students starting off studying linear algebra?

You need to understand singular value decomposition (SVD)! This will probably be towards the end of an introductory linear algebra course, but it really should be the take-home message of any such course. It lies at the heart of so many things, from computing the rank of a matrix, to solving least-square problems and computing the inverse or pseudoinverse of a matrix. I will quote Prof. Dianne O'Leary of the University of Maryland: SVD is the Swiss Army knife and the Rolls Royce of numerical linear algebra. It has numerous uses (like a Swiss Army knife), but it is computationally expensive (like a Rolls Royce). I hope that some of my own work on randomized algorithms for the SVD has helped make it a bit more affordable!

6 Determinants and the Inverse Matrix

SECTION 6.1 Determinant of a Matrix

By the end of this section you will be able to

- evaluate the determinant of 2 by 2 matrices
- understand the geometric interpretation of a determinant

Figure 6.1 T. Seki 1642–1708.

(*Source:* http://turnbull.mcs.st-and.ac.uk/~history/).

T. Seki (Fig. 6.1) is credited with being the first person to study determinants in 1683. Seki taught himself mathematics from an early age after being initially introduced to the subject by a household servant. He was from a family of Samurai warriors.

The famous author Lewis Carroll, better known for his popular book *Alice's Adventures in Wonderland*, wrote a book on determinants called *An Elementary Theory of Determinants* in 1867.

Lewis Carroll proposed that Oxford University set up a mathematical institute 65 years before it was eventually built. He also wrote to the Dean proposing his salary be lowered from £300 to £200 per year because his College was suffering a financial crisis.

In this chapter we will find that every square matrix has a unique value associated with it, called the **determinant**. We can use this value to establish whether the matrix has an inverse or *not*, as well as finding whether the linear system has a *unique* solution. The determinant of a matrix **A** is used like a pregnancy test to see whether the linear system $\mathbf{Ax} = \mathbf{b}$ has a unique solution or not. Hence the determinant is useful whenever linear systems appear on the scene.

6.1.1 The determinant of a 2 by 2 matrix

We find the determinant of a 2 by 2 matrix in this section and then expand to 3 by 3, ..., n by n size matrices in the next section.

You can only find the determinant of a *square* matrix.

The determinant of a matrix **A** is normally denoted by det(**A**) and is a scalar *not* a matrix.

The **determinant** of the general 2 by 2 matrix $\mathbf{A} = \begin{pmatrix} a & b \\ c & d \end{pmatrix}$ is defined as:

(6.1) $$\det(\mathbf{A}) = ad - bc$$

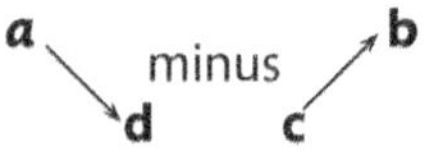

What does this formula (6.1) mean?

It means the determinant of a 2 by 2 matrix is the result of multiplying the entries of the leading diagonal and subtracting the product of the other diagonal.

Example 6.1

Find the determinant of $\mathbf{B} = \begin{pmatrix} 2 & 0 \\ 3 & 4 \end{pmatrix}$.

Solution

By applying the above formula (6.1) $\det\begin{pmatrix} a & b \\ c & d \end{pmatrix} = ad - bc$ we have

$$\det(\mathbf{B}) = \det\begin{pmatrix} 2 & 0 \\ 3 & 4 \end{pmatrix} = (2 \times 4) - (3 \times 0) = 8$$

Note that the determinant is a number (8), *not* a matrix.

Next, we look at the geometric interpretation of the determinant of matrix **B**.

6.1.2 Applications to transformations

Example 6.2

Let the matrix **A** represent the corners of the triangle *PQR* whose coordinates are given by $P(0, 0)$, $Q(2, 0)$ and $R(0, 3)$. Determine the image of this triangle under the transformation **BA** where once again, $\mathbf{B} = \begin{pmatrix} 2 & 0 \\ 3 & 4 \end{pmatrix}$.

By illustrating this transformation, determine the areas of the triangle *PQR* and the transformed triangle *P′Q′R′*.

How does this transformation change the size of the area?

Solution

We are given coordinates $P\ (0,\ 0)$, $Q(2,\ 0)$ and $R(0,\ 3)$, therefore $\mathbf{A} = \begin{pmatrix} 0 & 2 & 0 \\ 0 & 0 & 3 \end{pmatrix}$. Evaluating the matrix multiplication **BA**:

$$\mathbf{BA} = \begin{pmatrix} 2 & 0 \\ 3 & 4 \end{pmatrix}\overset{\begin{matrix} P & Q & R \end{matrix}}{\begin{pmatrix} 0 & 2 & 0 \\ 0 & 0 & 3 \end{pmatrix}} = \overset{\begin{matrix} P' & Q' & R' \end{matrix}}{\begin{pmatrix} 0 & 4 & 0 \\ 0 & 6 & 12 \end{pmatrix}}$$

The plot of this is shown in Fig. 6.2.

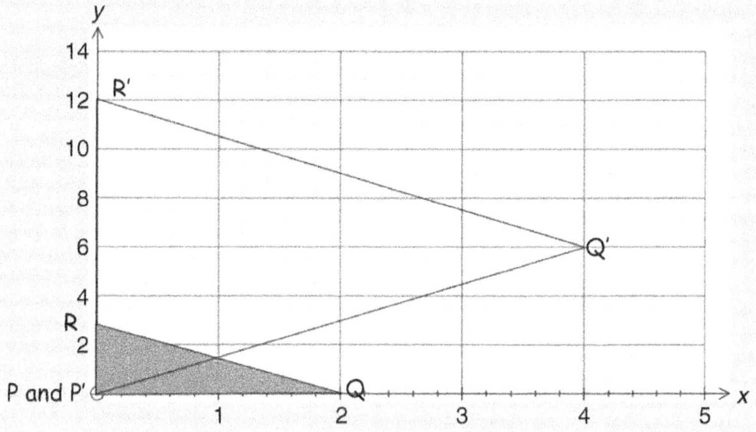

Figure 6.2

The area of shaded triangle $PQR = \frac{2 \times 3}{2} = 3$ and area of large triangle $P'Q'R' = \frac{12 \times 4}{2} = 24$.

The transformation **B** increases the original area by a factor of 8 because $24/3 = 8$.

This ratio 8 is the **area scale factor** and is equal to the *determinant* of matrix **B** which was evaluated in Example 6.1:

Determinant of matrix $\begin{pmatrix} 2 & 0 \\ 3 & 4 \end{pmatrix} = (2 \times 4) - (3 \times 0) = 8$

The determinant of a 2 by 2 matrix is the *area scale factor.*

The points $\begin{pmatrix} 0 \\ 0 \end{pmatrix}, \begin{pmatrix} 1 \\ 0 \end{pmatrix}, \begin{pmatrix} 0 \\ 1 \end{pmatrix}, \begin{pmatrix} 1 \\ 1 \end{pmatrix}$ form a unit square with an area of 1, as shown in Fig. 6.3(a).

What affect does multiplying the matrix $\mathbf{A} = \begin{pmatrix} 3 & 1 \\ 1 & 4 \end{pmatrix}$ *have on this square?*

$$\begin{pmatrix} 3 & 1 \\ 1 & 4 \end{pmatrix} \overset{\begin{matrix} P & Q & R & S \end{matrix}}{\begin{pmatrix} 0 & 1 & 0 & 1 \\ 0 & 0 & 1 & 1 \end{pmatrix}} = \begin{pmatrix} 0 & 3 & 1 & 3+1 \\ 0 & 1 & 4 & 1+4 \end{pmatrix} = \overset{\begin{matrix} P' & Q' & R' & S' \end{matrix}}{\begin{pmatrix} 0 & 3 & 1 & 4 \\ 0 & 1 & 4 & 5 \end{pmatrix}}$$

The transformation of the unit square to the parallelogram is shown in Fig. 6.3(b).

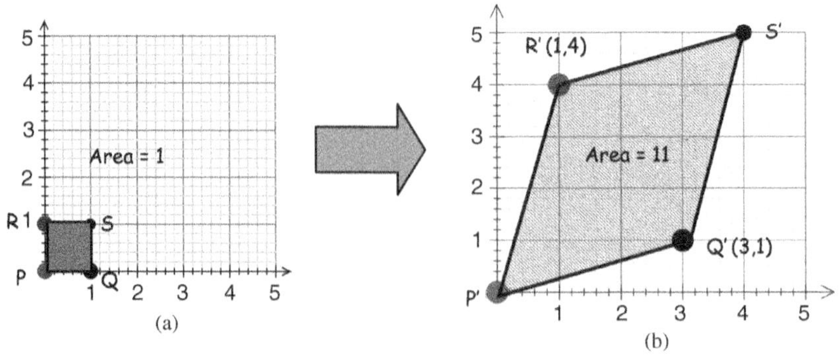

Figure 6.3

The matrix $\mathbf{A} = \begin{pmatrix} 3 & 1 \\ 1 & 4 \end{pmatrix}$ changes the area by a factor of 11 because

$$\frac{\text{Area of } P'Q'R'S'}{\text{Area of } PQRS} = \frac{11}{1} = 11$$

Do you notice any relationship between the matrix $\mathbf{A}$ *and the area of 11?*

11 is the determinant of the matrix $\mathbf{A}$ because $\det\begin{pmatrix} 3 & 1 \\ 1 & 4 \end{pmatrix} = (3 \times 4) - (1 \times 1) = 11$. Hence the area of the parallelogram, 11, is equal to the determinant of the matrix $\mathbf{A}$.

Consider the general 2 by 2 matrix $\mathbf{A} = \begin{pmatrix} a & b \\ c & d \end{pmatrix}$. Transforming the above unit square *PQRS* by multiplying by this matrix $\mathbf{A}$ gives :

$$\begin{pmatrix} a & b \\ c & d \end{pmatrix}\begin{pmatrix} 0 & 1 & 0 & 1 \\ 0 & 0 & 1 & 1 \end{pmatrix} = \begin{pmatrix} 0 & a & b & a+b \\ 0 & c & d & c+d \end{pmatrix}$$

We can illustrate this as shown in Fig. 6.4.

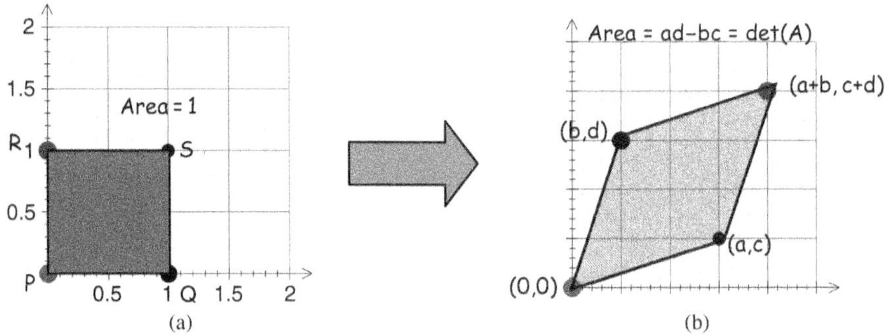

Figure 6.4

The matrix $\mathbf{A} = \begin{pmatrix} a & b \\ c & d \end{pmatrix}$ changes the area by a factor of $ad - bc$ because

$$\frac{\text{Area of parallelogram}}{\text{Area of square}} = \frac{ad - bc}{1} = ad - bc$$

This $ad - bc$ is the *determinant* of the matrix $\mathbf{A}$; depending on the values we choose for the symbols $a,\ b,\ c$ and d the shape may change but the area is always $ad - bc$. Hence $\det(\mathbf{A})$ represents the shaded area shown in Fig. 6.4(b).

What does a determinant equal to 1 mean in relation to transformations?
Applying a matrix with a determinant of 1 means the total area will *remain the same* under the transformation. For example, a simple rotation leaves the area unchanged, so any matrix that achieves a pure rotation will have a determinant of 1.

What does a negative determinant mean?
It changes the area over as shown in Fig. 6.5(c).

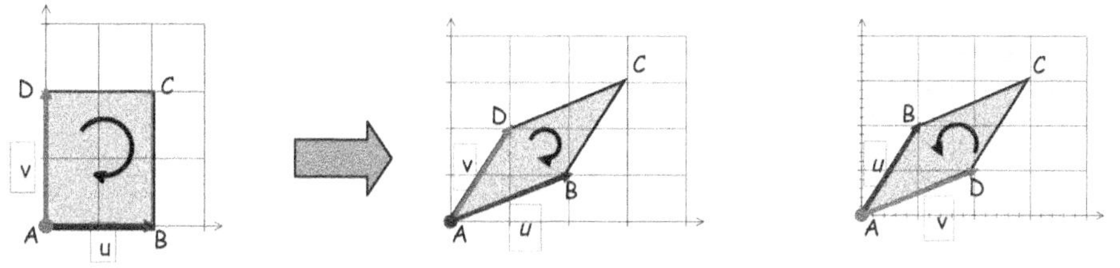

(a) (b) determinant is *positive* (c) determinant is *negative*

Figure 6.5

A negative determinant *changes* the *orientation* of the area. If matrix $\mathbf{A} = (\mathbf{u}\ \ \mathbf{v})$ and the determinant of matrix $\mathbf{A} = (\mathbf{u}\ \ \mathbf{v})$ is positive then *swapping* vectors gives a *negative* determinant, that is $\det(\mathbf{v}\ \ \mathbf{u})$ is negative.

6.1.3 Inverse of a matrix

Example 6.3

Let $\mathbf{A} = \begin{pmatrix} a & b \\ c & d \end{pmatrix}$ and $\mathbf{B} = \dfrac{1}{\det(\mathbf{A})}\begin{pmatrix} d & -b \\ -c & a \end{pmatrix}$. Evaluate the matrix multiplication $\mathbf{AB}$ provided $\det(\mathbf{A}) \neq 0$.
What do you notice about your result?

Solution

$$\mathbf{AB} = \begin{pmatrix} a & b \\ c & d \end{pmatrix} \times \frac{1}{\det(\mathbf{A})}\begin{pmatrix} d & -b \\ -c & a \end{pmatrix}$$

$$= \frac{1}{\det(\mathbf{A})}\begin{pmatrix} a & b \\ c & d \end{pmatrix}\begin{pmatrix} d & -b \\ -c & a \end{pmatrix} \qquad \left[\begin{matrix}\text{remember } \mathbf{A}(k\mathbf{B}) = k\,(\mathbf{AB}) \\ \text{where } k = 1/\det(\mathbf{A}) \text{ is a scalar}\end{matrix}\right]$$

$$= \frac{1}{\det(\mathbf{A})}\begin{pmatrix} ad - bc & 0 \\ 0 & ad - bc \end{pmatrix}$$

$$= \frac{1}{\det(\mathbf{A})}(ad - bc)\begin{pmatrix} 1 & 0 \\ 0 & 1 \end{pmatrix} = \frac{1}{\det(\mathbf{A})}\det(\mathbf{A})\mathbf{I} \qquad \left[\begin{matrix}\text{taking out the common} \\ \text{factor } ad - bc = \det(\mathbf{A})\end{matrix}\right]$$

$$= \mathbf{I} \qquad \left[\text{cancelling out } \det(\mathbf{A})\right]$$

Note that the matrix multiplication $\mathbf{AB} = \mathbf{I}$.

Since $\mathbf{AB} = \mathbf{I}$*, what conclusions can we draw about the matrices* $\mathbf{A}$ *and* $\mathbf{B}$*?*
$\mathbf{B}$ is the inverse of matrix $\mathbf{A}$, that is $\mathbf{B} = \mathbf{A}^{-1}$.

This means the inverse of the general matrix $\mathbf{A} = \begin{pmatrix} a & b \\ c & d \end{pmatrix}$ is defined by:

$$\mathbf{A}^{-1} = \frac{1}{\det(\mathbf{A})}\begin{pmatrix} d & -b \\ -c & a \end{pmatrix} \text{ provided } \det(\mathbf{A}) \neq 0 \qquad (6.2)$$

The inverse of a 2×2 matrix is calculated by interchanging entries along the leading diagonal and placing a negative sign next to the other entries, and then dividing this by the scalar $\det(\mathbf{A})$.

What does (6.2) imply?
It means that we can find the inverse matrix $\mathbf{A}^{-1}$, such that $\mathbf{A}^{-1}\mathbf{A} = \mathbf{I}$. Being able to find and use the inverse of a matrix can make solving some equations much easier. Furthermore, if a linear transformation $T(\mathbf{x}) = \mathbf{Ax}$ is applied to an object whose vertices (corners) are the vectors $\mathbf{x}$ then this transformation expands the area of the object by $\det(\mathbf{A})$. This means that $T^{-1}(\mathbf{x}) = \mathbf{A}^{-1}(\mathbf{x})$ must reverse this expansion, so we divide by $\det(\mathbf{A})$ as you can see in formula (6.2).

In the above we have described what is meant by a negative determinant and a determinant of 1.

What can we say if the determinant is zero, that is $\det(\mathbf{A}) = 0$*?*
If $\det(\mathbf{A}) = 0$ then the matrix $\mathbf{A}$ is non-invertible (singular) - it has *no inverse*. The geometric significance of this can be seen by examining transformations.

Consider the image of the triangle *PQR*, represented by matrix $\mathbf{A} = \begin{pmatrix} 0 & 2 & 0 \\ 0 & 0 & 3 \end{pmatrix}$ under the transformation $\mathbf{BA}$ where $\mathbf{B} = \begin{pmatrix} 3 & 2 \\ 6 & 4 \end{pmatrix}$. Multiplying the matrices, we have (Fig. 6.6):

$$\mathbf{BA} = \begin{pmatrix} 3 & 2 \\ 6 & 4 \end{pmatrix}\overset{\begin{matrix} P & Q & R \end{matrix}}{\begin{pmatrix} 0 & 2 & 0 \\ 0 & 0 & 3 \end{pmatrix}} = \overset{\begin{matrix} P' & Q' & R' \end{matrix}}{\begin{pmatrix} 0 & 6 & 6 \\ 0 & 12 & 12 \end{pmatrix}}$$

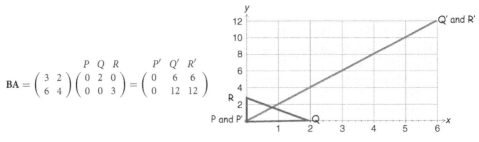

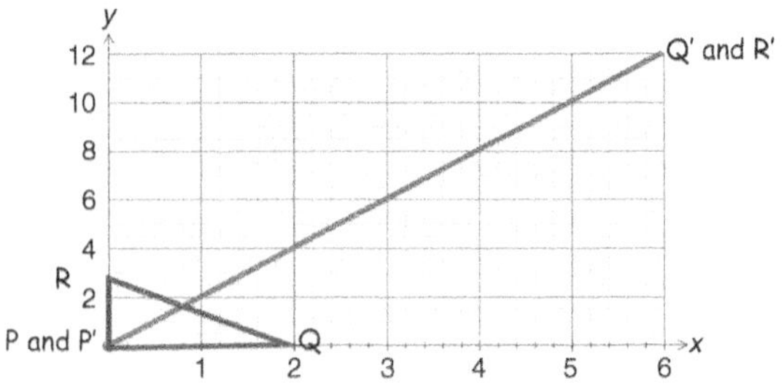

Figure 6.6

Under the transformation performed by **B**, the triangle represented by matrix **A** becomes a straight line $P'Q'R'$ which means that it has *no* area. The transformation **B** collapses the area of triangle to zero so that matrix **B** *cannot* have an inverse because there is *no* matrix that can transform the one-dimensional line back to the original two-dimensional object. The area is increased by a factor of *zero*, therefore we expect $\det(\mathbf{B}) = 0$ and we can confirm this by using the formula:

$$\det\begin{pmatrix} 3 & 2 \\ 6 & 4 \end{pmatrix} = (3 \times 4) - (6 \times 2) = 0$$

Example 6.4

Find the inverses of the following matrices:

(a) $\mathbf{A} = \begin{pmatrix} 2 & 3 \\ -1 & 5 \end{pmatrix}$ **(b)** $\mathbf{B} = \begin{pmatrix} \sqrt{2} & 1 \\ -1 & \sqrt{2} \end{pmatrix}$ **(c)** $\mathbf{C} = \begin{pmatrix} \pi & \pi \\ \pi & \pi \end{pmatrix}$

Solution

(a) When finding the inverse, we should evaluate the determinant first.

Why?

Because finding the determinant of a 2 by 2 matrix is simple, and if it turns out to be 0 then the matrix does *not* have an inverse and we can save ourselves a lot of unnecessary additional calculation.

Therefore by (6.1) we have

$$\det(\mathbf{A}) = \det\begin{pmatrix} 2 & 3 \\ -1 & 5 \end{pmatrix} = (2 \times 5) - (-1 \times 3) = 13$$

The inverse matrix $\mathbf{A}^{-1}$ is given by the above formula (6.2) with $\det(\mathbf{A}) = 13$:

$$\mathbf{A}^{-1} = \begin{pmatrix} 2 & 3 \\ -1 & 5 \end{pmatrix}^{-1} = \frac{1}{13}\begin{pmatrix} 5 & -3 \\ 1 & 2 \end{pmatrix} \quad \left[\text{By (6.2)} \quad \begin{pmatrix} a & b \\ c & d \end{pmatrix}^{-1} = \frac{1}{\det(\mathbf{A})}\begin{pmatrix} d & -b \\ -c & a \end{pmatrix}\right]$$

(b) We adopt the same procedure as part (a) to find $\mathbf{B}^{-1}$. Applying (6.1) $\det(\mathbf{B}) = ad - bc$:

$$\det(\mathbf{B}) = \det\begin{pmatrix} \sqrt{2} & 1 \\ -1 & \sqrt{2} \end{pmatrix} = \left(\sqrt{2} \times \sqrt{2}\right) - (-1 \times 1) = 2 + 1 = 3$$

By substituting $\det(\mathbf{B}) = 3$ into the inverse formula (6.2) we have

$$\mathbf{B}^{-1} = \begin{pmatrix} \sqrt{2} & 1 \\ -1 & \sqrt{2} \end{pmatrix}^{-1} = \frac{1}{3}\begin{pmatrix} \sqrt{2} & -1 \\ 1 & \sqrt{2} \end{pmatrix} \quad \left[\text{By (6.2)} \quad \begin{pmatrix} a & b \\ c & d \end{pmatrix}^{-1} = \frac{1}{\det(\mathbf{B})}\begin{pmatrix} d & -b \\ -c & a \end{pmatrix}\right]$$

(c) Similarly applying (6.1) $\det(\mathbf{C}) = ad - bc$ we have

$$\det(\mathbf{C}) = \det\begin{pmatrix} \pi & \pi \\ \pi & \pi \end{pmatrix} = (\pi \times \pi) - (\pi \times \pi) = 0$$

What can we conclude about the matrix **C**?

Since $\det(\mathbf{C}) = 0$, the matrix **C** does *not* have an inverse.

6.1.4 Wronskian determinant

The Wronskian determinant is used in differential equations to test whether solutions to the differential equation are *linearly independent*. If the Wronskian determinant is *non-zero* then the solutions are *linearly independent*. Let $f(x)$ and $g(x)$ be two solutions to a differential equation then the Wronskian $W(f, g)$ is defined by:

$$W(f, g) = \det \begin{pmatrix} f(x) & g(x) \\ f'(x) & g'(x) \end{pmatrix}$$

where $f'(x)$ and $g'(x)$ are the derivatives of $f(x)$ and $g(x)$ respectively. If $W(f,g) \neq 0$ then f and g are linearly independent.

For example, the Wronskian $W(\cos(x), \sin(x))$ is given by:

$$\begin{aligned} W\left(\cos(x), \sin(x)\right) &= \det \begin{pmatrix} \cos(x) & \sin(x) \\ -\sin(x) & \cos(x) \end{pmatrix} \qquad \begin{bmatrix} \text{because } \left(\cos(x)\right)' = -\sin(x) \\ \text{and } \left(\sin(x)\right)' = \cos(x) \end{bmatrix} \\ &= \cos^2(x) + \sin^2(x) = 1 \neq 0 \end{aligned}$$

Remember, the fundamental trigonometric identity $\cos^2(x) + \sin^2(x) = 1$. Hence sine and cosine are linearly independent solutions because $W\left(\cos(x), \sin(x)\right) \neq 0$ [Non-zero]

Summary

We can only find the determinant of a square matrix.

Let $\mathbf{A} = \begin{pmatrix} a & b \\ c & d \end{pmatrix}$ then the determinant and inverse of matrix $\mathbf{A}$ are given by

$$\det(\mathbf{A}) = ad - bc \quad \text{and} \quad \mathbf{A}^{-1} = \frac{1}{\det(\mathbf{A})} \begin{pmatrix} d & -b \\ -c & a \end{pmatrix} \quad \text{provided } \det(\mathbf{A}) \neq 0$$

EXERCISES 6.1

(Brief solutions at end of book. Full solutions available at <http://www.oup.co.uk/companion/singh>.)

You may like to check your numerical answers using MATLAB.

1. Compute the determinant of the following matrices.

   (a) $\mathbf{A} = \begin{pmatrix} 7 & 9 \\ 5 & 7 \end{pmatrix}$ (b) $\mathbf{B} = \begin{pmatrix} 9 & 2 \\ 13 & 3 \end{pmatrix}$ (c) $\mathbf{C} = \begin{pmatrix} 17 & 7 \\ 12 & 5 \end{pmatrix}$ (d) $\mathbf{D} = \begin{pmatrix} 7 & 1 \\ 14 & 2 \end{pmatrix}$

2. Compute $\det(\mathbf{A})$ and $\det(\mathbf{B})$ for the following:

   (a) $\mathbf{A} = \begin{pmatrix} 1 & 3 \\ 5 & 7 \end{pmatrix}$ and $\mathbf{B} = \begin{pmatrix} 1 & 5 \\ 3 & 7 \end{pmatrix}$ (b) $\mathbf{A} = \begin{pmatrix} -1 & 2 \\ 5 & 3 \end{pmatrix}$ and $\mathbf{B} = \begin{pmatrix} -1 & 5 \\ 2 & 3 \end{pmatrix}$

(c) $\mathbf{A} = \begin{pmatrix} \cos(-\pi) & \cos(\pi) \\ \sin(-\pi) & \sin(\pi) \end{pmatrix}$ and $\mathbf{B} = \begin{pmatrix} \cos(-\pi) & \sin(-\pi) \\ \cos(\pi) & \sin(\pi) \end{pmatrix}$

What do you notice about matrices **A** *and* **B** *and your results for* det(**A**) *and* det(**B**)?

3. For $\mathbf{A} = \begin{pmatrix} a & b \\ c & d \end{pmatrix}$. Show that $\det(\mathbf{A}) = \det\left(\mathbf{A}^T\right)$.

4. Let $\mathbf{A} = \begin{pmatrix} 2 & 1 \\ 0 & 3 \end{pmatrix}$. Show that det(**A**) is the area of the parallelogram which connects the column vectors, $\begin{pmatrix} 2 \\ 0 \end{pmatrix}$ and $\begin{pmatrix} 1 \\ 3 \end{pmatrix}$, of matrix **A**.

   What is $\det\begin{pmatrix} 1 & 2 \\ 3 & 0 \end{pmatrix}$ *equal to?*

5. By using determinants decide whether the following linear system has a unique solution? If so find it.

$$\begin{aligned} 2x + 3y &= -2 \\ 5x - 2y &= 14 \end{aligned}$$

6. Let $T: \mathbb{R}^2 \to \mathbb{R}^2$ be a linear transformation given by $T(\mathbf{x}) = \mathbf{Ax}$. Determine a formula for $\det\left(\mathbf{A}^2\right)$ in terms of det(**A**).
7. Show that the Wronskian $W\left(e^{-x}, e^{-3x}\right) \neq 0$.
8. Decide whether the transformation $T: M_{22} \to \mathbb{R}$ given by $T(\mathbf{A}) = \det(\mathbf{A})$ is linear or not.

SECTION 6.2 Determinant of Other Matrices

By the end of this section you will be able to

- understand what is meant by the terms cofactor, minor and adjoint of a matrix
- determine the inverse of an n by n matrix

In the last section we found the determinant of a 2 by 2 matrix, and used it to work out the inverse matrix. However, finding the determinant and inverse of a 3 by 3 matrix is not quite as straightforward as for a 2 by 2. First we need to examine what is meant by the terms 'minors', 'cofactors' and 'adjoint'. We need to define these terms before we can find the inverse.

6.2.1 Minors and cofactors

Consider the general 3 by 3 matrix $\mathbf{A} = \begin{pmatrix} a & b & c \\ d & e & f \\ g & h & i \end{pmatrix}$. The *determinant* of the *remaining* matrix after *deleting* the *row* and *column* of an entry is called the **minor** of that entry.

For example, in the case of the matrix **A** we have

$$\det\begin{pmatrix} e & f \\ h & i \end{pmatrix} \text{ is the } \textit{minor} \text{ of entry } a$$

$$\det\begin{pmatrix} d & f \\ g & i \end{pmatrix} \text{ is the } \textit{minor} \text{ of entry } b$$

What is the minor of entry e?

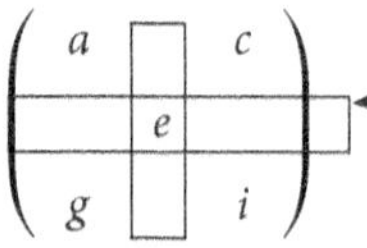

Hence $\det\begin{pmatrix} a & c \\ g & i \end{pmatrix}$ is the minor of entry e.

Example 6.5

Determine the minor of -1 in $\begin{pmatrix} 3 & 5 & 7 \\ -1 & 2 & 3 \\ -4 & 4 & -9 \end{pmatrix}$

Solution

After deleting the rows and columns containing -1,

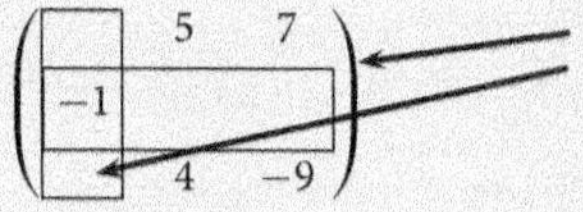

we obtain the matrix $\begin{pmatrix} 5 & 7 \\ 4 & -9 \end{pmatrix}$. The minor of -1 is the determinant of this matrix:

$$\det\begin{pmatrix} 5 & 7 \\ 4 & -9 \end{pmatrix} = (5 \times (-9)) - (4 \times 7) = -73 \qquad \left[\text{by (6.1) } \det\begin{pmatrix} a & b \\ c & d \end{pmatrix} = ad - bc\right]$$

Next we give the formal definition of **minor**.

Definition (6.3). Consider a square matrix **A**. Let a_{ij} be the entry in the ith row and jth column of matrix **A**. The **minor** M_{ij} of entry a_{ij} is the *determinant* of the remaining matrix after *deleting* the entries in the ith row and jth column.

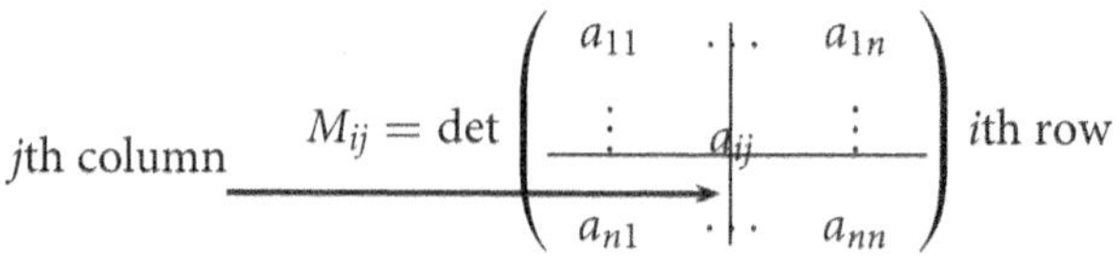

Note that when calculating the minor of an entry in a 3×3 matrix, we need to find the determinant of the remaining 2×2 matrix as seen in the above Example 6.5. When calculating the minor of an entry in a 4×4 matrix we would need to find the determinant of the remaining 3×3 matrix which we have *not* yet stated.

Next, we define the term **cofactor**, which is a sign associated with the minor M_{ij}.

Definition (6.4). Consider a square matrix **A**. Let a_{ij} be the entry in the ith row and jth column of matrix **A**. The **cofactor** C_{ij} of the entry a_{ij} is defined as

$$C_{ij} = (-1)^{i+j} M_{ij}$$

where M_{ij} is the minor of entry a_{ij}.

This definition may appear at first to be a little complicated, but it simply states that for any n by n matrix, the cofactors are just the minors with the following place signs:

$$\begin{pmatrix} + & - & + & \cdots \\ - & + & - & \cdots \\ + & - & + & \cdots \\ \vdots & \vdots & \vdots & \vdots \end{pmatrix}$$

What do you notice about the place signs?

The first entry of a matrix has a positive sign and then the place signs *alternate*.

Example 6.6

Determine the cofactor of 5 in $\begin{pmatrix} 3 & 5 & 7 \\ -1 & 2 & 3 \\ -4 & 4 & -9 \end{pmatrix}$.

Solution

After deleting the row and column containing 5 we find that the minor of 5 is

$$\det\begin{pmatrix} -1 & 3 \\ -4 & -9 \end{pmatrix} \underset{\text{by (6.1)}}{=} (-1 \times (-9)) - (-4 \times 3) = 21 \qquad \left[\text{(6.1) } \det\begin{pmatrix} a & b \\ c & d \end{pmatrix} = ad - bc\right]$$

According to the rule, the place sign of the central entry in the top row is negative, so the cofactor of 5 is -21.

Note that the *minor* of 5 is 21 but the *cofactor* is -21 because the position of 5 in the matrix means that the cofactor inherits a negative sign.

By the above Definition (6.4) the cofactor is given by

$C_{ij} = (-1)^{i+j} \det \begin{pmatrix} a_{11} & \cdots & a_{1n} \\ \vdots & a_{ij} & \vdots \\ a_{n1} & \cdots & a_{nn} \end{pmatrix}$. This definition might seem difficult to follow because of the complex ij notation, but this ij simply locates the entry of a matrix. There is *no* easier way to locate an entry. The cofactor is just the minor of an entry with a plus or minus sign depending on its position in the matrix. For example, the cofactor of the first entry a_{11} is equal to the minor, $C_{11} = M_{11}$ because $(-1)^{1+1} = (-1)^2 = 1$.

What is the cofactor of the entry a_{12}?
In this case, a_{12} means $i = 1, j = 2$ and $i + j = 1 + 2 = 3$, therefore $C_{12} = (-1)^3 M_{12} = -M_{12}$.

What is the cofactor of the entry a_{13}?
Since $i + j = 1 + 3 = 4$ therefore $C_{13} = (-1)^4 M_{13} = M_{13}$. If we carry on developing the cofactors we find that they are just the minors with a place sign.

We can write the determinant of a 3 by 3 matrix in terms of its cofactors. Let

$$\mathbf{A} = \begin{pmatrix} a & b & c \\ d & e & f \\ g & h & i \end{pmatrix} \quad \left[\text{remember that place signs are } \begin{pmatrix} + & - & + \\ - & + & - \\ + & - & + \end{pmatrix}\right]$$

then:

$$\det(\mathbf{A}) = a\left(\text{cofactor of } a\right) + b\left(\text{cofactor of } b\right) + c\left(\text{cofactor of } c\right) \tag{6.5}$$

Expanding out (6.5) gives:

$$\det(\mathbf{A}) = a\left[\det\begin{pmatrix} e & f \\ h & i \end{pmatrix}\right] - b\left[\det\begin{pmatrix} d & f \\ g & i \end{pmatrix}\right] + c\left[\det\begin{pmatrix} d & e \\ g & h \end{pmatrix}\right] \tag{6.6}$$

Why is there a minus sign in front of the b in formula (6.6)?
This is *no mistake*; the minus sign comes about because the place sign for b is minus.

We can find the determinant of a matrix by expanding along any of the rows. For example, the formula for expanding along the middle row is

$$\det(\mathbf{A}) = d\left(\text{cofactor of } d\right) + e\left(\text{cofactor of } e\right) + f\left(\text{cofactor of } f\right)$$

What is the formula for expanding along the bottom row?

$$\det(\mathbf{A}) = g\left(\text{cofactor of } g\right) + h\left(\text{cofactor of } h\right) + i\left(\text{cofactor of } i\right)$$

We can also expand along any of the columns to find the determinant of $\mathbf{A}$. The formula for expanding along the first column is

$$\det(\mathbf{A}) = a\left(\text{cofactor of } a\right) + d\left(\text{cofactor of } d\right) + g\left(\text{cofactor of } g\right)$$

If any of the rows or columns contain zeros then we choose to expand along that row or column because it simplifies the arithmetic.

A transformation $T : \mathbb{R}^3 \to \mathbb{R}^3$ means that T goes from three-dimensional space $\mathbb{R}^3$ to $\mathbb{R}^3$.

Example 6.7

Let $T : \mathbb{R}^3 \to \mathbb{R}^3$ be defined by $T(\mathbf{x}) = \mathbf{A}\mathbf{x}$ where $\mathbf{A} = \begin{pmatrix} -1 & 5 & -2 \\ -6 & 6 & 0 \\ 3 & -7 & 1 \end{pmatrix}$. Find $\det(\mathbf{A})$.

Solution

Since there is a 0 in the last column, it is easier to expand along this column:

$$\det(\mathbf{A}) = -2\,(\text{cofactor of } -2) + 0\,(\text{cofactor of } 0) + 1\,(\text{cofactor of } 1)$$

$$\det\begin{pmatrix} -1 & 5 & \boxed{-2} \\ -6 & 6 & \boxed{0} \\ 3 & -7 & \boxed{1} \end{pmatrix} = (-2)\det\begin{pmatrix} -6 & 6 \\ 3 & -7 \end{pmatrix} - 0\det\begin{pmatrix} -1 & 5 \\ 3 & -7 \end{pmatrix} + 1\det\begin{pmatrix} -1 & 5 \\ -6 & 6 \end{pmatrix}$$

expanding along this column

$$= -2\underbrace{(42-18)}_{\text{by (6.1)}} - 0 + \underbrace{(-6+30)}_{\text{by (6.1)}} = -24$$

We can illustrate the given transformation by examining how T acts on the unit cube determined by $\mathbf{e}_1 = (1,\ 0,\ 0)$, $\mathbf{e}_2 = (0,\ 1,\ 0)$ and $\mathbf{e}_3 = (0,\ 0,\ 1)$. Applying T to each of these vectors gives:

$$T(\mathbf{e}_1) = (-1,\ -6,\ 3),\quad T(\mathbf{e}_2) = (5,\ 6,\ -7) \text{ and } T(\mathbf{e}_3) = (-2,\ 0,\ 1)$$

These coordinates are the column vectors of matrix $\mathbf{A}$.

What does the determinant of -24 represent in geometric terms?

We have a 3 by 3 matrix $\mathbf{A}$, and the modulus (absolute value) of the determinant, -24, is $+24$ which is the volume of the three-dimensional parallelogram formed by the three column vectors of matrix $\mathbf{A}$; $(-1,\ -6,\ 3)$, $(5,\ 6,\ -7)$ and $(-2,\ 0,\ 1)$ (Fig. 6.7).

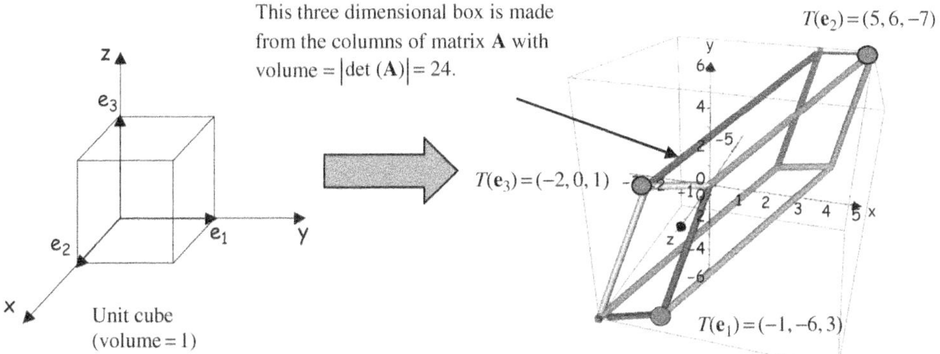

Figure 6.7

The volume scale factor of transformation T is $|\det(\mathbf{A})| = 24$.

The determinant measures the (signed) volume. Negative determinant means the column vectors of matrix **A** are left handed set of vectors. This means with our left hand the positive x axis is in the direction our forefinger, the positive y axis is in the direction of our middle finger and the positive z axis is in the direction of our thumb. This is called the **left handed orientation** – Fig. 6.8(a). Similarly we have the **right handed orientation** – Fig. 6.8(b).

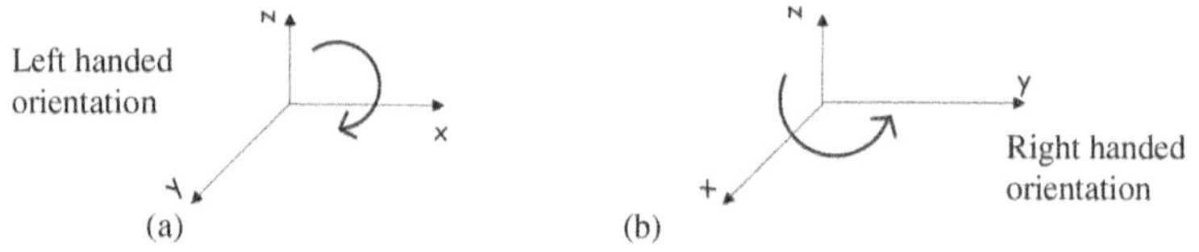

Figure 6.8

The minus sign for the determinant in three dimensions means that we are changing from a right handed set of vectors (unit cube) to a left handed set of vectors.

We can use determinants to find the area of a parallelogram or the volume of a parallelepiped (higher dimension parallelogram) given just the coordinates.

Next we look at a more formal definition of the determinant, with respect to n by n matrices.

The general formula for the determinant of an n by n matrix $\mathbf{A} = \begin{pmatrix} a_{11} & \cdots & a_{1n} \\ \vdots & \ddots & \vdots \\ a_{n1} & \cdots & a_{nn} \end{pmatrix}$ where $n \geq 3$ is:

$$\det(\mathbf{A}) = a_{11}C_{11} + a_{12}C_{12} + \cdots + a_{1n}C_{1n} = \sum_{k=1}^{n} a_{1k}C_{1k} \tag{6.7}$$

where the a's are the entries in the matrix **A**, and the C's are the corresponding cofactors.

Don't be put off by the sigma $\sum$ notation. This is just a shorthand way of writing the lengthy sum (6.7), developed by the great Swiss mathematician Euler, pronounced 'oiler', (1707–1783). The sigma $\sum$ notation is often used by mathematicians to write a sum in compact form. For example

$$\text{(a) } 1 + 2 + 3 + 4 + \cdots + n = \sum_{k=1}^{n} k \quad \text{(b) } 2 + 4 + 6 + 8 + \cdots + 2n = \sum_{k=1}^{n} 2k$$

$\sum_{k=1}^{n} a_{1k}C_{1k}$ means summing every value of $a_{1k}C_{1k}$, from $k = 1$ to $k = n$, which has the form given in (6.7).

(6.7) is the formula for finding the determinant by expanding along the first row of the matrix **A**.

What is the formula for expanding along the ith row?

For expanding along the *i*th row of the matrix, the formula is

$$\text{(6.8)} \qquad \det(\mathbf{A}) = a_{i1}C_{i1} + a_{i2}C_{i2} + \cdots + a_{in}C_{in} = \sum_{k=1}^{n} a_{ik}C_{ik}$$

6.2.2 Cofactor matrix

Let **C** be the new matrix consisting of the cofactors of the general matrix **A**. If

$$\mathbf{A} = \begin{pmatrix} a & b & c \\ d & e & f \\ g & h & i \end{pmatrix} \quad \text{then} \quad \mathbf{C} = \begin{pmatrix} A & B & C \\ D & E & F \\ G & H & I \end{pmatrix}$$

where A is the cofactor of entry a, B is the cofactor of entry b, C is the cofactor of entry c ... The matrix **C** is called the **cofactor matrix** and it is used in finding the *inverse* of **A**. Note that bold **C** represents the cofactor matrix and plain C is the cofactor of the entry c.

Example 6.8

Find the cofactor matrix **C** of

$$\mathbf{A} = \begin{pmatrix} 1 & -1 & 5 \\ 3 & 9 & 7 \\ -2 & 1 & 0 \end{pmatrix} \quad \left[\text{remember the place signs are } \begin{pmatrix} + & - & + \\ - & + & - \\ + & - & + \end{pmatrix}\right]$$

Solution

Cofactor of the first entry, 1, is

$$\det\begin{pmatrix} 9 & 7 \\ 1 & 0 \end{pmatrix} \underbrace{=}_{\text{by (6.1)}} [(9 \times 0) - (1 \times 7)] = -7$$

Cofactor of -1 is

$$\underbrace{=}_{\text{Minus place sign}} \det\begin{pmatrix} 3 & 7 \\ -2 & 0 \end{pmatrix} \underbrace{=}_{\text{By (6.1)}} -[(3 \times 0) - (-2 \times 7)] = -14$$

Cofactor of 5 is

$$\det\begin{pmatrix} 3 & 9 \\ -2 & 1 \end{pmatrix} \underbrace{=}_{\text{by (6.1)}} [(3 \times 1) - (-2 \times 9)] = 21$$

(continued...)

Cofactor of 3 is

$$\underbrace{-}_{\text{minus place sign}} \det\begin{pmatrix} -1 & 5 \\ 1 & 0 \end{pmatrix} \underbrace{=}_{\text{by (6.1)}} -[(-1 \times 0) - (1 \times 5)] = 5$$

Cofactor of 9 is

$$\det\begin{pmatrix} 1 & 5 \\ -2 & 0 \end{pmatrix} \underbrace{=}_{\text{by (6.1)}} [(1 \times 0) - (-2 \times 5)] = 10$$

Cofactor of 7 is

$$\underbrace{-}_{\text{minus place sign}} \det\begin{pmatrix} 1 & -1 \\ -2 & 1 \end{pmatrix} \underbrace{=}_{\text{by (6.1)}} -[(1 \times 1) - (-2 \times (-1))] = 1$$

Cofactor of −2 is

$$\det\begin{pmatrix} -1 & 5 \\ 9 & 7 \end{pmatrix} \underbrace{=}_{\text{by (6.1)}} [(-1 \times 7) - (9 \times 5)] = -52$$

Cofactor of 1 (the 1 on the bottom row of the given matrix) is

$$\underbrace{-}_{\text{minus place sign}} \det\begin{pmatrix} 1 & 5 \\ 3 & 7 \end{pmatrix} \underbrace{=}_{\text{by (6.1)}} -[(1 \times 7) - (3 \times 5)] = 8$$

Cofactor of the last entry 0 is

$$\det\begin{pmatrix} 1 & -1 \\ 3 & 9 \end{pmatrix} \underbrace{=}_{\text{by (6.1)}} [(1 \times 9) - (3 \times (-1))] = 12$$

Hence by collecting these together and placing them in the corresponding position we form the cofactor matrix:

$$\mathbf{C} = \begin{pmatrix} -7 & -14 & 21 \\ 5 & 10 & 1 \\ -52 & 8 & 12 \end{pmatrix}$$

As stated above, we use the cofactor matrix to find the inverse of an invertible matrix. The final term we need to define is ‘**adjoint**’.

Definition (6.9). Let **A** be a square matrix then the matrix consisting of the cofactors of each entry in **A** is called the cofactor matrix and is normally denoted by **C**. The *transpose* of this cofactor matrix is called the **adjoint** of **A** and is denoted by *adj*(**A**). That is

$$adj(\mathbf{A}) = \mathbf{C}^T$$

Remember, we discussed the transpose of a matrix in chapter 1 and it simply means *swapping* the rows and columns around.

Example 6.9

Find the adjoint of the matrix **A** given in the above Example 6.8.

Solution

We have already done all the hard work in evaluating the cofactor matrix **C** above. The adjoint is the *transpose* of this matrix **C**:

$$adj(\mathbf{A}) = \mathbf{C}^T = \begin{pmatrix} -7 & 5 & -52 \\ -14 & 10 & 8 \\ 21 & 1 & 12 \end{pmatrix} \quad \left[\text{because } \mathbf{C} = \begin{pmatrix} -7 & -14 & 21 \\ 5 & 10 & 1 \\ -52 & 8 & 12 \end{pmatrix} \right]$$

6.2.3 Inverse of a matrix

The remainder of this section might seem demanding because we use mathematical proofs to support the propositions. If you are struggling to understand the chain of arguments in the proof then come back and go over the proof a second time. It is *not* necessary that you understand every detail on the first reading.

The statement and proofs of the next three propositions are working towards the proof of the inverse formula given in Theorem (6.13).

Proposition (6.10). If a square matrix **A** consists of *two identical rows* then $\det(\mathbf{A}) = 0$.

Proof – Exercises 6.3.

■

This proposition may seem a bit random and out of context, but we need this (6.10) to prove the next proposition.

Proposition (6.11). Let **A** be an n by n matrix. If C_{jk} denotes the cofactor of the entry a_{jk} for $k = 1, 2, 3, \ldots$ and n then

$$a_{i1}C_{j1} + a_{i2}C_{j2} + a_{i3}C_{j3} + \cdots + a_{in}C_{jn} = \begin{cases} \det(\mathbf{A}) & \text{if } i = j \\ 0 & \text{if } i \neq j \end{cases}$$

Proof.

How do we prove this result?
We consider the two cases $i = j$ and $i \neq j$ then show the required result in each case.

Case 1: Let $i = j$ then substituting this into the left hand side of the above gives:

$$a_{i1} C_{i1} + a_{i2} C_{i2} + a_{i3} C_{i3} + \cdots + a_{in} C_{in} = \det(\mathbf{A}) \qquad [\text{by } (6.8)]$$

Case 2: Consider the case when $i \neq j$ (not equal):
Let $\mathbf{A}^*$ be the matrix obtained from matrix $\mathbf{A}$ by copying the entries of the ith row into the jth row of matrix $\mathbf{A}$. That is the matrix $\mathbf{A}^*$ is the matrix $\mathbf{A}$, apart from the jth row being replaced with a duplicate of the ith row.

$$
\text{ith row} \quad \mathbf{A} = \begin{pmatrix} a_{11} & a_{12} & \cdots & a_{1n} \\ \vdots & \vdots & \vdots & \vdots \\ a_{i1} & a_{i2} & \cdots & a_{in} \\ \vdots & \vdots & \vdots & \vdots \\ a_{j1} & a_{j2} & \cdots & a_{jn} \\ a_{n1} & a_{n2} & \cdots & a_{nn} \end{pmatrix} \quad \text{and} \quad \mathbf{A}^* = \begin{pmatrix} a_{11} & a_{12} & \cdots & a_{1n} \\ \vdots & \vdots & \vdots & \vdots \\ a_{i1} & a_{i2} & \cdots & a_{in} \\ \vdots & \vdots & \vdots & \vdots \\ a_{i1} & a_{i2} & \cdots & a_{in} \\ a_{n1} & a_{n2} & \cdots & a_{nn} \end{pmatrix} \quad \text{ith row = jth row}
$$

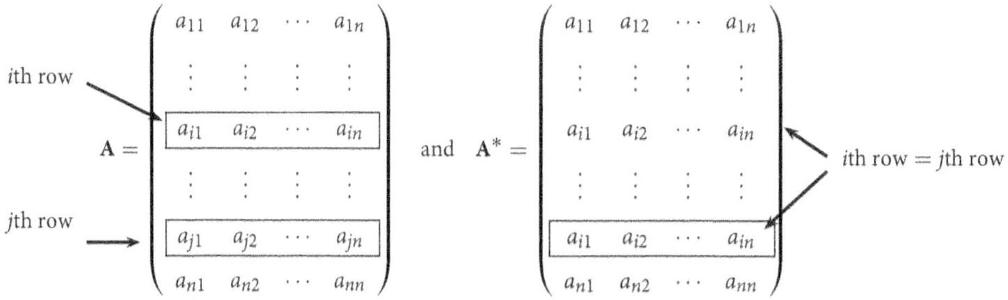

We have $\det(\mathbf{A}^*) = 0$.

Why?
Because we have *two identical rows*, i and j, in $\mathbf{A}^*$, therefore by the previous Proposition (6.10) the determinant is zero, that is $\det(\mathbf{A}^*) = 0$.

Expand along the jth row of $\mathbf{A}^*$ by using (6.8) which is given by

$$\det(\mathbf{A}) = a_{i1} C_{i1} + a_{i2} C_{i2} + \cdots + a_{in} C_{in}$$

In the right hand matrix $\mathbf{A}^*$ we have

$$0 = \det(\mathbf{A}^*) = a_{i1} C_{j1}^* + a_{i2} C_{j2}^* + \cdots + a_{in} C_{jn}^* \qquad (\dagger)$$

where C_{jk}^* is the cofactor of entry a_{ik} in the jth row of matrix $\mathbf{A}^*$.

Consider the cofactor C_{j1}, which is the place sign multiplied by the determinant of the remaining matrix after deleting the row and column containing the entry a_{j1} in the left hand matrix $\mathbf{A}$. Similarly, the cofactor C_{j1}^* is the place sign multiplied by the determinant of the remaining matrix after deleting the row and column containing the entry a_{i1} in the right hand matrix $\mathbf{A}^*$. We have

$$C_{j1} = (-1)^{j+1}\det\left(\begin{array}{c|ccc} & a_{12} & \cdots & a_{1n}\\ & \vdots & \vdots & \vdots\\ & a_{i2} & \cdots & a_{in}\\ & \vdots & \vdots & \vdots\\ \hline \cancel{a_{j1}} & \multicolumn{3}{c}{\rule[0.5ex]{6em}{1pt}}\\ \hline & a_{n2} & \cdots & a_{nn}\end{array}\right) \underset{\longleftarrow\ j\text{th row}\ \longrightarrow}{\text{and}\quad C^*_{j1} = (-1)^{j+1}\det}\left(\begin{array}{c|ccc} & a_{12} & \cdots & a_{1n}\\ & \vdots & \vdots & \vdots\\ & a_{i2} & \cdots & a_{in}\\ & \vdots & \vdots & \vdots\\ \hline \cancel{a_{i1}} & \multicolumn{3}{c}{\rule[0.5ex]{6em}{1pt}}\\ \hline & a_{n2} & \cdots & a_{nn}\end{array}\right)$$

What can you conclude about C_{j1} and C^*_{j1}?

$C^*_{j1} = C_{j1}$ because the cofactor is made up of the *same entries* of matrix $\mathbf{A}$ and $\mathbf{A}^*$. In both cases we delete the jth row and first column. Similarly, we have $C^*_{j2} = C_{j2}$, $C^*_{j3} = C_{j3}, \ldots$ and $C^*_{jn} = C_{jn}$. Substituting these, $C^*_{j1} = C_{j1}$, $C^*_{j2} = C_{j2}, \ldots$ and $C^*_{jn} = C_{jn}$, into (†) gives

$$\begin{aligned} 0 = \det\left(\mathbf{A}^*\right) &= a_{i1}C^*_{j1} + a_{i2}C^*_{j2} + \cdots + a_{in}C^*_{jn}\\ &= a_{i1}C_{j1} + a_{i2}C_{j2} + \cdots + a_{in}C_{jn}\end{aligned}$$

Hence in the case where $i \neq j$ we have

$$a_{i1}C_{j1} + a_{i2}C_{j2} + \cdots + a_{in}C_{jn} = 0$$

which is our required result.

■

The final piece of jigsaw we need for finding the inverse matrix is the following:

Proposition (6.12). Let $\mathbf{A}$ be a square matrix. Then

$$\mathbf{A}\, adj(\mathbf{A}) = \det(\mathbf{A})\mathbf{I}$$

Proof.

Writing out the entries of the general n by n matrix $\mathbf{A}$ and the transpose of the corresponding cofactors of each entry, because $adj(\mathbf{A})$ is the transpose of the cofactors, gives

$$\underset{i\text{th row}}{\mathbf{A} =}\left(\begin{array}{cccc} a_{11} & a_{12} & \cdots & a_{1n}\\ a_{21} & a_{22} & \cdots & a_{2n}\\ \vdots & \vdots & \vdots & \vdots\\ \hline a_{i1} & a_{i2} & \cdots & a_{in}\\ \hline \vdots & \vdots & \vdots & \vdots\\ a_{n1} & a_{n2} & \cdots & a_{nn}\end{array}\right) \quad\text{and}\quad adj(\mathbf{A}) = \left(\begin{array}{cccc|c|cc} C_{11} & C_{21} & \cdots & C_{j1} & \cdots & C_{n1}\\ C_{12} & C_{22} & \cdots & C_{j2} & \cdots & C_{n2}\\ \vdots & \vdots & \vdots & \vdots & \vdots & \vdots\\ C_{1n} & C_{2n} & \cdots & C_{jn} & \cdots & C_{nn}\end{array}\right)$$

$$j\text{th column}$$

Consider the ij entry of the left hand matrix multiplication in $\mathbf{A}\, adj(\mathbf{A})$.

How do we evaluate the ij entry in the matrix multiplication $\mathbf{A}\ adj(\mathbf{A})$*?*
Remember, matrix multiplication is row by column. So the ij entry of $\mathbf{A}\ adj(\mathbf{A})$ is the ith row times the jth column. For an ij entry we have

$$[\mathbf{A}\ adj(\mathbf{A})]_{ij} = a_{i1}C_{j1} + a_{i2}C_{j2} + \cdots + a_{in}C_{jn}$$
$$= \begin{cases} \det(\mathbf{A}) & \text{if } i = j \\ 0 & \text{if } i \neq j \end{cases} \qquad \left[\text{by the above proposition (6.11)}\right]$$

Repeating this for each ij entry and writing out the matrix $\mathbf{A}\ adj(\mathbf{A})$ means that we have $\det(\mathbf{A})$ when $i = j$, along the *leading diagonal*, and 0 *everywhere else* in the matrix $\mathbf{A}\ adj(\mathbf{A})$:

$$\mathbf{A}\ adj(\mathbf{A}) = \begin{pmatrix} \det(\mathbf{A}) & 0 & \cdots & 0 \\ 0 & \det(\mathbf{A}) & \cdots & 0 \\ \vdots & 0 & \ddots & \vdots \\ 0 & \vdots & \cdots & \det(\mathbf{A}) \end{pmatrix}$$

$$\underset{\text{taking out the common factor}}{=} \det(\mathbf{A}) \underbrace{\begin{pmatrix} 1 & 0 & \cdots & 0 \\ 0 & 1 & \cdots & 0 \\ \vdots & 0 & \ddots & \vdots \\ 0 & \vdots & \cdots & 1 \end{pmatrix}}_{=\mathbf{I}} = \det(\mathbf{A})\mathbf{I} \qquad \left[\text{Remember } \mathbf{I} \text{ is the identity matrix, which has 1's along the leading diagonal and 0's everywhere else.}\right]$$

Hence we have $\mathbf{A}\ adj(\mathbf{A}) = \det(\mathbf{A})\mathbf{I}$.

■

All the above propositions are an aid in proving the next result, which is an important theorem in linear algebra.

Theorem (6.13). If $\det(\mathbf{A}) \neq 0$ (not zero) then

$$\mathbf{A}^{-1} = \frac{1}{\det(\mathbf{A})} adj(\mathbf{A})$$

Proof.
This follows from the previous Proposition (6.12). Since $\det(\mathbf{A}) \neq 0$ we can divide the above formula given in Proposition (6.12) $\mathbf{A}\ adj(\mathbf{A}) = \det(\mathbf{A})\mathbf{I}$ by $\det(\mathbf{A})$:

$$\mathbf{A}\left[\frac{1}{\det(\mathbf{A})} adj(\mathbf{A})\right] = \mathbf{I} \qquad (\dagger)$$

As we have $\mathbf{A}$ times another matrix gives $\mathbf{I}$, therefore $\mathbf{A}^{-1} = \frac{1}{\det(\mathbf{A})} adj(\mathbf{A})$ provided $\det(\mathbf{A}) \neq 0$.

■

What does Theorem (6.13) mean?
The inverse of an invertible matrix $\mathbf{A}$ is given by the formula $\mathbf{A}^{-1} = \frac{1}{\det(\mathbf{A})} adj(\mathbf{A})$. To determine the inverse of a matrix you need to find the cofactors and the determinant of the given matrix.

What is the point of finding the inverse of a matrix?
We need the inverse to solve linear system of equations, although it is more efficient to use the elimination process discussed in chapter 1 for large systems. For example, to find the determinant of a 10 by 10 matrix requires 10! = 3,628,800 operations. From a computational point of view, a determinant is expensive so we try *not* to compute it for large matrices. However, determinants play an important role in many areas of mathematics.

Example 6.10

Find the inverse of the matrix given in Example 6.8 which is

$$\mathbf{A} = \begin{pmatrix} 1 & -1 & 5 \\ 3 & 9 & 7 \\ -2 & 1 & 0 \end{pmatrix}$$

Solution
We need to find $\mathbf{A}^{-1}$ which is given by the above Proposition: $\mathbf{A}^{-1} = \frac{1}{\det(\mathbf{A})} adj(\mathbf{A})$
What is $adj(\mathbf{A})$ equal to?
Remember, $adj(\mathbf{A})$ is the cofactor matrix transposed, and was found in Example 6.9 above:

$$adj(\mathbf{A}) = \mathbf{C}^T = \begin{pmatrix} -7 & 5 & -52 \\ -14 & 10 & 8 \\ 21 & 1 & 12 \end{pmatrix}$$

We only need to find $\det(\mathbf{A})$. Expanding along the *bottom row* of the given matrix $\mathbf{A}$, because it contains a 0:

$$\det(\mathbf{A}) = \det \begin{pmatrix} 1 & -1 & 5 \\ 3 & 9 & 7 \\ \boxed{-2 \quad 1 \quad 0} \end{pmatrix} = -2 \det \begin{pmatrix} -1 & 5 \\ 9 & 7 \end{pmatrix} - 1 \det \begin{pmatrix} 1 & 5 \\ 3 & 7 \end{pmatrix} + 0$$

$$= -2(-7 - 45) - (7 - 15) = 112$$

expanding along this row

Substituting these into formula (6.13) gives

$$\mathbf{A}^{-1} = \frac{1}{\det(\mathbf{A})} adj(\mathbf{A}) = \frac{1}{112} \begin{pmatrix} -7 & 5 & -52 \\ -14 & 10 & 8 \\ 21 & 1 & 12 \end{pmatrix}$$

Normally, we would find the determinant first and then the adjoint of the matrix, because if the determinant is zero, this tells us that the matrix has *no* inverse or is non-invertible (singular).

Check that the matrix found in Example 6.10 is indeed the inverse of $\mathbf{A}$.

How?

Check the matrix multiplication $\mathbf{A} \times \mathbf{A}^{-1} = \mathbf{I}$.

Summary

If we expand along the ith row of a matrix $\mathbf{A}$, then the formula for determinant is

$$\det(\mathbf{A}) = a_{i1}C_{i1} + a_{i2}C_{i2} + \cdots + a_{in}C_{in} \tag{6.8}$$

If a row or column vector of a matrix contains zero(s) then expand along that row or column.
The inverse of a square matrix is defined as

$$\mathbf{A}^{-1} = \frac{1}{\det(\mathbf{A})} adj(\mathbf{A}) \quad \text{provided } \det(\mathbf{A}) \neq 0. \tag{6.13}$$

EXERCISES 6.2

(Brief solutions at end of book. Full solutions available at <http://www.oup.co.uk/companion/singh>.)

You may like to check your numerical answers using MATLAB.

1. Calculate the determinants of the following matrices:

(a) $\mathbf{A} = \begin{pmatrix} 1 & 3 & -1 \\ 2 & 0 & 5 \\ -6 & 3 & 1 \end{pmatrix}$ (b) $\mathbf{B} = \begin{pmatrix} 2 & -10 & 11 \\ 5 & 3 & -4 \\ 7 & 9 & 12 \end{pmatrix}$ (c) $\mathbf{C} = \begin{pmatrix} -12 & 9 & -5 \\ 3 & 9 & 1 \\ -7 & 2 & -2 \end{pmatrix}$

2. Show that $\det \begin{pmatrix} \mathbf{i} & \mathbf{j} & \mathbf{k} \\ 7 & 3 & -2 \\ 4 & 2 & 7 \end{pmatrix} = 25\mathbf{i} - 57\mathbf{j} + 2\mathbf{k}$.

3. Find the values of x so that $\det \begin{pmatrix} 1 & 0 & -3 \\ 5 & x & -7 \\ 3 & 9 & x-1 \end{pmatrix} = 0$.

4. Find the cofactor matrices $\mathbf{C}$ and $\mathbf{C}^T$ of $\mathbf{A} = \begin{pmatrix} 1 & 0 & 5 \\ -2 & 3 & 7 \\ 6 & -1 & 0 \end{pmatrix}$. Also determine $\mathbf{A}^{-1}$.

5. Determine the inverse of the following matrices:

(a) $\mathbf{A} = \begin{pmatrix} 9 & 2 \\ 13 & 3 \end{pmatrix}$ (b) $\mathbf{B} = \begin{pmatrix} 17 & 7 \\ 12 & 5 \end{pmatrix}$

(c) $\mathbf{C} = \begin{pmatrix} 5 & 4 \\ 3 & 1 \end{pmatrix}$ (d) $\mathbf{D} = \begin{pmatrix} 3 & -5 & 3 \\ 2 & 1 & -7 \\ -10 & 4 & 5 \end{pmatrix}$

6. Find the determinants of the following:

(a) $\mathbf{A} = \begin{pmatrix} 2 & 3 & 5 \\ 0 & 0 & 6 \\ 1 & 5 & 3 \end{pmatrix}$ (b) $\mathbf{B} = \begin{pmatrix} 6 & 7 & 1 \\ 1 & 3 & 2 \\ 0 & 1 & 5 \end{pmatrix}$

(c) $\mathbf{C} = \begin{pmatrix} 1 & 5 & 1 \\ 0 & 3 & 7 \\ 0 & 2 & 9 \end{pmatrix}$ (d) $\mathbf{D} = \begin{pmatrix} 9 & 5 & 1 \\ 13 & 0 & 2 \\ 11 & 0 & 3 \end{pmatrix}$

7. The formula for the area of a triangle with coordinates (x_1, y_1), (x_2, y_2), (x_3, y_3) is given by Area $= \frac{1}{2}\left|\det\begin{pmatrix} x_1 & y_1 & 1 \\ x_2 & y_2 & 1 \\ x_3 & y_3 & 1 \end{pmatrix}\right|$. Determine the areas of triangles connecting:

(a) $(0,\ 0)$, $(3,\ 2)$, $(7,\ -4)$
(b) $(-3,\ 2)$, $(2,\ 6)$, $(8,\ -3)$
(c) $(-2,\ -1)$, $(1,\ 5)$ $(0.5,\ 4)$. What do you notice about this result?

8. Let $P_1 = (x_1, y_1)$ and $P_2 = (x_2, y_2)$ be two points in a plane. The equation of the line going through these two points is given by

$$\det\begin{pmatrix} x & y & 1 \\ x_1 & y_1 & 1 \\ x_2 & y_2 & 1 \end{pmatrix} = 0$$

Find the equation of the line through the points:
(a) $(1,\ 2)$ and $(5,\ 6)$ (b) $(-3,\ 7)$ and $(10,\ 10)$
(c) $(-3,\ 7)$ and $(9,\ -21)$

9. In a 6 by 6 matrix $\mathbf{A}$, decide the place sign of the following entries:

$$a_{31},\ a_{56},\ a_{62},\ a_{65} \text{ and } a_{71}$$

10. Show that in an n by n matrix the place sign of a_{mn} is equal to that of a_{nm}.

11. Find $\det(\mathbf{A})$ where $\mathbf{A} = \begin{pmatrix} a & b & c & d \\ 0 & 0 & 0 & 0 \\ e & f & g & h \\ i & j & k & l \end{pmatrix}$.

12. Find the values of k for which the following matrix is invertible:

$$\mathbf{A} = \begin{pmatrix} k & 1 & 2 \\ 0 & k & 2 \\ 5 & -5 & k \end{pmatrix}$$

13. Show that the 3 by 3 Vandermonde determinant is written as

$$\det\begin{pmatrix} 1 & 1 & 1 \\ x & y & z \\ x^2 & y^2 & z^2 \end{pmatrix} = (x-y)(y-z)(z-x)$$

14. The volume of a parallelepiped (three-dimensional parallelogram) which is spanned by the vectors $\mathbf{u}$, $\mathbf{v}$ and $\mathbf{w}$ is given by $\left|\det\begin{pmatrix}\mathbf{u} & \mathbf{v} & \mathbf{w}\end{pmatrix}\right|$.
Find the volume of the parallelepiped generated by the vectors

$$\mathbf{u} = (1\ \ 2\ \ 1)^T,\quad \mathbf{v} = (2\ \ 3\ \ 5)^T \text{ and } \mathbf{w} = (7\ \ 10\ \ -1)^T$$

15. Show that the determinant of the following rotational matrix $\mathbf{R}$ is 1.

$$\mathbf{R} = \begin{pmatrix} \cos(\theta) & \sin(\theta) & 0 \\ -\sin(\theta) & \cos(\theta) & 0 \\ 0 & 0 & 1 \end{pmatrix}$$

What does determinant equal to 1 mean in this context?

16. In multivariable calculus we have to transform from rectangular coordinates $(x,\ y)$ to polar coordinates $(r,\ \theta)$. The Jacobian matrix $\mathbf{J}$ given below is used for such a transformation. Show that $\det(\mathbf{J}) = r$.

$$\mathbf{J} = \begin{pmatrix} \cos(\theta) & -r\sin(\theta) \\ \sin(\theta) & r\cos(\theta) \end{pmatrix}$$

17. The Jacobian determinant J is defined as

$$J = \left|\det\begin{pmatrix} \cos(\theta)\sin(\phi) & -\rho\sin(\theta)\sin(\phi) & \rho\cos(\theta)\cos(\phi) \\ \sin(\theta)\sin(\phi) & \rho\cos(\theta)\sin(\phi) & \rho\sin(\theta)\cos(\phi) \\ \cos(\phi) & 0 & -\rho\sin(\phi) \end{pmatrix}\right|$$

This determinant is used in change-of-variable formulae when studying multivariable calculus. This is used to transform from $(x,\ y,\ z)$ coordinates to spherical coordinates $(\rho,\ \theta,\ \phi)$. Show that $J = \rho^2\sin(\phi)$.

18. The Wronskian $W(f,\ g,\ h) = \det\begin{pmatrix} f(x) & g(x) & h(x) \\ f'(x) & g'(x) & h'(x) \\ f''(x) & g''(x) & h''(x) \end{pmatrix}$. Determine

$$W(1,\ \cos(x),\ \sin(x))$$

19. Show that for an n by n identity matrix $\mathbf{I}_n$ we have $\det(\mathbf{I}_n) = 1$ for every natural number n.

20. Prove that $\det\left(\mathbf{A}\mathbf{A}^{-1}\right) = 1$ where $\mathbf{A}$ is an invertible matrix.

21. Prove that if a square matrix $\mathbf{A}$ contains a zero row or zero column then $\det(\mathbf{A}) = 0$.

22. Let $\mathbf{A}$ be an n by n matrix. Prove that $\det\left(\mathbf{A}^T\right) = \det(\mathbf{A})$.

23. Let $\mathbf{B}$ be an n by n matrix obtained from a matrix $\mathbf{A}$ by multiplying one row (or column) by a scalar k. Prove that $\det(\mathbf{B}) = k\det(\mathbf{A})$.

24. Prove that $\det\left(k\mathbf{A}\right) = k^n\det(\mathbf{A})$ where $\mathbf{A}$ is an n by n matrix.

SECTION 6.3 Properties of Determinants

By the end of this section you will be able to

- find the determinant of a triangular and diagonal matrix
- convert a given matrix into a triangular matrix
- establish certain properties of determinants of matrices

To calculate the determinant of matrices of size 4 by 4 or larger is a lengthy process. In this section we establish some properties of determinants that make evaluating such determinants a lot simpler. You will need to remember the definition of a determinant of a matrix and the technique to evaluate it by using cofactors. This is a challenging section because you will need to understand the abstract mathematics and recall some of the definitions of chapter 1 to prove certain results.

6.3.1 Revision of properties of a determinant

You have proved certain properties of the determinant of a matrix in Exercises 6.2 such as

Proposition (6.14). Let $\mathbf{A}$ be a square matrix then $\det\left(\mathbf{A}^T\right) = \det(\mathbf{A})$.

What does this mean?

The determinant of a transposed matrix is equal to the determinant of the initial matrix. Since $\det\left(\mathbf{A}^T\right) = \det(\mathbf{A})$, we can expand along a row *or* a column and achieve the same result. Propositions about the determinant of the matrix with the word *row* can be swapped with the word *column* because of this result. Another important property is:

Proposition (6.15). Let $\mathbf{B}$ be a matrix obtained from matrix $\mathbf{A}$ by multiplying *one* row (or column) of $\mathbf{A}$ by a non-zero scalar k then

$$\det(\mathbf{B}) = k\det(\mathbf{A})$$

We can visualize this proposition by looking at transformations. Multiplying one row of a matrix by k means one of the row vectors has its length multiplied by a factor of k.

For example, the unit square in Fig. 6.9(a) is transformed to the rectangle shown in Fig. 6.9(b):

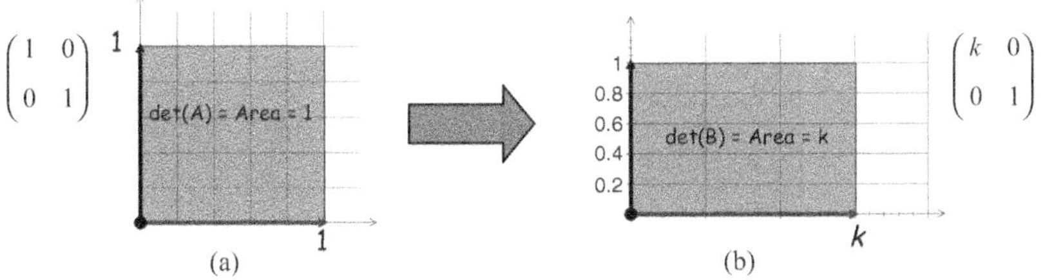

Figure 6.9

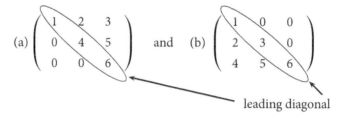

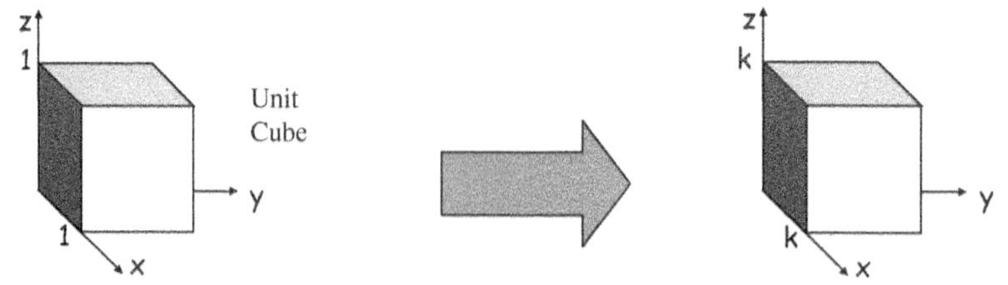

Figure 6.10

Proposition (6.16). Let $\mathbf{A}$ be any n by n matrix and k be a scalar then

$$\det\left(k\mathbf{A}\right) = k^n \det(\mathbf{A})$$

What does this mean?

The scalar multiplication $k\mathbf{A}$ means that *each* column vector of $\mathbf{A}$ has been multiplied by k. For a 2 by 2 matrix, $k\mathbf{A}$ means the length of each side of a parallelogram has been changed by k so the area has changed by a factor of $k \times k = k^2$.

For a 3 by 3 matrix, $k\mathbf{A}$ means the length of each side of a three-dimensional box has been multiplied by k so the volume has changed by $k \times k \times k = k^3$. If we double each column vector of a 3 by 3 matrix then the volume increase will be $2 \times 2 \times 2 = 2^3$. Fig. 6.10 shows the transformation from a unit cube with volume 1 to k^3.

For an n by n matrix the length of each side of the n-dimensional box has been multiplied by k for the scalar multiplication $k\mathbf{A}$ which means that the volume has changed by k^n. For example, if we double every side of an n-dimensional box or parallelepiped (higher-dimension version of a parallelogram) then the volume will increase by a factor of 2^n.

In general, we have $\det\left(k\mathbf{A}\right) = k^n \det(\mathbf{A})$.

6.3.2 Determinant properties of particular matrices

We can easily find determinants of particular matrices such as triangular matrices.

What are triangular matrices?

Definition (6.17). A triangular matrix is an n by n matrix where all entries to one side of the leading diagonal are zero.

For example, the following are triangular matrices:

$$\text{(a)} \begin{pmatrix} 1 & 2 & 3 \\ 0 & 4 & 5 \\ 0 & 0 & 6 \end{pmatrix} \quad \text{and} \quad \text{(b)} \begin{pmatrix} 1 & 0 & 0 \\ 2 & 3 & 0 \\ 4 & 5 & 6 \end{pmatrix}$$

leading diagonal

(a) is an example of an upper triangular matrix.

(b) is an example of a lower triangular matrix.

Another type of matrix is a diagonal matrix.

Definition (6.18). A diagonal matrix is an n by n matrix where *all* entries to both sides of the leading diagonal are zero.

Can you think of an example of a diagonal matrix?

The identity matrix $\mathbf{I} = \begin{pmatrix} 1 & 0 & 0 \\ 0 & 1 & 0 \\ 0 & 0 & 1 \end{pmatrix}$. Another example is $\begin{pmatrix} 1 & 0 & 0 \\ 0 & 2 & 0 \\ 0 & 0 & 3 \end{pmatrix}$.

A diagonal matrix is both an upper and lower triangular matrix.
Triangular and diagonal matrices have the following *useful* property.

Proposition (6.19). The determinant of a triangular or diagonal matrix is the *product* of the entries along the leading diagonal.

What does this proposition mean?

Let $\mathbf{A}$ be an upper triangle matrix then

leading diagonal

$$\det(\mathbf{A}) = \det \begin{pmatrix} a_{11} & & a_{1n} \\ 0 & \ddots & \\ \vdots & 0 \dots & a_{nn} \end{pmatrix} = a_{11}a_{22}a_{33} \times \cdots \times a_{nn}$$

This is a useful result because finding the determinant of such matrices is just a matter of multiplying the entries on the leading diagonal.

We prove this for an upper triangular matrix; the proof for the lower triangular matrix is similar. The diagonal matrix is a special case of an upper (or lower) triangular matrix, therefore the proof of the diagonal matrix follows from the upper triangular result.

Proof.

In each of the steps below we expand along the first column of the remaining matrix because only the first entry makes a contribution to the determinant as the remaining entries are zero.

Expanding along the first column of the above matrix $\mathbf{A}$ we have

$$\det(\mathbf{A}) = \det \begin{pmatrix} \boxed{a_{11}} & & a_{1n} \\ 0 & \ddots & \\ \vdots & 0 \dots & a_{nn} \end{pmatrix} = (a_{11}) \det \begin{pmatrix} \boxed{a_{22}} & \dots & a_{2n} \\ 0 & \ddots & \vdots \\ \vdots & 0 \dots & a_{nn} \end{pmatrix}$$

(continued)

$$\underbrace{=}_{\substack{\text{expanding along the}\\ \text{first column with } a_{22} \text{ entry}}} (a_{11}a_{22}) \det \begin{pmatrix} \boxed{a_{33}} & \cdots & a_{3n} \\ 0 & \ddots & \vdots \\ \vdots & 0 \cdots & a_{nn} \end{pmatrix}$$

$$\underbrace{=}_{\substack{\text{taking out } a_{33} \text{ and}\\ \text{deleting the rows}\\ \text{and columns containing } a_{33}}} (a_{11}a_{22}a_{33}) \det \begin{pmatrix} \boxed{a_{44}} & \cdots & a_{4n} \\ 0 & \ddots & \vdots \\ \vdots & 0 \cdots & a_{nn} \end{pmatrix}$$

$$= \left(a_{11}a_{22}a_{33} \cdots a_{(n-2)(n-2)}\right) \det \begin{pmatrix} a_{(n-1)(n-1)} & a_{(n-1)n} \\ 0 & a_{nn} \end{pmatrix}$$

$$= a_{11}a_{22}a_{33} \cdots a_{(n-2)(n-2)}a_{(n-1)(n-1)}a_{nn}$$

Hence the determinant of an upper triangular matrix is the product of the entries along the leading diagonal. This completes our proof.

■

Example 6.11

Find the determinants of the following matrices:

(a) $U = \begin{pmatrix} 1 & 2 & 3 & 4 \\ 0 & 5 & 6 & 7 \\ 0 & 0 & 8 & 9 \\ 0 & 0 & 0 & 10 \end{pmatrix}$ **(b)** $L = \begin{pmatrix} 1 & 0 & 0 & 0 \\ 3 & -2 & 0 & 0 \\ -11 & 8 & 5 & 0 \\ 89 & 9 & 3 & 2 \end{pmatrix}$ **(c)** $D = \begin{pmatrix} -2 & 0 & 0 & 0 & 0 \\ 0 & -6 & 0 & 0 & 0 \\ 0 & 0 & 4 & 0 & 0 \\ 0 & 0 & 0 & 3 & 0 \\ 0 & 0 & 0 & 0 & 10 \end{pmatrix}$

Solution

We use the above result, Proposition (6.19), because the three matrices are upper triangular, lower triangular and diagonal respectively.

In each case the determinant is the product of the leading diagonal entries.

(a) $\det(U) = 1 \times 5 \times 8 \times 10 = 400$

(b) $\det(L) = 1 \times (-2) \times 5 \times 2 = -20$

(c) $\det(D) = (-2) \times (-6) \times 4 \times 3 \times 10 = 1440$

You may like to check these answers by using MATLAB – the command is det(A) where A is the matrix you have to enter.

6.3.3 Determinant properties of elementary matrices

Do you remember what an elementary matrix is?

An *elementary matrix* is a matrix obtained by a *single* row operation on the identity matrix. Examples of 2 by 2 elementary matrices are

$$\mathbf{E}_1 = \begin{pmatrix} 5 & 0 \\ 0 & 1 \end{pmatrix}, \quad \mathbf{E}_2 = \begin{pmatrix} 1 & 0 \\ 0.5 & 1 \end{pmatrix} \text{ and } \mathbf{E}_3 = \begin{pmatrix} 0 & 1 \\ 1 & 0 \end{pmatrix}$$

Recall from chapter 1 that there are *three* different types of elementary matrices:

1. An elementary matrix **E** obtained from the identity matrix, **I**, by multiplying a row by a non-zero scalar k. For example, $\mathbf{E}_1$.
2. An elementary matrix **E** obtained from the identity matrix, **I**, by adding (or subtracting) a multiple of one row to another. For example, $\mathbf{E}_2$.
3. An elementary matrix **E** obtained from the identity matrix, **I**, by interchanging two rows (or columns). For example, $\mathbf{E}_3$.

Since $\mathbf{E}_1$ and $\mathbf{E}_2$ are triangular matrices we have $\det(\mathbf{E}_1) = 5 \times 1 = 5$ and $\det(\mathbf{E}_2) = 1 \times 1 = 1$. The row vectors of matrix $\mathbf{E}_3$ have swapped over so $\det(\mathbf{E}_3) = (0 \times 0) - (1 \times 1) = -1$, see Fig. 6.11(c).

Applying these transformation matrices $\mathbf{E}_1, \mathbf{E}_2$ and $\mathbf{E}_3$ to the unit square we have:

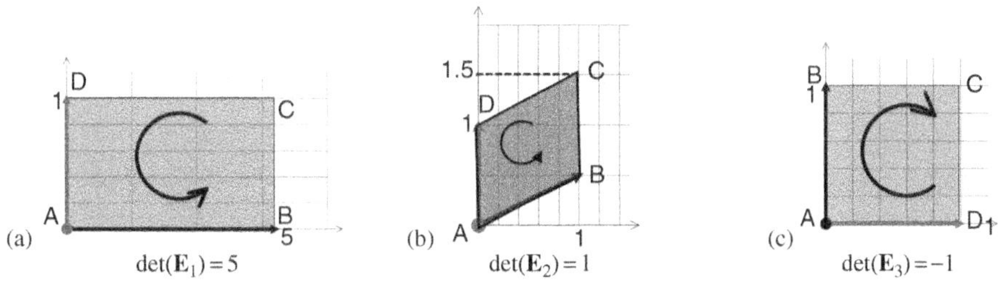

Figure 6.11

Example 6.12

Find the determinants of the following elementary matrices:

(a) $\mathbf{A} = \begin{pmatrix} 1 & 0 & 0 \\ 0 & -5 & 0 \\ 0 & 0 & 1 \end{pmatrix}$ **(b)** $\mathbf{B} = \begin{pmatrix} 0 & 0 & 1 \\ 0 & 1 & 0 \\ 1 & 0 & 0 \end{pmatrix}$ **(c)** $\mathbf{C} = \begin{pmatrix} 1 & 0 & -5 \\ 0 & 1 & 0 \\ 0 & 0 & 1 \end{pmatrix}$

Solution

(a) Matrix **A** is obtained from the identity matrix by multiplying the middle row by -5. Since **A** is a diagonal matrix we have $\det(\mathbf{A}) = 1 \times (-5) \times 1 = -5$.

(b) *How is matrix **B** obtained from the identity matrix?* By swapping top and bottom rows. Expanding along the top row gives $\det(\mathbf{B}) = -1$.

(c) We have added -5 times the bottom row to the top row to get matrix **C** from the identity. Since **C** is an upper triangular matrix, $\det(\mathbf{C}) = 1 \times 1 \times 1 = 1$.

We can write all these operations and their determinants in general terms.

Proposition (6.20). Let $\mathbf{E}$ be an elementary matrix.

(a) If the elementary matrix $\mathbf{E}$ is obtained from the identity matrix $\mathbf{I}$ by multiplying a row by a non-zero scalar k then $\det(\mathbf{E}) = k$. (Matrix $\mathbf{A}$ in Example 6.12)

(b) If the elementary matrix $\mathbf{E}$ is obtained from the identity matrix $\mathbf{I}$ by adding a multiple of one row to another then $\det(\mathbf{E}) = 1$. (Matrix $\mathbf{C}$ in Example 6.12)

(c) If the elementary matrix $\mathbf{E}$ is obtained from the identity matrix $\mathbf{I}$ by interchanging two rows then $\det(\mathbf{E}) = -1$. (Two row vectors have swapped over, matrix $\mathbf{B}$ in Example 6.12.)

Proof.

(a) With one row of the identity matrix multiplied by non-zero k we have a diagonal matrix with one entry equal to k on the leading diagonal and the remaining entries being 1. Hence

$$\det(\mathbf{E}) = 1 \times 1 \times \cdots \times k \times 1 \cdots \times 1 = k$$

(b) If we add a multiple of one row of the identity matrix to another then we have a triangular matrix $\mathbf{E} = \begin{pmatrix} 1 & k & 0 \\ 0 & \ddots & \vdots \\ 0 & \cdots & 1 \end{pmatrix}$ or $\mathbf{E} = \begin{pmatrix} 1 & 0 & 0 \\ 0 & \ddots & \vdots \\ \vdots & k & 1 \end{pmatrix}$. By (6.19): Determinant of a triangular matrix is a product of entries along the leading diagonal.

We have $\det(\mathbf{E}) = 1$ because $\mathbf{E}$ is a triangular matrix with 1's along the leading diagonal so

$$\det(\mathbf{E}) = 1 \times 1 \times 1 \times \cdots \times 1 = 1$$

(c) We have to prove it for the case when two rows (or columns) have been interchanged. This is a more complex proof and is by *induction*. Recall the three steps of induction:

1. Prove it for the simplest case, $n = 1$ or $n = 2$ (or for some other base $n = k_0$).
2. Assume that it is true for $n = k$.
3. Prove it for $n = k + 1$.

Step 1: We first prove the result for $n = 2$ which means a 2 by 2 elementary matrix:

$$\begin{aligned} \det(\mathbf{E}) &= \det\begin{pmatrix} 0 & 1 \\ 1 & 0 \end{pmatrix} \qquad \left[\begin{array}{l}\text{rows of } \mathbf{I}_2 \text{ have} \\ \text{been interchanged}\end{array}\right] \\ &= 0 - 1 = -1 \end{aligned}$$

Hence if we swap the two rows of the 2 by 2 identity matrix then the determinant is -1.

Step 2: Assume that the result is true for $n = k$, that is, for a k by k elementary matrix $\mathbf{E}_k$ with rows i and j interchanged:

$$\det(\mathbf{E}_k) = -1 \qquad (\dagger)$$

Step 3: We need to prove the result for $n = k + 1$. Let $\mathbf{E}_{k+1}$ be the $k + 1$ by $k + 1$ elementary matrix with rows i and j of the identity matrix interchanged.

$$\mathbf{E}_{k+1} = \begin{pmatrix} 1 & 0 & \cdots & \cdots & \cdots & \cdots & 0 \\ \vdots & \ddots & \vdots & \vdots & \vdots & \vdots & \vdots \\ 0 & \cdots & 0 & \cdots & 1 & 0 & \vdots \\ \vdots & \vdots & 0 & 1 & 0 & \vdots & \vdots \\ 0 & \cdots & 1 & 0 & 0 & \vdots & \vdots \\ \vdots & \vdots & 0 & \vdots & \vdots & 1 & \vdots \\ 0 & \cdots & \cdots & \cdots & \cdots & \cdots & 1 \end{pmatrix} \begin{matrix} \\ \\ \leftarrow i\text{th row} \\ \\ \leftarrow j\text{th row} \\ \\ \\ \end{matrix}$$

To find the determinant of this matrix we can expand along the kth row where the kth row is *not one* of ith or jth row. Note that in the kth row *all* the entries should be 0's apart from the diagonal item e_{kk} which is 1. Therefore the determinant of this matrix $\mathbf{E}_{k+1}$ is

$$\det\left(\mathbf{E}_{k+1}\right) = (-1)^{k+k} \det\left(\mathbf{E}_k\right) \qquad (*)$$

Why?

Because if you delete the elements containing the row and column containing the entry e_{kk} then the remaining matrix is the k by k elementary matrix with rows i and j interchanged, which is $\mathbf{E}_k$.

What is the determinant of $\mathbf{E}_k$?

By (†) we have $\det\left(\mathbf{E}_k\right) = -1$. Substituting this into (*) gives

$$\begin{aligned} \det\left(\mathbf{E}_{k+1}\right) &= (-1)^{k+k}(-1) \\ &= (-1)^{2k}(-1) = -1 \qquad \left[\text{because } (-1)^{2k} = 1\right] \end{aligned}$$

Hence we have proved that the determinant of an elementary matrix of size $k + 1$ by $k + 1$ which has two rows interchanged is -1.

Therefore by mathematical induction we have our result that if the elementary matrix $\mathbf{E}$ is obtained from the identity matrix $\mathbf{I}$ by swapping two rows then $\det(\mathbf{E}) = -1$.

■

Note that we can use this Proposition (6.20) to find the determinants of *elementary* matrices. Summarizing this Proposition (6.20) we have that the determinant of an elementary matrix $\mathbf{E}$ is given by:

$$(6.21) \qquad \det(\mathbf{E}) = \begin{cases} 1 & \text{if a multiple of one row is added to another} \\ -1 & \text{if two rows have been interchanged} \\ k & \text{if a row has been multiplied by non-zero } k \end{cases}$$

Hence the determinant of an elementary matrix can only have values 1, -1 or k (non-zero).

6.3.4 Determinant properties of other matrices

In the above, we restricted ourselves to *elementary* matrices. However, in this section we deal with general matrices and find that they have identical results.

We can extend Proposition (6.20) or result (6.21) to *any* square matrix $\mathbf{A}$ as the following:

Proposition (6.22). Let $\mathbf{B}$ be a matrix obtained from the square matrix $\mathbf{A}$ by:

(i) adding (or subtracting) a multiple of one row to another; we have $\det(\mathbf{B}) = \det(\mathbf{A})$

(ii) interchanging two rows; in this case $\det(\mathbf{B}) = -\det(\mathbf{A})$

(iii) multiplying a row by a non-zero scalar k; in this case $\det(\mathbf{B}) = k\det(\mathbf{A})$.

Proof.
Part (iii) is Proposition (6.15) stated earlier in this section. See Exercises 6.3 for proofs of parts (i) and (ii).

■

We can summarize this into the following result.

(6.23) $$\det(\mathbf{B}) = \begin{cases} \text{(i)} & \det(\mathbf{A}) & \text{if a multiple of one row is added to another} \\ \text{(ii)} & -\det(\mathbf{A}) & \text{if two rows have been interchanged} \\ \text{(iii)} & k\det(\mathbf{A}) & \text{if a row has been multiplied by non-zero } k \end{cases}$$

We use this result (6.23) to cut down on the arithmetic required to evaluate a determinant as the next example illustrates.

Example 6.13

Find the determinant of the following matrix:

$$\mathbf{A} = \begin{pmatrix} 1/23 & -2/23 & 1 \\ 1/2 & 1/6 & 5/6 \\ 1/11 & 3/11 & -1/11 \end{pmatrix}$$

Solution

How can we find the determinant of this matrix?

Expanding along a row would be a very tedious task because of the fractions involved. We can create a simpler matrix, which is matrix $\mathbf{B}$, with the top row multiplied by 23, second row multiplied by 6 and bottom row multiplied by 11.

How is the determinant of this new matrix, $\mathbf{B}$, related to the determinant of matrix $\mathbf{A}$?

By part (iii) of result (6.23), if a row has been multiplied by a non-zero scalar k, then $\det(\mathbf{B}) = k\det(\mathbf{A})$:

Taking out the multiple of each row (23, 6 and 11):

$$(23 \times 6 \times 11)\det(\mathbf{A}) = \det(\mathbf{B}) = \det\begin{pmatrix} 1 & -2 & 23 \\ 3 & 1 & 5 \\ 1 & 3 & -1 \end{pmatrix} \qquad (\dagger)$$

We can find the determinant of matrix $\mathbf{B}$:

$$\det(\mathbf{B}) = \det\begin{pmatrix} 1 & -2 & 23 \\ 3 & 1 & 5 \\ 1 & 3 & -1 \end{pmatrix} = \det\begin{pmatrix} 1 & 5 \\ 3 & -1 \end{pmatrix} - (-2)\det\begin{pmatrix} 3 & 5 \\ 1 & -1 \end{pmatrix} + 23\det\begin{pmatrix} 3 & 1 \\ 1 & 3 \end{pmatrix}$$

$$= (-1-15) + 2(-3-5) + 23(9-1)$$

$$= -16 - 16 + 23(8) = 152$$

Note that evaluating the determinant of matrix $\mathbf{B}$ is much easier than trying to find the determinant of matrix $\mathbf{A}$, because of the simple integer values of matrix $\mathbf{B}$. We would prefer to work with integers rather than fractions.

Substituting this, $\det(\mathbf{B}) = 152$, into (†) gives

$$(23 \times 6 \times 11)\det(\mathbf{A}) = 152$$

$$\det(\mathbf{A}) = \frac{152}{(23 \times 6 \times 11)} = \frac{152}{1518} = \frac{76}{759}$$

Hence $\det(\mathbf{A}) = \frac{76}{759}$. Note that we don't have to involve fractions until the end, which makes the evaluation a lot easier.

6.3.5 The determinant of larger matrices

Proposition (6.19) stated that: Determinant of a triangular matrix is the product of the entries along the leading diagonal.

This implies that if we can use row operations to convert a large matrix to a triangular or diagonal matrix, then evaluating its determinant is simply a case of finding the product of the entries along its leading diagonal. The next two examples demonstrate this.

Example 6.14

Find the determinant of the following 4 by 4 matrix by using row operations:

$$\mathbf{A} = \begin{pmatrix} \boxed{1} & 2 & 2 & 4 \\ 7 & \boxed{8} & 3 & 0 \\ 3 & 2 & \boxed{0} & 0 \\ 1 & 0 & 0 & \boxed{0} \end{pmatrix}$$

Solution

Can we convert this matrix $\mathbf{A}$ *into a triangular matrix?*

Yes, by using row operations. Note that the given matrix $\mathbf{A}$ is *not* a triangular matrix because both sides of the leading diagonal contain non-zero entries. You will need to recall your work from chapter 1 on row operations.

(continued...)

How can we convert the matrix **A** *into a triangular matrix?*
First we label the rows of matrix **A** and then we apply row operations:

$$\begin{matrix} R_1 \\ R_2 \\ R_3 \\ R_4 \end{matrix} \begin{pmatrix} 1 & 2 & 2 & 4 \\ 7 & 8 & 3 & 0 \\ 3 & 2 & 0 & 0 \\ 1 & 0 & 0 & 0 \end{pmatrix}$$

Swapping top and bottom rows R_1 and R_4, and middle rows R_2 and R_3 we have

$$\begin{matrix} R_4 \\ R_3 \\ R_2 \\ R_1 \end{matrix} \begin{pmatrix} \boxed{1} & 0 & 0 & 0 \\ 3 & \boxed{2} & 0 & 0 \\ 7 & 8 & \boxed{3} & 0 \\ 1 & 2 & 2 & \boxed{4} \end{pmatrix}$$

What is the determinant of this matrix?
We have a *lower triangular* matrix, so the determinant is the product of the entries on the leading diagonal, that is $1 \times 2 \times 3 \times 4 = 24$.
What is the determinant of the given matrix **A**?
The bottom matrix is obtained from matrix **A** by interchanging rows R_1 and R_4, R_2 and R_3.
How does interchanging rows affect the determinant of the matrix?
By result (6.23) part (ii):

$$\det(\mathbf{B}) = -\det(\mathbf{A}) \quad \text{if two rows have been interchanged}$$

Interchanging rows multiplies the determinant by -1. We have *two interchanges*, therefore the determinant of the matrix **A** is given by

$$(-1)(-1)\det(\mathbf{A}) = 24 \text{ which gives } \det(\mathbf{A}) = 24$$

Each time an *interchange* is made, the determinant is *multiplied by* -1. This is much simpler than trying to expand along a row or column of the 4 by 4 matrix **A**.

Example 6.15

Find the determinant of the following matrix:

$$\mathbf{A} = \begin{pmatrix} 5 & 1 & 1 & 1 & 1 \\ 1 & 5 & 1 & 1 & 1 \\ 5 & 1 & 5 & 1 & 1 \\ 1 & 5 & 1 & 5 & 1 \\ 5 & 1 & 5 & 1 & 5 \end{pmatrix}$$

Solution
Expanding along a row or column to find the determinant of this 5 by 5 matrix would be a tedious task. We aim to convert this into a triangular matrix and then find the determinant.

Labelling the rows of this matrix we have

$$\begin{matrix} R_1 \\ R_2 \\ R_3 \\ R_4 \\ R_5 \end{matrix} \begin{pmatrix} 5 & 1 & 1 & 1 & 1 \\ 1 & 5 & 1 & 1 & 1 \\ 5 & 1 & 5 & 1 & 1 \\ 1 & 5 & 1 & 5 & 1 \\ 5 & 1 & 5 & 1 & 5 \end{pmatrix}$$

Executing the following row operations:

$$\begin{matrix} R_1 \\ R_2^* = R_2 - R_1 \\ R_3^* = R_3 - R_1 \\ R_4^* = R_4 - R_2 \\ R_5^* = R_5 - R_3 \end{matrix} \begin{pmatrix} 5 & 1 & 1 & 1 & 1 \\ -4 & 4 & 0 & 0 & 0 \\ 0 & 0 & 4 & 0 & 0 \\ 0 & 0 & 0 & 4 & 0 \\ 0 & 0 & 0 & 0 & 4 \end{pmatrix}$$

Carrying out the row operation $5R_2^*$ gives

$$\begin{matrix} R_1 \\ R_2^{**} = 5R_2^* \\ R_3^* \\ R_4^* \\ R_5^* \end{matrix} \begin{pmatrix} 5 & 1 & 1 & 1 & 1 \\ -20 & 20 & 0 & 0 & 0 \\ 0 & 0 & 4 & 0 & 0 \\ 0 & 0 & 0 & 4 & 0 \\ 0 & 0 & 0 & 0 & 4 \end{pmatrix}$$

Executing the row operation $R_2^{**} + 4R_1$ yields

$$\begin{matrix} R_1 \\ R_2^{**} + 4R_1 \\ R_3^* \\ R_4^* \\ R_5^* \end{matrix} \begin{pmatrix} 5 & 1 & 1 & 1 & 1 \\ 0 & 24 & 4 & 4 & 4 \\ 0 & 0 & 4 & 0 & 0 \\ 0 & 0 & 0 & 4 & 0 \\ 0 & 0 & 0 & 0 & 4 \end{pmatrix} = \mathbf{B}$$

What is the determinant of this last matrix $\mathbf{B}$*?*

We have an upper triangular matrix so that the determinant is the product of all the entries on the leading diagonal, that is $\det(\mathbf{B}) = 5 \times 24 \times 4 \times 4 \times 4 = 7680$.

What is the determinant of the given matrix $\mathbf{A}$*?*

All the above row operations apart from $5R_2^*$ makes *no* difference to the determinant.

Why not?

Because by (6.23) (i) adding a multiple of one row to another has the same determinant:

$\det(\mathbf{B}) = \det(\mathbf{A})$ if a multiple of one row is added to another.

How does the row operation $5R_2^*$ *change the determinant?*

By (6.23) (iii) $\det(\mathbf{B}) = k\det(\mathbf{A})$ if a row has been multiplied by non-zero k.

We have the above determinant $\det(\mathbf{B}) = 7680$ is equal to $5 \times \det(\mathbf{A})$, that is

$$5\det(\mathbf{A}) = 7680 \text{ which gives } \det(\mathbf{A}) = \frac{7680}{5} = 1536$$

Generally, to find the determinant of matrices of size 4 by 4 or larger it is a lot easier to try to convert these into a triangular matrix. To do this, we apply the row operations as discussed above.

6.3.6 Further properties of matrices

Proposition (6.24). Let $\mathbf{E}$ be an elementary matrix. For any square matrix $\mathbf{A}$ we have

$$\det(\mathbf{EA}) = \det(\mathbf{E})\det(\mathbf{A}) \text{ where } \mathbf{EA} \text{ is a valid operation}$$

What does this proposition mean?

It means that we can find the determinant of each individual matrix, $\mathbf{E}$ and $\mathbf{A}$, and then multiply the scalars $\det(\mathbf{E})$ and $\det(\mathbf{A})$ to give $\det(\mathbf{EA})$.

Proof.

We consider the *three* different cases of elementary matrices separately.

Case 1. Let $\mathbf{E}$ be the elementary matrix obtained from the identity matrix by adding a multiple of one row to another. Then from chapter 1 we have $\mathbf{EA}$ performs the *same row operation* of adding one row to another of matrix $\mathbf{A}$. By result (6.23) (i):

$\det(\mathbf{B}) = \det(\mathbf{A})$ if a row has been multiplied by non-zero k.

We have

$$\det(\mathbf{EA}) = \det(\mathbf{A})$$

By the first line of result (6.21):

$\det(\mathbf{E}) = 1$ if a multiple of one row is added to another. We have $\det(\mathbf{E}) = 1$ therefore

$$\begin{aligned}\det(\mathbf{EA}) &= \det(\mathbf{A})\\ &= 1 \times \det(\mathbf{A}) = \det(\mathbf{E}) \times \det(\mathbf{A})\end{aligned}$$

Hence for this case we have $\det(\mathbf{EA}) = \det(\mathbf{E})\det(\mathbf{A})$.

Case 2. Let $\mathbf{E}$ be the elementary matrix obtained from the identity matrix by multiplying a row by a non-zero scalar k. By the bottom line of result (6.21):

$\det(\mathbf{E}) = k$ if a row has been multiplied by non-zero k

We have $\det(\mathbf{E}) = k$. The matrix multiplication $\mathbf{EA}$ performs the same row operation of multiplying a row by a scalar k on matrix $\mathbf{A}$. By

(6.23) (iii): $\det(\mathbf{B}) = k\det(\mathbf{A})$ if a row has been multiplied by non-zero k

We have $\det(\mathbf{EA}) = k\det(\mathbf{A})$.

$$\det(\mathbf{EA}) = \underbrace{k}_{=\det(\mathbf{E})}\det(\mathbf{A}) = \det(\mathbf{E})\det(\mathbf{A})$$

We have proved $\det(\mathbf{EA}) = \det(\mathbf{E})\det(\mathbf{A})$ for the second case.

Case 3. See website.

■

Proposition (6.25). Let $\mathbf{E}_1, \mathbf{E}_2, \ldots$ and $\mathbf{E}_k$ be elementary matrices and $\mathbf{B}$ be a square matrix of the same size. Then

$$\det(\mathbf{E}_1\mathbf{E}_2\cdots\mathbf{E}_k\mathbf{B}) = \det(\mathbf{E}_1) \times \det(\mathbf{E}_2) \times \cdots \times \det(\mathbf{E}_k) \times \det(\mathbf{B})$$

Proof – Exercises 6.3.

■

Next we prove an *important test* to identify invertible (non-singular) matrices.

Theorem (6.26). A square matrix $\mathbf{A}$ is invertible (has an inverse) $\Leftrightarrow$ $\det(\mathbf{A}) \neq 0$.

What does this proposition mean?

It means that if matrix $\mathbf{A}$ is *invertible* then the determinant of $\mathbf{A}$ does *not* equal zero. Also if the determinant is *not* equal to zero then the matrix $\mathbf{A}$ is invertible (non-singular). This result works both ways.

How do we prove this result?

($\Rightarrow$). First we assume matrix $\mathbf{A}$ is invertible and show that this leads to $\det(\mathbf{A}) \neq 0$.
($\Leftarrow$). Then we assume that $\det(\mathbf{A}) \neq 0$ and prove that matrix $\mathbf{A}$ is invertible.

Proof.

($\Rightarrow$). Assume the matrix $\mathbf{A}$ is invertible. By Theorem (1.29) part (d):

$\mathbf{A}$ is invertible $\Leftrightarrow$ $\mathbf{A}$ is a product of elementary matrices

We have matrix $\mathbf{A}$ is a product of elementary matrices. We can write the matrix $\mathbf{A}$ as

$$\mathbf{A} = \mathbf{E}_1\mathbf{E}_2\mathbf{E}_3\cdots\mathbf{E}_k$$

where $\mathbf{E}_1, \mathbf{E}_2, \mathbf{E}_3, \ldots$ and $\mathbf{E}_k$ are elementary matrices. By the above Proposition (6.25):

$\det(\mathbf{E}_1\mathbf{E}_2\mathbf{E}_3\cdots\mathbf{E}_k\mathbf{B}) = \det(\mathbf{E}_1)\det(\mathbf{E}_2)\det(\mathbf{E}_3)\cdots\det(\mathbf{E}_k)\det(\mathbf{B})$

We have

$$\begin{aligned}\det(\mathbf{A}) &= \det(\mathbf{E}_1\mathbf{E}_2\mathbf{E}_3\cdots\mathbf{E}_k)\\ &= \det(\mathbf{E}_1)\det(\mathbf{E}_2)\det(\mathbf{E}_3)\cdots\det(\mathbf{E}_k)\end{aligned}$$

By (6.21):

$$\det(\mathbf{E}) = \begin{cases} 1 & \text{if a multiple of one row is added to another} \\ -1 & \text{if two rows have been interchanged} \\ k & \text{if a row has been multiplied by non-zero } k \end{cases}$$

The determinant of an elementary matrix can only be 1, -1 or the non-zero k. Multiplying these non-zero real numbers, $\det(\mathbf{E}_1) \times \det(\mathbf{E}_2) \times \cdots \times \det(\mathbf{E}_k)$, *cannot* give 0. Therefore $\det(\mathbf{A}) \neq 0$.

Now we go the other way ($\Leftarrow$). Assume $\det(\mathbf{A}) \neq 0$ then by Proposition (6.13) $\mathbf{A}^{-1} = \frac{1}{\det(\mathbf{A})}adj(\mathbf{A})$ which means that matrix $\mathbf{A}$ has an inverse so it is invertible.

■

This result says that *invertible* matrix $\mathbf{A}$ is equivalent to $\det(\mathbf{A}) \neq 0$.

We can add this to the test for unique solutions of linear systems described in chapter 1: Proposition (1.37). The linear system $\mathbf{Ax} = \mathbf{b}$ has a unique solution $\Leftrightarrow$ $\mathbf{A}$ is invertible. Hence we have:

Proposition (6.27). The linear system $\mathbf{Ax} = \mathbf{b}$ has a unique solution $\Leftrightarrow$ $\det(\mathbf{A}) \neq 0$.

We can extend Proposition (6.24) to any two square matrices of the same size as the next proposition states.

Proposition (6.28). If $\mathbf{A}$ and $\mathbf{B}$ are square matrices of the same size then

$$\det(\mathbf{AB}) = \det(\mathbf{A})\det(\mathbf{B})$$

We can see this result by examining transformations.

Consider the transformations $T(\mathbf{x}) = \mathbf{Ax}$ and $S(\mathbf{x}) = \mathbf{Bx}$. The transformation S changes the volume (area for 2 by 2) by $\det(\mathbf{B})$ and T changes the volume (or area) by $\det(\mathbf{A})$. The composite transform $(T \circ S)(\mathbf{x}) = \mathbf{AB}(\mathbf{x})$ must change the volume first by $\det(\mathbf{B})$ and then by $\det(\mathbf{A})$ so the overall volume change is $\det(\mathbf{B}) \times \det(\mathbf{A})$. Hence

$$\det(\mathbf{AB}) = \det(\mathbf{A})\det(\mathbf{B})$$

Proof.

We consider the two cases of matrix $\mathbf{A}$. Case 1 is where the matrix $\mathbf{A}$ is invertible (has an inverse) and case 2 is where matrix $\mathbf{A}$ is non-invertible.

Case 1. Assume the matrix $\mathbf{A}$ is invertible. Then by Theorem (1.29) part (d):

$\mathbf{A}$ is invertible $\Leftrightarrow$ $\mathbf{A}$ is a product of elementary matrices

We have matrix $\mathbf{A}$ is a product of elementary matrices. We can write

$$\mathbf{A} = \mathbf{E}_1\mathbf{E}_2\mathbf{E}_3\cdots\mathbf{E}_k$$

where $\mathbf{E}_1, \mathbf{E}_2, \mathbf{E}_3, \ldots$ and $\mathbf{E}_k$ are elementary matrices. We have

$$\begin{aligned}
\det(\mathbf{AB}) &= \det(\mathbf{E}_1\mathbf{E}_2\mathbf{E}_3\cdots\mathbf{E}_k\mathbf{B}) \\
&= \det(\mathbf{E}_1)\det(\mathbf{E}_2)\det(\mathbf{E}_3)\cdots\det(\mathbf{E}_k)\det(\mathbf{B}) && \left[\text{by (6.25)}\right] \\
&= \det\Big(\underbrace{\mathbf{E}_1\mathbf{E}_2\mathbf{E}_3\cdots\mathbf{E}_k}_{=\mathbf{A}}\Big)\det(\mathbf{B}) = \det(\mathbf{A})\det(\mathbf{B})
\end{aligned}$$

Case 2. Assume matrix $\mathbf{A}$ is non-invertible. By the above Proposition (6.26) we conclude that $\det(\mathbf{A}) = 0$. Matrix $\mathbf{A}$ is non-invertible, therefore matrix multiplication $\mathbf{AB}$ is also non-invertible.

Why?

Because Proposition (1.27) says: $(\mathbf{AB})^{-1} = \mathbf{B}^{-1}\mathbf{A}^{-1}$

Since $\mathbf{AB}$ is also non-invertible, we have $\det(\mathbf{AB}) = 0$. Hence we have our result

$$\det(\mathbf{AB}) = \det(\mathbf{A})\det(\mathbf{B}) \text{ because } \det(\mathbf{A}) = 0 \text{ and } \det(\mathbf{AB}) = 0$$

■

Again, we can extend the result of Proposition (6.28) to n square matrices as the next proposition states.

Proposition (6.29). If $\mathbf{A}_1, \mathbf{A}_2, \mathbf{A}_3, \ldots$ and $\mathbf{A}_n$ are square matrices of the same size then

$$\det(\mathbf{A}_1\mathbf{A}_2\cdots\mathbf{A}_n) = \det(\mathbf{A}_1) \times \det(\mathbf{A}_2) \times \cdots \times \det(\mathbf{A}_n)$$

Proof – Exercises 6.3.

■

What use is this result?

Knowing that

$$\det(\mathbf{A}_1\mathbf{A}_2\cdots\mathbf{A}_n) = \det(\mathbf{A}_1) \times \det(\mathbf{A}_2) \times \cdots \times \det(\mathbf{A}_n)$$

allows us to test whether the matrix $\mathbf{A}_1\mathbf{A}_2\cdots\mathbf{A}_n$ is invertible or not. This is much faster than multiplying $\mathbf{A}_1, \mathbf{A}_2, \ldots, \mathbf{A}_n$ and then testing for invertibility.

Generally a function in mathematics which has the properties of Proposition (6.28) is called a **multiplicative function.**

In mathematics, we say a function f is multiplicative if

$$f(xy) = f(x)f(y)$$

The determinant is an example of a multiplicative function. Another example is the square root function, because $\sqrt{ab} = \sqrt{a}\sqrt{b}$.

Proposition (6.30). If $\mathbf{A}$ is an invertible (non-singular) matrix then $\det\left(\mathbf{A}^{-1}\right) = \dfrac{1}{\det(\mathbf{A})}$.

Proof – Exercises 6.3.

■

Summary

The determinant of a triangular or diagonal matrix is the product of the entries along the leading diagonal.

$$(6.23) \qquad \det(\mathbf{B}) = \begin{cases} \text{(i)} \ \det(\mathbf{A}) & \text{if a multiple of one row is added to another} \\ \text{(ii)} \ -\det(\mathbf{A}) & \text{if two rows have been interchanged} \\ \text{(iii)} \ k\det(\mathbf{A}) & \text{if a row has been multiplied by non-zero } k \end{cases}$$

If $\mathbf{A}$ and $\mathbf{B}$ are square matrices of the same size then $\det(\mathbf{AB}) = \det(\mathbf{A})\det(\mathbf{B})$.
A square matrix $\mathbf{A}$ is invertible $\Leftrightarrow$ the determinant of $\mathbf{A}$ does *not* equal zero.

EXERCISES 6.3

(Brief solutions at end of book. Full solutions available at <http://www.oup.co.uk/companion/singh>.)

1. Compute the determinants of the following matrices:

$$\text{(a) } \mathbf{A} = \begin{pmatrix} 1 & 0 & 0 \\ 0 & -10 & 0 \\ 0 & 0 & 1 \end{pmatrix} \quad \text{(b) } \mathbf{B} = \begin{pmatrix} 1 & 0 & 0 \\ 0 & 0 & 1 \\ 0 & 1 & 0 \end{pmatrix} \quad \text{(c) } \mathbf{C} = \begin{pmatrix} 1 & 0 & 0 \\ 0 & 1 & 1 \\ 0 & 0 & 1 \end{pmatrix}$$

$$\text{(d) } \mathbf{D} = \begin{pmatrix} 0 & 0 & 0 & 1 \\ 0 & 1 & 0 & 0 \\ 0 & 0 & 1 & 0 \\ 1 & 0 & 0 & 0 \end{pmatrix} \quad \text{(e) } \mathbf{E} = \begin{pmatrix} 1 & 0 & 0 & 0 \\ 0 & 1 & 0 & 0 \\ 0 & 0 & -0.6 & 0 \\ 0 & 0 & 0 & 1 \end{pmatrix} \quad \text{(f) } \mathbf{F} = \begin{pmatrix} 1 & 0 & 0 & 0 \\ 0 & 1 & -7 & 0 \\ 0 & 0 & 1 & 0 \\ 0 & 0 & 0 & 1 \end{pmatrix}$$

2. Find the determinants of the following matrices:

$$\text{(a) } \mathbf{A} = \begin{pmatrix} 1 & 0 & 0 \\ 0 & 2 & 0 \\ 0 & 0 & 3 \end{pmatrix} \quad \text{(b) } \mathbf{B} = \begin{pmatrix} 1 & -1 & 7 \\ 0 & 2 & 9 \\ 0 & 0 & 3 \end{pmatrix} \quad \text{(c) } \mathbf{C} = \begin{pmatrix} -2 & 0 & 0 \\ -98 & -3 & 0 \\ 67 & 17 & 1 \end{pmatrix}$$

$$\text{(d) } \mathbf{D} = \begin{pmatrix} 1 & -3 & 5 & 9 \\ 0 & -3 & 67 & 9 \\ 0 & 0 & 8 & 90 \\ 0 & 0 & 0 & 3 \end{pmatrix} \quad \text{(e) } \mathbf{E} = \begin{pmatrix} 10 & 0 & 0 & 0 \\ 0 & 20 & 0 & 0 \\ 0 & 0 & 30 & 0 \\ 0 & 0 & 0 & 40 \end{pmatrix}$$

$$\text{(f) } \mathbf{F} = \begin{pmatrix} -9 & 0 & 0 & 0 \\ 67 & 3 & 0 & 0 \\ 67 & 57 & 7 & 0 \\ 23 & 78 & 6 & 5 \end{pmatrix} \quad \text{(g) } \mathbf{G} = \begin{pmatrix} 1 & 0 & 0 \\ 2 & 2 & 0 \\ -5 & 3 & 4 \\ 9 & -8 & 1 \end{pmatrix}$$

3. Compute the determinants of the following matrices:

$$\text{(a) } \mathbf{A} = \begin{pmatrix} \alpha & -\beta & \delta \\ 0 & \beta & -\alpha \\ 0 & 0 & \gamma \end{pmatrix} \quad \text{(b) } \mathbf{B} = \begin{pmatrix} \sin(\theta) & 0 & 0 \\ 1 & \cos(\theta) & 0 \\ -2 & \sin(\theta) & 2 \end{pmatrix} \quad \text{(c) } \mathbf{C} = \begin{pmatrix} x & y & z \\ 0 & y & 0 \\ 0 & 0 & z \end{pmatrix}$$

4. Find the value(s) of x which make the following matrix non-invertible (singular):

$$\mathbf{A} = \begin{pmatrix} 1 & 2 & x \\ 3 & x & 4 \\ 5 & 5 & x \end{pmatrix}$$

5. Find the determinant of the following matrices:

$$\text{(a) } \mathbf{A} = \begin{pmatrix} 0 & 0 & 0 & 9 \\ 0 & 1 & 4 & 5 \\ 0 & 0 & 3 & 5 \\ 1 & 2 & 3 & 4 \end{pmatrix} \quad \text{(b) } \mathbf{A} = \begin{pmatrix} 1 & 2 & 3 & 4 \\ 1 & 3 & 5 & 6 \\ 1 & 4 & 3 & 7 \\ 1 & 6 & 1 & 9 \end{pmatrix} \quad \text{(c) } \mathbf{A} = \begin{pmatrix} 1 & 2 & 5 & 7 \\ 3 & 6 & 2 & 8 \\ -1 & -2 & 8 & 7 \\ -4 & -5 & 1 & 2 \end{pmatrix}$$

6. Evaluate the determinant of the following matrices:

$$\text{(a) } \mathbf{A} = \begin{pmatrix} 1 & 1 & 0 & 0 & 1 \\ -1 & 2 & 0 & 0 & 0 \\ 0 & 0 & 1 & 4 & 16 \\ 0 & 0 & 1 & 5 & 25 \\ 0 & 0 & 1 & 2 & 4 \end{pmatrix} \quad \text{(b) } \mathbf{A} = \begin{pmatrix} 1 & 2 & 3 & 4 & 5 \\ 2 & 5 & 8 & 11 & 1 \\ 7 & 6 & 1 & 9 & 8 \\ 4 & 10 & 16 & 22 & 2 \\ 2 & 3 & 7 & 9 & 5 \end{pmatrix}$$

7. (a) Show that $\det(\mathbf{A}+\mathbf{B}) \neq \det(\mathbf{A})+\det(\mathbf{B})$.

   (b) Show that $\det(\mathbf{u}\ \ \mathbf{v}\ \ \mathbf{w}) \neq \det(\mathbf{v}\ \ \mathbf{u}\ \ \mathbf{w})$.

   (c) Show that $\det(k\mathbf{A}) \neq k\det(\mathbf{A})$ where $\mathbf{A}$ is an n by n matrix and $n \geq 2$.

8. Evaluate the determinants of the following matrices:

   (a) $\mathbf{A} = \begin{pmatrix} 1/2 & 1/2 & -1/2 \\ 2 & 3 & 4 \\ 1 & -1 & 1 \end{pmatrix}$ (b) $\mathbf{B} = \begin{pmatrix} 1/2 & 1/3 & 1 \\ 1/7 & 2/21 & 1/21 \\ 2/3 & 1/3 & -4/3 \end{pmatrix}$

   (c) $\mathbf{C} = \begin{pmatrix} 10 & 20 & -30 \\ -4 & 5 & -6 \\ -70 & 80 & -90 \end{pmatrix}$

9. Let $\mathbf{A}$, $\mathbf{B}$ be 3 by 3 matrices with $\det(\mathbf{A}) = 3$, $\det(\mathbf{B}) = -4$. Determine
   (a) $\det(-2\mathbf{AB})$ (b) $\det(\mathbf{A}^5\mathbf{B}^6)$ (Hint: See question 19 below.)
   (c) $\det(\mathbf{A}^{-1}\mathbf{A}^T)$

10. Let $\mathbf{A} = \begin{pmatrix} 1 & 2 \\ 3 & 4 \end{pmatrix}$, $\mathbf{B} = \begin{pmatrix} 5 & 6 \\ 7 & 8 \end{pmatrix}$ and $\mathbf{C} = \begin{pmatrix} 9 & 10 \\ 11 & 12 \end{pmatrix}$. Decide whether the matrix $\mathbf{A} \times \mathbf{B} \times \mathbf{C}$ is invertible.

11. Without evaluating the determinants of the following matrices, decide whether they are positive, negative or zero:

    (a) $\begin{pmatrix} 0 & 0 & 0 & 61 \\ 0 & 89 & 0 & 0 \\ 0 & 0 & 98 & 0 \\ 21 & 0 & 0 & 0 \end{pmatrix}$ (b) $\begin{pmatrix} 2 & 5 & 7 & 8 \\ 7 & 8 & 3 & 1 \\ 12 & 30 & 42 & 48 \\ 10 & 51 & 41 & 44 \end{pmatrix}$ (c) $\begin{pmatrix} 0 & 0 & 25 & 1 \\ 0 & 0 & -5 & 0 \\ 0 & 2 & 0 & 0 \\ 23 & 0 & 0 & 0 \end{pmatrix}$

12. Prove that the determinant of a lower triangular matrix is the product of the entries along the leading diagonal.

13. Prove that the determinant of a diagonal matrix is the product of the entries along the leading diagonal
    (a) by using induction (b) without using induction

14. Prove parts (i) and (ii) of Proposition (6.22).

15. Prove Proposition (6.30).

16. Prove that the matrix $\mathbf{A}$ is invertible (non-singular) if and only if the matrix multiplication $\mathbf{A}^T\mathbf{A}$ is invertible (non-singular).

17. Prove Proposition (6.25).

18. Prove Proposition (6.29).

19. Prove that $\det(\mathbf{A}^n) = [\det(\mathbf{A})]^n$ where $\mathbf{A}$ is a square matrix.

20. Let $\mathbf{A}$ and $\mathbf{B}$ be square matrices. Prove that $\det(\mathbf{AB}) = \det(\mathbf{BA})$.

21. If a square matrix $\mathbf{A}$ satisfies $\mathbf{A}^T\mathbf{A} = \mathbf{I}$ ($\mathbf{A}$ is an orthogonal matrix) where $\mathbf{I}$ is the identity matrix, show that $\det(\mathbf{A}) = \pm 1$.

22. Let $\mathbf{A}$ be an invertible matrix. Show that $\det\left(\left(\mathbf{A}^T\right)^{-1}\right) = \dfrac{1}{\det(\mathbf{A})}$.

23. Let $\mathbf{A}$ be a matrix which contains a multiple of a row within the matrix. Show that $\det(\mathbf{A}) = 0$.
24. We say matrices $\mathbf{A}$ and $\mathbf{B}$ are similar if there exists an invertible matrix $\mathbf{P}$ such that $\mathbf{B} = \mathbf{P}^{-1}\mathbf{AP}$. Prove that if $\mathbf{A}$ and $\mathbf{B}$ are similar matrices then $\det(\mathbf{B}) = \det(\mathbf{A})$.
25. Prove that $\det\left(adj(\mathbf{A})\right) = \left[\det(\mathbf{A})\right]^{n-1}$ where matrix $\mathbf{A}$ is an n by n invertible matrix.
26. The linear system $\mathbf{Ax} = \mathbf{0}$ has an infinite number of solutions $\Leftrightarrow \det(\mathbf{A}) = 0$. Prove this proposition.

SECTION 6.4 LU Factorization

By the end of this section you will be able to

- find the **LU** factorization or decomposition of a matrix
- solve linear equations

6.4.1 Introduction to LU factorization

(**L** is a lower triangular matrix and **U** is an upper triangular matrix.)

Solving a system of equations of the form $\mathbf{Ax} = \mathbf{b}$, where $\mathbf{A}$ is a 4 by 4 or larger matrix, can be a tedious task. For example, to calculate the inverse of such a matrix is a lengthy process. We are interested in finding the inverse of matrix $\mathbf{A}$ because we may want to solve $\mathbf{Ax} = \mathbf{b}$ for various different $\mathbf{b}$ vectors.

In this section we establish a new numerical method to solve such linear systems.

Example 6.16

(i) Convert $\mathbf{A} = \begin{pmatrix} 1 & 1 \\ 3 & 2 \end{pmatrix}$ into an upper triangular matrix $\mathbf{U}$ by using row operations.

(ii) Find a lower triangular matrix $\mathbf{L}$ such that $\mathbf{LU} = \mathbf{A}$.

Solution

(i) *What is an upper triangular matrix?*

All the entries *below* the leading diagonal are zero. This means that we need to convert the 3 in the bottom row of matrix $\mathbf{A}$ into zero.

How?

By applying row operations. Subtracting three times the top row from the bottom, $R_2 - 3R_1$:

$$\begin{matrix} R_1 \\ R_2 \end{matrix} \begin{pmatrix} 1 & 1 \\ 3 & 2 \end{pmatrix} \Longrightarrow \begin{matrix} R_1 \\ R_2 - 3R_1 \end{matrix} \begin{pmatrix} 1 & 1 \\ 0 & -1 \end{pmatrix} = \mathbf{U}$$

(ii) A lower triangular matrix has the form $\mathbf{L} = \begin{pmatrix} a & 0 \\ b & c \end{pmatrix}$ where all the entries *above* the leading diagonal are zero. Substituting this $\mathbf{L}$ and $\mathbf{U}$ into $\mathbf{LU} = \mathbf{A}$ gives:

$$\mathbf{LU} = \begin{pmatrix} a & 0 \\ b & c \end{pmatrix}\begin{pmatrix} 1 & 1 \\ 0 & -1 \end{pmatrix}$$

$$\underset{\text{matrix multiplication}}{=} \begin{pmatrix} a & a \\ b & b-c \end{pmatrix} = \begin{pmatrix} 1 & 1 \\ 3 & 2 \end{pmatrix} = \mathbf{A} \quad \text{implies} \quad a = 1,\ b = 3 \text{ and } c = 1$$

Hence $\mathbf{L} = \begin{pmatrix} a & 0 \\ b & c \end{pmatrix} = \begin{pmatrix} 1 & 0 \\ 3 & 1 \end{pmatrix}$. We can check this by multiplying out the matrices:

$$\mathbf{LU} = \begin{pmatrix} 1 & 0 \\ 3 & 1 \end{pmatrix}\begin{pmatrix} 1 & 1 \\ 0 & -1 \end{pmatrix} = \begin{pmatrix} 1 & 1 \\ 3 & 2 \end{pmatrix} = \mathbf{A}$$

This result means that we can factorize the matrix $\mathbf{A}$ into $\mathbf{LU}$ where $\mathbf{L}$ and $\mathbf{U}$ are lower and upper triangular matrices respectively.

Why do we want to break a matrix into **LU**?

As mentioned above, finding the inverse of a matrix by using cofactors is very slow and cumbersome. If we wanted to invert a 30 by 30 matrix by using cofactors it would take more than the lifetime of the universe. We can use the $\mathbf{LU}$ factorization of a matrix $\mathbf{A}$ to solve linear systems $\mathbf{Ax} = \mathbf{b}$ and also to find the inverse of an invertible matrix $\mathbf{A}$ because it is a lot more efficient.

Solving linear systems of equations of the form $\mathbf{Ax} = \mathbf{b}$ is the fundamental concept of linear algebra. If the linear system is not too large, we can solve it by using a direct method such as *Gaussian elimination* which was described in chapter 1. This Gaussian elimination procedure gives us an upper triangular matrix, but computer software tends to use the $\mathbf{LU}$ factorization method, which is one of the fastest ways that computers can solve $\mathbf{Ax} = \mathbf{b}$. The numerical software MATLAB uses $\mathbf{LU}$ factorization.

This factorization is particularly useful if we need to solve many linear systems of the form $\mathbf{Ax} = \mathbf{b}$ for different $\mathbf{b}$ vectors but the same coefficient matrix $\mathbf{A}$.

But why don't we use Gaussian elimination?

We do use the Gaussian elimination procedure to get the upper triangular matrix $\mathbf{U}$, but this is not sufficient.

Why not?

We would have to repeat all the steps of Gaussian elimination for every different $\mathbf{b}$ vector, which would be tedious, especially if we had 100 or more different $\mathbf{b}$ vectors.

In this section, our aim is to factorize a given matrix $\mathbf{A}$ into $\mathbf{LU}$ where $\mathbf{L}$ is a lower triangular matrix and $\mathbf{U}$ is an upper triangular matrix. We can illustrate this as shown in Fig. 6.12

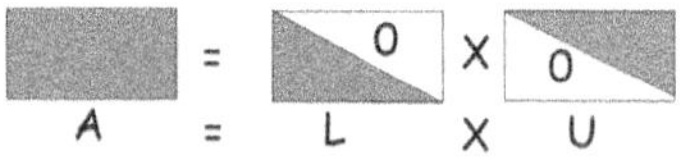

Figure 6.12

6.4.2 LU factorization procedure

Given a matrix $\mathbf{A}$*, can we form an upper triangular matrix* $\mathbf{U}$ *by using the following two row operations?*

1. Multiplying a row by a non-zero constant.
2. Adding a multiple of one row to another.

If the answer to this question is yes then we can form an **LU** factorization of the matrix **A**. Suppose these row operations are represented by the elementary matrices:

$$\mathbf{E}_1,\ \mathbf{E}_2,\ \mathbf{E}_3,\ \ldots,\ \mathbf{E}_k$$

Remember, an elementary matrix is a matrix with *one* row operation applied to an identity matrix. These are all examples of 2 by 2 elementary matrices:

$$\begin{pmatrix} 2 & 0 \\ 0 & 1 \end{pmatrix},\ \begin{pmatrix} 1 & 2 \\ 0 & 1 \end{pmatrix},\ \begin{pmatrix} 0 & 1 \\ 1 & 0 \end{pmatrix}$$

For **LU** factorization we do *not* allow the row operation of swapping rows. By applying Gaussian elimination with only the above two operations (1 and 2) we have $\left(\mathbf{E}_k\mathbf{E}_{k-1} \cdots \mathbf{E}_2\mathbf{E}_1\right)\mathbf{A} = \mathbf{U}$. Taking the inverse of $\left(\mathbf{E}_k\mathbf{E}_{k-1} \cdots \mathbf{E}_2\mathbf{E}_1\right)$ gives

$$\mathbf{A} = \left(\mathbf{E}_k\mathbf{E}_{k-1} \cdots \mathbf{E}_2\mathbf{E}_1\right)^{-1}\mathbf{U}$$

If we do *not* allow the third row operation of swapping rows then this $\left(\mathbf{E}_k\mathbf{E}_{k-1} \cdots \mathbf{E}_2\mathbf{E}_1\right)^{-1}$ is a lower triangular matrix **L**; thus

$$\begin{aligned} \mathbf{L} &= \left(\mathbf{E}_k\mathbf{E}_{k-1} \cdots \mathbf{E}_2\mathbf{E}_1\right)^{-1} \\ &= \mathbf{E}_1^{-1}\mathbf{E}_2^{-1} \cdots \mathbf{E}_{k-1}^{-1}\mathbf{E}_k^{-1} && \left[\text{by (1.27) } (\mathbf{ABC})^{-1} = \mathbf{C}^{-1}\mathbf{B}^{-1}\mathbf{A}^{-1}\right] \\ &= \left(\mathbf{E}_1^{-1}\mathbf{E}_2^{-1} \cdots \mathbf{E}_{k-1}^{-1}\mathbf{E}_k^{-1}\right)\mathbf{I} \end{aligned}$$

This means that if a given matrix **A** can be converted into an upper triangular matrix **U** by only using the above row operations (1 and 2) then we can factorize matrix **A** into **LU** where **L** is a lower triangular matrix. This matrix **L** is obtained from the identity **I** by using the *reverse* row operations used to find **U**.

In the above Example 6.16 we carried out the single row operation of $R_2 - 3R_1$ to find the upper triangular matrix **U**. To find the lower triangular matrix **L** we carry out the reverse row operation of $R_2 + 3R_1$ on the identity **I** to give us the lower triangular matrix $\mathbf{L} = \begin{pmatrix} 1 & 0 \\ 3 & 1 \end{pmatrix}$.

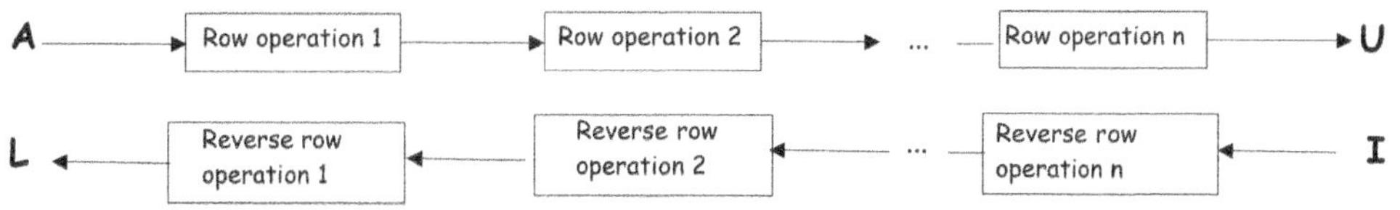

Figure 6.13

The flow chart in Fig. 6.13 shows how to obtain the upper **U** and lower **L** triangular matrices of a given matrix **A** by applying the above two row operations (1 and 2). We can apply these operations as many times as we want.

We obtain the matrix **L** by using the *reverse row* operations that were applied to **A** in finding **U**. Also we reverse the order as you can observe from the flow chart above. (Go back from **U** to **A**, or if we start with the identity **I** then we obtain the matrix **L**.)

Example 6.17

Find an **LU** factorization of $\mathbf{A} = \begin{pmatrix} 1 & 2 & 3 \\ 4 & 5 & 6 \\ 3 & -3 & 5 \end{pmatrix}$.

Solution

We apply row operations to find an upper triangular matrix **U**, which means all the entries *below* the leading diagonal are zero. Labelling rows

$$\begin{matrix} R_1 \\ R_2 \\ R_3 \end{matrix} \begin{pmatrix} 1 & 2 & 3 \\ 4 & 5 & 6 \\ 3 & -3 & 5 \end{pmatrix}$$

Steps 1 and 2: Executing the row operations $R_2 - 4R_1$ and $R_3 - 3R_1$ gives

$$\begin{matrix} R_1 \\ R_2^* = R_2 - 4R_1 \\ R_3^* = R_3 - 3R_1 \end{matrix} \begin{pmatrix} 1 & 2 & 3 \\ 0 & -3 & -6 \\ 0 & -9 & -4 \end{pmatrix}$$

Step 3: Carrying out the row operation $R_3^* - 3R_2^*$ yields

$$\begin{matrix} R_1 \\ R_2^* \\ R_3^* - 3R_2^* \end{matrix} \begin{pmatrix} 1 & 2 & 3 \\ 0 & -3 & -6 \\ 0 & 0 & 14 \end{pmatrix} = \mathbf{U}$$

We have got our upper triangular matrix **U**.

How do we obtain the lower triangular matrix **L**?

Starting with the identity matrix **I** we *reverse* the above row operations and the order:

$$\begin{matrix} R_1 \\ R_2 \\ R_3 \end{matrix} \begin{pmatrix} 1 & 0 & 0 \\ 0 & 1 & 0 \\ 0 & 0 & 1 \end{pmatrix} = \mathbf{I}$$

The *reverse* row operation of step 3 above is $R_3 + 3R_2$. Executing $R_3 + 3R_2$ on the identity matrix:

$$\begin{matrix} R_1 \\ R_2 \\ R_3^\dagger = R_3 + 3R_2 \end{matrix} \begin{pmatrix} 1 & 0 & 0 \\ 0 & 1 & 0 \\ 0 & 3 & 1 \end{pmatrix}$$

(continued...)

The reverse row operation of steps 2 and 1 are $R_3^{\dagger} + 3R_1$ and $R_2 + 4R_1$. Carrying out these row operations $R_3^{\dagger} + 3R_1$ and $R_2 + 4R_1$ gives

$$
\begin{array}{c} R_1 \\ R_2 + 4R_1 \\ R_3^{\dagger} + 3R_1 \end{array}
\begin{pmatrix} 1 & 0 & 0 \\ 4 & 1 & 0 \\ 3 & 3 & 1 \end{pmatrix} = L
$$

Hence we have our lower triangular matrix L.
By multiplying matrices we can check that $LU = A$, that is

$$
LU = \begin{pmatrix} 1 & 0 & 0 \\ 4 & 1 & 0 \\ 3 & 3 & 1 \end{pmatrix} \begin{pmatrix} 1 & 2 & 3 \\ 0 & -3 & -6 \\ 0 & 0 & 14 \end{pmatrix} = \begin{pmatrix} 1 & 2 & 3 \\ 4 & 5 & 6 \\ 3 & -3 & 5 \end{pmatrix} = A
$$

There is a slightly easier way of finding the lower triangular matrix L which is described next.

If we carry out the row operation Row $k + c$ (Row j) then c is called the **multiplier**. The negative multiplier of $+c$ is $-c$ which is used for obtaining matrix L.

Observe in Example 6.17 that the entries in the lower triangular matrix L are the *negatives* of the *multipliers* used in the row operations to obtain the upper triangular matrix U.

Here are the steps carried out above in the derivation of the upper triangular matrix U and the resulting operations on the identity matrix I to get the lower triangular matrix L:

Step 1
The first row operation was, $R_2 - 4R_1$, so our multiplier is -4. We replace the zero in $(2, 1)$ (second row, first column) position of the identity by $+4$.

Step 2
Our second row operation was $R_3 - 3R_1$ which means our multiplier is -3. We replace the zero in the $(3, 1)$ (third row, first column) position of the identity matrix with $+3$.

Step 3
Our third step was $R_3^* - 3R_2^*$. We replace the zero in the $(3, 2)$ (third row, second column) position of the identity matrix with $+3$.

Carrying out these three steps we have our lower triangular matrix L:

$$
I = \begin{pmatrix} 1 & 0 & 0 \\ 0 & 1 & 0 \\ 0 & 0 & 1 \end{pmatrix} \implies \begin{pmatrix} 1 & 0 & 0 \\ \boxed{4} & 1 & 0 \\ \boxed{3} & \boxed{3} & 1 \end{pmatrix} = L
$$

(2, 1) position

(3, 1) position

(3, 2) position

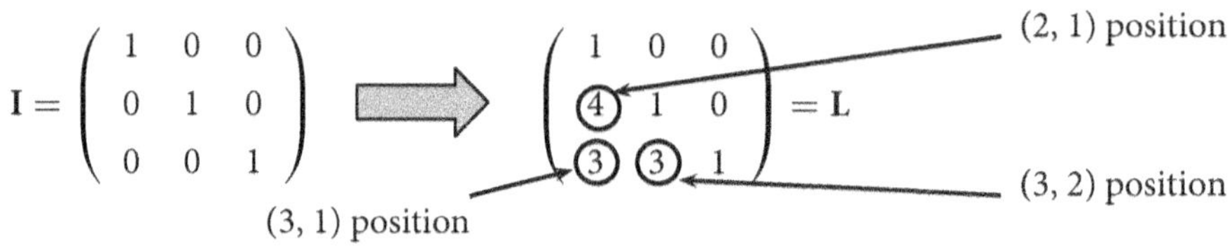

The general row operation Row $k + c$(Row j) means that we add c times row j to row k. The reverse of this operation is Row $k - c$(Row j). In order to find matrix L we carry out the reverse row operations on the identity matrix I, as shown in Fig. 6.14.

If we carry out the row operation Row $k + c$(Row j) on the given matrix A to get the upper triangular matrix U, then in the identity matrix I we replace the zero in the (k, j)

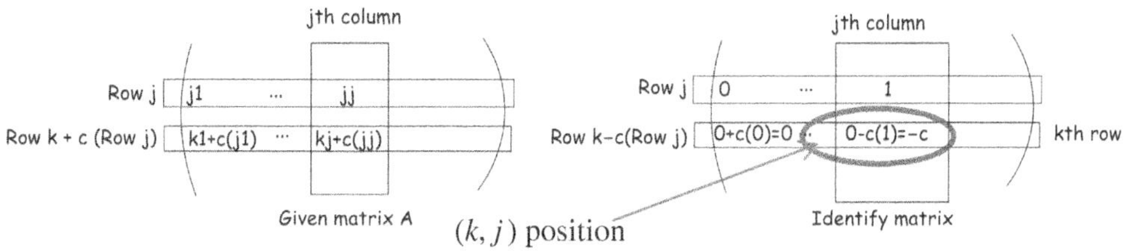

Figure 6.14

position which is row k and column j by the negative multiplier, $-c$. If we continue to do this then we obtain the lower triangular matrix **L** from the identity.

If we carry out the row operation $c \times (\text{Row } k)$ to produce a 1 on the leading diagonal in matrix **U** then we place the reciprocal $1/c$ in the same position on the leading diagonal in the matrix **L**.

This is the easiest way to find the lower triangular matrix **L** because all the work is done in finding the upper triangular matrix **U**. We do *not* need to carry out extra row operations to find **L**. However, we need to record the row operations executed in evaluating **U**.

Example 6.18

Find an **LU** factorization of $\mathbf{A} = \begin{pmatrix} 1 & 4 & 5 & 3 \\ 5 & 22 & 27 & 11 \\ 6 & 19 & 27 & 31 \\ 5 & 28 & 35 & -8 \end{pmatrix}$.

Solution

We record our row operations to find the upper triangular matrix **U**, and we use the negatives of the multipliers to get the lower triangular matrix **L** from the identity.

$$\begin{matrix} R_1 \\ R_2 \\ R_3 \\ R_4 \end{matrix} \begin{pmatrix} 1 & 4 & 5 & 3 \\ 5 & 22 & 27 & 11 \\ 6 & 19 & 27 & 31 \\ 5 & 28 & 35 & -8 \end{pmatrix} = \mathbf{A} \quad \begin{pmatrix} 1 & 0 & 0 & 0 \\ 0 & 1 & 0 & 0 \\ 0 & 0 & 1 & 0 \\ 0 & 0 & 0 & 1 \end{pmatrix} = \mathbf{I}$$

Carrying out $R_2 - 5R_1$, $R_3 - 6R_1$ and $R_4 - 5R_1$ on matrix **A** and the *negative multipliers* (+5, +6 and +5) on the identity gives:

$$\begin{matrix} R_1 \\ R_2^* = R_2 - 5R_1 \\ R_3^* = R_3 - 6R_1 \\ R_4^* = R_4 - 5R_1 \end{matrix} \begin{pmatrix} 1 & 4 & 5 & 3 \\ 0 & 2 & 2 & -4 \\ 0 & -5 & -3 & 13 \\ 0 & 8 & 10 & -23 \end{pmatrix} \begin{matrix} \\ \longleftarrow \text{multiplier} = -5 \\ \longleftarrow \text{multiplier} = -6 \\ \longleftarrow \text{multiplier} = -5 \end{matrix} \quad \begin{pmatrix} 1 & 0 & 0 & 0 \\ 5 & 1 & 0 & 0 \\ 6 & 0 & 1 & 0 \\ 5 & 0 & 0 & 1 \end{pmatrix}$$

Carrying out $R_2^*/2$ and $1/(1/2) = 2$ in position (2, 2) in the right hand matrix:

$$\begin{matrix} R_1 \\ R_2^\dagger = R_2^*/2 \\ R_3^* \\ R_4^* \end{matrix} \begin{pmatrix} 1 & 4 & 5 & 3 \\ 0 & \textcircled{1} & 1 & -2 \\ 0 & -5 & -3 & 13 \\ 0 & 8 & 10 & -23 \end{pmatrix} \longleftarrow \text{multiplier} = 1/2 \quad \begin{pmatrix} 1 & 0 & 0 & 0 \\ 5 & \textcircled{2} & 0 & 0 \\ 6 & 0 & 1 & 0 \\ 5 & 0 & 0 & 1 \end{pmatrix}$$

(continued...)

Executing $R_3^* + 5R_2^\dagger$ and $R_4^* - 8R_2^\dagger$ gives

$$\begin{matrix} R_1 \\ R_2^\dagger \\ R_3^\dagger = R_3^* + 5R_2^\dagger \\ R_4^\dagger = R_4^* - 8R_2^\dagger \end{matrix} \begin{pmatrix} 1 & 4 & 5 & 3 \\ 0 & 1 & 1 & -2 \\ 0 & \boxed{0} & 2 & 3 \\ 0 & \boxed{0} & 2 & -7 \end{pmatrix} \begin{matrix} \\ \\ \longleftarrow \text{multiplier} = +5 \\ \longleftarrow \text{multiplier} = -8 \end{matrix} \qquad \begin{pmatrix} 1 & 0 & 0 & 0 \\ 5 & 2 & 0 & 0 \\ 6 & \boxed{-5} & 1 & 0 \\ 5 & \boxed{8} & 0 & 1 \end{pmatrix}$$

Executing $R_4^\dagger - R_3^\dagger$ gives

$$\begin{matrix} R_1 \\ R_2^\dagger \\ R_3^\dagger \\ R_4^\dagger - R_3^\dagger \end{matrix} \begin{pmatrix} 1 & 4 & 5 & 3 \\ 0 & 1 & 1 & -2 \\ 0 & 0 & 2 & 3 \\ 0 & 0 & \boxed{0} & -10 \end{pmatrix} = \mathbf{U} \longleftarrow \text{multiplier} = -1 \qquad \begin{pmatrix} 1 & 0 & 0 & 0 \\ 5 & 2 & 0 & 0 \\ 6 & -5 & 1 & 0 \\ 5 & 8 & \boxed{1} & 1 \end{pmatrix} = \mathbf{L}$$

Once we can place linear systems into triangular forms, such as **L** and **U**, the linear system becomes easy to solve. Doing this involves throwing away a lot of the matrix, as you can see in the above example.

6.4.3 Solving linear systems

We want to solve a linear system $\mathbf{Ax} = \mathbf{b}$ and assume that we can break matrix **A** into **LU**. The linear system $\mathbf{Ax} = \mathbf{b}$ becomes

$$(\mathbf{LU})\,\mathbf{x} = \mathbf{b} \quad \text{implies} \quad \mathbf{L}\,(\mathbf{Ux}) = \mathbf{b}$$

Suppose applying the upper triangular matrix **U** to the vector **x** gives the vector **y**, that is $\mathbf{Ux} = \mathbf{y}$. Substituting this into the above gives

$$\mathbf{L}\,(\mathbf{Ux}) = \mathbf{b} \quad \text{implies} \quad \mathbf{Ly} = \mathbf{b}$$

We illustrate this in Fig. 6.15.

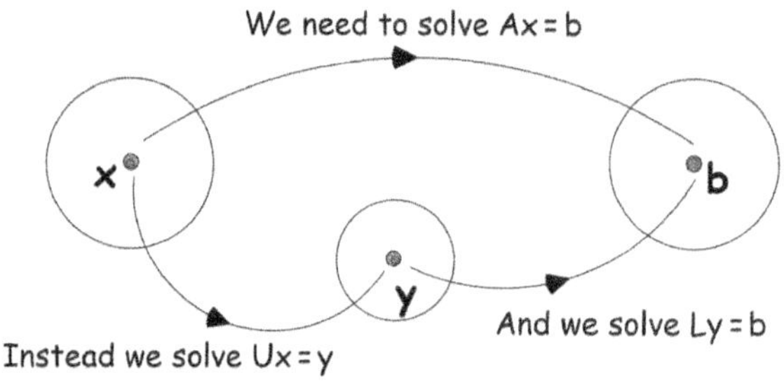

Figure 6.15

This factorization of the matrix **A** in the linear system $\mathbf{Ax} = \mathbf{b}$ into **LU** means that we need to solve two equations; $\mathbf{Ux} = \mathbf{y}$ and $\mathbf{Ly} = \mathbf{b}$.

Why should we solve two equations rather than just the given one?
Because both these $\mathbf{Ux} = \mathbf{y}$ and $\mathbf{Ly} = \mathbf{b}$ are much easier to solve than the given $\mathbf{Ax} = \mathbf{b}$.

Why is this?
Because $\mathbf{Ly} = \mathbf{b}$ means that we have a lower triangular matrix $\mathbf{L}$ times the given vector $\mathbf{b}$.

We can use a method called *forward substitution* because we have

$$\mathbf{Ly} = \begin{pmatrix} a_{11} & & O \\ \vdots & \ddots & \\ a_{m1} & \cdots & a_{mn} \end{pmatrix} \begin{pmatrix} y_1 \\ \vdots \\ y_m \end{pmatrix} = \begin{pmatrix} b_1 \\ \vdots \\ b_m \end{pmatrix} = \mathbf{b} \qquad (\dagger)$$

Expanding the top row of (†) gives

$$a_{11}y_1 = b_1 \quad \text{which implies} \quad y_1 = \frac{b_1}{a_{11}} \quad \text{provided } a_{11} \neq 0$$

Similarly we can find y_2 from the second row, and so on for the rest of the y values. By applying this we find all the entries in the vector $\mathbf{y}$.

Now we use this vector $\mathbf{y}$ to find our unknown vector $\mathbf{x}$ from $\mathbf{Ux} = \mathbf{y}$.

How can we do this?
Since $\mathbf{U}$ is an upper triangular matrix, writing out $\mathbf{Ux} = \mathbf{y}$ we have

$$\mathbf{Ux} = \begin{pmatrix} a_{11} & \cdots & a_{1n} \\ & \ddots & \\ O & & a_{mn} \end{pmatrix} \begin{pmatrix} x_1 \\ \vdots \\ x_m \end{pmatrix} = \begin{pmatrix} y_1 \\ \vdots \\ y_m \end{pmatrix} = \mathbf{y} \qquad (*)$$

This time, we use *back substitution* to find our unknowns $x_1, x_2, \ldots, x_m$.

Expanding the bottom row of (*) gives

$$a_{mn}x_m = y_m \quad \text{implies} \quad x_m = \frac{y_m}{a_{mn}} \quad \text{provided } a_{mn} \neq 0$$

Substituting these x_m into the penultimate row of (*) will produce the value for x_{m-1}. Repeating this process we can find all the unknown x's, which is what we were looking for.

Example 6.19

Solve the linear system $\mathbf{Ax} = \mathbf{b}$ where $\mathbf{A}$ is the matrix of Example 6.18 and $\mathbf{b} = (7 \;\; 13 \;\; 106 \;\; -94)^T$.

Solution

We first need to find the vector $\mathbf{y}$ such that $\mathbf{Ly} = \mathbf{b}$ where $\mathbf{L}$ is the lower triangular matrix that we evaluated in the previous example. Substituting this $\mathbf{L}$ and the given $\mathbf{b}$ into $\mathbf{Ly} = \mathbf{b}$:

$$\mathbf{Ly} = \begin{pmatrix} 1 & 0 & 0 & 0 \\ 5 & 2 & 0 & 0 \\ 6 & -5 & 1 & 0 \\ 5 & 8 & 1 & 1 \end{pmatrix} \begin{pmatrix} y_1 \\ y_2 \\ y_3 \\ y_4 \end{pmatrix} = \begin{pmatrix} 7 \\ 13 \\ 106 \\ -94 \end{pmatrix} \qquad (\dagger)$$

(continued...)

From the first row we have $y_1 = 7$. Substituting this $y_1 = 7$ into the second row gives

$$\begin{aligned} 5y_1 + 2y_2 &= 13 \\ 5(7) + 2y_2 &= 13 \text{ gives } y_2 = -11 \end{aligned}$$

Using these values $y_1 = 7$ and $y_2 = -11$ in the third row of (†) yields

$$\begin{aligned} 6y_1 - 5y_2 + y_3 &= 106 \\ 6(7) - 5(-11) + y_3 &= 106 \Rightarrow y_3 = 9 \end{aligned}$$

Repeating this process with $y_1 = 7, y_2 = -11$ and $y_3 = 9$ on the bottom row of (†) gives

$$\begin{aligned} 5y_1 + 8y_2 + y_3 + y_4 &= -94 \\ 5(7) + 8(-11) + 9 + y_4 &= -94 \Rightarrow y_4 = -50 \end{aligned}$$

We have $y_1 = 7, y_2 = -11, y_3 = 9$ and $y_4 = -50$, or in vector form:

$$\mathbf{y} = \begin{pmatrix} 7 & -11 & 9 & -50 \end{pmatrix}^T$$

Now we use our upper triangular matrix $\mathbf{U}$ to find the required unknown x's from $\mathbf{Ux} = \mathbf{y}$. We found the upper triangular matrix $\mathbf{U}$ in the above Example 6.18. Substituting that matrix $\mathbf{U}$ and the above vector $\mathbf{y}$ into $\mathbf{Ux} = \mathbf{y}$ gives

$$\begin{pmatrix} 1 & 4 & 5 & 3 \\ 0 & 1 & 1 & -2 \\ 0 & 0 & 2 & 3 \\ 0 & 0 & 0 & -10 \end{pmatrix} \begin{pmatrix} x_1 \\ x_2 \\ x_3 \\ x_4 \end{pmatrix} = \begin{pmatrix} 7 \\ -11 \\ 9 \\ -50 \end{pmatrix} \qquad (*)$$

From the bottom row of (*) we have $x_4 = 5$. Substituting this $x_4 = 5$ into the penultimate row of (*) gives

$$\begin{aligned} 2x_3 + 3x_4 &= 9 \\ 2x_3 + 3(5) &= 9 \Rightarrow x_3 = -3 \end{aligned}$$

Using the second row and x values already found, $x_3 = -3$ and $x_4 = 5$ gives

$$\begin{aligned} x_2 + x_3 - 2x_4 &= -11 \\ x_2 - 3 - 2(5) &= -11 \Rightarrow x_2 = 2 \end{aligned}$$

Using these values $x_2 = 2, x_3 = -3$ and $x_4 = 5$ in the top row of (*) yields

$$\begin{aligned} x_1 + 4x_2 + 5x_3 + 3x_4 &= 7 \\ x_1 + 4(2) + 5(-3) + 3(5) &= 7 \Rightarrow x_1 = -1 \end{aligned}$$

Hence our unknowns are $x_1 = -1, x_2 = 2, x_3 = -3$ and $x_4 = 5$.

Remember, we apply this **LU** factorization method if we want to solve a linear system $\mathbf{Ax} = \mathbf{b}$ for different $\mathbf{b}$ vectors.

This **LU** factorization method is especially useful if you have a lot of **b**'s for which you want to find solutions, because once you have found **L** and **U**, then the remaining work is pretty straightforward.

We can also use **LU** factorization to find the determinant of a matrix, as the next example demonstrates.

Example 6.20

Find the determinant of $\mathbf{A}$ where $\mathbf{A}$ is the matrix of Example 6.18.

Solution

From Example 6.18 we have

$$\mathbf{A} = \mathbf{LU} \text{ where } \mathbf{L} = \begin{pmatrix} 1 & 0 & 0 & 0 \\ 5 & 2 & 0 & 0 \\ 6 & -5 & 1 & 0 \\ 5 & 8 & 1 & 1 \end{pmatrix}, \quad \mathbf{U} = \begin{pmatrix} 1 & 4 & 5 & 3 \\ 0 & 1 & 1 & -2 \\ 0 & 0 & 2 & 3 \\ 0 & 0 & 0 & -10 \end{pmatrix}$$

Since we have a lower triangular $\mathbf{L}$ and an upper triangular matrix $\mathbf{U}$, the determinant is the product of the diagonal entries, because we have the following proposition from the last section: Proposition (6.19). The determinant of a triangular matrix is the *product* of the entries along the leading diagonal.

Using this proposition we have

$$\det(\mathbf{L}) = 1 \times 2 \times 1 \times 1 = 2 \text{ and } \det(\mathbf{U}) = 1 \times 1 \times 2 \times (-10) = -20$$

By the multiplicative property of determinants: Proposition (6.28). $\det(\mathbf{AB}) = \det(\mathbf{A}) \times \det(\mathbf{B})$.

We have

$$\det(\mathbf{A}) = \det(\mathbf{LU}) = \det(\mathbf{L}) \times \det(\mathbf{U}) \\ = 2 \times (-20) = -40$$

Hence $\det(\mathbf{A}) = -40$.

Once we have an **LU** factorization then evaluating the determinant is simple multiplication.

We can also use **LU** factorization of an invertible matrix $\mathbf{A}$ to find the inverse of $\mathbf{A}$.

How?

We have $\mathbf{A} = \mathbf{LU}$ so taking the inverse of both sides gives

$$\mathbf{A}^{-1} = (\mathbf{LU})^{-1} = \mathbf{U}^{-1}\mathbf{L}^{-1} \qquad \left[\text{By (1.27) } (\mathbf{XY})^{-1} = \mathbf{Y}^{-1}\mathbf{X}^{-1}\right]$$

Since **L** and **U** are triangular matrices, their inverses are a lot less demanding to find. We use row operations to find the inverse rather than using cofactors – see Exercises 6.4.

Summary

We can break a given matrix $\mathbf{A}$ into a lower triangular matrix $\mathbf{L}$ and an upper triangular matrix $\mathbf{U}$ such that $\mathbf{A} = \mathbf{LU}$. This approach is used by computers to solve linear systems $\mathbf{Ax} = \mathbf{b}$ for the same coefficient matrix $\mathbf{A}$ but many different vectors $\mathbf{b}$ because it is much less computationally expensive.

EXERCISES 6.4

(Brief solutions at end of book. Full solutions available at <http://www.oup.co.uk/companion/singh>.)

1. Solve the following linear systems $\mathbf{Ax} = \mathbf{b}$ by using $\mathbf{LU}$ factorization for:

   (a) $\mathbf{A} = \begin{pmatrix} 1 & 2 & 2 \\ 3 & -3 & -2 \\ 4 & -1 & -5 \end{pmatrix}$, $\mathbf{b} = \begin{pmatrix} 5 \\ 0 \\ -10 \end{pmatrix}$

   (b) $\mathbf{A} = \begin{pmatrix} 1 & 5 & 6 \\ 2 & 11 & 19 \\ 3 & 19 & 47 \end{pmatrix}$, $\mathbf{b} = \begin{pmatrix} 1 \\ -4 \\ -22 \end{pmatrix}$

   (c) $\mathbf{A} = \begin{pmatrix} 2 & 1 & 3 \\ 4 & -1 & 0 \\ 10 & 12 & 34 \end{pmatrix}$, $\mathbf{b} = \begin{pmatrix} 5 \\ -11 \\ 84 \end{pmatrix}$

   (d) $\mathbf{A} = \begin{pmatrix} 1 & 2 & 3 \\ 3 & -9 & -12 \\ -1 & 8 & 18 \end{pmatrix}$, $\mathbf{b} = \begin{pmatrix} -5 \\ 9 \\ -4 \end{pmatrix}$

2. Solve the following linear system $\mathbf{Ax} = \mathbf{b}$ by using $\mathbf{LU}$ factorization:

   $$\mathbf{A} = \begin{pmatrix} 1 & 2 & 3 & 4 \\ 17 & 22 & 27 & 8 \\ 77 & 44 & 47 & -494 \\ -10 & 1 & 7 & 63 \end{pmatrix}, \quad \mathbf{b} = \begin{pmatrix} -10 \\ 22 \\ 2106 \\ -243 \end{pmatrix}$$

3. Find the determinant of the matrix $\mathbf{A}$ given in question 2.

4. Decompose the matrix $\mathbf{A} = \begin{pmatrix} 1 & 2 & 3 \\ 0 & 4 & 5 \\ 0 & 0 & 6 \end{pmatrix}$ into $\mathbf{LU}$.

5. Let $\mathbf{A} = \begin{pmatrix} 1 & 2 & 3 \\ 3 & 7 & 14 \\ 4 & 13 & 38 \end{pmatrix}$.

   (a) Convert the matrix $\mathbf{A}$ into $\mathbf{LU}$ form.

   (b) Find $\mathbf{A}^{-1}$.

   (c) Solve $\mathbf{Ax} = \mathbf{b}$ for: (i) $\mathbf{b} = (1\ \ 2\ \ 3)^T$ (ii) $\mathbf{b} = (-1\ \ 3\ \ 1)^T$

6. Explain why we cannot convert the following $\begin{pmatrix} 0 & 1 \\ 2 & 3 \end{pmatrix}$ into $\mathbf{LU}$ factorization.

MISCELLANEOUS EXERCISES 6

(Brief solutions at end of book. Full solutions available at <http://www.oup.co.uk/companion/singh>.)

In this exercise you may check your numerical answers using MATLAB.

Cramer's Method is described on the book's website.

6.1. The determinant of the matrix

$$\mathbf{A} = \begin{bmatrix} 0 & 1 & 5 \\ 3 & -6 & 9 \\ 2 & 6 & 1 \end{bmatrix}$$

is

A. 165 B. 0 C. -35 D. -75 E. 68

University of Ottawa, Canada

6.2. Let $\mathbf{E} = \begin{pmatrix} -1 & 7 & 8 \\ 0 & 5 & 6 \\ 0 & 0 & 4 \end{pmatrix}$ and $\mathbf{F} = \begin{pmatrix} 3 & 0 & 0 \\ -1 & -5 & 0 \\ 0 & 1 & -2 \end{pmatrix}$.

Find $\det(\mathbf{E})$, $\det(\mathbf{F})$, $\det(\mathbf{EF})$ and $\det(\mathbf{E}+\mathbf{F})$.

University of Western Ontario, Canada

6.3. Evaluate the determinant of

$$\begin{bmatrix} 3 & 1 & 1 & 1 & 1 \\ 1 & 3 & 1 & 1 & 1 \\ 3 & 1 & 3 & 1 & 1 \\ 1 & 3 & 1 & 3 & 1 \\ 3 & 1 & 3 & 1 & 3 \end{bmatrix}$$

University of Western Ontario, Canada

6.4. Let $\mathbf{A}$ and $\mathbf{B}$ be $n \times n$ matrices. Complete the following formulas with the simplest possible expressions. If a formula does not exist, write 'No Formula'.

(a) $\det(\mathbf{AB}) =$

(b) $\det(\mathbf{A}^{-1}) =$

(c) $\det(\mathbf{A}+\mathbf{B}) =$

(d) $\det(3\mathbf{A}) =$

(e) $\det(\mathbf{A}^{T}) =$

Mount Holyoke College, Massachusetts, USA

6.5. Consider the matrix $\mathbf{B} = \begin{pmatrix} 1 & 2 & 0 & 0 \\ 0 & 1 & 2 & 0 \\ 0 & 0 & 1 & 2 \\ -3 & 4 & -5 & 6 \end{pmatrix}$.

Use row operations or straightforward calculation to find $\det(\mathbf{B})$.

Mount Holyoke College, Massachusetts, USA

6.6. Suppose

$$\begin{aligned} 2x_1 + x_2 + x_3 &= 8 \\ 3x_1 - 2x_2 - x_3 &= 1 \\ 4x_1 - 7x_2 + 3x_3 &= 10 \end{aligned}$$

By Cramer's rule we find (fill in determinants without doing the computation):

$$x_2 =$$

Mount Holyoke College, Massachusetts, USA

6.7. Let $\mathbf{A} = \begin{bmatrix} 2 & 123 & -1 \\ 1 & 456 & 1 \\ 2 & 789 & 1 \end{bmatrix}$. If you know that $\det(\mathbf{A}) = -420$, then find the value of x_2 in the solution of the linear system $\mathbf{A}\begin{bmatrix} x_1 \\ x_2 \\ x_3 \end{bmatrix} = \begin{bmatrix} 1 \\ 0 \\ 1 \end{bmatrix}$.

Purdue University, USA

6.8. Using Cramer's method, solve the system

$$\begin{aligned} 5x_2 + x_3 &= -8 \\ x_1 - 2x_2 - x_3 &= 2 \\ 6x_1 + x_2 - 3x_3 &= -8 \end{aligned}$$

University of Wisconsin, USA (part question)

6.9. Calculate the determinant of the matrix

$$\begin{pmatrix} 2 & 1 & 3 & 1 \\ 0 & 8 & -2 & 5 \\ -6 & -3 & -5 & -5 \\ 2 & 5 & 3 & -4 \end{pmatrix}$$

National University of Ireland, Galway (part question)

6.10. Compute the determinant of the following matrix

$$\mathbf{A} = \begin{pmatrix} 10 & -2 & 3 & 16 \\ 1 & 1 & 1 & 1 \\ 9 & -3 & 2 & 15 \\ 110 & 23 & 12 & -15 \end{pmatrix}$$

RWTH Aachen University, Germany

6.11. Let $\mathbf{A}$ be the 3×3 matrix

$$\mathbf{A} = \begin{pmatrix} 0 & 1 & 3 \\ 1 & 2 & 0 \\ 4 & 5 & 1 \end{pmatrix}$$

(a) Calculate $\det(\mathbf{A})$ using the cofactor expansion.

(b) Calculate $\mathbf{A}^{-1}$ using elementary row operations.

(c) Use the calculation you performed in (b) to find $\det(\mathbf{A})$.

University of Maryland, Baltimore County, USA

6.12. (i) Compute the inverse of the matrix

$$\mathbf{A} = \begin{bmatrix} 2 & -1 & -4 \\ -1 & 1 & 2 \\ -1 & 1 & 3 \end{bmatrix}$$

(ii) Use the result of part (i) to find the inverses of the three matrices $\mathbf{A}^t$, $3\mathbf{A}$ and $\mathbf{A}^2$.

University of New Brunswick, Canada

6.13. Let $\mathbf{A}$ be an upper triangular $n \times n$ matrix with determinant equal to 3. Multiply by 5 all terms in the matrix above and to the right of the diagonal (but not on the diagonal). What is the determinant of the new matrix?

Johns Hopkins University, USA

6.14. Find $adj\mathbf{A}$, $\det \mathbf{A}$ and hence $\mathbf{A}^{-1}$ when

$$\mathbf{A} = \begin{bmatrix} 1 & -1 \\ -1 & 1 \end{bmatrix}$$

and when

$$\mathbf{A} = \begin{bmatrix} 3 & -4 & 1 \\ 1 & -1 & 3 \\ 2 & -2 & 5 \end{bmatrix}$$

University of Sussex, UK (part question)

6.15. Let $\mathbf{A} = \begin{bmatrix} 2 & 0 & 0 & 0 \\ 1 & -7 & -5 & 0 \\ 3 & 8 & 6 & 0 \\ 0 & 7 & 5 & 4 \end{bmatrix}$.

(a) Use determinants to decide if $\mathbf{A}$ is invertible or not.

(b) What is $\det\left(\mathbf{A}^T\right)$?

(c) What is $\det\left(\mathbf{A}^{-1}\right)$?

Washington State University, USA (part question)

6.16. Consider the following determinants:

$$D_1 = \begin{vmatrix} 1 & 0 & 2 & 0 \\ 0 & 2 & 0 & -1 \\ 2 & 0 & 3 & 0 \\ 0 & -1 & 0 & 4 \end{vmatrix}, \quad D_2 = \begin{vmatrix} 1 & 0 & 2 & 0 \\ 2 & 0 & 3 & 0 \\ 0 & 2 & 0 & -1 \\ 0 & -1 & 0 & 4 \end{vmatrix}, \quad D_3 = \begin{vmatrix} 1 & 0 & 2 & 0 \\ 0 & 2 & 0 & -1 \\ 2 & 0 & 3 & 0 \\ 3 & 2 & 5 & -1 \end{vmatrix}$$

Evaluate D_1 by direct expansion. Evaluate D_2 by relating it to D_1. Evaluate D_3 by using standard properties of determinants. In each case, explain your method briefly.

Queen Mary, University of London, UK

6.17. Find the value of the determinant of each of the following matrices.

(a) $\mathbf{A} = \begin{bmatrix} 3 & 0 & 0 & 0 \\ x & -2 & 0 & 0 \\ 17 & y & 1 & 0 \\ \pi & 42 & -6 & 4 \end{bmatrix}$ (b) $\mathbf{B} = \begin{bmatrix} 1 & 2 & 3 & 4 & 5 \\ 4 & 6 & 1 & -9 & 8 \\ 12 & 10 & 2 & 7 & 0 \\ 4 & 6 & 1 & -9 & 8 \\ 1 & 5 & 2 & 4 & 3 \end{bmatrix}$

(c) $\mathbf{C} = \begin{bmatrix} 4 & -2 & 1 & 5 \\ 2 & 9 & 0 & 3 \\ 0 & 2 & 0 & 0 \\ 5 & 8 & 0 & 7 \end{bmatrix}$

Memorial University, Canada

6.18. (a) Consider the matrices

$$\mathbf{A} = \begin{pmatrix} -1 & -2 & -3 & 1 \\ 1 & -2 & 0 & 3 \\ 2 & -3 & 1 & -1 \\ 0 & 1 & 0 & 0 \end{pmatrix} \text{ and } \mathbf{B} = \begin{pmatrix} 2 & 0 & 0 & 0 \\ 5 & 1 & 0 & 0 \\ -1 & 1 & 2 & 0 \\ 2 & -3 & 1 & -3 \end{pmatrix}$$

Find the determinants of **A**, **B**, **AB** and $\mathbf{A}^3$.

(b) Let **A** be an invertible $n \times n$ matrix. Prove that $\det(\mathbf{A}) \neq 0$.

University of Southampton, UK (part question)

6.19. (a) By elimination or otherwise, find the determinant of **A**:

$$\mathbf{A} = \begin{bmatrix} 1 & 0 & 0 & u_1 \\ 0 & 1 & 0 & u_2 \\ 0 & 0 & 1 & u_3 \\ v_1 & v_2 & v_3 & 0 \end{bmatrix}$$

(b) If that zero in the lower right hand corner of **A** changes to 100, what is the change (if any) in the determinant of **A**? (You can consider its cofactors.)

Massachusetts Institute of Technology, USA (part question)

6.20. Use row operations to calculate the determinant

$$\begin{vmatrix} 1 & a^2 & b+c \\ 1 & b^2 & a+c \\ 1 & c^2 & a+b \end{vmatrix}$$

University of Maryland, Baltimore County, USA

6.21. Let **A** be an invertible 3×3 matrix. Suppose it is known that

$$\mathbf{A} = \begin{bmatrix} u & v & w \\ 3 & 3 & -2 \\ x & y & z \end{bmatrix} \text{ and that } adj(\mathbf{A}) = \begin{bmatrix} a & 3 & b \\ -1 & 1 & 2 \\ c & -2 & d \end{bmatrix}$$

Find $\det(\mathbf{A})$. (Give an answer not involving any of the unknown variables.)

McGill University, Canada (part question)

6.22. Evaluate $\det\left((2\mathbf{A})^{-1}(3\mathbf{A})^T\right)$, where **A** is any invertible (2×2) matrix.

Memorial University, Canada

6.23. (a) Evaluate the determinants:

$$\text{(i)} \begin{vmatrix} 1 & 2 & 2 \\ 3 & 1 & 3 \\ 1 & 3 & 1 \end{vmatrix}; \qquad \text{(ii)} \begin{vmatrix} 3 & 0 & 0 & 2 \\ 0 & 0 & 3 & 0 \\ 0 & 0 & 2 & 2 \\ 3 & 1 & 0 & 0 \end{vmatrix}$$

(b) Assume that **A**, **B** and **C** are 4×4 matrices with determinants

$$\det(\mathbf{A}) = 2, \qquad \det(\mathbf{B}) = -3 \text{ and } \det(\mathbf{C}) = 5$$

Find the following determinants:

(i) $\det(3\mathbf{A})$; (ii) $\det\left(\mathbf{C}^{-1}\mathbf{B}\right)$; (iii) $\det\left(\mathbf{A}^2\mathbf{C}^{-1}\mathbf{B}^T\right)$

University of Manchester, UK

6.24. (a) Evaluate the determinants:

$$\text{(i)} \begin{vmatrix} 0 & 0 & 1 & 3 \\ 0 & 0 & 3 & 1 \\ 0 & 1 & a & b \\ 3 & 1 & c & d \end{vmatrix} \qquad \text{(ii)} \begin{vmatrix} 0 & 0 & a & 0 & 0 \\ 0 & b & 0 & 0 & 0 \\ c & 0 & 0 & 0 & 0 \\ 0 & 0 & 0 & 0 & 1 \\ 0 & 0 & 0 & 2 & 0 \end{vmatrix} \qquad \text{(iii)} \begin{vmatrix} 0 & 0 & 3 & 0 & 0 \\ 0 & 2 & 4 & 0 & 0 \\ 1 & 0 & 5 & 0 & 0 \\ 0 & 0 & 6 & 0 & 9 \\ 0 & 0 & 7 & 8 & 0 \end{vmatrix}$$

(b) Assume that **A** and **B** are 3×3 matrices with determinants $\det \mathbf{A} = 3$ and $\det \mathbf{B} = -2$. Find the following determinants:

(i) $\det\left(\mathbf{B}\mathbf{B}^T\right)$ (ii) $\det\left(\mathbf{A}^{-1}\mathbf{B}\mathbf{A}\right)$ (iii) $\det\left(\mathbf{A}\mathbf{B}^{-1}\right)$

University of Manchester, UK

6.25. (a) Let **A**, **B** be 2×2 matrices with $\det(\mathbf{A}) = 2$, $\det(\mathbf{B}) = 3$. Find

(i) $\det\left(-\mathbf{A}^3\mathbf{B}^{-2}\right)$ (ii) $\det\left(2\mathbf{A}^{-1}\mathbf{B}\mathbf{A}\right)$ (iii) $\det\left(\mathbf{A}^{-1}\mathbf{A}^T\right)$

(b) If $\begin{vmatrix} a & d & g \\ b & e & h \\ c & f & i \end{vmatrix} = 1$ find $\begin{vmatrix} a+d & d+g & g+a \\ b+e & e+h & h+b \\ c+f & f+i & i+c \end{vmatrix}$.

McGill University, Canada

6.26. We have a 3×3 matrix $\mathbf{A} = \begin{bmatrix} a & 1 & 2 \\ b & 3 & 4 \\ c & 5 & 6 \end{bmatrix}$ with $\det(\mathbf{A}) = 3$. Compute the determinants of the following matrices:

(a) $\begin{bmatrix} a-2 & 1 & 2 \\ b-4 & 3 & 4 \\ c-6 & 5 & 6 \end{bmatrix}$ (b) $\begin{bmatrix} 7a & 7 & 14 \\ b & 3 & 4 \\ c & 5 & 6 \end{bmatrix}$ (c) $2\mathbf{A}^{-1}\mathbf{A}^T$ (d) $\begin{bmatrix} a-2 & 1 & 2 \\ b & 3 & 4 \\ c & 5 & 6 \end{bmatrix}$

Purdue University, USA 2006

In the following question, consider the field F to be a set of real numbers.

6.27. Let F be a field, and $a,\ b,\ c,\ d,\ e,\ f,\ p,\ q,\ r,\ s,\ t,\ u \in F$. Consider the matrix

$$\mathbf{A} = \begin{pmatrix} a & b & c \\ 0 & d & e \\ 0 & 0 & f \end{pmatrix} \begin{pmatrix} p & 0 & 0 \\ q & r & 0 \\ s & t & u \end{pmatrix}$$

Find $\det(\mathbf{A})$. (Advice: you can do this the hard way, by multiplying out and calculating, or more easily, using facts about determinants.)

University of California, Berkeley, USA

6.28. Use Cramer's rule to solve for x in the system of equations

$$\begin{bmatrix} a & 1 & 1 \\ 1 & 2 & 3 \\ 1 & 3 & 6 \end{bmatrix} \begin{bmatrix} x \\ y \\ z \end{bmatrix} = \begin{bmatrix} 1 \\ 1 \\ 1 \end{bmatrix}$$

Columbia University, New York, USA

6.29. Supply a short proof for the following statement:
If $\mathbf{A}$ is a $n \times n$ matrix satisfying $\mathbf{A}^5 = \mathbf{O}$, then $\det(\mathbf{A}) = 0$.

University of California, Berkeley, USA (part question)

6.30. Let $\mathbf{A}_n$ denote the following $n \times n$ matrix.

$$\mathbf{A}_n = \begin{bmatrix} 1 & 2 & 3 & 4 & \cdots & n \\ -1 & 0 & 3 & 4 & \cdots & n \\ -1 & -2 & 0 & 4 & \cdots & n \\ \vdots & \vdots & \vdots & \vdots & \ddots & \vdots \\ -1 & -2 & -3 & -4 & \cdots & 0 \end{bmatrix}$$

(a) Determine $\mathbf{A}_2, \mathbf{A}_3$ and $\mathbf{A}_4$. Use elementary row operations and properties of determinants to compute the determinants of these matrices.

(b) Based on your work on part (a), use elementary row operations and properties of determinants to compute $\mathbf{A}_n$, the determinant of the matrix $\mathbf{A}_n$ for an integer $n \geq 2$.

Illinois State University, USA

6.31. What is the formula for the inverse to the following matrix?

$$\begin{bmatrix} A & B & D \\ 0 & C & E \\ 0 & 0 & F \end{bmatrix}$$

Columbia University, New York, USA

6.32. Using Cramer's rule, solve for y in the following system of equations:

$$\begin{bmatrix} A & B & D \\ 0 & C & E \\ 0 & 0 & F \end{bmatrix} \begin{bmatrix} x \\ y \\ z \end{bmatrix} = \begin{bmatrix} 1 \\ 1 \\ 0 \end{bmatrix}$$

Columbia University, New York, USA

6.33. Evaluate the determinant of the matrix $\begin{bmatrix} 1 & 0 & 0 & 2 & 4 & 6 & 8 \\ 0 & 1 & 0 & 5 & 12 & 13 & 9 \\ 0 & 0 & 1 & -1 & 31 & 5 & 23 \\ 0 & 0 & 0 & 4 & 2 & 7 & 1 \\ 0 & 0 & 0 & -2 & 1 & 3 & -2 \\ 0 & 0 & 0 & 0 & 1 & 0 & 0 \\ 0 & 0 & 0 & -1 & 2 & 5 & 3 \end{bmatrix}$.

University of California, Berkeley, USA

6.34. For which real numbers k is the following matrix invertible?

$$\begin{pmatrix} 1 & 1 & 0 & 0 & 1 \\ -1 & k & 0 & 0 & 0 \\ 0 & 0 & 1 & 3 & 9 \\ 0 & 0 & 1 & 4 & 16 \\ 0 & 0 & 1 & k & k^2 \end{pmatrix}$$

Stanford University, USA

6.35. Let $\mathbf{B}$ be the invertible matrix given below, where ? means that the value of the entry does not affect the answer to this problem. The second matrix $\mathbf{C}$ is the adjoint of $\mathbf{B}$. Find the value of $\det\left(2\mathbf{B}^{-1}\left(\mathbf{C}^T\right)^{-2}\right)$.

$$\mathbf{B} = \begin{pmatrix} ? & ? & ? & 0 \\ 0 & -1 & 2 & 0 \\ 1 & 1 & 0 & 0 \\ ? & ? & ? & -3 \end{pmatrix}, \quad \mathbf{C} = \begin{pmatrix} 6 & 3 & 9 & 0 \\ -6 & -3 & 6 & 0 \\ -3 & 6 & 3 & 0 \\ 2 & 1 & 3 & -5 \end{pmatrix}$$

University of Utah, USA

Sample questions

6.36. Find the determinant of

$$\mathbf{A} = \begin{pmatrix} 1 & 2 & 3 \\ 4 & 5 & 6 \end{pmatrix}$$

6.37. Let $\mathbf{P}$ be an invertible n by n matrix and $\mathbf{A}$ be a n by n matrix. Show that

$$\det\left(\mathbf{P}^{-1}\mathbf{AP}\right) = \det(\mathbf{A})$$

6.38. Let $\mathbf{A}$ be a n by n matrix given by $\mathbf{A} = \begin{pmatrix} 0 & 0 & \cdots & a_{1n} \\ 0 & \cdots & a_{2(n-1)} & a_{2n} \\ \vdots & .\cdot^{\cdot} & .\cdot^{\cdot} & \vdots \\ a_{n1} & a_{n2} & \cdots & a_{nn} \end{pmatrix}$.

Prove that $\det(\mathbf{A}) = (-1)^{\lfloor n/2 \rfloor} a_{1n} \cdots a_{2(n-1)} a_{n1}$ where $\lfloor\ \rfloor$ is the floor function.

[The floor function $\lfloor\ \rfloor$ is defined on the set of real numbers. Let x be a real number such that $x = n + y$ where $0 \le y < 1$

We define the floor function to be $\lfloor x \rfloor = n$

That is $\lfloor x \rfloor = n$ gives the largest integer less than or equal to x. For example

$$\lfloor 2.4 \rfloor = 2, \quad \lfloor 6.99 \rfloor = 6, \quad \lfloor 10 \rfloor = 10, \quad \lfloor \pi \rfloor = 3, \quad \lfloor -3.4 \rfloor = -4]$$

Françoise Tisseur

is Professor of Numerical Analysis in the School of Mathematics, University of Manchester, UK.

Tell us about yourself and your work.

I am a numerical analyst specializing in numerical linear algebra. My work spans the full range from developing fundamental theory to deriving algorithms and implementing them in software.

I am particularly interested in the theory and numerical solution of algebraic eigenvalue problems, that is, to find scalars λ and non-zero vectors x and y satisfying $N(\lambda)x = 0$, where $N : \Omega \to \mathbb{C}nxn$ is an analytic function on an open set $\Omega \subseteq \mathbb{C}$, the standard eigenvalue problem $(A - \lambda I)x = 0$ studied in Chapter 7 being the simplest case. However, in practical applications, such as in the dynamic analysis of mechanical systems (where the eigenvalues represent vibrational frequencies), in the linear stability of flows in fluid mechanics, or in the stability analysis of time-delay systems, the elements of $N(\lambda)$ can be polynomial, rational or exponential functions of λ and the corresponding non-linear eigenvalue problems can be very difficult to solve.

I am also interested in *tropical algebra* (also known as max-plus algebra), where matrices and vectors have entries in $\mathbb{R} \cup \{-\infty\}$ and where the addition $a + b$ is replaced by a maximization $\max(a, b)$ and the multiplication ab is replaced by an addition $a + b$. As in section 1.4 of this book, we can add two tropical matrices and also multiply them. In fact, many of the tools of linear algebra described in this book are available in tropical algebra. A major difference with classical linear algebra is that the maximum operation lacks inverses. Tropical algebra allows us to describe, in a linear way, a phenomenon that is non-linear in the conventional algebra. These include, for example, parallel computation, transportation networks and scheduling.

What are the challenges connected with the subject?

Non-linear eigenvalue problems present many mathematical challenges. For some there is a lack of underlying theory. For others, numerical methods struggle to provide any accuracy or to solve very large problems in a reasonable time.

7 Eigenvalues and Eigenvectors

SECTION 7.1 Introduction to Eigenvalues and Eigenvectors

By the end of this section you will be able to

- determine eigenvalues and eigenvectors
- prove properties of eigenvalues and eigenvectors

Eigenvector/value problems crop up frequently in the physical sciences and engineering. They take the form $\mathbf{Av} = (\text{scalar}) \times \mathbf{v}$ where $\mathbf{v}$ is a non-zero vector and $\mathbf{A}$ is a square matrix. By knowing the eigenvalues and eigenvectors of a matrix we can easily find its determinant, decide whether the matrix has an inverse and determine the powers of the matrix. For an example of linear algebra at work, one needs to look no further than Google's search engine, which relies upon eigenvalues and eigenvectors to rank pages with respect to relevance.

7.1.1 Definition of eigenvalues and eigenvectors

Before we define what is meant by an eigenvalue and an eigenvector let's do an example which involves them.

Example 7.1

Let $\mathbf{A} = \begin{pmatrix} 4 & -2 \\ 1 & 1 \end{pmatrix}$ and $\mathbf{u} = \begin{pmatrix} 2 \\ 1 \end{pmatrix}$ then evaluate $\mathbf{Au}$.

Solution

Multiplying the matrix $\mathbf{A}$ and vector $\mathbf{u}$ we have

$$\mathbf{Au} = \begin{pmatrix} 4 & -2 \\ 1 & 1 \end{pmatrix}\begin{pmatrix} 2 \\ 1 \end{pmatrix} = \begin{pmatrix} 6 \\ 3 \end{pmatrix} = 3\begin{pmatrix} 2 \\ 1 \end{pmatrix}$$

What do you notice about the result?

We have $\mathbf{Au} = 3\mathbf{u}$. The matrix $\mathbf{A}$ scalar multiplies the vector $\mathbf{u}$ by 3, as shown in Fig. 7.1.

In general terms, this can be written as

(7.1) $$\mathbf{Au} = \lambda\mathbf{u} \quad (\text{matrix } \mathbf{A} \text{ scalar multiplies vector } \mathbf{u})$$

where $\mathbf{A}$ is a square matrix, $\mathbf{u}$ is a vector and the Greek letter λ (lambda) is a scalar.

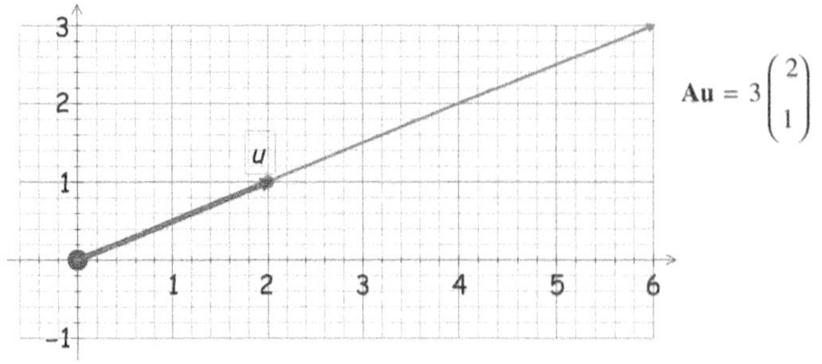

Figure 7.1

This is an important result which is used throughout this chapter and well worth becoming familiar with.

Why is formula (7.1) important?

Because the matrix $\mathbf{A}$ transforms the vector $\mathbf{u}$ by scalar multiplying it, which means that the transformation only changes the length of the vector $\mathbf{u}$ unless $\lambda = \pm 1$ (in which case the length remains unchanged). Note that (7.1) says that the matrix $\mathbf{A}$ applied to $\mathbf{u}$ gives a vector in the same or opposite (negative λ) direction of $\mathbf{u}$.

Can you think of a vector, $\mathbf{u}$, which satisfies equation (7.1)?

The zero vector $\mathbf{u} = \mathbf{O}$ because $\mathbf{AO} = \lambda\mathbf{O} = \mathbf{O}$. In this case we say that we have the trivial solution $\mathbf{u} = \mathbf{O}$. In this chapter we consider the non-trivial solutions, $\mathbf{u} \neq \mathbf{O}$ (not zero), and these solutions are powerful tools in linear algebra.

For a *non-zero* vector $\mathbf{u}$ the scalar λ is called an **eigenvalue** of the matrix $\mathbf{A}$ and the vector $\mathbf{u}$ is called an **eigenvector** belonging to or corresponding to λ, which satisfies $\mathbf{Au} = \lambda\mathbf{u}$.

In most linear algebra literature the Greek letter lambda, λ, is used for eigenvalues. These terms eigenvalue and eigenvector are derived from the German word 'Eigenwert' which means 'proper value'. The word eigen is pronounced 'i-gun'.

Eigenvalues were initially developed in the field of differential equations by Jean d' Alembert.

Figure 7.2 Jean d'Alembert 1717 to 1783.

Jean d'Alembert 1717–1783 (Fig. 7.2) was a French mathematician and the illegitimate son of Madam Tencin and an army officer, Louis Destouches. His mother left him on the steps of a local church and he was consequently sent to a home for orphans. His father recognised his son's difficulties and placed him under the care of Madam Rousseau, wife of a wealthy architect.

However, d'Alembert's father died when he was only nine years old and his father's family looked after his financial situation so that he could continue his education.

In 1735, Alembert graduated, and he thought that a career in law would suit him, but his real thirst and enthusiasm was for mathematics, and he studied this in his spare time. For most of his life he worked for the Paris Academy of Science and the French Academy.

Example 7.2

Let $\mathbf{A} = \begin{pmatrix} 1 & 1 \\ -2 & 4 \end{pmatrix}$. Verify the following:

(a) $\mathbf{u} = \begin{pmatrix} 1 \\ 1 \end{pmatrix}$ is an eigenvector of matrix $\mathbf{A}$ belonging to the eigenvalue $\lambda_1 = 2$.

(b) $\mathbf{v} = \begin{pmatrix} 1 \\ 2 \end{pmatrix}$ is an eigenvector of matrix $\mathbf{A}$ belonging to the eigenvalue $\lambda_2 = 3$.

Solution

(a) Multiplying the given matrix $\mathbf{A}$ and vector $\mathbf{u}$ we have

$$\mathbf{Au} = \begin{pmatrix} 1 & 1 \\ -2 & 4 \end{pmatrix}\begin{pmatrix} 1 \\ 1 \end{pmatrix} = \begin{pmatrix} 2 \\ 2 \end{pmatrix} = 2\begin{pmatrix} 1 \\ 1 \end{pmatrix}$$

Thus $\mathbf{u} = (1 \quad 1)^T$ is an eigenvector of the matrix $\mathbf{A}$ belonging to $\lambda_1 = 2$ because $\mathbf{Au} = 2\mathbf{u}$. Matrix $\mathbf{A}$ doubles the vector $\mathbf{u}$.

(b) Similarly we have

$$\mathbf{Av} = \begin{pmatrix} 1 & 1 \\ -2 & 4 \end{pmatrix}\begin{pmatrix} 1 \\ 2 \end{pmatrix} = \begin{pmatrix} 3 \\ 6 \end{pmatrix} = 3\begin{pmatrix} 1 \\ 2 \end{pmatrix}$$

Thus $\mathbf{v} = (1 \ 2)^T$ is an eigenvector of the matrix $\mathbf{A}$ belonging to $\lambda_2 = 3$ because $\mathbf{Av} = 3\mathbf{v}$. This $\mathbf{Av} = 3\mathbf{v}$ means matrix $\mathbf{A}$ triples the vector $\mathbf{v}$.

What do you notice about your results?
A 2 by 2 matrix can have *more than* one eigenvalue and eigenvector.

We have eigenvalues λ and eigenvectors $\mathbf{u}$ for any *square* matrix $\mathbf{A}$ such that $\mathbf{Au} = \lambda\mathbf{u}$.

Example 7.3

Let $\mathbf{A} = \begin{pmatrix} 5 & 0 & 0 \\ -9 & 4 & -1 \\ -6 & 2 & 1 \end{pmatrix}$ and $\mathbf{u} = \begin{pmatrix} 0 \\ 1 \\ 2 \end{pmatrix}$. Show that the matrix $\mathbf{A}$ scalar multiplies the vector $\mathbf{u}$ and find the value of this scalar, λ, the eigenvalue.

(continued...)

Solution

Applying the matrix $\mathbf{A}$ to the vector $\mathbf{u}$ we have

$$\mathbf{Au} = \begin{pmatrix} 5 & 0 & 0 \\ -9 & 4 & -1 \\ -6 & 2 & 1 \end{pmatrix} \begin{pmatrix} 0 \\ 1 \\ 2 \end{pmatrix} = \begin{pmatrix} 0 \\ 2 \\ 4 \end{pmatrix} = 2 \begin{pmatrix} 0 \\ 1 \\ 2 \end{pmatrix}$$

We have $\mathbf{Au} = 2\mathbf{u}$ so $\lambda = 2$. Hence $\lambda = 2$ is an eigenvalue of the matrix $\mathbf{A}$ with an eigenvector $\mathbf{u}$. Matrix $\mathbf{A}$ transforms the vector $\mathbf{u}$ by a scalar multiple of 2 because $\mathbf{Au} = 2\mathbf{u}$.

7.1.2 Characteristic equation

From the above formula (7.1) $\mathbf{Au} = \lambda\mathbf{u}$ we have

$$\mathbf{Au} = \lambda\mathbf{Iu}$$

[$\lambda\mathbf{Iu} = \lambda\mathbf{u}$ – multiplying by the identity keeps it the same]
where $\mathbf{I}$ is the identity matrix. We can rewrite this as

$$\mathbf{Au} - \lambda\mathbf{Iu} = \mathbf{O}$$
$$(\mathbf{A} - \lambda\mathbf{I})\mathbf{u} = \mathbf{O}$$

Under what condition is the non-zero vector $\mathbf{u}$ a solution of this equation?

By question 26 of Exercises 6.3:

$$\mathbf{Ax} = \mathbf{O} \text{ has an infinite number of solutions} \Leftrightarrow \det(\mathbf{A}) = 0.$$

Applying this result to $(\mathbf{A} - \lambda\mathbf{I})\mathbf{u} = \mathbf{O}$ means that we must have a non-zero vector $\mathbf{u}$ (because there are an infinite number of solutions which satisfy this equation) $\Leftrightarrow$

$$\det(\mathbf{A} - \lambda\mathbf{I}) = 0$$

This is an important equation because we use this to find the eigenvalues and it is called the **characteristic equation**:

$$\det(\mathbf{A} - \lambda\mathbf{I}) = 0 \tag{7.2}$$

The procedure for determining eigenvalues and eigenvectors is:

1. Solve the characteristic equation (7.2) for the scalar λ.
2. For the eigenvalue λ determine the corresponding eigenvector $\mathbf{u}$ by solving the system $(\mathbf{A} - \lambda\mathbf{I})\mathbf{u} = \mathbf{O}$.

Let's follow this procedure for the next example.

Note that eigenvalues and eigenvectors come in pairs. You *cannot* have one without the other. It is a relationship like mother and child because eigenvalues give birth to eigenvectors.

Example 7.4

Determine the eigenvalues and corresponding eigenvectors of $\mathbf{A} = \begin{pmatrix} 2 & 0 \\ 1 & 3 \end{pmatrix}$. Also sketch the effect of multiplying the eigenvectors by matrix $\mathbf{A}$.

Solution

What do we find first, the eigenvalues or eigenvectors?

Eigenvalues, because they produce eigenvectors.

We carry out the above procedure:

Step 1.

We need to find the values of λ which satisfy $\det(\mathbf{A} - \lambda\mathbf{I}) = 0$. First we obtain $\mathbf{A} - \lambda\mathbf{I}$:

$$\mathbf{A} - \lambda\mathbf{I} = \begin{pmatrix} 2 & 0 \\ 1 & 3 \end{pmatrix} - \lambda \begin{pmatrix} 1 & 0 \\ 0 & 1 \end{pmatrix}$$

$$= \begin{pmatrix} 2 & 0 \\ 1 & 3 \end{pmatrix} - \begin{pmatrix} \lambda & 0 \\ 0 & \lambda \end{pmatrix} = \begin{pmatrix} 2-\lambda & 0 \\ 1 & 3-\lambda \end{pmatrix}$$

Substituting this into $\det(\mathbf{A} - \lambda\mathbf{I})$ gives

$$\det(\mathbf{A} - \lambda\mathbf{I}) = \det \begin{pmatrix} 2-\lambda & 0 \\ 1 & 3-\lambda \end{pmatrix}$$

To find the determinant, we use formula (6.1), $\det \begin{pmatrix} a & b \\ c & d \end{pmatrix} = ad - bc$, thus:

$$\det(\mathbf{A} - \lambda\mathbf{I}) = \det \begin{pmatrix} 2-\lambda & 0 \\ 1 & 3-\lambda \end{pmatrix} = (2-\lambda)(3-\lambda) - 0$$

For eigenvalues we equate this determinant to zero:

$$(2-\lambda)(3-\lambda) = 0 \quad \text{implies } \lambda_1 = 2 \text{ or } \lambda_2 = 3$$

Step 2.

For each eigenvalue, λ determine the corresponding eigenvector $\mathbf{u}$ by solving the system $(\mathbf{A} - \lambda\mathbf{I})\mathbf{u} = \mathbf{O}$.

Let $\mathbf{u}$ be the eigenvector corresponding to $\lambda_1 = 2$. Substituting $\mathbf{A} = \begin{pmatrix} 2 & 0 \\ 1 & 3 \end{pmatrix}$ and $\lambda_1 = \lambda = 2$ into $(\mathbf{A} - \lambda\mathbf{I})\mathbf{u} = \mathbf{O}$ gives

$$(\mathbf{A} - \lambda\mathbf{I})\mathbf{u} = \left[\begin{pmatrix} 2 & 0 \\ 1 & 3 \end{pmatrix} - \begin{pmatrix} 2 & 0 \\ 0 & 2 \end{pmatrix}\right]\mathbf{u} = \mathbf{O}$$

$$\begin{pmatrix} 0 & 0 \\ 1 & 1 \end{pmatrix}\mathbf{u} = \mathbf{O}$$

Remember, $\mathbf{O} = \begin{pmatrix} 0 \\ 0 \end{pmatrix}$ and let $\mathbf{u} = \begin{pmatrix} x \\ y \end{pmatrix}$, so we have

$$\begin{pmatrix} 0 & 0 \\ 1 & 1 \end{pmatrix}\begin{pmatrix} x \\ y \end{pmatrix} = \begin{pmatrix} 0 \\ 0 \end{pmatrix}$$

Multiplying out gives

$$0 + 0 = 0$$
$$x + y = 0$$

(continued...)

Remember, the eigenvector *cannot* be the zero vector, therefore at least one of the values, x or y, must be non-zero. From the bottom equation we have $x = -y$.

The simplest solution is $x = 1, y = -1$ but we could have $\begin{pmatrix} 2 \\ -2 \end{pmatrix}, \begin{pmatrix} -3 \\ 3 \end{pmatrix}, \begin{pmatrix} \pi \\ -\pi \end{pmatrix}, \begin{pmatrix} 5 \\ -5 \end{pmatrix}, \ldots$

Hence we have an infinite number of eigenvectors belonging to $\lambda = 2$. We can write down the general eigenvector $\mathbf{u}$.

How?

Let $x = s$ then $y = -s$ where $s \neq 0$ and is a real number. Thus the eigenvectors belonging to $\lambda = 2$ are

$$\mathbf{u} = \begin{pmatrix} s \\ -s \end{pmatrix} = s\begin{pmatrix} 1 \\ -1 \end{pmatrix} \text{ where } s \neq 0 \qquad \left[\begin{pmatrix} 1 \\ -1 \end{pmatrix} \text{ is one of the simplest eigenvectors}\right]$$

Similarly, we find the general eigenvector $\mathbf{v}$ belonging to the other eigenvalue $\lambda_2 = 3$. Putting $\lambda_2 = \lambda = 3$ into $[\mathbf{A} - \lambda\mathbf{I}]\,\mathbf{v} = \mathbf{O}$ gives

$$(\mathbf{A} - \lambda\mathbf{I})\mathbf{v} = \left[\begin{pmatrix} 2 & 0 \\ 1 & 3 \end{pmatrix} - \begin{pmatrix} 3 & 0 \\ 0 & 3 \end{pmatrix}\right]\mathbf{v} = \mathbf{O} \text{ simplifies to } \begin{pmatrix} -1 & 0 \\ 1 & 0 \end{pmatrix}\mathbf{v} = \mathbf{O}$$

By writing $\mathbf{v} = \begin{pmatrix} x \\ y \end{pmatrix}$ [different x and y from those above] and $\mathbf{O} = \begin{pmatrix} 0 \\ 0 \end{pmatrix}$ we obtain

$$\begin{pmatrix} -1 & 0 \\ 1 & 0 \end{pmatrix}\begin{pmatrix} x \\ y \end{pmatrix} = \begin{pmatrix} 0 \\ 0 \end{pmatrix}$$

Multiplying out:

$$-x + 0 = 0, \quad x + 0 = 0$$

From these equations we must have $x = 0$.

What is y equal to?

We can choose y to be any real number apart from zero because the eigenvector cannot be zero. Thus

$$y = s \text{ where } s \neq 0$$

The general eigenvector belonging to $\lambda_2 = 3$ is

$$\mathbf{v} = \begin{pmatrix} x \\ y \end{pmatrix} = \begin{pmatrix} 0 \\ s \end{pmatrix} = s\begin{pmatrix} 0 \\ 1 \end{pmatrix} \text{ where } s \neq 0 \qquad \left[\begin{pmatrix} 0 \\ 1 \end{pmatrix} \text{ is one of the simplest eigenvectors}\right]$$

Summarizing the above we have:

Eigenvector $\mathbf{u} = s\begin{pmatrix} 1 \\ -1 \end{pmatrix}$ belonging to $\lambda_1 = 2$ and eigenvector $\mathbf{v} = s\begin{pmatrix} 0 \\ 1 \end{pmatrix}$ belonging to $\lambda_2 = 3$.

What does all this mean?

The given matrix $\mathbf{A}$ scalar multiplies the eigenvector $\mathbf{u}$ by 2 and $\mathbf{v}$ by 3 because

$$\mathbf{Au} = 2\mathbf{u} \text{ and } \mathbf{Av} = 3\mathbf{v}$$

Plotting these eigenvectors, the effect of multiplying by the matrix $\mathbf{A}$ is shown in Fig. 7.3.

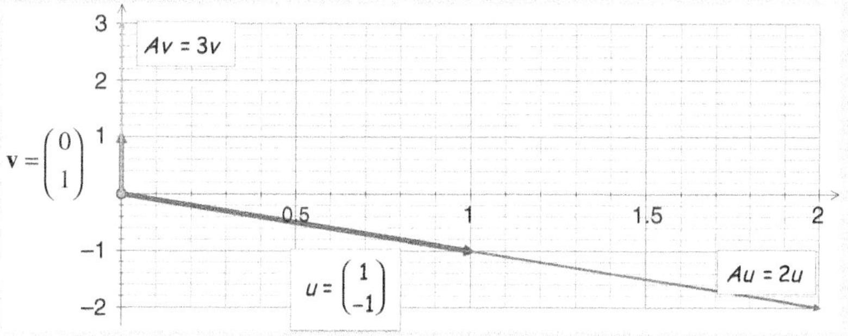

Figure 7.3

Matrix $\mathbf{A}$ doubles ($\lambda_1 = 2$) the eigenvector $\mathbf{u}$ and triples ($\lambda_2 = 3$) the eigenvector $\mathbf{v}$ as you can see in Fig. 7.3. Matrix $\mathbf{A}$ does *not* change the direction of the eigenvectors.

Eigenvectors are non-zero vectors which are transformed by the matrix $\mathbf{A}$ to a scalar multiple λ of itself.

Next, we find the eigenvalues and eigenvectors of a 3 by 3 matrix. Follow the algebra carefully because you will have to expand brackets like $(1-\lambda)(-3-\lambda)$.

To expand this it is usually easier to take out two minus signs and then expand, that is:

$$(1-\lambda)(-3-\lambda) = --(-1+\lambda)(3+\lambda) = (\lambda-1)(3+\lambda) \quad [\text{Because } -- = +]$$

Example 7.5

Determine the eigenvalues of $\mathbf{A} = \begin{pmatrix} 1 & 0 & 4 \\ 0 & 4 & 0 \\ 3 & 5 & -3 \end{pmatrix}$

Solution

We have

$$\mathbf{A} - \lambda\mathbf{I} = \begin{pmatrix} 1 & 0 & 4 \\ 0 & 4 & 0 \\ 3 & 5 & -3 \end{pmatrix} - \begin{pmatrix} \lambda & 0 & 0 \\ 0 & \lambda & 0 \\ 0 & 0 & \lambda \end{pmatrix} = \begin{pmatrix} 1-\lambda & 0 & 4 \\ 0 & 4-\lambda & 0 \\ 3 & 5 & -3-\lambda \end{pmatrix}$$

It is easier to remember that $\mathbf{A} - \lambda\mathbf{I}$ is actually matrix $\mathbf{A}$ with $-\lambda$ along the leading diagonal (from top left to bottom right). We need to evaluate $\det(\mathbf{A} - \lambda\mathbf{I})$.

What is the simplest way to find $\det(\mathbf{A} - \lambda\mathbf{I})$?

From the properties of determinants of the last chapter, we know that it will be easier to evaluate the determinant along the middle row, containing the elements 0, $4-\lambda$ and 0.

Why?

Because it has two zeros we do *not* have to evaluate the 2 by 2 determinants associated with these zeros. [Spending a second or two in choosing an easy way forward can really help save on the arithmetic later on.] From above we have

(continued...)

$$\det(\mathbf{A}-\lambda\mathbf{I}) = \det\begin{pmatrix} 1-\lambda & 0 & 4 \\ 0 & 4-\lambda & 0 \\ 3 & 5 & -3-\lambda \end{pmatrix} \longleftarrow \text{middle row}$$

$$= (4-\lambda)\left[\det\begin{pmatrix} 1-\lambda & 4 \\ 3 & -3-\lambda \end{pmatrix}\right] \quad \left[\begin{matrix}\text{expanding the}\\ \text{middle Row}\end{matrix}\right]$$

$$= (4-\lambda)\left[(1-\lambda)(-3-\lambda)-(3\times 4)\right] \quad [\text{by determinant of 2 by 2}]$$

$$= (4-\lambda)\left[(\lambda-1)(3+\lambda)-12\right] \quad [\text{taking out minus signs}]$$

$$= (4-\lambda)\left[3\lambda+\lambda^2-3-\lambda-12\right] \quad [\text{opening brackets}]$$

$$= (4-\lambda)\left[\lambda^2+2\lambda-15\right] \quad [\text{simplifying}]$$

$$= (4-\lambda)(\lambda+5)(\lambda-3) \quad [\text{factorizing}]$$

By the characteristic equation (7.2), $\det(\mathbf{A}-\lambda\mathbf{I}) = 0$, we equate all the above to zero:

$$(4-\lambda)(\lambda+5)(\lambda-3) = 0$$

Solving this equation gives the eigenvalues $\lambda_1 = 4$, $\lambda_2 = -5$ and $\lambda_3 = 3$.

Example 7.6

Determine the eigenvectors associated with $\lambda_3 = 3$ for the matrix $\mathbf{A}$ given in Example 7.5.

Solution

Substituting the eigenvalue $\lambda_3 = \lambda = 3$ and the matrix $\mathbf{A} = \begin{pmatrix} 1 & 0 & 4 \\ 0 & 4 & 0 \\ 3 & 5 & -3 \end{pmatrix}$ into $(\mathbf{A}-\lambda\mathbf{I})\mathbf{u} = \mathbf{O}$ (subtract 3 from the leading diagonal) gives:

$$(\mathbf{A}-3\mathbf{I})\,\mathbf{u} = \begin{pmatrix} 1-3 & 0 & 4 \\ 0 & 4-3 & 0 \\ 3 & 5 & -3-3 \end{pmatrix}\mathbf{u} = \mathbf{O}$$

where $\mathbf{u}$ is the eigenvector corresponding to $\lambda_3 = 3$.

What is the zero vector, $\mathbf{O}$, equal to?

Remember, this zero vector is $\mathbf{O} = \begin{pmatrix} 0 \\ 0 \\ 0 \end{pmatrix}$. Let $\mathbf{u} = \begin{pmatrix} x \\ y \\ z \end{pmatrix}$. Substituting these into the above and simplifying gives

$$\begin{pmatrix} -2 & 0 & 4 \\ 0 & 1 & 0 \\ 3 & 5 & -6 \end{pmatrix}\begin{pmatrix} x \\ y \\ z \end{pmatrix} = \begin{pmatrix} 0 \\ 0 \\ 0 \end{pmatrix}$$

Expanding this yields the linear system

$$\begin{aligned} -2x + 0 + 4z &= 0 \quad (1) \\ 0 + y + 0 &= 0 \quad (2) \\ 3x + 5y - 6z &= 0 \quad (3) \end{aligned}$$

From the middle equation (2) we have $y = 0$. From the top equation (1) we have

$$2x = 4z \text{ which gives } x = 2z$$

If $z = 1$ then $x = 2$; or more generally if $z = s$ then $x = 2s$ where $s \neq 0$ [not zero].

The general eigenvector $\mathbf{u} = \begin{pmatrix} x \\ y \\ z \end{pmatrix} = \begin{pmatrix} 2s \\ 0 \\ s \end{pmatrix} = s\begin{pmatrix} 2 \\ 0 \\ 1 \end{pmatrix}$ where $s \neq 0$ and corresponds to $\lambda_3 = 3$.

The given matrix $\mathbf{A}$ triples the eigenvector $\mathbf{u}$ because $\mathbf{Au} = 3\mathbf{u}$.

You are asked to find the eigenvectors belonging to $\lambda_1 = 4$ and $\lambda_2 = -5$ in Exercises 7.1.

7.1.3 Eigenspace

Note that for $\lambda_3 = 3$ in the above Example 7.6 we have an infinite number of eigenvectors by substituting various non-zero values of s:

$$\mathbf{u} = \begin{pmatrix} 2 \\ 0 \\ 1 \end{pmatrix} \text{ or } \mathbf{u} = \begin{pmatrix} 4 \\ 0 \\ 2 \end{pmatrix} \text{ or } \mathbf{u} = \begin{pmatrix} 1 \\ 0 \\ 1/2 \end{pmatrix} \text{ or } \mathbf{u} = \begin{pmatrix} -4 \\ 0 \\ -2 \end{pmatrix} \ldots$$

Check that the matrix $\mathbf{A}$ triples each of these eigenvectors by verifying $\mathbf{Au} = 3\mathbf{u}$. The above solutions $\mathbf{u}$ are given by *all* the points (apart from $x = y = z = 0$) on the line shown in Fig. 7.4:

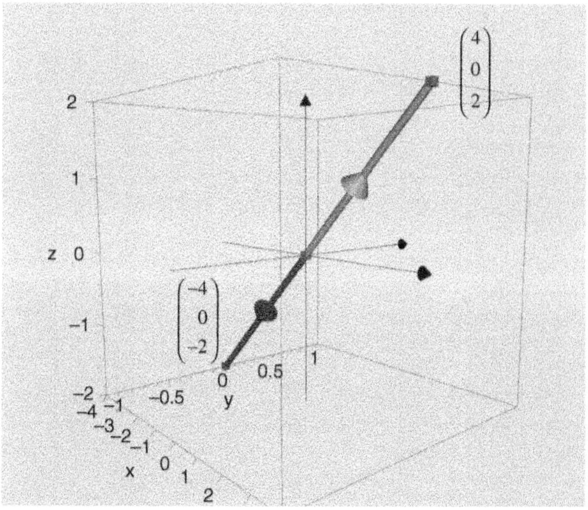

Figure 7.4

In general, if $\mathbf{A}$ is a square matrix and λ is an eigenvalue of $\mathbf{A}$ with an eigenvector $\mathbf{u}$ then every scalar multiple (apart from 0) of the vector $\mathbf{u}$ is also an eigenvector belonging to the eigenvalue λ. For example, if we have $\mathbf{Au} = 3\mathbf{u}$ then $666\mathbf{u}$ is also an eigenvector because

$$\mathbf{A}(666\mathbf{u}) = 666(\mathbf{A}\mathbf{u}) = 666 \underbrace{(3\mathbf{u})}_{\text{because } \mathbf{Au}=3\mathbf{u}} = 3(666\mathbf{u})$$

Since the matrix **A** *triples* the vector 666**u** so 666**u** is an eigenvector with eigenvalue 3. Thus we have the general proposition:

Proposition (7.3). If λ is an eigenvalue of a square matrix **A** with an eigenvector **u** then *every* non-zero scalar multiplication of **u**, such as $k\mathbf{u}$, is also an eigenvector belonging to λ.

This means that if **u** is an eigenvector belonging to λ then so is 2**u**, 0.53**u**, −666**u**, ...

Proof.
Consider an arbitrary non-zero scalar k, then

$$\begin{aligned} \mathbf{A}(k\mathbf{u}) &= k(\mathbf{A}\mathbf{u}) && [\text{by rules of matrices}] \\ &= k(\lambda\mathbf{u}) && [\text{by (7.1) } \mathbf{A}\mathbf{u} = \lambda\mathbf{u}] \\ &= \lambda(k\mathbf{u}) \end{aligned}$$

Thus we have $\mathbf{A}(k\mathbf{u}) = \lambda(k\mathbf{u})$, which means that the matrix **A** acting on the vector $k\mathbf{u}$ produces a scalar multiple λ of $k\mathbf{u}$. Hence $k\mathbf{u}$ is an eigenvector belonging to the eigenvalue λ. Since k was arbitrary, every non-zero scalar multiple of **u** is an eigenvector of the matrix **A** belonging to the eigenvalue λ.

Hence the scalar λ produces an infinite number of eigenvectors.

■

Proposition (7.4). If **A** is an n by n matrix with an eigenvalue of λ, then the set S of *all* eigenvectors of **A** belonging to λ together with the zero vector, **O**, is a subspace of $\mathbb{R}^n$:

$$S = \{\mathbf{O}\} \cup \{\mathbf{u} \mid \mathbf{u} \text{ is an eigenvector belonging to } \lambda\}$$

How do we prove the given set is a subspace of $\mathbb{R}^n$?
We can use result (3.7) of chapter 3:

Proposition (3.7). A non-empty subset S containing vectors **u** and **v** is a subspace of a vector space $V \Leftrightarrow$ any linear combination $k\mathbf{u} + c\mathbf{v}$ is also in S (k and c are scalars).

This means that we need to show:

If vectors **u** and **v** are in S then for any scalars k and c we have $k\mathbf{u} + c\mathbf{v}$ is also in S. This means that S is a subspace $\Leftrightarrow$ S is closed so the vector $k\mathbf{u} + c\mathbf{v}$ cannot escape from S, as shown in Fig. 7.5.

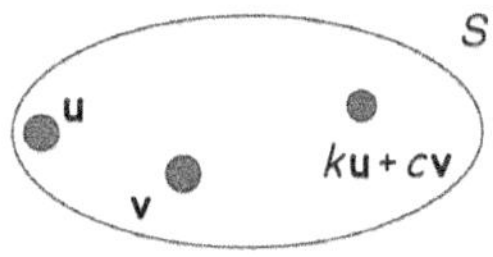

Figure 7.5

Proof.

What do we need to prove?

Required to prove that if $\mathbf{u}$ and $\mathbf{v}$ are eigenvectors belonging to the eigenvalue λ then $k\mathbf{u} + c\mathbf{v}$ is also an eigenvector belonging to λ.

Let $\mathbf{u}$ and $\mathbf{v}$ be eigenvectors belonging to the same eigenvalue λ, and k and c be any non-zero scalars. Then by the above Proposition (7.3):

If λ is an eigenvalue of a square matrix $\mathbf{A}$ with an eigenvector $\mathbf{u}$ then $k\mathbf{u}$ is also an eigenvector belonging to λ.

We have $k\mathbf{u}$ and $c\mathbf{v}$ are eigenvectors belonging to λ, therefore by (7.1):

$$\mathbf{A}(k\mathbf{u}) = \lambda(k\mathbf{u}) \text{ and } \mathbf{A}(c\mathbf{v}) = \lambda(c\mathbf{v}) \qquad (*)$$

We need to show that $\mathbf{A}(k\mathbf{u} + c\mathbf{v}) = \lambda(k\mathbf{u} + c\mathbf{v})$:

$$\begin{aligned} \mathbf{A}(k\mathbf{u} + c\mathbf{v}) &= \mathbf{A}(k\mathbf{u}) + \mathbf{A}(c\mathbf{v}) && \left[\text{applying the rules of matrices}\right] \\ &= \lambda(k\mathbf{u}) + \lambda(c\mathbf{v}) && \left[\text{by the above } (*)\right] \\ &= \lambda(k\mathbf{u} + c\mathbf{v}) && \left[\text{factorizing}\right] \end{aligned}$$

Since $\mathbf{A}(k\mathbf{u} + c\mathbf{v}) = \lambda(k\mathbf{u} + c\mathbf{v})$ so the matrix $\mathbf{A}$ scalar (λ) multiplies the vector $k\mathbf{u} + c\mathbf{v}$ so it is an eigenvector belonging to λ which means it is a member of the set S. By Proposition (3.7) we conclude that the set S is a subspace of $\mathbb{R}^n$.

■

This subspace S of Proposition (7.4)

$$S = \{\mathbf{O}\} \cup \left\{\mathbf{u} \mid \mathbf{u} \text{ is an eigenvector belonging to } \lambda\right\}$$

is called an **eigenspace** of λ and is denoted by E_λ, that is $E_\lambda = S$.

For example, the eigenspace associated with Example 7.4 for the eigenvalue $\lambda_1 = 2$ is the eigenvector $\mathbf{u} = s\begin{pmatrix} 1 \\ -1 \end{pmatrix}$ and for $\lambda_2 = 3$ the eigenvector $\mathbf{v} = s\begin{pmatrix} 0 \\ 1 \end{pmatrix}$ which are shown in Fig. 7.6.

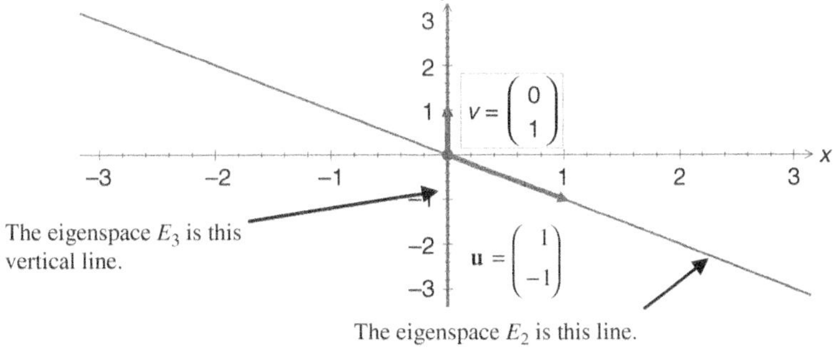

Figure 7.6

E_2 and E_3 denote the eigenspaces given by $\lambda = 2$ and $\lambda = 3$ respectively.

Note that the vector $\begin{pmatrix} 0 \\ 1 \end{pmatrix}$ is a basis (axis) for the eigenspace E_3 and $\begin{pmatrix} 1 \\ -1 \end{pmatrix}$ is a basis (axis) for the eigenspace E_2. These *eigenvectors* are a basis (axis) for each eigenspace.

We can also use the numerical software MATLAB to find eigenvalues and eigenvectors:

As an example, we'll find consider the matrix in Example 7.6: $\mathbf{A} = \begin{pmatrix} 1 & 0 & 4 \\ 0 & 4 & 0 \\ 3 & 5 & -3 \end{pmatrix}$.

In MATLAB, we enter the matrix **A** by typing: `A=[1 0 4 ; 0 4 0 ; 3 5 -3]` where the semicolon denotes the start of the new row. We'll let the matrix containing the eigenvectors be called V and the matrix containing the eigenvalues as d. We then use the following MATLAB command. `[V,d]=eig(A,'nobalance')`. The 'nobalance' prevents MATLAB from normalizing the vector. The result of this command is:

```
V =
1.0000    0.6667   -1.0000
     0         0   -0.4500
0.5000   -1.0000   -0.7500
```

Reading down each column gives the eigenvectors.

```
d =
3    0   0
0   -5   0
0    0   4
```

The leading diagonal entries give the eigenvalues.

By reading the above MATLAB output, the eigenvectors are

$$\begin{pmatrix} 1 \\ 0 \\ 1/2 \end{pmatrix}, \begin{pmatrix} 2/3 \\ 0 \\ -1 \end{pmatrix}, \begin{pmatrix} 1 \\ 9/20 \\ 3/4 \end{pmatrix}$$

So the general eigenvectors are

$$\mathbf{u} = r\begin{pmatrix} 2 \\ 0 \\ 1 \end{pmatrix}, \mathbf{v} = s\begin{pmatrix} 2 \\ 0 \\ -3 \end{pmatrix}, \mathbf{w} = t\begin{pmatrix} 20 \\ 9 \\ 15 \end{pmatrix}$$

where $r, s, t \neq 0$. Note that the eigenvector **u** is the eigenvector found in Example 7.6. You are asked to verify the other two by hand in Exercises 7.1.

Summary

The eigenvector **u** belonging to eigenvalue λ satisfies:

(7.1) $\mathbf{Au} = \lambda\mathbf{u}$ (matrix **A** scalar multiplies eigenvector **u** by λ)

The following equation is used to find the eigenvalues:

(7.2) $\det(\mathbf{A} - \lambda\mathbf{I}) = 0$

Eigenvectors **u** are found using $(\mathbf{A} - \lambda\mathbf{I})\mathbf{u} = \mathbf{O}$.

EXERCISES 7.1

(Brief solutions at end of book. Full solutions available at <http://www.oup.co.uk/companion/singh>.)

In this exercise you may check your numerical answers using MATLAB.

1. Find the eigenvalues and particular eigenvectors of the following matrices:
   (a) $\mathbf{A} = \begin{pmatrix} 7 & 3 \\ 0 & -4 \end{pmatrix}$ (b) $\mathbf{A} = \begin{pmatrix} 5 & -2 \\ 4 & -1 \end{pmatrix}$ (c) $\mathbf{A} = \begin{pmatrix} -1 & 4 \\ 2 & 1 \end{pmatrix}$
2. Obtain the general eigenvectors of $\lambda_1 = 4$ and $\lambda_2 = -5$ for the matrix in Example 7.5.
3. Find the eigenvalues and eigenvectors of $\mathbf{A} = \begin{pmatrix} 3 & 1 \\ 1 & 3 \end{pmatrix}$. Plot the eigenspaces E_λ.
4. Find the eigenvalues and eigenvectors of $\mathbf{A} = \begin{pmatrix} 5 & -2 \\ 7 & -4 \end{pmatrix}$. State the effect of multiplying the eigenvector by the matrix **A**. Plot the eigenspaces E_λ and write down a basis vector for each of the eigenspaces.
5. Let $\mathbf{A} = \begin{pmatrix} -2 & 8 \\ 5 & 1 \end{pmatrix}$ and $\mathbf{B} = \begin{pmatrix} -4 & 16 \\ 10 & 2 \end{pmatrix}$.
   (a) Determine the eigenvalues of **A**.
   (b) Determine the eigenvalues of **B**.
   (c) State a relationship between the eigenvalues of **A** and **B** and predict a general relationship.
6. Let **A** be a square matrix and $\mathbf{B} = r\mathbf{A}$, where r is a real number. Prove that if λ is the eigenvalue of matrix **A** then the eigenvalue of **B** is $r\lambda$.
7. Prove that the zero $n \times n$ matrix, $\mathbf{O} = \begin{pmatrix} 0 & \cdots & 0 \\ \vdots & \ddots & \vdots \\ 0 & \cdots & 0 \end{pmatrix}$, only has *zero* eigenvalues.
   [Hint: Use the Proposition of chapter 6 which gives the determinant of a diagonal matrix.]
8. Determine the eigenvalues and eigenvectors of $\mathbf{A} = \begin{pmatrix} 1 & 2 & 1 \\ 2 & 1 & 1 \\ 1 & 1 & 2 \end{pmatrix}$. By using appropriate software plot the eigenspaces and write down a basis for each eigenspace.

SECTION 7.2 Properties of Eigenvalues and Eigenvectors

By the end of this section you will be able to

- determine eigenvalues and eigenvectors of particular matrices
- prove some properties of eigenvalues and eigenvectors
- apply the Cayley–Hamilton theorem

As we discovered in chapter 6, finding the inverse of a matrix can be a lengthy process. In this section we will show an easier way to find the inverse. We will also show that the Cayley–Hamilton theorem significantly reduces the workload in finding the powers of matrices.

It is worth reminding ourselves that eigenvalues and eigenvectors always come in pairs. You *cannot* have an eigenvalue on its own; it must have an associated eigenvector. But they are like chalk and cheese because an eigenvalue is a scalar and an eigenvector is a vector.

First we look at multiple eigenvalues and their corresponding eigenvectors.

7.2.1 Multiple eigenvalues

What is the characteristic equation of a square matrix $\mathbf{A}$*?*

The characteristic equation was defined in the last section 7.1 as:

(7.2) $\det(\mathbf{A}-\lambda\mathbf{I})=0$ where $\mathbf{A}$ is a square matrix, $\mathbf{I}$ is the identity matrix and λ is the eigenvalue which is a scalar. By expanding this we have

$$\det(\mathbf{A}-\lambda\mathbf{I})=\det\left[\begin{pmatrix} a_{11} & \cdots & a_{1n} \\ \vdots & \ddots & \vdots \\ a_{n1} & \cdots & a_{nn}\end{pmatrix}-\begin{pmatrix} \lambda & 0 & \cdots \\ 0 & \ddots & 0 \\ 0 & \cdots & \lambda\end{pmatrix}\right]$$

$$=\det\begin{pmatrix} a_{11}-\lambda & \cdots & a_{1n} \\ \vdots & \ddots & \vdots \\ a_{n1} & \cdots & a_{nn}-\lambda\end{pmatrix}=0 \qquad \left[\begin{array}{l}\text{taking away } \lambda \text{ along} \\ \text{the leading diagonal}\end{array}\right]$$

Evaluating the determinant of this matrix results in a polynomial equation of degree n. An example of a polynomial $p(\lambda)$ of degree 3 is

$$5\lambda^3-2\lambda^2+3\lambda+69$$

An example of a polynomial $p(\lambda)$ of degree 4 is

$$7\lambda^4-5\lambda^3-0.1\lambda^2+34\lambda+5$$

An example of a polynomial $p(\lambda)$ of degree n is

$$333\lambda^n+2\lambda^{n-1}+\cdots+6\lambda^2+\lambda+2.7\,1828$$

In general terms, we can write a polynomial $p(\lambda)$ of degree n as

$$p(\lambda)=c_n\lambda^n+c_{n-1}\lambda^{n-1}+\cdots+c_1\lambda+c_0$$

where the c's are the coefficients.

What is the difference between a characteristic polynomial and an equation?

A characteristic polynomial is an expression $p(\lambda)$ while an equation is $p(\lambda)=0$.

How many roots does the equation $p(\lambda) = 0$ have?

The impressively titled 'Fundamental Theorem of Algebra' tells us that the polynomial equation $p(\lambda) = 0$ of degree n has *exactly* n roots. Don't worry if you have *not* heard of the 'Fundamental Theorem of Algebra'. It is a known theorem in algebra which claims that a polynomial equation of degree n has *exactly* n roots. For example, $x^2 - 2x + 1 = 0$ has exactly *two* roots because it is a polynomial equation of degree 2 (quadratic equation).

Other examples are:

Equation	Number of roots
$x^5 - 1 = 0$	5
$2x^{12} - 2x^3 + 1 = 0$	12
$-5x^{101} - x^{100} - \cdots - 1 = 0$	101

We will *not* prove the fundamental theorem of algebra but assume it is true. Thus by the fundamental theorem of algebra we conclude that

$$p(\lambda) = c_n\lambda^n + c_{n-1}\lambda^{n-1} + c_{n-2}\lambda^{n-2} + \cdots + c_2\lambda^2 + c_1\lambda + c_0 = 0$$

has n eigenvalues (roots). There might be n distinct eigenvalues or they might be *repeated* such as

$$(\lambda - 1)^3(\lambda - 2) = 0 \text{ which gives } \lambda_1 = 1, \lambda_2 = 1, \lambda_3 = 1 \text{ and } \lambda_4 = 2$$

Normally we write the first three roots in compact form as $\lambda_{1,2,3} = 1$ rather than as above.

We distinguish between simple and multiple eigenvalues in the next definition.

Definition (7.5). Let $\mathbf{A}$ be an n by n matrix and have the eigenvalues λ_1, λ_2, $\lambda_3, \ldots$ and λ_n. If λ occurs *only once* then we say λ is a **simple eigenvalue**, otherwise it is called a **multiple eigenvalue**. If λ occurs m times where $m > 1$ then we say λ is an eigenvalue with **multiplicity** of m or λ has *multiplicity m*.

In the above equation $(\lambda - 1)^3(\lambda - 2) = 0$ we have $\lambda_{1,2,3} = 1$ is an eigenvalue of multiplicity 3 and $\lambda_4 = 2$ is a simple eigenvalue.

Example 7.7

Determine the eigenvalues and eigenspaces of $\mathbf{A} = \begin{pmatrix} 2 & 1 & 3 \\ 0 & 2 & 0 \\ 0 & 0 & 2 \end{pmatrix}$.

Solution

What do we determine first?

The eigenvalues, because they produce the eigenvectors. Using the characteristic equation (7.2) $\det(\mathbf{A} - \lambda\mathbf{I}) = 0$ we have

(continued...)

$$\det(\mathbf{A}-\lambda\mathbf{I}) = \det\begin{pmatrix} \boxed{2-\lambda} & 1 & 3 \\ 0 & 2-\lambda & 0 \\ 0 & 0 & 2-\lambda \end{pmatrix}$$

$$= (2-\lambda)\det\begin{pmatrix} 2-\lambda & 0 \\ 0 & 2-\lambda \end{pmatrix} \qquad \begin{bmatrix}\text{expanding along} \\ \text{the first column}\end{bmatrix}$$

$$= (2-\lambda)\left[(2-\lambda)(2-\lambda)-0\right] = (2-\lambda)^3 = 0$$

Thus we *only* have one repeated eigenvalue $\lambda = \lambda_{1,2,3} = 2$. We say that $\lambda = 2$ has multiplicity 3.
How do we find the corresponding eigenvector?
By substituting this, $\lambda = 2$, into $(\mathbf{A}-\lambda\mathbf{I})\mathbf{u} = \mathbf{O}$ where $\mathbf{u}$ is the eigenvector:

$$(\mathbf{A}-2\mathbf{I})\,\mathbf{u} = \begin{pmatrix} 2-2 & 1 & 3 \\ 0 & 2-2 & 0 \\ 0 & 0 & 2-2 \end{pmatrix}\mathbf{u} = \mathbf{O}$$

Simplifying this and substituting unknowns x, y, z for the eigenvector $\mathbf{u}$ and zeros into the zero vector, $\mathbf{O}$, gives

$$\begin{pmatrix} 0 & 1 & 3 \\ 0 & 0 & 0 \\ 0 & 0 & 0 \end{pmatrix}\begin{pmatrix} x \\ y \\ z \end{pmatrix} = \begin{pmatrix} 0 \\ 0 \\ 0 \end{pmatrix}$$

Note that this matrix is already in reduced row echelon form (rref). There is only one non-zero equation and three unknowns, therefore there are $3-1=2$ free variables (x and z, because none of the equations begin with these). The term 'free variables' was defined in chapter 1.

By expanding the first row we have $y+3z=0$, which gives $y=-3z$. Let $z=s$ where $s \neq 0$ then $y=-3s$. Clearly x can be any real number, that is $x=t$. Hence, we write the eigenvector ($x=t$, $y=-3s$ and $z=s$) in terms of two separate vectors which are a basis for the eigenspace E_2:

$$\mathbf{u} = \begin{pmatrix} x \\ y \\ z \end{pmatrix} = \begin{pmatrix} t \\ -3s \\ s \end{pmatrix} = \begin{pmatrix} t \\ 0 \\ 0 \end{pmatrix} + \begin{pmatrix} 0 \\ -3s \\ s \end{pmatrix} = t\begin{pmatrix} 1 \\ 0 \\ 0 \end{pmatrix} + s\begin{pmatrix} 0 \\ -3 \\ 1 \end{pmatrix}$$

where s and t are *not both zero*. If s and t were *both* zero then $\mathbf{u}$ would be the zero vector but $\mathbf{u}$ is an eigenvector so $\neq \mathbf{O}$. This $\mathbf{u}$ is our general eigenvector and we can write our eigenspace as

$E_2 = \left\{ t\begin{pmatrix} 1 \\ 0 \\ 0 \end{pmatrix} + s\begin{pmatrix} 0 \\ -3 \\ 1 \end{pmatrix} \right\}$ and plot this in $\mathbb{R}^3$ as shown in Fig. 7.7.

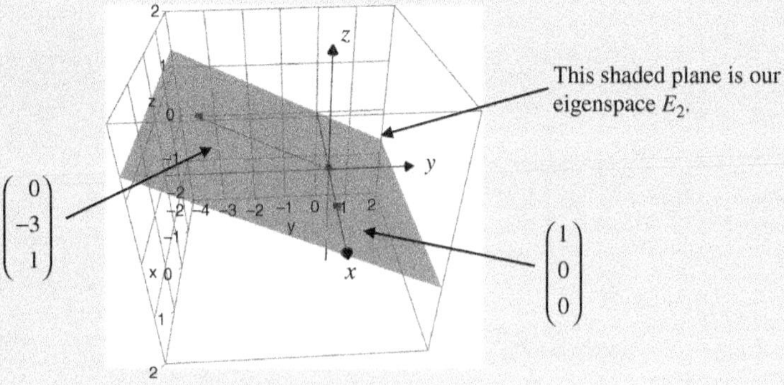

Figure 7.7

Note that we have a plane rather than just a line because we have two basis vectors (as shown above) which span a plane. A set of basis (axes) vectors B of the eigenspace E_2 are given by

$$B = \left\{ \begin{pmatrix} 1 \\ 0 \\ 0 \end{pmatrix}, \begin{pmatrix} 0 \\ -3 \\ 1 \end{pmatrix} \right\}$$

The matrix given in the above example is a type of matrix called a triangular matrix which was defined in the previous chapter. A smarter way to evaluate the eigenvalues of such matrices is described next.

7.2.2 Eigenvalues of diagonal and triangular matrices

In Exercises 7.1 you proved that the eigenvalues of the zero matrix, $\mathbf{O}$, are 0.

What sort matrix is the zero matrix?
A diagonal matrix.

What is a diagonal or triangular matrix?
By definition (6.17), a *triangular* matrix is an n by n matrix where *all* the entries to *one* side of the leading diagonal are zero.

By definition (6.18) we have a *diagonal* matrix is an n by n matrix where *all* the entries to *both* sides of the leading diagonal are zero.

The following are examples:

$$\mathbf{A} = \underbrace{\begin{pmatrix} 1 & 0 & 0 & 0 \\ 0 & 2 & 0 & 0 \\ 0 & 0 & 5 & 0 \\ 0 & 0 & 0 & 4 \end{pmatrix}}_{\text{diagonal matrix}}, \mathbf{B} = \underbrace{\begin{pmatrix} 8 & -1 & 5 & 9 \\ 0 & -8 & 3 & -6 \\ 0 & 0 & 5 & 7 \\ 0 & 0 & 0 & 4 \end{pmatrix}}_{\text{triangular matrix}} \text{ and } \mathbf{C} = \underbrace{\begin{pmatrix} -9 & 0 & 0 & 0 \\ -8 & 3 & 0 & 0 \\ -2 & 4 & -1 & 0 \\ 1 & 0 & 5 & 10 \end{pmatrix}}_{\text{triangular matrix}}$$

B is actually called an *upper* triangular matrix and **C** a *lower* triangular matrix.

We prove that for a diagonal or triangular matrix the eigenvalues are given by the entries along the leading diagonal which goes from top left to bottom right of the matrix. For example, the above diagonal matrix **A** has eigenvalues 1, 2, 5 and 4.

What are the eigenvalues of the triangular matrices **B** *and* **C**?
B has eigenvalues 8, − 8, 5 and 4. **C** has eigenvalues −9, 3, −1 and 10.

Proposition (7.6). If an n by n matrix **A** is a diagonal or triangular matrix then the eigenvalues of **A** are the entries along the leading diagonal.

How do we prove this?
We apply Proposition (6.19) to the matrix **A**: The determinant of a triangular or diagonal matrix is the product of the entries along the leading diagonal.

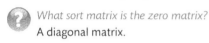

The proofs for the three different cases of upper, lower triangular and diagonal are very similar, so we only proof this for one case. In mathematical proof, we say 'without loss of generality (WLOG)' meaning that we prove it for one case, and the proof for the other case is identical.

Proof of (7.6).

Without loss of generality (WLOG), let $\mathbf{A} = \begin{pmatrix} a_{11} & a_{12} & \cdots & a_{1n} \\ 0 & a_{22} & \cdots & a_{2n} \\ \vdots & 0 & \ddots & \vdots \\ 0 & \vdots & 0 & a_{nn} \end{pmatrix}$ be an upper triangular matrix. The characteristic equation $\det(\mathbf{A} - \lambda\mathbf{I}) = 0$ is given by

$$\det(\mathbf{A} - \lambda\mathbf{I}) = \det \begin{pmatrix} a_{11} - \lambda & a_{12} & \cdots & a_{1n} \\ 0 & a_{22} - \lambda & \cdots & a_{2n} \\ \vdots & 0 & \ddots & \vdots \\ 0 & \vdots & 0 & a_{nn} - \lambda \end{pmatrix}$$

$$\underbrace{=}_{\substack{\text{multiplying the leading} \\ \text{diagonal entries.} \\ \text{(by Proposition (6.19))}}} (a_{11} - \lambda)(a_{22} - \lambda) \cdots (a_{nn} - \lambda) = 0 \text{ implies } \lambda_1 = a_{11}, \lambda_2 = a_{22}, \ldots, \lambda_n = a_{nn}$$

The roots or eigenvalues of this equation are a_{11}, a_{22}, $a_{33}, \ldots$ and a_{nn} which are the leading diagonal entries of the given matrix **A**. This completes our proof. ■

Example 7.8

Determine the eigenvalues of $\mathbf{A} = \begin{pmatrix} 5 & 3 & 5 & 8 \\ 0 & 9 & 7 & -9 \\ 0 & 0 & 6 & -5 \\ 0 & 0 & 0 & -17 \end{pmatrix}$.

Solution

*What do you notice about matrix **A**?*

Matrix **A** is an (upper) triangular matrix, therefore we can apply the above Proposition (7.6) to find the eigenvalues of matrix **A**.

*What are the eigenvalues of **A**?*

The eigenvalues are the entries on the leading diagonal which runs from top left to bottom right. Thus the eigenvalues are

$$\lambda_1 = 5,\ \lambda_2 = 9,\ \lambda_3 = 6 \text{ and } \lambda_4 = -17$$

Note that the eigenvalues of the matrix **A** given in Example 7.7 are

$$\lambda_{1,\,2,\,3} = 2$$

because *all* the entries along the leading diagonal of the triangular matrix **A** are 2.

7.2.3 Properties of eigenvalues and eigenvectors

Proposition (7.7). A square matrix **A** is invertible (has an inverse) $\Leftrightarrow \lambda = 0$ is *not* an eigenvalue of the matrix **A**.

Proof – Exercises 7.2.

■

This proposition means that if $\lambda = 0$ is an eigenvalue of matrix **A** then **A** has no inverse.

Proposition (7.8). Let **A** be a square matrix with eigenvector **u** belonging to eigenvalue λ.

(a) If m is a natural number then λ^m is an eigenvalue of the matrix $\mathbf{A}^m$ with the *same* eigenvector **u**.

(b) If the matrix **A** is invertible (has an inverse) then the eigenvalue of the inverse matrix $\mathbf{A}^{-1}$ is $\frac{1}{\lambda} = \lambda^{-1}$ with the *same* eigenvector **u**.

What does this proposition mean?

Matrix	Eigenvector	Eigenvalue
$\mathbf{A}$	$\mathbf{u}$	λ
(a) $\mathbf{A}^m$ (power)	$\mathbf{u}$	λ^m
(b) $\mathbf{A}^{-1}$ (inverse)	$\mathbf{u}$	λ^{-1}

We illustrate this for $\lambda = 2$ (the matrix **A** doubles the eigenvector **u**) as shown in Fig. 7.8.

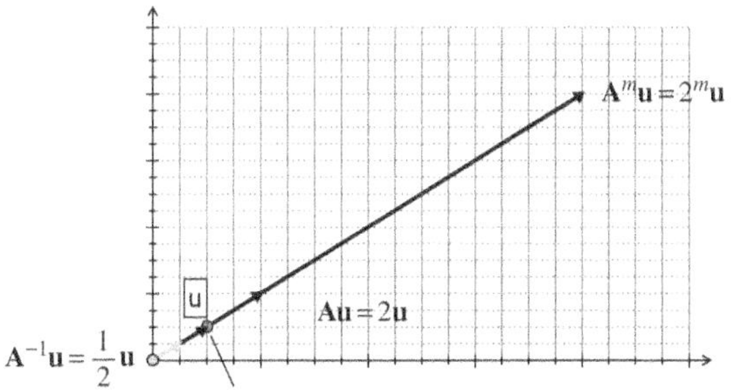

Figure 7.8

How do we prove proposition (a)?

By using mathematical induction. The three steps of mathematical induction are:

Step 1: Check the result for some base case $m = m_0$.

Step 2: Assume that the result is true for $m = k$.

Step 3: Prove the result for $m = k + 1$.

Proof.
Step 1: Using the definition of eigenvalues and eigenvectors (7.1) we have $\mathbf{Au} = \lambda\mathbf{u}$ which means the result holds for $m = 1$:

$$\mathbf{Au} = \lambda\mathbf{u} \qquad (*)$$

Step 2: Assume that the result is true for $m = k$:

$$\mathbf{A}^k\mathbf{u} = \lambda^k\mathbf{u} \qquad (\dagger)$$

Step 3: Required to prove the case $m = k + 1$; that is, we need to prove

$$\mathbf{A}^{k+1}\mathbf{u} = \lambda^{k+1}\mathbf{u}$$

Expanding the left hand side:

$$\begin{aligned}\mathbf{A}^{k+1}\mathbf{u} &= \mathbf{A}\left(\mathbf{A}^k\mathbf{u}\right) \qquad \left[\text{writing } \mathbf{A}^{k+1} = \mathbf{A}\mathbf{A}^k\right] \\ &= \mathbf{A}\underbrace{\left(\lambda^k\mathbf{u}\right)}_{\text{by } (\dagger)} = \lambda^k(\mathbf{Au}) = \lambda^k \underbrace{(\lambda\mathbf{u})}_{\text{by } (*)} = \underbrace{\lambda^k\lambda}_{=\lambda^{k+1}}\mathbf{u} = \lambda^{k+1}\mathbf{u}\end{aligned}$$

Thus $\mathbf{A}^{k+1}\mathbf{u} = \lambda^{k+1}\mathbf{u}$, therefore by mathematical induction we have our result that λ^m is an eigenvalue of the matrix $\mathbf{A}^m$ with the eigenvector $\mathbf{u}$.

■

Proof of (b).
Using formula (7.1) we have $\mathbf{Au} = \lambda\mathbf{u}$. Left multiplying both sides of this by $\mathbf{A}^{-1}$ gives

$$\underbrace{\left(\mathbf{A}^{-1}\mathbf{A}\right)}_{=\mathbf{I}}\mathbf{u} = \mathbf{A}^{-1}(\lambda\mathbf{u}) = \lambda\mathbf{A}^{-1}\mathbf{u}$$

Remember, multiplying by the identity $\mathbf{I}$ keeps it the same, that is $\mathbf{Iu} = \mathbf{u}$, so we have

$$\mathbf{u} = \lambda\mathbf{A}^{-1}\mathbf{u}$$

Dividing both sides by λ gives

$$\frac{1}{\lambda}\mathbf{u} = \mathbf{A}^{-1}\mathbf{u} \text{ or writing this the other way, we have } \mathbf{A}^{-1}\mathbf{u} = \frac{1}{\lambda}\mathbf{u}$$

Since $\mathbf{A}^{-1}$ scalar multiplies $\mathbf{u}$ by $\frac{1}{\lambda}$ so the eigenvalue of $\mathbf{A}^{-1}$ is $\frac{1}{\lambda}$ with the eigenvector $\mathbf{u}$.

■

Example 7.9

Find the eigenvalues of $\mathbf{A}^7$ where $\mathbf{A} = \begin{pmatrix} 1 & 4 & 5 \\ 0 & 2 & 6 \\ 0 & 0 & 3 \end{pmatrix}$.

Solution

Because $\mathbf{A}$ is an upper triangular matrix, the eigenvalues are the entries on the leading diagonal, that is $\lambda_1 = 1, \lambda_2 = 2$ and $\lambda_3 = 3$. By the above Proposition (7.8) (a) we can see that the eigenvalues of $\mathbf{A}^7$ are

$$\lambda_1^7 = 1^7 = 1, \quad \lambda_2^7 = 2^7 = 128 \quad \text{and} \quad \lambda_3^7 = 3^7 = 2187$$

Proposition (7.9). Let $\mathbf{A}$ be any n by n matrix with eigenvalues $\lambda_1, \lambda_2, \lambda_3, \ldots \lambda_n$. We have:

(a) The determinant of the matrix $\mathbf{A}$ is given by $\det(\mathbf{A}) = \lambda_1 \times \lambda_2 \times \lambda_3 \times \cdots \times \lambda_n$.

(b) The trace of the matrix $\mathbf{A}$ is given by $tr(\mathbf{A}) = \lambda_1 + \lambda_2 + \lambda_3 + \cdots + \lambda_n$.

What is the trace of a matrix?

The trace (tr) of a matrix is the addition of all the entries on the leading diagonal:

$$tr \begin{pmatrix} a_{11} & \cdots & a_{1n} \\ \vdots & \ddots & \vdots \\ a_{n1} & \cdots & a_{nn} \end{pmatrix} = a_{11} + a_{22} + a_{33} + \cdots + a_{nn}$$

What does part (b) mean?

Adding *all* the eigenvalues of a matrix is equal to the *trace* of the matrix.

Can you see any use for part (b)?

We can use part (b) as a rough check to see if we have the correct eigenvalues.

Why is (b) a rough rather than an exact check for eigenvalues?

Because there are so many different ways of adding to the same value.

What does the first part (a) mean?

It means that we can find the determinant of a matrix by *multiplying* all the eigenvalues of that matrix.

How do we prove part (a)?

By using the characteristic equation $\det(\mathbf{A} - \lambda\mathbf{I}) = 0$.

Proof of (a).

We are given that matrix $\mathbf{A}$ has eigenvalues $\lambda_1, \lambda_2, \lambda_3, \ldots$ and λ_n.

What does this mean?

It means that these, $\lambda_1, \lambda_2, \ldots$ and λ_n, are roots of the characteristic equation, $\det(\mathbf{A} - \lambda\mathbf{I}) = 0$, which implies that we have

$$\det(\mathbf{A} - \lambda\mathbf{I}) = (\lambda_1 - \lambda)(\lambda_2 - \lambda)(\lambda_3 - \lambda)\cdots(\lambda_n - \lambda) \quad (\dagger)$$

[For example, if the eigenvalues of a matrix **B** are $\lambda_1 = 2$, $\lambda_2 = 4$ and $\lambda_3 = 5$ then the characteristic equation would be given by $\det(\mathbf{B} - \lambda\mathbf{I}) = (2 - \lambda)(4 - \lambda)(5 - \lambda)$.]

If we substitute $\lambda = 0$ into ($\dagger$) we get

$$\begin{aligned}\det(\mathbf{A}) &= (\lambda_1 - 0)(\lambda_2 - 0)(\lambda_3 - 0)\cdots(\lambda_n - 0)\\ &= (\lambda_1)(\lambda_2)(\lambda_3)\cdots(\lambda_n) = \lambda_1 \times \lambda_2 \times \lambda_3 \times \cdots \times \lambda_n\end{aligned}$$

This is our required result.

■

Proof of (b) – See website.

■

Summarizing this proposition we have:

Matrix A	$\lambda_1, \lambda_2, \lambda_3, \ldots, \lambda_n$	Eigenvalues
Determinant of **A**	$\lambda_1 \times \lambda_2 \times \lambda_3 \times \cdots \times \lambda_n$	Product of eigenvalues
Trace of **A**	$\lambda_1 + \lambda_2 + \lambda_3 + \cdots + \lambda_n$	Addition of eigenvalues

The above proposition states that we can evaluate the determinant (or trace) of any n by n matrix by finding the eigenvalues and multiplying (or adding) them.

Example 7.10

Find the determinant and trace of $\mathbf{A} = \begin{pmatrix} 1 & 2 & 1 \\ 2 & 1 & 1 \\ 1 & 1 & 2 \end{pmatrix}$ given that the eigenvalues of **A** are 1, 4 and -1.

Solution

By the above Proposition (7.9) we have

$$\det(\mathbf{A}) = 1 \times 4 \times (-1) = -4 \quad \text{[multiplying the eigenvalues]}$$

$$Tr(\mathbf{A}) = 1 + 4 - 1 = 4 \quad \text{[adding \textit{all} eigenvalues]}$$

Remember, the trace of the matrix is found by adding all the entries in the leading diagonal

$$Tr(\mathbf{A}) = 1 + 1 + 2 = 4$$

Hence we have illustrated for matrix **A** that the trace of the matrix is the same as the sum of *the* eigenvalues.

Proposition (7.10). Let $\mathbf{A}$ be an n by n matrix with *distinct* eigenvalues $\lambda_1, \lambda_2, \lambda_3, \ldots, \lambda_m$ and corresponding eigenvectors $\mathbf{u}_1, \mathbf{u}_2, \mathbf{u}_3, \ldots, \mathbf{u}_m$ where $1 \leq m \leq n$. Then these eigenvectors $\mathbf{u}_1, \mathbf{u}_2, \mathbf{u}_3, \ldots$ and $\mathbf{u}_m$ are *linearly independent*.

We need this proposition to prove the Cayley–Hamilton theorem which is given in the next subsection.

Proof – Exercises 7.2.

■

7.2.4 Cayley-Hamilton theorem

The biography of Arthur Cayley was given in section 1.4.1. Here we give a brief profile of Sir William Rowan Hamilton.

Figure 7.9

William Hamilton (Fig. 7.9) was born in Dublin, Ireland in 1805 and became one of the greatest Irish mathematicians. Initially he took an interest in languages, but during his early school days he found an affection for mathematics. At the age of 18 he entered Trinity College, Dublin and spent the rest of his life there. In 1827, Hamilton was appointed Professor of Astronomy at Trinity College, but he did not take much interest in astronomy and devoted all his time to mathematics. Hamilton is best known for his work on *quaternions* which is a vector space of four dimensions. In fact, in 1843 while he was walking along the local canal in Dublin with his wife he had a flash of inspiration and discovered the formula for quaternion multiplication.

At present there is plaque at the bridge stating this formula. He also invented the dot and cross product of vectors.

Throughout his life he had problems with alcohol and love. He fell in love with Catherine but due to unfortunate circumstances he ended up marrying Helen which he regretted for the rest of his life.

In general, we have *not* evaluated powers of matrices such as $\mathbf{A}^3, \mathbf{A}^4, \ldots$. This is because trying to determine $\mathbf{A}^n$ is a tedious task for almost all matrices. Finding $\mathbf{A}^2$ by $\mathbf{A} \times \mathbf{A}$ is simple enough for small size matrices but $\mathbf{A}^3$ is *not* so elementary. The Cayley–Hamilton theorem simplifies evaluation of $\mathbf{A}^n$ for a given matrix $\mathbf{A}$.

The statement of the Cayley–Hamilton theorem is straightforward.

Cayley–Hamilton (7.11). Every square matrix $\mathbf{A}$ is a root of the characteristic equation, that is $p(\mathbf{A}) = \mathbf{O}$ where p represents the characteristic polynomial.

Proof – This is a difficult proof and is on the book's website.

■

Example 7.11

Find the characteristic polynomial $p(\lambda)$ of the matrix $\mathbf{A} = \begin{pmatrix} 3 & 2 \\ 3 & 4 \end{pmatrix}$ and illustrate the Cayley-Hamilton theorem for this matrix $\mathbf{A}$.

Solution

We use $p(\lambda) = \det(\mathbf{A} - \lambda\mathbf{I})$ and substitute the given matrix $\mathbf{A}$ into this:

$$p(\lambda) = \det(\mathbf{A} - \lambda\mathbf{I}) = \det\begin{pmatrix} 3-\lambda & 2 \\ 3 & 4-\lambda \end{pmatrix} \qquad \left[\begin{array}{l}\text{taking away } \lambda \text{ along} \\ \text{the leading diagonal}\end{array}\right]$$

$$= (3-\lambda)(4-\lambda) - 6$$

$$= 12 - 7\lambda + \lambda^2 - 6 = \lambda^2 - 7\lambda + 6$$

Thus the characteristic polynomial is $p(\lambda) = \lambda^2 - 7\lambda + 6$.

What is $p(\mathbf{A})$ equal to?

Substituting matrix $\mathbf{A}$ into this equation yields:

$$p(\mathbf{A}) = \mathbf{A}^2 - 7\mathbf{A} + 6\mathbf{I}$$

$$= \begin{pmatrix} 3 & 2 \\ 3 & 4 \end{pmatrix}^2 - 7\begin{pmatrix} 3 & 2 \\ 3 & 4 \end{pmatrix} + 6\begin{pmatrix} 1 & 0 \\ 0 & 1 \end{pmatrix} \qquad [\text{substituting } \mathbf{A} \text{ and } \mathbf{I}]$$

$$= \begin{pmatrix} 3 & 2 \\ 3 & 4 \end{pmatrix}\begin{pmatrix} 3 & 2 \\ 3 & 4 \end{pmatrix} - \begin{pmatrix} 21 & 14 \\ 21 & 28 \end{pmatrix} + \begin{pmatrix} 6 & 0 \\ 0 & 6 \end{pmatrix} \qquad \left[\begin{array}{l}\text{carrying out scalar} \\ \text{multiplication}\end{array}\right]$$

$$= \begin{pmatrix} 15 & 14 \\ 21 & 22 \end{pmatrix} - \begin{pmatrix} 21 & 14 \\ 21 & 28 \end{pmatrix} + \begin{pmatrix} 6 & 0 \\ 0 & 6 \end{pmatrix}$$

$$= \begin{pmatrix} 15-21+6 & 14-14+0 \\ 21-21+0 & 22-28+6 \end{pmatrix} = \begin{pmatrix} 0 & 0 \\ 0 & 0 \end{pmatrix} = \mathbf{O}$$

Thus $p(\mathbf{A}) = \mathbf{O}$.

What does this mean?

This means that matrix $\mathbf{A}$ satisfies its characteristic equation $p(\mathbf{A}) = \mathbf{A}^2 - 7\mathbf{A} + 6\mathbf{I} = \mathbf{O}$ which is what the Cayley-Hamilton theorem states.

We can use this result $\mathbf{A}^2 - 7\mathbf{A} + 6\mathbf{I} = \mathbf{O}$ to find higher powers of matrix $\mathbf{A}$.

We can apply the Cayley–Hamilton theorem to find powers of matrices. For example, if

$$\mathbf{A} = \begin{pmatrix} 0.9 & 0.2 \\ 0.3 & 0.6 \end{pmatrix} \text{ then } \mathbf{A}^{100} = \begin{pmatrix} 29.4 & 13.5 \\ 20.2 & 9.2 \end{pmatrix} \text{ (1dp)}$$

$\mathbf{A}^{100}$ is *not* evaluated by multiplying 100 copies of matrix $\mathbf{A}$. The evaluation of $\mathbf{A}^{100}$ is found by using the eigenvalues of matrix $\mathbf{A}$. Calculating $\mathbf{A}^{100}$ is a very tedious task which can be significantly reduced by applying the Cayley–Hamilton theorem. The powers of a matrix can be written in terms of a polynomial of lower degree in matrix $\mathbf{A}$. This is illustrated in the next example.

Example 7.12

Let $\mathbf{A} = \begin{pmatrix} -2 & -4 \\ 1 & 3 \end{pmatrix}$. Determine $\mathbf{A}^4$.

Solution

Working out the determinant of $\mathbf{A} - \lambda\mathbf{I}$ gives

$$\begin{aligned}\det(\mathbf{A} - \lambda\mathbf{I}) = \det\begin{pmatrix} -2-\lambda & -4 \\ 1 & 3-\lambda \end{pmatrix} &= (-2-\lambda)(3-\lambda) + 4 \\ &= (\lambda+2)(\lambda-3) + 4 \\ &= \lambda^2 - \lambda - 2 = p(\lambda)\end{aligned}$$

By the Cayley-Hamilton theorem we have $p(\mathbf{A}) = \mathbf{A}^2 - \mathbf{A} - 2\mathbf{I} = \mathbf{O}$. Transposing this gives

$$\begin{aligned}\mathbf{A}^2 &= \mathbf{A} + 2\mathbf{I} \qquad (\dagger) \\ \mathbf{A}^3 &= \mathbf{A}^2 + 2\mathbf{A} \qquad \left[\text{multiplying by } \mathbf{A}\right] \\ &= \underbrace{\mathbf{A} + 2\mathbf{I}}_{=\mathbf{A}^2 \text{ by above } (\dagger)} + 2\mathbf{A} = 3\mathbf{A} + 2\mathbf{I}\end{aligned}$$

Multiplying the last equation $\mathbf{A}^3 = 3\mathbf{A} + 2\mathbf{I}$ by $\mathbf{A}$ gives

$$\begin{aligned}\mathbf{A}^4 &= 3\mathbf{A}^2 + 2\mathbf{A} \\ &= 3\underbrace{\left[\mathbf{A} + 2\mathbf{I}\right]}_{=\mathbf{A}^2 \text{ by above } (\dagger)} + 2\mathbf{A} = 5\mathbf{A} + 6\mathbf{I}\end{aligned}$$

Thus $\mathbf{A}^4 = 5\mathbf{A} + 6\mathbf{I} = 5\begin{pmatrix} -2 & -4 \\ 1 & 3 \end{pmatrix} + 6\begin{pmatrix} 1 & 0 \\ 0 & 1 \end{pmatrix} = \begin{pmatrix} -4 & -20 \\ 5 & 21 \end{pmatrix}$.

Note that we can evaluate $\mathbf{A}^4$ without working out $\mathbf{A}^3$ and $\mathbf{A}^2$ because $\mathbf{A}^4 = 5\mathbf{A} + 6\mathbf{I}$ is a polynomial of degree 1 in matrix $\mathbf{A}$.

In MATLAB we can find the characteristic polynomial of a matrix $\mathbf{A}$ by using the command poly(A). We can also use the Cayley–Hamilton theorem to find the *inverse* of a matrix as the next example demonstrates. Remember, determining the inverse of a 3 by 3 or larger matrix is a laborious job. However, by applying the Cayley–Hamilton theorem we find it becomes much simpler.

Example 7.13

Determine $\mathbf{A}^{-1}$ where $\mathbf{A} = \begin{pmatrix} 10 & 15 & 0 \\ 2 & 4 & 0 \\ 3 & 6 & 6 \end{pmatrix}$ given that the characteristic polynomial of this matrix is:

$$p(\lambda) = \lambda^3 - 20\lambda^2 + 94\lambda - 60$$

Solution

By the Cayley-Hamilton theorem (7.11) we have

$$p(\mathbf{A}) = \mathbf{A}^3 - 20\mathbf{A}^2 + 94\mathbf{A} - 60\mathbf{I} = \mathbf{O} \qquad \left[\text{because } p(\lambda) = \lambda^3 - 20\lambda^2 + 94\lambda - 60\right]$$

(continued...)

Adding $60\mathbf{I}$ to both sides gives

$$\mathbf{A}^3 - 20\mathbf{A}^2 + 94\mathbf{A} = 60\mathbf{I}$$
$$\mathbf{A}(\mathbf{A}^2 - 20\mathbf{A} + 94\mathbf{I}) = 60\mathbf{I} \qquad \text{[factorizing out the matrix } \mathbf{A}\text{]}$$

By the definition of the inverse matrix we have $\mathbf{A}\mathbf{A}^{-1} = \mathbf{I}$, which means that dividing both sides by 60 in the above gives

$$\mathbf{A}\underbrace{\frac{1}{60}\left(\mathbf{A}^2 - 20\mathbf{A} + 94\mathbf{I}\right)}_{=\mathbf{A}^{-1}} = \mathbf{I}$$

Hence $\mathbf{A}^{-1} = \frac{1}{60}\left(\mathbf{A}^2 - 20\mathbf{A} + 94\mathbf{I}\right)$. Evaluating the components of this yields

$$\mathbf{A}^2 = \begin{pmatrix} 10 & 15 & 0 \\ 2 & 4 & 0 \\ 3 & 6 & 6 \end{pmatrix}\begin{pmatrix} 10 & 15 & 0 \\ 2 & 4 & 0 \\ 3 & 6 & 6 \end{pmatrix} = \begin{pmatrix} 130 & 210 & 0 \\ 28 & 46 & 0 \\ 60 & 105 & 36 \end{pmatrix}$$

$$20\mathbf{A} = 20\begin{pmatrix} 10 & 15 & 0 \\ 2 & 4 & 0 \\ 3 & 6 & 6 \end{pmatrix} = \begin{pmatrix} 200 & 300 & 0 \\ 40 & 80 & 0 \\ 60 & 120 & 120 \end{pmatrix} \quad \text{and} \quad 94\mathbf{I} = \begin{pmatrix} 94 & 0 & 0 \\ 0 & 94 & 0 \\ 0 & 0 & 94 \end{pmatrix}$$

Putting these into $\mathbf{A}^{-1} = \frac{1}{60}\left(\mathbf{A}^2 - 20\mathbf{A} + 94\mathbf{I}\right)$ gives

$$\mathbf{A}^{-1} = \frac{1}{60}\left(\mathbf{A}^2 - 20\mathbf{A} + 94\mathbf{I}\right) = \frac{1}{60}\left[\begin{pmatrix} 130 & 210 & 0 \\ 28 & 46 & 0 \\ 60 & 105 & 36 \end{pmatrix} - \begin{pmatrix} 200 & 300 & 0 \\ 40 & 80 & 0 \\ 60 & 120 & 120 \end{pmatrix} + \begin{pmatrix} 94 & 0 & 0 \\ 0 & 94 & 0 \\ 0 & 0 & 94 \end{pmatrix}\right]$$

$$= \frac{1}{60}\begin{pmatrix} 130 - 200 + 94 & 210 - 300 + 0 & 0 - 0 + 0 \\ 28 - 40 + 0 & 46 - 80 + 94 & 0 - 0 + 0 \\ 60 - 60 + 0 & 105 - 120 + 0 & 36 - 120 + 94 \end{pmatrix}$$

$$= \frac{1}{60}\begin{pmatrix} 24 & -90 & 0 \\ -12 & 60 & 0 \\ 0 & -15 & 10 \end{pmatrix}$$

Finding $\mathbf{A}^{-1}$ still involves a fair amount of calculation but is generally easier than finding the cofactors of each entry.

Summary

Proposition (7.6). If $\mathbf{A}$ is a diagonal or triangular matrix then the eigenvalues of $\mathbf{A}$ are the entries along the leading diagonal.

Let $\mathbf{A}$ be a square matrix with an eigenvalue λ and eigenvector $\mathbf{u}$ belonging to λ.

Then we have the following:

Matrix		Eigenvector	Eigenvalue
(7.8) (a)	$\mathbf{A}^m$	$\mathbf{u}$	λ^m
(7.8) (b)	$\mathbf{A}^{-1}$	$\mathbf{u}$	λ^{-1}

Cayley-Hamilton theorem (7.11). Every square matrix $\mathbf{A}$ is a root of the characteristic equation, that is $p(\mathbf{A}) = \mathbf{O}$.

EXERCISES 7.2

(Brief solutions at end of book. Full solutions available at <http://www.oup.co.uk/companion/singh>.)

In this exercise check your numerical answers using any appropriate software.

1. Determine the eigenvalues and eigenvectors of $\mathbf{A} = \begin{pmatrix} 5 & 2 \\ -2 & 1 \end{pmatrix}$ and plot the eigenspace E_λ and write down a set of basis vectors for E_λ.
2. Determine the eigenvalues and eigenvectors of $\mathbf{A} = \begin{pmatrix} -12 & 7 \\ -7 & 2 \end{pmatrix}$ and plot the eigenspace E_λ and write down a set of basis vectors for E_λ.
3. Let $\mathbf{A}$ be a 2 by 2 matrix. Show that the characteristic polynomial $p(\lambda)$ is given by
$$p(\lambda) = \lambda^2 - tr(\mathbf{A})\lambda + \det(\mathbf{A})$$
where $tr(\mathbf{A})$ is the trace of the matrix $\mathbf{A}$.
4. Let $\mathbf{A}$ be a 2 by 2 matrix with $tr(\mathbf{A}) = 2a$ and $\det(\mathbf{A}) = a^2$ where a is a real number. Show that this matrix has an eigenvalue of a with *multiplicity* of 2.
5. Determine the eigenvalues and eigenvectors of $\mathbf{A} = \begin{pmatrix} 1 & 0 & 0 \\ -3 & 1 & 0 \\ 7 & 9 & 1 \end{pmatrix}$. By using appropriate software or otherwise, plot the eigenspace E_λ and write down a set of basis vectors for this eigenspace.
6. Determine the eigenvalues and eigenvectors of the following matrices and write down a set of basis vectors for the eigenspace E_λ.

   (a) $\mathbf{A} = \begin{pmatrix} 5 & 0 & 0 \\ 0 & 5 & 0 \\ 0 & 0 & 2 \end{pmatrix}$ (b) $\mathbf{B} = \begin{pmatrix} 1 & 2 & 3 \\ 0 & 5 & 1 \\ 0 & 0 & 9 \end{pmatrix}$ (c) $\mathbf{C} = \begin{pmatrix} -2 & 0 & 0 \\ 2 & -2 & 0 \\ 4 & 10 & -2 \end{pmatrix}$
7. Determine the eigenvalues and eigenvectors of the following matrices:

   (a) $\mathbf{A} = \begin{pmatrix} 7 & 0 & 2 & 3 \\ 0 & 7 & 4 & 6 \\ 0 & 0 & 5 & -3 \\ 0 & 0 & 0 & 5 \end{pmatrix}$ (b) $\mathbf{B} = \begin{pmatrix} 1 & 0 & 0 & 0 \\ 0 & 1 & 0 & 0 \\ 0 & 0 & 1 & 0 \\ 0 & 0 & 0 & 3 \end{pmatrix}$ (c) $\mathbf{C} = \begin{pmatrix} 3 & 0 & 0 & 0 \\ 0 & 3 & 0 & 0 \\ 0 & 0 & 3 & 0 \\ 0 & 0 & 0 & 3 \end{pmatrix}$

8. Prove Proposition (7.7), which says a matrix is invertible $\Leftrightarrow \lambda = 0$ is *not* an eigenvalue.
9. Let $\mathbf{A}$ be a square matrix and λ be an eigenvalue with the corresponding eigenvector $\mathbf{u}$. Prove that the eigenvalue λ is unique for the eigenvector $\mathbf{u}$.
10. Let $\mathbf{A}$ be a 2 by 2 matrix. By using the result of question 3, the characteristic polynomial $p(\lambda)$ is given by

$$p(\lambda) = \lambda^2 - tr(\mathbf{A})\lambda + \det(\mathbf{A})$$

Show that if $[tr(\mathbf{A})]^2 > 4\det(\mathbf{A})$ then $\mathbf{A}$ has distinct eigenvalues.
State under what conditions we have equal and complex eigenvalues.

11. For each of the following matrices

(a) $\mathbf{A} = \begin{pmatrix} 1 & 0 & 0 & 0 \\ 3 & 2 & 0 & 0 \\ 5 & 2 & 3 & 0 \\ 9 & 8 & 1 & 4 \end{pmatrix}$ (b) $\mathbf{A} = \begin{pmatrix} -1 & 3 & 4 & 7 \\ 0 & 6 & -3 & 5 \\ 0 & 0 & -8 & 9 \\ 0 & 0 & 0 & 3 \end{pmatrix}$ (c) $\mathbf{A} = \begin{pmatrix} 2 & 0 & 0 & 0 \\ 0 & -4 & 0 & 0 \\ 0 & 0 & -7 & 0 \\ 0 & 0 & 0 & 0 \end{pmatrix}$

Determine
(i) eigenvalues of $\mathbf{A}$ (ii) eigenvalues of $\mathbf{A}^5$ (iii) eigenvalues of $\mathbf{A}^{-1}$
(iv) $\det(\mathbf{A})$ (v) $tr(\mathbf{A})$

12. Find the characteristic polynomial $p(\lambda)$ of the matrix $\mathbf{A} = \begin{pmatrix} 2 & 1 \\ -2 & 1 \end{pmatrix}$ and determine the inverse of this matrix by using the Cayley–Hamilton theorem.
13. Let $\mathbf{A} = \begin{pmatrix} 6 & 5 \\ 3 & 4 \end{pmatrix}$. By using the Cayley–Hamilton theorem determine $\mathbf{A}^2$ and $\mathbf{A}^3$.
14. Let $\mathbf{A} = \begin{pmatrix} 1 & 2 & 1 \\ 2 & 1 & 1 \\ 1 & 1 & 2 \end{pmatrix}$ and the characteristic polynomial of this matrix be given by

$$p(\lambda) = \lambda^3 - 4\lambda^2 - \lambda + 4$$

Determine expressions for $\mathbf{A}^{-1}$ and $\mathbf{A}^4$ in terms of the matrices $\mathbf{A}$, $\mathbf{I}$ and $\mathbf{A}^2$.

15. Prove Proposition (7.10).
16. Prove that the eigenvalues of the transposed matrix, $\mathbf{A}^T$, are exactly the eigenvalues of the matrix $\mathbf{A}$.

SECTION 7.3 Diagonalization

By the end of this section you will be able to

- understand what is meant by similar matrices
- diagonalize a matrix
- find powers of matrices

If $\mathbf{A} = \begin{pmatrix} 1 & 2 \\ 3 & 4 \end{pmatrix}$ *then how do we find* $\mathbf{A}^{100}$*?*

We could apply the Cayley-Hamilton theorem of the last section. However, using Cayley-Hamilton requires us to find a formula for $\mathbf{A}^2, \mathbf{A}^3, \ldots, \mathbf{A}^{99}$ and then use this to determine $\mathbf{A}^{100}$. Clearly this is a very laborious way of finding $\mathbf{A}^{100}$. In this section, we examine an easier method to find powers of matrices such as $\mathbf{A}^{100}$. We factorize a given matrix $\mathbf{A}$ into three matrices, one of which is a diagonal matrix. It is a lot simpler to deal with a diagonal matrix because *any* matrix calculation is easier with diagonal matrices. For example, it is easy to find the inverse of a diagonal matrix. Also, proving results about diagonal matrices is simpler than proving results about general matrices.

In this section we aim to answer the following question:

For a square matrix $\mathbf{A}$*, is there an invertible matrix* $\mathbf{P}$ *(has an inverse) such that* $\mathbf{P}^{-1}\mathbf{AP}$ *produces a diagonal matrix?*

The goal of this section is to convert an n by n matrix into a diagonal matrix. The process of converting any n by n matrix into a diagonal matrix is called **diagonalization** (Fig. 7.10).

n by n matrix Diagonal matrix

Figure 7.10

What has this got to do with eigenvalues and eigenvectors?

This section explores how to diagonalize a matrix, for which we need to find the eigenvalues and the corresponding eigenvectors. These eigenvalues are the leading diagonal entries in the diagonal matrix.

First, we define **similar** matrices and some of their properties.

7.3.1 Similar matrices

Example 7.14

Let $\mathbf{A} = \begin{pmatrix} 1 & 0 \\ 1 & 2 \end{pmatrix}$ and $\mathbf{P} = \begin{pmatrix} 0 & -1 \\ 1 & 1 \end{pmatrix}$. Determine $\mathbf{P}^{-1}\mathbf{AP}$.

Solution

We first find the inverse of the matrix $\mathbf{P}$, denoted $\mathbf{P}^{-1}$:

$$\mathbf{P}^{-1} = \begin{pmatrix} 0 & -1 \\ 1 & 1 \end{pmatrix}^{-1} = \frac{1}{0-(-1)}\begin{pmatrix} 1 & 1 \\ -1 & 0 \end{pmatrix} = \begin{pmatrix} 1 & 1 \\ -1 & 0 \end{pmatrix} \quad \left[\begin{pmatrix} a & b \\ c & d \end{pmatrix}^{-1} = \frac{1}{ad-bc}\begin{pmatrix} d & -b \\ -c & a \end{pmatrix}\right]$$

Carrying out the matrix multiplication

$$\mathbf{P}^{-1}\mathbf{AP} = \begin{pmatrix} 1 & 1 \\ -1 & 0 \end{pmatrix}\begin{pmatrix} 1 & 0 \\ 1 & 2 \end{pmatrix}\begin{pmatrix} 0 & -1 \\ 1 & 1 \end{pmatrix}$$

$$= \begin{pmatrix} 2 & 2 \\ -1 & 0 \end{pmatrix}\begin{pmatrix} 0 & -1 \\ 1 & 1 \end{pmatrix} = \begin{pmatrix} 2 & 0 \\ 0 & 1 \end{pmatrix} \quad \left[\begin{array}{l}\text{first multiplying the} \\ \text{left and centre matrices.}\end{array}\right]$$

Definition (7.12). A square matrix $\mathbf{B}$ is **similar** to a matrix $\mathbf{A}$ if there exists an invertible matrix $\mathbf{P}$ such that $\mathbf{P}^{-1}\mathbf{AP} = \mathbf{B}$.

In Example 7.14, the final matrix $\begin{pmatrix} 2 & 0 \\ 0 & 1 \end{pmatrix}$ is *similar* to matrix $\mathbf{A}$ because $\mathbf{P}^{-1}\mathbf{AP} = \begin{pmatrix} 2 & 0 \\ 0 & 1 \end{pmatrix}$.

Similar matrices have the following properties (equivalence relation):

Proposition (7.13). Let $\mathbf{A}$, $\mathbf{B}$ and $\mathbf{C}$ be square matrices. Then

(a) Matrix $\mathbf{A}$ is similar to matrix $\mathbf{A}$.

(b) If matrix $\mathbf{B}$ is similar to matrix $\mathbf{A}$ then the other way round is also true, that is matrix $\mathbf{A}$ is similar to matrix $\mathbf{B}$.

(c) If matrix $\mathbf{A}$ is similar to $\mathbf{B}$ and $\mathbf{B}$ is similar to matrix $\mathbf{C}$ then matrix $\mathbf{A}$ is similar to matrix $\mathbf{C}$.

Proof – Exercises 7.3.

■

By property (b) we can say matrices $\mathbf{A}$ and $\mathbf{B}$ are similar. The following proposition gives another important property of similar matrices.

Proposition (7.14). Let $\mathbf{A}$ and $\mathbf{B}$ be similar matrices. The eigenvalues of these matrices are identical.

Proof.

We are given that matrices $\mathbf{A}$ and $\mathbf{B}$ are similar.

What does this mean?

By (7.12) there exists an invertible matrix $\mathbf{P}$ such that $\mathbf{P}^{-1}\mathbf{AP} = \mathbf{B}$. Let $\det(\mathbf{B} - \lambda\mathbf{I})$ be the characteristic polynomial for the matrix $\mathbf{B}$ and $\det(\mathbf{A} - \lambda\mathbf{I})$ be the characteristic polynomial for the matrix $\mathbf{A}$.

Required to prove that the polynomials given by these determinants are equal:

$$\det(\mathbf{B} - \lambda\mathbf{I}) = \det(\mathbf{A} - \lambda\mathbf{I})$$

We have

$$\begin{aligned}
\det(\mathbf{B} - \lambda\mathbf{I}) &= \det\left(\mathbf{P}^{-1}\mathbf{AP} - \lambda\mathbf{I}\right) && \left[\text{replacing } \mathbf{B} = \mathbf{P}^{-1}\mathbf{AP}\right] \\
&= \det\left(\mathbf{P}^{-1}\mathbf{AP} - \lambda\mathbf{P}^{-1}\mathbf{P}\right) && \left[\text{substituting } \mathbf{I} = \mathbf{P}^{-1}\mathbf{P}\right] \\
&= \det\left(\mathbf{P}^{-1}\mathbf{AP} - \mathbf{P}^{-1}\lambda\mathbf{P}\right) && \left[\text{moving the scalar } \lambda\right] \\
&= \det\left(\mathbf{P}^{-1}\mathbf{AP} - \mathbf{P}^{-1}\lambda\mathbf{IP}\right) && \left[\text{rewriting } \lambda\mathbf{P} = \lambda\mathbf{IP}\right] \\
&= \det\left[\mathbf{P}^{-1}(\mathbf{A} - \lambda\mathbf{I})\mathbf{P}\right] && \left[\text{factorizing out } \mathbf{P}^{-1} \text{ and } \mathbf{P}\right] \\
&= \det\left(\mathbf{P}^{-1}\right)\det(\mathbf{A} - \lambda\mathbf{I})\det(\mathbf{P}) && \left[\text{using } \det(\mathbf{ABC}) = \det(\mathbf{A})\det(\mathbf{B})\det(\mathbf{C})\right] \\
&= \det(\mathbf{A} - \lambda\mathbf{I}) && \left[\text{because } \det\left(\mathbf{P}^{-1}\right)\det(\mathbf{P}) = 1\right]
\end{aligned}$$

Similar matrices **A** and **B** have the *same* eigenvalues λ because λ satisfies the equation:

$$\det(\mathbf{B}-\lambda\mathbf{I})=\det(\mathbf{A}-\lambda\mathbf{I})=0$$

■

Proposition (7.14) means that similar matrices carry out an identical transformation – they scalar multiply each eigenvector by the same scalar λ.

7.3.2 Introduction to diagonalization

What do we mean when we say that a matrix is diagonalizable?

Definition (7.15). An n by n matrix **A** is diagonalizable if it is *similar* to a diagonal matrix **D**.

The matrix $\mathbf{A}=\begin{pmatrix}1&0\\1&2\end{pmatrix}$ in the above Example 7.14 is diagonalizable because the matrix $\mathbf{P}=\begin{pmatrix}0&-1\\1&1\end{pmatrix}$ gives $\mathbf{P}^{-1}\mathbf{AP}=\mathbf{D}$ where $\mathbf{D}=\begin{pmatrix}2&0\\0&1\end{pmatrix}$ is a *diagonal* matrix.

What does this definition (7.15) mean?
$\mathbf{P}^{-1}\mathbf{AP}=\mathbf{D}$ which means that we can convert a matrix **A** into a diagonal matrix by left multiplying by $\mathbf{P}^{-1}$ and right multiplying by **P**. We say the matrix **P** **diagonalizes** the matrix **A**.

Why diagonalize a matrix?
In general, a diagonal matrix is easier to work with because if you are multiplying, solving a system of equations or finding eigenvalues, it is always preferable to have a diagonal matrix. The diagonal matrix significantly reduces the amount of numerical calculations needed to find powers of a matrix, for example. We will show later in this section that it is easy to evaluate the 100th power of a diagonal matrix. Remember, we aim to convert a given matrix into a diagonal matrix.

Next, we give a trivial example of a matrix which is diagonalizable.

Example 7.15

Show that a diagonal matrix $\mathbf{A}=\begin{pmatrix}3&0\\0&2\end{pmatrix}$ is diagonalizable.

Solution
We need to find a matrix **P** such that $\mathbf{P}^{-1}\mathbf{AP}=\mathbf{D}$ where **D** is a diagonal matrix.
*What matrix **P** should we consider?*
The identity matrix $\mathbf{P}=\mathbf{I}$ because the given matrix **A** is already a diagonal matrix:

$$\mathbf{I}^{-1}\mathbf{AI}=\mathbf{A}$$

Thus the given matrix **A** is diagonalizable.

Example 7.16

Show that any diagonal matrix is diagonalizable.

Solution

Let $\mathbf{D}$ be any diagonal matrix. The diagonalizing matrix is the identity matrix $\mathbf{P} = \mathbf{I}$ because $\mathbf{I}^{-1}\mathbf{DI} = \mathbf{D}$.

How do we know which matrices are diagonalizable?

The next theorem states a test for establishing whether a matrix is diagonalizable or not.

Theorem (7.16). An n by n matrix $\mathbf{A}$ is diagonalizable $\Leftrightarrow$ it has n *linearly independent* eigenvectors.

Proof – Exercises 7.3.

■

We must have n linearly *independent* eigenvectors for the matrix to be *diagonalizable.* The procedure for diagonalizing an n by n matrix $\mathbf{A}$ can be derived from the proof of (7.16). In a nutshell it is the following procedure:

1. Using $\det(\mathbf{A} - \lambda\mathbf{I})\mathbf{u} = \mathbf{O}$ for each eigenvalue $\lambda_1, \lambda_2, \ldots, \lambda_n$, we find the eigenvectors belonging to these $\lambda_1, \lambda_2, \ldots, \lambda_n$. Call these eigenvectors $\mathbf{p}_1, \mathbf{p}_2, \mathbf{p}_3, \ldots$ and $\mathbf{p}_n$. If matrix $\mathbf{A}$ does *not* have n linearly independent eigenvectors then it is *not* diagonalizable.
2. Form the matrix $\mathbf{P}$ by having these eigenvectors $\mathbf{p}_1, \mathbf{p}_2, \ldots$ and $\mathbf{p}_n$, as its columns. That is matrix $\mathbf{P}$ contains the eigenvectors:

$$\mathbf{P} = (\mathbf{p}_1 \;\; \mathbf{p}_2 \;\; \mathbf{p}_3 \;\; \cdots \;\; \mathbf{p}_n)$$

3. The diagonal matrix $\mathbf{D} = \mathbf{P}^{-1}\mathbf{AP}$ will have the eigenvalues $\lambda_1, \lambda_2, \ldots$ and λ_n of $\mathbf{A}$ along its leading diagonal, that is

The eigenvector $\mathbf{p}_j$ belongs to the eigenvalue λ_j.

4. It is good practice to check that matrices $\mathbf{P}$ and $\mathbf{D}$ actually work. For matrices of size *greater than* 2 by 2, the evaluation of the inverse matrix $\mathbf{P}^{-1}$ can be lengthy so prone to calculation errors. To bypass this evaluation of the inverse, simply check that $\mathbf{PD} = \mathbf{AP}$.

Why?

Because left multiplying the above $\mathbf{D} = \mathbf{P}^{-1}\mathbf{AP}$ by $\mathbf{P}$ gives

$$\begin{aligned}\mathbf{PD} &= \mathbf{P}(\mathbf{P}^{-1}\mathbf{AP})\\ &= (\mathbf{PP}^{-1})\mathbf{AP} = \mathbf{IAP} = \mathbf{AP} \qquad \left[\text{remember } \mathbf{PP}^{-1} = \mathbf{I}\right]\end{aligned}$$

Hence it is enough to check that $\mathbf{PD} = \mathbf{AP}$.

Note that since matrices **A** and **D** are similar, they have the same eigenvalues.

The matrix **P** is called the **eigenvector matrix** and the diagonal matrix **D** the **eigenvalue matrix**.

Example 7.17

Determine the eigenvector matrix **P** which diagonalizes the matrix $\mathbf{A} = \begin{pmatrix} 1 & 4 \\ 2 & 3 \end{pmatrix}$ given that the eigenvalues of this matrix are $\lambda_1 = -1$ and $\lambda_2 = 5$ with corresponding eigenvectors $\mathbf{u} = \begin{pmatrix} -2 \\ 1 \end{pmatrix}$ and $\mathbf{v} = \begin{pmatrix} 1 \\ 1 \end{pmatrix}$ respectively.

Solution

Steps 1 and 2.

We have been given the eigenvectors, **u** and **v**, of matrix **A** so the eigenvector matrix **P** is

$$\mathbf{P} = (\mathbf{u} \quad \mathbf{v}) = \begin{pmatrix} -2 & 1 \\ 1 & 1 \end{pmatrix}$$

Step 3.

Since matrices **A** and **D** are similar, so our diagonal matrix **D** contains the eigenvalues of **A**, $\lambda_1 = -1$ and $\lambda_2 = 5$:

$$\mathbf{D} = \begin{pmatrix} -1 & 0 \\ 0 & 5 \end{pmatrix}$$

eigenvalues of **A**

Step 4.

We need to confirm that this matrix **P** does indeed diagonalize the given matrix **A**.

How?

By checking that $\mathbf{PD} = \mathbf{AP}$:

$$\mathbf{PD} = \begin{pmatrix} -2 & 1 \\ 1 & 1 \end{pmatrix}\begin{pmatrix} -1 & 0 \\ 0 & 5 \end{pmatrix} = \begin{pmatrix} 2 & 5 \\ -1 & 5 \end{pmatrix}$$

$$\mathbf{AP} = \begin{pmatrix} 1 & 4 \\ 2 & 3 \end{pmatrix}\begin{pmatrix} -2 & 1 \\ 1 & 1 \end{pmatrix} = \begin{pmatrix} 2 & 5 \\ -1 & 5 \end{pmatrix}$$

Thus the eigenvector matrix **P** does indeed diagonalize the given matrix **A**.

Notice that the eigenvalues, $\lambda_1 = -1$ and $\lambda_2 = 5$, of the given matrix **A** (and **D**) are entries along the leading diagonal in the matrix **D**. They occur in the order λ_1 and then λ_2

because the matrix $\mathbf{P}$ is created by $\mathbf{P} = (\mathbf{u} \quad \mathbf{v})$ where $\mathbf{u}$ is the eigenvector belonging to λ_1 and $\mathbf{v}$ is the eigenvector belonging to the other eigenvalue λ_2.

What would be the diagonal matrix if the matrix $\mathbf{P}$ *was created by swapping* $\mathbf{u}$ *and* $\mathbf{v}$, *that is* $\mathbf{P} = (\mathbf{v} \ \ \mathbf{u})$?

Our diagonal matrix would be $\mathbf{D}' = \begin{pmatrix} 5 & 0 \\ 0 & -1 \end{pmatrix}$. [See Exercises 7.3].

Note that the diagonal eigenvalue matrix $\mathbf{D}$ contains the eigenvalues, the eigenvector matrix $\mathbf{P}$ contains the corresponding eigenvectors.

Example 7.18

Show that the matrix $\mathbf{A} = \begin{pmatrix} 1 & -2 & 3 \\ 0 & 2 & 5 \\ 0 & 0 & 2 \end{pmatrix}$ with eigenvalues and eigenvectors given by

$$\lambda_1 = 1, \mathbf{u} = \begin{pmatrix} 1 \\ 0 \\ 0 \end{pmatrix}; \lambda_2 = 2, \mathbf{v} = \begin{pmatrix} -2 \\ 1 \\ 0 \end{pmatrix} \text{ and } \lambda_3 = 2, \mathbf{w} = \begin{pmatrix} 2 \\ -1 \\ 0 \end{pmatrix}$$

is *not* diagonalizable.

Solution

Why can't we diagonalize the given matrix $\mathbf{A}$?

According to the procedure outlined above we *cannot* diagonlize matrix $\mathbf{A}$ if the three eigenvectors are linearly dependent. (Linear dependency occurs when we can write one vector in terms of the others.)

Note that vectors $\mathbf{v}$ and $\mathbf{w}$ are linearly *dependent* because $\mathbf{v} = -\mathbf{w}$ or $\mathbf{v} + \mathbf{w} = \mathbf{O}$. Thus the matrix $\mathbf{A}$ is *not* diagonalizable.

7.3.3 Distinct eigenvalues

Next, we state a proposition that if an n by n matrix has n *distinct* eigenvalues then the eigenvectors belonging to these are linearly *independent*.

Proposition (7.17). Let $\mathbf{A}$ be an n by n matrix with n *distinct* eigenvalues, $\lambda_1, \lambda_2, \ldots$ and λ_n with the corresponding eigenvectors $\mathbf{u}_1$, $\mathbf{u}_2$, $\mathbf{u}_3$, ... and $\mathbf{u}_n$. These eigenvectors are *linearly independent.*

Proof.

By Proposition (7.10):

Let $\mathbf{A}$ have *distinct* eigenvalues $\lambda_1, \lambda_2, \ldots, \lambda_m$ with corresponding eigenvectors $\mathbf{u}_1, \mathbf{u}_2, \ldots, \mathbf{u}_m$ where $1 \leq m \leq n$. Then these eigenvectors are *linearly independent.*

Using (7.10) with $n = m$ gives us our required result.

■

Proposition (7.18). (a) If an n by n matrix $\mathbf{A}$ has n *distinct* eigenvalues then the matrix $\mathbf{A}$ is diagonalizable. (b) $\mathbf{A}$ is diagonalizable $\Leftrightarrow$ the dimension of each eigenspace is equal to the multiplicity of the corresponding eigenvalue.

Proof.

The n by n matrix $\mathbf{A}$ has n distinct eigenvalues, therefore by the above proposition (7.17) the corresponding n eigenvectors are linearly independent. Thus by the above result (7.16):

An n by n matrix $\mathbf{A}$ is diagonalizable $\Leftrightarrow$ it has n *independent* eigenvectors.

We conclude that the matrix $\mathbf{A}$ is diagonalizable.

Proof of (b) see website.

■

Example 7.19

Determine whether the matrix $\mathbf{A} = \begin{pmatrix} 1 & -6 & 2 \\ 0 & 4 & 25 \\ 0 & 0 & 9 \end{pmatrix}$ is diagonalizable.

Solution

What type of matrix is $\mathbf{A}$*?*

Matrix $\mathbf{A}$ is an upper triangular matrix, therefore by Proposition (7.6): If an n by n matrix $\mathbf{A}$ is a diagonal or triangular matrix then the eigenvalues of $\mathbf{A}$ are the entries along the leading diagonal.

The eigenvalues are the entries along the leading diagonal; $\lambda_1 = 1, \lambda_2 = 4$ and $\lambda_3 = 9$.

How do we determine whether the given matrix $\mathbf{A}$ *is diagonalizable or not?*

$\mathbf{A}$ is a 3 by 3 matrix and it has three distinct eigenvalues; $\lambda_1 = 1, \lambda_2 = 4$ and $\lambda_3 = 9$,

Therefore by Proposition (7.18) the matrix $\mathbf{A}$ is diagonalizable.

An n by n matrix might be diagonalizable even if it does *not* have n distinct eigenvalues.

For example, the identity matrix $\mathbf{I}$ is diagonalizable even though it has n copies of the same eigenvalue 1. (If we have n distinct eigenvalues for an n by n matrix then we are guaranteed that the matrix is diagonalizable.)

The process of finding the eigenvalues, eigenvectors and the matrix $\mathbf{P}$ which diagonalizes a given matrix can be a tedious task if you are not given the eigenvalues and eigenvectors; you have to go through the whole process.

Example 7.20

For the matrix $\mathbf{A} = \begin{pmatrix} 1 & -6 & 2 \\ 0 & 4 & 25 \\ 0 & 0 & 9 \end{pmatrix}$, determine the eigenvector matrix $\mathbf{P}$ which diagonalizes matrix $\mathbf{A}$ given that the eigenvalues of $\mathbf{A}$ are $\lambda_1 = 1, \lambda_2 = 4$ and $\lambda_3 = 9$ with corresponding eigenvectors

$\mathbf{u} = \begin{pmatrix} 1 \\ 0 \\ 0 \end{pmatrix}$, $\mathbf{v} = \begin{pmatrix} -2 \\ 1 \\ 0 \end{pmatrix}$ and $\mathbf{w} = \begin{pmatrix} -7 \\ 10 \\ 2 \end{pmatrix}$ respectively.

Check that $\mathbf{P}$ does indeed diagonalize the given matrix $\mathbf{A}$.

Solution

Step 1 and Step 2:

We have been given the eigenvalues and corresponding eigenvectors of the matrix $\mathbf{A}$.

(continued...)

What is our eigenvector matrix $\mathbf{P}$ *equal to?*

Eigenvector matrix $\mathbf{P}$ contains the eigenvectors:

$$\mathbf{P} = \begin{pmatrix} \mathbf{u} & \mathbf{v} & \mathbf{w} \end{pmatrix} = \begin{pmatrix} 1 & -2 & -7 \\ 0 & 1 & 10 \\ 0 & 0 & 2 \end{pmatrix} \left[\text{given } \mathbf{u} = \begin{pmatrix} 1 \\ 0 \\ 0 \end{pmatrix}, \mathbf{v} = \begin{pmatrix} -2 \\ 1 \\ 0 \end{pmatrix}, \mathbf{w} = \begin{pmatrix} -7 \\ 10 \\ 2 \end{pmatrix} \right]$$

Step 3:

The diagonal eigenvalue matrix $\mathbf{D}$ has the eigenvalues $\lambda_1 = 1$, $\lambda_2 = 4$ and $\lambda_3 = 9$ along the leading diagonal, that is

$$\mathbf{P}^{-1}\mathbf{AP} = \mathbf{D} = \begin{pmatrix} 1 & 0 & 0 \\ 0 & 4 & 0 \\ 0 & 0 & 9 \end{pmatrix}$$

eigenvalues of matrix **A**.

Step 4:

Checking that we have the correct $\mathbf{P}$ and $\mathbf{D}$ matrices by showing $\mathbf{PD} = \mathbf{AP}$:

$$\mathbf{PD} = \begin{pmatrix} 1 & -2 & -7 \\ 0 & 1 & 10 \\ 0 & 0 & 2 \end{pmatrix} \begin{pmatrix} 1 & 0 & 0 \\ 0 & 4 & 0 \\ 0 & 0 & 9 \end{pmatrix} = \begin{pmatrix} 1 & -8 & -63 \\ 0 & 4 & 90 \\ 0 & 0 & 18 \end{pmatrix}$$

$$\mathbf{AP} = \begin{pmatrix} 1 & -6 & 2 \\ 0 & 4 & 25 \\ 0 & 0 & 9 \end{pmatrix} \begin{pmatrix} 1 & -2 & -7 \\ 0 & 1 & 10 \\ 0 & 0 & 2 \end{pmatrix} = \begin{pmatrix} 1 & -8 & -63 \\ 0 & 4 & 90 \\ 0 & 0 & 18 \end{pmatrix}$$

Hence, this confirms that the matrix $\mathbf{P}$ does indeed diagonalize the given matrix $\mathbf{A}$.

7.3.4 Powers of matrices

We discussed the above diagonalization process so that we can find powers of matrices. For example, to find $\mathbf{A}^{100}$ is a difficult task.

What does diagonalization have to do with powers of matrices?

If $\mathbf{A}$ is a square matrix which is diagonalizable, so that there exists a matrix $\mathbf{P}$ such that $\mathbf{P}^{-1}\mathbf{AP} = \mathbf{D}$ where $\mathbf{D}$ is a diagonal matrix then

$$\mathbf{A}^m = \mathbf{PD}^m\mathbf{P}^{-1} \text{ where } m \text{ is any real number}$$

We will show this result in the next proposition.

In the meantime, using this formula, we can find the inverse of matrix $\mathbf{A}$ by substituting $m = -1$. To find $\mathbf{A}^{100}$ it is much easier to use this formula, $\mathbf{A}^m = \mathbf{PD}^m\mathbf{P}^{-1}$, rather than multiplying a 100 copies of the matrix $\mathbf{A}$. We can use this formula to find $\mathbf{A}^m$ if we first determine $\mathbf{D}^m$.

How?

The matrix $\mathbf{D}^m$ is simply a diagonal matrix with its leading diagonal entries raised to the power m, that is

$$\text{If } \mathbf{D} = \begin{pmatrix} d_1 & & O \\ & \ddots & \\ O & & d_n \end{pmatrix} \quad \text{then } \mathbf{D}^m = \begin{pmatrix} {d_1}^m & & O \\ & \ddots & \\ O & & d_n^m \end{pmatrix}$$

You are asked to show this result in Exercises 7.3. $\mathbf{A}^m$ might be hard to calculate but because $\mathbf{D}$ is a diagonal matrix, $\mathbf{D}^m$ is simply $\mathbf{D}$ with the entries on the leading diagonal raised to the power m.

The eigenvalue matrix $\mathbf{D}$ consists of the eigenvalues on the leading diagonal, therefore $\mathbf{D}^m$ has these *eigenvalues* to the *power m* on the leading diagonal.

Proposition (7.19). If an n by n matrix $\mathbf{A}$ is diagonalizable with $\mathbf{P}^{-1}\mathbf{AP} = \mathbf{D}$ where $\mathbf{D}$ is a diagonal matrix then

$$\mathbf{A}^m = \mathbf{PD}^m\mathbf{P}^{-1}$$

How do we prove this result?
By using mathematical induction.

What is the mathematical induction procedure?

Step 1: Check for $m = 1$.

Step 2: Assume that the result is true for $m = k$.

Step 3: Prove it for $m = k + 1$.

Proof.
Step 1: Check for $m = 1$, that is we need to show $\mathbf{PDP}^{-1} = \mathbf{A}$.

We have $\mathbf{D} = \mathbf{P}^{-1}\mathbf{AP}$. Left multiplying this by matrix $\mathbf{P}$ and right multiplying it by $\mathbf{P}^{-1}$:

$$\begin{aligned} \mathbf{PDP}^{-1} &= \mathbf{P}\underbrace{\left(\mathbf{P}^{-1}\mathbf{AP}\right)}_{=\mathbf{D}}\mathbf{P}^{-1} \\ &= \left(\mathbf{PP}^{-1}\right)\mathbf{A}\left(\mathbf{PP}^{-1}\right) = \mathbf{IAI} = \mathbf{A} \end{aligned}$$

Thus we have our result for $m = 1$ which is $\mathbf{A} = \mathbf{PDP}^{-1}$. This means that we can factorize matrix $\mathbf{A}$ into three matrices $\mathbf{P}$, $\mathbf{D}$ and $\mathbf{P}^{-1}$.

Step 2: Assume that the result is true for $m = k$, that is

$$\mathbf{A}^k = \mathbf{PD}^k\mathbf{P}^{-1} \qquad (*)$$

Step 3: We need to prove the result for $m = k + 1$, that is we need to prove

$$\mathbf{A}^{k+1} = \mathbf{PD}^{k+1}\mathbf{P}^{-1}$$

Starting with the left hand side of this we have

$$\begin{aligned}
\mathbf{A}^{k+1} &= \mathbf{A}^k\mathbf{A} && \left[\text{applying the rules of indices}\right] \\
&= \underbrace{\left(\mathbf{PD}^k\mathbf{P}^{-1}\right)}_{=\mathbf{A}^k \text{ by } (*)}\underbrace{\left(\mathbf{PDP}^{-1}\right)}_{=\mathbf{A} \text{ by Step 1}} \\
&= \mathbf{PD}^k\left(\mathbf{P}^{-1}\mathbf{P}\right)\mathbf{DP}^{-1} && \left[\text{using the rules of matrices } (\mathbf{AB})\mathbf{C} = \mathbf{A}(\mathbf{BC})\right] \\
&= \mathbf{P}\underbrace{\mathbf{D}^k(\mathbf{I})\,\mathbf{D}}_{=\mathbf{D}^k\mathbf{D}}\mathbf{P}^{-1} && \left[\text{because } \mathbf{P}^{-1}\mathbf{P} = \mathbf{I}\right] \\
&= \mathbf{PD}^k\mathbf{DP}^{-1} = \mathbf{PD}^{k+1}\mathbf{P}^{-1} && \left[\text{because } \mathbf{D}^k\mathbf{D} = \mathbf{D}^{k+1}\right]
\end{aligned}$$

This is our required result. Hence by mathematical induction we have $\mathbf{A}^m = \mathbf{PD}^m\mathbf{P}^{-1}$. ■

In general, to find $\mathbf{A}^m$ we have to multiply m copies of the matrix $\mathbf{A}$ which is a laborious task. It is much easier if we factorize $\mathbf{A}^m$ into $\mathbf{A}^m = \mathbf{PD}^m\mathbf{P}^{-1}$ (even though we need to find $\mathbf{P}$, $\mathbf{P}^{-1}$ and $\mathbf{D}$, which is no easy task in itself). This formula means that if you want to evaluate $\mathbf{A}^m$ without working out lower powers then using the diagonal matrix is more efficient than the Cayley–Hamilton method discussed in the previous section.

Note that in the above subsection when we diagonalized a matrix we could avoid the calculation of $\mathbf{P}^{-1}\mathbf{AP}$.

Why?

Because we check that $\mathbf{PD} = \mathbf{AP}$. This means that $\mathbf{P}^{-1}\mathbf{AP} = \mathbf{D}$ is the diagonal eigenvalue matrix. However, in evaluating $\mathbf{A}^m$ we need to find $\mathbf{P}^{-1}$ because $\mathbf{A}^m = \mathbf{PD}^m\mathbf{P}^{-1}$.

Example 7.21

Let $\mathbf{A} = \begin{pmatrix} 1 & -2 & 3 \\ 0 & 2 & 5 \\ 0 & 0 & 3 \end{pmatrix}$. Find $\mathbf{A}^5$ given that $\mathbf{P} = \begin{pmatrix} 1 & -2 & -7 \\ 0 & 1 & 10 \\ 0 & 0 & 2 \end{pmatrix}$ and $\mathbf{P}^{-1} = \frac{1}{2}\begin{pmatrix} 2 & 4 & -13 \\ 0 & 2 & -10 \\ 0 & 0 & 1 \end{pmatrix}$.

Solution

The diagonal matrix $\mathbf{D} = \mathbf{P}^{-1}\mathbf{AP} = \begin{pmatrix} 1 & 0 & 0 \\ 0 & 2 & 0 \\ 0 & 0 & 3 \end{pmatrix}$. Because $\mathbf{A}$ is an upper triangular matrix, its eigenvalues are the entries on the leading diagonal of $\mathbf{A}$, that is $\lambda_1 = 1$, $\lambda_2 = 2$ and $\lambda_3 = 3$.

How do we find $\mathbf{A}^5$?

By applying the above result (7.19), $\mathbf{A}^m$ factorizes into $\mathbf{A}^m = \mathbf{PD}^m\mathbf{P}^{-1}$ with $m = 5$:

$$\mathbf{A}^5 = \mathbf{PD}^5\mathbf{P}^{-1}$$

Substituting matrices $\mathbf{P}$, $\mathbf{D}$ and $\mathbf{P}^{-1}$ into this $\mathbf{A}^5 = \mathbf{PD}^5\mathbf{P}^{-1}$ gives

$$\mathbf{A}^5 = \mathbf{PD}^5\mathbf{P}^{-1}$$

eigenvalues of $\mathbf{A}$.

$$= \begin{pmatrix} 1 & -2 & -7 \\ 0 & 1 & 10 \\ 0 & 0 & 2 \end{pmatrix} \begin{pmatrix} 1 & 0 & 0 \\ 0 & 2 & 0 \\ 0 & 0 & 3 \end{pmatrix}^5 \frac{1}{2} \begin{pmatrix} 2 & 4 & -13 \\ 0 & 2 & -10 \\ 0 & 0 & 1 \end{pmatrix}$$

$$= \frac{1}{2} \begin{pmatrix} 1 & -2 & -7 \\ 0 & 1 & 10 \\ 0 & 0 & 2 \end{pmatrix} \begin{pmatrix} 1^5 & 0 & 0 \\ 0 & 2^5 & 0 \\ 0 & 0 & 3^5 \end{pmatrix} \begin{pmatrix} 2 & 4 & -13 \\ 0 & 2 & -10 \\ 0 & 0 & 1 \end{pmatrix}$$ [taking $\frac{1}{2}$ to the front and e.values to the power 5]

$$= \frac{1}{2} \begin{pmatrix} 1 & -2 & -7 \\ 0 & 1 & 10 \\ 0 & 0 & 2 \end{pmatrix} \begin{pmatrix} 1 & 0 & 0 \\ 0 & 32 & 0 \\ 0 & 0 & 243 \end{pmatrix} \begin{pmatrix} 2 & 4 & -13 \\ 0 & 2 & -10 \\ 0 & 0 & 1 \end{pmatrix}$$ [replacing the e.values $1^5 = 1, 2^5 = 32$ and $3^5 = 243$]

$$= \frac{1}{2} \begin{pmatrix} 1 & -64 & -1701 \\ 0 & 32 & 2430 \\ 0 & 0 & 486 \end{pmatrix} \begin{pmatrix} 2 & 4 & -13 \\ 0 & 2 & -10 \\ 0 & 0 & 1 \end{pmatrix}$$ [multiplying the first two matrices on the left]

$$= \frac{1}{2} \begin{pmatrix} 2 & -124 & -1074 \\ 0 & 64 & 2110 \\ 0 & 0 & 486 \end{pmatrix} = \begin{pmatrix} 1 & -62 & -537 \\ 0 & 32 & 1055 \\ 0 & 0 & 243 \end{pmatrix}$$ [multiplying by the scalar 1/2]

You may wish to check this final result by using appropriate software.

Note that in the above example the diagonal matrix $\mathbf{D}$ has eigenvalues 1, 2 and 3 on the leading diagonal, and $\mathbf{D}^5$ has 1^5, 2^5 and 3^5 on the leading diagonal. 1^5, 2^5 and 3^5 are the eigenvalues of $\mathbf{A}^5$.

7.3.5 Application of powers of matrices

Matrix powers are particularly useful in **Markov chains** – these are based on matrices whose entries are probabilities. Many real life systems have an element of uncertainty which develops over time, and this can be explained through Markov chains.

Example 7.22

The transition matrix $\mathbf{T}$ below gives the percentage of people involved in accidents who were either injured (I) or were killed (K) on urban (U) and rural (R) roads. The entries in the first column of matrix $\mathbf{T}$ indicate that 60% of road injuries on urban roads and 40% on rural roads. The second column of $\mathbf{T}$ represents 50% of road accident deaths occured on urban roads and 50% on rural roads. Out of a sample of 100 accidents this year the number of accidents on urban roads was 90 and rural roads 10 and this is represented by the vector $\mathbf{x}$.

$$\mathbf{T} = \begin{matrix} & \text{I} & \text{K} & \\ & \begin{pmatrix} 0.6 & 0.5 \\ 0.4 & 0.5 \end{pmatrix} & & \begin{matrix} \text{U} \\ \text{R} \end{matrix} \end{matrix} \quad \text{and} \quad \mathbf{x} = \begin{pmatrix} 90 \\ 10 \end{pmatrix}$$

The vector $\mathbf{x}_n$ given by $\mathbf{x}_n = \mathbf{T}^n\mathbf{x}$ gives us the number of accidents on urban and rural roads out of a sample of 100 accidents after n number of years. Determine to 2sf
(i) $\mathbf{x}_n$ for $n = 10$. (ii) $\mathbf{x}_n$ as $n \to \infty$

(continued...)

(This gives us our long term number of injuries and deaths on urban and rural roads out of a sample of 100 accidents.) For a Markov chain, we are interested in the long-term behaviour of $\mathbf{x}_n$.

Solution

(i) This means that we need to find $\mathbf{x}_n = \mathbf{T}^n\mathbf{x}$ when $n = 10$, that is $\mathbf{x}_{10} = \mathbf{T}^{10}\mathbf{x}$. To evaluate $\mathbf{T}^{10}$, we diagonalize the matrix $\mathbf{T}$ by finding the eigenvalue $\mathbf{D}$ and eigenvector $\mathbf{P}$ matrices. Verify that the eigenvalues and the corresponding eigenvectors are given by

$$\lambda_1 = 1,\ \mathbf{u} = \begin{pmatrix} 5 \\ 4 \end{pmatrix} \text{ and } \lambda_2 = 0.1, \mathbf{v} = \begin{pmatrix} -1 \\ 1 \end{pmatrix}$$

Our eigenvector matrix $\mathbf{P} = (\mathbf{u}\ \ \mathbf{v}) = \begin{pmatrix} 5 & -1 \\ 4 & 1 \end{pmatrix}$ and eigenvalue matrix $\mathbf{D} = \begin{pmatrix} 1 & 0 \\ 0 & 0.1 \end{pmatrix}$.

By result (7.19) $\mathbf{A}^m = \mathbf{P}\mathbf{D}^m\mathbf{P}^{-1}$ with $m = 10$ and $\mathbf{A} = \mathbf{T}$ we have

$$\mathbf{T}^{10} = \mathbf{P}\mathbf{D}^{10}\mathbf{P}^{-1} \quad (\dagger)$$

To evaluate $\mathbf{T}^{10}$ we need to find $\mathbf{P}^{-1}$, which is given by

$$\mathbf{P}^{-1} = \frac{1}{9}\begin{pmatrix} 1 & 1 \\ -4 & 5 \end{pmatrix}$$

Substituting $\mathbf{P} = \begin{pmatrix} 5 & -1 \\ 4 & 1 \end{pmatrix}$, $\mathbf{D} = \begin{pmatrix} 1 & 0 \\ 0 & 0.1 \end{pmatrix}$ and $\mathbf{P}^{-1} = \frac{1}{9}\begin{pmatrix} 1 & 1 \\ -4 & 5 \end{pmatrix}$ into (†) gives

$$\begin{aligned}
\mathbf{T}^{10} = \mathbf{P}\mathbf{D}^{10}\mathbf{P}^{-1} &= \begin{pmatrix} 5 & -1 \\ 4 & 1 \end{pmatrix}\begin{pmatrix} 1 & 0 \\ 0 & 0.1 \end{pmatrix}^{10}\frac{1}{9}\begin{pmatrix} 1 & 1 \\ -4 & 5 \end{pmatrix} \\
&= \frac{1}{9}\begin{pmatrix} 5 & -1 \\ 4 & 1 \end{pmatrix}\begin{pmatrix} 1 & 0 \\ 0 & 0 \end{pmatrix}\begin{pmatrix} 1 & 1 \\ -4 & 5 \end{pmatrix} \quad \begin{bmatrix} \text{because } 1^{10} = 1 \text{ and} \\ 0.1^{10} = 1 \times 10^{-10} = 0\text{(3dp)} \end{bmatrix} \\
&= \frac{1}{9}\begin{pmatrix} 5 & 0 \\ 4 & 0 \end{pmatrix}\begin{pmatrix} 1 & 1 \\ -4 & 5 \end{pmatrix} = \frac{1}{9}\begin{pmatrix} 5 & 5 \\ 4 & 4 \end{pmatrix}
\end{aligned}$$

Substituting $\mathbf{T}^{10} = \frac{1}{9}\begin{pmatrix} 5 & 5 \\ 4 & 4 \end{pmatrix}$ and $\mathbf{x} = \begin{pmatrix} 90 \\ 10 \end{pmatrix}$ into $\mathbf{x}_{10} = \mathbf{T}^{10}\mathbf{x}$ gives

$$\mathbf{x}_{10} = \frac{1}{9}\begin{pmatrix} 5 & 5 \\ 4 & 4 \end{pmatrix}\begin{pmatrix} 90 \\ 10 \end{pmatrix} \underset{\substack{\text{taking out a} \\ \text{factor of 10}}}{=} \frac{10}{9}\begin{pmatrix} 5 & 5 \\ 4 & 4 \end{pmatrix}\begin{pmatrix} 9 \\ 1 \end{pmatrix} = \frac{10}{9}\begin{pmatrix} 50 \\ 40 \end{pmatrix} = \begin{pmatrix} 55.5 \\ 44.4 \end{pmatrix} \begin{matrix} \text{U} \\ \text{R} \end{matrix}$$

This means that after 10 years the number of people likely to be injured or killed on an urban road is 56 (2sf) and on a rural road is 44 (2sf) out of a sample of 100 accidents.

(ii) We have $\mathbf{D}^n = \begin{pmatrix} 1 & 0 \\ 0 & 0.1 \end{pmatrix}^n = \begin{pmatrix} 1^n & 0 \\ 0 & 0.1^n \end{pmatrix}$.

How does this change as $n \to \infty$?

As $n \to \infty$ we have $1^n \to 1$ and $0.1^n \to 0$. This means that $\mathbf{D}^n = \mathbf{D}^{10}$ correct to 2sf which gives the same results as part (i).

Summary

If $\mathbf{A}$ is diagonalizable then we can convert $\mathbf{A}$ into a diagonal matrix $\mathbf{D}$.
If an n by n matrix $\mathbf{A}$ is diagonalizable with $\mathbf{P}^{-1}\mathbf{AP} = \mathbf{D}$ where $\mathbf{D}$ is a diagonal matrix then

$$\mathbf{A}^m = \mathbf{PD}^m\mathbf{P}^{-1}$$

EXERCISES 7.3

(Brief solutions at end of book. Full solutions available at <http://www.oup.co.uk/companion/singh>.)

In this exercise check your numerical answers using MATLAB.

1. For the following matrices find:

   (i) The eigenvalues and corresponding eigenvectors.

   (ii) Eigenvector matrix $\mathbf{P}$ and eigenvalue matrix $\mathbf{D}$.

   (a) $\mathbf{A} = \begin{pmatrix} 1 & 0 \\ 0 & 2 \end{pmatrix}$ (b) $\mathbf{A} = \begin{pmatrix} 1 & 1 \\ 1 & 1 \end{pmatrix}$ (c) $\mathbf{A} = \begin{pmatrix} 3 & 0 \\ 4 & 4 \end{pmatrix}$ (d) $\mathbf{A} = \begin{pmatrix} 2 & 2 \\ 1 & 3 \end{pmatrix}$

2. (i) For the matrices in question 1 find $\mathbf{A}^5$.

   (ii) For the matrix in question 1 part (c) find $\mathbf{A}^{-1/2}$.

3. For the following matrices find:

   (i) The eigenvalues and corresponding eigenvectors.

   (ii) Matrices $\mathbf{P}$ and $\mathbf{D}$ where $\mathbf{P}$ is the invertible (non-singular) matrix and $\mathbf{D} = \mathbf{P}^{-1}\mathbf{AP}$ is the diagonal matrix. To find $\mathbf{P}^{-1}$ you may use MATLAB.

   (iii) Determine $\mathbf{A}^4$ in each case by using the results of parts (i) and (ii).

   (a) $\mathbf{A} = \begin{pmatrix} 1 & 0 & 0 \\ 0 & 2 & 0 \\ 0 & 0 & 3 \end{pmatrix}$ (b) $\mathbf{A} = \begin{pmatrix} -1 & 4 & 0 \\ 0 & 4 & 3 \\ 0 & 0 & 5 \end{pmatrix}$ (c) $\mathbf{A} = \begin{pmatrix} 2 & 0 & 0 \\ 1 & 5 & 0 \\ 1 & 2 & 6 \end{pmatrix}$

4. For the following matrices determine whether they are diagonalizable.

   (a) $\mathbf{A} = \begin{pmatrix} 1 & 0 & 0 \\ 0 & 1 & 0 \\ 0 & 0 & 1 \end{pmatrix}$ (b) $\mathbf{A} = \begin{pmatrix} -1 & 2 & 3 \\ 0 & 2 & 5 \\ 0 & 0 & 8 \end{pmatrix}$ (c) $\mathbf{A} = \begin{pmatrix} \sqrt{2} & 0 & 0 & 0 \\ 1 & \sqrt{3} & 0 & 0 \\ 6 & 7 & 1/2 & 0 \\ 2 & 9 & 7 & -5 \end{pmatrix}$

5. In Example 7.17 take $\mathbf{P} = (\mathbf{v} \quad \mathbf{u})$ and determine the diagonal matrix $\mathbf{D}$ given by $\mathbf{P}^{-1}\mathbf{AP}$ where $\mathbf{A} = \begin{pmatrix} 1 & 4 \\ 2 & 3 \end{pmatrix}$.

   What do you notice about your diagonal matrix D?

6. Let **A** be a 3 by 3 matrix with the following eigenvalues and eigenvectors:

$$\lambda_1 = -2,\ \mathbf{u} = \begin{pmatrix} 1 \\ 2 \\ 0 \end{pmatrix}, \quad \lambda_2 = -5,\ \mathbf{v} = \begin{pmatrix} 5 \\ 4 \\ 0 \end{pmatrix} \text{ and } \lambda_3 = -1,\ \mathbf{w} = \begin{pmatrix} 0 \\ 0 \\ 1 \end{pmatrix}$$

Is the matrix **A** diagonalizable? If it is then find the diagonal eigenvalue matrix **D** which is similar to the matrix **A** and also determine the invertible matrix **P** such that $\mathbf{P}^{-1}\mathbf{AP} = \mathbf{D}$.

Find $\mathbf{A}^3$. [Note that you do *not* need to know the elements of matrix **A**.]

7. Show that the following matrices are *not* diagonalizable:

(a) $\mathbf{A} = \begin{pmatrix} 2 & -1 \\ 1 & 4 \end{pmatrix}$ (b) $\mathbf{A} = \begin{pmatrix} -2 & 4 \\ -1 & -6 \end{pmatrix}$ (c) $\mathbf{A} = \begin{pmatrix} 1 & 2 & 3 \\ 0 & 1 & 3 \\ 0 & 0 & 1 \end{pmatrix}$

8. Let $\mathbf{A} = \begin{pmatrix} -4 & 2 \\ -9 & 5 \end{pmatrix}$. Determine (i) $\mathbf{A}^{11}$ (ii) $\mathbf{A}^{-1}$

9. Prove that if **D** is a diagonal matrix then the matrix $\mathbf{D}^m$ is simply a diagonal matrix with its leading diagonal entries raised to the power m.

10. Prove Proposition (7.13).

11. Prove that if **A** is diagonalizable then the transpose of **A**, that is $\mathbf{A}^T$, is also diagonalizable.

12. In a differential equations course, the matrix $\exp(\mathbf{A}t)$ is defined as

$$\exp(\mathbf{A}t) = \mathbf{I} + \mathbf{A}t + \mathbf{A}^2\frac{t^2}{2!} + \mathbf{A}^3\frac{t^3}{3!} + \mathbf{A}^4\frac{t^4}{4!} + \cdots$$

Let $\mathbf{A} = \begin{pmatrix} 3 & 5 \\ 0 & 2 \end{pmatrix}$ and find an expression for $\exp(\mathbf{A}t)$ up to and including the term t^4 by diagonalizing matrix **A**.

13. Let $\mathbf{F} = \begin{pmatrix} 1 & 1 \\ 1 & 0 \end{pmatrix}$ [**F** is known as the **Fibonacci matrix**]. Evaluate the matrices **P**, $\mathbf{P}^{-1}$ and **D**, where **P** is an invertible matrix and $\mathbf{D} = \mathbf{P}^{-1}\mathbf{AP}$ is a diagonal matrix.

14. Let **A** be a 2 by 2 matrix with $[tr(\mathbf{A})]^2 > 4\det(\mathbf{A})$ where tr represents the trace of the matrix. Show that **A** is diagonalizable.

15. Let **A** be an invertible matrix which is diagonalizable. Prove that $\mathbf{A}^{-1}$ is also diagonalizable.

16. Prove that if **A** is diagonalizable then $\mathbf{A}^m$ (where $m \in \mathbb{N}$) is diagonalizable.

17. Let **A** and **B** be invertible matrices. Prove that **AB** is similar to **BA**.

18. If matrices **A** and **B** are similar, prove that

(i) $tr(\mathbf{A}) = tr(\mathbf{B})$ where tr is trace.

(ii) $\det(\mathbf{A}) = \det(\mathbf{B})$.

19. Let $\mathbf{A}$ be a diagonal matrix such that the modulus of each eigenvalue is less than 1. Evaluate the matrix $\mathbf{A}^m$ as $m \to \infty$.
    [You may assume that if $|x| < 1$ then $\lim_{m\to\infty} (x^m) = 0$.]
20. Let $\mathbf{A}$ be a diagonalizable matrix with eigenvalues $\lambda_1, \lambda_2, \ldots$ and λ_n. Prove that the eigenvalues of $\mathbf{A}^m$ are $(\lambda_1)^m, (\lambda_2)^m, \ldots$ and $(\lambda_n)^m$.
21. Prove Theorem (7.16).

SECTION 7.4 Diagonalization of Symmetric Matrices

By the end of this section you will be able to

- prove properties of symmetric matrices
- orthogonally diagonalize symmetric matrices

In this section we continue the diagonalization process. Diagonalization was described in the previous section – we found a matrix $\mathbf{P}$ which diagonalized a given matrix; this allowed us to find the matrix $\mathbf{D}$:

$$\mathbf{P}^{-1}\mathbf{A}\mathbf{P} = \mathbf{D} \text{ where } \mathbf{D} \text{ is a diagonal matrix.}$$

Left multiplying this by $\mathbf{P}$ and right multiplying by $\mathbf{P}^{-1}$ gives the factorization of matrix $\mathbf{A}$:

$$\mathbf{A} = \mathbf{P}\mathbf{D}\mathbf{P}^{-1}$$

Eigenvector matrix $\mathbf{P}$ contains the eigenvectors of $\mathbf{A}$, and $\mathbf{D}$ contains the eigenvalues of $\mathbf{A}$.

From this we deduced (result (7.19)) that the powers of matrix $\mathbf{A}$ can be found by factorizing $\mathbf{A}^m$ into three matrices:

$$\mathbf{A}^m = \mathbf{P}\mathbf{D}^m\mathbf{P}^{-1}$$

If you want to find $\mathbf{A}^{10}$ then $\mathbf{A}^{10} = \mathbf{P}\mathbf{D}^{10}\mathbf{P}^{-1}$. This can still be a tedious task even for a small size matrix such as 3 by 3.

Why?

Because we need to find the inverse matrix $\mathbf{P}^{-1}$, which will involve putting the matrix $\mathbf{P}$ into reduced row echelon form or using cofactors. Either way, a cumbersome task.

Is there a type of matrix for which we can easily find the inverse?
Yes, the orthogonal matrix $\mathbf{Q}$, described in chapter 4, for which $\mathbf{Q}^{-1} = \mathbf{Q}^T$.

What is an orthogonal matrix?
It's a square matrix whose columns are orthonormal (perpendicular unit) vectors.

In this section we aim to find a diagonalizing matrix $\mathbf{Q}$ which is an *orthogonal* matrix. Eigenvector matrix $\mathbf{Q}$ is the diagonalizing matrix which is made up by writing its columns as the eigenvectors of the given matrix $\mathbf{A}$.

However, when we find eigenvectors, they are usually *not* perpendicular (orthogonal) to each other. In this section we obtain eigenvectors which are perpendicular and normalized. We aim to get *orthonormal* (perpendicular unit) eigenvectors as columns of the diagonalizing matrix. Once we have achieved unit perpendicular eigenvectors as columns of the diagonalizing matrix then we will find working with the diagonal matrix even easier than the previous section. The columns of $\mathbf{Q}$ are also an orthonormal *basis* for the eigenspace E_λ. Remember, orthonormal bases (axes) are one of the simplest bases to work with.

We cannot *guarantee* that the diagonalizing matrix will be an orthogonal matrix. However, in this section we will show that if the given matrix is *symmetric* then we can always find an orthogonal diagonalizing matrix.

7.4.1 Symmetric matrices

Can you recall what a symmetric matrix is?
A square matrix $\mathbf{A}$ is a symmetric matrix if $\mathbf{A}^T = \mathbf{A}$ ($\mathbf{A}$ transpose equals $\mathbf{A}$).

Examples are

$$\mathbf{A} = \begin{pmatrix} 1 & 0 \\ 0 & 1 \end{pmatrix}, \quad \mathbf{B} = \begin{pmatrix} 1 & \sqrt{2} \\ \sqrt{2} & 3 \end{pmatrix}, \quad \mathbf{C} = \begin{pmatrix} 1 & -1 & \sqrt{5} \\ -1 & 2 & \pi \\ \sqrt{5} & \pi & 3 \end{pmatrix}, \quad \mathbf{D} =$$

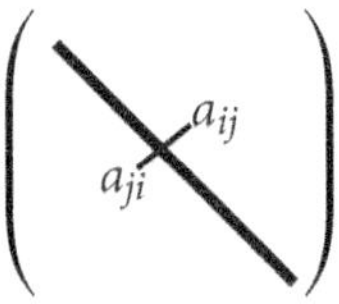

What do you notice?
We get a reflection of the entries by placing a mirror on the leading diagonal as highlighted.

Why are symmetric matrices important?
We will show later in this section that *all* symmetric matrices are *diagonalizable* by an orthogonal matrix. This is *not* the case for non-symmetric matrices.

Example 7.23

Let $\mathbf{A} = \begin{pmatrix} -3 & 4 \\ 4 & 3 \end{pmatrix}$. Diagonalize matrix $\mathbf{A}$.

Solution

The characteristic equation is given by

$$\det(\mathbf{A}-\lambda\mathbf{I}) = \det\begin{pmatrix} -3-\lambda & 4 \\ 4 & 3-\lambda \end{pmatrix} = \lambda^2 - 25 = 0 \text{ which gives } \lambda_1 = 5 \text{ and } \lambda_2 = -5$$

The corresponding eigenvectors are $\mathbf{u} = \begin{pmatrix} 1 \\ 2 \end{pmatrix}$ for $\lambda_1 = 5$ and $\mathbf{v} = \begin{pmatrix} 2 \\ -1 \end{pmatrix}$ for $\lambda_2 = -5$.

The eigenvectors $\mathbf{u}$ and $\mathbf{v}$ are linearly independent (*not* multiples of each other) therefore

$$\mathbf{P} = (\mathbf{u} \quad \mathbf{v}) = \begin{pmatrix} 1 & 2 \\ 2 & -1 \end{pmatrix} \text{ and } \mathbf{D} = \mathbf{P}^{-1}\mathbf{A}\mathbf{P} = \begin{pmatrix} 5 & 0 \\ 0 & -5 \end{pmatrix}$$

The eigenvector matrix $\mathbf{P} = (\mathbf{u} \quad \mathbf{v})$ contains the eigenvectors of $\mathbf{A}$ and the eigenvalue matrix $\mathbf{D}$ contains the eigenvalues of $\mathbf{A}$.

Remember, matrices $\mathbf{A}$ and $\mathbf{D}$ are similar, so they have the same eigenvalues.

What do you notice about the eigenvectors $\mathbf{u} = \begin{pmatrix} 1 \\ 2 \end{pmatrix}$ *and* $\mathbf{v} = \begin{pmatrix} 2 \\ -1 \end{pmatrix}$*?*

The inner (dot) product of eigenvectors $\mathbf{u}$ and $\mathbf{v}$ is zero:

$$\mathbf{u}\cdot\mathbf{v} = \begin{pmatrix} 1 \\ 2 \end{pmatrix}\cdot\begin{pmatrix} 2 \\ -1 \end{pmatrix} = (1\times 2) + (2\times(-1)) = 0$$

What does $\mathbf{u}\cdot\mathbf{v} = 0$ *mean?*

Eigenvectors $\mathbf{u}$ and $\mathbf{v}$ are *orthogonal* which means that they are perpendicular to each other. (See Fig. 7.11 overleaf.) We can normalize these eigenvectors (that is make their length 1).

How?

By dividing by its length : (2.16) $\widehat{\mathbf{u}} = \frac{1}{\|\mathbf{u}\|}\mathbf{u}$ where $\widehat{\mathbf{u}}$ is the normalized (unit) vector and $\|\mathbf{u}\|$ is the norm (length) of $\mathbf{u}$.

For the eigenvector $\mathbf{u}$ *what is length* $\|\mathbf{u}\|$ *equal to?*

$$\|\mathbf{u}\|^2 = \mathbf{u}\cdot\mathbf{u} = \begin{pmatrix} 1 \\ 2 \end{pmatrix}\cdot\begin{pmatrix} 1 \\ 2 \end{pmatrix} = 1^2 + 2^2 = 5 \quad \text{[by Pythagoras]}$$

We have $\|\mathbf{u}\|^2 = 5$, therefore taking the square root gives $\|\mathbf{u}\| = \sqrt{5}$. Thus the normalized eigenvector $\widehat{\mathbf{u}} = \dfrac{1}{\|\mathbf{u}\|}\mathbf{u} = \dfrac{1}{\sqrt{5}}\begin{pmatrix} 1 \\ 2 \end{pmatrix}$.

Similarly, the other normalized eigenvector $\widehat{\mathbf{v}} = \dfrac{1}{\|\mathbf{v}\|}\mathbf{v} = \dfrac{1}{\sqrt{5}}\begin{pmatrix} 2 \\ -1 \end{pmatrix}$.

Note that $\widehat{\mathbf{u}}$ and $\widehat{\mathbf{v}}$ are unit eigenvectors, which means that they have norm (length) of 1. Plotting these in $\mathbb{R}^2$ is shown in Fig. 7.11.

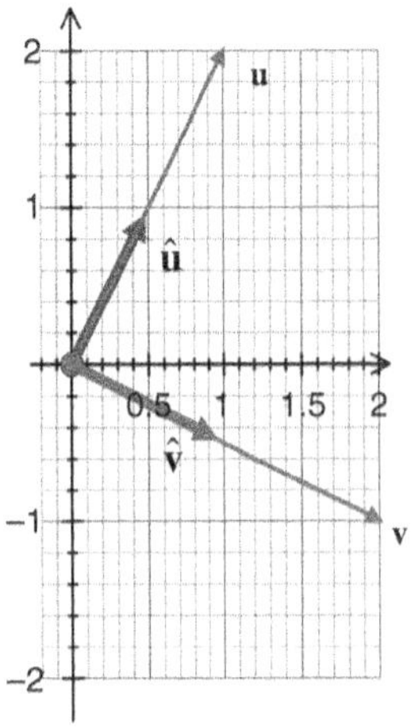

Figure 7.11

These vectors $\hat{\mathbf{u}}$ and $\hat{\mathbf{v}}$ form an orthonormal (perpendicular unit) basis for $\mathbb{R}^2$.

In the above Example 7.23 the diagonalizing matrix for $\mathbf{A}$ was $\mathbf{P} = (\mathbf{u} \quad \mathbf{v})$. However, we can show that the matrix $\mathbf{Q} = (\hat{\mathbf{u}} \quad \hat{\mathbf{v}})$ also diagonalizes the matrix $\mathbf{A}$ because $\hat{\mathbf{u}}$ and $\hat{\mathbf{v}}$ are the same vectors $\mathbf{u}$ and $\mathbf{v}$ but normalized, which means that they have the same direction as $\mathbf{u}$ and $\mathbf{v}$ but have length 1-see Fig. 7.11.

Example 7.24

Show that $\mathbf{Q} = (\hat{\mathbf{u}} \quad \hat{\mathbf{v}})$ diagonalizes the matrix $\mathbf{A}$ of Example 7.23.

Solution

What is the matrix $\mathbf{Q}$ *equal to?*

$$\mathbf{Q} = (\hat{\mathbf{u}} \quad \hat{\mathbf{v}}) = \left(\frac{1}{\sqrt{5}}\begin{pmatrix} 1 \\ 2 \end{pmatrix} \quad \frac{1}{\sqrt{5}}\begin{pmatrix} 2 \\ -1 \end{pmatrix} \right) \quad \left[\text{because } \hat{\mathbf{u}} = \frac{1}{\sqrt{5}}\begin{pmatrix} 1 \\ 2 \end{pmatrix} \text{ and } \hat{\mathbf{v}} = \frac{1}{\sqrt{5}}\begin{pmatrix} 2 \\ -1 \end{pmatrix}\right]$$

$$= \frac{1}{\sqrt{5}}\begin{pmatrix} 1 & 2 \\ 2 & -1 \end{pmatrix} = \frac{1}{\sqrt{5}}\mathbf{P} \quad \left[\text{because } \mathbf{P} = \begin{pmatrix} 1 & 2 \\ 2 & -1 \end{pmatrix}\right]$$

How do we show that matrix $\mathbf{Q}$ *diagonalizes matrix* $\mathbf{A}$*?*

We need to verify that $\mathbf{Q}^{-1}\mathbf{A}\mathbf{Q} = \mathbf{D}$, where $\mathbf{D}$ is the diagonal eigenvalue matrix. From the above we have $\mathbf{Q} = \frac{1}{\sqrt{5}}\mathbf{P}$, and taking the inverse of this gives

$$\mathbf{Q}^{-1} = \left(\frac{1}{\sqrt{5}}\mathbf{P}\right)^{-1} = \left(\frac{1}{\sqrt{5}}\right)^{-1}\mathbf{P}^{-1} = \sqrt{5}\,\mathbf{P}^{-1} \quad \left[\text{because } (k\mathbf{A})^{-1} = k^{-1}\mathbf{A}^{-1}\right]$$

Substituting these into $\mathbf{Q}^{-1}\mathbf{A}\mathbf{Q}$ yields

$$\mathbf{Q}^{-1}\mathbf{A}\mathbf{Q} = \left(\sqrt{5}\,\mathbf{P}^{-1}\right)\mathbf{A}\left(\frac{1}{\sqrt{5}}\mathbf{P}\right) \underset{\text{cancelling } \sqrt{5}}{=} \mathbf{P}^{-1}\mathbf{A}\mathbf{P} \underset{\text{by Example 7.23}}{=} \mathbf{D}$$

Thus matrix $\mathbf{Q}$ diagonalizes the matrix $\mathbf{A}$ because $\mathbf{Q}^{-1}\mathbf{A}\mathbf{Q} = \mathbf{D}$.

7.4.2 Orthogonal matrices

The matrix **Q** which diagonalizes **A** in the above Example 7.24 is an *orthogonal* matrix.

Can you remember what an orthogonal matrix is?
By chapter 4 Definition (4.18): A square matrix $\mathbf{Q} = (\mathbf{v}_1\ \mathbf{v}_2\ \ldots\ \mathbf{v}_n)$ the columns of which $\mathbf{v}_1, \mathbf{v}_2, \ldots, \mathbf{v}_n$ are *orthonormal* (perpendicular unit) vectors is called an *orthogonal matrix.*

*Why is **Q** an orthogonal matrix in the above Example 7.24?*
That's because the columns of $\mathbf{Q} = (\widehat{\mathbf{u}}\ \ \widehat{\mathbf{v}})$ are perpendicular unit vectors, $\widehat{\mathbf{u}}$ and $\widehat{\mathbf{v}}$, as illustrated in Fig. 7.11 above.

We use an orthogonal matrix **Q** to diagonalize a given matrix **A**. One critical application of diagonalization is the evaluation of powers of a matrix, which we found in the previous section and was given by formula (7.19): $\mathbf{A}^m = \mathbf{P}\mathbf{D}^m\mathbf{P}^{-1}$

From chapter 4, Proposition (4.20) we have: If **Q** is an orthogonal matrix then $\mathbf{Q}^{-1} = \mathbf{Q}^T$.

In this case, the diagonalizing matrix is the orthogonal matrix **Q** therefore

$$\mathbf{A}^m = \mathbf{Q}\mathbf{D}^m\mathbf{Q}^{-1} = \mathbf{Q}\mathbf{D}^m\mathbf{Q}^T \qquad \left[\text{because for an orthogonal matrix } \mathbf{Q}^{-1} = \mathbf{Q}^T\right]$$

This means that calculating the power of a matrix is even simpler because we don't have to evaluate the inverse of matrix **Q** by some tedious method but just transpose matrix **Q**. This is the great advantage of using an orthogonal matrix to diagonalize a matrix because

$$\mathbf{A}^m = \mathbf{Q}\mathbf{D}^m\mathbf{Q}^T \tag{7.20}$$

Evaluation of all the matrices on the right hand side; **Q**, $\mathbf{D}^m$ and $\mathbf{Q}^T$, is straightforward.

Example 7.25

Find $\mathbf{A}^6$ for the matrix **A** given in Example 7.23.

Solution

To find $\mathbf{A}^6$, we use the above formula $\mathbf{A}^m = \mathbf{Q}\mathbf{D}^m\mathbf{Q}^T$ with $m = 6$:

$$\mathbf{A}^6 = \mathbf{Q}\mathbf{D}^6\mathbf{Q}^T \qquad (*)$$

By Examples 7.23 and 7.24 we have

$$\mathbf{D} = \begin{pmatrix} 5 & 0 \\ 0 & -5 \end{pmatrix}, \quad \mathbf{Q} = \frac{1}{\sqrt{5}}\begin{pmatrix} 1 & 2 \\ 2 & -1 \end{pmatrix} \text{ and taking the transpose } \mathbf{Q}^T = \frac{1}{\sqrt{5}}\begin{pmatrix} 1 & 2 \\ 2 & -1 \end{pmatrix} = \mathbf{Q}$$

(continued...)

Substituting these into (*) yields

$$\mathbf{A}^6 = \mathbf{Q}\mathbf{D}^6\mathbf{Q}^T = \frac{1}{\sqrt{5}}\begin{pmatrix} 1 & 2 \\ 2 & -1 \end{pmatrix}\begin{pmatrix} 5 & 0 \\ 0 & -5 \end{pmatrix}^6 \frac{1}{\sqrt{5}}\begin{pmatrix} 1 & 2 \\ 2 & -1 \end{pmatrix}$$

$$= \frac{1}{\sqrt{5}}\frac{1}{\sqrt{5}}\begin{pmatrix} 1 & 2 \\ 2 & -1 \end{pmatrix}\begin{pmatrix} 5^6 & 0 \\ 0 & (-5)^6 \end{pmatrix}\begin{pmatrix} 1 & 2 \\ 2 & -1 \end{pmatrix}$$

$$= \frac{5^6}{5}\begin{pmatrix} 1 & 2 \\ 2 & -1 \end{pmatrix}\begin{pmatrix} 1 & 0 \\ 0 & 1 \end{pmatrix}\begin{pmatrix} 1 & 2 \\ 2 & -1 \end{pmatrix} \quad \left[\text{because } (-5)^6 = 5^6\right]$$

$$= 5^5\begin{pmatrix} 1 & 2 \\ 2 & -1 \end{pmatrix}\begin{pmatrix} 1 & 2 \\ 2 & -1 \end{pmatrix} = 5^5\begin{pmatrix} 5 & 0 \\ 0 & 5 \end{pmatrix} = 5^6\begin{pmatrix} 1 & 0 \\ 0 & 1 \end{pmatrix} = 5^6\mathbf{I}$$

7.4.3 Properties of symmetric matrices

Proposition (7.21). Let $\mathbf{A}$ be a real symmetric matrix. Then *all* the eigenvalues of $\mathbf{A}$ are real.

Proof.
We omit the proof because we need to use complex numbers, which are not covered.

■

Proposition (7.22). Let $\mathbf{A}$ be a symmetric matrix. If λ_1 and λ_2 are *distinct* eigenvalues of matrix $\mathbf{A}$ then their corresponding eigenvectors $\mathbf{u}$ and $\mathbf{v}$ respectively are orthogonal (perpendicular).

How do we prove this result?
We use the dot product result of chapter 2:

$$\mathbf{u} \cdot \mathbf{v} = \mathbf{u}^T\mathbf{v}, \text{ and show that } \mathbf{u} \cdot \mathbf{v} = \mathbf{u}^T\mathbf{v} = 0.$$

Proof.
Let $\mathbf{u}$ and $\mathbf{v}$ be eigenvectors belonging to distinct eigenvalues λ_1 and λ_2 respectively.

What do we need to prove?
Required to prove that the eigenvectors are orthogonal, which means that we need to show $\mathbf{u} \cdot \mathbf{v} = 0$. Since $\mathbf{u}$ and $\mathbf{v}$ are eigenvectors belonging to distinct eigenvalues λ_1 and λ_2, matrix $\mathbf{A}$ scalar multiplies each of the eigenvectors by λ_1 and λ_2 respectively:

$$\mathbf{A}\mathbf{u} = \lambda_1\mathbf{u} \quad \text{and} \quad \mathbf{A}\mathbf{v} = \lambda_2\mathbf{v} \qquad (*)$$

Taking the transpose of both sides of $\lambda_1\mathbf{u} = \mathbf{A}\mathbf{u}$ gives

$$(\lambda_1\mathbf{u})^T = (\mathbf{A}\mathbf{u})^T$$

$$\lambda_1\mathbf{u}^T = \mathbf{u}^T\mathbf{A}^T \qquad \left[\text{by (1.19) (b) } (k\mathbf{C})^T = k\mathbf{C}^T \text{ and (d) } (\mathbf{A}\mathbf{B})^T = \mathbf{B}^T\mathbf{A}^T\right]$$

$$= \mathbf{u}^T\mathbf{A} \qquad \left[\text{because } \mathbf{A} \text{ is symmetric therefore } \mathbf{A}^T = \mathbf{A}\right]$$

Right-multiplying the last line $\lambda_1 \mathbf{u}^T = \mathbf{u}^T \mathbf{A}$ by the eigenvector $\mathbf{v}$ gives

$$\begin{aligned} \lambda_1 \mathbf{u}^T \mathbf{v} &= \mathbf{u}^T \mathbf{A} \mathbf{v} \\ &= \mathbf{u}^T \lambda_2 \mathbf{v} \qquad [\text{by } (*)] \\ \lambda_1 \mathbf{u}^T \mathbf{v} - \lambda_2 \mathbf{u}^T \mathbf{v} &= 0 \\ (\lambda_1 - \lambda_2)\, \mathbf{u}^T \mathbf{v} &= 0 \qquad [\text{factorizing}] \end{aligned}$$

λ_1 and λ_2 are distinct eigenvalues therefore $\lambda_1 - \lambda_2 \neq 0$ [not zero] so we have

$$\mathbf{u}^T \mathbf{v} = 0 \text{ which means that } \mathbf{u} \cdot \mathbf{v} = 0$$

The dot product of the eigenvectors $\mathbf{u}$ and $\mathbf{v}$ is *zero*, therefore they are *orthogonal*.

■

We can extend Proposition (7.22) to n distinct eigenvalues:

Proposition (7.23). Let $\mathbf{A}$ be a symmetric matrix with distinct eigenvalues $\lambda_1, \lambda_2, \ldots, \lambda_n$ and corresponding eigenvectors $\mathbf{v}_1, \mathbf{v}_2, \ldots$ and $\mathbf{v}_n$. Then these eigenvectors are orthogonal.

Proof – Exercises 7.3.

■

The next two examples show applications of this Proposition (7.23).

Example 7.26

Show that the eigenvectors of matrix $\mathbf{A} = \begin{pmatrix} 1 & 2 & 1 \\ 2 & 1 & 1 \\ 1 & 1 & 2 \end{pmatrix}$ are orthogonal.

Solution

If you have reached this point then you should be able to find the eigenvalues and eigenvectors of the given matrix. Verify that the characteristic equation $p(\lambda)$ is given by

$$\begin{aligned} p(\lambda) &= \lambda^3 - 4\lambda^2 - \lambda + 4 = 0 \\ &(\lambda + 1)(\lambda - 1)(\lambda - 4) = 0 \text{ yields } \lambda_1 = -1, \lambda_2 = 1 \text{ and } \lambda_3 = 4 \end{aligned}$$

Let $\mathbf{u}$, $\mathbf{v}$ and $\mathbf{w}$ be the eigenvectors belonging to $\lambda_1 = -1$, $\lambda_2 = 1$ and $\lambda_3 = 4$ respectively. We have (verify)

$$\mathbf{u} = \begin{pmatrix} 1 \\ -1 \\ 0 \end{pmatrix}, \mathbf{v} = \begin{pmatrix} 1 \\ 1 \\ -2 \end{pmatrix} \text{ and } \mathbf{w} = \begin{pmatrix} 1 \\ 1 \\ 1 \end{pmatrix}$$

(continued...)

How do we check that the eigenvectors $\mathbf{u}$, $\mathbf{v}$ and $\mathbf{w}$ are orthogonal to each other?
We need to confirm that the dot product is zero: $\mathbf{u}\cdot\mathbf{v}=0$, $\mathbf{u}\cdot\mathbf{w}=0$ and $\mathbf{v}\cdot\mathbf{w}=0$.

$$\mathbf{u}\cdot\mathbf{v}=\begin{pmatrix}1\\-1\\0\end{pmatrix}\cdot\begin{pmatrix}1\\1\\-2\end{pmatrix}=(1\times 1)+((-1)\times 1)+(0\times(-2))=0$$

$$\mathbf{u}\cdot\mathbf{w}=\begin{pmatrix}1\\-1\\0\end{pmatrix}\cdot\begin{pmatrix}1\\1\\1\end{pmatrix}=0 \text{ and } \mathbf{v}\cdot\mathbf{w}=\begin{pmatrix}1\\1\\-2\end{pmatrix}\cdot\begin{pmatrix}1\\1\\1\end{pmatrix}=0$$

Thus the eigenvectors $\mathbf{u}$, $\mathbf{v}$ and $\mathbf{w}$ are orthogonal (perpendicular) to each other.

Example 7.27

Show that the eigenvectors belonging to *distinct* eigenvalues of $\mathbf{A}=\begin{pmatrix}2&2&-2\\2&-1&4\\-2&4&-1\end{pmatrix}$ are orthogonal.

Solution

By solving the determinant you can verify that the characteristic equation $p(\lambda)$ is

$$p(\lambda)=\lambda^3-27\lambda+54=0$$
$$(\lambda-3)^2(\lambda+6)=0 \text{ gives } \lambda_1=3, \lambda_2=3 \text{ and } \lambda_3=-6$$

Let $\mathbf{u}$, $\mathbf{v}$ be the eigenvectors belonging to $\lambda_1=3$, $\lambda_2=3$ and $\mathbf{w}$ be the eigenvector belonging to $\lambda_3=-6$. You can verify the following in your own time:

$$\left\{\mathbf{u}=\begin{pmatrix}2\\1\\0\end{pmatrix},\mathbf{v}=\begin{pmatrix}-2\\0\\1\end{pmatrix}\right\} \text{ belong to the } \textit{same} \text{ e.value } \lambda_1=3,\ \lambda_2=3 \text{ and } \mathbf{w}=\begin{pmatrix}-1\\2\\-2\end{pmatrix} \text{ to } \lambda_3=-6$$

We need to check that eigenvectors $\mathbf{u}$, $\mathbf{w}$ and $\mathbf{v}$, $\mathbf{w}$ are orthogonal (dot product is zero):

$$\mathbf{u}\cdot\mathbf{w}=\begin{pmatrix}2\\1\\0\end{pmatrix}\cdot\begin{pmatrix}-1\\2\\-2\end{pmatrix}=0 \text{ and } \mathbf{v}\cdot\mathbf{w}=\begin{pmatrix}-2\\0\\1\end{pmatrix}\cdot\begin{pmatrix}-1\\2\\-2\end{pmatrix}=0$$

However, note that the eigenvectors $\mathbf{u}$ and $\mathbf{v}$ belonging to the *same* eigenvalue $\lambda_1=3$ and $\lambda_2=3$ need *not* be orthogonal to each other. Actually

$$\mathbf{u}\cdot\mathbf{v}=\begin{pmatrix}2\\1\\0\end{pmatrix}\cdot\begin{pmatrix}-2\\0\\1\end{pmatrix}=(2\times(-2))+(1\times 0)+(0\times 1)=-4 \quad (\dagger)$$

Thus the eigenvectors $\mathbf{u}$ and $\mathbf{v}$ belonging to the same eigenvalue $\lambda_1=3$ and $\lambda_2=3$ are *not* orthogonal because $\mathbf{u}\cdot\mathbf{v}=-4\neq 0$ [not zero].

7.4.4 Orthogonal diagonalization

Definition (7.24). In general, a matrix $\mathbf{A}$ is **orthogonally diagonalizable** if there is an orthogonal matrix $\mathbf{Q}$ such that

$$\mathbf{Q}^{-1}\mathbf{A}\mathbf{Q} = \mathbf{Q}^T\mathbf{A}\mathbf{Q} = \mathbf{D} \text{ where } \mathbf{D} \text{ is a diagonal matrix.}$$

Eigenvector matrix $\mathbf{Q}$ contains the eigenvectors of $\mathbf{A}$ and eigenvalue matrix $\mathbf{D}$ contains the eigenvalues of $\mathbf{A}$.

In the above Examples 7.23 and 7.24, the matrix $\mathbf{A} = \begin{pmatrix} -3 & 4 \\ 4 & 3 \end{pmatrix}$ is orthogonally diagonalizable because with $\mathbf{Q} = \frac{1}{\sqrt{5}}\begin{pmatrix} 1 & 2 \\ 2 & -1 \end{pmatrix}$ we have $\mathbf{Q}^{-1}\mathbf{A}\mathbf{Q} = \begin{pmatrix} 5 & 0 \\ 0 & -5 \end{pmatrix}$ which is a diagonal matrix.

Theorem (7.25). Let $\mathbf{A}$ be a square matrix. If the matrix $\mathbf{A}$ is orthogonally diagonalizable then $\mathbf{A}$ is a symmetric matrix.

How do we prove this result?
We assume that $\mathbf{A}$ is orthogonally diagonalizable and deduce that $\mathbf{A}$ is symmetric, which means that we need to show that $\mathbf{A}^T = \mathbf{A}$.

Why?
Because $\mathbf{A}^T = \mathbf{A}$ means that the matrix $\mathbf{A}$ is symmetric.

Proof.
Assume that the matrix $\mathbf{A}$ is orthogonally diagonalizable. This means that there is an orthogonal matrix $\mathbf{Q}$ such that $\mathbf{Q}^{-1}\mathbf{A}\mathbf{Q} = \mathbf{D}$, where $\mathbf{D}$ is a diagonal matrix. Left multiplying this $\mathbf{Q}^{-1}\mathbf{A}\mathbf{Q} = \mathbf{D}$ by $\mathbf{Q}$ and right multiplying by $\mathbf{Q}^{-1}$ gives

$$\mathbf{A} = \mathbf{Q}\mathbf{D}\mathbf{Q}^{-1} \qquad (\dagger)$$

Taking the transpose of both sides gives

$$\begin{aligned}
\mathbf{A}^T = \left(\mathbf{Q}\mathbf{D}\mathbf{Q}^{-1}\right)^T &= \left(\mathbf{Q}^{-1}\right)^T\mathbf{D}^T\mathbf{Q}^T && \left[\text{by } (\mathbf{ABC})^T = \mathbf{C}^T\mathbf{B}^T\mathbf{A}^T\right] \\
&= \left(\mathbf{Q}^T\right)^{-1}\mathbf{D}\mathbf{Q}^T && \left[\text{using } \left(\mathbf{A}^{-1}\right)^T = \left(\mathbf{A}^T\right)^{-1} \text{ and } \mathbf{D}^T = \mathbf{D}\right] \\
&= \left(\mathbf{Q}^{-1}\right)^{-1}\mathbf{D}\mathbf{Q}^{-1} && \left[\text{because } \mathbf{Q} \text{ is orthogonal, } \mathbf{Q}^T = \mathbf{Q}^{-1}\right] \\
&= \mathbf{Q}\mathbf{D}\mathbf{Q}^{-1} && \left[\text{because } \left(\mathbf{Q}^{-1}\right)^{-1} = \mathbf{Q}\right]
\end{aligned}$$

We have $\mathbf{A}^T = \mathbf{Q}\mathbf{D}\mathbf{Q}^{-1}$.

What do you notice?
By (†) we can see that this is equal to matrix **A**. Thus

$$\mathbf{A}^T = \mathbf{QDQ}^{-1} = \mathbf{A} \text{ which means we have } \mathbf{A}^T = \mathbf{A}$$

Hence **A** is a symmetric matrix because $\mathbf{A}^T = \mathbf{A}$, which is our required result.

■

Now we will show that the other way round is also true.

What does this mean?
If **A** is a symmetric matrix then **A** is orthogonally diagonalizable. This means that there is at least one type of matrix, *symmetric matrices*, which can be *diagonalized orthogonally*.

Lemma (7.26). If **A** is a real symmetric matrix with an eigenvalue λ of multiplicity m then λ has m linearly independent eigenvectors.

Proof – See <http://www.oup.co.uk/companion/singh>.

■

Theorem (7.27). If **A** is an n by n symmetric matrix then **A** is orthogonally diagonalizable.

How do we prove this?
We consider two cases:
(i) **A** has distinct eigenvalues (ii) **A** does *not* have distinct eigenvalues

Proof.
Case (i): Let the symmetric matrix **A** have distinct eigenvalues $\lambda_1, \lambda_2, \ldots$ and λ_n. Then by Proposition (7.23): Let **A** be a symmetric matrix with distinct eigenvalues $\lambda_1, \lambda_2, \ldots, \lambda_n$ and eigenvectors $\mathbf{v}_1, \mathbf{v}_2, \ldots$ and $\mathbf{v}_n$. Then these eigenvectors are orthogonal.

So the eigenvectors $\mathbf{v}_1, \mathbf{v}_2, \ldots$ and $\mathbf{v}_n$ belonging to distinct eigenvalues $\lambda_1, \lambda_2, \ldots$ and λ_n are orthogonal. Because they are orthogonal, they are linearly independent and so we have n linearly independent eigenvectors. By Theorem (7.16):

An n by n matrix **A** is diagonalizable $\Leftrightarrow$ it has n independent eigenvectors.

We have that the matrix **A** is diagonalizable. Let $\mathbf{Q} = (\widehat{\mathbf{v}_1} \;\; \widehat{\mathbf{v}_2} \;\; \cdots \;\; \widehat{\mathbf{v}_n})$, then the columns of the matrix **Q** are orthonormal, which means it is an orthogonal matrix. Thus we have $\mathbf{Q}^{-1}\mathbf{AQ} = \mathbf{D}$, so matrix **A** is orthogonally diagonalizable.

Case (ii): Let the symmetric matrix **A** have an eigenvalue λ with multiplicity $m > 1$. By the above Lemma (7.26) λ has m linearly independent eigenvectors $\mathbf{u}_1, \mathbf{u}_2, \ldots$ and $\mathbf{u}_m$. These are a basis for the eigenspace E_λ. By the Gram–Schmidt process we can convert these m vectors into orthonormal basis vectors for E_λ.

We can repeat this process for any other eigenvectors belonging to eigenvalues of **A** which have a multiplicity of more than 1.

All the remaining eigenvalues are *distinct*, so the eigenvectors are orthogonal.

Thus *all* the vectors are orthogonal. By repeating the procedure outlined in case (i), we conclude that matrix **A** is orthogonally diagonalizable.

■

By combining Theorems (7.25) and (7.27) what can we conclude?
If $\mathbf{A}$ is orthogonally diagonalizable then $\mathbf{A}$ is a symmetric matrix and the other way round; that is, if $\mathbf{A}$ is symmetric matrix then $\mathbf{A}$ is orthogonally diagonalizable. We have the main result of this section which has a special name – **spectral theorem**:

Spectral theorem (7.28). Matrix $\mathbf{A}$ is orthogonally diagonalizable $\Leftrightarrow$ $\mathbf{A}$ is a symmetric matrix.

Proof – By the above Theorems (7.25) and (7.27).

■

This means that if we have a symmetric matrix then we can orthogonally diagonalize it. This is the spectral theorem for real matrices.

Example 7.28

Determine an orthogonal matrix $\mathbf{Q}$ which orthogonally diagonalizes $\mathbf{A} = \begin{pmatrix} 3 & 2 \\ 2 & 0 \end{pmatrix}$.

Solution

Since matrix $\mathbf{A}$ is symmetric, we can find an orthogonal matrix $\mathbf{Q}$ which diagonalizes $\mathbf{A}$. Verify that the characteristic polynomial is given by:

$$\lambda^2 - 3\lambda - 4 = 0 \Rightarrow (\lambda + 1)(\lambda - 4) = 0$$
$$\lambda_1 = -1 \text{ and } \lambda_2 = 4$$

Let $\mathbf{u}$ and $\mathbf{v}$ be eigenvectors (verify) belonging to $\lambda_1 = -1$ and $\lambda_2 = 4$ respectively:

$$\mathbf{u} = \begin{pmatrix} 1 \\ -2 \end{pmatrix} \text{ and } \mathbf{v} = \begin{pmatrix} 2 \\ 1 \end{pmatrix}$$

The given matrix $\mathbf{A}$ is symmetric and we have distinct eigenvalues, therefore $\mathbf{u}$ and $\mathbf{v}$ are orthogonal (that is $\mathbf{u} \cdot \mathbf{v} = 0$). Remember, the question says that we have to find an orthogonal matrix $\mathbf{Q}$.
What is an orthogonal matrix?
A square matrix the columns of which form an orthonormal set.
What is an orthonormal set?
A set that is *orthogonal* and *normalized*. The eigenvectors $\mathbf{u}$ and $\mathbf{v}$ are orthogonal but we need to normalize them, which means make their norm (length) to be 1.
How?
Divide by the norm (length) of the eigenvector, which in both cases is $\sqrt{5}$ because

$$\|\mathbf{u}\| = \sqrt{1^2 + (-2)^2} = \sqrt{5} \text{ and } \|\mathbf{v}\| = \sqrt{1^2 + 2^2} = \sqrt{5}$$

Normalizing gives

$$\widehat{\mathbf{u}} = \frac{1}{\|\mathbf{u}\|}\mathbf{u} = \frac{1}{\sqrt{5}}\begin{pmatrix} 1 \\ -2 \end{pmatrix} \text{ and } \widehat{\mathbf{v}} = \frac{1}{\|\mathbf{v}\|}\mathbf{v} = \frac{1}{\sqrt{5}}\begin{pmatrix} 2 \\ 1 \end{pmatrix}$$

(continued...)

Thus our orthogonal matrix **Q** is given by

$$\mathbf{Q} = (\widehat{\mathbf{u}} \quad \widehat{\mathbf{v}}) = \frac{1}{\sqrt{5}}\begin{pmatrix} 1 & 2 \\ -2 & 1 \end{pmatrix}$$

Q is an orthogonal matrix, therefore $\mathbf{Q}^{-1} = \mathbf{Q}^T$. Verify that $\mathbf{Q}^T\mathbf{A}\mathbf{Q} = \mathbf{D}$ or $\mathbf{Q}\mathbf{D} = \mathbf{A}\mathbf{Q}$ where **D** is a diagonal matrix with eigenvalues along the leading diagonal:

$$\mathbf{Q}^T\mathbf{A}\mathbf{Q} = \mathbf{D} = \begin{pmatrix} -1 & 0 \\ 0 & 4 \end{pmatrix} \quad \left[\begin{array}{l} -1,\ 4 \text{ are the eigenvalues} \\ \text{of the given matrix } \mathbf{A} \end{array}\right]$$

7.4.5 Procedure for orthogonal diagonalization

We can work through a procedure to find an orthogonal matrix which diagonalizes a given matrix.

The procedure for orthogonal diagonalization of a symmetric matrix **A** is as follows:

1. Determine the eigenvalues of **A**.
2. Find the corresponding eigenvectors.
3. If any of the eigenvalues are repeated then check that the associated eigenvectors are orthogonal. If they are *not* orthogonal then place them into an orthogonal set by using the Gram–Schmidt process described in chapter 4.
4. Normalize all the eigenvectors.
5. Form the orthogonal matrix **Q** whose columns are the orthonormal eigenvectors.
6. Check that $\mathbf{QD} = \mathbf{AQ}$, where **D** is the diagonal matrix whose entries along the leading diagonal are the eigenvalues of matrix **A**.

Example 7.29

Determine an orthogonal matrix **Q** which diagonalizes $\mathbf{A} = \begin{pmatrix} 2 & 2 & -2 \\ 2 & -1 & 4 \\ -2 & 4 & -1 \end{pmatrix}$.

Solution

Steps 1 and 2:

This is the same matrix as Example 7.27. Thus we have

$$\left\{\mathbf{u} = \begin{pmatrix} 2 \\ 1 \\ 0 \end{pmatrix}, \mathbf{v} = \begin{pmatrix} -2 \\ 0 \\ 1 \end{pmatrix}\right\} \text{ belong to } \lambda_{1,2} = 3 \text{ and } \mathbf{w} = \begin{pmatrix} -1 \\ 2 \\ -2 \end{pmatrix} \text{ to } \lambda_3 = -6$$

Step 3:

We need to check that eigenvectors **u** and **v** belonging to repeated eigenvalue $\lambda_{1,2} = 3$ are orthogonal.

Remember that in Example 7.27 we showed that $\mathbf{u}\cdot\mathbf{v}=-4$, therefore eigenvectors $\mathbf{u}$ and $\mathbf{v}$ are *not* orthogonal. We need to convert these $\mathbf{u}$ and $\mathbf{v}$ into an orthogonal set, say $\mathbf{q}_1$ and $\mathbf{q}_2$ respectively.

How do we convert $\mathbf{u}$ and $\mathbf{v}$ into the orthogonal set $\mathbf{q}_1$ and $\mathbf{q}_2$?

By using the Gram–Schmidt process (4.16):

$$\mathbf{q}_1=\mathbf{u} \quad \text{and} \quad \mathbf{q}_2=\mathbf{v}-\frac{\mathbf{v}\cdot\mathbf{q}_1}{\|\mathbf{q}_1\|^2}\mathbf{q}_1 \qquad (*)$$

How do we evaluate $\mathbf{q}_1$ and $\mathbf{q}_2$?

By substituting $\mathbf{u}$ and $\mathbf{v}$ and evaluating $\mathbf{v}\cdot\mathbf{q}_1$ and $\|\mathbf{q}_1\|^2$. We have

$$\mathbf{q}_1=\mathbf{u}=\begin{pmatrix}2\\1\\0\end{pmatrix} \text{ and } \mathbf{v}\cdot\mathbf{q}_1=\mathbf{v}\cdot\mathbf{u}=\mathbf{u}\cdot\mathbf{v}=-4 \text{ [already evaluated]}$$

$$\|\mathbf{q}_1\|^2=\left\|\begin{pmatrix}2\\1\\0\end{pmatrix}\right\|^2=\begin{pmatrix}2\\1\\0\end{pmatrix}\cdot\begin{pmatrix}2\\1\\0\end{pmatrix}=2^2+1^2+0^2=5 \quad (**)$$

Substituting the above $\mathbf{v}=\begin{pmatrix}-2\\0\\1\end{pmatrix}$, $\mathbf{q}_1=\begin{pmatrix}2\\1\\0\end{pmatrix}$, $\mathbf{v}\cdot\mathbf{q}_1=-4$ and $\|\mathbf{q}_1\|^2=5$ into $\mathbf{q}_2=\mathbf{v}-\frac{\mathbf{v}\cdot\mathbf{q}_1}{\|\mathbf{q}_1\|^2}\mathbf{q}_1$:

$$\begin{aligned}\mathbf{q}_2=\mathbf{v}-\frac{\mathbf{v}\cdot\mathbf{q}_1}{\|\mathbf{q}_1\|^2}\mathbf{q}_1&=\begin{pmatrix}-2\\0\\1\end{pmatrix}-\frac{(-4)}{5}\begin{pmatrix}2\\1\\0\end{pmatrix}\\&=\begin{pmatrix}-2+8/5\\0+4/5\\1-0\end{pmatrix}=\begin{pmatrix}-2/5\\4/5\\1\end{pmatrix}=\frac{1}{5}\begin{pmatrix}-2\\4\\5\end{pmatrix}=\mathbf{q}_2\end{aligned}$$

Thus we have $\mathbf{q}_1=\begin{pmatrix}2\\1\\0\end{pmatrix}$, $\mathbf{q}_2=\frac{1}{5}\begin{pmatrix}-2\\4\\5\end{pmatrix}$ and $\mathbf{w}=\begin{pmatrix}-1\\2\\-2\end{pmatrix}$, which are orthogonal to each other.

What else do we need to do?

Step 4:

We need to normalize these eigenvectors by dividing by the norm (length) of each. We have already established above in (**) that $\|\mathbf{q}_1\|^2=5$. Taking the square root gives $\|\mathbf{q}_1\|=\sqrt{5}$.

Remember, we can ignore any fractions (scalars) because vectors are orthogonal independent of scalars. Thus for the eigenvector $\mathbf{q}_2$, we can ignore the fraction 1/5 and call this $\mathbf{q}_2^*$:

$$\|\mathbf{q}_2^*\|^2=\begin{pmatrix}-2\\4\\5\end{pmatrix}\cdot\begin{pmatrix}-2\\4\\5\end{pmatrix}=(-2)^2+4^2+5^2=45$$

Taking the square root gives $\|\mathbf{q}_2^*\|=\sqrt{45}=3\sqrt{5}$. Similarly we have

$$\|\mathbf{w}\|^2=\begin{pmatrix}-1\\2\\-2\end{pmatrix}\cdot\begin{pmatrix}-1\\2\\-2\end{pmatrix}=(-1)^2+2^2+(-2)^2=9$$

(continued...)

Taking the square root gives the norm $\|\mathbf{w}\| = 3$. Normalizing means we divide each vector by its norm (length):

$$\widehat{\mathbf{q}_1} = \frac{1}{\sqrt{5}}\begin{pmatrix} 2 \\ 1 \\ 0 \end{pmatrix}, \quad \widehat{\mathbf{q}_2^*} = \frac{1}{3\sqrt{5}}\begin{pmatrix} -2 \\ 4 \\ 5 \end{pmatrix} \quad \text{and} \quad \widehat{\mathbf{w}} = \frac{1}{3}\begin{pmatrix} -1 \\ 2 \\ -2 \end{pmatrix}$$

Step 5:
What is the orthogonal matrix $\mathbf{Q}$ *equal to?*

$$\mathbf{Q} = \begin{pmatrix} \widehat{\mathbf{q}_1} & \widehat{\mathbf{q}_2^*} & \widehat{\mathbf{w}} \end{pmatrix} = \begin{pmatrix} 2/\sqrt{5} & -2/3\sqrt{5} & -1/3 \\ 1/\sqrt{5} & 4/3\sqrt{5} & 2/3 \\ 0 & 5/3\sqrt{5} & -2/3 \end{pmatrix}$$

Step 6:
Check $\mathbf{QD} = \mathbf{AQ}$, where $\mathbf{D}$ is a diagonal matrix with entries on the leading diagonal given by the eigenvalues of the matrix $\mathbf{A}$.
What is $\mathbf{D}$ *equal to?*

$$\mathbf{D} = \begin{pmatrix} 3 & 0 & 0 \\ 0 & 3 & 0 \\ 0 & 0 & -6 \end{pmatrix} \quad \left[\begin{array}{l}\text{eigenvalues of the given matrix} \\ \mathbf{A} \text{ are } \lambda_1 = \lambda_2 = 3 \text{ and } \lambda_3 = -6\end{array}\right]$$

Summary

Definition (7.23). In general, a matrix $\mathbf{A}$ is *orthogonally diagonalizable* if there is an orthogonal matrix $\mathbf{Q}$ such that $\mathbf{Q}^{-1}\mathbf{AQ} = \mathbf{Q}^T\mathbf{AQ} = \mathbf{D}$, where $\mathbf{D}$ is a diagonal matrix.

Spectral Theorem (7.28).

Matrix $\mathbf{A}$ is orthogonally diagonalizable $\Leftrightarrow$ $\mathbf{A}$ is a symmetric matrix.

EXERCISES 7.4

(Brief solutions at end of book. Full solutions available at <http://www.oup.co.uk/companion/singh>.)

In this exercise check your numerical answers using MATLAB.

1. For the following matrices find an orthogonal matrix $\mathbf{Q}$ which diagonalizes the given matrix. Also check that $\mathbf{Q}^T\mathbf{AQ} = \mathbf{D}$ where $\mathbf{D}$ is a diagonal matrix.

   (a) $\mathbf{A} = \begin{pmatrix} 1 & 0 \\ 0 & 2 \end{pmatrix}$ (b) $\mathbf{A} = \begin{pmatrix} 1 & 1 \\ 1 & 1 \end{pmatrix}$ (c) $\mathbf{A} = \begin{pmatrix} 2 & 1 \\ 1 & 2 \end{pmatrix}$ (d) $\mathbf{A} = \begin{pmatrix} 5 & 12 \\ 12 & -5 \end{pmatrix}$

2. For the following matrices find an orthogonal matrix $\mathbf{Q}$ which diagonalizes the given matrix. Also check that $\mathbf{Q}^T\mathbf{AQ} = \mathbf{D}$ where $\mathbf{D}$ is a diagonal matrix.

   (a) $\mathbf{A} = \begin{pmatrix} 9 & 3 \\ 3 & 1 \end{pmatrix}$ (b) $\mathbf{A} = \begin{pmatrix} 3 & \sqrt{2} \\ \sqrt{2} & 2 \end{pmatrix}$

   (c) $\mathbf{A} = \begin{pmatrix} -5 & \sqrt{3} \\ \sqrt{3} & -3 \end{pmatrix}$ (d) $\mathbf{A} = \begin{pmatrix} 5 & \sqrt{12} \\ \sqrt{12} & 1 \end{pmatrix}$

3. For the following matrices find an orthogonal matrix $\mathbf{Q}$ which diagonalizes the given matrix. By using MATLAB or otherwise check that $\mathbf{Q}^T\mathbf{A}\mathbf{Q} = \mathbf{D}$ where $\mathbf{D}$ is a diagonal matrix.

(a) $\mathbf{A} = \begin{pmatrix} 1 & 0 & 0 \\ 0 & 2 & 0 \\ 0 & 0 & 3 \end{pmatrix}$ (b) $\mathbf{A} = \begin{pmatrix} 2 & 2 & 2 \\ 2 & 2 & 2 \\ 2 & 2 & 2 \end{pmatrix}$ (c) $\mathbf{A} = \begin{pmatrix} 0 & 0 & 0 \\ 0 & 1 & 1 \\ 0 & 1 & 1 \end{pmatrix}$

4. For the following matrices find an orthogonal matrix $\mathbf{Q}$ which diagonalizes the given matrix. By using MATLAB or otherwise check that $\mathbf{Q}^T\mathbf{A}\mathbf{Q} = \mathbf{D}$.

(a) $\mathbf{A} = \begin{pmatrix} 1 & 2 & 2 \\ 2 & 1 & 2 \\ 2 & 2 & 1 \end{pmatrix}$ (b) $\mathbf{A} = \begin{pmatrix} 2 & 1 & 1 \\ 1 & 2 & 1 \\ 1 & 1 & 2 \end{pmatrix}$ (c) $\mathbf{A} = \begin{pmatrix} -5 & 4 & 2 \\ 4 & -5 & 2 \\ 2 & 2 & -8 \end{pmatrix}$

5. Let $\mathbf{A} = \begin{pmatrix} 1 & 1 \\ 1 & 1 \end{pmatrix}$. Show that $\mathbf{A}^{10} = 2^9\mathbf{A}$. Also prove that $\mathbf{A}^m = 2^{m-1}\mathbf{A}$ where m is a positive integer.

6. Show that if $\mathbf{A}$ is a diagonal matrix then orthogonal diagonalising matrix $\mathbf{Q} = \mathbf{I}$.

7. Prove that (a) the zero matrix $\mathbf{O}$ and (b) the identity matrix $\mathbf{I}$ are orthogonally diagonalisable.

8. Prove that $\mathbf{A} = \begin{pmatrix} a & b \\ b & c \end{pmatrix} \neq \mathbf{O}$ is orthogonally diagonalisable and find the orthogonal matrix $\mathbf{Q}$ which diagonalizes the matrix $\mathbf{A}$.

   [Hint: If the quadratic $x^2 + px + q = 0$ has roots a and b then $a + b = -p$.]

9. Let $\mathbf{A}$ be a symmetric invertible matrix. If $\mathbf{Q}$ orthogonally diagonalizes the matrix $\mathbf{A}$ show that $\mathbf{Q}$ also diagonalizes the matrix $\mathbf{A}^{-1}$.

10. Prove Proposition (7.23).

SECTION 7.5 Singular Value Decomposition

By the end of this section you will be able to

- understand what is meant by SVD
- find a triple factorization of any matrix

The singular value decomposition (SVD) is one of the most important factorizations of a matrix. SVD factorization breaks the matrix down into useful parts such as orthogonal matrices, and the method can be applied to any matrix; it does *not* need to be a square or symmetric matrix.

The SVD of a matrix gives us an orthogonal basis (axes) for the row and column space of the matrix. If we consider a matrix as a transformation then SVD factorization gives an orthogonal basis (axes) for the start and arrival vector spaces (Fig. 7.12).

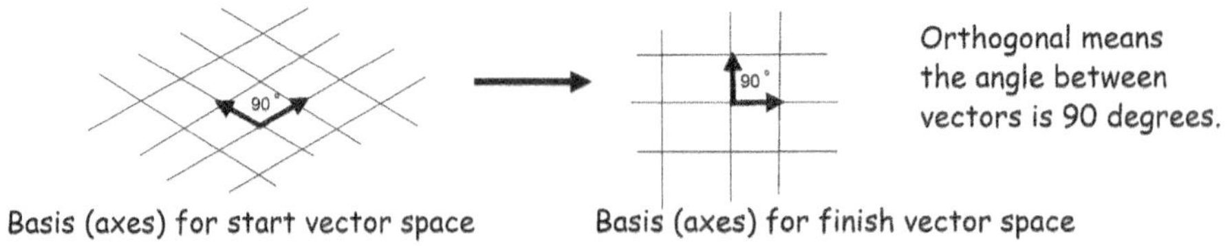

Figure 7.12

A very good application of SVD is given in the article by David Austin in the Monthly Essays on Mathematical Topics of August 2009:

> Netflix, the online movie rental company, is currently offering a \$1 million prize for anyone who can improve the accuracy of its movie recommendation system by 10%. Surprisingly, this seemingly modest problem turns out to be quite challenging, and the groups involved are now using rather sophisticated techniques. At the heart of all of them is the singular value decomposition.

First, we look at the geometric significance of SVD factorization.

7.5.1 Geometric interpretation of singular value decomposition (SVD)

To find the SVD of any matrix **A** we use the matrix $\mathbf{A}^T\mathbf{A}$ because $\mathbf{A}^T\mathbf{A}$ is a symmetric matrix, as we will show later in this section. Remember, symmetric matrices can be orthogonally diagonalized.

First we define the **singular values** of a matrix **A**.

Definition (7.29). Let **A** be any m by n matrix and $\lambda_1, \lambda_2, \ldots, \lambda_n$ be the eigenvalues of $\mathbf{A}^T\mathbf{A}$, then the **singular values** of **A** denoted by $\sigma_1, \sigma_2, \ldots, \sigma_n$ are the numbers:

$$\sigma_1 = \sqrt{\lambda_1},\ \sigma_2 = \sqrt{\lambda_2}, \ldots, \sigma_n = \sqrt{\lambda_n} \quad \text{[positive root only]}$$

Example 7.30

Find the eigenvalues and eigenvectors of $\mathbf{A}^T\mathbf{A}$ where $\mathbf{A} = \begin{pmatrix} 2 & 1 \\ 1 & 2 \end{pmatrix}$.

Solution

Since the given matrix **A** is a symmetric matrix so $\mathbf{A}^T = \mathbf{A}$. We have $\mathbf{A}^T\mathbf{A} = \mathbf{A}\mathbf{A} = \mathbf{A}^2$.

We found the eigenvalues t_1 and t_2 with the normalized eigenvectors $\mathbf{v}_1$ and $\mathbf{v}_2$ of this matrix **A** in Exercises 7.4 question 1(c):

$$t_1 = 3,\ \mathbf{v}_1 = \frac{1}{\sqrt{2}}\begin{pmatrix} 1 \\ 1 \end{pmatrix} \quad \text{and} \quad t_2 = 1,\ \mathbf{v}_2 = \frac{1}{\sqrt{2}}\begin{pmatrix} 1 \\ -1 \end{pmatrix}$$

What are the eigenvalues and eigenvectors of matrix $\mathbf{A}^2$*?*

By Proposition (7.8)(a): If m is a natural number then λ^m is an eigenvalue of the matrix $\mathbf{A}^m$ with the *same* eigenvector **u**.

Let λ_1 and λ_2 be the eigenvalues of matrix $\mathbf{A}^2$, then by using this proposition we have

$$\lambda_1 = 3^2 = 9, \mathbf{v}_1 = \frac{1}{\sqrt{2}}\begin{pmatrix} 1 \\ 1 \end{pmatrix} \quad \text{and} \quad \lambda_2 = 1^2 = 1, \mathbf{v}_2 = \frac{1}{\sqrt{2}}\begin{pmatrix} 1 \\ -1 \end{pmatrix}$$

Example 7.31

(i) Find the singular values $\sigma_1 = \sqrt{\lambda_1}$ and $\sigma_2 = \sqrt{\lambda_2}$ of the matrix $\mathbf{A}$ given in the above Example 7.30.

(ii) Determine $\sigma_1\mathbf{u}_1 = \mathbf{A}\mathbf{v}_1$ and $\sigma_2\mathbf{u}_2 = \mathbf{A}\mathbf{v}_2$, where $\mathbf{v}_1$ and $\mathbf{v}_2$ are the normalized eigenvectors belonging to the eigenvalues of matrix $\mathbf{A}^T\mathbf{A}$.

Solution

(i) From Example 7.30 we have the eigenvalues $\lambda_1 = 9$ and $\lambda_2 = 1$. Taking the square root:

$$\sigma_1 = \sqrt{9} = 3 \quad \text{and} \quad \sigma_2 = \sqrt{1} = 1 \qquad \text{[positive root only]}$$

The singular values of matrix $\mathbf{A}$ are 3 and 1. Since $\mathbf{A}$ is a symmetric matrix, the singular values of $\mathbf{A}$ are the eigenvalues of $\mathbf{A}$. This would *not* be the case if $\mathbf{A}$ was a non-symmetric matrix.

(ii) Substituting $\mathbf{A} = \begin{pmatrix} 2 & 1 \\ 1 & 2 \end{pmatrix}$, $\mathbf{v}_1 = \frac{1}{\sqrt{2}}\begin{pmatrix} 1 \\ 1 \end{pmatrix}$ and $\sigma_1 = 3$ into $\sigma_1\mathbf{u}_1 = \mathbf{A}\mathbf{v}_1$ gives

$$3\mathbf{u}_1 = \mathbf{A}\mathbf{v}_1 = \begin{pmatrix} 2 & 1 \\ 1 & 2 \end{pmatrix}\frac{1}{\sqrt{2}}\begin{pmatrix} 1 \\ 1 \end{pmatrix} = \frac{1}{\sqrt{2}}\begin{pmatrix} 3 \\ 3 \end{pmatrix}$$

Similarly $\sigma_2\mathbf{u}_2 = \mathbf{A}\mathbf{v}_2$ is

$$(1)\,\mathbf{u}_2 = \mathbf{A}\mathbf{v}_2 = \begin{pmatrix} 2 & 1 \\ 1 & 2 \end{pmatrix}\frac{1}{\sqrt{2}}\begin{pmatrix} 1 \\ -1 \end{pmatrix} = \frac{1}{\sqrt{2}}\begin{pmatrix} 1 \\ -1 \end{pmatrix}$$

We have

$$\mathbf{A}\mathbf{v}_1 = 3\mathbf{u}_1 \quad \text{and} \quad \mathbf{A}\mathbf{v}_2 = \mathbf{u}_2$$

What does this mean?

It means that the transformation $T : \mathbb{R}^2 \to \mathbb{R}^2$ given by $T(\mathbf{v}) = \mathbf{A}\mathbf{v}$ transforms the vector $\mathbf{v}_1$ to 3 times the vector $\mathbf{u}_1$ in the same direction. The vector $\mathbf{v}_2$ is transformed under the matrix $\mathbf{A}$ to the same vector, $\mathbf{v}_2 = \mathbf{u}_2$ (Fig. 7.13).

The unit circle in Fig. 7.13(a) is transformed to an ellipse in Fig. 7.13(b) under the matrix $\mathbf{A}$. Note that the unit circle in Fig. 7.13(a) is stretched by the factors $\sigma_1 = 3$ and $\sigma_2 = 1$ in the direction of $\mathbf{u}_1$ and $\mathbf{u}_2$ respectively. Remember, these factors $\sigma_1 = 3$ and $\sigma_2 = 1$ are the singular values of matrix $\mathbf{A}$. Also observe that vectors $\mathbf{u}_1$ and $\mathbf{u}_2$ are orthogonal (perpendicular) and normalized (length equals 1). In SVD factorization, orthogonal vectors get transformed to orthogonal vectors – this is why this factorization is the most useful.

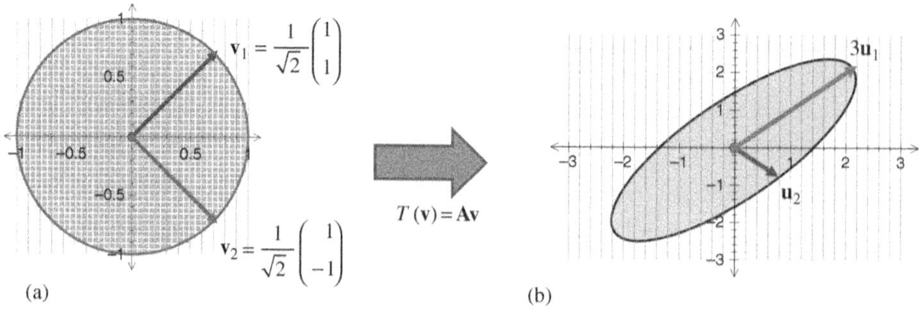

Figure 7.13

Actually the vectors $\mathbf{u}_1$ and $\mathbf{u}_2$ form an orthonormal (perpendicular unit) basis (axes) for the arrival vector space $\mathbb{R}^2$, similarly $\mathbf{v}_1$ and $\mathbf{v}_2$ form an orthonormal basis (axes) for the start vector space $\mathbb{R}^2$.

The eigenvectors $\mathbf{u}_1$ and $\mathbf{u}_2$ give us the direction of the semi-axes and the singular values σ_1 and σ_2 give us the length of the semi-axes.

What do we mean by semi-axes?

We illustrate the semi-axes for an ellipse in Fig. 7.14.

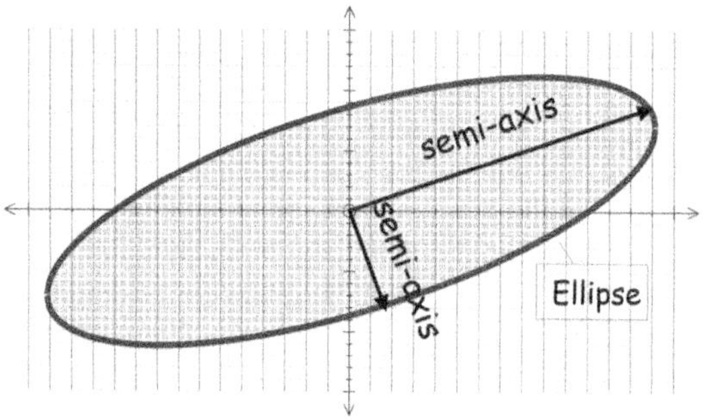

Figure 7.14

We can write $\mathbf{Av}_1 = 3\mathbf{u}_1$ and $\mathbf{Av}_2 = \mathbf{u}_2$ in matrix form as

$$\mathbf{A}(\mathbf{v}_1 \ \ \mathbf{v}_2) = (3\mathbf{u}_1 \ \ \mathbf{u}_2) = (\mathbf{u}_1 \ \ \mathbf{u}_2)\begin{pmatrix}3 & 0\\0 & 1\end{pmatrix} \qquad (*)$$

If we let $\mathbf{V} = (\mathbf{v}_1 \ \ \mathbf{v}_2)$, $\mathbf{U} = (\mathbf{u}_1 \ \ \mathbf{u}_2)$ and $\mathbf{D} = \begin{pmatrix}3 & 0\\0 & 1\end{pmatrix}$ then (*) can be written as

$$\mathbf{AV} = \mathbf{UD} \quad (\dagger)$$

Matrices $\mathbf{U}$ and $\mathbf{V}$ are orthogonal because $\mathbf{V} = (\mathbf{v}_1 \ \ \mathbf{v}_2)$ and $\mathbf{U} = (\mathbf{u}_1 \ \ \mathbf{u}_2)$ contain the orthonormal vectors $\mathbf{v}_1$, $\mathbf{v}_2$, $\mathbf{u}_1$ and $\mathbf{u}_2$ which are illustrated in the above Fig. 7.13.

Right multiplying (†) by the inverse of $\mathbf{V}$ (remember for an orthogonal matrix $\mathbf{V}^{-1} = \mathbf{V}^T$) which is $\mathbf{V}^T$:

$$(\mathbf{AV})\,\mathbf{V}^T = \mathbf{A}\underbrace{\left(\mathbf{V}\mathbf{V}^T\right)}_{=\mathbf{I}} = \mathbf{A} = \mathbf{U}\mathbf{D}\mathbf{V}^T$$

Hence we have factorized matrix $\mathbf{A} = \mathbf{U}\mathbf{D}\mathbf{V}^T$. Matrix $\mathbf{A}$ is broken into a *triple* matrix with the diagonal matrix $\mathbf{D}$ sandwiched between the two orthogonal matrices $\mathbf{U}$ and $\mathbf{V}^T$.

This is a singular value decomposition (SVD) of matrix $\mathbf{A}$.

Example 7.32

Check that $\mathbf{A} = \mathbf{U}\mathbf{D}\mathbf{V}^T$ for $\mathbf{A}$ of Example 7.30, that is the matrix $\mathbf{A} = \begin{pmatrix} 2 & 1 \\ 1 & 2 \end{pmatrix}$.

Solution

What are the orthogonal matrices U, V and the diagonal matrix D equal to?

By the results of Example 7.31 we have $\mathbf{u}_1 = \frac{1}{3\sqrt{2}}\begin{pmatrix} 3 \\ 3 \end{pmatrix} = \frac{1}{\sqrt{2}}\begin{pmatrix} 1 \\ 1 \end{pmatrix}$, $\mathbf{u}_2 = \frac{1}{\sqrt{2}}\begin{pmatrix} 1 \\ -1 \end{pmatrix}$, $\mathbf{v}_1 = \frac{1}{\sqrt{2}}\begin{pmatrix} 1 \\ 1 \end{pmatrix}$, $\mathbf{v}_2 = \frac{1}{\sqrt{2}}\begin{pmatrix} 1 \\ -1 \end{pmatrix}$, $\sigma_1 = 3$ and $\sigma_2 = 1$:

$$\mathbf{U} = (\mathbf{u}_1\ \mathbf{u}_2) = \frac{1}{\sqrt{2}}\begin{pmatrix} 1 & 1 \\ 1 & -1 \end{pmatrix},\ \mathbf{D} = \begin{pmatrix} \sigma_1 & 0 \\ 0 & \sigma_2 \end{pmatrix} = \begin{pmatrix} 3 & 0 \\ 0 & 1 \end{pmatrix} \text{ and } \mathbf{V} = (\mathbf{v}_1\ \mathbf{v}_2) = \frac{1}{\sqrt{2}}\begin{pmatrix} 1 & 1 \\ 1 & -1 \end{pmatrix}$$

Carrying out the matrix multiplication $\mathbf{U} \times \mathbf{D} \times \mathbf{V}^T$:

$$\begin{aligned} \mathbf{U}\mathbf{D}\mathbf{V}^T &= \frac{1}{\sqrt{2}}\begin{pmatrix} 1 & 1 \\ 1 & -1 \end{pmatrix}\begin{pmatrix} 3 & 0 \\ 0 & 1 \end{pmatrix}\frac{1}{\sqrt{2}}\begin{pmatrix} 1 & 1 \\ 1 & -1 \end{pmatrix}^T \\ &= \frac{1}{2}\begin{pmatrix} 3 & 1 \\ 3 & -1 \end{pmatrix}\begin{pmatrix} 1 & 1 \\ 1 & -1 \end{pmatrix} = \frac{1}{2}\begin{pmatrix} 4 & 2 \\ 2 & 4 \end{pmatrix} = \begin{pmatrix} 2 & 1 \\ 1 & 2 \end{pmatrix} = \mathbf{A} \end{aligned}$$

Note, the similarity to orthogonal diagonalization from the last section, $\mathbf{A} = \mathbf{Q}\mathbf{D}\mathbf{Q}^T$. As the given matrix $\mathbf{A}$ is a symmetric matrix, then $\mathbf{U} = \mathbf{Q}$, $\mathbf{D} = \mathbf{D}$ and $\mathbf{Q}^T = \mathbf{V}^T$.

What use is this section on SVD if we can simply apply the orthogonal diagonalization technique of the last section?

Well, for SVD you do *not* need square or symmetric matrix. (Recall that symmetric matrices must be *square* matrices because if the number of rows does *not* equal the number of columns then transposing changes the shape of the matrix.) SVD can be applied any matrix.

7.5.2 Introduction to singular value decomposition (SVD)

In the previous section, and again above, we factorized only symmetric matrices. In this section we extend the factorization or decomposition to any matrix. We factorize the matrix $\mathbf{A}$ where $\mathbf{A}$ is m by n and $m \geq n$. The results in this section are also true if $m < n$ but we have chosen $m \geq n$ for convenience. Note that the results in this section are valid for ANY matrix **A**.

Singular value decomposition theorem (7.30).

We can decompose any given matrix **A** of size m by n with positive singular values $\sigma_1 \geq \sigma_2 \geq \cdots \geq \sigma_k > 0$ where $k \leq n$, into $\mathbf{UDV}^T$, that is

$$\mathbf{A} = \mathbf{UDV}^T$$

where **U** is an m by m *orthogonal* matrix, **D** is an m by n matrix and **V** is an n by n *orthogonal* matrix. The values of m (rows) and n (columns) is the size of the given matrix **A**.

We have the situation shown in Fig. 7.15.

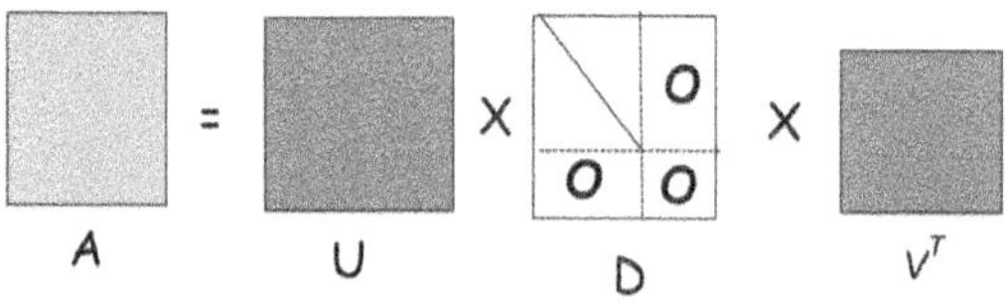

Figure 7.15

The matrix **D** looks a bit odd but it is a diagonal-like matrix. The matrix **D** has the positive singular values $\sigma_1 \geq \sigma_2 \geq \cdots \geq \sigma_k > 0$ of matrix **A** starting from the top left hand corner of the matrix and working diagonally towards the bottom right. The symbol **O** represents the zero matrix of an appropriate size. Matrix **D** is of shape (Fig. 7.16).

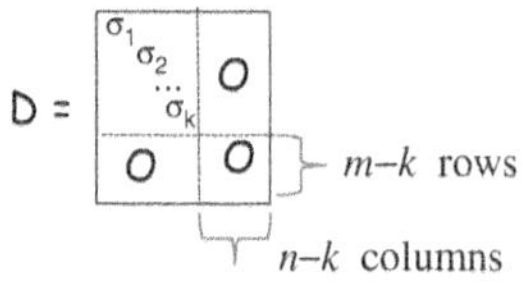

Figure 7.16

We need to be careful with the size of the matrices in Theorem (7.30).

*If the given matrix **A** is a 3 by 2 then what size are the matrices **U**, **D** and **V** in the above formula (7.30)?*

U is a 3 by 3 matrix, **D** is a 3 by 2 matrix and **V** is a 2 by 2 matrix.

Example 7.33

Find the eigenvalues and eigenvectors of $\mathbf{A}^T\mathbf{A}$ where $\mathbf{A} = \begin{pmatrix} 1 & 1 \\ 1 & 1 \\ -2 & 1 \end{pmatrix}$.

Solution

First, we carry out the matrix multiplication:

$$\mathbf{A}^T\mathbf{A} = \begin{pmatrix} 1 & 1 & -2 \\ 1 & 1 & 1 \end{pmatrix} \begin{pmatrix} 1 & 1 \\ 1 & 1 \\ -2 & 1 \end{pmatrix} = \begin{pmatrix} 6 & 0 \\ 0 & 3 \end{pmatrix}$$

Since $\mathbf{A}^T\mathbf{A}$ is a diagonal matrix, the entries on the leading diagonal are the eigenvalues

$$\lambda_1 = 6 \text{ and } \lambda_2 = 3.$$

Verify that the eigenvectors $\mathbf{v}_1$ and $\mathbf{v}_2$ belonging to these eigenvalues $\lambda_1 = 6$ and $\lambda_2 = 3$ are

$$\mathbf{v}_1 = \begin{pmatrix} 1 \\ 0 \end{pmatrix} \text{ and } \mathbf{v}_2 = \begin{pmatrix} 0 \\ 1 \end{pmatrix}$$

Note that we *cannot* find eigenvalues of matrix $\mathbf{A}$ because $\mathbf{A}$ is a non-square matrix. This is why the question says find the eigenvalues of $\mathbf{A}^T\mathbf{A}$.

Note that in the above example $\mathbf{A}^T\mathbf{A}$ is a symmetric matrix.

Is this always the case?

Yes.

Proposition (7.30). Let $\mathbf{A}$ be any matrix. Then $\mathbf{A}^T\mathbf{A}$ is a symmetric matrix.

Proof.

Remember, a matrix $\mathbf{X}$ is symmetric if $\mathbf{X}^T = \mathbf{X}$.

To show that $\mathbf{A}^T\mathbf{A}$ is a symmetric matrix, we need to prove $\left(\mathbf{A}^T\mathbf{A}\right)^T = \mathbf{A}^T\mathbf{A}$:

$$\begin{aligned} \left(\mathbf{A}^T\mathbf{A}\right)^T &= \mathbf{A}^T\left(\mathbf{A}^T\right)^T && \left[\text{by (1.19) (d) } (\mathbf{XY})^T = \mathbf{Y}^T\mathbf{X}^T\right] \\ &= \mathbf{A}^T\mathbf{A} && \left[\text{by (1.19) (a) } \left(\mathbf{X}^T\right)^T = \mathbf{X}\right] \end{aligned}$$

Hence $\mathbf{A}^T\mathbf{A}$ is a symmetric matrix.

■

What do we know about diagonalizing a symmetric matrix?

From the previous section, we have the spectral theorem (7.28):

Matrix $\mathbf{A}$ is orthogonally diagonalizable $\Leftrightarrow$ $\mathbf{A}$ is a symmetric matrix.

This means that we can orthogonally diagonalize the matrix $\mathbf{A}^T\mathbf{A}$ because it is a symmetric matrix. This is why we examine the matrix $\mathbf{A}^T\mathbf{A}$ for the purposes of finding the SVD of $\mathbf{A}$.

We show that the eigenvalues of this matrix $\mathbf{A}^T\mathbf{A}$ are positive or zero:

Proposition (7.31). Let $\mathbf{A}$ be any matrix. Then the eigenvalues of $\mathbf{A}^T\mathbf{A}$ are positive or zero.

Proof – Exercises 7.5.

■

Example 7.34

(i) Find the singular values $\sigma_1 = \sqrt{\lambda_1}$ and $\sigma_2 = \sqrt{\lambda_2}$ of the matrix $\mathbf{A}$ given in the above Example 7.33.

(ii) Determine $\sigma_1 \mathbf{u}_1 = \mathbf{A}\mathbf{v}_1$ and $\sigma_2 \mathbf{u}_2 = \mathbf{A}\mathbf{v}_2$, where $\mathbf{v}_1$ and $\mathbf{v}_2$ are the normalized eigenvectors belonging to the eigenvalues of matrix $\mathbf{A}^T\mathbf{A}$.

Solution

(i) From Example 7.33 we have the eigenvalues $\lambda_1 = 6$ and $\lambda_2 = 3$. Taking the square root:

$$\sigma_1 = \sqrt{6} \quad \text{and} \quad \sigma_2 = \sqrt{3}$$

The singular values of matrix $\mathbf{A}$ are $\sqrt{6}$ and $\sqrt{3}$. Note that matrix $\mathbf{A}$ does *not* have eigenvalues because it is a non-square matrix but has singular values which are the square roots of the eigenvalues of $\mathbf{A}^T\mathbf{A}$.

(ii) Substituting $\mathbf{A} = \begin{pmatrix} 1 & 1 \\ 1 & 1 \\ -2 & 1 \end{pmatrix}$, eigenvector $\mathbf{v}_1 = \begin{pmatrix} 1 \\ 0 \end{pmatrix}$ and $\sigma_1 = \sqrt{6}$ into $\sigma_1 \mathbf{u}_1 = \mathbf{A}\mathbf{v}_1$:

$$\sqrt{6}\mathbf{u}_1 = \mathbf{A}\mathbf{v}_1 = \begin{pmatrix} 1 & 1 \\ 1 & 1 \\ -2 & 1 \end{pmatrix}\begin{pmatrix} 1 \\ 0 \end{pmatrix} = \begin{pmatrix} 1 \\ 1 \\ -2 \end{pmatrix}$$

Similarly for $\sigma_2 \mathbf{u}_2 = \mathbf{A}\mathbf{v}_2$ we have

$$\sqrt{3}\mathbf{u}_2 = \mathbf{A}\mathbf{v}_2 = \begin{pmatrix} 1 & 1 \\ 1 & 1 \\ -2 & 1 \end{pmatrix}\begin{pmatrix} 0 \\ 1 \end{pmatrix} = \begin{pmatrix} 1 \\ 1 \\ 1 \end{pmatrix}$$

The matrix $\mathbf{A}$ transforms the two-dimensional eigenvectors $\mathbf{v}_1 = \begin{pmatrix} 1 \\ 0 \end{pmatrix}$ and $\mathbf{v}_2 = \begin{pmatrix} 0 \\ 1 \end{pmatrix}$ to three-dimensional vectors $(1\ 1\ {-2})^T$ and $(1\ \ 1\ \ 1)^T$ respectively. This transformation is $T : \mathbb{R}^2 \to \mathbb{R}^3$, such that $T(\mathbf{v}) = \mathbf{A}\mathbf{v}$ where $\mathbf{A}$ is the given 3 by 2 matrix.

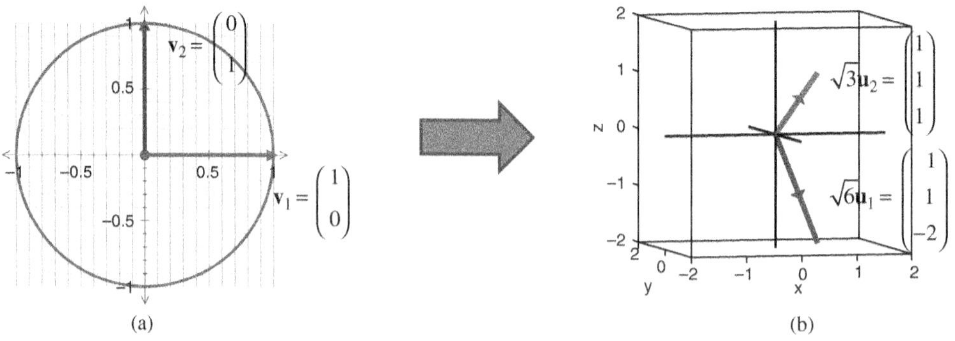

Figure 7.17

The transformation of the unit circle in Fig. 7.17(a) under the given matrix **A** is a two-dimensional ellipse in 3d space $\mathbb{R}^3$, but not illustrated above because of the limitation of software available.

Since the given **A** is a 3 by 2 matrix, by using the above formula (7.30) we find that **U** is a 3 by 3 matrix. This means that $\mathbf{U} = (\mathbf{u}_1 \ \ \mathbf{u}_2 \ \ \mathbf{u}_3)$.

In the above Example 7.34 we have found $\mathbf{u}_1$ and $\mathbf{u}_2$ but what is $\mathbf{u}_3$ equal to?

The vector $\mathbf{u}_3$ needs to be orthogonal (perpendicular) to both vectors $\mathbf{u}_1$ and $\mathbf{u}_2$ because **U** is an orthogonal matrix. This means vector $\mathbf{u}_3$ must satisfy both $\mathbf{u}_1 \cdot \mathbf{u}_3 = 0$ and $\mathbf{u}_2 \cdot \mathbf{u}_3 = 0$. Let $\mathbf{u}_3 = (x \ \ y \ \ z)^T$. We can ignore the scalars because the vectors will be orthogonal (perpendicular) independently of their scalars. We need to solve

$$\begin{pmatrix} 1 \\ 1 \\ -2 \end{pmatrix} \cdot \begin{pmatrix} x \\ y \\ z \end{pmatrix} = 0 \quad \text{and} \quad \begin{pmatrix} 1 \\ 1 \\ 1 \end{pmatrix} \cdot \begin{pmatrix} x \\ y \\ z \end{pmatrix} = 0$$

[Remember, for orthogonal (perpendicular) vectors, the dot product is zero]

In matrix form, and solving by inspection, yields

$$\begin{pmatrix} 1 & 1 & -2 \\ 1 & 1 & 1 \end{pmatrix} \begin{pmatrix} x \\ y \\ z \end{pmatrix} = \begin{pmatrix} 0 \\ 0 \end{pmatrix} \text{ gives } x = 1, \ y = -1 \text{ and } z = 0$$

Normalizing the vector gives $\mathbf{u}_3 = \dfrac{1}{\sqrt{2}}(1 \ -1 \ 0)^T$. Hence

$$\mathbf{U} = \left(\mathbf{u}_1 \ \ \mathbf{u}_2 \ \ \mathbf{u}_3 \right) = \begin{pmatrix} 1/\sqrt{6} & 1/\sqrt{3} & 1/\sqrt{2} \\ 1/\sqrt{6} & 1/\sqrt{3} & -1/\sqrt{2} \\ -2/\sqrt{6} & 1/\sqrt{3} & 0 \end{pmatrix}$$

Note that the column vectors $\left\{ \mathbf{u}_1, \mathbf{u}_2, \mathbf{u}_3 \right\}$ of matrix **U** is a basis (axes) for $\mathbb{R}^3$ (Fig. 7.18).

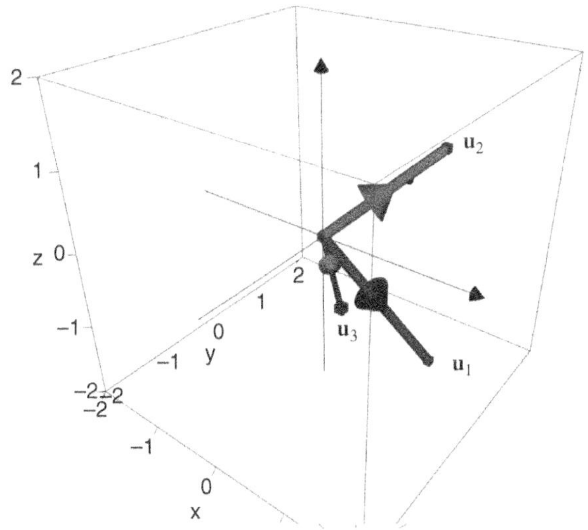

Figure 7.18

Example 7.35

Check that $\mathbf{A} = \mathbf{UDV}^T$ for the matrix $\mathbf{A} = \begin{pmatrix} 1 & 1 \\ 1 & 1 \\ -2 & 1 \end{pmatrix}$ given in the above Example 7.33.

Solution

We need to break the given matrix **A** into $\mathbf{A} = \mathbf{UDV}^T$ where **U** is 3 by 3, **D** is 3 by 2 and **V** is 2 by 2.

What are the matrices $\mathbf{U}$, $\mathbf{D}$ *and* $\mathbf{V}$ *equal to?*

By the results of Example 7.34 and above we have:

$$\mathbf{U} = (\mathbf{u}_1 \quad \mathbf{u}_2 \quad \mathbf{u}_3) = \begin{pmatrix} 1/\sqrt{6} & 1/\sqrt{3} & 1/\sqrt{2} \\ 1/\sqrt{6} & 1/\sqrt{3} & -1/\sqrt{2} \\ -2/\sqrt{6} & 1/\sqrt{3} & 0 \end{pmatrix}, \mathbf{D} = \begin{pmatrix} \sigma_1 & 0 \\ 0 & \sigma_2 \\ 0 & 0 \end{pmatrix} = \begin{pmatrix} \sqrt{6} & 0 \\ 0 & \sqrt{3} \\ 0 & 0 \end{pmatrix}$$

$$\text{and } \mathbf{V} = (\mathbf{v}_1 \quad \mathbf{v}_2) = \begin{pmatrix} 1 & 0 \\ 0 & 1 \end{pmatrix} = \mathbf{I} \quad [\text{identity matrix}]$$

By transposing the last matrix $\mathbf{V}^T = \mathbf{I}^T = \mathbf{I}$, let us check that this triple factorization actually works; that is $\mathbf{UDV}^T = \mathbf{UDI} = \mathbf{UD} = \mathbf{A}$:

$$\mathbf{UDV}^T = \mathbf{UD} = \begin{pmatrix} 1/\sqrt{6} & 1/\sqrt{3} & 1/\sqrt{2} \\ 1/\sqrt{6} & 1/\sqrt{3} & -1/\sqrt{2} \\ -2/\sqrt{6} & 1/\sqrt{3} & 0 \end{pmatrix} \times \begin{pmatrix} \sqrt{6} & 0 \\ 0 & \sqrt{3} \\ 0 & 0 \end{pmatrix} = \begin{pmatrix} 1 & 1 \\ 1 & 1 \\ -2 & 1 \end{pmatrix} = \mathbf{A}$$

Hence a non-square matrix such as **A** can be broken into $\mathbf{UDV}^T$.

7.5.3 Proof of the singular value decomposition (SVD)

Singular value decomposition can be applied to any matrix **A**.

Proposition (7.32). Let $\mathbf{v}_1, \mathbf{v}_2, \ldots, \mathbf{v}_k$ be the eigenvectors of $\mathbf{A}^T\mathbf{A}$ such that they belong to the positive eigenvalues $\lambda_1 \geq \lambda_2 \geq \lambda_3 \geq \cdots \geq \lambda_k > 0$. Then

(i) for $j = 1, 2, 3, \ldots, k$ we have $\|\mathbf{A}\mathbf{v}_j\| = \sigma_j$ where $\sigma_j = \sqrt{\lambda_j}$ is the singular value of **A**.

(ii) $\{\mathbf{A}\mathbf{v}_1, \mathbf{A}\mathbf{v}_2, \ldots, \mathbf{A}\mathbf{v}_k\}$ is an *orthogonal* set of vectors.

Note, the following:

(i) σ_j gives the size of the vector $\mathbf{A}\mathbf{v}_j$, or the length of the semi-axis ($\mathbf{u}_j$) of the ellipse.

(ii) This part means that orthogonal (perpendicular) vectors $\{\mathbf{v}_1, \mathbf{v}_2, \ldots, \mathbf{v}_k\}$ are transformed to orthogonal (perpendicular) vectors $\{\mathbf{A}\mathbf{v}_1, \mathbf{A}\mathbf{v}_2, \ldots, \mathbf{A}\mathbf{v}_k\}$ under the matrix **A**.

Proof of (i).

By the above proposition (7.30): Let **A** be any matrix. Then $\mathbf{A}^T\mathbf{A}$ is a symmetric matrix.

$\mathbf{A}^T\mathbf{A}$ is a symmetric matrix, so we can orthogonally diagonalize this because of the spectral theorem (7.28): **A** is orthogonally diagonalizable $\Leftrightarrow$ **A** is a symmetric matrix.

This means that the eigenvectors belonging to the positive eigenvalues of $\mathbf{A}^T\mathbf{A}$, given by $\mathbf{v}_1, \mathbf{v}_2, \ldots, \mathbf{v}_k$ are orthonormal (perpendicular unit) vectors. Consider $\|\mathbf{Av}_1\|^2$. Then

$$\begin{aligned}
\|\mathbf{Av}_1\|^2 = \mathbf{Av}_1 \cdot \mathbf{Av}_1 = (\mathbf{Av}_1)^T \mathbf{Av}_1 \quad & \left[\text{by (2.4) } \mathbf{u}\cdot\mathbf{v} = \mathbf{u}^T\mathbf{v}\right] \\
= \mathbf{v}_1^T \mathbf{A}^T \mathbf{Av}_1 \quad & \left[\text{because } (\mathbf{XY})^T = \mathbf{Y}^T\mathbf{X}^T\right] \\
= \mathbf{v}_1^T (\lambda_1 \mathbf{v}_1) \quad & \left[\lambda_1 \text{ and } \mathbf{v}_1 \text{ are e.values and e.vectors of } \mathbf{A}^T\mathbf{A} \right. \\
& \left. \text{so } \mathbf{A}^T\mathbf{Av}_1 = \lambda_1\mathbf{v}_1\right] \\
= \lambda_1 \mathbf{v}_1^T \mathbf{v}_1 \quad & \\
= \lambda_1 (\mathbf{v}_1 \cdot \mathbf{v}_1) = \lambda_1 \quad & \left[\text{because } \mathbf{v}_1 \text{ is normalized so } (\mathbf{v}_1 \cdot \mathbf{v}_1) = \|\mathbf{v}_1\|^2 = 1\right]
\end{aligned}$$

Taking the square root of this result $\|\mathbf{Av}_1\|^2 = \lambda_1$ gives

$$\|\mathbf{Av}_1\| = \sqrt{\lambda_1} = \sigma_1$$

Similarly for $j = 2, 3, \ldots, n$ we have $\|\mathbf{Av}_j\| = \sigma_j$. This completes our proof for part (i).

■

Proof of (ii).
Required to prove that the vectors in the set $\{\mathbf{Av}_1, \mathbf{Av}_2, \ldots, \mathbf{Av}_k\}$ are orthogonal (perpendicular). We prove that any two arbitrary *different* vectors $\mathbf{Av}_i$ and $\mathbf{Av}_j$ where $i \neq j$ in the set are orthogonal, which means that we need to show that $\mathbf{Av}_i \cdot \mathbf{Av}_j = 0$:

$$\begin{aligned}
\mathbf{Av}_i \cdot \mathbf{Av}_j = (\mathbf{Av}_i)^T \mathbf{Av}_j = \mathbf{v}_i^T \mathbf{A}^T \mathbf{Av}_j & \\
= \mathbf{v}_i^T \lambda_j \mathbf{v}_j \quad & \left[\lambda_j \text{ and } \mathbf{v}_j \text{ are e.values and e.vectors of } \mathbf{A}^T\mathbf{A}\right. \\
& \left. \text{so } \mathbf{A}^T\mathbf{Av}_j = \lambda_j\mathbf{v}_j\right] \\
= \lambda_j \left(\mathbf{v}_i^T \mathbf{v}_j\right) = \lambda_j \left(\mathbf{v}_i \cdot \mathbf{v}_j\right) = 0 \quad & \left[\mathbf{v}_i \text{ and } \mathbf{v}_j \text{ are orthogonal so}\right. \\
& \left. \mathbf{v}_i \cdot \mathbf{v}_j = 0\right]
\end{aligned}$$

We have $\mathbf{Av}_i \cdot \mathbf{Av}_j = 0$, therefore $\mathbf{Av}_i$ and $\mathbf{Av}_j$ are orthogonal (perpendicular). Hence the set of vectors $\{\mathbf{Av}_1, \mathbf{Av}_2, \ldots, \mathbf{Av}_k\}$ are orthogonal to each other.

■

Next we prove the main result in this section, which was stated above and also repeated here:

Singular Value Decomposition Theorem (7.30).

We can decompose any given matrix $\mathbf{A}$ of size m by n with positive singular values $\sigma_1 \geq \sigma_2 \geq \cdots \geq \sigma_k > 0$ where $k \leq n$, into $\mathbf{UDV}^T$, that is

$$\mathbf{A} = \mathbf{UDV}^T$$

where $\mathbf{U}$ is an m by m *orthogonal* matrix, $\mathbf{D}$ is an m by n matrix and $\mathbf{V}$ is an n by n *orthogonal* matrix.

Proof.
The size of matrix $\mathbf{A}$ is m by n. This means that the transformation T given by matrix $\mathbf{A}$ is $T : \mathbb{R}^n \to \mathbb{R}^m$ such that $T(\mathbf{v}) = \mathbf{Av}$ (Fig. 7.19).

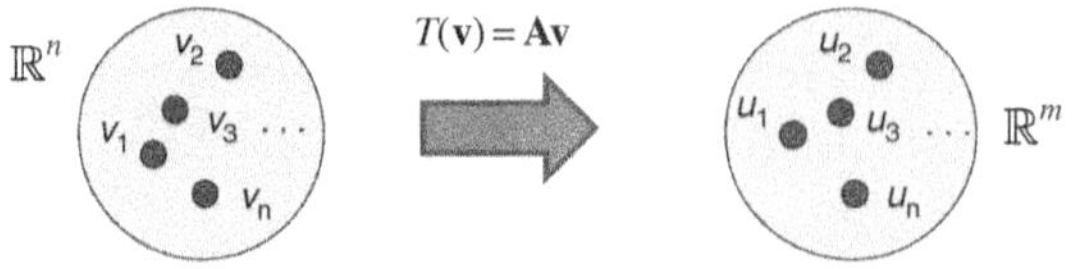

Figure 7.19

Let $\mathbf{v}_1, \mathbf{v}_2, \ldots, \mathbf{v}_k$ be the eigenvectors of $\mathbf{A}^T\mathbf{A}$ belonging to the positive eigenvalues $\lambda_1 \geq \lambda_2 \geq \cdots \geq \lambda_k > 0$ respectively. Then by the above Proposition (7.32) part (ii)

$$\mathbf{Av}_1, \mathbf{Av}_2, \ldots, \mathbf{Av}_k \quad (^*)$$

is an orthogonal set of vectors. We want to convert this into an orthonormal set which means we need to normalize each of the vectors in the list (*). Convert the first vector $\mathbf{Av}_1$ into a unit vector $\mathbf{u}_1$ say, (length of 1):

$$\mathbf{u}_1 = \frac{1}{\|\mathbf{Av}_1\|}\mathbf{Av}_1 = \frac{1}{\sigma_1}\mathbf{Av}_1 \qquad \left[\text{by (7.32) part(i) } \|\mathbf{Av}_j\| = \sigma_j\right]$$

Similarly converting the remaining vectors $\mathbf{Av}_2, \ldots, \mathbf{Av}_k$ into unit vectors we have

$$\mathbf{u}_j = \frac{1}{\sigma_j}\mathbf{Av}_j \text{ for } j = 2, 3, 4, \ldots, k$$

The vectors in $S = \{\mathbf{u}_1, \mathbf{u}_2, \ldots, \mathbf{u}_k\}$ constitute an orthonormal set of vectors. We need to produce a matrix $\mathbf{U}$ which is of size m by m. If $k < m$ then the vectors in this set $S = \{\mathbf{u}_1, \mathbf{u}_2, \ldots, \mathbf{u}_k\}$ are the first k vectors of the matrix $\mathbf{U}$. However, we need m vectors because

$$\mathbf{U} = (\mathbf{u}_1 \ \ \mathbf{u}_2 \ \ \cdots \ \ \mathbf{u}_k \ \ \mathbf{u}_{k+1} \ \ \cdots \ \ \mathbf{u}_m)$$

We extend the above set S to an orthonormal set $S' = \left\{\mathbf{u}_1, \mathbf{u}_2, \ldots, \mathbf{u}_k, \mathbf{u}_{k+1}, \ldots, \mathbf{u}_m\right\}$. This S' is an orthonormal (perpendicular unit) basis for R^m. Let

$$\mathbf{U} = (\mathbf{u}_1 \ \ \mathbf{u}_2 \ \ \cdots \ \ \mathbf{u}_m)$$

Multiplying the above result $\mathbf{u}_j = \frac{1}{\sigma_j}\mathbf{Av}_j$ by σ_j for $j = 1, 2, 3, \ldots, k$ gives

$$\sigma_j\mathbf{u}_j = \mathbf{Av}_j \quad (\dagger)$$

The remaining singular values are zero, that is $\sigma_{k+1} = \sigma_{k+2} = \cdots = \sigma_n = 0$. We have

$$\mathbf{Av}_j = \sigma_j \mathbf{u}_j = 0\mathbf{u}_j \text{ for } j = k+1, \ldots, n \qquad (\dagger\dagger)$$

Collecting all these together we have

$$\mathbf{Av}_1 = \sigma_1 \mathbf{u}_1,\ \mathbf{Av}_2 = \sigma_2 \mathbf{u}_2, \ldots, \mathbf{Av}_k = \sigma_k \mathbf{u}_k,\ \mathbf{Av}_{k+1} = 0\mathbf{u}_{k+1}, \ldots, \mathbf{Av}_n = 0\mathbf{u}_n$$

In matrix form we have

$$\begin{aligned}
\mathbf{AV} &= \mathbf{A}\left(\mathbf{v}_1 \cdots \mathbf{v}_k \quad \mathbf{v}_{k+1} \cdots \mathbf{v}_n\right) && \left[\mathbf{V} \text{ is an n by n orthogonal matrix}\right] \\
&= \left(\mathbf{Av}_1 \cdots \mathbf{Av}_k \quad \mathbf{Av}_{k+1} \cdots \mathbf{Av}_n\right) \\
&= \left(\sigma_1 \mathbf{u}_1 \cdots \sigma_k \mathbf{u}_k \quad 0\mathbf{u}_{k+1} \cdots 0\mathbf{u}_n\right) && \left[\text{by } (\dagger) \text{ and } (\dagger\dagger)\right]
\end{aligned}$$

The scalars $\sigma_1, \sigma_2, \ldots$ in front of the **u**'s can be placed in the diagonal-like matrix **D** and the vectors **u**'s into the above matrix $\mathbf{U} = \left(\mathbf{u}_1\ \mathbf{u}_2\ \cdots\ \mathbf{u}_m\right)$.

Hence the above derivation

$$\mathbf{AV} = \left(\ \sigma_1 \mathbf{u}_1 \ \cdots \ \sigma_k \mathbf{u}_k \ \ 0\mathbf{u}_{k+1} \ \cdots \ 0\mathbf{u}_n\ \right)$$

can be written as:

$$\left(\sigma_1 \mathbf{u}_1 \cdots \sigma_k \mathbf{u}_k \quad 0\mathbf{u}_{k+1} \cdots 0\mathbf{u}_n\right) = \underbrace{\left(\mathbf{u}_1 \quad \mathbf{u}_2 \cdots \mathbf{u}_m\right)}_{=\mathbf{U}} \underbrace{\left(\begin{array}{ccc|c} \sigma_1 & & O & \\ & \ddots & & O \\ O & & \sigma_k & \\ \hline & O & & O \end{array}\right)}_{=\mathbf{D}}$$

$$= \mathbf{UD} = \mathbf{AV}$$

Since **V** is an orthogonal matrix so $\mathbf{V}^{-1} = \mathbf{V}^T$. Right multiplying both sides of $\mathbf{UD} = \mathbf{AV}$ by $\mathbf{V}^T$ gives us our required result $\mathbf{UDV}^T = \mathbf{A}$.

■

Proposition (7.33). Let **A** be an m by n matrix with factorization given by $\mathbf{A} = \mathbf{UDV}^T$. Let matrix **A** have $k \leq n$ positive singular values. Then we have the following:

(a) The set of vectors $\{\mathbf{u}_1, \mathbf{u}_2, \ldots, \mathbf{u}_k\}$ form an orthonormal basis for the column space of matrix **A**.

(b) The set of vectors $\{\mathbf{v}_1, \mathbf{v}_2, \ldots, \mathbf{v}_k\}$ form an orthonormal basis for the row space of matrix **A**.

(c) The set of vectors $\left\{\mathbf{v}_{k+1}, \mathbf{v}_{k+2}, \ldots, \mathbf{v}_n\right\}$ form an orthonormal basis for the null space of matrix **A**.

Proof – Exercises 7.5.

■

Summary

We can break any matrix $\mathbf{A}$ into a triple factorization $\mathbf{A} = \mathbf{U} \times \mathbf{D} \times \mathbf{V}^T$.

EXERCISES 7.5

(Brief solutions at end of book. Full solutions available at <http://www.oup.co.uk/companion/singh>.)

1. Determine the matrices $\mathbf{U}$, $\mathbf{D}$ and $\mathbf{V}$ such that $\mathbf{A} = \mathbf{UDV}^T$ for the following:

(a) $\mathbf{A} = \begin{pmatrix} 1 & 0 \\ 0 & 2 \end{pmatrix}$ (b) $\mathbf{A} = \begin{pmatrix} 1 & 4 \\ 4 & 7 \end{pmatrix}$ (c) $\mathbf{A} = \begin{pmatrix} 1 & 0 \\ 0 & 1 \\ 1 & 2 \end{pmatrix}$

(d) $\mathbf{A} = \begin{pmatrix} 1 & 0 & 1 \\ 0 & 1 & 2 \end{pmatrix}$ (e) $\mathbf{A} = \begin{pmatrix} 1 & 1 \\ 3 & 3 \end{pmatrix}$ (f) $\mathbf{A} = \begin{pmatrix} 1 & 1 & 1 \\ 3 & 3 & 3 \end{pmatrix}$

2. Prove Proposition (7.31).
3. Let matrix $\mathbf{A}$ have k positive singular values. Show that the rank of matrix $\mathbf{A}$ is k.
   Hint: You may use the following result:
   If matrix $\mathbf{X}$ is invertible then $rank(\mathbf{XA}) = rank(\mathbf{A})$ and $rank(\mathbf{AX}) = rank(\mathbf{A})$.
4. Prove Proposition (7.33).
5. Prove that the singular values $\sigma_1, \sigma_2, \ldots, \sigma_n$ of an m by n matrix $\mathbf{A}$ are unique.
6. Prove that the column vectors of the orthogonal matrix $\mathbf{U}$ in $\mathbf{A} = \mathbf{UDV}^T$ are the eigenvectors of $\mathbf{AA}^T$.
7. Prove that the singular values of $\mathbf{A}$ and $\mathbf{A}^T$ are identical.
8. Let $T : \mathbb{R}^n \to \mathbb{R}^m$ be a linear transformation given by $T(\mathbf{x}) = \mathbf{Ax}$, where $\mathbf{A}$ is an m by n matrix. Let the singular value decomposition $\mathbf{A} = \mathbf{UDV}^T$ with $k \le n$ positive singular values of matrix $\mathbf{A}$. Prove the following results:

   (a) The set of vectors $\{\mathbf{u}_1, \mathbf{u}_2, \ldots, \mathbf{u}_k\}$ form an orthonormal basis for the range of T.

   (b) The set of vectors $\{\mathbf{v}_{k+1}, \mathbf{v}_{k+2}, \ldots, \mathbf{v}_n\}$ form an orthonormal basis for the kernel of T.

MISCELLANEOUS EXERCISES 7

(Brief solutions at end of book. Full solutions available at <http://www.oup.co.uk/companion/singh>.)

In this exercise you may check your numerical answers using MATLAB.

7.1. If $\mathbf{A} = \begin{pmatrix} 1 & 1 \\ 0 & 3 \end{pmatrix}$.

   (a) Find all the eigenvalues of $\mathbf{A}$.

   (b) Find a non-singular matrix $\mathbf{Q}$ and a diagonal matrix $\mathbf{D}$ such that $\mathbf{Q}^{-1}\mathbf{AQ} = \mathbf{D}$ (that is $\mathbf{A} = \mathbf{QDQ}^{-1}$).

(c) For the matrix $\mathbf{A}$ find $\mathbf{A}^5$.

Purdue University, USA

7.2. Let $\mathbf{A} = \begin{pmatrix} 4 & 5 \\ -3 & -4 \end{pmatrix}$. Compute $\mathbf{A}^{1\,000\,001}$.

Harvey Mudd College, California, USA

7.3. Find the eigenvalues and bases for the corresponding eigenspaces for the matrix

$$\mathbf{A} = \begin{pmatrix} 0 & 2 & 2 \\ 2 & 0 & 2 \\ 2 & 2 & 0 \end{pmatrix}$$

Harvey Mudd College, California, USA

7.4. Answer each of the following by filling in the blank. No explanation is necessary.

(a) Let $\mathbf{A}$ be an invertible $n \times n$ matrix with eigenvalue λ. Then ______ is an eigenvalue of $\mathbf{A}^{-1}$.

(b) An $n \times n$ matrix $\mathbf{A}$ is diagonalizable if and only if $\mathbf{A}$ has n ___.

Illinois State University, USA

7.5. Let $\mathbf{A}$ be the matrix

$$\begin{pmatrix} 7 & 5 \\ 3 & -7 \end{pmatrix}$$

(a) Find matrices S and Λ such that $\mathbf{A}$ has factorization of the form

$$\mathbf{A} = \mathbf{S}\Lambda\mathbf{S}^{-1}$$

where $\mathbf{S}$ is invertible and Λ is diagonal: $\Lambda = \text{diag}(\lambda_1, \lambda_2)$.

(b) Find a matrix $\mathbf{B}$ such that $\mathbf{B}^3 = \mathbf{A}$: (Hint: first find such a matrix for Λ. Then use the formula above.)

Massachusetts Institute of Technology, USA

7.6. (a) Consider the matrix

$$\mathbf{A} = \begin{pmatrix} 4 & -3 \\ 1 & 0 \end{pmatrix}$$

(i) Find the eigenvalues and eigenvectors of $\mathbf{A}$.

(ii) Find a matrix $\mathbf{P}$ such that $\mathbf{P}^{-1}\mathbf{A}\mathbf{P}$ is diagonal.

(iii) Find the eigenvalues and the determinant of $\mathbf{A}^{2008}$.

(b) Consider the matrix

$$\mathbf{B} = \begin{pmatrix} 1 & 1 & 0 \\ 0 & 1 & 0 \\ 0 & 0 & 0 \end{pmatrix}$$

Is $\mathbf{B}$ diagonalisable? Justify your answer.

Loughborough University, UK

7.7. Do **one** of the following.

(a) Is the matrix $\mathbf{A} = \begin{bmatrix} 1 & 0 \\ 10 & 2 \end{bmatrix}$ diagonalizable? If not, explain why not. If so, find an invertible matrix $\mathbf{S}$ for which $\mathbf{S}^{-1}\mathbf{AS}$ is diagonal.

(b) The matrices $\mathbf{A} = \begin{bmatrix} 2 & 0 \\ 0 & 3 \end{bmatrix}$ and $\mathbf{B} = \begin{bmatrix} 3 & 0 \\ 0 & 2 \end{bmatrix}$ are similar. Exhibit a matrix $\mathbf{S}$ for which $\mathbf{B} = \mathbf{S}^{-1}\mathbf{AS}$.

University of Puget Sound, USA

7.8. Consider the matrix

$$\mathbf{B} = \begin{pmatrix} 1 & 1 \\ 1 & -1 \end{pmatrix}$$

Do there exist matrices $\mathbf{P}$ and $\mathbf{D}$ such that $\mathbf{B} = \mathbf{PDP}^T$ where $\mathbf{P}^{-1} = \mathbf{P}^T$ and $\mathbf{D}$ is a diagonal matrix? Why? If these matrices exist then write down a possible $\mathbf{P}$ and the corresponding $\mathbf{D}$.

New York University, USA

7.9. Consider the 3×3 matrix $\mathbf{A} = \begin{pmatrix} 4 & 2 & -2 \\ 0 & 3 & 1 \\ 0 & 1 & 3 \end{pmatrix}$.

Already computed are the eigenpairs $\left(2, \begin{pmatrix} 2 \\ -1 \\ 1 \end{pmatrix}\right), \left(4, \begin{pmatrix} 1 \\ 0 \\ 0 \end{pmatrix}\right)$.

(a) Find the remaining eigenpairs of $\mathbf{A}$.

(b) Display an invertible matrix $\mathbf{P}$ and a diagonal matrix $\mathbf{D}$ such that $\mathbf{AP} = \mathbf{PD}$.

University of Utah, USA (part question)

7.10. Prove that, if λ_1 and λ_2 are *distinct* eigenvalues of a symmetric matrix $\mathbf{A}$, then the corresponding eigenspaces are orthogonal.

Harvey Mudd College, California, USA

7.11. (a) Find the eigenvalues and corresponding eigenvectors of $\mathbf{A} = \begin{bmatrix} 6 & 2 \\ 2 & 3 \end{bmatrix}$.

(b) Determine which of the following vectors

$$\vec{v} = \begin{bmatrix} 1 \\ 1 \\ 1 \\ -1 \end{bmatrix}, \vec{w} = \begin{bmatrix} 1 \\ -1 \\ -1 \\ 2 \end{bmatrix}$$

is an eigenvector of the matrix $\begin{bmatrix} 1 & 1 & 1 & 1 \\ 1 & -1 & 1 & -1 \\ 1 & 1 & -1 & -1 \\ 1 & -1 & -1 & 1 \end{bmatrix}$ and find the corresponding eigenvalue.

University of New Brunswick, Canada

7.12. (a) Find the eigenvalues and corresponding eigenvectors for

$$\mathbf{A} = \begin{bmatrix} 1 & 1 & -1 \\ 1 & 1 & 1 \\ 1 & 1 & 1 \end{bmatrix}$$

(b) Is **A** diagonalizable? Give reasons for your answer.

University of New Brunswick, Canada

7.13. (a) (i) Show that the eigenvalues and corresponding eigenvectors of

$$\mathbf{A} = \begin{pmatrix} 1 & 1 & 1 \\ 1 & 1 & 1 \\ 1 & 1 & 1 \end{pmatrix}$$

are given by

$$\lambda_1 = 0, \mathbf{u} = \begin{pmatrix} 1 \\ 1 \\ -2 \end{pmatrix}, \lambda_2 = 0, \mathbf{v} = \begin{pmatrix} 1 \\ -1 \\ 0 \end{pmatrix} \text{ and } \lambda_3 = 3, \mathbf{w} = \begin{pmatrix} 1 \\ 1 \\ 1 \end{pmatrix}$$

(ii) Find an orthogonal matrix **Q** which diagonalises the matrix **A**.

(iii) Use the Cayley–Hamilton theorem or otherwise to find $\mathbf{A}^3$.

(b) Let **B** be an n by n real matrix. Prove that **B** and $\mathbf{B}^T$ have the same eigenvalues.

University of Hertfordshire, UK

7.14. (a) Define what is meant by an eigenvector and an eigenvalue for a real $n \times n$ matrix.

(b) Let $\mathbf{A} = \begin{pmatrix} 0 & 2 & 1 \\ 1 & 1 & 1 \\ 1 & 2 & 0 \end{pmatrix}$. Show that the vector $\begin{pmatrix} 1 \\ 1 \\ 1 \end{pmatrix}$ is an eigenvector for **A**. What is the corresponding eigenvalue?

(c) Show that the matrix **A** given in (b) is diagonalizable and hence find an invertible 3×3 matrix **P** (and $\mathbf{P}^{-1}$) such that $\mathbf{P}^{-1}\mathbf{A}\mathbf{P}$ is diagonal.

City University, London, UK

7.15. (a) Let **A** be the matrix

$$\mathbf{A} = \begin{pmatrix} 1 & -2 & 2 \\ 8 & 11 & -8 \\ 4 & 4 & -1 \end{pmatrix}$$

By finding a basis of eigenvectors, determine an invertible matrix **P** and its inverse $\mathbf{P}^{-1}$ such that $\mathbf{P}^{-1}\mathbf{A}\mathbf{P}$ is diagonal.

(b) State the Cayley–Hamilton theorem and verify it for the above matrix.

City University, London, UK

7.16. Let $\mathbf{A} \in M_4(\mathbb{R})$ be the following matrix

$$\mathbf{A} = \begin{pmatrix} 0 & -3 & 0 & 0 \\ 1 & -1 & 0 & 0 \\ 1 & -1 & 0 & -9 \\ 1 & -1 & 1 & 10 \end{pmatrix}$$

(a) Show that the characteristic polynomial of $\mathbf{A}$ is given by

$$P_A(t) = \left(t^2 + t + 3\right)\left(t^2 - 10t + 9\right)$$

(b) Compute all real eigenvalues of $\mathbf{A}$. Choose one real eigenvalue λ of $\mathbf{A}$ and find a basis of its eigenspace E_λ.

Jacobs University, Germany

7.17. Find the eigenvalues of the matrix $\mathbf{A} = \begin{pmatrix} 0 & 1 & 2 & 0 & 0 \\ -1 & -1 & 1 & 0 & 1 \\ 0 & 0 & 1 & 0 & 0 \\ 0 & 0 & 2 & -3 & 0 \\ 1 & 2 & 3 & 4 & 0 \end{pmatrix}$. To save time, **do not** find the eigenvectors.

University of Utah, USA

7.18. (Calculators are **not allowed** for this question). Let $\mathbf{A} = \begin{pmatrix} 1 & 1 & 0 \\ 1 & 2 & 1 \\ 0 & 1 & 1 \end{pmatrix}$.

(a) Find the eigenvalues for $\mathbf{A}$.

(b) For $\mathbf{A}$, find an eigenvector for each of the eigenvalues. To make it easier to grade, choose eigenvectors with integer coordinates where the integers are as small as possible.

(c) Use your eigenvectors to make a basis of $\mathbb{R}^3$. Choose the first basis vector to be the eigenvector associated with the largest eigenvalue and the third basis vector to be the eigenvector associated with the smallest eigenvalue. Call this basis β. We have a linear transformation given to us by $\mathbf{A}$. What is the matrix $\mathbf{D}$ when we use coordinates from this new basis, $\mathbf{D} : [x]_\beta \to [\mathbf{A}\mathbf{x}]_\beta$?

(d) We know that there is a matrix $\mathbf{S}$ such that $\mathbf{S}^{-1}\mathbf{A}\mathbf{S} = \mathbf{D}$. Find $\mathbf{S}$.

(e) Find $\mathbf{S}^{-1}$.

(f) Modify the basis β so that you can orthogonally diagonalize $\mathbf{A}$.

(g) Find the new $\mathbf{S}$ needed to orthogonally diagonalize $\mathbf{A}$.

(h) Find the inverse of this last $\mathbf{S}$.

Johns Hopkins University, USA

7.19. (a) If $a \neq c$, find the eigenvalue matrix Λ and eigenvector matrix $\mathbf{S}$ in

$$\mathbf{A} = \begin{bmatrix} a & b \\ 0 & c \end{bmatrix} = \mathbf{S}\Lambda\mathbf{S}^{-1}$$

(b) Find the *four entries* in the matrix $\mathbf{A}^{1000}$.

Massachusetts Institute of Technology, USA

7.20. (a) Define what is meant by saying that an $n \times n$ matrix $\mathbf{A}$ is *diagonalizable.*

(b) Define what is meant by an eigenvector and an eigenvalue of an $n \times n$ matrix $\mathbf{A}$.

(c) Define the algebraic multiplicity and the geometric multiplicity of an eigenvalue λ_0 of a matrix $\mathbf{A}$.

[Note that geometric multiplicity of an eigenvalue is **not** discussed in this book but have a look at other sources to find the definition of this.]

(d) Suppose that a 3×3 matrix $\mathbf{A}$ has characteristic polynomial $(\lambda - 1)(\lambda - 2)^2$. What is the algebraic multiplicity of each eigenvalue of $\mathbf{A}$? State a necessary and sufficient condition for the diagonalizability of $\mathbf{A}$ using geometric multiplicity.

(e) Each of the following matrices has characteristic polynomial $(\lambda - 1)(\lambda - 2)^2$.

(i) $\begin{bmatrix} 1 & -2 & 1 \\ 1 & 1 & 2 \\ 1 & 0 & 3 \end{bmatrix}$ (ii) $\begin{bmatrix} 1 & -1 & 1 \\ -1 & 1 & 1 \\ -1 & -1 & 3 \end{bmatrix}$

Determine whether each of the matrices is diagonalizable. In each case, if the matrix is diagonalizable, find a diagonalizing matrix.

University of Manchester, UK

7.21. (a) Define what it means for vectors $\mathbf{w}_1, \mathbf{w}_2, \mathbf{w}_3$ in $\mathbb{R}^3$ to be orthonormal.

(b) Apply the Gram–Schmidt process to the vectors

$$\mathbf{v}_1 = \begin{bmatrix} 0 \\ 1 \\ 1 \end{bmatrix}, \mathbf{v}_2 = \begin{bmatrix} 1 \\ 1 \\ 0 \end{bmatrix}, \mathbf{v}_3 = \begin{bmatrix} 5 \\ 4 \\ 6 \end{bmatrix}$$

in $\mathbb{R}^3$ to find an orthonormal set in $\mathbb{R}^3$.

(c) Consider the matrix

$$\mathbf{A} = \begin{pmatrix} 3 & 2 & 4 \\ 2 & 0 & 2 \\ 4 & 2 & 3 \end{pmatrix}$$

(i) Obtain the characteristic polynomial and find the eigenvalues of $\mathbf{A}$.

(ii) Find a complete set of linearly independent eigenvectors corresponding to each eigenvalue.

(d) Let the non-zero vectors $\mathbf{v}_1, \mathbf{v}_2, \mathbf{v}_3 \in \mathbb{R}^3$ be an orthogonal set. Prove that they are linearly independent.

University of Southampton, UK

7.22. (a) Define the terms orthogonal and orthonormal applied to a set of vectors in a vector space on which an inner product is defined.

(b) State the relationship between an *orthogonal matrix* and its transpose. Prove that the set of columns of an orthogonal matrix forms an orthonormal set of vectors.

(c) (i) Show that $(1, 0, 1, 0)$ and $(1, 0, -1, 0)$ are eigenvectors of the matrix

$$\mathbf{A} = \begin{pmatrix} 5 & 0 & -1 & 0 \\ 0 & 1 & 0 & -1 \\ -1 & 0 & 5 & 0 \\ 0 & -1 & 0 & 1 \end{pmatrix}$$

and find the corresponding eigenvalues.

(ii) Find the two other eigenvalues and corresponding eigenvectors of **A**.

(iii) Find matrices $\mathbf{P}, \mathbf{Q}, \Lambda$ such that $\mathbf{PQ} = \mathbf{I}$ and $\mathbf{PAQ} = \Lambda$ is diagonal.

Queen Mary, University of London, UK

[The wording has been modified for the next two questions so that it is compatible with the main text. You will need to look at the website material for chapter 7 to attempt these questions.]

7.23. Express the following quadratic in its diagonal form:

$$3x^2 + 4xy + 6y^2 = aX^2 + bY^2$$

Write X and Y in terms of x and y.

Columbia University, USA

7.24. Express the following quadratic in its diagonal form:

$$2xy + 4xz + 4yz + 3z^2 = aX^2 + bY^2 + cZ^2$$

Write X, Y and Z in terms of x, y and z.

Columbia University, USA

Brief Solutions

SOLUTIONS TO EXERCISES 1.1

1. (a) Linear (b) Not linear (c) Not linear (d) Not linear
   (e) Linear (f) Linear (g) Linear (h) Linear
   (i) Linear (j) Not linear (k) Linear (l) Linear
   (m) Linear

2. (a) $x = 1,\ y = 1$ (b) $x = 1,\ y = -1$ (c) $x = -29,\ y = -31$
   (d) $x = \frac{3}{4},\ y = \frac{1}{4}$ (e) $x = \frac{3}{4\pi},\ y = -\frac{1}{4}$ (f) $x = \frac{1}{e},\ y = -\frac{1}{e}$

3. (a) $x = y = z = 1$ (b) $x = -2,\ y = 1$ and $z = -3$
   (c) $x = 3,\ y = -4$ and $z = \frac{1}{2}$ (d) $x = 3,\ y = -3$ and $z = 2$

4. (a) Unique (b) Infinite (c) No solution (d) No solution
   (e) No solution (f) Unique

5. (a) Unique (b) Unique (c) No solution

SOLUTIONS TO EXERCISES 1.2

1. (a) $x = 6$ and $y = 1$ (b) $x = 2,\quad y = -4$ and $z = -3$
   (c) $x = 1,\ y = 3$ and $z = 2$ (d) $x = 1/2,\quad y = 1/8$ and $z = 1/4$
   (e) $x = -6.2,\quad y = 30$ and $z = -10$

2. (a) $x = 1,\quad y = 4, z = 1$ (b) $x = -1, y = 2, z = -1$ (c) $x = y = z = 1$

3. (a) $x = 1,\quad y = 2$ and $z = 3$ (b) $x = -2,\quad y = 1$ and $z = -1$
   (c) $x = -1/2,\quad y = 1$ and $z = -2$ (d) $x = -3,\quad y = 1$ and $z = 1/2$

SOLUTIONS TO EXERCISES 1.3

2. (e) 1 (f) 1 (g) 2 (h) 5

3. (a) 15 (b) 15 (c) 14 (d) 30

5. (a) $\begin{pmatrix} 2 \\ 0 \end{pmatrix}$ (b) $\begin{pmatrix} -4 \\ 2 \end{pmatrix}$ (c) $\begin{pmatrix} 1/2 \\ 1/2 \end{pmatrix}$ (d) $\begin{pmatrix} -5/2 \\ 3/2 \end{pmatrix}$ (e) $\begin{pmatrix} 0 \\ 2/3 \end{pmatrix}$

6. (a) $\begin{pmatrix} 0 \\ 3 \end{pmatrix}$ (b) $\begin{pmatrix} 0 \\ 3/2 \end{pmatrix}$ (c) $\begin{pmatrix} 2 \\ -1/2 \end{pmatrix}$ (d) $\begin{pmatrix} -2 \\ 1/2 \end{pmatrix}$

7. $x = -\frac{1}{2}, y = -\frac{9}{2}$

9. (a) $\begin{pmatrix} 4 \\ 4 \\ 10 \end{pmatrix}$ (b) $\begin{pmatrix} -15 \\ 25 \\ 40 \end{pmatrix}$ (c) $\begin{pmatrix} 36 \\ 4 \\ 28 \end{pmatrix}$ (d) $\begin{pmatrix} -24 \\ 8 \\ 2 \end{pmatrix}$ (e) $\begin{pmatrix} -13 \\ -21 \\ -48 \end{pmatrix}$

10. (a) $\begin{pmatrix} -10 \\ 1 \\ 12 \end{pmatrix}$ (b) $\begin{pmatrix} -8 \\ 3 \\ -4 \end{pmatrix}$ (c) $\begin{pmatrix} -21 \\ 7 \\ 6 \end{pmatrix}$ (d) $\begin{pmatrix} 17 \\ -11 \\ 26 \end{pmatrix}$

11. $x = 7$, $y = -11$ and $z = 1$

13. (a) $\begin{pmatrix} 2 \\ 0 \\ 8 \\ 3 \end{pmatrix}$ (b) $\begin{pmatrix} -4 \\ 6 \\ -4 \\ -3 \end{pmatrix}$ (c) $\begin{pmatrix} -7 \\ 4 \\ -5 \\ 4 \end{pmatrix}$ (d) $\begin{pmatrix} x-1 \\ y+6 \\ z-1 \\ a-6 \end{pmatrix}$

$x = -2, y = 0, z = -8$ and $a = -3$

SOLUTIONS TO EXERCISES 1.4

1. (a) $\begin{pmatrix} 7 & 1 \\ 8 & 2 \end{pmatrix}$ (b) $\begin{pmatrix} 7 & 1 \\ 8 & 2 \end{pmatrix}$ (c) $\begin{pmatrix} 18 & -3 \\ 15 & 9 \end{pmatrix}$

(d) $\begin{pmatrix} 18 & -3 \\ 15 & 9 \end{pmatrix}$ (e) $\begin{pmatrix} 15 & 4 \\ 19 & 3 \end{pmatrix}$ (h) $\begin{pmatrix} 1 \\ -4 \end{pmatrix}$

(i) $\begin{pmatrix} -7 \\ -2 \end{pmatrix}$ (k) $\begin{pmatrix} 17 \\ -8 \end{pmatrix}$. Parts (f), (g) and (j) **cannot** be evaluated.

2. (a) $\begin{pmatrix} 0 & 0 & 0 \\ 0 & 0 & 0 \end{pmatrix}$ (b) **A** (c) $\begin{pmatrix} 42 \\ 73 \\ 81 \end{pmatrix}$

(e) $\begin{pmatrix} 12 & 10 & 8 \\ 9 & 5 & 2 \\ 3 & 3 & 6 \end{pmatrix}$ (f) $\begin{pmatrix} 12 & 10 & 8 \\ 9 & 5 & 2 \\ 3 & 3 & 6 \end{pmatrix}$ (h) $\begin{pmatrix} 1 \\ 2 \\ 3.5 \end{pmatrix}$

(i) $\begin{pmatrix} 44 & 44 & 3 \\ 72 & 71 & 1 \\ 29 & 12 & -43 \end{pmatrix}$ (j) $\begin{pmatrix} 111 & 41 & 127 \\ 24 & -13 & -20 \\ -1 & -12 & -26 \end{pmatrix}$ (k) $\begin{pmatrix} -67 & 3 & -124 \\ 48 & 84 & 21 \\ 30 & 24 & -17 \end{pmatrix}$

(d) (g) and (l) Impossible

3. (a) $\begin{pmatrix} 2 & 4 \\ 3 & 9 \end{pmatrix}$ (b) $\begin{pmatrix} 6 & 7 \\ 2 & 3 \end{pmatrix}$ (c) $\begin{pmatrix} a & b \\ c & d \end{pmatrix}$ (d) $\begin{pmatrix} 2 & 3 & 6 \\ 1 & 4 & 5 \\ 0 & 9 & 7 \end{pmatrix}$

The result is always the first (or left hand) matrix.

4. (a) $\begin{pmatrix} 1 & 0 \\ 0 & 1 \end{pmatrix}$ (b) $\begin{pmatrix} 1 & 0 \\ 0 & 1 \end{pmatrix}$ (c) $\begin{pmatrix} 1 & 0 \\ 0 & 1 \end{pmatrix}$

5. and 6. $\begin{pmatrix} 0 & 0 \\ 0 & 0 \end{pmatrix}$ and $\begin{pmatrix} 0 & 0 & 0 \\ 0 & 0 & 0 \\ 0 & 0 & 0 \end{pmatrix}$. Multiplying two non-zero matrices gives a zero matrix.

7. $\begin{pmatrix} a_{\boxed{11}} & a_{\boxed{12}} & a_{\boxed{13}} & a_{\boxed{14}} \\ a_{\boxed{21}} & a_{\boxed{22}} & a_{\boxed{23}} & a_{\boxed{24}} \end{pmatrix}$

8. (a) $\mathbf{A}^2 = \mathbf{A}^3 = \mathbf{A}^4 = \begin{pmatrix} 1 & 0 \\ 0 & 1 \end{pmatrix}$ and $\mathbf{x}_n = \begin{pmatrix} 1 & 0 \\ 0 & 1 \end{pmatrix}\mathbf{x} = \mathbf{x}$

(b) $\mathbf{A}^2 = \begin{pmatrix} -1 & 0 \\ 0 & -1 \end{pmatrix}$, $\mathbf{A}^3 = \begin{pmatrix} 0 & -1 \\ 1 & 0 \end{pmatrix}$, $\mathbf{A}^4 = \begin{pmatrix} 1 & 0 \\ 0 & 1 \end{pmatrix}$ and $\mathbf{x}_n = \mathbf{A}^r\mathbf{x}$ where r is the reminder after dividing n by 4. If the reminder $r = 0$ then $\mathbf{A}^r\mathbf{x} = \mathbf{x}$.

(c) $\mathbf{A}^2 = \mathbf{A}^3 = \mathbf{A}^4 = \mathbf{A}$ and $\mathbf{x}_n = \mathbf{A}\mathbf{x}$.

9. $\mathbf{A} = \begin{pmatrix} 1 & 2 & 2 & 1 \\ 2 & 2 & 4 & 4 \end{pmatrix}$ (a) $\begin{pmatrix} -3 & -2 & -2 & -3 \\ 2 & 2 & 4 & 4 \end{pmatrix}$ (b) $\begin{pmatrix} 3 & 6 & 6 & 3 \\ 6 & 6 & 12 & 12 \end{pmatrix}$

(c) $\begin{pmatrix} 2 & 2 & 4 & 4 \\ -1 & -2 & -2 & -1 \end{pmatrix}$ (d) $\begin{pmatrix} -1 & -2 & -2 & -1 \\ 2 & 2 & 4 & 4 \end{pmatrix}$

10. $\mathbf{AF} = \begin{pmatrix} 1.2 & 1.6 & 2.6 & 2.52 & 1.92 & 1.8 & 2.4 & 2.32 & 1.72 & 1.6 \\ 1 & 3 & 3 & 2.6 & 2.6 & 2 & 2 & 1.6 & 1.6 & 1 \end{pmatrix}$

11. (a) $x = 0,\quad y = 0$ (b) $x = 0,\quad y = 0$

(c) $x = -4r$, $y = r$ where r is any real number.

12. (i) $\mathbf{u} - 2\mathbf{v}$ (ii) $\frac{1}{4}(5\mathbf{v} - 2\mathbf{u})$ (iii) No (iv) $2\mathbf{u} - 7\mathbf{v}$

SOLUTIONS TO EXERCISES 1.5

1. (a) $\begin{pmatrix} 1 & 1 & 12 \\ 2 & -10 & 13 \end{pmatrix}$ (b) $\begin{pmatrix} 10 & 6 & 20 \\ -4 & -11 & 19 \end{pmatrix}$ (c) $\begin{pmatrix} 10 & 6 & 20 \\ -4 & -11 & 19 \end{pmatrix}$

(d) $\begin{pmatrix} 10 & 6 & 20 \\ -4 & -11 & 19 \end{pmatrix}$ (e) $\begin{pmatrix} 9 & 5 & 8 \\ -6 & -1 & 6 \end{pmatrix}$ (f) $\begin{pmatrix} -5 & 15 & 35 \\ 10 & -45 & 30 \end{pmatrix}$

(g) Same as part (f). (h) $\begin{pmatrix} 0 & 0 & 0 \\ 0 & 0 & 0 \end{pmatrix}$

2. (a) $\begin{pmatrix} -5 & 1 & 2 \\ -1 & 3 & -2 \end{pmatrix}$ (b) $\begin{pmatrix} 4 & -3 & -5 \\ -3 & -8 & -4 \end{pmatrix}$

(c) $\begin{pmatrix} 9 & -4 & -7 \\ -2 & -11 & -2 \end{pmatrix}$ and $\begin{pmatrix} 11 & 0 & -1 \\ 6 & -1 & 10 \end{pmatrix}$

3. (a) $\begin{pmatrix} -8 & 8 & 0 \\ 12 & 5 & -26 \end{pmatrix}$ (b) Impossible (c) $\begin{pmatrix} -53 & 52 & -41 \\ -77 & 98 & -116 \end{pmatrix}$

(d) $\begin{pmatrix} -53 & 52 & -41 \\ -77 & 98 & -116 \end{pmatrix}$ (e) Impossible (f) Impossible (g) $\mathbf{O}_{33}$ (h) $\mathbf{O}_{33}$

(i) $\mathbf{O}_{33}$ (j) Impossible (k) Impossible

4. We have $\mathbf{AI} = \mathbf{IA} = \begin{pmatrix} 1 & 2 & 3 \\ 4 & 5 & 6 \\ 7 & 8 & 9 \end{pmatrix}$. Note, that $\mathbf{AI} = \mathbf{IA} = \mathbf{A}$.

5. $(\mathbf{AB})_{11} = 27$, $(\mathbf{AB})_{12} = 30$, $(\mathbf{AB})_{21} = 60$, $(\mathbf{AB})_{22} = 66$ and $\mathbf{AB} = \begin{pmatrix} 27 & 30 \\ 60 & 66 \end{pmatrix}$.

6. (a) $\begin{pmatrix} 72 & -216 & -360 \\ -288 & -72 & 504 \end{pmatrix}$ (b) $\begin{pmatrix} 72 & -216 & -360 \\ -288 & -72 & 504 \end{pmatrix}$

(c) $\mathbf{O}_{23}$ (d) $\mathbf{O}_{23}$ (e) $\mathbf{B}$ (f) $\mathbf{B}$ (g) $\mathbf{A}$ (h) $\mathbf{A}$

7. $\lambda = 3$

8. $\mathbf{A} = \mathbf{A}^2 = \mathbf{A}^3 = \mathbf{A}^4 = \frac{1}{3}\begin{pmatrix} 1 & 1 & 1 \\ 1 & 1 & 1 \\ 1 & 1 & 1 \end{pmatrix}$, $\mathbf{A}^n = \mathbf{A}$. Also $\mathbf{x}_n = \mathbf{Ax}$.

9. (a) $\mathbf{A} = \begin{pmatrix} 1 & 1 \\ 1 & 1 \end{pmatrix}$, $\mathbf{B} = \begin{pmatrix} 1 & 2 \\ -1 & -2 \end{pmatrix}$; (b) $\mathbf{A} = \mathbf{O}$, $\mathbf{B} \neq \mathbf{O}$

10. $\begin{pmatrix} 0.65 \\ 0.35 \end{pmatrix}$, $\begin{pmatrix} 0.635 \\ 0.365 \end{pmatrix}$, $\begin{pmatrix} 0.636 \\ 0.364 \end{pmatrix}$, $\begin{pmatrix} 0.636 \\ 0.364 \end{pmatrix}$, and $\begin{pmatrix} 0.636 \\ 0.364 \end{pmatrix}$.

For large k we have $\mathbf{p}_k = \mathbf{T}^k\mathbf{p} = \begin{pmatrix} 0.636 \\ 0.364 \end{pmatrix}$.

SOLUTIONS TO EXERCISES 1.6

1. (a) $\begin{pmatrix} 1 & 3 \\ 2 & 4 \end{pmatrix}$ (b) $\begin{pmatrix} 1 & -1 \\ 2 & -2 \\ 3 & -3 \end{pmatrix}$ (c) $\begin{pmatrix} -1 \\ 5 \\ 9 \\ 100 \end{pmatrix}$

(d) $\begin{pmatrix} a & c \\ b & d \end{pmatrix}$ (e) $\begin{pmatrix} 0 & 0 & 0 \\ 0 & 0 & 0 \end{pmatrix}$ (f) $\begin{pmatrix} 0 & 0 & 0 & 0 \\ 1 & 0 & 0 & 0 \\ 0 & 1 & 0 & 0 \\ 0 & 0 & 1 & 0 \end{pmatrix}$

2. (a) $\begin{pmatrix} 35 & -35 \\ 16 & -66 \end{pmatrix}$ (b) $\begin{pmatrix} 28 & -36 & 16 \\ 45 & -30 & 35 \end{pmatrix}$ (c) $\begin{pmatrix} 0 & -5 \\ 5 & 0 \end{pmatrix}$

(d) $\begin{pmatrix} 0 & -5 & -5 \\ 5 & 0 & -1 \\ 5 & 1 & 0 \end{pmatrix}$ (e) $\begin{pmatrix} -6 & 3 & 1 \\ 8 & 9 & -2 \\ 6 & -1 & 5 \end{pmatrix}$ (f) $\begin{pmatrix} -8 & 12 \\ 2 & 10 \end{pmatrix}$

(g) $\begin{pmatrix} 4 & -1 \\ 4 & -5 \\ 13 & 8 \end{pmatrix}$ (h) $\begin{pmatrix} 4 & 4 & 13 \\ -1 & -5 & 8 \end{pmatrix}$ (i) $\begin{pmatrix} 4 & 4 & 13 \\ -1 & -5 & 8 \end{pmatrix}$

(j) $\begin{pmatrix} -27 & 8 & -9 \\ -58 & 32 & -26 \end{pmatrix}$ (k) $\begin{pmatrix} 6 & -8 & -6 \\ -3 & -9 & 1 \\ -1 & 2 & -5 \end{pmatrix}$ (l) $\begin{pmatrix} 6 & -8 & -6 \\ -3 & -9 & 1 \\ -1 & 2 & -5 \end{pmatrix}$

(m) $\begin{pmatrix} 22 & 6 \\ 1 & 31 \end{pmatrix}$ (n) $\begin{pmatrix} 22 & 6 \\ 1 & 31 \end{pmatrix}$

3. (a) 32 (b) 32 (c) 32

15. $x = -1, y = -1$ and $z = 1$

16. $x = \sin(\theta) + \cos(\theta), y = \cos(\theta) - \sin(\theta)$

SOLUTIONS TO EXERCISES 1.7

1. (a) $x = 3/4,\ y = 13/4$ and $z = -11/4$
   (b) $x = 0,\ y = 0$ and $z = 0$
   (c) No Solution (d) $x = 3t,\ y = 2 - 2t$ and $z = t$
   (e) $x = -7t,\ y = -6t,\ z = -t,\ w = t$
   (f) $x = 11t - 20/3, y = 7 - 8t, z = -1/3, w = t$
   (g) $x = \frac{1}{20}(33 - 144t - 20s), y = \frac{1}{5}(19t - 3), z = t$ and $w = s$.
   (h) No Solution
   (i) $x = \frac{1}{9}(71t - 18s - 47), y = 5t - 3, z = \frac{1}{3}(26 - 32t), w = s$ and $u = t$
   (j) $x = -(1 + 2t + 3s), y = \frac{1}{9}(15t - 6s - 13), z = \frac{1}{27}(14 + 6t + 3s)$, $w = t$ and $u = s$.
2. (a) $x = -9 + 10t - s, y = s, z = -7 + 7t$ and $w = t$
   (b) $x_1 = -p - q - r,\quad x_2 = p,\ x_3 = q,\ x_4 = r,\ x_5 = 0$ and $x_6 = s$
   (c) $x_1 = 2 - 3r - 6s - 2t, x_2 = r,\quad x_3 = 1 - 3s,\ x_4 = 0,\quad x_5 = s,\ x_6 = t$
3. $x = 4 + 2s, y = 6$ and $z = s$
8. $x_1 = -10 + s + t, x_2 = 7 - s, x_3 = 9 - t, x_4 = 15 - s - t, x_5 = s$ and $x_6 = t$
9. $x_1 = -10 + s + t, x_2 = 7 - s, x_3 = 9 - t, x_4 = 15 - s - t, x_5 = s$ and $x_6 = t$, $x_7 = 8 - s - t, x_8 = s$ and $x_9 = t$.
10. $x_1 = -2 + p + r - t,\ x_2 = -7 - p + s + t,\ x_3 = p,\ x_4 = 3 + q - r + t$, $x_5 = 25 - q - s - t,\ x_6 = q,\ x_7 = r,\ x_8 = s$ and $x_9 = t$
11. $x = 5.41 - 0.095t, y = 5.41 - 0.095t, z = 3.67 - 0.067t$ and t is free.

SOLUTIONS TO EXERCISES 1.8

1. Matrices **B**, **C**, **D** and **G** are elementary matrices.

2. (a) $\mathbf{E}_1^{-1} = \begin{pmatrix} -1 & 0 \\ 0 & 1 \end{pmatrix}$ (b) $\mathbf{E}_2^{-1} = \begin{pmatrix} 0 & 1 \\ 1 & 0 \end{pmatrix}$

   (c) $\mathbf{E}_3^{-1} = \begin{pmatrix} 1 & 0 \\ 0 & -1/2 \end{pmatrix}$ (d) $\mathbf{E}_4^{-1} = \begin{pmatrix} -1/5 & 0 & 0 \\ 0 & 1 & 0 \\ 0 & 0 & 1 \end{pmatrix}$

   (e) $\mathbf{E}_5^{-1} = \begin{pmatrix} 1 & 0 & 0 \\ 0 & -1/\sqrt{2} & 0 \\ 0 & 0 & 1 \end{pmatrix}$ (f) $\mathbf{E}_6^{-1} = \begin{pmatrix} 1 & 0 & 0 \\ 0 & 1 & 0 \\ 0 & 0 & 1/\pi \end{pmatrix}$

3. (a) $\begin{pmatrix} a & b & c \\ -d & -e & -f \\ g & h & i \end{pmatrix}$ (b) $\begin{pmatrix} g & h & i \\ d & e & f \\ a & b & c \end{pmatrix}$

(c) $\begin{pmatrix} ka & kb & kc \\ d & e & f \\ g & h & i \end{pmatrix}$ (d) $\begin{pmatrix} a & b & c \\ d & e & f \\ -\frac{g}{k} & -\frac{h}{k} & -\frac{i}{k} \end{pmatrix}$

4. (a) $\frac{1}{6}\begin{pmatrix} 4 & -2 \\ 1 & 1 \end{pmatrix}$ (b) $-\frac{1}{28}\begin{pmatrix} 1 & 5 \\ 6 & 2 \end{pmatrix}$

(d) $\begin{pmatrix} 0 & 1 & -1 \\ -1 & 1 & -2 \\ 0 & 0 & -1 \end{pmatrix}$ (e) $\frac{1}{4}\begin{pmatrix} 3 & 2 & 1 \\ 2 & 4 & 2 \\ 1 & 2 & 3 \end{pmatrix}$

(f) $\frac{1}{15}\begin{pmatrix} 3 & -10 & 1 \\ 0 & 10 & 5 \\ 3 & -5 & -4 \end{pmatrix}$ (i) $\frac{1}{5}\begin{pmatrix} 131 & -2 & 8 & 15 \\ -43 & 1 & -9 & 0 \\ -25 & 0 & -5 & 0 \\ -30 & 0 & 0 & -5 \end{pmatrix}$

(c) Non-invertible (g) Non-invertible
(h) Non-invertible

5. (a) $x = 1/3,\ y = 4/3$ (b) $x = 1/14$ and $y = -4/7$
(c) No unique solution (d) $x = -2,\ y = -17$ and $z = -5$
(e) $x = 8,\ y = 11$ and $z = 7$ (f) $x = -2,\ y = 4$ and $z = -3$
(g) $x = 222.6,\ y = -66.8,\ z = -39$ and $w = -53$

7. $\mathbf{p} = (220.04\ \ 277.58\ \ 235.40)^T$

SOLUTIONS TO MISCELLANEOUS EXERCISES 1

1.1. (a) $-16\begin{pmatrix} 4 & 5 \\ 4 & 5 \end{pmatrix}$ (b) $-4\begin{pmatrix} 15 & 17 \\ 19 & 21 \end{pmatrix}$ This is because **AB** does not equal **BA**.

1.2. Error in first line because $(\mathbf{AB})^{-1} \neq \mathbf{A}^{-1}\mathbf{B}^{-1}$ and error in line 2 because matrix multiplication is **not commutative**.

1.3. (a) $\mathbf{A}^2 = \mathbf{I}$ and $\mathbf{A}^3 = \mathbf{A}$ (b) $\mathbf{A}^{-1} = \mathbf{A}$ and $\mathbf{A}^{2004} = \mathbf{I}$

1.4. $\mathbf{A}^t = \begin{pmatrix} 1 & 1 & 1 \\ 0 & 1 & 2 \\ -1 & 1 & 3 \end{pmatrix}$. The operations **AB**, **B** + **C**, **A** − **B** and $\mathbf{BC}^t$ are **not** valid.

$\mathbf{CB} = \begin{pmatrix} 0 & -2 \\ -1 & 1 \end{pmatrix}$, $\mathbf{A}^2 = \begin{pmatrix} 0 & -2 & -4 \\ 3 & 3 & 3 \\ 6 & 8 & 10 \end{pmatrix}$

1.5. Prove that $(\mathbf{A}^T\mathbf{A})^T = \mathbf{A}^T\mathbf{A}$ using $(\mathbf{XY})^T = \mathbf{Y}^T\mathbf{X}^T$ and $(\mathbf{X}^T)^T = \mathbf{X}$.

1.6. $a = c$ and $b = d = \frac{2}{3}c$

1.7. An example is $\mathbf{Au} = \mathbf{Av} = \begin{pmatrix} 0 \\ 0 \end{pmatrix} = \mathbf{O}$ with $\mathbf{A} = \begin{pmatrix} 1 & 2 \\ 2 & 4 \end{pmatrix}$ and $\mathbf{u} = \begin{pmatrix} -2 \\ 1 \end{pmatrix} \neq \begin{pmatrix} 2 \\ -1 \end{pmatrix} = \mathbf{v}$.

1.8. (a) $\mathbf{A}^2 = \begin{pmatrix} 7 & 10 \\ 15 & 22 \end{pmatrix}$, $\mathbf{B}^2 = \begin{pmatrix} -1 & 0 \\ 0 & -1 \end{pmatrix}$, $\mathbf{AB} = \begin{pmatrix} -2 & 1 \\ -4 & 3 \end{pmatrix}$ and $\mathbf{BA} = \begin{pmatrix} 3 & 4 \\ -1 & -2 \end{pmatrix}$

(b) $\mathbf{AB} - \mathbf{BA} = \begin{pmatrix} bg - cf & af + bh - be - df \\ ce + dg - ag - ch & cf - bg \end{pmatrix}$

1.9. $\mathbf{M}^2 = 2\mathbf{M}$, $\mathbf{M}^3 = 4\mathbf{M}$ and $\mathbf{M}^4 = 8\mathbf{M}$. $c(n) = 2^{n-1}$.

1.10. (i) $\mathbf{A}^2 = \frac{1}{2}\left(\frac{2}{3}\right)^2 \mathbf{A}$ (ii) $\mathbf{A}^3 = \frac{1}{2}\left(\frac{2}{3}\right)^3 \mathbf{A}$
To prove the required result use mathematical induction.

1.11. Matrix **B** has four rows.

1.12. (a) The given matrix is **non-invertible (singular)**.
(b) $\mathbf{B}^{-1} = \mathbf{CA}$ so the matrix **B** is invertible.

1.13. (b) $\mathbf{A}^{-1} = \begin{bmatrix} 7/2 & 0 & -3 \\ -1 & 1 & 0 \\ 0 & -1 & 1 \end{bmatrix}$

1.14. $x = 3$, $y = 1$ and $z = 2$.

1.15. (a) Reduced row echelon form is $\left[\begin{array}{ccc|c} 1 & 0 & 3 & 1 \\ 0 & 1 & 2 & 1 \\ 0 & 0 & 0 & 0 \end{array}\right]$.

(b) $\begin{bmatrix} x \\ y \\ z \end{bmatrix} = \begin{bmatrix} 1 - 3t \\ 1 - 2t \\ t \end{bmatrix} = \begin{bmatrix} 1 \\ 1 \\ 0 \end{bmatrix} + t \begin{bmatrix} -3 \\ -2 \\ 1 \end{bmatrix}$

1.16. $\begin{bmatrix} x_1 \\ x_2 \\ x_3 \\ x_4 \end{bmatrix} = \begin{bmatrix} -2 + t \\ 3 - 3t \\ -1 - 2t \\ t \end{bmatrix} = \begin{bmatrix} -2 \\ 3 \\ -1 \\ 0 \end{bmatrix} + t \begin{bmatrix} 1 \\ -3 \\ -2 \\ 1 \end{bmatrix}$

1.17. $\begin{bmatrix} x_1 \\ x_2 \\ x_3 \\ x_4 \end{bmatrix} = \begin{bmatrix} -1 + 11s + 16t \\ 1 - 6s - 6t \\ s \\ t \end{bmatrix} = \begin{bmatrix} -1 \\ 1 \\ 0 \\ 0 \end{bmatrix} + s \begin{bmatrix} 11 \\ -6 \\ 1 \\ 0 \end{bmatrix} + t \begin{bmatrix} 16 \\ -6 \\ 0 \\ 1 \end{bmatrix}$

1.18. (i) $\begin{bmatrix} 1 & 2 & 5 & 6 \\ 0 & 1 & -12 & -16 \\ 0 & 0 & 13 & 26 \end{bmatrix}$ (ii) $\begin{bmatrix} 1 & 2 & 5 & 6 \\ 0 & 1 & -12 & -16 \\ 0 & 0 & 1 & 2 \end{bmatrix}$ (iii) $\begin{bmatrix} 1 & 0 & 0 & -20 \\ 0 & 1 & 0 & 8 \\ 0 & 0 & 1 & 2 \end{bmatrix}$

1.19. The system is inconsistent because if you carry our row operations you end up with something like $0x_1 + 0x_2 + 0x_3 + 0x_4 = -33$ which means that $0 = -33$

1.20. $\mathbf{A} = \begin{pmatrix} 1 & 0 & -1 \\ 0 & 1 & -1 \\ 1 & 1 & 1 \end{pmatrix}$

1.21. $rref\,(\mathbf{A}) = \begin{pmatrix} 1 & 1 & 0 & 1 \\ 0 & 0 & 1 & 1 \\ 0 & 0 & 0 & 0 \\ 0 & 0 & 0 & 0 \end{pmatrix}$

1.22. $\begin{pmatrix} x_1 \\ x_2 \\ x_3 \\ x_4 \\ x_5 \end{pmatrix} = \begin{pmatrix} 3+2t-2s \\ 1+s-t \\ s \\ 2t+2 \\ t \end{pmatrix} = \begin{pmatrix} 3 \\ 1 \\ 0 \\ 2 \\ 0 \end{pmatrix} + s\begin{pmatrix} -2 \\ 1 \\ 1 \\ 0 \\ 0 \end{pmatrix} + t\begin{pmatrix} 2 \\ -1 \\ 0 \\ 2 \\ 1 \end{pmatrix}$

1.23. (a) $\mathbf{E}_2\mathbf{E}_1 = \begin{pmatrix} 1 & 0 \\ 0 & 1/2 \end{pmatrix}\begin{pmatrix} 1 & 0 \\ 5 & 1 \end{pmatrix} = \begin{pmatrix} 1 & 0 \\ 5/2 & 1/2 \end{pmatrix} = \mathbf{A}^{-1}$

(b) $\mathbf{E}_1^{-1}\mathbf{E}_2^{-1} = \begin{pmatrix} 1 & 0 \\ -5 & 1 \end{pmatrix}\begin{pmatrix} 1 & 0 \\ 0 & 2 \end{pmatrix} = \begin{pmatrix} 1 & 0 \\ -5 & 2 \end{pmatrix} = \mathbf{A}$

1.24. $\mathbf{E} = \begin{bmatrix} 0 & 0 & 1 & 0 \\ 0 & 1 & 0 & 0 \\ 1 & 0 & 0 & 0 \\ 0 & 0 & 0 & 1 \end{bmatrix}$

1.25. Option D which is $\begin{bmatrix} 4 & -4 \\ 5 & 11 \end{bmatrix}$.

1.26. $\mathbf{X} = \begin{bmatrix} -6 & -2 & 2 \\ -5 & 4 & 3 \\ -2 & -3 & 0 \end{bmatrix}$

1.27. (a) $x = -4$, $y = -24$ and $z = 33$ (b) Inverse matrix is $\begin{pmatrix} -1 & 0 & 1 \\ -5 & 1 & 3 \\ 7 & -1 & -4 \end{pmatrix}$.

(c) If our z coefficient is 5/4 then the linear system has no solutions.

1.28. Prove $\left(\mathbf{A}^{-1}\right)^T = \mathbf{A}^{-1}$ using matrix operations.

1.29. Pre- and post-multiply **A** by **I** − **A** and in each case the result should be **I**.

1.30. (a) To show uniqueness, assume that we have two matrices which are inverses of a matrix **A** and then show that they are equal.

(b) Use mathematical induction.

(c) $\mathbf{A}^{-1} = \frac{1}{2}\begin{bmatrix} 1 & -1 & 1 \\ 0 & 0 & 2 \\ 1 & 1 & -1 \end{bmatrix}$

1.31. Pre-multiply $\mathbf{AXA}^{-1} = \mathbf{B}$ by $\mathbf{A}^{-1}$ and post-multiply by **A**. You should get $\mathbf{X} = \mathbf{A}^{-1}\mathbf{BA}$. Take the inverse of both sides to find $\mathbf{X}^{-1} = \mathbf{A}^{-1}\mathbf{B}^{-1}\mathbf{A}$.

1.32. $\begin{bmatrix} a \\ b \\ c \\ d \\ e \\ f \\ g \end{bmatrix} = \begin{bmatrix} p \\ 7-q-2r-4t \\ q \\ 8-3r-5t \\ r \\ 9-6t \\ t \end{bmatrix} = \begin{bmatrix} 0 \\ 7 \\ 0 \\ 8 \\ 0 \\ 9 \\ 0 \end{bmatrix} + p\begin{bmatrix} 1 \\ 0 \\ 0 \\ 0 \\ 0 \\ 0 \\ 0 \end{bmatrix} + q\begin{bmatrix} 0 \\ -1 \\ 1 \\ 0 \\ 0 \\ 0 \\ 0 \end{bmatrix} + r\begin{bmatrix} 0 \\ -2 \\ 0 \\ -3 \\ 1 \\ 0 \\ 0 \end{bmatrix} + t\begin{bmatrix} 0 \\ -4 \\ 0 \\ -5 \\ 0 \\ -6 \\ 1 \end{bmatrix}$

1.33. Go both ways ($\Rightarrow$) and ($\Leftarrow$) to show the required result.

1.34. Prove $(\mathbf{AB})\left(\mathbf{B}^{-1}\mathbf{A}^{-1}\right) = \mathbf{I}$ by using matrix operations and then apply result of question (33) to show that $\mathbf{B}^{-1}\mathbf{A}^{-1}$ is the inverse of $\mathbf{AB}$.

1.35. (i) Use mathematical induction to show the required result.
(ii) Multiply out the two matrices to show that $\mathbf{DD}^{-1} = \mathbf{I}$.

SOLUTIONS TO EXERCISES 2.1

1. (a) 1 (b) 1 (c) 10 (d) 5 (e) 10 (f) 5
(g) 3.16 (2 dp) (h) 2.24 (2dp) (i) 17 (j) 3.61 (2dp)

2. (a) 15 (b) 15 (c) 14 (d) 30 (e) 14 (f) 30
(g) 3.74 (2dp) (h) 5.48 (2dp) (i) 74 (j) 3.74 (2dp)

3. (a) −28 (b) −28 (c) 39 (d) 39 (e) 39 (f) 39
(g) 6.25 (2dp) (h) 6.25 (2dp) (i) 22 (j) 11.58 (2dp)

5. $x = 2s,\ y = 3s$ where s is any real number

6. (a) $\sqrt{5}$ (b) 13

9. (a) $\dfrac{1}{\sqrt{53}}(2\ \ -7)^T$ (b) $\dfrac{1}{\sqrt{139}}(-9\ \ 3\ \ 7)^T$ (c) $\dfrac{1}{\sqrt{134}}(-3\ \ 5\ \ 8\ \ 6)^T$
(d) $\dfrac{1}{\sqrt{138}}(-6\ \ 2\ \ 8\ \ 3\ \ 5)^T$

SOLUTIONS TO EXERCISES 2.2

1. (a) $\theta = 45°$ (b) $\theta = 90°$ (c) $\theta = 168.69°$

2. (a) $\theta = 55.90°$ (b) $\theta = 90°$ (c) $\theta = 115.38°$

3. (a) $\theta = 41.98°$ (b) $\theta = 135°$ (c) $\theta = 56.56°$

4. (a) $k = -13/7$ (b) $k = -5/3$ (c) $k = 0$

5. (a) $\hat{\mathbf{u}} = \dfrac{1}{\sqrt{13}}\begin{pmatrix}2\\3\end{pmatrix}$ (b) $\hat{\mathbf{u}} = \dfrac{1}{\sqrt{14}}\begin{pmatrix}1\\2\\3\end{pmatrix}$ (c) $\hat{\mathbf{u}} = \dfrac{1}{3}\begin{pmatrix}2\\-2\\1\end{pmatrix}$
(d) $\hat{\mathbf{u}} = \dfrac{1}{\sqrt{5}}(1\ \ \sqrt{2}\ \ -1\ \ 1)^T$ (e) $\hat{\mathbf{u}} = \dfrac{1}{\sqrt{205}}(-2\ \ 10\ \ -10\ \ 1\ \ 0)^T$

6. $k = \dfrac{1}{2}$ or $k = -\dfrac{1}{2}$

7. (d) $\theta = 90°$

11. $(1\ \ 1\ \ -2)$ and $(-s-t\ \ s\ \ t)^T$

12. (a) 0.71 (b) 0.89 (c) 6.12 (d) 0.76

SOLUTIONS TO EXERCISES 2.3

1. (a) independent. (b), (c), (d), (e) dependent
2. (a), (b) and (d) independent (c) dependent
3. (a) independent (b) and (c) dependent

SOLUTIONS TO EXERCISES 2.4

1. (a) Span $\mathbb{R}^2$ (b) Span $\mathbb{R}^2$ (c) Does not span $\mathbb{R}^2$ (d) Span $\mathbb{R}^2$
2. (a) Span $\mathbb{R}^3$ (b) Does not span $\mathbb{R}^3$ (c) Span $\mathbb{R}^3$ (d) Span $\mathbb{R}^3$
3. (a) Forms a basis (b) No basis (c) Forms a basis (d) Forms a basis
5. (a) **b** is in the space spanned by columns of **A**.
   (b) **b** is not spanned by the columns of **A**.
9. (a) independent. (b) $(x\ \ y\ \ 0)^T$ (c) $\mathbf{w} = (0\ \ 0\ \ z)^T\ (z \neq 0)$

SOLUTIONS TO MISCELLANEOUS EXERCISES 2

2.1. Linearly dependent.

2.2. No such example exists because to span $\mathbb{R}^3$ you need three vectors.

2.3. (a) Linearly independent (b) $2\mathbf{v}_1 + \mathbf{v}_2 = (3,\ 2,\ 1)$ (c) Yes

2.4. $(3,\ 1,\ 1)$ **cannot** be expressed as a linear combination of the vectors $(2,\ 5,\ -1)$, $(1,\ 6,\ 0)$, $(5,\ 2,\ -4)$.

2.5. (b) $-3\begin{bmatrix}1\\2\\-1\end{bmatrix} + \begin{bmatrix}2\\-1\\3\end{bmatrix} + \begin{bmatrix}1\\7\\-6\end{bmatrix} = \begin{bmatrix}0\\0\\0\end{bmatrix}$

2.6. (a) $\left\{\begin{pmatrix}1\\0\end{pmatrix}, \begin{pmatrix}0\\1\end{pmatrix}\right\}$ and $\left\{\begin{pmatrix}1\\0\end{pmatrix}, \begin{pmatrix}1\\1\end{pmatrix}\right\}$

(b) (i) No because the set is linearly dependent but it does span $\mathbb{R}^3$.
(ii) Yes the given set of vectors is a basis for $\mathbb{R}^3$.

2.7. A basis is $\left\{\begin{bmatrix}a\\0\\0\\0\\0\end{bmatrix}, \begin{bmatrix}b\\b\\0\\0\\b\end{bmatrix}, \begin{bmatrix}0\\0\\c\\0\\c\end{bmatrix}\right\}$.

2.8. The first three vectors are linearly independent.

2.9. (a) Use definition (3.22) and show that **all** scalars are equal to zero.
(b) For all real values of λ providing $\lambda \neq -4$.

2.10. $c = 17$

2.11. By showing $c_1\mathbf{v}_1 + c_2(\mathbf{v}_1 + \mathbf{v}_2) = \mathbf{O} \Rightarrow c_1 = c_2 = 0$

2.12. None of the given sets form a basis for $\mathbb{R}^3$.

(a) We only have two vectors. (b) Linear dependence $2\begin{pmatrix}1\\2\\3\end{pmatrix} + \begin{pmatrix}4\\5\\6\end{pmatrix} = \begin{pmatrix}6\\9\\12\end{pmatrix}$

(c) We are given four vectors.

2.13. a. $\mathbf{v} + \mathbf{w} = \mathbf{u}$ b. Yes, they are linearly dependent because $\mathbf{v} + \mathbf{w} = \mathbf{u}$.

2.14. (a) Since the dot product is zero therefore $\{\mathbf{u}_1, \mathbf{u}_2, \mathbf{u}_3\}$ is orthogonal.
(b) $\frac{5}{2}\mathbf{u}_1 - \frac{3}{2}\mathbf{u}_2 + 2\mathbf{u}_3 = \mathbf{x}$

2.15. (a) $\mathbf{u} \cdot \mathbf{u} = 13$ (b) $\|\mathbf{u}\| = \sqrt{13}$ (c) $\widehat{\mathbf{u}} = \frac{1}{\sqrt{13}}\begin{bmatrix}2\\0\\3\end{bmatrix}$ (d) $\mathbf{u} \cdot \mathbf{v} = 0$ (e) Yes

(f) $\|\mathbf{v} - \mathbf{x}\| = 6$

2.16. (c) $\theta = \cos^{-1}\left(\frac{1}{2}\right) = \frac{\pi}{3}$

2.17. (c) $\mathbf{v} = \begin{pmatrix}x\\y\\z\end{pmatrix} = s\begin{pmatrix}-1\\0\\1\end{pmatrix} + t\begin{pmatrix}-1\\1\\0\end{pmatrix}$ where $s, t \in \mathbb{R}$. A particular $\mathbf{v} = \begin{pmatrix}-2\\1\\1\end{pmatrix}$.

$\mathbf{w} = \begin{pmatrix}0\\-1\\1\end{pmatrix}$. Also $\widehat{\mathbf{u}} = \frac{1}{\sqrt{3}}\begin{pmatrix}1\\1\\1\end{pmatrix}$, $\widehat{\mathbf{v}} = \frac{1}{\sqrt{6}}\begin{pmatrix}-2\\1\\1\end{pmatrix}$ and $\widehat{\mathbf{w}} = \frac{1}{\sqrt{2}}\begin{pmatrix}0\\-1\\1\end{pmatrix}$ where $\{\widehat{\mathbf{u}}, \widehat{\mathbf{v}}, \widehat{\mathbf{w}}\}$ is an orthonormal set of vectors in $\mathbb{R}^3$.

2.18. (a) $\mathbf{u} \cdot \mathbf{v} = 0$, $(\mathbf{u} - \mathbf{v}) \cdot (\mathbf{u} - \mathbf{w}) = 20$ and $(2\mathbf{u} - 3\mathbf{v}) \cdot (\mathbf{u} + 4\mathbf{w}) = -20$.
(b) $|\mathbf{u} \cdot \mathbf{v}| = 2$, $\|\mathbf{u}\|\,\|\mathbf{v}\| = 3\sqrt{2}$ therefore $|\mathbf{u} \cdot \mathbf{v}| = 2 \le 3\sqrt{2} = \|\mathbf{u}\|\,\|\mathbf{v}\|$

2.19. Consider the linear combination $k_1\mathbf{v}_1 + k_2\mathbf{v}_2 + k_3\mathbf{v}_3 = \mathbf{O}$ then show that the dot product of $\mathbf{O} \cdot \mathbf{v}_1 = \mathbf{O} \cdot \mathbf{v}_2 = \mathbf{O} \cdot \mathbf{v}_3$ gives $k_1 = k_2 = k_3 = 0$.

2.20. Show that $k_1(\mathbf{A}\mathbf{u}_1) + k_2(\mathbf{A}\mathbf{u}_2) + \cdots + k_n(\mathbf{A}\mathbf{u}_n) = \mathbf{O}$ implies $k_1 = k_2 = \cdots k_n = 0$.

2.21. Prove that $\|\mathbf{u} + \mathbf{v}\|^2 = \|\mathbf{u} - \mathbf{v}\|^2$.

2.22. From $\mathbf{u} \cdot (\mathbf{v} - \mathbf{w}) = 0$ we have

$$u_1(v_1 - w_1) + u_2(v_2 - w_2) + \cdots + u_n(v_n - w_n) = 0$$

Thus $v_1 = w_1, v_2 = w_2, \ldots, v_n = w_n$.

2.23. ($\Leftarrow$). Show that $k_1\mathbf{u} + k_2\mathbf{v} = \mathbf{O} \Rightarrow k_1 = k_2 = 0$.
($\Rightarrow$). By assuming $k_1\mathbf{u} + k_2\mathbf{v} = \mathbf{O}$ and $k_1 = k_2 = 0$ show that $ad - bc \neq 0$.

2.24. Use proof by contradiction. Suppose m linearly independent vectors span $\mathbb{R}^n$.

2.25. Consider $k_1\mathbf{v}_1 + k_2\mathbf{v}_2 + k_3\mathbf{v}_3 + \cdots + k_n\mathbf{v}_n = \mathbf{O}$ and prove that

$$k_1 = k_2 = k_3 = \cdots = k_n = 0$$

Then $\{\mathbf{v}_1, \mathbf{v}_2, \mathbf{v}_3, \ldots, \mathbf{v}_n\}$ are linearly independent which means they form a basis for $\mathbb{R}^n$.

SOLUTIONS TO EXERCISES 3.2

7. S is a subspace of $\mathbb{R}^3$.
18. (a), (b) and (c) are in span$\{\mathbf{f},\ \mathbf{g}\}$ but part (d) is not in the span$\{\mathbf{f},\ \mathbf{g}\}$.
19. Does not span.

SOLUTIONS TO EXERCISES 3.3

1. (a) and (c) independent (b) and (d) dependent
2. (a) dependent (b), (c), (d) and (e) independent
3. (a), (b), (d) and (e) dependent. Others independent.
6. $2 + 2(t-1) + (t-1)^2$.

SOLUTIONS TO EXERCISES 3.4

1. (a) 5 (b) 7 (c) 11 (d) 13 (e) 9 (f) 16 (g) 6 (h) 4 (i) 6 (j) 0
2. $\dim(S) = 1$
3. $\dim(S) = 2$
4. $\dim(S) = 2$
5. $\dim(S) = 3$
6. (a) mn (b) 3 (c) 4

SOLUTIONS TO EXERCISES 3.5

1. (a) Row vectors – $\begin{pmatrix}1\\2\end{pmatrix}, \begin{pmatrix}3\\4\end{pmatrix}$ and column vectors – $\begin{pmatrix}1\\3\end{pmatrix}$ and $\begin{pmatrix}2\\4\end{pmatrix}$.

   (b) Row vectors $(1\ \ 2\ \ 3\ \ 4)^T$, $(5\ \ 6\ \ 7\ \ 8)^T$ and $(9\ \ 10\ \ 11\ \ 12)^T$.
   Column vectors $(1\ \ 5\ \ 9)^T$, $(2\ \ 6\ \ 10)^T$, $(3\ \ 7\ \ 11)^T$ and $(4\ \ 8\ \ 12)^T$.

   (c) Row vectors $(1\ \ 2)^T$, $(3\ \ 4)^T$, $(5\ \ 6)^T$. Column vectors $(1\ 3\ 5)^T$ and $(2\ 4\ 6)^T$.

   (d) Row vectors $(1\ 2\ 3)^T$ and $(4\ 5\ 6)^T$. Column vectors $(1\ 4)^T$, $(2\ 5)^T$ and $(3\ 6)^T$.

   (e) Row vectors $\begin{pmatrix}-1\\2\\5\end{pmatrix}, \begin{pmatrix}-3\\7\\0\end{pmatrix}$ and $\begin{pmatrix}-8\\1\\3\end{pmatrix}$.
   Column vectors $\begin{pmatrix}-1\\-3\\-8\end{pmatrix}, \begin{pmatrix}2\\7\\1\end{pmatrix}$ and $\begin{pmatrix}5\\0\\3\end{pmatrix}$.

(f) Row vectors $\begin{pmatrix} -5 \\ 2 \\ 3 \end{pmatrix}$, $\begin{pmatrix} 7 \\ 1 \\ 0 \end{pmatrix}$, $\begin{pmatrix} -7 \\ 6 \\ 1 \end{pmatrix}$ and $\begin{pmatrix} -2 \\ 5 \\ 2 \end{pmatrix}$.

Column vectors $\begin{pmatrix} -5 \\ 7 \\ -7 \\ -2 \end{pmatrix}$, $\begin{pmatrix} 2 \\ 1 \\ 6 \\ 5 \end{pmatrix}$ and $\begin{pmatrix} 3 \\ 0 \\ 1 \\ 2 \end{pmatrix}$.

2. (a) $\left\{\begin{pmatrix} 1 \\ 0 \end{pmatrix}, \begin{pmatrix} 0 \\ 1 \end{pmatrix}\right\}$, $rank\,(\mathbf{A}) = 2$ (b) $\left\{\begin{pmatrix} 1 \\ 0 \\ -1 \\ -2 \end{pmatrix}, \begin{pmatrix} 0 \\ 1 \\ 2 \\ 3 \end{pmatrix}\right\}$, $rank\,(\mathbf{B}) = 2$

(c) $\left\{\begin{pmatrix} 1 \\ 0 \end{pmatrix}, \begin{pmatrix} 0 \\ 1 \end{pmatrix}\right\}$, $rank\,(\mathbf{C}) = 2$ (d) $\left\{\begin{pmatrix} 1 \\ 0 \\ -1 \end{pmatrix}, \begin{pmatrix} 0 \\ 1 \\ 2 \end{pmatrix}\right\}$, $rank\,(\mathbf{D}) = 2$

(e) and (f) $\left\{\begin{pmatrix} 1 \\ 0 \\ 0 \end{pmatrix}, \begin{pmatrix} 0 \\ 1 \\ 0 \end{pmatrix}, \begin{pmatrix} 0 \\ 0 \\ 1 \end{pmatrix}\right\}$ with $rank\,(\mathbf{E}) = 3$ and $rank\,(\mathbf{F}) = 3$

3. (a) $\left\{\begin{pmatrix} 1 \\ 0 \end{pmatrix}, \begin{pmatrix} 0 \\ 1 \end{pmatrix}\right\}$ (b) $\left\{\begin{pmatrix} 1 \\ 3 \end{pmatrix}\right\}$

(c) and (d) $\left\{\begin{pmatrix} 1 \\ 0 \\ 0 \end{pmatrix}, \begin{pmatrix} 0 \\ 1 \\ 0 \end{pmatrix}, \begin{pmatrix} 0 \\ 0 \\ 1 \end{pmatrix}\right\}$ (e) $\left\{\begin{pmatrix} 1 \\ 0 \\ -0.8 \\ -1.6 \end{pmatrix}, \begin{pmatrix} 0 \\ 1 \\ 1.4 \\ 1.8 \end{pmatrix}\right\}$

4. $\left\{\frac{1}{7}\begin{pmatrix} 7 \\ 0 \\ 8 \\ -4 \end{pmatrix}, \frac{1}{7}\begin{pmatrix} 0 \\ 7 \\ 1 \\ 3 \end{pmatrix}\right\}$

5. For rank, see solution to question 2.

(a) $\left\{\begin{pmatrix} 1 \\ 0 \end{pmatrix}, \begin{pmatrix} 0 \\ 1 \end{pmatrix}\right\}$ (b) $\left\{\begin{pmatrix} 1 \\ 0 \\ -1 \end{pmatrix}, \begin{pmatrix} 0 \\ 1 \\ 2 \end{pmatrix}\right\}$

(c) $\left\{\begin{pmatrix} 1 \\ 0 \\ -1 \end{pmatrix}, \begin{pmatrix} 0 \\ 1 \\ 2 \end{pmatrix}\right\}$ (d) $\left\{\begin{pmatrix} 1 \\ 0 \end{pmatrix}, \begin{pmatrix} 0 \\ 1 \end{pmatrix}\right\}$

(e) $\left\{\begin{pmatrix} 1 \\ 0 \\ 0 \end{pmatrix}, \begin{pmatrix} 0 \\ 1 \\ 0 \end{pmatrix}, \begin{pmatrix} 0 \\ 0 \\ 1 \end{pmatrix}\right\}$ (f) $\left\{\begin{pmatrix} 128 \\ 0 \\ 0 \\ 61 \end{pmatrix}, \begin{pmatrix} 0 \\ 128 \\ 0 \\ 80 \end{pmatrix}, \begin{pmatrix} 0 \\ 0 \\ 128 \\ 73 \end{pmatrix}\right\}$

SOLUTIONS TO EXERCISES 3.6

1. (a) $\{\mathbf{O}\}$ (b) $\left\{ s\begin{pmatrix} -2 \\ 1 \end{pmatrix} \;\middle|\; s \in \mathbb{R} \right\}$ (c) $\{\mathbf{O}\}$
   (d) $\{\mathbf{O}\}$
2. (a) $\begin{pmatrix} x \\ y \\ z \end{pmatrix} = s\begin{pmatrix} -1 \\ -2 \\ 1 \end{pmatrix}$ (b) $\begin{pmatrix} x \\ y \\ z \end{pmatrix} = s\begin{pmatrix} 1 \\ 1 \\ 0 \end{pmatrix} + t\begin{pmatrix} 1 \\ 0 \\ 1 \end{pmatrix}$
   (c) and (d) $\mathbf{x} = \mathbf{O}$
3. (a) $s\begin{pmatrix} -1 \\ -2 \\ 1 \end{pmatrix}$, $rank(\mathbf{A}) = 2$, $nullity(\mathbf{A}) = 1$

   (b) $s\begin{pmatrix} 1 \\ 1 \\ 0 \end{pmatrix} + t\begin{pmatrix} 1 \\ 0 \\ 1 \end{pmatrix}$, $rank(\mathbf{B}) = 1$, $nullity(\mathbf{B}) = 2$

   (c) and (d) $\{\mathbf{O}\}$, $rank(\mathbf{C}) = rank(\mathbf{D}) = 3$, $nullity(\mathbf{C}) = nullity(\mathbf{D}) = 0$
4. $(s + 2t + 3p + 4q + 5r \;\; -2s - 3t - 4p - 5q - 6r \;\; s \;\; t \;\; p \;\; q \;\; r)^T$
5. (a) $\left\{ \begin{pmatrix} 13 \\ 0 \\ -73 \end{pmatrix}, \begin{pmatrix} 0 \\ 13 \\ 11 \end{pmatrix} \right\}, \left\{ \begin{pmatrix} 1 \\ 0 \end{pmatrix}, \begin{pmatrix} 0 \\ 1 \end{pmatrix} \right\}, \left\{ \begin{pmatrix} 73 \\ -11 \\ 13 \end{pmatrix} \right\}$, $rank(\mathbf{A}) = 2$, $nullity(\mathbf{A}) = 1$

   (b) $\left\{ \begin{pmatrix} 1 \\ 0 \end{pmatrix}, \begin{pmatrix} 0 \\ 1 \end{pmatrix} \right\}, \left\{ \begin{pmatrix} 1 \\ 0 \\ -4 \end{pmatrix}, \begin{pmatrix} 0 \\ 1 \\ -5 \end{pmatrix} \right\}$, $\{\mathbf{O}\}$, $rank(\mathbf{B}) = 2$, $nullity(\mathbf{B}) = 0$

   (c) $\left\{ \begin{pmatrix} 1 \\ 3 \\ 0 \\ 0 \end{pmatrix}, \begin{pmatrix} 0 \\ 0 \\ 1 \\ 0 \end{pmatrix}, \begin{pmatrix} 0 \\ 0 \\ 0 \\ 1 \end{pmatrix} \right\}, \left\{ \begin{pmatrix} 1 \\ 0 \\ 0 \end{pmatrix}, \begin{pmatrix} 0 \\ 1 \\ 0 \end{pmatrix}, \begin{pmatrix} 0 \\ 0 \\ 1 \end{pmatrix} \right\}, \left\{ \begin{pmatrix} -3 \\ 1 \\ 0 \\ 0 \end{pmatrix} \right\}$, $rank(\mathbf{C}) = 3$, $nullity(\mathbf{C}) = 1$
6. (a) $$\begin{pmatrix} x \\ y \\ z \\ w \end{pmatrix} = \underbrace{s\begin{pmatrix} -1 \\ -3 \\ 1 \\ 0 \end{pmatrix} + t\begin{pmatrix} -5 \\ 0 \\ 0 \\ 1 \end{pmatrix}}_{=\mathbf{x}_H} + \underbrace{\begin{pmatrix} 9 \\ -3 \\ 0 \\ 0 \end{pmatrix}}_{=\mathbf{x}_P}$$

   (b) $\begin{pmatrix} x \\ y \\ z \end{pmatrix} = \begin{pmatrix} 2 \\ -1 \\ -2 \end{pmatrix}$
7. (a) infinite (b) infinite (c) no solution (d) unique solution
8. (a) and (b) in the null space. (c) and (d) not in the null space.

SOLUTIONS TO MISCELLANEOUS EXERCISES 3

3.1. $N = \left\{ \begin{bmatrix} 2 \\ 1 \\ 0 \\ 0 \end{bmatrix}, \begin{bmatrix} 1 \\ 0 \\ -2 \\ 1 \end{bmatrix} \right\}$

3.2. B

3.3. F

3.4. (a) Use $\mathbf{Av} = \mathbf{b}$ and $\mathbf{Aw} = \mathbf{O}$ (b) $\mathbf{Au} = \mathbf{b}$ and $\mathbf{Ax} = \mathbf{b}$

3.5. (a) Matrix $\mathbf{A}$ is invertible (non-singular) (b) linear independent rows
(c) $rank(\mathbf{A}) = n$

3.6. (a) false (b) true (c) false (d) true

3.7. (a) true (b) true (c) false

3.8. All three statements are true.

3.9. (a) $rank(\mathbf{A}) = 3$ (b) $nullity(\mathbf{A}) = 2$

(c) $\left\{ \begin{pmatrix} 1 \\ 0 \\ 0 \\ 0 \end{pmatrix}, \begin{pmatrix} 0 \\ 1 \\ 0 \\ -1 \end{pmatrix}, \begin{pmatrix} 0 \\ 0 \\ 1 \\ 1 \end{pmatrix} \right\}$ (d) $\left\{ \begin{pmatrix} 1 \\ 0 \\ -2 \\ 1 \\ 0 \end{pmatrix}, \begin{pmatrix} 0 \\ 1 \\ 3 \\ 0 \\ 0 \end{pmatrix}, \begin{pmatrix} 0 \\ 0 \\ 0 \\ 0 \\ 1 \end{pmatrix} \right\}$

(e) $\left\{ \begin{bmatrix} 2 \\ -3 \\ 1 \\ 0 \\ 0 \end{bmatrix}, \begin{bmatrix} -1 \\ 0 \\ 0 \\ 1 \\ 0 \end{bmatrix} \right\}$

3.10. (b) $\left\{ s \begin{pmatrix} 1 \\ 1 \\ 1 \end{pmatrix} \middle| \; s \in \mathbb{R} \right\}$ (c) $\mathbf{v}_1$, $\mathbf{v}_2$ and $\mathbf{v}_3$ are linearly independent.

3.11. (a) $\{\mathbf{O}\}$ and the space $\mathbb{R}^n$ (c) $-4\mathbf{v}_1 + 3\mathbf{v}_2 + \mathbf{v}_3 = \mathbf{O}$ and dimension is 2.

3.12. (a) $\left[\begin{array}{ccc|c} 1 & 0 & -2 & -3 \\ 0 & 1 & 0 & 1 \\ 0 & 0 & 0 & 0 \end{array}\right] = \mathbf{R}$ (b) The system is consistent. (c) $\vec{x} = \underbrace{s \begin{bmatrix} 2 \\ 0 \\ 1 \end{bmatrix}}_{=\mathbf{x}_H} + \underbrace{\begin{bmatrix} -3 \\ 1 \\ 0 \end{bmatrix}}_{=\mathbf{x}_P}$

3.13. $\left\{ \begin{pmatrix} 0 \\ 0 \\ 0 \\ -2 \\ 1 \\ 0 \end{pmatrix}, \begin{pmatrix} -2 \\ 1 \\ 0 \\ 0 \\ 0 \\ 0 \end{pmatrix}, \begin{pmatrix} -3 \\ 0 \\ 1 \\ 0 \\ 0 \\ 0 \end{pmatrix} \right\}$

3.14. (b) $\begin{bmatrix} 1 & -1 & 0 & 2 \\ 0 & 0 & 1 & -1 \\ 0 & 0 & 0 & 0 \end{bmatrix} = \mathbf{R}$, $\left\{ \begin{bmatrix} 1 \\ 1 \\ 0 \\ 0 \end{bmatrix}, \begin{bmatrix} -2 \\ 0 \\ 1 \\ 1 \end{bmatrix} \right\}$, nullity and rank of matrix **A** is 2.

Two linearly independent and linear combination of first and last columns of the matrix **A**.

3.15. (a) (i) Not a subspace (ii) Subspace (iii) Subspace (b) $\begin{pmatrix} 1 \\ 0 \\ 0 \end{pmatrix}, \begin{pmatrix} 0 \\ 1 \\ 0 \end{pmatrix}, \begin{pmatrix} 3 \\ 5 \\ -4 \end{pmatrix}$

(c) (i) Not a basis, linearly dependent, span $\mathbb{R}^3$. (ii) Is a basis

3.16. Apply the definition of subspace.

3.17. (a) **v** is in the span of S (b) **v** is in the span of S

3.18. $\begin{bmatrix} 1 & 0 & 0 & 0 \\ 0 & 1 & 1 & 0 \\ 0 & 0 & 0 & 1 \end{bmatrix} = \mathbf{R}$ (a) $\begin{bmatrix} 0 \\ -1 \\ 1 \\ 0 \end{bmatrix}$ (b) $\left\{ \begin{pmatrix} 1 \\ 0 \\ 0 \end{pmatrix}, \begin{pmatrix} 0 \\ 1 \\ 0 \end{pmatrix}, \begin{pmatrix} 0 \\ 0 \\ 1 \end{pmatrix} \right\}$

(c) *Nullity* $(\mathbf{B}) = 1$ (d) *rank* $(\mathbf{B}) = 3$

3.19. Use the definition of subspace.

3.20. (a) Yes (b) Dim is 3 and a set of basis vectors are $\mathbf{u}_1 = (2 \ \ 0 \ \ 5 \ \ 3)^T$, $\mathbf{u}_2 = (0 \ \ 1 \ \ 0 \ \ 0)^T$ and $\mathbf{u}_3 = (0 \ \ 0 \ \ 2 \ \ 1)^T$

(c) Yes is in V. (d) 1 (e) $\begin{pmatrix} -0.25 \\ 0 \\ -0.5 \\ 1 \end{pmatrix}$

3.21. Use Proposition (4.34).

3.22. Use Proposition (4.34).

3.23. (a) $N = \left\{ t \begin{bmatrix} -3 \\ 2 \\ 0 \\ 0 \end{bmatrix} + s \begin{bmatrix} -1 \\ 0 \\ 2 \\ 0 \end{bmatrix} \middle| \ s, t \in \mathbb{R} \right\}$

(b) $\mathbf{x} = \mathbf{x}_P + \mathbf{x}_H = \begin{bmatrix} 0 \\ 0 \\ 0 \\ -1 \end{bmatrix} + \underbrace{t \begin{bmatrix} -3 \\ 2 \\ 0 \\ 0 \end{bmatrix} + s \begin{bmatrix} -1 \\ 0 \\ 2 \\ 0 \end{bmatrix}}_{\mathbf{x}_H \text{ solution of part (a)}}$

(c) $n - m$ solutions.

3.24. $c = 17$

3.25. $\mathbf{A} = \begin{pmatrix} 4 & 7 & -6 \\ 5 & 8 & -7 \\ 6 & 9 & -8 \end{pmatrix}$

3.26. (a) Use the trigonometric identities of $\cos(x)$, $\cos(2x)$, $\cos(3x)$.
(b) Expand each of these to show that they are in the *span* $\{1, x, x^2, x^3\}$.

3.27. (a) Write two matrices whose rows are each of the given vectors and then show that they have the same basis which means they are equal.
(b) $k = 0, \quad k = -1$

3.28. All three statements (a), (b) and (c) are false.

3.29. Suppose $\mathbf{v}_1, \ldots, \mathbf{v}_n$ are linearly dependent and derive a contradiction. Because these are n linearly independent vectors and dimension of $\mathbb{R}^n$ is n therefore they form a basis.
To show that $\mathbf{A}$ is invertible write an arbitrary vector $\mathbf{u}$ uniquely as

$$c_1\mathbf{A}\mathbf{v}_1 + c_2\mathbf{A}\mathbf{v}_2 + \cdots + c_n\mathbf{A}\mathbf{v}_n = \mathbf{u}$$

Rewrite this as $\mathbf{Ax} = \mathbf{u}$ where $\mathbf{x}$ is unique then $\mathbf{A}$ is invertible.

SOLUTIONS TO EXERCISES 4.1

1. (a) $1/4$ (b) $1/4$ (c) $3/4$ (d) $1/3$ (e) $1/\sqrt{3}$ (f) $1/5$ (g) $1/\sqrt{5}$
2. (a) $2/5$ (b) $2/5$ (c) $6/5$ (d) $2/3$ (e) $\sqrt{2/3}$ (f) $2/7$ (g) $\sqrt{2/7}$
3. (a) -21 (b) -21 (c) 63 (d) 38 (e) $\sqrt{38}$ (f) 90 (g) $\sqrt{90}$
5. (a) 70 (b) 350 (c) 70 (d) $\sqrt{30}$ (e) $\sqrt{174}$ (f) 27 (g) 55 (h) 82 (i) 97
7. (a) $\frac{7}{12}$ (b) $-\frac{1}{12}$ (c) $-\frac{2}{3}$ (d) $-\frac{3}{4}$ (e) $-\frac{1}{12}$ (f) $-\frac{11}{12}$
(g) $-\frac{5}{12}$ (h) $-\frac{7}{12}$ (i) $-10\frac{1}{2}$ (j) $-15\frac{5}{6}$
11. (a) 4 (b) 12 (c) $\sqrt{22}$ (d) $\sqrt{67}$

SOLUTIONS TO EXERCISES 4.2

4. (b) 17.23 (2dp)
5. (a) $k = -8$ (b) $k = -1$
8. (d) $\frac{\mathbf{f}}{\|\mathbf{f}\|} = \frac{\cos(x)}{\sqrt{\pi}}$ and $\frac{\mathbf{g}}{\|\mathbf{g}\|} = \frac{\sin(x)}{\sqrt{\pi}}$
9. The orthonormal set is $\left\{\frac{1}{\sqrt{\pi}}, \sqrt{\frac{2}{\pi}}\sin(2t), \sqrt{\frac{2}{\pi}}\sin(4t), \sqrt{\frac{2}{\pi}}\sin(6t), \cdots\right\}$
11. 1.713
14. (a) $\|\mathbf{u}+\mathbf{v}\| = \sqrt{2}$ (b) $\|\mathbf{u}-\mathbf{v}\| = \sqrt{2}$
15. (ii) $\sqrt{n}$

SOLUTIONS TO EXERCISES 4.3

1. (a) $\mathbf{w}_1 = \begin{pmatrix} 1 \\ 0 \end{pmatrix}, \mathbf{w}_2 = \begin{pmatrix} 0 \\ 1 \end{pmatrix}$ (b) $\mathbf{u}_1 = \frac{1}{\sqrt{10}}\begin{pmatrix} 1 \\ 3 \end{pmatrix}$ and $\mathbf{u}_2 = \frac{1}{\sqrt{10}}\begin{pmatrix} 3 \\ -1 \end{pmatrix}$

   (c) $\mathbf{u}_1 = \frac{1}{\sqrt{13}}\begin{pmatrix} 2 \\ 3 \end{pmatrix}$ and $\mathbf{u}_2 = \frac{1}{\sqrt{13}}\begin{pmatrix} 3 \\ -2 \end{pmatrix}$

   (d) $\mathbf{u}_1 = \frac{1}{\sqrt{29}}\begin{pmatrix} -2 \\ -5 \end{pmatrix}$ and $\mathbf{u}_2 = \frac{1}{\sqrt{29}}\begin{pmatrix} -5 \\ 2 \end{pmatrix}$

2. $\mathbf{u}_1 = \frac{1}{\sqrt{2}}\begin{pmatrix} 1 \\ 1 \end{pmatrix}$ and $\mathbf{u}_2 = \frac{1}{\sqrt{2}}\begin{pmatrix} 1 \\ -1 \end{pmatrix}$

3. $\mathbf{u}_1 = \frac{1}{\sqrt{3}}\begin{pmatrix} 1 \\ 1 \\ 1 \end{pmatrix}, \mathbf{u}_2 = \frac{1}{\sqrt{6}}\begin{pmatrix} 1 \\ 1 \\ -2 \end{pmatrix}$ and $\mathbf{u}_3 = \frac{1}{\sqrt{2}}\begin{pmatrix} 1 \\ -1 \\ 0 \end{pmatrix}$

4. (a) $\mathbf{u}_1 = \frac{1}{\sqrt{2}}\begin{pmatrix} 1 \\ 0 \\ 1 \end{pmatrix}, \mathbf{u}_2 = \frac{1}{\sqrt{3}}\begin{pmatrix} 1 \\ 1 \\ -1 \end{pmatrix}$ and $\mathbf{u}_3 = \frac{1}{\sqrt{6}}\begin{pmatrix} 1 \\ -2 \\ -1 \end{pmatrix}$

   (b) $\mathbf{u}_1 = \frac{1}{\sqrt{3}}\begin{pmatrix} 1 \\ 1 \\ 1 \end{pmatrix}, \mathbf{u}_2 = \frac{1}{\sqrt{6}}\begin{pmatrix} 1 \\ -2 \\ 1 \end{pmatrix}$ and $\mathbf{u}_3 = \frac{1}{\sqrt{2}}\begin{pmatrix} -1 \\ 0 \\ 1 \end{pmatrix}$

   (c) $\mathbf{u}_1 = \frac{1}{\sqrt{5}}\begin{pmatrix} 1 \\ 2 \\ 0 \end{pmatrix}, \mathbf{u}_2 = \frac{1}{\sqrt{45}}\begin{pmatrix} 4 \\ -2 \\ 5 \end{pmatrix}$, and $\mathbf{u}_3 = \frac{1}{3}\begin{pmatrix} -2 \\ 1 \\ 2 \end{pmatrix}$

5. (a) $\mathbf{u}_1 = \frac{1}{\sqrt{10}}\begin{pmatrix} 1 \\ 0 \\ 3 \\ 0 \end{pmatrix}, \mathbf{u}_2 = \frac{1}{\sqrt{110}}\begin{pmatrix} 3 \\ 10 \\ -1 \\ 0 \end{pmatrix}$ and $\mathbf{u}_3 = \frac{1}{\sqrt{11}}\begin{pmatrix} 3 \\ -1 \\ -1 \\ 0 \end{pmatrix}$

   (b) $\mathbf{w}_1 = \frac{1}{\sqrt{31}}\begin{pmatrix} 1 \\ 1 \\ 5 \\ 2 \end{pmatrix}, \mathbf{w}_2' = \frac{1}{\sqrt{28582}}\begin{pmatrix} -109 \\ 77 \\ 44 \\ -94 \end{pmatrix}$ and $\mathbf{w}_3' = \frac{1}{\sqrt{15555062}}\begin{pmatrix} -2929 \\ -1289 \\ -78 \\ 2304 \end{pmatrix}$

   (c) $\mathbf{u}_1 = \frac{1}{\sqrt{30}}\begin{pmatrix} 1 \\ 2 \\ 3 \\ 4 \end{pmatrix}, \mathbf{u}_2 = \frac{1}{\sqrt{3930}}\begin{pmatrix} 53 \\ 16 \\ 9 \\ -28 \end{pmatrix}$ and $\mathbf{u}_3 = \frac{1}{\sqrt{224665}}\begin{pmatrix} 224 \\ -140 \\ -308 \\ 245 \end{pmatrix}$

6. $\widehat{\mathbf{p}_1} = \frac{1}{\sqrt{2}}, \widehat{\mathbf{p}_2} = \sqrt{\frac{3}{2}}x$ and $\widehat{\mathbf{p}_3} = \sqrt{\frac{45}{8}}\left(x^2 - \frac{1}{3}\right)$

7. $\mathbf{p}_1 = x^2, \quad \mathbf{p}_2 = x$ and $\mathbf{p}_3 = 1 - \frac{5}{3}x^2$

8. (a) $\frac{4}{3}\mathbf{p}_1 + \mathbf{p}_2 + \frac{2}{3}\mathbf{p}_3$ (b) $-\frac{1}{3}\mathbf{p}_1 + \frac{4}{3}\mathbf{p}_3$ (c) $3\mathbf{p}_1$ (d) $\frac{1}{3}\mathbf{p}_1 + \frac{2}{3}\mathbf{p}_3$

   (e) $2\mathbf{p}_1 + 5\mathbf{p}_2$

SOLUTIONS TO EXERCISES 4.4

2. In each case $\mathbf{Q}^{-1} = \mathbf{Q}$.
3. (a) and (b) not orthogonal.

   (c) Orthogonal and inverse is $\begin{pmatrix} 1/\sqrt{2} & -1/\sqrt{2} & 0 \\ 1/\sqrt{3} & 1/\sqrt{3} & 1/\sqrt{3} \\ 1/\sqrt{6} & 1/\sqrt{6} & -2/\sqrt{6} \end{pmatrix}$.
4. Lie on the unit circle and perpendicular.
5. Two matrices $\begin{pmatrix} 1/\sqrt{3} & 1/\sqrt{2} & 1/\sqrt{6} \\ 1/\sqrt{3} & 0 & -2/\sqrt{6} \\ 1/\sqrt{3} & -1/\sqrt{2} & 1/\sqrt{6} \end{pmatrix}$ or $\begin{pmatrix} 1/\sqrt{3} & 1/\sqrt{2} & -1/\sqrt{6} \\ 1/\sqrt{3} & 0 & 2/\sqrt{6} \\ 1/\sqrt{3} & -1/\sqrt{2} & -1/\sqrt{6} \end{pmatrix}$.
6. $\begin{pmatrix} \cos(\theta) \\ \sin(\theta) \end{pmatrix}$. Matrix $\mathbf{Q}$ rotates the vector.
7. (a) $\mathbf{Q} = \dfrac{1}{\sqrt{6}} \begin{pmatrix} \sqrt{2} & 1 & -\sqrt{3} \\ \sqrt{2} & -2 & 0 \\ \sqrt{2} & 1 & \sqrt{3} \end{pmatrix}$, $\quad \mathbf{R} = \dfrac{1}{\sqrt{6}} \begin{pmatrix} 6\sqrt{2} & -2\sqrt{2} & 4\sqrt{2} \\ 0 & -2 & -2 \\ 0 & 0 & 4\sqrt{3} \end{pmatrix}$

   (b) $x = 1,\ y = 2,\ z = 3$ (c) $x = 2,\ y = 5$ and $z = 4$

SOLUTIONS TO MISCELLANEOUS EXERCISES 4

4.1. False.

4.2. $\mathbf{w}_1 = \begin{bmatrix} 1 \\ 1 \\ 1 \\ 1 \end{bmatrix}$, $\mathbf{w}_2' = \begin{bmatrix} 1 \\ -1 \\ -1 \\ 1 \end{bmatrix}$ and $\mathbf{w}_3' = \begin{bmatrix} 1 \\ 1 \\ -1 \\ -1 \end{bmatrix}$

4.3. (a) $\left\{ \begin{bmatrix} 1 \\ 0 \\ 1 \\ 0 \end{bmatrix}, \begin{bmatrix} 0 \\ 1 \\ 1 \\ 1 \end{bmatrix}, \begin{bmatrix} 1 \\ 1 \\ 2 \\ 2 \end{bmatrix} \right\}$ (b) $\left\{ \dfrac{1}{\sqrt{2}} \begin{bmatrix} 1 \\ 0 \\ 1 \\ 0 \end{bmatrix}, \dfrac{1}{\sqrt{10}} \begin{bmatrix} -1 \\ 2 \\ 1 \\ 2 \end{bmatrix}, \dfrac{1}{\sqrt{15}} \begin{bmatrix} 1 \\ -2 \\ -1 \\ 3 \end{bmatrix} \right\}$

4.4. (b) Show $\mathbf{f}_1$, $\mathbf{f}_2$ and $\mathbf{f}_3$ are linearly independent. (c) $\left\{ \begin{pmatrix} -1 \\ 0 \\ 2 \end{pmatrix}, \begin{pmatrix} 6 \\ -10 \\ 3 \end{pmatrix}, \begin{pmatrix} 4 \\ 3 \\ 2 \end{pmatrix} \right\}$

4.5. (a) Expanding $\langle \mathbf{v}_1, \mathbf{O} \rangle = \langle \mathbf{v}_2, \mathbf{O} \rangle = \langle \mathbf{v}_3, \mathbf{O} \rangle = \cdots = \langle \mathbf{v}_m, \mathbf{O} \rangle = 0$.

(b) Consider $\mathbf{w} = k_1\mathbf{v}_1 + k_2\mathbf{v}_2 + \cdots + k_n\mathbf{v}_n$ and show that $k_1 = (\mathbf{v}_1 \cdot \mathbf{w}), \ldots, k_n = (\mathbf{v}_n \cdot \mathbf{w})$

4.6. (b) $\left\{ \dfrac{1}{\sqrt{2}} \begin{bmatrix} 0 \\ 1 \\ 1 \end{bmatrix}, \dfrac{1}{\sqrt{6}} \begin{bmatrix} 2 \\ 1 \\ -1 \end{bmatrix}, \dfrac{1}{\sqrt{3}} \begin{bmatrix} 1 \\ -1 \\ 1 \end{bmatrix} \right\}$

4.7. (a) $\sqrt{39}$ (b) No

(c) Yes (d) $\left\{\begin{pmatrix}1\\0\\0\\0\end{pmatrix}, \frac{1}{\sqrt{2}}\begin{pmatrix}0\\1\\0\\1\end{pmatrix}, \begin{pmatrix}0\\0\\1\\0\end{pmatrix}, \frac{1}{\sqrt{2}}\begin{pmatrix}0\\1\\0\\-1\end{pmatrix}\right\}$

4.8. (a) $\sqrt{39}$ (b) No (c) Yes (d) $\frac{1}{\sqrt{2}}\begin{pmatrix}0 & 1\\0 & -1\end{pmatrix}$

(e) $\lambda_1 = 1,\ \lambda_2 = \frac{1}{\sqrt{2}},\ \lambda_3 = 0,\ \lambda_4 = -\frac{1}{\sqrt{2}}$

4.9. (a) $\sqrt{17}$

(b) $\widehat{\mathbf{w}} = \frac{\mathbf{w}}{\|\mathbf{w}\|}$ where $\mathbf{w} = \mathbf{u} - \langle \mathbf{u}, \mathbf{q}_1\rangle \mathbf{q}_1 - \langle \mathbf{u}, \mathbf{q}_2\rangle \mathbf{q}_2 - \langle \mathbf{u}, \mathbf{q}_3\rangle \mathbf{q}_3$

(c) Because $\mathbf{u}$ is linearly dependent on $\mathbf{q}_1, \mathbf{q}_2, \mathbf{q}_3$.

4.10. $\mathbf{w}_1 = \begin{bmatrix}0\\1\\0\\1\\0\end{bmatrix}, \mathbf{w}_2' = \begin{bmatrix}0\\1\\2\\-1\\0\end{bmatrix}$ and $\mathbf{w}_3 = \begin{bmatrix}0\\0\\0\\0\\1\end{bmatrix}$

4.11. (a) and (b). We have $\mathbf{v} = \begin{bmatrix}2, & -1\end{bmatrix}$ and $\mathbf{w} = \begin{bmatrix}1, & 3\end{bmatrix}$.

4.12. $\left\{\frac{1}{\sqrt{2}}, \frac{\sqrt{3}\mathbf{x}}{\sqrt{2}}\right\}$

4.13. Check definitions (i) to (iv) of (4.1). (b) $\|\mathbf{A} - \mathbf{B}\| = \sqrt{9} = 3$ (c) $31.65°$

4.14. $\{4 - 10t + 4t^2\}$

4.15. $\left\{1, \mathbf{x} - \frac{1}{2}, \mathbf{x}^2 - \mathbf{x} + \frac{1}{6}\right\}$

4.16. Show that condition (iv) of definition (4.1) fails.

4.17. Let $\mathbf{u} \in V$ where $\mathbf{u} = k_1\mathbf{v}_1 + k_2\mathbf{v}_2 + k_3\mathbf{v}_3 + \cdots + k_n\mathbf{v}_n$. Show that $k_j = \langle \mathbf{u}, \mathbf{v}_j\rangle$ for $j = 1, 2, 3, \ldots, n$.

4.18. $\left\{\frac{1}{\sqrt{2}}, \sqrt{\frac{3}{2}}\mathbf{x}, \sqrt{\frac{5}{8}}\left(3\mathbf{x}^2 - 1\right), \sqrt{\frac{7}{8}}\left(5\mathbf{x}^3 - 3\mathbf{x}\right)\right\}$

4.19. $\left\{\begin{pmatrix}1\\0\\2\\0\end{pmatrix}, \begin{pmatrix}-2\\1\\1\\4\end{pmatrix}, \begin{pmatrix}2\\5\\-1\\0\end{pmatrix}, \begin{pmatrix}-4\\2\\2\\-3\end{pmatrix}\right\}$

4.20. Let $\mathbf{u} = \mathbf{v}$ then $\langle \mathbf{u}, \mathbf{v}\rangle = \langle \mathbf{u}, \mathbf{u}\rangle = 0$ gives $\mathbf{u} = \mathbf{O}$.

4.21. $\langle k\mathbf{u}, \mathbf{v}\rangle = k\langle \mathbf{u}, \mathbf{v}\rangle = 0$

4.22. Use the definition of orthonormal vectors.

4.23. (a) Use Proposition (4.5).

(b) Apply Minkowski's inequality (4.7).

SOLUTIONS TO EXERCISES 5.1

1. (a) $(3 \;\; 1)^T$ (b) $(5 \;\; -1)^T$ (c) $(1 \;\; \sqrt{2})^T$ (d) $(-3 \;\; -2)^T$
2. (a) $(3 \;\; 5)^T$ (b) $(4 \;\; -4)^T$ (c) $(4 \;\; 6)^T$ (d) $(6 \;\; 8)^T$
3. (a) $\begin{pmatrix} 6 \\ 11 \\ 9 \end{pmatrix}$ (b) $\begin{pmatrix} 5 \\ 2 \\ -9 \end{pmatrix}$ (c) $\begin{pmatrix} 3\pi \\ 7\pi \\ 6\pi \end{pmatrix}$ (d) $\begin{pmatrix} 7/6 \\ 17/12 \\ 5/4 \end{pmatrix}$
5. (a) and (d) linear (b), (c) and (e) not linear
6. (a) linear (b) not linear
7. (a) linear (b) linear (c) not linear

SOLUTIONS TO EXERCISES 5.2

1. (a) $\{\mathbf{O}\}$ (b) $\left\{ r\begin{pmatrix} -1 \\ 1 \end{pmatrix} \,\middle|\, r \in \mathbb{R} \right\}$ (c) $\mathbb{R}^2$

   (d) $\left\{ \begin{pmatrix} 0 \\ a \\ b \end{pmatrix} \,\middle|\, a \in \mathbb{R},\ b \in \mathbb{R} \right\}$ (e) $\left\{ r\begin{pmatrix} 1 \\ 1 \\ 1 \end{pmatrix} \,\middle|\, r \in \mathbb{R} \right\}$
2. $\ker(T) = \{\mathbf{O}\}$ and $\text{range}(T) = M_{nn}$
3. $\ker(T) = \{c \mid c \in \mathbb{R}\} = P_0$ and $\text{range}(T) = P_1$

SOLUTIONS TO EXERCISES 5.3

1. (a) (i) $\left\{ r\begin{pmatrix} -2 \\ 1 \end{pmatrix} \,\middle|\, r \in \mathbb{R} \right\}$ (ii) 1 (iii) $\left\{ \begin{pmatrix} a \\ a \end{pmatrix} \,\middle|\, a \in \mathbb{R} \right\}$ (iv) 1

   (b) (i) $\{\mathbf{O}\}$ (ii) 0 (iii) $\mathbb{R}^3$ (iv) 3

   (c) (i) $\left\{ s\begin{pmatrix} -3 \\ 1 \\ 0 \end{pmatrix} + t\begin{pmatrix} -5 \\ 0 \\ 1 \end{pmatrix} \,\middle|\, s \in \mathbb{R} \text{ and } t \in \mathbb{R} \right\}$ (ii) 2 (iii) $\mathbb{R}$ (iv) 1

   (d) (i) P_0 (ii) 1 (iii) $\left\{ dx^3 + ex^2 + fx \,\middle|\, d,\ e \text{ and } f \in \mathbb{R} \right\}$ (iv) 3

   (e) $\{c \mid c \in \mathbb{R}\} = P_0$ (ii) 1 (iii) P_2 (iv) 3

   (f) (i) $\left\{ ax^3 + bx^2 + cx + d \,\middle|\, a = -\frac{4}{3}b - 2c - 4d \right\}$ (ii) 3 (iii) $\mathbb{R}$ (iv) 1

   (g) (i) $\ker(T) = \left\{ \begin{pmatrix} a & b \\ -a & -b \end{pmatrix} \right\}$ (ii) 2 (iii) P_1 (iv) 2

2. Substitute the values obtained in question 1 into nullity (T) + rank $(T) = n$.
The values of n for each part are as follows:
(a) $n = 2$ (b) $n = 3$ (c) $n = 3$ (d) $n = 4$ (e) $n = 4$ (f) $n = 4$
(g) $n = 4$

3. (i) $B = \{(-3\ \ 1\ \ 0\ \ 0\ \ 0)^T, (3.2\ \ 0\ \ -0.4\ \ 1\ \ 0)^T, (-3.6\ \ 0\ \ 0.2\ \ 0\ \ 1)^T\}$
(ii) $B' = \{(1\ \ 0\ \ 8/3)^T, (0\ \ 1\ \ 2/3)^T\}$

SOLUTIONS TO EXERCISES 5.4

3. (a) and (b) T is not one-to-one nor onto.
(c) T is one-to-one and onto.
(d) T is one-to-one but not onto.

17. $T^{-1}\left[\begin{pmatrix} a \\ b \end{pmatrix}\right] = \frac{1}{3}\begin{pmatrix} a+b \\ a-2b \end{pmatrix}$

SOLUTIONS TO EXERCISES 5.5

1. (a) $\mathbf{A} = \begin{pmatrix} 1 & 1 \\ 2 & 2 \end{pmatrix}$ (b) $\mathbf{A} = \begin{pmatrix} 1 & 1 \\ -1 & -1 \end{pmatrix}$

(c) $\mathbf{A} = \begin{pmatrix} 2 & 3 \\ 1 & -5 \end{pmatrix}$ (d) $\mathbf{A} = \begin{pmatrix} 1 & 1 & 1 \\ 1 & -1 & -1 \\ 2 & 1 & -1 \end{pmatrix}$

(e) $\mathbf{A} = \begin{pmatrix} 2 & -1 & 0 \\ 4 & -1 & 3 \\ 7 & -1 & -1 \end{pmatrix}$ (f) $\mathbf{A} = \begin{pmatrix} 1 & -1 & 1 & -3 \\ -1 & 3 & -7 & -1 \\ 9 & 5 & 6 & 12 \\ 1 & 0 & 0 & 0 \end{pmatrix}$

(g) $\mathbf{A} = \mathbf{O}_4$

2. (a) $\mathbf{A} = \begin{pmatrix} 1 & -1 \\ 1 & 2 \end{pmatrix}$ (b) $\mathbf{A} = \begin{pmatrix} 3 & -2 \\ -1 & 5 \end{pmatrix}$

(c) $\mathbf{A} = \begin{pmatrix} 1 & -1 & -1 \\ 1 & 1 & 1 \end{pmatrix}$ (d) $\mathbf{A} = \mathbf{O}_3$

(e) $\mathbf{A} = \begin{pmatrix} -3 & -5 & -6 \\ -2 & 7 & 5 \\ 0 & 0 & 0 \end{pmatrix}$

3. $\mathbf{A} = \begin{pmatrix} 1 & 4 \\ 2 & 5 \\ 3 & 6 \end{pmatrix}$

4. $\mathbf{A} = \begin{pmatrix} 0 & -1 & -2 \\ 0 & 0 & 2 \end{pmatrix}$ and $T\left(2x^2 + 3x + 1\right) = 3 + 4x$

5. (i) $\mathbf{A} = \begin{pmatrix} 0 & 1 & 0 & 0 \\ 0 & 0 & 2 & 0 \\ 0 & 0 & 0 & 3 \end{pmatrix}$ (ii) $\mathbf{A} = \begin{pmatrix} 0 & 0 & 0 & 3 \\ 0 & 0 & 2 & 0 \\ 0 & 1 & 0 & 0 \end{pmatrix}$ (iii) $\mathbf{A} = \begin{pmatrix} 0 & 0 & 1 & 0 \\ 0 & 2 & 0 & 0 \\ 3 & 0 & 0 & 0 \end{pmatrix}$

$T\left(-1 + 3x - 7x^2 - 2x^3\right) = 3 - 14x - 6x^2.$

6. $\mathbf{A} = \begin{pmatrix} 3 & -1 \\ 7 & -3 \end{pmatrix}$ and $T\left(\begin{pmatrix} -3 \\ 1 \end{pmatrix}\right) = \begin{pmatrix} -2 \\ -4 \end{pmatrix}$

7. $\left\{\mathbf{v}_1 = \begin{pmatrix} 1/\sqrt{2} \\ 1/\sqrt{2} \end{pmatrix}, \mathbf{v}_2 = \begin{pmatrix} -1/\sqrt{2} \\ 1/\sqrt{2} \end{pmatrix}\right\}$ and $\frac{1}{\sqrt{2}}\begin{pmatrix} 3 \\ -1 \end{pmatrix}$

8. $\mathbf{A} = \begin{pmatrix} 7/3 & 1 \\ -10/3 & -2 \\ 14/3 & 3 \end{pmatrix}$ and $T\left(\begin{pmatrix} 2 \\ 1 \end{pmatrix}\right) = \begin{pmatrix} -2 \\ -1 \\ 5 \end{pmatrix}$

9. $\mathbf{A} = \begin{pmatrix} 1 & 0 & 0 & 0 \\ 0 & 0 & 1 & 0 \\ 0 & 1 & 0 & 0 \\ 0 & 0 & 0 & 1 \end{pmatrix}$ and $\begin{pmatrix} 1 & 3 \\ 2 & 4 \end{pmatrix}$

10. $\mathbf{A} = \begin{pmatrix} 0 & -1 \\ 1 & 0 \end{pmatrix}$ (i) $-2\sin(x) + 5\cos(x)$ (ii) $-n\sin(x) + m\cos(x)$

11. $\mathbf{A} = \begin{pmatrix} 1 & 3 & 9 \\ 0 & 1 & 6 \\ 0 & 0 & 1 \end{pmatrix}$ and $T\left(q + nx + mx^2\right) = q + 3n + 9m + (n + 6m)\,x + mx^2$

12. $\mathbf{A} = \begin{pmatrix} 0 & -1 & 0 \\ 1 & 0 & 0 \\ 0 & 0 & 1 \end{pmatrix}$ (i) $T\left(-\sin(x) + 4\cos(x) - 2e^x\right) = -4\sin(x) - \cos(x) - 2e^x$

(ii) $T\left(m\sin(x) + n\cos(x) + pe^x\right) = -n\sin(x) + m\cos(x) + pe^x$

13. $\mathbf{A} = \begin{pmatrix} 2 & 1 & 0 \\ 0 & 2 & 2 \\ 0 & 0 & 2 \end{pmatrix}$ and

$T\left(ae^{2x} + bxe^{2x} + cx^2e^{2x}\right) = \left(2a + b\right)e^{2x} + \left(2b + 2c\right)xe^{2x} + 2cx^2e^{2x}$

SOLUTIONS TO EXERCISES 5.6

1. (i) $\begin{pmatrix} 1 \\ 0 \end{pmatrix}$ and $\begin{pmatrix} 0 \\ 2 \end{pmatrix}$ (ii) $\begin{pmatrix} 1 \\ 0 \end{pmatrix}$ and $\begin{pmatrix} 0 \\ 1 \end{pmatrix}$

2. (a) $2ax + b$ (b) $2ax$ (c) $2a$ (d) $ax^2 + bx$

3. $\begin{pmatrix} 3 & 1 \\ -2 & 1 \end{pmatrix}$ and $\begin{pmatrix} 0 & -1 \\ 5 & 4 \end{pmatrix}$.

(i) $\begin{pmatrix} 8 \\ 3 \end{pmatrix}$ (ii) $\begin{pmatrix} -5 \\ 25 \end{pmatrix}$ (iii) $\begin{pmatrix} -17 \\ 53 \end{pmatrix}$ (iv) $\begin{pmatrix} 8 \\ -1 \end{pmatrix}$

4. $\begin{pmatrix} -6 & 3 & -7 \\ 7 & 0 & -2 \\ 11 & -2 & -5 \end{pmatrix}$ and $\begin{pmatrix} 7 & 17 & 26 \\ -2 & 4 & 5 \\ -7 & -16 & -22 \end{pmatrix}$

(i) $\begin{pmatrix} -21 \\ 1 \\ -8 \end{pmatrix}$ (ii) $\begin{pmatrix} 119 \\ 21 \\ -105 \end{pmatrix}$ (iii) $\begin{pmatrix} -151 \\ -85 \\ 243 \end{pmatrix}$ (iv) $\begin{pmatrix} -17 \\ 31 \\ 54 \end{pmatrix}$

5. (a) (i) $\begin{pmatrix} 0 & 2 & 1 \\ 1 & 1 & 0 \\ 1 & 0 & 0 \end{pmatrix}$ (ii) $\begin{pmatrix} 0 & 0 & 1 \\ 0 & 1 & 1 \\ 1 & 2 & 0 \end{pmatrix}$ (iii) $\begin{pmatrix} 1 & 0 & 0 \\ 2 & 1 & 0 \\ 2 & 4 & 1 \end{pmatrix}$ (iv) $\begin{pmatrix} 1 & 0 & 0 \\ 0 & 1 & 0 \\ 0 & 0 & 1 \end{pmatrix}$

$S^{-1}(\mathbf{p}) = S(\mathbf{p}) = ce^x + bxe^x + ax^2e^x$ (b) $2e^x - 2xe^x + x^2e^x + C$

6. (i) $\begin{pmatrix} 1 & 0 & 0 \\ 0 & 1 & 0 \\ 0 & 0 & 0 \end{pmatrix}$ (ii) $\begin{pmatrix} 1 & 0 & 0 \\ 0 & 1 & 0 \\ 0 & 0 & 0 \end{pmatrix}$ (iii) $\begin{pmatrix} 1 & 0 & 0 \\ 0 & 1 & 0 \\ 0 & 0 & 1 \end{pmatrix}$ (iv) $\begin{pmatrix} 1 & 0 & 0 \\ 0 & 1 & 0 \\ 0 & 0 & 0 \end{pmatrix}$

$(S \circ T)(1 + 2x + 3x^2) = 1 + 2x$, $(T \circ S)(1 + 2x + 3x^2) = 1 + 2x$,
$(T \circ T)(1 + 2x + 3x^2) = 1 + 2x + 3x^2$ and $(S \circ S)(1 + 2x + 3x^2) = 1 + 2x$

7. $\mathbf{A} = \begin{pmatrix} 1 & -1 \\ 1 & 1 \end{pmatrix}$ and $\mathbf{A}^{-1} = \frac{1}{2}\begin{pmatrix} 1 & 1 \\ -1 & 1 \end{pmatrix}$.

8. $T^{-1}\left(\begin{bmatrix} x \\ y \\ z \end{bmatrix}\right) = \frac{1}{2}\begin{pmatrix} x + z \\ y - z \\ x - y \end{pmatrix}$

9. $T^{-1}(\mathbf{p}) = T^{-1}(ax^3 + bx^2 + cx + d) = \frac{a}{4}x^3 + \frac{b}{3}x^2 + \frac{c}{2}x + d$

10. $T^{-1}\left(\begin{bmatrix} x \\ y \\ z \end{bmatrix}\right) = \begin{pmatrix} x - z \\ 2x + y + z \\ 2x - y \end{pmatrix}$

11. (a) Not Invertible (b) Not Invertible (c) Invertible, $T^{-1}(\mathbf{p}) = \mathbf{p}$

12. (i) No change.

13. (i) $\mathbf{x}_1 = \begin{pmatrix} 0.3333 \\ 0.6667 \end{pmatrix}$, $\mathbf{x}_2 = \begin{pmatrix} 0.5556 \\ 0.4444 \end{pmatrix}$, $\mathbf{x}_3 = \begin{pmatrix} 0.4815 \\ 0.5185 \end{pmatrix}$, $\mathbf{x}_4 = \begin{pmatrix} 0.5062 \\ 0.4938 \end{pmatrix}$

and $\mathbf{x}_5 = \begin{pmatrix} 0.4979 \\ 0.5021 \end{pmatrix}$.

(ii) $\begin{pmatrix} 0.5 \\ 0.5 \end{pmatrix}$

SOLUTIONS TO MISCELLANEOUS EXERCISES 5

5.2. $\mathbf{S} = \begin{bmatrix} 2 & -3 & 4 \\ -1 & 1 & 0 \end{bmatrix}$

5.3. $f\left(\begin{bmatrix} x \\ y \end{bmatrix}\right) = \begin{pmatrix} x^2 \\ xy \end{pmatrix}$

5.4. (a) $\mathbf{S} = \begin{pmatrix} 0 & 3 & 2 \\ 3 & -4 & 0 \end{pmatrix}$ (b) T is **not** one-to-one. (c) T is **onto**

5.5. (a) S is a subspace of $\mathbb{R}^2$.

(b) (i) $T(\vec{e}_1) = \begin{bmatrix} 1 \\ -2 \end{bmatrix}$, $T(\vec{e}_2) = \begin{bmatrix} -1 \\ -3 \end{bmatrix}$ (ii) $\mathbf{A} = \begin{bmatrix} 1 & -1 \\ -2 & -3 \end{bmatrix}$ (iii) $\mathbf{w} = \begin{bmatrix} -1 \\ -2 \end{bmatrix}$

5.6. (a) $\mathbf{A} = \begin{pmatrix} 1 & -4 & 2 \\ 2 & 7 & -1 \\ -1 & -8 & 2 \\ 2 & 1 & 1 \end{pmatrix}$ (b) $\left\{ \begin{pmatrix} 1 \\ 2 \\ -1 \\ 2 \end{pmatrix}, \begin{pmatrix} 0 \\ 5 \\ -4 \\ 3 \end{pmatrix} \right\}$

5.7. (d) $\ker\ T = t \begin{pmatrix} -1 \\ 1 \\ 0 \end{pmatrix}$

5.8. (a) $B = \left\{ \begin{pmatrix} 1 \\ 0 \\ 0 \end{pmatrix}, \begin{pmatrix} 0 \\ 1 \\ 0 \end{pmatrix}, \begin{pmatrix} 0 \\ 0 \\ 1 \end{pmatrix} \right\}$ (b) T is onto

(c) $\left\{ \begin{pmatrix} 1 \\ -1 \\ 0 \\ 0 \end{pmatrix} \right\}$ (d) T is **not** one-to-one

5.9. $\begin{bmatrix} 9 \\ -3 \\ 2 \end{bmatrix}$

5.10. $\begin{bmatrix} 1 \\ -2 \\ -4 \end{bmatrix}$

5.11. (a) We have $T\ :\ \mathbb{R}^5 \to \mathbb{R}^6$ which gives $m = 6$ and $n = 5$. (b) 6
(c) T is **not** onto (d) T is one-to-one

5.12. For **all** real h and k provided $h \neq 0$ and $k \neq \dfrac{3}{2}$.

5.13. (a) zero function (c) $\ker(T) = Ae^{2x}$ and dim is 1.

5.14. (c) $\dim\big(\ker(T)\big) = 1$ and $\dim\big(image\,(T)\big) = 2$

5.15. Consider $T\left(\begin{bmatrix} x \\ y \end{bmatrix}\right) = \begin{pmatrix} x - y \\ x - y \end{pmatrix}$ with $\mathbf{v}_1 = \begin{pmatrix} 1 \\ 1 \end{pmatrix}$ and $\mathbf{v}_2 = \begin{pmatrix} 2 \\ 1 \end{pmatrix}$.

5.16. (b) $\mathbf{A} = \begin{pmatrix} 0 & 3 & 0 \\ 0 & 0 & 2 \\ 1 & 0 & 0 \\ 0 & 1 & 1 \end{pmatrix}$ (c) $15 - 2t + 2t^2 + 4t^3$ (d) yes

5.17. (b) (ii) $\mathbf{S} = \begin{bmatrix} 1 & 2 & 0 \\ 0 & 2 & -3 \\ 1 & 0 & 1 \\ 1 & 0 & 0 \end{bmatrix}$

5.18. (b) $\begin{bmatrix} 1/2 \\ 1/2 \\ 1 \\ -1 \end{bmatrix}$ (c) $\begin{bmatrix} 1/2 & -1/2 & 0 & 0 \\ 1/2 & -1/2 & 0 & 0 \\ 0 & 0 & 1 & 1 \\ 0 & 0 & 0 & 1 \end{bmatrix}$ (d) $\begin{bmatrix} 0 \\ 0 \\ 0 \\ -1 \end{bmatrix}$

5.19. (a) (i) f is a linear map (ii) f is **not** a linear map

(b) $Rank\,(f) + Nullity\,(f) = n$

(c) f is **not** injective nor surjective

(d) f is **not** injective nor surjective.

A basis for $\ker(f)$ is $\left\{ \begin{pmatrix} -1 \\ 1 \\ 0 \\ 0 \end{pmatrix}, \begin{pmatrix} 0 \\ 0 \\ -1 \\ 1 \end{pmatrix} \right\}$ and basis for image is $\left\{ \begin{pmatrix} 1 \\ 0 \\ 0 \end{pmatrix}, \begin{pmatrix} 0 \\ 0 \\ 1 \end{pmatrix} \right\}$.

5.20. Option C.

5.21. $\mathbf{C} = \mathbf{BA} = \begin{pmatrix} 9 & -6 & 15 \\ 20 & -1 & 0 \\ 15 & 5 & 5 \end{pmatrix}$

5.22. (a) $\begin{pmatrix} 1 \\ 1 \\ 1 \\ -1 \end{pmatrix}$ (b) $\det(\mathbf{A}) = 27$

(c) A basis for $\mathrm{Im}(\phi)$ is $\left\{ \begin{pmatrix} 1 \\ 0 \\ 0 \\ 0 \end{pmatrix}, \begin{pmatrix} 0 \\ 1 \\ 0 \\ 0 \end{pmatrix}, \begin{pmatrix} 0 \\ 0 \\ 1 \\ 0 \end{pmatrix}, \begin{pmatrix} 0 \\ 0 \\ 0 \\ 1 \end{pmatrix} \right\}$. No basis for $\ker(\phi)$.

(d) ϕ is invertible since $\det(\mathbf{A}) = 27 \neq 0$

5.23. (b) $[T]_B = \begin{pmatrix} -1 & -1 \\ 1 & -1 \end{pmatrix}$ (c) $[T]_B^{-1} = \frac{1}{2}\begin{pmatrix} -1 & 1 \\ -1 & -1 \end{pmatrix}$ (d) $f(x) = \frac{1}{2}\sin x - \frac{5}{2}\cos x$

5.26. True.

SOLUTIONS TO EXERCISES 6.1

1. (a) 4 (b) 1 (c) 1 (d) 0
2. (a) -8 and -8 (b) -13 and -13 (c) $\det(\mathbf{A}) = 0$ and $\det(\mathbf{B}) = 0$
3. In both cases $\det(\mathbf{A}) = \det(\mathbf{B}) = ad - bc$.
4. 6 and -6.
5. $x = 2,\ y = -2$.
6. $\det(\mathbf{A}^2) = \det(\mathbf{A}) \times \det(\mathbf{A})$.
8. not linear

SOLUTIONS TO EXERCISES 6.2

1. (a) -117 (b) 1288 (c) -114
3. $x = -13.38, 5.38$
4. $C = \begin{pmatrix} 7 & 42 & -16 \\ -5 & -30 & 1 \\ -15 & -17 & 3 \end{pmatrix}$, $C^T = \begin{pmatrix} 7 & -5 & -15 \\ 42 & -30 & -17 \\ -16 & 1 & 3 \end{pmatrix}$,

   $\mathbf{A}^{-1} = -\frac{1}{73}\begin{pmatrix} 7 & -5 & -15 \\ 42 & -30 & -17 \\ -16 & 1 & 3 \end{pmatrix}$
5. (a) $\mathbf{A}^{-1} = \begin{pmatrix} 3 & -2 \\ -13 & 9 \end{pmatrix}$ (b) $\mathbf{B}^{-1} = \begin{pmatrix} 5 & -7 \\ -12 & 17 \end{pmatrix}$

   (c) $\mathbf{C}^{-1} = -\frac{1}{7}\begin{pmatrix} 1 & -4 \\ -3 & 5 \end{pmatrix}$ (d) $\mathbf{D}^{-1} = -\frac{1}{147}\begin{pmatrix} 33 & 37 & 32 \\ 60 & 45 & 27 \\ 18 & 38 & 13 \end{pmatrix}$
6. (a) -42 (b) 44 (c) 13 (d) -85
7. (a) 13 (b) 34.5 (c) 0
8. The place sign for a_{31}, a_{56}, a_{62} and a_{65} is 1, -1, 1 and -1 respectively. There is **no** a_{71} entry in a 6 by 6 matrix.
10. $\det(\mathbf{A}) = 0$
11. All real values of k provided $k \neq \sqrt[3]{-10}$.
14. 7
18. 1

SOLUTIONS TO EXERCISES 6.3

1. (a) $\det(\mathbf{A}) = -10$ (b) $\det(\mathbf{B}) = -1$ (c) $\det(\mathbf{C}) = 1$ (d) $\det(\mathbf{D}) = -1$
   (e) $\det(\mathbf{E}) = -0.6$ (f) $\det(\mathbf{F}) = 1$
2. (a) 6 (b) 6 (c) 6 (d) -72 (e) 240 000 (f) -945
   (g) impossible
3. (a) $\det(\mathbf{A}) = \alpha\beta\gamma$ (b) $\det(\mathbf{B}) = \sin(2\theta)$ (c) $\det(\mathbf{C}) = xyz$
4. -1.38(2 dp) and 3.63 (2 dp).
5. (a) -27 (b) 2 (c) -39
6. (a) 18 (b) 0
8. (a) 7 (b) $-5/378$ (c) 600
9. (a) 96 (b) 995 328 (c) 1
10. invertible
11. (a) and (c) are negative. (b) is zero.

SOLUTIONS TO EXERCISES 6.4

1. (a) $x_1 = 1,\ x_2 = -1$ and $x_3 = 3$ (b) $x_1 = 2,\ x_2 = 1$ and $x_3 = -1$
   (c) $x_1 = -2,\ x_2 = 3$ and $x_3 = 2$ (d) $x_1 = -2,\ x_2 = -3,\ x_3 = 1$
2. $x_1 = 1,\ x_2 = -2,\ x_3 = 3,\ x_4 = -4$
3. 15 984
4. $\mathbf{L} = \mathbf{I},\ \mathbf{U} = \mathbf{A}.$
5. (a) $\mathbf{L} = \begin{pmatrix} 1 & 0 & 0 \\ 3 & 1 & 0 \\ 4 & 5 & 1 \end{pmatrix},\ \mathbf{U} = \begin{pmatrix} 1 & 2 & 3 \\ 0 & 1 & 5 \\ 0 & 0 & 1 \end{pmatrix}$ (b) $\mathbf{A}^{-1} = \begin{pmatrix} 84 & -37 & 7 \\ -58 & 26 & -5 \\ 11 & -5 & 1 \end{pmatrix}$
   (c) (i) $x_1 = 31,\ x_2 = -21$ and $x_3 = 4$ (ii) $x_1 = -181,\ x_2 = 131$ and $x_3 = -25$

SOLUTIONS TO MISCELLANEOUS EXERCISES 6

6.1. $\det(\mathbf{A}) = 165$

6.2. $\det(\mathbf{E}) = -20,\ \det(\mathbf{F}) = 30,\ \det(\mathbf{EF}) = -600$ and $\det(\mathbf{E}+\mathbf{F}) = -6$

6.3. 64

6.4. (a) $\det(\mathbf{AB}) = \det(\mathbf{A})\det(\mathbf{B})$ (b) $\det\left(\mathbf{A}^{-1}\right) = \dfrac{1}{\det(\mathbf{A})}$ provided $\det(\mathbf{A}) \neq 0$
(c) $\det(\mathbf{A}+\mathbf{B}) =$ No Formula (d) $\det(3\mathbf{A}) = 3^n \det(\mathbf{A})$
(e) $\det\left(\mathbf{A}^T\right) = \det(\mathbf{A})$

6.5. 56

6.6. $x_2 = \det\begin{pmatrix} 2 & 8 & 1 \\ 3 & 1 & -1 \\ 4 & 10 & 3 \end{pmatrix} \Big/ \det\begin{pmatrix} 2 & 1 & 1 \\ 3 & -2 & -1 \\ 4 & -7 & 3 \end{pmatrix}$

6.7. $x_2 = \dfrac{1}{210}$

6.8. $x_1 = 0,\ x_2 = -2$ and $x_3 = 2$

6.9. -448

6.10. 0

6.11. (a) $\det(\mathbf{A}) = -10$ (b) $\mathbf{A}^{-1} = \dfrac{1}{10}\begin{pmatrix} -2 & -14 & 6 \\ 1 & 12 & -3 \\ 3 & -4 & 1 \end{pmatrix}$

(c) $\det(\mathbf{A}) = -10$

6.12. (i) $\mathbf{A}^{-1} = \begin{pmatrix} 1 & -1 & 2 \\ 1 & 2 & 0 \\ 0 & -1 & 1 \end{pmatrix}$

(ii) $\left(\mathbf{A}^t\right)^{-1} = \begin{pmatrix} 1 & 1 & 0 \\ -1 & 2 & -1 \\ 2 & 0 & 1 \end{pmatrix},\ (3\mathbf{A})^{-1} = \dfrac{1}{3}\begin{pmatrix} 1 & -1 & 2 \\ 1 & 2 & 0 \\ 0 & -1 & 1 \end{pmatrix},\ \left(\mathbf{A}^2\right)^{-1} = \begin{pmatrix} 0 & -5 & 4 \\ 3 & 3 & 2 \\ -1 & -3 & 1 \end{pmatrix}$

6.13. 3

6.14. For the 2 by 2 matrix: $adj\mathbf{A} = \begin{bmatrix} 1 & 1 \\ 1 & 1 \end{bmatrix}$, $\det(\mathbf{A}) = 0$ which means that $\mathbf{A}^{-1}$ does not exist. 3 by 3 gives $adj\mathbf{A} = \begin{pmatrix} 1 & 18 & -11 \\ 1 & 13 & -8 \\ 0 & -2 & 1 \end{pmatrix}$, $\det \mathbf{A} = -1$ and

$$\mathbf{A}^{-1} = \begin{pmatrix} -1 & -18 & 11 \\ -1 & -13 & 8 \\ 0 & 2 & -1 \end{pmatrix}.$$

6.15. (a) -16 means that $\mathbf{A}$ is invertible. (b) $\det\left(\mathbf{A}^T\right) = -16$ (c) $\det\left(\mathbf{A}^{-1}\right) = -\dfrac{1}{16}$

6.16. $D_1 = -7, D_2 = 7, D_3 = 0$

6.17. (a) $\det(\mathbf{A}) = -24$ (b) $\det(\mathbf{B}) = 0$ (c) $\det(\mathbf{C}) = -2$

6.18. (a) $\det(\mathbf{A}) = -17$, $\det(\mathbf{B}) = -12$, $\det(\mathbf{AB}) = 204$, $\det\left(\mathbf{A}^3\right) = -4913$
(b) Take the determinants of $\mathbf{AA}^{-1} = \mathbf{I}$ to show the required result.

6.19. (a) $-(u_1v_1 + u_2v_2 + u_3v_3)$ (b) $100 - (u_1v_1 + u_2v_2 + u_3v_3)$

6.20. $(a-b)(a-c)(c-b)$

6.21. 16

6.22. $\dfrac{9}{4}$

6.23. (a) (i) 8 (ii) 18 (b) (i) 162 (ii) $-\dfrac{3}{5}$ (iii) $-\dfrac{12}{5}$

6.24. (a) (i) 24 (ii) $2abc$ (iii) 432 (b) (i) 4 (ii) -2 (iii) $-\dfrac{3}{2}$

6.25. (a) (i) 8/9 (ii) 12 (iii) 1 (b) 2

6.26. (a) 3 (b) 21 (c) 8 (d) 7

6.27. *adfpru*

6.28. $x = \dfrac{1}{3a-2}, \quad y = \dfrac{3a-3}{3a-2}, \quad z = \dfrac{1-a}{3a-2}$

6.29. Apply this $\det(\mathbf{A}_1\mathbf{A}_2\mathbf{A}_3\cdots\mathbf{A}_n) = \det(\mathbf{A}_1)\det(\mathbf{A}_2)\det(\mathbf{A}_3)\cdots\det(\mathbf{A}_n)$ to get $\det\left(\mathbf{A}^5\right) = \det(\mathbf{A})\det(\mathbf{A})\cdots\det(\mathbf{A}) = 0$

6.30. (a) $\det(\mathbf{A}_2) = 2!$, $\det(\mathbf{A}_3) = 3! = 6$ and $\det(\mathbf{A}_4) = 4! = 24$ (b) $\det(\mathbf{A}_n) = n!$

6.31. $\mathbf{X}^{-1} = \dfrac{1}{ACF}\begin{bmatrix} CF & -BF & BE-CD \\ 0 & AF & -AE \\ 0 & 0 & AC \end{bmatrix}$

6.32. $y = \dfrac{1}{C}$

6.33. -125

6.34. $k \neq 3,\ k \neq 4$ or $k \neq -1$

6.35. $2^4/15^7$

6.36. Not possible

6.37. Use properties of determinants.

6.38. Convert the given matrix into an upper triangular matrix by swapping rows. The number of swaps is $\lfloor n/2 \rfloor$, which means we multiply the resulting determinant which is the product of the leading diagonal by $(-1)^{\lfloor n/2 \rfloor}$.

SOLUTIONS TO EXERCISES 7.1

1. (a) $\lambda_1 = -4$, $\mathbf{u} = \begin{pmatrix} -3/11 \\ 1 \end{pmatrix}$ and $\lambda_2 = 7$, $\mathbf{v} = \begin{pmatrix} 1 \\ 0 \end{pmatrix}$

   (b) $\lambda_1 = 1$, $\mathbf{u} = \begin{pmatrix} 1 \\ 2 \end{pmatrix}$ and $\lambda_2 = 3$, $\mathbf{v} = \begin{pmatrix} 1 \\ 1 \end{pmatrix}$

   (c) $\lambda_1 = -3$, $\mathbf{u} = \begin{pmatrix} -2 \\ 1 \end{pmatrix}$ and $\lambda_2 = 3$, $\mathbf{v} = \begin{pmatrix} 1 \\ 1 \end{pmatrix}$

2. (a) $\lambda_1 = 4$, $\mathbf{u} = s\begin{pmatrix} 20 \\ 9 \\ 15 \end{pmatrix}$ and $\lambda_2 = -5$, $\mathbf{v} = s\begin{pmatrix} -2/3 \\ 0 \\ 1 \end{pmatrix}$

3. $\lambda_1 = 2$, $\mathbf{u} = s\begin{pmatrix} -1 \\ 1 \end{pmatrix}$, $E_2 = \left\{ s\begin{pmatrix} -1 \\ 1 \end{pmatrix} \right\}$ and $\lambda_2 = 4$ $\mathbf{v} = s\begin{pmatrix} 1 \\ 1 \end{pmatrix}$, $E_4 = \left\{ s\begin{pmatrix} 1 \\ 1 \end{pmatrix} \right\}$.

4. $\lambda_1 = -2$, $\mathbf{u} = s\begin{pmatrix} 2 \\ 7 \end{pmatrix}$, $E_{-2} = \left\{ s\begin{pmatrix} 2 \\ 7 \end{pmatrix} \right\}$ and $\lambda_2 = 3$, $\mathbf{v} = s\begin{pmatrix} 1 \\ 1 \end{pmatrix}$, $E_3 = \left\{ s\begin{pmatrix} 1 \\ 1 \end{pmatrix} \right\}$

   A basis vector for E_{-2} is $\begin{pmatrix} 2 \\ 7 \end{pmatrix}$ and a basis vector for E_3 is $\begin{pmatrix} 1 \\ 1 \end{pmatrix}$.

5. (a) The eigenvalues of $\mathbf{A}$ are $\lambda_1 = -7$, $\lambda_2 = 6$.
   (b) The eigenvalues of $\mathbf{B}$ are $\lambda_3 = -14$, $\lambda_4 = 12$.
   (c) If t is the eigenvalue of matrix $\mathbf{B}$ and λ is the eigenvalue of matrix $\mathbf{A}$ then $t = 2\lambda$.

8. $\lambda_1 = 1$, $\mathbf{u} = s\begin{pmatrix} 1 \\ 1 \\ -2 \end{pmatrix}$, basis vector is $\begin{pmatrix} 1 \\ 1 \\ -2 \end{pmatrix}$. $\lambda_2 = 4$, $\mathbf{v} = s\begin{pmatrix} 1 \\ 1 \\ 1 \end{pmatrix}$,

   basis vector is $\begin{pmatrix} 1 \\ 1 \\ 1 \end{pmatrix}$. $\lambda_3 = -1$, $\mathbf{w} = s\begin{pmatrix} -1 \\ 1 \\ 0 \end{pmatrix}$, basis vector is $\begin{pmatrix} -1 \\ 1 \\ 0 \end{pmatrix}$.

SOLUTIONS TO EXERCISES 7.2

The variables are non-zero in the following answers.

1. A basis vector for E_3 is $\begin{pmatrix} 1 \\ -1 \end{pmatrix}$ which corresponds to $\lambda_{1,\,2} = 3$.

2. A basis vector for E_{-5} is $\begin{pmatrix} 1 \\ 1 \end{pmatrix}$ which corresponds to $\lambda_{1,\,2} = -5$.

5. A basis vector for E_1 is $\begin{pmatrix} 0 \\ 0 \\ 1 \end{pmatrix}$ which correspond to $\lambda_{1,\,2,\,3} = 1$.

6. (a) Basis vectors for E_5 is $B = \left\{ \begin{pmatrix} 1 \\ 0 \\ 0 \end{pmatrix}, \begin{pmatrix} 0 \\ 1 \\ 0 \end{pmatrix} \right\}$ which corresponds to $\lambda_{1,\,2} = 5$ and a basis vector for E_2 is $B = \left\{ \begin{pmatrix} 0 \\ 0 \\ 1 \end{pmatrix} \right\}$ which corresponds to $\lambda_3 = 2$.

(b) For $\lambda_1 = 1$, $B = \left\{ \begin{pmatrix} 1 \\ 0 \\ 0 \end{pmatrix} \right\}$. For $\lambda_2 = 5$, $B = \left\{ \begin{pmatrix} 1 \\ 2 \\ 0 \end{pmatrix} \right\}$ and for $\lambda_3 = 9$, $B = \left\{ \begin{pmatrix} 7 \\ 4 \\ 16 \end{pmatrix} \right\}$

(c) For $\lambda_{1,\,2,\,3} = -2$ and a basis is $B = \left\{ \begin{pmatrix} 0 \\ 0 \\ 1 \end{pmatrix} \right\}$

7. (a) $\lambda_{1,\,2} = 7$, $\mathbf{u} = s\begin{pmatrix} 1 \\ 0 \\ 0 \\ 0 \end{pmatrix} + t\begin{pmatrix} 0 \\ 1 \\ 0 \\ 0 \end{pmatrix}$ and $\lambda_{3,\,4} = 5$, $\mathbf{v} = s\begin{pmatrix} -1 \\ -2 \\ 1 \\ 0 \end{pmatrix}$

(b) $\lambda_{1,\,2,\,3} = 1$, $\mathbf{u} = \begin{pmatrix} s \\ t \\ r \\ 0 \end{pmatrix} = s\begin{pmatrix} 1 \\ 0 \\ 0 \\ 0 \end{pmatrix} + t\begin{pmatrix} 0 \\ 1 \\ 0 \\ 0 \end{pmatrix} + r\begin{pmatrix} 0 \\ 0 \\ 1 \\ 0 \end{pmatrix}$ and $\lambda_4 = 3$, $\mathbf{v} = s\begin{pmatrix} 0 \\ 0 \\ 0 \\ 1 \end{pmatrix}$

(c) $\lambda_{1,\,2,\,3,\,4} = 3$, $\mathbf{u} = s\begin{pmatrix} 1 \\ 0 \\ 0 \\ 0 \end{pmatrix} + t\begin{pmatrix} 0 \\ 1 \\ 0 \\ 0 \end{pmatrix} + r\begin{pmatrix} 0 \\ 0 \\ 1 \\ 0 \end{pmatrix} + q\begin{pmatrix} 0 \\ 0 \\ 0 \\ 1 \end{pmatrix}$

11. (a) (i) Eigenvalues $\lambda_1 = 1$, $\lambda_2 = 2$, $\lambda_3 = 3$ and $\lambda_4 = 4$.

(ii) Eigenvalues of $\mathbf{A}^5$ are $(\lambda_1)^5 = 1$, $(\lambda_2)^5 = 32$, $(\lambda_3)^5 = 243$ and $(\lambda_4)^5 = 1024$

(iii) Eigenvalues of $\mathbf{A}^{-1}$ are $(\lambda_1)^{-1} = 1$, $(\lambda_2)^{-1} = \frac{1}{2}$, $(\lambda_3)^{-1} = \frac{1}{3}$ and $(\lambda_4)^{-1} = \frac{1}{4}$

(iv) $\det(\mathbf{A}) = 24$ (v) $tr(\mathbf{A}) = 10$

(b) (i) $\lambda_1 = -1$, $\lambda_2 = 6$, $\lambda_3 = -8$ and $\lambda_4 = 3$.

(ii) $(\lambda_1)^5 = -1$, $(\lambda_2)^5 = 7776$, $(\lambda_3)^5 = -32768$ and $(\lambda_4)^5 = 243$

(iii) Eigenvalues of $\mathbf{A}^{-1}$ are $(\lambda_1)^{-1} = -1$, $(\lambda_2)^{-1} = \frac{1}{6}$, $(\lambda_3)^{-1} = -\frac{1}{8}$ and $(\lambda_4)^{-1} = \frac{1}{3}$

(iv) $\det(\mathbf{A}) = 144$ (v) $tr(\mathbf{A}) = 0$

(c) (i) $\lambda_1 = 2$, $\lambda_2 = -4$, $\lambda_3 = -7$ and $\lambda_4 = 0$.

(ii) Eigenvalues of $\mathbf{A}^5$ $(\lambda_1)^5 = 32,\ (\lambda_2)^5 = -1024,\ (\lambda_3)^5 = -16807$ and $(\lambda_4)^5 = 0$

(iii) $\mathbf{A}^{-1}$ does *not* exist.

(iv) $\det(\mathbf{A}) = 0$ (v) $tr(\mathbf{A}) = -9$

12. $p(\mathbf{A}) = \mathbf{A}^2 - 3\mathbf{A} + 4\mathbf{I} = \mathbf{O}$ and $\mathbf{A}^{-1} = \frac{1}{4}\begin{pmatrix} 1 & -1 \\ 2 & 2 \end{pmatrix}$

13. $\mathbf{A}^2 = \begin{pmatrix} 51 & 50 \\ 30 & 31 \end{pmatrix}$ and $\mathbf{A}^3 = \begin{pmatrix} 456 & 455 \\ 273 & 274 \end{pmatrix}$

14. $\mathbf{A}^{-1} = -\frac{1}{4}(\mathbf{A}^2 - 4\mathbf{A} - \mathbf{I})$ and $\mathbf{A}^4 = 17\mathbf{A}^2 - 16\mathbf{I}$

SOLUTIONS TO EXERCISES 7.3

1. (i) $\lambda_1 = 1, \mathbf{u} = \begin{pmatrix} 1 \\ 0 \end{pmatrix}$ and $\lambda_2 = 2, \mathbf{v} = \begin{pmatrix} 0 \\ 1 \end{pmatrix}$ (ii) $\mathbf{P} = \begin{pmatrix} 1 & 0 \\ 0 & 1 \end{pmatrix} = \mathbf{I}, \mathbf{D} = \begin{pmatrix} 1 & 0 \\ 0 & 2 \end{pmatrix} = \mathbf{A}$

(b) (i) $\lambda_1 = 0, \mathbf{u} = \begin{pmatrix} -1 \\ 1 \end{pmatrix}$ and $\lambda_2 = 2, \mathbf{v} = \begin{pmatrix} 1 \\ 1 \end{pmatrix}$ (ii) $\mathbf{P} = \begin{pmatrix} -1 & 1 \\ 1 & 1 \end{pmatrix}, \mathbf{D} = \begin{pmatrix} 0 & 0 \\ 0 & 2 \end{pmatrix}$

(c) (i) $\lambda_1 = 3, \mathbf{u} = \begin{pmatrix} 1 \\ -4 \end{pmatrix}$ and $\lambda_2 = 4, \mathbf{v} = \begin{pmatrix} 0 \\ 1 \end{pmatrix}$ (ii) $\mathbf{P} = \begin{pmatrix} 1 & 0 \\ -4 & 1 \end{pmatrix}, \mathbf{D} = \begin{pmatrix} 3 & 0 \\ 0 & 4 \end{pmatrix}$

(d) (i) $\lambda_1 = 1, \mathbf{u} = \begin{pmatrix} 2 \\ -1 \end{pmatrix}$ and $\lambda_2 = 4, \mathbf{v} = \begin{pmatrix} 1 \\ 1 \end{pmatrix}$ (ii) $\mathbf{P} = \begin{pmatrix} 2 & 1 \\ -1 & 1 \end{pmatrix}, \mathbf{D} = \begin{pmatrix} 1 & 0 \\ 0 & 4 \end{pmatrix}$

2. (i) (a) $\begin{pmatrix} 1 & 0 \\ 0 & 32 \end{pmatrix}$ (b) $\begin{pmatrix} 16 & 16 \\ 16 & 16 \end{pmatrix}$ (c) $\begin{pmatrix} 243 & 0 \\ 3124 & 1024 \end{pmatrix}$ (d) $\begin{pmatrix} 342 & 682 \\ 341 & 683 \end{pmatrix}$

(ii) $\begin{pmatrix} 1/\sqrt{3} & 0 \\ 2 - 4/\sqrt{3} & 1/2 \end{pmatrix}$

3. (a) (i) $\lambda_1 = 1, \mathbf{u} = \begin{pmatrix} 1 \\ 0 \\ 0 \end{pmatrix}, \lambda_2 = 2, \mathbf{v} = \begin{pmatrix} 0 \\ 1 \\ 0 \end{pmatrix}$ and $\lambda_3 = 3, \mathbf{w} = \begin{pmatrix} 0 \\ 0 \\ 1 \end{pmatrix}$

(ii) $\mathbf{P} = \mathbf{I}, \mathbf{D} = \begin{pmatrix} 1 & 0 & 0 \\ 0 & 2 & 0 \\ 0 & 0 & 3 \end{pmatrix}$ (iii) $\mathbf{A}^4 = \begin{pmatrix} 1 & 0 & 0 \\ 0 & 16 & 0 \\ 0 & 0 & 81 \end{pmatrix}$

(b) (i) $\lambda_1 = -1, \mathbf{u} = \begin{pmatrix} 1 \\ 0 \\ 0 \end{pmatrix}, \lambda_2 = 4, \mathbf{v} = \begin{pmatrix} 4 \\ 5 \\ 0 \end{pmatrix}$ and $\lambda_3 = 5, \mathbf{w} = \begin{pmatrix} 2 \\ 3 \\ 1 \end{pmatrix}$

(ii) $\mathbf{P} = \begin{pmatrix} 1 & 4 & 2 \\ 0 & 5 & 3 \\ 0 & 0 & 1 \end{pmatrix}, \mathbf{D} = \begin{pmatrix} -1 & 0 & 0 \\ 0 & 4 & 0 \\ 0 & 0 & 5 \end{pmatrix}$ (iii) $\mathbf{A}^4 = \begin{pmatrix} 1 & 204 & 636 \\ 0 & 256 & 1107 \\ 0 & 0 & 625 \end{pmatrix}$

(c) (i) $\lambda_1 = 2, \mathbf{u} = \begin{pmatrix} -12 \\ 4 \\ 1 \end{pmatrix}, \lambda_2 = 5, \mathbf{v} = \begin{pmatrix} 0 \\ 1 \\ -2 \end{pmatrix}$ and $\lambda_3 = 6, \mathbf{w} = \begin{pmatrix} 0 \\ 0 \\ 1 \end{pmatrix}$

(ii) $\mathbf{P} = \begin{pmatrix} -12 & 0 & 0 \\ 4 & 1 & 0 \\ 1 & -2 & 1 \end{pmatrix}$, $\mathbf{D} = \begin{pmatrix} 2 & 0 & 0 \\ 0 & 5 & 0 \\ 0 & 0 & 6 \end{pmatrix}$ (iii) $\mathbf{A}^4 = \begin{pmatrix} 16 & 0 & 0 \\ 203 & 625 & 0 \\ 554 & 1342 & 1296 \end{pmatrix}$

4. All (a) (b) and (c) are diagonalizable.

5. $\mathbf{D} = \mathbf{P}^{-1}\mathbf{AP} = \begin{pmatrix} 5 & 0 \\ 0 & -1 \end{pmatrix}$

6. $\mathbf{A}$ is diagonalizable, $\mathbf{D} = \begin{pmatrix} -2 & 0 & 0 \\ 0 & -5 & 0 \\ 0 & 0 & -1 \end{pmatrix}$, $\mathbf{P} = \begin{pmatrix} 1 & 5 & 0 \\ 2 & 4 & 0 \\ 0 & 0 & 1 \end{pmatrix}$,

$\mathbf{A}^3 = \begin{pmatrix} -203 & 97.5 & 0 \\ -156 & 70 & 0 \\ 0 & 0 & -1 \end{pmatrix}$

7. (a) Only one eigenvalue $\lambda = 3$ and one independent eigenvector $\mathbf{u} = \begin{pmatrix} 1 \\ -1 \end{pmatrix}$.

(b) Only one independent e.vector $\mathbf{u} = \begin{pmatrix} -2 \\ 1 \end{pmatrix}$ to the only eigenvalue $\lambda = -4$.

(c) Only 1 linearly independent eigenvector $\begin{pmatrix} 1 & 0 & 0 \end{pmatrix}^T$

8. (i) $\mathbf{A}^{11} = \begin{pmatrix} -2050 & 1366 \\ -6147 & 4097 \end{pmatrix}$ (ii) $\dfrac{1}{2}\begin{pmatrix} -5 & 2 \\ -9 & 4 \end{pmatrix}$

12. $\begin{pmatrix} 1 & 0 \\ 0 & 1 \end{pmatrix} + \begin{pmatrix} 3 & 5 \\ 0 & 2 \end{pmatrix} t + \begin{pmatrix} 9 & 25 \\ 0 & 4 \end{pmatrix} \dfrac{t^2}{2!} + \begin{pmatrix} 27 & 95 \\ 0 & 8 \end{pmatrix} \dfrac{t^3}{3!} + \begin{pmatrix} 81 & 325 \\ 0 & 16 \end{pmatrix} \dfrac{t^4}{4!} + \cdots$

13. $\mathbf{P} = \begin{pmatrix} \lambda_1 & \lambda_2 \\ 1 & 1 \end{pmatrix}$, $\mathbf{P}^{-1} = \dfrac{1}{\lambda_1 - \lambda_2}\begin{pmatrix} 1 & -\lambda_2 \\ -1 & \lambda_1 \end{pmatrix}$, $\mathbf{D} = \begin{pmatrix} \lambda_1 & 0 \\ 0 & \lambda_2 \end{pmatrix}$

where $\lambda_1 = \dfrac{1+\sqrt{5}}{2}$ and $\lambda_2 = \dfrac{1-\sqrt{5}}{2}$.

19. $\mathbf{O}$

SOLUTIONS TO EXERCISES 7.4

1. (a) $\mathbf{Q} = \mathbf{I}$ and $\mathbf{D} = \mathbf{A}$ (b) $\mathbf{Q} = \dfrac{1}{\sqrt{2}}\begin{pmatrix} 1 & 1 \\ -1 & 1 \end{pmatrix}$, $\mathbf{D} = \begin{pmatrix} 0 & 0 \\ 0 & 2 \end{pmatrix}$

(c) $\mathbf{Q} = \dfrac{1}{\sqrt{2}}\begin{pmatrix} 1 & 1 \\ -1 & 1 \end{pmatrix}$, $\mathbf{D} = \begin{pmatrix} 1 & 0 \\ 0 & 3 \end{pmatrix}$ (d) $\mathbf{Q} = \dfrac{1}{\sqrt{13}}\begin{pmatrix} 3 & -2 \\ 2 & 3 \end{pmatrix}$, $\mathbf{D} = \begin{pmatrix} 13 & 0 \\ 0 & -13 \end{pmatrix}$

2. (a) $\mathbf{Q} = \dfrac{1}{\sqrt{10}}\begin{pmatrix} -1 & 3 \\ 3 & 1 \end{pmatrix}$, $\mathbf{D} = \begin{pmatrix} 0 & 0 \\ 0 & 10 \end{pmatrix}$ (b) $\mathbf{Q} = \dfrac{1}{\sqrt{6}}\begin{pmatrix} 2 & \sqrt{2} \\ \sqrt{2} & -2 \end{pmatrix}$, $\mathbf{D} = \begin{pmatrix} 4 & 0 \\ 0 & 1 \end{pmatrix}$

(c) $\mathbf{Q} = \dfrac{1}{\sqrt{12}}\begin{pmatrix} -3 & \sqrt{3} \\ \sqrt{3} & 3 \end{pmatrix}$, $\mathbf{D} = \begin{pmatrix} -6 & 0 \\ 0 & -2 \end{pmatrix}$ (d) $\mathbf{Q} = \dfrac{1}{2}\begin{pmatrix} -1 & \sqrt{3} \\ \sqrt{3} & 1 \end{pmatrix}$,

$\mathbf{D} = \begin{pmatrix} -1 & 0 \\ 0 & 7 \end{pmatrix}$

3. (a) $\mathbf{Q} = \mathbf{I}$ and $\mathbf{D} = \mathbf{A}$ (b) $\mathbf{Q} = \begin{pmatrix} 1/\sqrt{6} & 1/\sqrt{2} & 1/\sqrt{3} \\ 1/\sqrt{6} & -1/\sqrt{2} & 1/\sqrt{3} \\ -2/\sqrt{6} & 0 & 1/\sqrt{3} \end{pmatrix}$, $\mathbf{D} = \begin{pmatrix} 0 & 0 & 0 \\ 0 & 0 & 0 \\ 0 & 0 & 6 \end{pmatrix}$

(c) $\mathbf{Q} = \frac{1}{\sqrt{2}} \begin{pmatrix} \sqrt{2} & 0 & 0 \\ 0 & 1 & 1 \\ 0 & -1 & 1 \end{pmatrix}$, $\mathbf{D} = \begin{pmatrix} 0 & 0 & 0 \\ 0 & 0 & 0 \\ 0 & 0 & 2 \end{pmatrix}$

4. (a) $\mathbf{Q} = \begin{pmatrix} -2/\sqrt{6} & 0 & 1/\sqrt{3} \\ 1/\sqrt{6} & -1/\sqrt{2} & 1/\sqrt{3} \\ 1/\sqrt{6} & 1/\sqrt{2} & 1/\sqrt{3} \end{pmatrix}$, $\mathbf{D} = \begin{pmatrix} -1 & 0 & 0 \\ 0 & -1 & 0 \\ 0 & 0 & 5 \end{pmatrix}$

(b) $\mathbf{Q} = \begin{pmatrix} -1/\sqrt{2} & -1/\sqrt{6} & 1/\sqrt{3} \\ 1/\sqrt{2} & -1/\sqrt{6} & 1/\sqrt{3} \\ 0 & 2/\sqrt{6} & 1/\sqrt{3} \end{pmatrix}$, $\mathbf{D} = \begin{pmatrix} 1 & 0 & 0 \\ 0 & 1 & 0 \\ 0 & 0 & 4 \end{pmatrix}$

(c) $\mathbf{Q} = \begin{pmatrix} -1/\sqrt{2} & 1/3\sqrt{2} & 2/3 \\ 1/\sqrt{2} & 1/3\sqrt{2} & 2/3 \\ 0 & -4/3\sqrt{2} & 1/3 \end{pmatrix}$, $\mathbf{D} = \begin{pmatrix} -9 & 0 & 0 \\ 0 & -9 & 0 \\ 0 & 0 & 0 \end{pmatrix}$

8. $\mathbf{Q} = \begin{pmatrix} \dfrac{b}{\sqrt{b^2 + (\lambda_1 - a)^2}} & \dfrac{\lambda_2 - c}{\sqrt{(\lambda_2 - c)^2 + b^2}} \\ \dfrac{\lambda_1 - a}{\sqrt{b^2 + (\lambda_1 - a)^2}} & \dfrac{b}{\sqrt{(\lambda_2 - c)^2 + b^2}} \end{pmatrix}$ where λ_1 and λ_2 are eigenvalues of $\mathbf{A}$.

9. By taking the inverse of $\mathbf{Q}^T\mathbf{A}\mathbf{Q} = \mathbf{D}$ show that $\mathbf{D}^{-1} = \mathbf{Q}^{-1}\mathbf{A}^{-1}\mathbf{Q}$.

SOLUTIONS TO EXERCISES 7.5

1. (a) $\mathbf{U} = \mathbf{I}$, $\mathbf{D} = \begin{pmatrix} 1 & 0 \\ 0 & 2 \end{pmatrix}$, $\mathbf{V} = \mathbf{I}$ (b) $\mathbf{U} = \frac{1}{\sqrt{5}} \begin{pmatrix} 1 & 2 \\ 2 & -1 \end{pmatrix}$, $\mathbf{D} = \begin{pmatrix} 9 & 0 \\ 0 & 1 \end{pmatrix}$,

$\mathbf{V}^T = \frac{1}{\sqrt{5}} \begin{pmatrix} 1 & 2 \\ -2 & 1 \end{pmatrix}$

(c) $\mathbf{U} = \begin{pmatrix} 1/\sqrt{30} & -2/\sqrt{5} & 1/\sqrt{6} \\ 2/\sqrt{30} & 1/\sqrt{5} & 2/\sqrt{6} \\ 5/\sqrt{30} & 0 & -1/\sqrt{6} \end{pmatrix}$, $\mathbf{D} = \begin{pmatrix} \sqrt{6} & 0 \\ 0 & 1 \\ 0 & 0 \end{pmatrix}$, $\mathbf{V}^T = \frac{1}{\sqrt{5}} \begin{pmatrix} 1 & 2 \\ -2 & 1 \end{pmatrix}$

(d) $\mathbf{U} = \begin{pmatrix} 6/\sqrt{180} & -2/\sqrt{5} \\ 12/\sqrt{180} & 1/\sqrt{5} \end{pmatrix}$, $\mathbf{D} = \begin{pmatrix} \sqrt{6} & 0 & 0 \\ 0 & 1 & 0 \end{pmatrix}$,

$\mathbf{V}^T = \begin{pmatrix} 1/\sqrt{30} & 2/\sqrt{30} & 5/\sqrt{30} \\ -2/\sqrt{5} & 1/\sqrt{5} & 0 \\ 1/\sqrt{6} & 2/\sqrt{6} & -1/\sqrt{6} \end{pmatrix}$

(e) $\mathbf{U} = \frac{1}{\sqrt{10}} \begin{pmatrix} 1 & 3 \\ 3 & -1 \end{pmatrix}$, $\mathbf{D} = \begin{pmatrix} \sqrt{20} & 0 \\ 0 & 0 \end{pmatrix}$, $\mathbf{V}^T = \frac{1}{\sqrt{2}} \begin{pmatrix} 1 & 1 \\ 1 & -1 \end{pmatrix}$

(f) $\mathbf{U} = \frac{1}{\sqrt{10}} \begin{pmatrix} 1 & 3 \\ 3 & -1 \end{pmatrix}$, $\mathbf{D} = \begin{pmatrix} \sqrt{30} & 0 & 0 \\ 0 & 0 & 0 \end{pmatrix}$, $\mathbf{V}^T = \begin{pmatrix} 1/\sqrt{3} & 1/\sqrt{3} & 1/\sqrt{3} \\ 1/\sqrt{2} & -1/\sqrt{2} & 0 \\ 1/\sqrt{6} & 1/\sqrt{6} & -2/\sqrt{6} \end{pmatrix}$

SOLUTIONS TO MISCELLANEOUS EXERCISES 7

7.1. (a) $\lambda_1 = 1$ and $\lambda_2 = 3$ (b) $\mathbf{Q} = \begin{pmatrix} 1 & 1 \\ 0 & 2 \end{pmatrix}$, $\mathbf{D} = \begin{pmatrix} 1 & 0 \\ 0 & 3 \end{pmatrix}$

(c) $\mathbf{A}^5 = \begin{pmatrix} 1 & 121 \\ 0 & 243 \end{pmatrix}$

7.2. $\mathbf{A} = \begin{pmatrix} 4 & 5 \\ -3 & -4 \end{pmatrix}$

7.3. $\lambda_{1,2} = -2$ and $\lambda_3 = 4$, E_{-2} is $\left\{ \begin{pmatrix} -1 \\ 0 \\ 1 \end{pmatrix}, \begin{pmatrix} -1 \\ 1 \\ 0 \end{pmatrix} \right\}$, E_4 is $\left\{ \begin{pmatrix} 1 \\ 1 \\ 1 \end{pmatrix} \right\}$

7.4. λ^{-1} and linearly independent eigenvectors.

7.5. (a) $\mathbf{S} = \begin{pmatrix} 5 & 1 \\ 1 & -3 \end{pmatrix}$, $\Lambda = \begin{pmatrix} 8 & 0 \\ 0 & -8 \end{pmatrix}$ (b) $\mathbf{B} = \frac{1}{4}\begin{pmatrix} 7 & 5 \\ 3 & -7 \end{pmatrix} = \frac{1}{4}\mathbf{A}$

7.6. (a) (i) $\lambda_1 = 3$, $\mathbf{u} = \begin{pmatrix} 3 \\ 1 \end{pmatrix}$ and $\lambda_2 = 1$, $\mathbf{v} = \begin{pmatrix} 1 \\ 1 \end{pmatrix}$ (ii) $\mathbf{P} = \begin{pmatrix} 3 & 1 \\ 1 & 1 \end{pmatrix}$

(iii) The eigenvalues of $\mathbf{A}^{2008}$ are $\lambda_1 = 1$, $\lambda_2 = 3^{2008}$ and $\det\left(\mathbf{A}^{2008}\right) = 3^{2008}$.

(b) We have three linearly independent eigenvectors for a 3 by 3 matrix so the matrix is diagonalisable.

7.7. (a) $\mathbf{S} = \begin{pmatrix} -1 & 0 \\ 10 & 1 \end{pmatrix}$ (b) $\mathbf{S} = \begin{pmatrix} 0 & 1 \\ 1 & 0 \end{pmatrix}$

7.8. Eigenvalues are distinct $\lambda_1 = \sqrt{2}$ and $\lambda_2 = -\sqrt{2}$ therefore the matrix **B** is diagonalisable. The matrix $\mathbf{P} = \begin{pmatrix} 1+\sqrt{2} & 1-\sqrt{2} \\ 1 & 1 \end{pmatrix}$ and $\mathbf{D} = \begin{pmatrix} \sqrt{2} & 0 \\ 0 & -\sqrt{2} \end{pmatrix}$.

7.9. (a) $\left(4, \begin{pmatrix} 0 \\ 1 \\ 1 \end{pmatrix}\right)$ (b) $\mathbf{P} = \begin{pmatrix} 2 & 1 & 0 \\ -1 & 0 & 1 \\ 1 & 0 & 1 \end{pmatrix}$ and $\mathbf{D} = \begin{pmatrix} 2 & 0 & 0 \\ 0 & 4 & 0 \\ 0 & 0 & 4 \end{pmatrix}$

7.10. Show that $\mathbf{u}^T\mathbf{v} = 0$ by taking the transpose of $\mathbf{Au} = \lambda_1\mathbf{u}$.

7.11. (a) $\lambda_1 = 2$, $\mathbf{u} = \begin{pmatrix} 1 \\ -2 \end{pmatrix}$ and $\lambda_2 = 7$, $\mathbf{v} = \begin{pmatrix} 2 \\ 1 \end{pmatrix}$ (b) $\begin{bmatrix} 1 \\ 1 \\ 1 \\ -1 \end{bmatrix}$ with eigenvalue 2

7.12. (a) $\lambda_1 = 1$, $\mathbf{u} = \begin{pmatrix} -1 \\ 1 \\ 1 \end{pmatrix}$, $\lambda_2 = 2$, $\mathbf{v} = \begin{pmatrix} 0 \\ 1 \\ 1 \end{pmatrix}$ and $\lambda_3 = 0$, $\mathbf{w} = \begin{pmatrix} 1 \\ -1 \\ 0 \end{pmatrix}$

(b) We have distinct eigenvalues therefore **A** is diagonalizable.

7.13. (a) (i) Substitute these into $\mathbf{Ax} = \lambda\mathbf{x}$. (ii) $\mathbf{Q} = \begin{pmatrix} 1/\sqrt{6} & 1/\sqrt{2} & 1/\sqrt{3} \\ 1/\sqrt{6} & -1/\sqrt{2} & 1/\sqrt{3} \\ -2/\sqrt{6} & 0 & 1/\sqrt{3} \end{pmatrix}$

(iii) $\mathbf{A}^3 = \begin{pmatrix} 9 & 9 & 9 \\ 9 & 9 & 9 \\ 9 & 9 & 9 \end{pmatrix}$

(b) Show that they have the same $p_{B^T}(\lambda) = p_B(\lambda)$.

7.14. (b) $\lambda_1 = 3$ (c) $\mathbf{P} = \begin{pmatrix} 1 & -1 & -2 \\ 1 & 0 & 1 \\ 1 & 1 & 0 \end{pmatrix}$, $\mathbf{P}^{-1} = \frac{1}{4}\begin{pmatrix} 1 & 2 & 1 \\ -1 & -2 & 3 \\ -1 & 2 & -1 \end{pmatrix}$

7.15. (a) $\mathbf{P} = \begin{pmatrix} -1 & 1 & -1 \\ 4 & 0 & 1 \\ 2 & 1 & 0 \end{pmatrix}$, $\mathbf{P}^{-1} = \begin{pmatrix} 1 & 1 & -1 \\ -2 & -2 & 3 \\ -4 & -3 & 4 \end{pmatrix}$, $\mathbf{P}^{-1}\mathbf{AP} = \begin{pmatrix} 5 & 0 & 0 \\ 0 & 3 & 0 \\ 0 & 0 & 3 \end{pmatrix}$

(b) Check $p(\mathbf{A}) = (\mathbf{A} - 5\mathbf{I})(\mathbf{A} - 3\mathbf{I})^2 = \mathbf{O}$

7.16. (b) Real eigenvalues of $\mathbf{A}$ are $t_1 = 1$ and $t_2 = 9$. A basis for E_1 is $\begin{pmatrix} 0 \\ 0 \\ 9 \\ -1 \end{pmatrix}$.

7.17. $\lambda_{1,2} = 1$, $\lambda_{3,4} = -1$ and $\lambda_5 = -3$

7.18. (a) $\lambda_1 = 0$, $\lambda_2 = 1$ and $\lambda_3 = 3$

(b) $\mathbf{u} = \begin{pmatrix} -1 \\ 1 \\ -1 \end{pmatrix}$, $\mathbf{v} = \begin{pmatrix} 1 \\ 0 \\ -1 \end{pmatrix}$, $\mathbf{w} = \begin{pmatrix} 1 \\ 2 \\ 1 \end{pmatrix}$

(c) $\beta = \left\{ \begin{pmatrix} 1 \\ 2 \\ 1 \end{pmatrix}, \begin{pmatrix} 1 \\ 0 \\ -1 \end{pmatrix}, \begin{pmatrix} -1 \\ 1 \\ -1 \end{pmatrix} \right\}$, $\mathbf{D} = \begin{pmatrix} 3 & 0 & 0 \\ 0 & 1 & 0 \\ 0 & 0 & 0 \end{pmatrix}$

(d) $\mathbf{S} = (\mathbf{w} \ \mathbf{v} \ \mathbf{u}) = \begin{pmatrix} 1 & 1 & -1 \\ 2 & 0 & 1 \\ 1 & -1 & -1 \end{pmatrix}$

(e) $\mathbf{S}^{-1} = \frac{1}{6}\begin{pmatrix} 1 & 2 & 1 \\ 3 & 0 & -3 \\ -2 & 2 & -2 \end{pmatrix}$

(f) $\beta' = \left\{ \frac{1}{\sqrt{6}}\begin{pmatrix} 1 \\ 2 \\ 1 \end{pmatrix}, \frac{1}{\sqrt{2}}\begin{pmatrix} 1 \\ 0 \\ -1 \end{pmatrix}, \frac{1}{\sqrt{3}}\begin{pmatrix} -1 \\ 1 \\ -1 \end{pmatrix} \right\}$

(g) $\mathbf{S}' = \begin{pmatrix} 1/\sqrt{6} & 1/\sqrt{2} & -1/\sqrt{3} \\ 2/\sqrt{6} & 0 & 1/\sqrt{3} \\ 1/\sqrt{6} & -1/\sqrt{2} & -1/\sqrt{3} \end{pmatrix}$

(h) $(\mathbf{S}')^{-1} = \begin{pmatrix} 1/\sqrt{6} & 2/\sqrt{6} & 1/\sqrt{6} \\ 1/\sqrt{2} & 0 & -1/\sqrt{2} \\ -1/\sqrt{3} & 1/\sqrt{3} & -1/\sqrt{3} \end{pmatrix}$

7.19. (a) $\mathbf{S} = \begin{pmatrix} 1 & b \\ 0 & c - a \end{pmatrix}$, $\Lambda = \begin{pmatrix} a & 0 \\ 0 & c \end{pmatrix}$

(b) $\mathbf{A}^{1000} = \begin{pmatrix} a^{1000} & b\left(c^{999} + c^{998}a + c^{997}a^2 + \cdots + ca^{998} + a^{999}\right) \\ 0 & c^{1000} \end{pmatrix}$

7.20. (d) The algebraic multiplicity of $\lambda_1 = 1$ is one and $\lambda_{2,3} = 2$ is two.
Matrix **A** is diagonalizable if and only if $\lambda_1 = 1$ has geometric multiplicity of 1 and $\lambda_{2,3} = 2$ has geometric multiplicity of 2.

(e) (i) is **not** diagonalizable (ii) $\mathbf{P} = \begin{pmatrix} 1 & 1 & -1 \\ 1 & 0 & 1 \\ 1 & 1 & 0 \end{pmatrix}$

7.21. (b) $\widehat{\mathbf{w}_1} = \frac{1}{\sqrt{2}}\begin{bmatrix} 0 \\ 1 \\ 1 \end{bmatrix}$, $\widehat{\mathbf{w}_2} = \frac{1}{\sqrt{6}}\begin{bmatrix} 2 \\ 1 \\ -1 \end{bmatrix}$ and $\widehat{\mathbf{w}_3} = \frac{1}{\sqrt{3}}\begin{bmatrix} 1 \\ -1 \\ 1 \end{bmatrix}$

(c) (i) $\lambda^3 - 6\lambda^2 - 15\lambda - 8 = (\lambda + 1)^2(\lambda - 8) = 0$. The eigenvalues are $\lambda_{1,2} = -1$ and $\lambda_3 = 8$.

(ii) $\left\{ \begin{pmatrix} 1 \\ -4 \\ 1 \end{pmatrix}, \begin{pmatrix} 1 \\ 0 \\ -1 \end{pmatrix}, \begin{pmatrix} 2 \\ 1 \\ 2 \end{pmatrix} \right\}$

(c) Consider $k_1\mathbf{v}_1 + k_2\mathbf{v}_2 + k_3\mathbf{v}_3 = \mathbf{O}$. Prove that $k_1 = k_2 = k_3 = 0$.

7.22. (b) See chapter 4.

(c) (i) $\lambda_1 = 4$ and $\lambda_2 = 6$ (ii) $\lambda_3 = 0$, $\mathbf{w} = \begin{pmatrix} 0 \\ 1 \\ 0 \\ 1 \end{pmatrix}$ and $\lambda_4 = 2$, $\mathbf{x} = \begin{pmatrix} 0 \\ 1 \\ 0 \\ -1 \end{pmatrix}$

(iii) $\mathbf{P} = \frac{1}{\sqrt{2}}\begin{pmatrix} 1 & 1 & 0 & 0 \\ 0 & 0 & 1 & 1 \\ 1 & -1 & 0 & 0 \\ 0 & 0 & 1 & -1 \end{pmatrix}$, $\mathbf{Q} = \mathbf{P}^T = \frac{1}{\sqrt{2}}\begin{pmatrix} 1 & 0 & 1 & 0 \\ 1 & 0 & -1 & 0 \\ 0 & 1 & 0 & 1 \\ 0 & 1 & 0 & -1 \end{pmatrix}$,

$\Lambda = \begin{pmatrix} 4 & 0 & 0 & 0 \\ 0 & 6 & 0 & 0 \\ 0 & 0 & 0 & 0 \\ 0 & 0 & 0 & 2 \end{pmatrix}$

7.23. $X = \frac{1}{\sqrt{5}}(2x - y)$ and $Y = \frac{1}{\sqrt{5}}(x + 2y)$. The diagonal form is
$3x^2 + 4xy + 6y^2 = 2X^2 + 7Y^2$

7.24. $X = \frac{1}{\sqrt{3}}(x + y - z)$, $Y = \frac{1}{\sqrt{2}}(y - x)$ and $Z = \frac{1}{\sqrt{6}}(x + y + 2z)$.
The diagonal form is $2xy + 4xz + 4yz + 3z^2 = -X^2 - Y^2 + 5Z^2$.

Index

Ax 49
$\mathbf{Ax} = \mathbf{b}$ 48
$\mathbf{A}^T$ 76
$\mathbf{A}^{-1}$ 82
$(\mathbf{AB})^{-1}$ 87
$|$ 80
$\mathbf{u} \cdot \mathbf{v}$ 37
$\mathbf{O}_{mn}$ 59
$P \Leftrightarrow Q$ 108
$P \Rightarrow Q$ 108
$||\mathbf{u}||$ 137
$\hat{\mathbf{u}}$ 154
$\langle \mathbf{u}, \mathbf{v} \rangle$ 278
$d(\mathbf{u}, \mathbf{v})$ 285
$P(t)$ 193
M_{23} 193
$F[a, b]$ 197
$T : U \rightarrow V$ 340
$\ker(T)$ 354
$[\mathbf{u}]_B$ 399
$S \circ T$ 407
$T^{-1} : W \rightarrow V$ 414
$\det(A)$ 432
$adj(A)$ 447
E_λ 501
σ_n 548
$\mathbf{Au} = \lambda\mathbf{u}$ 491
$\det(\mathbf{A} - \lambda\mathbf{I}) = 0$ 494
$d(\mathbf{u}, \mathbf{v})$ 285
T^{-1} 382
$1 - 1$ 374

A

additive inverse 59, 131, 192
adjoint 447
Alembert, Jean d' 492
approximation 2, 319, 329
area scale factor 433
associative 59, 67, 130–1, 192
augmented matrix 14–26, 92, 97, 99, 114, 248, 264–6
axiom 191
axioms of inner product 278
axioms of vector spaces 191–2

B

back substitution 16, 329, 479
basis 171, 175, 394
basis for Fourier series 307
bijective transformation 381

C

calculus 389, 403
Carroll, Lewis 431
Cauchy, Augustin 151
Cauchy–Schwarz inequality 149–50, 153, 291–4
Cayley, Arthur 42, 77, 504, 513
Cayley–Hamilton 513
characteristic equation 494
characteristic polynomial 504, 513
Chebyshev polynomials 319
closure 191, 205
codomain 340
cofactor 441
cofactor formula for determinants 445
cofactors 439, 442, 444
column space 241–2
column vectors 43, 48, 159, 240, 555
commutative 64–5, 67, 83, 130, 136, 191, 278, 412
composite transformation 409–10
compress 250
computer graphics 43, 249, 339, 356
consistent 9, 99, 102, 104
coordinates 159, 391, 394
coordinates of a polynomial 385
coordinates with respect to non-standard basis 395–402
cosine rule 145–6
cryptography 115, 373

D

decompose 552, 557
degree of a polynomial 193
derivative 377, 389
derivative matrix 401–3
determinant 431–72
  by LU factorization 481
  by row operations 463–5
  formula of a 2 by 2 matrix 432
  formula of a 3 by 3 matrix 442
  formula of a n by n matrix 444–5
  formula of triangular or diagonal matrix 457
  geometric interpretation 432–3, 443
  negative 435, 444
  properties 462
diagonalization 519, 526
  illustration 519
differential equations 438, 492, 532
differential operator 412
dimension 229, 231
  dimension infinite 232
  dimension n 231
  dimension theorem of linear transforms 364
  dimension theorem of matrices 259
  dimension zero 230
  direction 27–30, 129–30, 144–7
distance function 139, 143, 290, 319
distributive law 67, 131, 136, 192, 278
document term 295
domain 340
donkey theorem 152, 294

E

eigenspace 501
eigenvalue 492
  addition of 512
  eigenvector 492
  the inverse matrix 509
  powers of matrices 509
  product of 512
elementary matrices 106–10, 459, 466–8
elementary row operations 14–5, 19, 92, 107, 243–45
elimination 5, 12–27, 91–105, 451, 473
ellipse 549–50
equivalence 108, 520
Euclidean norm 139–40
Euclidean space 36, 129, 220, 230, 287, 319

F

Fermat, Pierre de 58
finite dimensional vector space 230
forward substitution 479
Fourier series 225, 296–7, 301

free variables 98, 101–3, 260
full rank 249, 251, 265–6
function 340
functions 196–7
Fundamental Theorem of Algebra
505

G

Gauss 15, 181
Gaussian elimination 12–27,
91–105, 451, 473
Gauss–Jordan 23
general solution 96, 262
general solution in vector form 103
generate 171–5
Google 491
Gram–Schmidt 314–5
graph 3, 5, 10–11

H

Hilbert space 35
homogeneous 94, 97
homogeneous solution 261
homogeneous system 254–5
hyperplane 155

I

identity operator 412
identity transformation 414–5
if and only if 108
image of a transformation 340, 359
inconsistent 9, 85, 92–4, 212, 225
independent variables 130
inequality 149–53, 291–4
infinite dimensional vector space
301
infinite number of solutions 92, 265
infinity norm 287
information carried by a transform
365
information retrieval 295
injective transformation 374
inner product 277–90, 292, 295
integers 196
interchange rows 15
intersection 216
interval 196–7
inverse by determinants 447–51
inverse by LU factorization 481
inverse by row operations 110
inverse transformation 372, 381–2
illustration 373
invertible linear transform 414–5
isomorphic 319, 384
isomorphic to Euclidean space 387

K

kernel of a transformation 352–4

L

largest independent set 236
leading coefficient 23
leading diagonal 80, 281
leading one 23
left handed orientation 444
Legendre polynomials 295, 319
lemma 181
Leontief input-output 119
linear algebra 3, 108
linear combination 37, 47–8, 160,
208–10
linear dependence 163, 219
linear dependent 183
linear equations 2
linear independence 163, 168, 175,
217, 223, 308
linear mapping 342
linear operator 344
linear system 2, 3
linear transformation 342
LU factorization 472–81
LU procedure 474

M

map 340
MAPLE 52, 102
Markov chain 69, 75, 529–30
MATHEMATICA 52
MATLAB 52, 473
angle between vectors 156
converting to rationals 253
determinant 458
dot product 133
eigenvalues, eigenvectors 502
entering a matrix 53
matrix dimensions must agree 53
matrix multiplication 53
no balance 502
norm 141
polynomial 515
power 56
rref 115
transpose 79
MATLAB commands 53, 56, 79,
115, 133, 141, 156, 253, 458,
502, 515
matrix 42, 70, 80, 251, 321, 431, 449,
491, 519, 526
cofactor 445
diagonal 328
eigenvalue 523, 541
eigenvector 523, 533–5
elementary 106
Fibonacci 532
identity 80–2, 106, 108, 110
inverse 85, 114, 436, 451
invertible 83, 85, 88, 114
Jacobian 454
lower triangular 328, 457, 476
matrix representation of a
transformation 390
non-invertible 85, 114–5, 436,
451
non-singular 83, 85, 88, 114
orthogonal 321–9, 534
similar 520–1
singular 85
skew-symmetric 273
square 42, 70
standard 391
symmetric 273
upper triangular 328–9, 457, 476
zero 59
matrix times matrix 15, 19
matrix vector multiplication
49–51
maximum independent set 236
metric 139, 285
Miller indices 394
minimum spanning set 236
Minkowski, Hermann 153
Minkowski inequality 151, 153,
293–4
minor 439–42
modulus 140
multiple eigenvalue 505
multiplicative function 469
multiplicity 505
multiplier 476

N

natural basis 176–7, 228
natural basis for polynomials 224
negative determinant 435, 444
Netflix challenge 429, 548
neutral element 131, 192
no solution 9, 11, 92, 94, 265
non-homogeneous 99, 261
non-standard bases 394
non-standard basis 176, 403
non-trivial solutions 182, 492
non-zero row 98, 243, 249
norm 137–43, 152, 155–6, 282,
286–8, 298
norm of a matrix 282–3
normalized 154–5, 298–300
normalized set 299
null space of a matrix 255
null space of a transformation 357

nullity of a matrix 255
nullity of a transformation 357

O

one norm 287
one-to-one transformations 373–7
one-to-one test 373–5, 377
onto transformations 377
onto test 379
optimization 2, 338
ordered basis 416
orthogonal 134, 155, 183, 290, 294–321
orthogonal diagonalization 541, 544, 551
orthogonal set 299, 307–9, 544, 558
orthogonality 134, 290, 295, 308, 310
orthogonally diagonalizable 541–3
orthonormal 155, 299
orthonormal basis 300, 306–8, 315, 534, 550
orthonormal vectors 155, 329
overdetermined system 263

P

parallel 177
parallel planes 94
parallelepiped 444, 454, 456
parallelogram 29, 31, 434
parallelogram higher dimensional 444
parallelogram three dimensional 443
parameter 96, 98
particular solution 96, 262
periodic function 225, 301
perpendicular unit vectors 155, 299, 537
place holders 4
place sign 441
plane 3, 8, 31, 98–9, 155–6, 249
point(s) of intersection 3, 5
polar basis 394
polar coordinates 454
polynomial approximation 2, 319
polynomials 193
powers of a matrix 70, 514, 537
procedure for diagonalization 522
procedure for eigenvalues and eigenvectors 494
procedure for orthogonal basis 314
procedure for orthogonal diagonalization 544
procedure for basis of a subspace 247
projection 312–3
proof 58
proof by contradiction 181
proof by induction 88–9
puzzles 101
Pythagoras 58, 137–8, 282, 287, 297, 321
Pythagoras' Theorem 297

Q

query term 295
QR factorization 328–30

R

range of a transformation 340, 359
rank
  column 249–51
  low 249
  of a matrix 250
  of a transformation 364
  row 249–51
rank-nullity theorem 259, 364
real inner product space 278
reduced row echelon form 13, 23, 97–9, 108–9, 114, 245
redundant rows 251
reverse order 317, 418
reverse row operations 474–6
right handed orientation 444
rms value 284
row by column 47, 50
row echelon form 23
row equivalent 19, 92, 94, 97, 99, 107–10, 243–4, 254
row equivalent matrix 243
row space 241–2
row vectors 43, 51, 240–4
rref 23, 115, 245, 253

S

scalar 27–30, 33
scalar multiplication 30, 33–4, 37, 43–5, 62, 130, 191–2
scalar product 277
Schwarz, Hermann 149–51
Seki, T. 431
semi-axes 550
set 168
shortest distance 155
sigma notation 65
signal processing 295
simple eigenvalue 505
singleton set 204
singular value decomposition (SVD) 430, 547–8
singular values 548
size 42
smallest spanning set 236
solution set 5, 19, 92, 97
solution test 265
span 171–5
spanning set 172, 235–6
spectral theorem 543
spherical coordinates 454
spherical polar coordinates 394
standard or natural basis 176, 183, 224, 231, 385, 391
standard unit vectors 154, 159–61, 171, 177
subset 203
subspace 203–7, 209–14
subspace test 204, 209
Sudoku 101
support vector machines 155
surjective transformation 378
SVD factorization 547
swapping vectors 435
Sylvester, James 77

T

three-dimensional space 30, 34, 130, 210
trace of a matrix 281, 511
transformation matrix for non-standard bases 394–6
transition matrix for Markov chain 529
transpose of a matrix 75–9, 447, 534
triangular inequality 151, 293
triple factorization 547
trivial solution 96, 256, 260
two norm 287
types of solution 8, 91–105

U

underdetermined system 263
union 216
unique solution 10–11, 92, 265
usual inner product 277
u hat 154

V

Vandermonde 453
vector space 191–3, 198
vector space over the real numbers 196
vectors 143, 145, 149
  angle between 156
  dot product 37

vectors (*continued*)
  length – Euclidean space 137
  length – non-Euclidean space 282
  negative 147
  parallel 145
  perpendicular 134, 147, 155
  positive 146
  unit 153
  zero 147
volume scale factor 444

W

Wiles, Andrew 58
without loss of generality (WLOG) 508
work done 144–6
Wronskian 438

X

x direction 171, 175, 391

Y

y direction 171, 175, 391

Z

zero determinant 436
zero function 197
zero polynomial 193
zero row 23, 97–8, 242–3, 248